26th Association for the Advancement of Animal Breeding and Genetics Conference 2025

Genetic Frontiers: Innovations in Livestock Breeding for a Sustainable Future

Queenstown, New Zealand
24 - 26 June 2025

ISBN: 979-8-3313-2668-5

Printed from e-media with permission by:

Curran Associates, Inc.
57 Morehouse Lane
Red Hook, NY 12571

Some format issues inherent in the e-media version may also appear in this print version.

AAABG – Association for the Advancement of Animal Breeding and Genetics
University of New England
Armidale NSW 2351
Australia

Phone: 61 2 677 320 55
Fax: 61 2 677 332 66

AGBU@une.edu.au

Additional copies of this publication are available from:

Curran Associates, Inc.
57 Morehouse Lane
Red Hook, NY 12571 USA
Phone: 845-758-0400
Fax: 845-758-2633
Email: curran@proceedings.com
Web: www.proceedings.com

Association for the Advancement of Animal Breeding and Genetics

AAABG

Association for the Advancement of Animal Breeding and Genetics

Proceedings of the Twenty-sixth Conference

26th AAABG Conference

24 - 26 June 2025

Queenstown, New Zealand

Genetic Frontiers:
Innovations in Livestock Breeding for a Sustainable Future

Millennium Hotel
Tahuna Queenstown, New Zealand

24 – 26 June 2025

ISSN Number: 1328-3227

Produced by:
Association for the Advancement of Animal Breeding and Genetics
C/- AGBU
University of New England
Armidale NSW 2351

http://www.aaabg.org

PRESIDENT'S MESSAGE

On behalf of the organising committee, it gives me much pleasure to welcome members, Fellows and Honorary members and delegates to Tāhuna Queenstown and the 26th Conference of the Association for the Advancement of Animal Breeding and Genetics (AAABG). Our conference theme is "Genetic Frontiers: Innovations in Animal Breeding for a Sustainable Future". We are looking forward to three days of learning, exchange of knowledge and ideas, and of meeting friends and colleagues, old and new, with a shared interest in the science and practice of modern animal breeding.

AAABG has been an incredibly important organisation in my career and professional development, and I know many of our members would express similar views. It has provided opportunity to look outwards and to engage with others working in animal breeding, to understand the latest developments and ideas in animal breeding, and to form long-standing connections and lifelong friendships. To those of you who are regular contributors, thanks for your presence here again. And for those of you who might be attending AAABG for the first time, welcome to the Association and please make the most of the opportunity to connect with others, to share your knowledge and to learn.

Animal breeding, and the scientific endeavours associated with it, has played an integral part in the economies of both Australia and New Zealand. Both countries have significant and long-standing livestock industries, and over the past few decades the aquaculture industries in both countries have also grown substantially. Compared to other parts of the world, both countries punch well above their weight when it comes to animal breeding, whether measured on an industry size basis or a population basis. What we do matters – it generates revenue and profit for our animal-based industries including farming families, it maintains their international competitiveness, it helps feed the world, it improves the lives of small-holding farmers in less wealthy countries and importantly, if done correctly, it can produce a better outcome for the health and welfare of the animals which we work with. It is also increasingly a platform that can be used to improve environmental outcomes.

We are grateful for the generous support of our sponsor organisations whose contributions make conferences sustainable. Thanks also to our conference organisers (ICMS Australasia), and to all the members of the conference organising committee whose enthusiasm and commitment to the cause is what makes a conference come together. We also have a debt of gratitude to Dr Sue Hatcher, who combines the dual roles of AAABG Editor and AAABG Executive Officer so well and keeps us on track.

Enjoy your time in this unique and beautiful place and take the opportunity to renew acquaintances and expand your networks with other delegates from the animal breeding community.

Jason Archer
President

ASSOCIATION FOR THE ADVANCEMENT OF ANIMAL BREEDING AND GENETICS

2025

TWENTY SIXTH CONFERENCE COMMITTEE

CITATION OF PAPERS

Papers in this publication should be cited as appearing in the Proceedings of the Association for the Advancement of Animal Breeding and Genetics.

(Abbreviation: Proc. Assoc. Advmt. Anim. Breed. Genet.)

Example citation:
Bowley F.E., Amer P.R. and Meier S. (2013) New approaches to genetic analysis of fertility traits in New Zealand dairy cattle. *Proc. Assoc. Advmt. Anim. Breed. Genet*. 20: 37-40.

REVIEWERS

All papers, invited and contributed, were subject to peer review by two referees. We acknowledge and thank those listed below for reviewing the AAABG papers contained in these proceedings.

Michael Aldridge
Panoraia Alexandri
Hassan Aliloo
Fazel Almasi
Peter Amer
Hamed Amirpour
Jason Archer
Michelle Axford
Robert Banks
Amy Bell
Timothy Bilton
Hugh Blair
Julie Blommaert
Vinzent Boerner
Clara Bradford
Peta Bradley
Forbes Brien
Annika Bunz
Andrew Byrne
Linda Cafe
Mark Camara
John Cavalieri
Amanda Chamberlain
Yizhou Chen
Evans Cheruiyot
Sam Clark
Bronwyn Clarke
Schalk Cloete
Roy Costilla
Brad Crook
Sean Crowley
Neil Cullen
Peter Deardon
Elena Dehnavi
Sara de las Heras Saldana
Ken Dodds
Kath Donoghue
Jigme Dorji
Yaser Fazel
Mohammad Ferdosi
Marina Fortes
Amber Garber
Daniel Garrick
Dorian Garrick
Arthur Gilmour
Tom Granleese
Johan Greeff
Philip Gurman
Bruce Hancock
Rhiannon Handcock
Jo-Anne Harper
Sue Hatcher
Ben Hayes
Susann Hermesch
Brad Hine
Dirk Hinrichs
Marnie Hodge
Patricia Johnson
Neville Jopson
Jie Kang
Tom Kelly
Hamideh Keshavarzi
Majid Khansefid
Mehar Khatkar
Brian Kinghorn
Li Li
Yuhao Li
Zhi Loh
Nicolas Lopez-Villalobos
Wai Low
Iona MacLeod
Mette Madsen
Graeme Martin
Rudi McEwin
Lorna McNaughton
Suzanne Mortimer
Marina Naval Sanchez
Cornelius Nel
Loan Nguyen
Thuy Nguyen
Tuan Nguyen
Mohammad Nilforooshan
Babatunde Olasege
Chian Teng Ong
Brendon O'Rourke
Beth Paganoni
Sameer Pant
Peter Parnell
Natalie Pickering
Wayne Pitchford
Laercio Porto-Neto
Jennie Pryce
Cheryl Quinton
Imtiaz Randhawa
Toni Reverter
Jess Rickard
Elizabeth Ross
Suzanne Rowe
Malshani Samaraweera
Boris Sepulveda
Ric Sherlock
Erin Smith
Jen Smith
Nick Sneddon
Melissa Stephen
Bolormaa Sunduimijid
Imke Tammen
Carel Teseling
Peter Thurn
Kathryn Tiplady
Hong Quan Tran
Irene van den Berg
Anouk van Breukelen
Christy Vander Jagt
Julius van der Werf
Alison Van Eenennaam
Laura Vargovic
Peter Wahinya
Sam Walkom
Bradley Walmsley
Yu Wang
Christie Warburton
Franziska Weik
Cameron Whistler
Matt Wolcott
Ruidong Xiang
Luna Zhang

SPONSORS OF THE 26TH AAABG CONFERENCE 2025

The financial assistance of the following organisations is gratefully acknowledged.

Platinum Sponsor

genomNZ from agresearch

Gold Sponsor

agbu

Australia's genetics institute for agriculture

Bronze Sponsors

decode science enabling discovery

TE KUNENGA KI PŪREHUROA MASSEY UNIVERSITY UNIVERSITY OF NEW ZEALAND

MOL BREEDING

Thermo Fisher SCIENTIFIC

Industry Session

mla
MEAT & LIVESTOCK AUSTRALIA

Field Trip Sponsor

beef+lamb
new zealand

Conference App

abacusbio.
BRIDGING SCIENCE & BUSINESS

THE ASSOCIATION FOR THE ADVANCEMENT OF ANIMAL BREEDING AND GENETICS[1] INCORPORATED[2]

OBJECTIVES

(i) to promote scientific research on the genetics of animals;
(ii) to foster the application of genetics in animal production;
(iii) to promote communication among all those interested in the application of genetics to animal production, particularly breeders and their organisations, consultants, extension workers, educators and geneticists.

To meet these objectives, the Association will:

(i) hold regular conferences to provide a forum for:
  (a) presentation of papers and in-depth discussions of general and industry-specific topics concerning the application of genetics in commercial animal production;
  (b) scientific discussions and presentation of papers on completed research and on proposed research projects;
(ii) publish the proceedings of each Regular Conference and circulate them to all financial members;
(iii) use any such other means as may from time to time be deemed appropriate.

MEMBERSHIP

Any person interested in the application of genetics to animal production may apply for membership of the Association and, at the discretion of the Committee, be admitted to membership as an Ordinary Member.

Any organisations interested in the application of genetics to animal production may apply for membership and, at the discretion of the Committee, be admitted to membership as a Corporate member. Each such Corporate Member shall have the privilege of being represented at any meeting of the Association by one delegate appointed by the Corporate Member.

Benefits to Individual Members.

- While it is not possible to produce specific recommendations or "recipes" for breeding plans that are applicable for all herd/flock sizes and management systems, principles for the development of breeding plans can be specified. Discussion of these principles, consideration of particular case studies, and demonstration of breeding programs that are in use will all be of benefit to breeders.
- Geneticists will benefit from the continuing contact with other research workers in refreshing and updating their knowledge.
- The opportunity for contact and discussions between breeders and geneticists in individual members' programs, and for geneticists in allowing for detailed discussion and appreciation of the practical management factors that often restrict application of optimum breeding programs.

Benefits to Member Organisations.

- Many of the benefits to individual breeders will also apply to breeding organisations. In addition, there are benefits to be gained through coordination and integration of their efforts.

---

[1] AAABG was formerly known as the Australian Association for Animal Breeding and Genetics. Following the 1995 OGM the name was changed when it became an organisation with a joint Australian and New Zealand membership.
[2] The Association for the Advancement of Animal Breeding and Genetics is incorporated in South Australia.

Recognition of this should follow from understanding of common problems, and would lead to increased effectiveness of action and initiatives.

- Corporate members can use the Association as a forum to float ideas aimed at improving and/or increasing service to their members.

General Benefits.

- Membership of the Association may be expected to provide a variety of benefits and, through the members, indirect benefits to all the animal industries.
- All members should benefit through increased recognition of problems, both at the level of research and of application, and increased understanding of current approaches to their solution.
- Well-documented communication of gains to be realised through effective breeding programs will stimulate breeders and breeding organisations, allowing increased effectiveness of application and, consequently, increased efficiency of operation.
- Increased recognition of practical problems and specific areas of major concern to individual industries should lead to increased relevance of applied research.
- All breeders will benefit indirectly because of improved services offered by the organisations which service them.
- The existence of the Association will increase appreciably the amount and use of factual information in public relations in the animal industries.
- Association members will comprise a pool of expertise – at both the applied and research levels – and, as such, individual members and the Association itself must have an impact on administrators at all levels of the animal industries and on Government organisations, leading to wiser decisions on all aspects of livestock improvement, and increased efficiency of animal production.

CONFERENCES

One of the main activities of the Association is the Conference. These Conferences will be structured to provide a forum for discussion of research problems and for breeders to discuss their problems with each other, with extension specialists and with geneticists.

ASSOCIATION FOR THE ADVANCEMENT OF ANIMAL BREEDING AND GENETICS

FELLOWS OF THE ASSOCIATION

"Persons who have rendered eminent service to animal breeding in Australia and/or New Zealand or elsewhere in the world, may be elected to Fellowship of the Association…"

Elected February 1990
R.B.M Dun
F.H.W. Morley
A.L. Rae
H.N Turner

Elected September 1992
K. Hammond

Elected July 1995
C.H.S. Dolling
J.R. Hawker
J. Litchfield

Elected February 1997
J.S.F. Barker
R.E. Freer

Elected June 1999
J. Gough
J.W. James

Elected July 2001
J.N. Clarke
A.R. Gilmour
L.R. Piper

Elected September 2005
B.M. Bindon
M.E. Goddard
H.-U. Graser
F.W. Nicholson

Elected September 2007
K.D. Atkins
R.G. Banks
G.H. Davis

Elected September 2009
N.M. Fogarty
A.R. Fyfe
J.C. McEwan
R. Mortimer
R.W. Ponzoni

Elected September 2011
B.P. Kinghorn
A. McDonald

Elected October 2013
H.M. Burrow
P.F. Fennessy
G. Nicol
P. Parnell

Elected October 2015
P.F. Arthur
D. Johnson
K. Meyer
B. Tier
R. Woolaston

Elected October 2019
S.A. Barwick
H.T. Blair
S.W.P. Cloete
I.W. Purvis

Elected November 2021
F.D. Brien
D. Garrick
J. Greeff
B.J. Hayes
J.E. Pryce
J.H.J van der Werf

Elected July 2023
K.G. Dodds
W.S. Pitchford
H.W. Raadsma
C.W. (Bill) Sandilands
A.A. Swan

Elected June 2025
P.R. Amer
K.L. Bunter
M. Haile-Mariam
S.I. Mortimer
M.Shaffer
S. Skinner
A.L. Van Eenennaam

PETER R. AMER

Peter grew up on a Dairy Farm in North Canterbury in New Zealand. He studied Agricultural Science at Lincoln College with his Honours dissertation titled "Methods of calculating breeding values for rams in flocks genetically linked by reference sires" completed in 1988. The following summer he joined the Genetics Section at Ruakura Agriculture Research Centre where he worked on researching and implementing the statistical analysis for the National Sire Referencing Scheme. He then moved to University of Guelph for his PhD "Economic evaluation of beef cattle genotypes using neoclassical production theory" awarded in 1992. His thesis was selected to represent the Department of Agricultural Economics and Business in a competition among North American Universities.

Peter had postdoctoral fellowships at the Swiss Federal Institute of Technology, Zurich (studied economic aspects of dual cattle production) in and Scottish Agricultural College, Edinburgh (working on economic selection indexes in beef cattle breeding). In 1996 he was appointed as a Geneticist with AgResearch at Invermay, Mosgiel NZ. He worked on optimising industry use of major genes; supported industry genetic improvements programmes for sheep and deer; analysed livestock data sets to detect evidence of gene segregation; and had contract work in aquaculture breeding programmes, predominantly salmon.

In 2001 he was one of the early scientists to join Abacus BioTech where he consulted on indexes for industry groups which has had continued. During the early days he also did some venture investment work for the original Abacus Biotech Investors. In 2006 Peter was a founding partner in the establishment of AbacusBio and has been Managing Director from 2022.

Peter has set an outstanding example of how to both effectively serve industry and continue to be scientifically active. He has published 341 articles (100 on Scopus) with a h-index of 26. He is highly collaborative publishing with 150 co-authors. He has been able to work across countries and production systems with his top three cited papers include developing breeding objectives for beef cattle in Ireland, sheep in New Zealand and dairy cattle in Australia.

Peter is extremely good at listening to farmers and understanding challenges in their production systems. He is a highly effective mentor of young scientists and consultants, and has been an early adopter of novel tools such as 1000minds for decision making and conjoint analysis. For his outstanding contribution to developing genetic improvement programs across multiple species, the Association for the Advancement of Animal Breeding and Genetics is pleased to elect Peter Amer as a Fellow of the Association.

KIM L. BUNTER

Dr Kim Bunter is a Principal Scientist at the Animal Genetics and Breeding Unit (AGBU). Since 2022 she has held the position of Deputy Director where she contributes leadership to a team of close to 40 scientists, staff and students that develop and build the genetic infrastructure that underpins genetic improvement programs for agriculture industries in Australia, including sheep, beef, swine, goats, bees and plants. Kim has been a part of the AGBU and by extension the AAABG communities for over 30 years. A product of the University of New England education pathway, Kim completed her PhD in 2002. Her foray into Ostrich breeding for her PhD, an unexplored frontier at the time, would foreshadow Kim's career as someone who tackled new problems, species and traits. As a result of her desire to answer practical industry questions, Kim developed important knowledge for implementation across many species including pigs, sheep, beef cattle, Sydney Rock oysters, ostrich, and prawns. Recently Kim provided pivotal leadership for the development of a genetic evaluation system for Australian honeybees called "Plan Bee".

Kim's largest impact on the Australian genetics industry is certainly in the pig industry. Kim collaborated widely with researchers in universities and breeding companies and has contributed to the long-term impact of genetics expertise at these breeding companies where former PhD students now hold influential roles. Kim led many projects in pigs including development of mate selection strategies, and projects on immune responsiveness, piglet mortality, heat stress, use of juvenile IGF-1 in pig breeding, development of selection criteria to improve welfare in group housed sows, the genetics of reproductive performance, sow longevity and feed efficiency, and determination of meat and eating quality differences between Australian genotypes. Her work contributed to several publications and developments in PIGBLUP software.

Kim has also applied her extensive knowledge in the genetics of reproduction to other species. Kim contributed to the development of the "Days to Calving" trait at AGBU, which became the primary reproductive trait in BREEDPLAN for beef cattle. In recent years Kim, alongside the team at AGBU, has developed the component reproduction analysis for the Sheep Genetics LAMBPLAN Maternal and Merino evaluations. This was a significant body of work with huge benefit to the Australian sheep industry as it enabled breeders to select on the components of reproduction and focus on improving weaning rates through improved survival of lambs, not just larger litters. Kim has continued to provide guidance on improving welfare in the sheep industry through her work with lamb survival and the development of footrot breeding values.

Kim has (co)authored over 200 publications across a wide variety of species and topics. Kim's first contributing paper to AAABG was in 1995 at the Roseworthy Conference. Since then, Kim has authored and co-authored over 50 AAABG papers and was President of the 2019 conference in Armidale. This conference demonstrated Kim's dedication to the organisation as she pulled together a great conference against the odds with the notorious black summer bush fires as a backdrop. She backed this up leading a AAABG conference hub in Armidale in 2021, following Covid and post-tornado. For her outstanding contributions to the science of genetics and animal improvement, the Association for the Advancement of Animal Breeding and Genetics is pleased to elect Kim Bunter as a Fellow of the Association.

MEKONNEN HAILE-MARIAM

Agriculture Victoria Senior Research Scientist and Senior Research Fellow of School of Applied Systems Biology at La Trobe University, Dr. Mekonnen Haile-Mariam has made significant contributions to the field of animal breeding and genetics over the past three decades. Mekonnen has been at the forefront of improving the fertility, health and animal welfare of Australian dairy cattle through genetic improvement. He specialises in developing and updating Australian Breeding Values (ABVs) which are used by dairy farmers to increase productivity through genetic gain.

Mekonnen grew up around animals in Ethiopia, where there are more than 11 million dairy cows, compared to just 1.5 million in Australia. He completed his BSc in Animal Science from Addis Ababa University in Ethiopia in 1982 and went on to do an MSc in Animal Production at Alemaya University of Agriculture. In the early 1990s he swapped warmth for snow and headed to Uppsala in Sweden. He earned his PhD in Animal Breeding and Genetics from the Swedish University of Agricultural Sciences in 1994 and has since focused on the genetic improvement of dairy cattle. During his MSc and PhD studies and after completing his PhD Mekonnen has taught animal breeding at Alemaya University.

A post-doctoral opportunity prompted a move to Australia to join Professor Michael Goddard's research team at Agriculture Victoria in 1999. His early work involved genetically evaluating the somatic cell count as an indicator of udder health (mastitis) to develop and provide a ranking of bulls to dairy farmers, the first research of its kind in Australia. A similar genetic evaluation was developed for fertility, at the time there were limited records to identify the most fertile cows. Mekonnen continued his research focusing on dairy cattle fitness, type, fertility, heat tolerance and methane emission traits and their genetic relationships with milk production traits.

In 2016, Mekonnen was involved in the development of a new gestation length ABV which was released for industry use in 2020 through DataGene. This breeding value allows dairy farmers to select cows that calve within a specific time, as longer gestational length can increase calving difficulty and disrupt the seasonal calving season. Mekonnen also played a key role on updating the calving ease ABV, which has helped to improve animal welfare and reduce veterinary costs, as the calving difficulties can cause cow and calf deaths. His research emphasises profitability, sustainability, and animal welfare, with a particular interest in functional traits such as reproduction and health.

Over the last 20 years, Dr. Haile-Mariam has actively participated in various dairy genetics projects, including "Improved Genetic Evaluation System for the Dairy Industry" and "Validating and Increasing Profit from Dairy Bull Selection in Australia". He has extensively contributed to the DFCRC Fertility Project and the development of the DairyBio Animals' Program (2016-current). These projects aimed to improve and expand breeding values for dairy cattle in Australia.

Dr. Haile-Mariam has extensive experience working with key stakeholders such as Holstein Australia, Jersey Australia, and Dairy Australia. While in his native Ethiopia, he secured funding for the genetic improvement of indigenous livestock in Ethiopia through collaborations with the Swedish University of Agricultural Sciences, funded by the Swedish International Development

Cooperation Agency (SIDA). Mekonnen's research interest and focus while in Ethiopia was genetic improvement of indigenous and crossbred cattle in production environments where genetic antagonistic between adaptation and production is important.

In Oceania and abroad, Dr. Haile-Mariam has been at the forefront of major dairy genetics innovations and insights over the past 25 years. It is fair to say that he had a hand in all the main genetic innovations implemented by ADHIS and now DataGene over the past decade, significantly contributing to the doubling of the genetic gain in the Australian national herds.

Mekonnen is one of the leading experts in using ASReml and is a go to person for statistical analysis in the Computational Biology group of Agriculture Victoria Research. He is a well-liked colleague and an admired supervisor of his PhD students. He is always happy to invest time in showing students how to prepare the input data for ASReml as well as trouble-shooting models. Always done in a kind way with plenty of empathy and understanding.

His work has not only advanced the genetic improvement of livestock in Australia but also demonstrated the importance of genetic and genomic selection globally. He has published more than 100 journal articles and numerous conference papers, which have been cited more than 3,000 times. Dr. Haile-Mariam's dedication to integrating phenomics and other 'omics data into livestock improvement illustrates his innovative approach and commitment to the field. For his extensive contribution to animal breeding and genetics and impactful research in dairy genetic improvement the Association for the Advancement of Animal Breeding and Genetics is pleased to elect Mekonnen Haile-Mariam as a Fellow of the Association.

SUZANNE I. MORTIMER

Sue Mortimer is a Research Scientist, NSW Department of Primary Industries and Regional Development (NSW DPIRD) within Livestock Systems. She joined NSW Department of Agriculture (now NSW DPIRD) on the 14th February 1985 and was based at the Trangie Agricultural Research Centre.

Sue's appetite for agricultural science stemmed from her childhood interest in Australian history, particularly the importance of the wool industry and its contribution to the economy. Sue grew up in Sydney with no real connection to the land, so she studied agriculture, choosing Wool and Pastoral Sciences at the University of New South Wales where she undertook genetics subjects and graduated with first class honours. Sue went on to do her PhD under the supervision of John James, looking at the relative efficiency of index selection and independent culling levels for two traits under a range of conditions and their responses in breeding objective traits, with and without restrictions on genetic gain in one trait. In 1985, at the completion of her PhD Sue was appointed to a position at Trangie Agriculture Research Centre. At the time, the Trangie site had a substantial mix of young and old scientists which Sue found to be exciting. Sue is now based at the NSW DPIRD's Armidale Livestock Industry Centre located at the University of New England, Armidale.

Over the past 40 years, Sue's research has applied quantitative genetics to sheep breeding, particularly by estimating genetic parameters for economically important traits for use in genetic evaluation and breeding program design to improve the profitability of sheep production enterprises. Working with Kevin Atkins, her early focus was on evaluating the relative amounts of genetic and environmental variation in Merino wool production traits. From there her research program has included: the genetics of sheep meat production and eating quality, particularly the relationships of those traits with wool production and quality traits in the Merino breed; genetics of flystrike in non-seasonal rainfall environments; and genetic benchmarking of Merinos using industry data (central test sire evaluation, wether trials). Sue has supervised the research programs of many of the genetic resource flocks conducted at Trangie, as well as being involved in managing several of the Merino central test sire evaluations hosted there. Her involvement in central test sire evaluation continued with the establishment of the Merino Lifetime Productivity site at Trangie. Her extensive involvement in the QPLU$ selection lines and earlier resource flocks at Trangie has had a large impact on Merino genetic improvement in Australia.

During her career, Sue has had an extensive publication contribution (over 150 papers). As senior author or a co-author, she has contributed more than 40 papers to the association's proceedings since 1985. Sue has served as an Associate Editor for Animal Production Science and as a member of the BREEDPLAN and Sheep Genetics Technical Committee.

For her excellent contribution to the genetic improvement of the Australian Merino industry, the Association for the Advancement of Animal Breeding and Genetics is pleased to elect Sue Mortimer as a Fellow of the Association.

MATTHEW SHAFFER

Dr Matthew Shaffer is the Chief Executive Officer of DataGene, Australia's national genetic evaluation centre that provides leadership in herd improvement and data-related services on behalf of dairy farmers. Matt led the transformational change from the former Australian Dairy Herd Improvement Scheme to a contemporary organisation with broad industry support as demonstrated by its membership base of 27 cross-sectoral organisations. At a time when the genetic improvement sector was at cross-roads, Matt was instrumental in establishing the Herd Improvement Industry Strategic Steering Committee, a whole-of-industry approach to tackling significant challenges. Matt's gravitas and personal approach made it work. Matt delivered a well-supported business case that resulted in the formation of DataGene through his leadership, astute ability to manage inter-organisational relationships and unwavering personal commitment. In creating an environment where consultation and collaboration were his team's priorities, he increased resource availability, and this has significantly improved service delivery to farmers and industry.

A pig farm in Pennsylvania, USA was Matt's early home. His first paid job was milking cows for a neighbour, signalling a long career in agriculture. He earned his BA degree in International Relations at Penn State University (USA) in 1993 before completing both a Masters (1994) and PhD (1998) in International Relations at Cambridge University (UK). Matt started his career working for Hancock Agricultural Investment Group as Director of International Operations. While working for Hancock, Matt was transferred to Australia to open a subsidiary and later immigrated with his family. Matt was first introduced to the Australian dairy industry as Chief Executive Officer of the Holstein-Friesian Association of Australia where he was appointed to the board of the Australian Dairy Herd Improvement Scheme and then became Program Manager of Genetics and Data Management at Dairy Australia.

Matt's significant influence on the genetic and herd improvement industries has stretched beyond Australian shores. He is a past chair of the Interbull steering committee of the International Committee for Animal Recording and was instrumental in re-shaping the governance of Interbull. He was a founding innovator in establishing the International Dairy Data Exchange Network. He is also a past president of the World Holstein-Friesian Federation where he was the first president from Oceania and the youngest to ever hold the position. He has been generous with his commitment to global-scale collaboration to advance dairy genetics and is very highly respected across the world for his contribution to the genetics and herd improvement sectors.

The Australian dairy herd improvement industry has been heavily influenced by Matt's leadership over the past two decades. His breadth of experience, networks and focus on delivering for farmers has benefited research and industry organisations and lead to ground-breaking initiatives. Examples include Ginfo (national reference herd of genomic information), modernising the infrastructure that supports the centralised data repository and genetic evaluation system, ImProving Herds, and the rapid acceleration of genetic gain towards an agreed national breeding objective. The successful delivery of DataGene, and its programs, is one of his greatest professional achievements.

Matt is respected by his peers in Australia and overseas for the way he works with people to listen, promote diversity and find common ground, especially when the issues are complicated. He is a leader, mentor and friend. For his outstanding contributions to the genetic and dairy herd improvement industries in Australia, the Association for the Advancement of Animal Breeding and Genetics is pleased to elect Matthew Shaffer as a Fellow of the Association.

STEVEN SKINNER

Steven Skinner completed a Bachelor of Agricultural Economics at the University of New England in 1980, before undertaking a seven year appointment in applied agricultural economic research with the Bureau of Agricultural Economics in Canberra. Steve then spent five years managing a sheep and cattle property, before joining the Agricultural Business Research Institute (ABRI) in 1993.

Throughout his 30 year span at ABRI, Steve played an integral role in the delivery of ABRI's products and services to the beef cattle breeding industry around the world, including the BREEDPLAN genetic evaluation service, Breed Secretariat services and breed registry software. Initially employed as a BREEDPLAN processor, which overlapped with a three year stint as Executive Officer of the Australian Brangus Cattle Association Ltd, Steve was promoted to numerous roles, including manager of the Pedigree/Performance section (appointed 1997), manager of the International Beef Recording Scheme (appointed 2011) and Chief Operational Officer (appointed 2015). Steve also spent eight years as Executive Officer of the Australian Registered Cattle Breeder's Association (ARCBA), representing the interests of ARCBA's breed society members and their breeders.

Throughout his career, Steve worked extensively with ABRI's Australian and international beef breed society clients to maintain the integrity of their studbooks through their use of the Breed Secretariat service and/or breed registry software, and to facilitate genetic improvement within their breeds through the use of the BREEDPLAN genetic evaluation. Always outcome focused, Steve's appreciation for and understanding of the commercial realities of beef breeding was bolstered by his previous experience in managing beef properties. Throughout Steve's career, there have been numerous enhancements made to the BREEDPLAN genetic evaluation, and to the use of DNA technologies in cattle breeding. These include, but are not limited to, the introduction of new traits into the BREEDPLAN genetic evaluation, revisions to selection indexes, and the emergence of DNA technology with applications for parentage verification, management of genetic conditions and Single-Step BREEDPLAN. Steve led ABRI through these changes, ensuring that both he and ABRI staff were ready to assist breed society personnel and seedstock producers understand these new technologies and how they can best be utilised in their own businesses.

In addition, Steve has been a mentor to numerous others in the cattle breeding industry. Throughout his career, Steve has been an advocate for young people in agriculture. He facilitated work experience opportunities for those completing university degrees in agriculture and employed numerous personnel who were newly graduated and/or in the early stage of their career. Many of those employed by Steve early in their careers have gone on to have their own successful careers in the cattle breeding industry (for example, moving on to positions with Meat and Livestock Australia, Australian beef cattle breed societies, state Agricultural Departments and/or service provider roles).

In 2024, Steve was awarded the RW Vincent award by ARCBA for his outstanding service to the Australian cattle breeding industry. For his outstanding contribution to genetic improvement of the Australian cattle industry, the Association for the Advancement of Animal Breeding and Genetics is pleased to elect Steve Skinner as a Fellow of the Association.

ALISON L. VAN EENENNAAM

Professor Alison Louise Van Eenennaam was born in 1963 in Melbourne, Australia. Although her passion for animal genetics was piqued through lectures given by Mike Goddard during her BAgSc (Hons) degree at the University of Melbourne, she initially worked as a Development Scientist for Cooper's Animal Health in Camden, NSW. She then undertook a Master of Science at the University of California, Davis (UCD) in the United States, following which she worked for three years as a University of California Cooperative Extension Dairy and Livestock Extension Advisor in the central valley of California, before returning to UCD to complete a PhD in Genetics with Professor Juan Medrano. Upon graduation, she worked as a Research Scientist and then a Project Leader at Calgene, the small Davis-based company that commercialized the first genetically engineered plant food product, the Flavr Savr tomato.

In 2002 she returned to academia after accepting a position as a Professor of Cooperative Extension in Animal Biotechnology and Genomics at UCD. In 2023, she was promoted to Distinguished Professor, the highest UC campus-level faculty title that can be bestowed and is reserved for those whose work "*has been internationally recognized and acclaimed, and who have achieved the highest level of distinction in their scholarship, teaching, and university/public service*".

Alison has over 100 peer-reviewed publications, and her current research focuses on technical aspects of producing genome-edited livestock (especially cattle and sheep), and policy relating to the regulation of genetically modified and genome-edited organisms. She is an elected Fellow of the American Association for the Advancement of Science (AAAS), the recipient of the 2019 American Society of Animal Science (ASAS) National Rockefeller Prentice Award in Animal Breeding and Genetics, the 2019 University of California Davis James H. Meyer Distinguished Career Achievement Award, and the 2021 U.S. Beef Improvement Federation (BIF) Pioneer Award.

Alison is a gifted communicator, and has been a leading voice and occasional YouTube star in the field of animal biotechnology, in particular the validation of the early generation of DNA marker tests, and the development and application of genetic modification in livestock. Alison presents widely and interacts frequently with a variety of media and has delivered more than 800 oral presentations to global audiences, ranging from cattle producers to Congress. She received the American Association of Public and Land-grant Universities (APLU) National Award for Excellence in Extension; and the 2014 Council for Agricultural Science and Technology (CAST) Borlaug Communication Award.

Alison served as the USDA National Animal Genome Research Program (NRSP8) Bovine Genome coordinator (2018-2023), wrote the BIF guidelines for genome edited animals, and has served on several international advisory committees including the Strategic Steering Committee of the Australian dairy project entitled "ImProving Herds"; the International Advisory Committee of the "Center for Tropical Livestock Genetics and Health" (CTLGH); International Research Advisory Committee for the Genome Canada project "Increasing feed efficiency and reducing methane emissions through genomics"; and as Chair of the Climate Smart Cattle Technical Advisory Group for Ag Research in New Zealand. She has also served on several World Congress Genetics Applied

Livestock Production (WCGALP) Scientific Advisory Committees, and is on the Executive Committee organizing the 2026 WCGALP in Madison, Wisconsin.

She first attended AAABG in 2009 in South Australia, and has since attended almost all conferences. It is fitting that at the 26th AAABG in Queenstown NZ, Alison has been invited to give a keynote presentation. For her contributions to research in animal biotechnology and providing a trusted voice and evidence-based materials on the importance of genetic innovations to varied stakeholders globally, the Association for the Advancement of Animal Breeding and Genetics is pleased to elect Alison Van Eenennaam as a Fellow of the Association.

HONORARY MEMBERS OF THE ASSOCIATION

"Members who have rendered eminent service to the Association may be elected to Honorary Membership…"

Elected September 2009
W.A. Pattie
J.R.W. Walkley

HELEN NEWTON TURNER MEDAL TRUST

The Helen Newton Turner Medal Trust was established in 1993 following an anonymous donation to the Animal Genetics and Breeding Unit to perpetuate the memory of Helen Newton Turner and to encourage and inspire those engaged in animal genetics.

Helen Newton Turner AO

THE HELEN NEWTON TURNER MEDAL

The Medal is named after Dr Helen Newton Turner whose career with CSIRO was dedicated to research into the genetic improvement of sheep for wool production.

The Helen Newton Turner Medal was first awarded in 1994 to Associate Professor John James and a list of all recipients to date is given below. The Helen Newton Turner Medallist is chosen by Trustees from the ranks of those persons who have made an outstanding contribution to genetic improvement of Australian livestock.

The recipient of the Medal is invited to deliver an Oration on a topical subject of their choice.

Medallists

1994	J.W. James	2003	F.W. Nicholas	2015	A.R. Gilmour
1995	L.R. Piper	2005	K. Hammond	20147	A. Collins
1997	J. Litchfield	2007	L. Corrigan	2019	K.D. Atkins
1998	J.S.F. Barker	2009	R. Hawker	2021	J.H.J van der Werf
1999	C.W. Sandilands	2011	R. Banks	2023	M. Mortimer
2001	G.A. Carnaby	2013	M. Goddard	2025	A.A. Swan

The Oration of the 2023 Medal recipient, Mark Mortimer, is reproduced in these proceedings.

THE HELEN NEWTON TURNER BRIGHT FUTURES AWARD

In 2021, the Trust established a new award, the Helen Newton Turner Bright Futures Award to recognise the achievements of an up-and-coming individual who is showing evidence of establishing a reputation for excellence in the field of animal genetics within Australia.

2021 S.A. Clark
2023 B.J. Walmsley

2023 HELEN NEWTON TURNER MEDALIST ORATION

Mark Mortimer

PHONE CALL FROM JAMES

I was out in the paddock one morning moving a mob of sheep and I received a phone call from James Rowe (Emeritus Professor James Rowe, CEO of the sheep CRC 2000 to 2019). I'd had a long involvement with James through the sheep CRC and after the usual discussions about Farm, Family and Sheep James asked if it would be OK to nominate me for the Helen Newton-Turner Medal. After a pause for processing, I had a lot of reasons why that would not be a good idea. James was his usual persistent self, and I realised that this was not the first time James had tried to talk me into doing something that I was not comfortable with. The previous time was to go on a national TV show by the ABC called The New Inventors. This was due to my work with the Sheep CRC in developing the Pedigree Match Maker system. I look back at that time with good memories, so I said yes to James and thought nothing more about it. I mean, it was only a nomination, there would be plenty of other more worthy people besides myself.

HIDDEN MEANING IN SHEEP DATA

For me playing with sheep data started as a teenager. Starting in 1980 Robert (My Father) started a ram breeding cooperative as part of the AMS (Australian Merino Society). To be a member of the group you had to fleece weigh and body weigh all maiden ewes for selection each year. Most of the members at that stage didn't have computers and it become my job to type up pages of hand recorded data and run simple indexes on them. Computers at that stage had very few distractions and to pad out the time I got spend in the office and not in the yards I would build hand drawn histograms for each drop of animals. One member had a drop of animals that had no tail on the body weight histogram. It was clear that these animals had been removed before fleece weighing. When asked about this at a group meeting this member said, "How could you possibly know that, I didn't even tell my wife". This really highlighted to me that there were hidden secrets just waiting to be unlocked (Granted not always secrets that had to be hidden from your wife!). I have been playing with sheep data looking for hidden secrets ever since.

MENTORS

I certainly wouldn't be writing this if it wasn't for my parents, Robert and Pam Mortimer. My siblings and I were often coopted into data gathering in the sheep yards. It just became natural that if you were working with sheep someone was working with pen and paper and later computers.

Dr Rob Banks (2007 Fellow of AAABG, 2011 HNT Medallist). The first time I met Rob was around our kitten table, he'd come to stay the night and talk with Dad about this new system called BLUP that would produce breeding values. I can clearly remember his excitement about the data (There may have been red wine involved), there was lots of hand draw graphs explaining the concepts involved. My involvement with Rob lasted for the rest of his career and he was always very generous with his time and in sharing his reading lists, thank you Rob.

Dr Kevin Atkins (2007 Fellow of AAABG, 2019 HNT Medallist). While away doing a three-year stint at Orange Ag College I had a subject on research and I chose sheep (I know you didn't see that coming), Dad suggested I meet Kevin Atkins at the NSW DPI. Sitting in Kevin's office piled with books he taught me how to do correlations on sheep data in a spread sheet (Lotus 123) long hand (No simple formulas back then). I still remember the excitement of finding what appeared to me to be hidden relationships in the data. I would later go on to have a long involvement with Kevin through both phases of the Sheep CRC (19 years)

Professor Julius van der Werf (2021 Fellow of AAABG, 2021 HNT Medallist). In order to get a better understanding of genetics I came up with the idea to ring the Armidale UNE book shop and convince them that even though I wasn't a student they could sell and post me the textbook on genetics (Genetic Improvement of Cattle and Sheep, by Geoff Simm). Now I'll be honest, there was a lot in that book I didn't fully understand. But it certainly enabled me to ask the researchers much better questions. The first target of this newfound, half understood knowledge was Julius at a Sheep CRC workshop on genomics. To this day I am still on committees with Julius, and he is still patiently answering my questions.

There are many other people who have influenced my genetics knowledge over the years both in and out of academia, and I would like to thank you all. With special thanks to Daniel Brown, Kim Bunter and the late Andew Swan.

SHEEP CRC & INDIVIDUAL ANIMAL MANAGEMENT ADOPTION.

In 2003 I became involved with the Sheep CRC under their precision sheep management program. The goal was to move industry from mob-based management to individual animal management. This program helped changed how we handle and measure our sheep. I used to really dislike visually reading ear tags and if you couldn't keep your error rate at or under 0.2% you would get a talking to from the old fella. When I heard about this organisation that was offering co-funding and technical support for producers to trial RFID (Radio Frequency Identification) tags I was in. All I had to do was write the software that made the hardware talk to each other because in 2003 no off the shelf products existed. 22 years later and I'm still updating and improving this software. During one of many conversations with James Rowe over this time. James said we are not getting the traction we had hope for on RFID; I need something more, I need something that RFID can do that visual tags cannot. I came back to James with the idea of pedigree matching. Tracking ewes and lambs in the paddock with a panel RFID reader and using this data to determine pedigree. The idea worked! Obviouslly, I thought it was really cool, but also a very niche product that not many people would be interested in. We ended up on the ABC LandLine and The New Inventors shows and won a CRC Excellence in Innovation award. It was also my first look at "Big Data". We had tens of thousands of records from a small number of sheep. Daniel Brown (Brown *et al.* 2011) took this raw data and found hidden meanings one of which is that sheep that turn up to the water site more often have a positive genetic correlation to worm egg count.

If I ask myself the hard question of did the Pedigree Match Maker system help with adoption of individual animal management? History tells us that the answer is no. After several decades of giving presentations on individual animal management across the country, hears my take on why it hasn't been adopted and won't be even though in Australia we now have mandatory RFID.

Three key factors that new tech needs for adoption.

1. Make what I already do easier.
2. Make my decision-making data a byproduct of management.
3. Improve labour efficiency.

Make what I already do easier. For those that were already practicing individual animal management the adoption of the new tech that the CRC was pushing was a game changer. It made life so much easier, but for the average sheep producer to get a pay off they needed to do extra work, and this is a hard sell when they are already time poor.

Make my decision-making data a byproduct of management. An example of this is milk yields in the dairy industry. With the right tech you can gather milk yields through your normal daily routines. Another is yield mapping in cropping, we get all our yield data simply by harvesting our crop. If I want individual fleece data on my sheep it's an extra job. This would be akin to if I wanted cropping yields going out before harvest and hand cutting and weighing small samples through the paddock. An example of this working in sheep is having a pregnancy scanner with RFID capabilities emailing you a file with the pregnancy status of all your sheep. The problem is we still have low uptake of pregnancy scanning, so for your average producer this would be just another job.

Improve labour efficiency. I have lots of sheep handling equipment that enables old me to do what young me used to do wrestling the sheep down the race, but I'm not more efficient. As a sector this means we fall prey to the Baumol effect (https://en.wikipedia.org/wiki/Baumol_effect). This is the tendency for wages in industries with no increases in labour productivity to rise in response to rising wages in other industries. In our area this is livestock wages having to compete with cropping and mining, two industries that have experienced labour productivity increases. This has seen a swing to cropping where land supports it. In sheep according to MLA (2025) the fastest growing breed type is shedding sheep jumping from 4% in 2022 to 8% in 2024. They need less infrastructure and labour to run. I see this as a one-off jump in labour productivity; it is not driven by continual technological adaptation. The Livestock industry needs technologies that deliver labour productivity gains, the Sheep CRC individual management program was delivering the opposite. Yes, it made your sheep more profitable, but this didn't offset the labour productivity gains achieved by other land use sectors.

DISPLAYING DATA & INSIGHTS

As a ram breeder I work with both scientists/researchers and commercial breeders. Over the years I have spent a lot of time thinking about how to display and communicate all the detailed genetic information that's currently available. Your average commercial breeder is only thinking about about genetics once a year when buying their rams. This information overload can be daunting. I built into my database a graphical reporting system so that commercial breeders could read in plain English the strengths and weaknesses of each animal. Figure 1 was my first attempt, displaying 20 breeding values and their percentile bands. It was at this at this stage I found a book dedicated to just this topic. "The Visual Display of Quantitative Information, by Edward R. Tufte" (If you get the chance, it's a cracking read). One of the takeaway concepts I picked up was the ink to information ratio of your graphic. More information and less distractions.

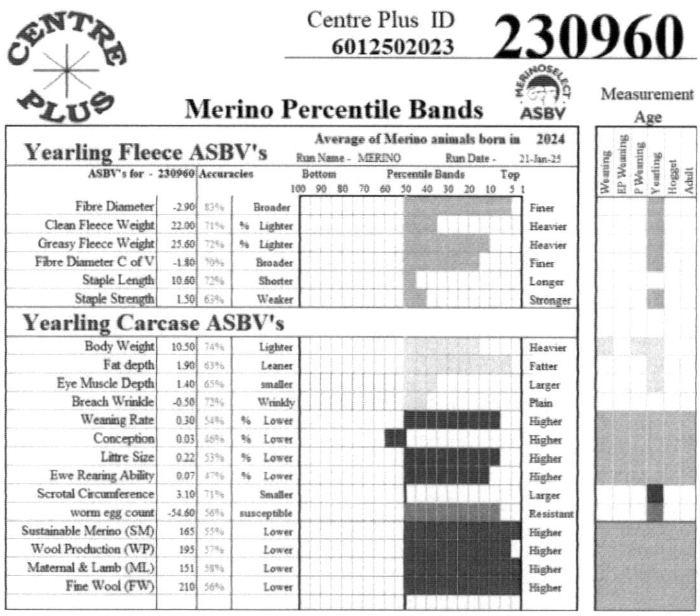

Figure 1. Breeding values and percentile band initial display

This led to the development of Figure 2 with 38 breeding values. Red is below industry average and blue is above. The position of the darker shading is how far above or below the average for each trait. To my surprise this changed how some breeders were selecting their animals.

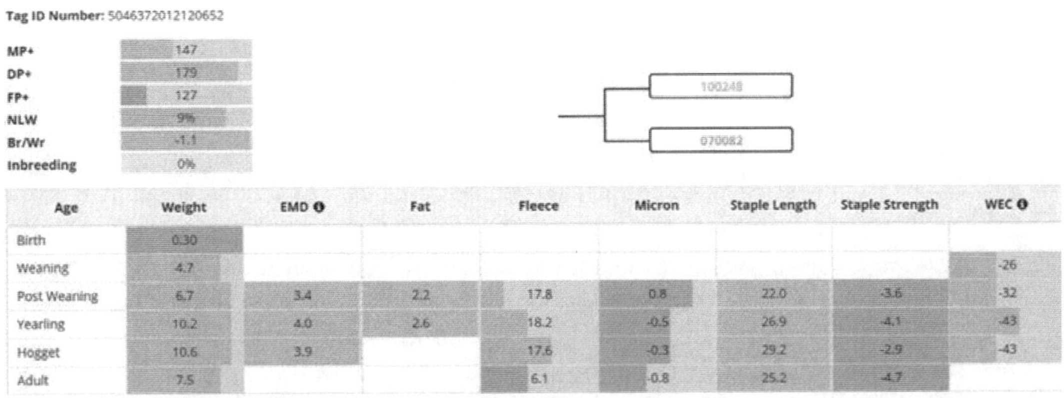

Figure 2. Revised display of breeding values

It was instantly obvious where each animals' weakness was. I found that breeders would move past animals that filled their core requirements to find an animal that just didn't have a weakness even though it would make them less money. This was even more obvious when I made an interactive version of the graphic. I would have customers asking for combinations of traits that rarely existed in one animal. To best show this, I built an interactive version so that customers could set their targets for core traits and the graphic would show them the trade-offs for all other traits. Customers would tweak the selection tool until they had a ram team with the least amount weaknesses. They were trading off gains in core economic traits to avoid losses in lower economic traits. It turns out this is a well understood concept in behavioural economics and psychology call loss aversion. Kahneman and Tversky (1979) showed that the psychological impact of a loss is roughly twice as powerful as a gain of the same size. With this new understanding of loss aversion, I re-worked my selection tool with the customer choosing one of the Sheep Genetics indexes that best matched their production system. Because the Sheep Genetics indexes are build using an economic model, I am able to show the customer how many dollars per breeding ewe their selections were costing them. I now find that customers are not willing to give up more than a few dollars per breeding ewe to get the mix of traits they are looking for.

REFERENCES

Brown D.J.; Swan A.A and Mortimer M.M. (2011) *Proc. Assoc. Advmt. Anim. Breet. Genet.* **19**: 231.

Kahneman D. and Tversky A. (1979) *Econometrica* **47**: 263.

Meat & Livestock Australia (2025) Strong growth in the uptake of sheepmeat genetics https://www.mla.com.au/news-and-events/industry-news/strong-growth-in-the-uptake-sheepmeat-genetics/

JOHN VERCOE MEMORIAL LECTURE

Initiated in 2007 by the 17th AAABG Committee, in honour of Dr J.E. (John) Vercoe (1936 – 2005), AM, BSc(Agric)(Hons)(Melb), MAgriSc (Hons)(Melb), PhD(Melb), AATS&E, FAIAS, FASAP. John was an Honorary Professor at the University of Central Queensland and an Honorary CSIRO Research Fellow.

John's research contribution to the breeding and management of cattle production in tropical and other stressful environments and belief in the value of interdisciplinary research to solving industry problems significantly improved the productivity of the Northen Australian beef industry. His other passion was international agriculture and John was honoured with executive roles with international research organisations for his contributions to international efforts aimed at reducing world poverty by exploiting the opportunities available from improved livestock production in developing countries. He also mentored many early career scientists both within Australia and internationally.

The Vercoe Lecturers have been:

2007	C.P. Van Tassell
2011	A.L. Van Eenennaam
2015	K. Marshall
2017	H.M. Burrow
2019	D.J. Johnston
2021	B.J. Hayes
2023	P. Boettcher
2025	R.G. Banks

TABLE OF CONTENTS

Genomic Prediction 2

Sustainability Methane Prediction

Proc. Assoc. Advmt. Anim. Breed. Genet. 26: 1-5

GLOBAL REGULATIONS GOVERNING GENOME EDITING IN FOOD ANIMALS

A.L. Van Eenennaam

Department of Animal Science, University of California -Davis, California, 95616
United States of America

SUMMARY

With the emergence of genome editing (GnEd), many countries are considering regulatory approaches that allow GnEd organisms that could have been developed through conventional breeding to be regulated under the same rules as conventionally-bred organisms, rather than as genetically modified organisms (GMO). However, this approach has not been universally adopted, and given the importance of international trade in agricultural products, the most risk-adverse policy (e.g. all GnEd animals are *a priori* regulated as GMOs, irrespective of the nature of the edit) may end up becoming the de facto global standard. This is a form of "risk colonialism", whereby precaution-based regulations to a biotechnology in one country effectively precludes or disincentives the adoption of that biotechnology by trading partners for fear of losing market access.

INTRODUCTION

The purported purpose of animal biotechnology regulations is to protect human and animal health and the environment. The high cost and unpredictable timeline for the approval of genetically modified organisms (GMOs) have effectively forestalled the commercialization of animal biotechnology products, especially in Europe (Lubieniechi *et al.* 2025). Despite active research on introducing valuable traits via recombinant DNA technologies in the late 1990s and early 2000s, only two genetically engineered food animals have been approved for food use in the world; the AquAdvantage Atlantic salmon, and the GalSafe pig, and neither can be grown on conventional farms. There are substantial opportunity and economic costs associated with delaying the adoption of safe breeding innovations in animal breeding programs (Van Eenennaam *et al.* 2021).

GnEd refers to making a targeted alteration to the genome using site-directed nucleases. Typically, this involves introducing a double stranded break (DSB) in the DNA helix which can then be repaired by the non-homologous end joining (NHEJ) pathway, or in the presence of a nucleic acid repair template flanked by DNA sequences homologous to the sequence on either side of the DSB, using the homology-directed repair (HDR) pathway. The latter can involve a template that contains DNA sequences from sexually-compatible organisms (cisgenic), or from other species (transgenic). Animals that are descended from a GnEd parent that do not inherit the genetic edited sequence themselves are called null segregants (e.g. 50% of offspring from a hemizygous animal).

The current status of global regulations of GnEd animals is summarized in Table 1 and was largely drawn from Wray-Cahen *et al.* (2024). Some regions are using an arbitrary classification of genome edits categorized as site-directed nuclease (SDN) -1,2,3 differentiated by the size and nature of the edit. SDN-1 edits are defined as those that involve simple NHEJ deletions and insertions and no nucleic acid repair template. Distinctions between the remaining SDN classes are not clear-cut, notably between "short" and "long" nucleic acid repair templates, and "novel" and "foreign" DNA.

Table 1. Global approaches to regulation of genome editing (GnEd) animals, and whether GnEd food animal applications have been received. Modified from Wray-Cahen *et al.* (2024)

Country	Policy	No nucleic acid template	Short[a] nucleic acid template	Long nucleic acid template	Foreign DNA Synthetic/ Transgenic	Null Segregant	Decisions made for GnEd Animal?
		Central and South America					
Argentina	Resolution 21/2021	Not GMO	Not GMO	Likely not GMO[b]	GMO	Not GMO	Yes
Brazil	Normative Resolution 16	Not GMO	Likely not GMO	Likely not GMO	GMO	Not GMO	Yes
Colombia	Resolution No. 22991	Not GMO	Not GMO	Likely not GMO	GMO	Not GMO	Yes
Uruguay	Decree No. 84/024	Not GMO	Likely not GMO	Likely not GMO	GMO	Not GMO	No
		North America					
Canada	Novel Foods Laws (product based)	Method of genetic modification does not determine whether safety assessment is required; "novelty" of product is regulatory trigger for pre-market assessment					No
USA	Coordinated Framework (GFI #187 A/B)	Intentional Genomic Alterations (IGAs) evaluated on case-by-case basis. May grant "enforcement discretion" for low-risk products				Not GMO	Yes
		Asia and Oceania					
Australia	Environmental release	Not GMO	GMO	GMO	GMO	Not GMO	No
Australia/ New Zealand	Proposal P0155 (FSANZ)	Not GMO	Likely not GMO	Likely not GMO	GMO	Not GMO	No
New Zealand	Current legislation	GMO	GMO	GMO	GMO	Not GMO	No
	Proposed legislation	Not GMO	Risk-tiering system proposed		GMO	Not GMO	
Indonesia	Draft proposal	Not GMO	Not GMO	Likely not GMO	GMO	Not GMO	
Japan	MAFF – environment	Not GMO	GMO	GMO	GMO	Not GMO	
	MAFF- Animal Products	Not GMO	Likely not GMO	Likely GMO	GMO	Not GMO	Yes
	Ministry of Health, Labor and Welfare	Not GMO	Likely not GMO	Likely GMO	GMO	Not GMO	
		Africa and Middle East					
Ghana, Kenya, Malawi		Not GMO	Not GMO	Not GMO	GMO	Not GMO	No
South Africa	GMO Act of 1997	GMO	GMO	GMO	GMO	Undetermined	No
		Europe					
EU	Directive 2001/18/EC	GMO	GMO	GMO	GMO	Not GMO	No
	Proposed	Not GMO	GMO	GMO	GMO	Not GMO	No
Norway	Proposed	Not GMO	Expeditated Assessment		GMO		
UK	Awaiting establishment of animal welfare provisions	Not GMO	Not GMO	Not GMO	GMO	Not GMO	No

[a] Defined by some countries as no more than 20 bp [b] Perfect allele replacement

The GnEd food animal applications that have undergone a regulatory process in at least one country are listed in Table 2 (Retrieved from https://www.isaaa.org/animalbiotechdatabase, 4/1/2025). These applications do not include a "new combination of genetic material", also known as transgenic DNA. Food products from three GnEd fish species have already reached the consumer marketplace in Japan. Kyoto University-based start-up, Regional Fish Co., Ltd., started selling GnEd red sea bream, tiger pufferfish, and olive flounder shortly after these applications were first reported in the scientific literature in 2018 and 2019, following determination by Japanese regulatory authorities that the deletions that occurred in the target gene in these GnEd fish were 'non-GMO'.

Table 2. GnEd animals that have undergone a regulatory determination in different countries

Country	Common name	Trait	Developer	Gene Targeted	Year
Argentina	Nile Tilapia	Increased yield	AquaBounty	Myostatin	2018
	Beef cattle	Heat tolerance	Acceligen	Prolactin receptor	2020
	Dairy cattle	Heat tolerance /Polled	Acceligen	Prolactin receptor /Pc Polled allele	2020
	Cattle	Increased yield	Acceligen	Myostatin	2021
	Other species (?)	Unknown as disclosure is not needed for non-GMO products			
Brazil	Nile Tilapia	Increased yield	AquaBounty	Myostatin	2019
	Beef cattle	Heat tolerance	Acceligen	Prolactin receptor	2021
	Dairy cattle	Heat tolerance	Acceligen	Prolactin receptor	2023
	Cattle	Increased yield	Acceligen	Myostatin	2021
	Pig	PRRS[a]-resistance	Genus, plc	CD-163	2024
Colombia	Pig	PRRS-resistance	Genus, plc	CD-163	2023
Japan	Red Sea Bream	Increased yield	Regional Fish	Myostatin	2021/22
	Tiger Pufferfish	Faster growth	Regional Fish	Leptin receptor	2022
	Olive Flounder	Faster growth	Regional Fish	Leptin receptor	2023
USA	Beef Cattle	Heat tolerance	Acceligen	Prolactin receptor	2022

[a] Porcine Reproductive and Respiratory Syndrome (PRRS) virus.

DISCUSSION

The introduction of useful GnEd alleles into livestock breeding programs will require a regulatory approach that is fit-for-purpose, meaning that safe products can come to market in a timely fashion and the regulatory evaluation timeline is compatible with the pace of the breeding program. As can be seen in Table 2, countries that are treating simple edits with no template guided repair as analogous to conventionally-bred animals, have already made regulatory decisions that allow products to come to market. Interestingly, the commercialization of the three GnEd fish species in Japan did not result in the anticipated consumer pushback. Efforts by groups opposed to GnEd did not gain traction, and media coverage was mostly positive (Matsuo and Tachikawa 2022).

Genus, plc has announced plans for its subsidiary the Pig Improvement Company, to introduce the PRRS-resistance allele lacking exon 7 of CD163 into its four elite grandparent swine breeding lines (two maternal & two paternal) in the homozygous condition, with the ultimate objective of producing commercial pigs that are resistant to Porcine Reproductive and Respiratory Syndrome (PRRS) virus (Burger *et al.* 2024). A requirement of any breeding program is to maintain sufficient genetic diversity for future genetic progress, meaning that in this case the edit had to be introduced into multiple founder breeding boars and gilts (10–15 per line). Illumina short amplicon sequencing was employed to screen 435 founder animals to identify the 90 (35 boars, 55 gilts) that had the intended edit, as well as screen for other unintended alterations within, or immediately adjacent to, the targeted CD163 locus. This was complemented by Oxford Nanopore Technologies sequencing of 2.8 kb PCR amplicons to evaluate the structural integrity around the target. Characterization of the intended edit, along with robust phenotyping of GnEd animals for production traits and in this specific example the PRRS disease resistance phenotype, was undertaken by the breeding company.

Unique to GnEd animal evaluations is this requirement to identify potential off-target mutations, defined as an edit at a location on the genome that is not the intended target. Even though Genus, plc is a large multinational company, using whole genome sequencing (WGS) on this large number of founder animals was not seen to be cost-effective off-target screening tool for large numbers of founders. Therefore, the SITE-Seq® assay was used to identify a total of 182 putative off-target sites associated with the dual guide RNAs used in to delete exon 7 of CD163. This was validated with WGS on a small number (n=24) of edited pigs and their progeny. WGS identified 63 *de novo* mutations in GnEd animals, compared with 80 in wild type animals. Given the GnEd pigs were otherwise healthy, it is unclear what purpose is served by requiring the identification of de novo mutations in GnEd animals, when such analyses are not required for conventionally-bred animals.

The value and purpose of de novo variation identification in GnEd animals needs to be given serious consideration, given it is difficult to envisage a path to harm associated with de novo mutations that do not manifest as a phenotype. Genetic variation *per se* in an otherwise healthy animal is not a food safety hazard. Further, animal breeders have been safely making genetic improvement by selecting on uncharacterized de novo mutations for centuries. We reported that unedited commercial rams in our campus flock had on average 5.1 million naturally-occurring SNP and indel mutations relative to the Rambouillet sheep reference genome (Mahdi *et al.* 2025). Similarly, the 1000 bull genome project sequenced 2,703 cattle of different breeds and found 84 million SNPs and 2.5 million naturally-occurring indels (Hayes and Daetwyler 2019). This variation drives genetic progress in breeding programs, and animals produced by conventional breeding methods are not routinely evaluated for de novo mutations at the molecular level. The food safety benefit resulting from expensive and time-consuming off-target analysis on the genomes of healthy GnEd animals is therefore unclear against this well-documented background of genetic variation.

The livestock breeding community is uniquely positioned to provide substantive and evidence-based comments regarding proposed regulatory approaches and data collection requirements for GnEd animals for agricultural purposes. Necessitating the documentation of all potential off-target edits uniquely for GnEd animals in the absence of a path to harm will increase the cost of using this technology in commercial breeding programs, with no clear benefit to public health and safety.

CONCLUSION

Many countries are currently developing regulatory frameworks for GnEd animals and their products. History shows that the expensive and overly precautionary regulatory approaches that were associated with GMO animals were not fit-for-purpose and had a chilling effect on investment and adoption of this technology in animal breeding programs. The animal breeding community has an opportunity to provide evidenced-based input to help ensure that emergent regulatory frameworks for GnEd animals are science-based, risk-proportionate, and focused on novel hazards associated with GnEd. Well-considered policies should enable safe products to reach the market, and permit the cost-effective incorporation of biotechnologies like GnEd into breeding programs and timelines.

ACKNOWLEDGEMENTS

I would like to acknowledge the support of AAABG in providing travel funding to allow me to attend the 26th AAABG conference in New Zealand. My program receives funding from USDA Foreign Agricultural Service Cooperative Agreement FX23TA-10960C036 “Virtual and in-Person Workshops on Regulatory Aspects of Genome Editing in Animal Biotechnology Development”

REFERENCES

Burger B.T., Beaton B.P., Campbell M.A., *et al.* (2024) *CRISPR J.* **7**: 12.

Hayes B.J. and Daetwyler, H.D. (2019) *Annu Rev Anim Biosci.* **7**: 89.
Lubieniechi S.A., Van Eenennaam A.L. and Smyth S.J. (2025) *Trends Biotechnol.* **43**: 511.
Mahdi A.K., Fitzpatrick D.S., Hagen D.E., *et al.* (2025) *CRISPR J.* **8**: 13.
Matsuo M. and Tachikawa M. (2022) *Front. Genome Ed.* **4**: 899154.
Van Eenennaam A.L., De Figueiredo Silva F., Trott J.F. and Zilberman D. (2021) *Annu. Rev. Anim. Biosci.* **16**: 453.
Wray-Cahen D., Hallerman E. and Tizard M. (2024) *Front. Genome Ed.* **6**: 1467080.

GROWING AOTEAROA NEW ZEALAND'S AQUACULTURE INDUSTRY THROUGH BREEDING AND GENOMICS: PROGRESS, CHALLENGES AND FUTURE OPPORTUNITIES

J.E. Symonds[1], M.R. Scholtens[1], K.G Dodds[2], S.M. Clarke[2], R. Costilla[1], S.P. Walker[1], N. King[1], N.L.C. Ragg[1], M. Camara[3], J.A. Ericson[1], J. Vignier[1], M. Wellenreuther[4,5], Z. Waddington[6], J. Bailey[7], R.D. Roberts[8], E. Malpot[9], M.L. Tate[10] and T. Slattery[11].

[1] Cawthron Institute, Nelson 7010, New Zealand
[2] AgResearch, Invermay Agricultural Centre, Mosgiel 9053, New Zealand
[3] National Institute of Water & Atmospheric Research Ltd (NIWA), Northland Marine Research Centre, Ruakaka 0116, New Zealand
[4] The Plant and Food Research Institute, Nelson, New Zealand
[5] School of Biological Sciences, University of Auckland, Auckland, New Zealand
[6] The New Zealand King Salmon Co. Ltd, Picton 7220, New Zealand
[7] Mount Cook Alpine Salmon, Twizel, New Zealand
[8] Shellfish Production and Technology New Zealand Ltd. (SPATnz), Nelson, New Zealand
[9] Moana New Zealand, Glenduan, Nelson 7071, New Zealand
[10] Natural Food Innovation, Sefton 7445, New Zealand
[11] Sanford Limited, Kaitangata 9281, New Zealand

SUMMARY

The aquaculture industry in Aotearoa New Zealand recognises that the application of selective breeding is an essential part of its business and is an important way to generate cumulative production and product quality gains. Many of the larger finfish and shellfish companies have established in-house breeding programmes and selected juveniles can also be accessed from breeding programmes established by research organisations. Three salmon farming companies are applying genomic selection, including a smaller company where genomic selection based on genotyping-by-sequencing is cost-effective. Within these programmes, significant genetic gains have been demonstrated across multiple generations. New traits such as thermotolerance and summer survival, needed to combat climate change, have been identified as new priorities for selection in finfish and shellfish. Families have been phenotyped and estimated breeding values generated to accelerate selection for climate change resilience. Two new finfish species, kingfish and snapper, are also being developed for aquaculture with multi-generation breeding programmes already in place. Selective breeding enhanced by genomic selection provides the industry with an essential climate change adaptation tool to ensure future survival and growth.

INTRODUCTION

Selective breeding (i.e. artificial selection), has made a significant contribution to the growth of Aotearoa New Zealand's aquaculture industry (Camara and Symonds 2014; Symonds *et al.* 2019). The first family-based breeding programmes were established in the 1990s and have been the foundation for improved stock management and significant production gains in all three key farmed species (Pacific oysters (*Crassostrea gigas)*, Greenshell™ mussels (*Perna canaliculus)* and King (Chinook) salmon (*Oncorhynchus tshawytscha*)). A genomic-based selective breeding programme for recently commercialised yellowtail kingfish (haku in te reo Māori) (*Seriola lalandi*) has also been established, and pre-commercialisation selective breeding of Australasian snapper (tāmure in te reo Māori) (*Chrysophrys auratus*) is underway. In contrast to terrestrial livestock, fish and shellfish have very high fecundity, shorter generation intervals (2-3 years), and relatively high heritabilities for economically important traits. These are key factors which define the design of breeding programmes in aquaculture. Breeding objectives have become more complex over the years with harvest quality traits and survival, as well as growth, all now being part of the selection process. Marine heatwaves are occurring more frequently around coastal Aotearoa New Zealand (Montie *et al.* 2024) and are often considered to be an important causal factor for summer mortality (Copedo *et al.* 2023; Ericson *et al.* 2023). These warming waters have generated new challenges for the aquaculture industry (Cummings *et al.* 2021; Greenhough *et al.* 2025) and for some companies heat tolerance is now a key trait in their breeding objectives. To enhance selection and provide more accurate gains for critical climate resilience traits, genomics resources are now available for all dominant farmed species and are applied by multiple salmon farming companies. This paper summarises the history of aquaculture selective breeding in Aotearoa New Zealand, discusses the successes and future challenges, and shows how selective breeding and genomics can help overcome these to enable future growth.

MATERIALS AND METHODS

Programme design, breeding objectives and selection strategies. For salmon, the first breeding programme was established in 1994 and was based on the production of full and half-sib families. Since then, two additional commercial salmon breeding programmes have been established, based on family or individual evaluation and genomic selection. A typical family-based breeding programme design for salmon is shown in Figure 1.

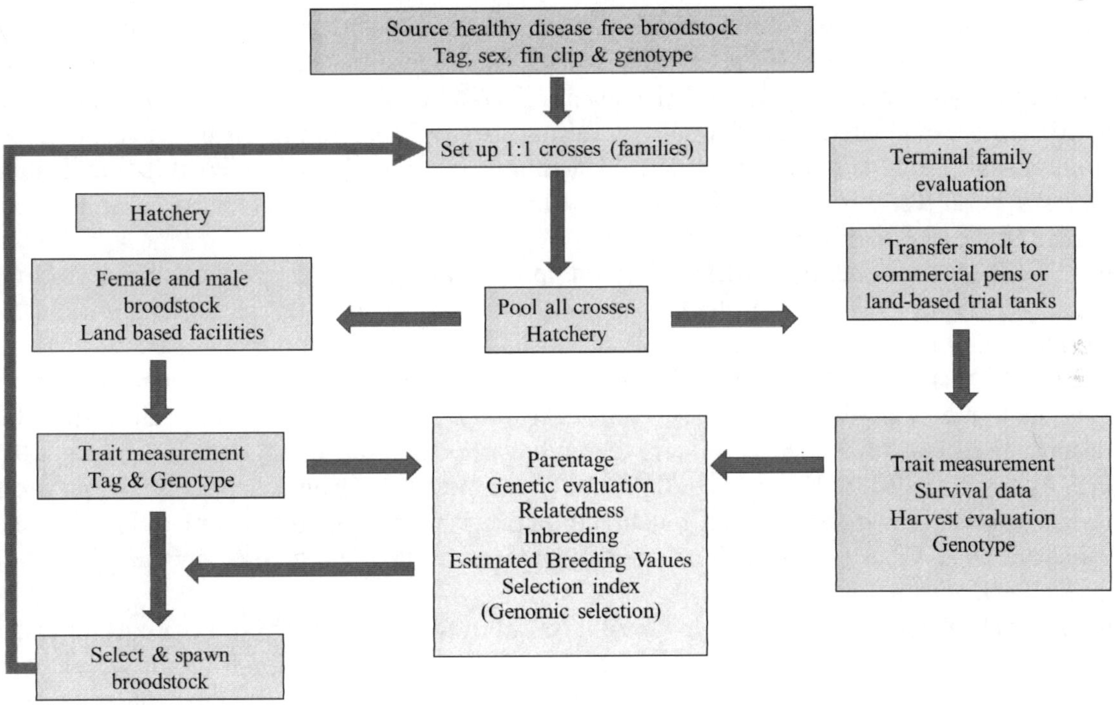

Figure 1. Example outline of a salmon breeding programme

Growth, usually based on weight at harvest or weight prior to spawning, is a priority trait for all three salmon programmes. In the family-based programmes quality traits are also included such as fillet colour and fat. More recently survival traits and thermotolerance have been introduced as priority breeding objectives. Thermotolerance (specifically, the ability to endure elevated water temperatures) is assessed in tank-based temperature challenges using families, and/or by evaluation of the survival of the same families over summer in a farm-based sea pen (Symonds *et al.* 2025).

Greenshell™ mussels (GSM), have been selectively bred since 2002 using family-based selection. The initial traits focussed on yields: shell length, meat weight and meat condition score, and were associated with significant heritabilities and substantial genetic gain. Traits were often evaluated at multiple farm sites to account for genetic x environment interactions. More recently, resilience traits, including thermotolerance to improve survival and growth during periods of elevated temperature, have been investigated and found to be significantly influenced by genetics (Delorme *et al.* 2024, Ericson *et al.* 2024; Scholtens *et al.* 2025). Phenotypic selection for GSM thermotolerance is possible utilising tank-based challenges and/or field assessments of summer survival.

The Cawthron Institute initiated a family-based Pacific oyster breeding programme in 1999 with the initial objectives of breeding for fast-growing oysters with good shell shape and a high meat-to-shell ratio using a combination of between- and within-family selection (Camara and Symonds 2014). In 2010, sudden mass mortalities of both cultured and wild oyster populations on the North Island and in Cawthron's oyster hatchery on the South Island, were caused by a highly pathogenic variant of the oyster herpes virus (OsHV-1 μ-var). This led to a shift in the programme's objectives and strategies, with surviving broodstock selected from field trials at OsHV-1infected sites. While OsHV-1 has become naturalised in most New Zealand oyster populations, commercial survival rates have rebounded thanks, substantially, to selective breeding. The current breeding strategy now combines survival, growth and quality traits, such as shell shape. Families are typically evaluated on multiple sites to account for genetic x environment interactions.

Genetic parameter estimation. Methods for evaluating key traits for the three major species have previously been published (Camara and Symonds 2014; Camara *et al.* 2017; Symonds *et al.* 2018, 2019, 2025; Gutierrez *et al.* 2020; Ericson *et al.* 2023; Scholtens *et al.* 2022, 2023, 2025). The number of families evaluated typically ranges from 20 to 135 depending on the species. Genetic models are usually fitted using Restricted Maximum Likelihood as implemented in standard software such as ASReml (Gilmour *et al.* 2009). More details of the statistics and animal models can be found in the publications above.

Genotyping-by-sequencing (GBS) for commercial and research evaluations is routinely conducted by GenomNZ (AgResearch) for salmon (Symonds *et al.* 2019; Scholtens *et al.* 2022). These evaluations include families and/or broodstock in two commercial salmon breeding programmes (Scholtens *et al.* 2023). The application of GBS includes parentage assignment, estimating genomic relatedness matrices, inbreeding, and measures of genetic diversity. GBS methods have also been developed for selective breeding programmes in kingfish (Jacob *et al.* 2025), for GSM (Ashby *et al.* 2018; Hess *et al.* 2018) and Pacific oysters (Vignier *et al.* 2025) but are not yet routinely applied in commercial shellfish breeding programmes. All these applications make use of the cost-effective GBS methods developed by AgResearch, implemented in the open-source software KGD (Dodds *et al.* 2015).

Genomic selection (GS) is employed by all salmon commercial breeding programmes, the snapper and the kingfish programmes. This allows for within family selection for traits which cannot be measured on selection candidates (e.g. slaughter traits). Because kingfish can only be group-spawned, selection is based on individual genomic breeding values (GEBVs). Optimal contribution methods are used to manage long-term inbreeding and genetic diversity in genomic and pedigree-based selection programmes. The benefits of genomic selection have also been assessed in Pacific oysters using a medium-density SNP array for oysters for improvement of resistance to OsHV-1 (Gutierrez *et al.* 2020).

Selection indices. Estimated breeding values (EBVs) are calculated for all candidate broodstock using either pedigree-based Best Linear Unbiased Prediction (BLUP) or Genomic-BLUP methodologies. The analysis is based on individual data recorded on previously selected parents and selection candidates (e.g. harvest weight, deformities), family-level survival based on mortality during field or tank challenges, and family-level quality traits based on measurements taken from sacrificed animals. Trait prioritisation and economic weightings are used to develop selection indices determined by the commercial companies and their genetics providers.

Kingfish. After over a decade of R&D to develop reliable hatchery and production systems, the National Institute of Water and Atmospheric Research (NIWA) recently initiated demonstration-scale commercial production and a genomics-based genetic improvement programme to improve production in yellowtail kingfish. The programme initially used a selection index that included growth rate and skeletal deformities and aims to supply genetically elite larvae, fry, and broodstock to both the fledgling NZ industry and the rapidly growing, global industry. Currently the focus is on performance in land-based recirculating aquaculture systems (RAS). RAS provides increased environmental control in the face of climate change, enhanced biosecurity, and absolute escape prevention that eliminates genetic interactions between selectively bred and wild stocks. The programme has also assessed the potential for genotype-by-environment interactions (Jacob *et al.* 2025). The mating design is based on a rotational breeding scheme using four-year cycles in which four-year-old males are mated with five-year old females to minimise the generation interval and conserve genetic diversity.

Snapper. Due to its commercial, recreational, and cultural importance, a breeding programme was initiated by Plant & Food Research using broodstock collected between 1994–2006. F1 populations were produced from 2004–2009, followed by F2 in 2013, which underwent domestication selection for survival and reproduction in captivity. A major shift came in 2016 with genomics-informed breeding, enabling full pedigree reconstruction and genomic breeding values for moderately heritable traits like length and weight (Ashton *et al.* 2019a). To support breeding, genomic resources were developed including a genome (Catanach *et al.* 2019, Blommaert, *et al.* 2024), linkage map (Ashton *et al.* 2019b), transcriptome (Wellenreuther *et al.* 2019), SNP chip (Montanari *et al.* 2023) and QTL identification (Ashton *et al.* 2019b; Ruigrok, *et al.* 2022, Sandoval *et al.* 2022). The F3 (2018) and F4 (2021) generations were selected using genomic data, marking the onset of precision breeding focused on growth and production traits.

RESULTS AND DISCUSSION

Salmon. Most of the key production traits for salmon are moderately to highly heritable (Symonds *et al.* 2000, 2019; Scholtens *et al.* 2022, 2023) and have shown improvement after generations of selection since the mid-1990s. For example, growth is no longer production limiting for The New Zealand King Salmon Co. Ltd. (NZKS) (Zac Waddington, *unpublished data*). Some unfavourable genetic correlations have been uncovered, for example between harvest weight and body fat, requiring incorporation of fat into the selection index to avoid increased fat content (Symonds *et al.* 2019). Higher body condition has also been found to be a risk factor for spinal curvature (Scholtens *et al.* 2022). Recent evaluation of thermotolerance (Symonds *et al.* 2025) has shown that survival and time to death during a temperature challenge are both moderately heritable, demonstrating the potential to improve thermotolerance through selection. The aim is to apply between and within family genomic selection to increase selection accuracy and improve gains for these thermotolerance traits, now a priority for NZKS due to increasing ocean temperatures.

Greenshell mussel family evaluation also demonstrated the potential for genetic gain in multiple production traits (Camara and Symonds 2014). Selective breeding combined with the development of improved hatchery technologies for juvenile production has dramatically improved production for one of the largest GSM farming companies (SPATnz, 2020). There is now significant interest by the wider industry in investing in new GSM hatcheries enabling the development of new breeding programmes and improving industry growth and resilience, and reducing reliance on wild caught spat.

With the challenge of increasing summer mortality in GSM, the recently determined estimated high heritability (0.48±0.10) for survival under simulated severe heatwave conditions was very encouraging, suggesting that if heat tolerance is included in a selection index, there is adequate genetic variation to allow genetic improvement in this trait (Scholtens *et al.* 2025).

Pacific oysters. Selective breeding to improve resistance to OsHV-1 has been successful (Camara *et al.* 2017) and survival is no longer the major threat that it was in 2010. Increasing emphasis is being placed on selecting traits such as tolerance to low salinity and applying genomic selection. The oyster family programme has developed from small scale research to a commercial programme utilised by New Zealand's largest oyster farming company to produce superior farmed product. There is increasing interest and investment in building new land-based oyster hatcheries that is also enabling further establishment of family-based breeding programmes.

Kingfish. The heritabilities of weight, length, and morphological development in juvenile yellowtail kingfish exposed to current and projected water conditions representing future global warming and ocean acidification is low but significant (Munday *et al.* 2019). Table 1 (Camara, *unpublished data*) presents estimates of genetic parameters for three production traits. All traits are sufficiently genetically variable to support a response to selection, but there is an unfavourable genetic correlation between growth (weight) and the probability of deformities that must be taken into account in the selection index to prevent an undesirable correlated response to selection on growth.

Table 1. Genetic parameter estimates (se) for kingfish traits grown in RAS. Heritability (bold) on the diagonal, genetic correlations below the diagonal, and phenotypic correlations above the diagonal.

Trait	Tagging weight	Harvest weight	Deformities at harvest
Tagging weight	**0.433** ± 0.039	0.302 ± 0.028	-0.048 ± 0.023
Harvest weight	0.325 ± 0.069	**0.560** ± 0.033	0.021 ± 0.023
Deformities at harvest	-0.020 ± 0.115	0.165 ± 0.103	**0.102** ± 0.020

Snapper. In 2024, the F5 generation of snapper was created, which shows faster growth, higher survivorship, and improved FCRs (Moran *et al.* 2023; Samuels *et al.* 2024). Heritabilities for the three growth-related traits, fork length, peduncle length, and weight, were similar, averaging 0.27 in the first year and declining to 0.10 by the third year (Ashton *et al.* 2019a). Snapper exhibit rapid growth at optimal temperatures between 18–21 °C, while growth slows significantly and nearly ceases at temperatures below 14 °C. The overarching aim is to develop a fast-growing, robust snapper strain, mirroring gains achieved in red sea bream (*Pagrus/Chrysophrus* major) and gilthead sea bream (*Sparus auratus*), to support a sustainable and commercially viable warm water resilient aquaculture industry in New Zealand.

Challenges and future opportunities. The three main aquaculture species encountered significant challenges during the first 20 to 30 years as the industry developed. The emerging RAS-based kingfish industry and anticipated snapper industries will likely follow this pattern. Selective breeding helped overcome these challenges, from improving OsHV-1 resistance to enhanced growth and product quality despite some unfavourable genetic correlations. More recently new challenges have emerged as climate change has exacerbated the frequency of marine heat waves, reduced salinity due to freshwater run off during floods and increased pathogen risks, accelerating the need to breed for climate change resilience, including thermotolerance and disease resistance. Researchers and industry are developing relevant tank and farm-based challenges to define and measure the accurate resilience phenotypes required for successful selection. The increasing application of genomic selection over the next few years will improve genetic gains across the industry.

CONCLUSIONS

Successful selective breeding programmes have demonstrated genetic gains for all three main commercial aquaculture species either through centralised programmes or in-house programmes operated by individual companies. The priority traits that are selected continue to evolve, especially as new challenges emerge such as increasing temperatures and new diseases. Breeding programmes are seen as key to the future adaptation of the industry to climate change. Genomic selection will become more widespread over the next few years as the additional gains and accuracy it provides become evident. However, breeders must stay ahead of the game, by selecting for the uncertain future and not just for business as usual.

ACKNOWLEDGEMENTS

Salmon thermotolerance evaluations were funded by the NZ Government Ministry for Business Innovation and Employment (MBIE) (CAWX2304), SFFF (project 220041) and NZKS. Xelect Ltd. helped with the salmon family thermal challenge evaluation as part of SFFF project 220041. Much of the shellfish work was funded by Cawthron's MBIE Shellfish Aquaculture Research Platform (CAWX1801) and also by the commercial companies operating the breeding programmes. Thanks to the GenomNZ lab (www.genomnz.co.nz) for their genotyping services. The AgResearch MBIE funded programme C10X1306 funded the development of GBS in a variety of species, including NZ aquaculture species. Snapper breeding work was supported through MBIE C11X1603 and thermotolerance evaluation through CAWX2304. GSM breeding research was supported by the Ministry for Primary Industries through the SPATnz Primary Growth Partnership Programme. The kingfish work was supported through MBIE's Strategic Science Investment Fund administered by NIWA.

REFERENCES

Ashby R., Hess A., Roberts R., King N., Symonds J., Gemmell N., Dodds K. and Clarke S. (2018) *Proc. World Congress Genet. Appl. Livest. Prod.* **11**: 463.

Ashton D.T., Hilario E., Jaksons P., Ritchie P.A. and Wellenreuther M. (2019a).*Aquaculture* **505**: 190.

Ashton D.T., Ritchie P.A. and Wellenreuther M. (2019b) *G3 Genes Genom. Genet.* **9**: 1027.

Blommaert J., Sandoval-Castillo J., Beheregaray L. and Wellenreuther M. (2024) *Genomics*: **110929**.

Camara M.D. and Symonds J.E. (2014) *N.Z.J. Mar. Freshw. Res.* **48**: 466.

Camara M.D., Yen S., Kaspar H.F., Kesarcodi-Watson A., King N., Jeffs A.G. and Tremblay L.A. (2017) *Aquaculture* **469**.

Catanach A., Crowhurst R., Deng C., Bernatchez C.D.L. and Wellenreuther M. (2019) *Mol Ecol* **28**: 1210.

Copedo J., Webb S.C., Ragg N.L.C., Ericson J.A., Venter L., Schmidt A.J., Delorme N.J. and Alfaro A.C. (2023) *J. Therm. Biol.* **117**.

Cummings V.J., Lundquist C.J., Dunn M.R., Francis M., Horn P., Law C., Pinkerton M.H., Sutton P., Tracey D., Hansen L. and Mielbrecht E. (2021) NZ Aquatic Environment and Biodiversity Report No. **261**.

Delorme N.J., King N., Cervantes-Loreto A., South P.M., Baettig C.G., Zamora L.N., Knight B.R., Ericson J.A., Smith K.F. and Ragg N.L.C. (2024) *Sci Rep* **14**: 19130.

Dodds K.G., McEwan J.C., Brauning R., Anderson R.A., Van Stijn T.C., Kristjánsson T. and Clarke S.M. (2015) *BMC Genomics* **16**: 1047.

Ericson J.A., Delorme N.J. and Ragg N.L.C. (2023) Cawthron Institute Report No. 3914.
Ericson J.A., Laroche O., Biessy L., Delorme N.J., Pochon X., Thomson-Laing J., Ragg N.L.C. and Smith K.F. (2024) *Front. Physiol.* **14**
Ericson J.A., Venter L., Copedo J.S., Nguyen V.T., Alfaro A.C. and Ragg N.L.C. (2023) *Aquaculture* **564**.
Gilmour A.R., Gogel B., Cullis B., Thompson R. and Butler D. (2009) ASReml User Guide Release 3.0. VSN International Ltd, Hemel Hempstead, UK.
Greenhough H., Vignier J., Smith K.F., Brown C.M., Kenny N. J. and Rolton A. (2025) *Sci Total Env.* **964**.
Hess A.S., Dodds K.G., Roberts R.D., King N., Symonds J. and Clarke S.M. (2018) *Proc. New Zeal. Soc. Anim. Prod.* **78**: 66.
Jacob N.K., Camara M., Setiawan A. and Bastiaansen J.W.M. (2025) *Aquaculture* **606**: 742592.
Montanari S., Jibran R., Deng C., David C., Koot E., Kirk C., Bassil N.V., Morrison-Whittle P., Worthington M. L., Pradelles J., Wellenreuther M. and Chagne M. (2023) *G3 Genes Genom. Genet.* **13**: jkad170.
Moran D., Schleyken J., Flammensbeck C., Fantham W., Ashton D. and Wellenreuther M. (2023) *Aquaculture* **563**.
Munday P.L., Schunter C., Allan Bridie J.M., Simon N., Parsons D.M., Pether S.M.J., Pope S., Ravasi T., Setiawan A.N., Smith N. and Domingos Jose A. (2019) *Front. Ecol. Evol.* **7**.
Ruigrok M., Xue B., Catanach A., Zhang M., Jesson L., Davy M. and Wellenreuther M. (2022) *Genes* **13**: 1129.
Samuels G., Hegarty L., Fantham W., Ashton D., Blommaert J., Wylie M.J., Moran D. and Wellenreuther M. (2024) *Aquaculture* **586**: 740782
Sandoval J., Beheregaray L. and Wellenreuther M. (2022) *G3* **12**: jkac015.
Scholtens M., Dodds K.G., Clarke S.M., Walker S.P., Tate M.L., Miller R., Slattery, T. and Symonds J.E. (2022) *Proc. World Congress Genet. Appl. Livest. Prod.***11**: 44.
Scholtens M., Dodds K., Walker S., Clarke S., Tate M., Slattery T., Preece M., Arratia L. and Symonds J. (2023) *Aquaculture* **563**: 738936.
Scholtens M.R., Roberts R., Delorme N.J., King N., Ragg N.L.C. Costilla R. and Ericson J.A. (2025) *Proc. Assoc. Advmt Anim. Breed. Genet.* **26**: *These proceedings.*
SPATnz. (2020) End of Programme Review report.
Symonds J.E., King N., Camara M.D., Ragg N.L.C., Hilton Z., Walker S.P., Roberts R., Malpot E., Preece M., Amer P.R., Helt F.S., Clarke S.M., Dodds K.G, Tate M. and Buxton P. (2018) *Proc. World Congress Genet. Appl. Livest. Prod.* **10**:1035.
Symonds J.E., Scholtens M.R., Dodds K.G., Costilla R., Clarke S.M., Arinez J., Kenny N., Walker S.P. and Waddington Z. (2025) *Proc. Assoc. Advmt Anim. Breed. Genet.* **26**: *These proceedings.*
Symonds J.E., Walker S.P., Amer P.R. and Dodds K.G. (2000). *Proc. 7th International Symp. Genetics in Aquaculture:* 115.
Vignier J., Reardon M., Exton M., Delisle L., Rolton A., Malpot E., Scholtens M., Welford M., Zamora Z., Delorme N., Dunphy B. and Adams S. (2025) *Aquaculture* **596**: 741894.
Wellenreuther M., Le Luyer J., Cook D., Ritchie P.A. and Bernatchez L. (2019). *G3 Genes Genom. Genet.* **9**: 105.

Proc. Assoc. Advmt. Anim. Breed. Genet: 26: 13-17

CLIMATE CHANGE MITIGATION THROUGH GENETIC IMPROVEMENT – A DEEPLY WORTHWHILE CHALLENGE IN THE SPIRIT OF JOHN VERCOE

R.G. Banks

Animal Genetics and Breeding Unit*, University of New England, Armidale, 2351 Australia

INTRODUCTION

It is an unexpected and considerable privilege to have been asked to deliver the John Vercoe Memorial Lecture at this Conference. John Vercoe contributed to livestock development both in Australia and globally, through research that was rigorous, insightful, and practical, and more broadly through his commitment to effective multi-disciplinary collaboration and leadership. That commitment meant that problems that were complex or even wicked (Wittel *et al.* 1973) at the time became tractable and yielded insights that could be implemented by breeders and farmers, and established solid foundations for further progress. Further, he did not shy away from challenging orthodoxy – the push to introduce and use *Bos indicus* genetics in Northern Australia in the face of considerable industry opposition being an outstanding example. Similar courage will be required in tackling climate change, given how deeply embedded in human activity the drivers are of that change and the inevitable extent of vested interests confronted.

This paper addresses the challenge of climate change mitigation through genetic improvement, focussing on animals (but acknowledging that genetic improvement of plants is also targeting methanogenic properties) attempting to highlight elements of how we can tackle that challenge drawing on inspiration from John Vercoe's practical philosophy. The paper does not attempt a review of all activity, research, publications etc with any detail; rather, to consider how genetics can contribute to tackling climate change, to outline key challenges in doing that, and suggest how our genetics community can contribute – in very positive ways.

THE CHALLENGE

Climate change arising from global warming is widely accepted as an existential challenge to humanity and indeed life on this planet (IPCC 2022). The underlying driver of the warming is increasing CO2 concentrations in the atmosphere, which result from a range of sources including burning fossil fuels, deforestation, construction, and others. Contributing also to warming, is methane. Indeed, almost half of recent warming can be attributed to rising emissions of methane from energy, waste and agriculture. Sources of methane emissions include those from enteric fermentation in the rumen, which are estimated to contribute nearly 30% of anthropogenic methane.

Methane emission from enteric fermentation in the rumen are potentially able to be reduced, since methane represents unused energy consumed by the animal. The questions for focus then become:

- How might methane emissions be reduced i.e. by what methods or approaches?
- How much reduction might be possible?
- What costs are likely to be involved?
- What are the prospects of success, and what is needed to ensure success?

Together, the threats generated by climate change, the complexity of the economic or activity systems contributing to it – including the ubiquitous dependence of human activity on conversion of energy, and the uncertainty around the range of possible approaches, mean that climate change

---

* A joint venture of NSW Department of Primary Industries and Regional Development and the University of New England

mitigation can be rightly considered a complex, or more probably, a "wicked" problem (Wittel *et al.* 1973).

Reducing emissions from enteric fermentation must focus on the anaerobic microbes in the rumen – the Archaea, and in principle either reducing their abundance and/or activity, and/or preventing/inhibiting them from generating methane. The main approaches to achieving these outcomes are:

- Supplementation of diets with molecules, compounds or feedstuffs that have these effects
- Developing vaccines against the specific microbial species
- Breeding animals that generate less methane

Assuming that one or more of these can be developed and deployed, they must be able to be deployed at very wide scale, and it is almost inevitable that interactions and synergies between them will arise and need to be understood.

Immediately we can see links back to themes of John Vercoe's life – the willingness to tackle complex or even wicked problems, the need for rigorous and insightful science coupled with the imperative for practical outcomes, and the need for genuine multi-disciplinary approaches.

GENETICS AS AN APPROACH TO METHANE MITIGATION

The strategy for deploying genetic change as a means of tackling a problem is well-known (e.g. Cunningham 1979; de Haas *et al.* 2021), consisting – adapted to use of genomics - of:

- Define the breeding objective, including both the traits being targeted and the value of changing each
- Measurement of the trait(s) or correlated predictor traits, which may include developing methods of measurement
- Estimating genetic parameters for the trait(s) and any correlated criterion and objective traits in the population, utilising both pedigree and genomic information
- Predicting breeding values for candidates, potentially in both the nucleus or seedstock sector and the commercial sector, relying on pedigree and phenotype, pedigree and phenotype and genotype, or genotype alone
- Selecting parents, and in parallel selecting animals for commercial production, both based on breeding values (which may be expressed in different format for different sectors of the value chain)

Applying this strategy may involve nuances such as utilisation of mate selection and coupling with reproductive technologies, and underlying the implementation is the question relevant to all genomic selection, of how much and where (what traits) to invest in ongoing phenotyping – especially relevant where the trait(s) is hard-to-measure (Banks 2022).

While this algorithm is applicable to any trait or combination of traits, considering its use in tackling methane mitigation brings some challenges:

- There is considerable debate around what the breeding objective trait is – absolute methane emission (i.e. gm per day), methane yield (i.e. per unit feed intake), or methane intensity (i.e. per unit product). Consequences of selection for each of these have been examined (e.g. van der Werf 2023), but as yet there is no consensus on the best approach, and the debates around the question are often emotionally charged, (and hindered by limited knowledge of the relevant genetic parameters), in part because many stakeholders either fear or cannot comprehend a situation where the outcome would be to reduce the rate of improvement of food (or fibre), and because mechanisms for credit or debit for emission level are absent or embryonic in most jurisdictions.
- There is no clear market signal in place, in any species or region, in the sense of a price of unit change in the trait leading to a payment or cost to the farmer. Options for dealing with this fact include:

- Using current market price
- Conducting sensitivity analysis, varying price from zero to some large value
- Using such sensitivity analysis to develop desired gain options

Interestingly, it is straightforward to derive an implicit carbon price from responses under consideration such as herd reduction, and those prices are orders of magnitude higher than any current market price.

These uncertainties around the price feed directly into challenges around funding for phenotyping: if methane is considered an externality of the whole food production system, the most efficient mechanism economically is a Pigouvian tax (analogous to an end-point royalty, or levy), usually requiring some form of government intervention. Individual breeder and/or producer stakeholders may invest in phenotyping intending to create a point of difference, but in the absence of a widespread carbon price, will likely have to either invest in additional marketing to maintain that point of difference, and/or underinvest in phenotyping.

- There is no single or obviously superior measurement technique. Methods in use or under development for individual animal measurement include respiration chambers, Greenfeed, sniffers, portable accumulation chambers (PACs), MIR-prediction, wearable devices. These vary widely in cost and practicality and will have different genetic parameters (ICAR Feed and Gas https://wiki.icar.org/index.php/Section_20_%E2%80%93_Methane_Emission_for_Genetic_Evaluation#Sub-sections, Van Breukelen, 2023).

  The fact that different measurement methods have different suitability in terms of where they can be deployed means that methane phenotypes will inevitably be collected in different production systems – ranging from barns with TMR feeding, through to extensive grazing, which means that there will inevitably be a range of criteria (measurement method x diet (quality and quantity)), which could be an issue even within a species in a region, but will certainly impact efforts to combine data across systems.

  In "normal" animal breeding where the focus is largely on selection within a population, such diversity of criteria is not a major problem, but for a trait that is currently expensive to measure and for which there would potentially be benefit in combining data from different sources, the challenge of building sufficiently robust genetic parameter sets will be relevant. The INTERBULL approach to this challenge provides an example of a solution (INTERBULL, https://interbull.org/index).

This list is likely not complete, but highlights aspects which require careful thought and analysis of options, and clear communication of those options and their consequences: the contributions from the community of animal breeders will be pivotal.

At the same time, initial modelling work suggests that valuable reductions in methane emission can be achieved, with 1% reduction per year without significant loss of gain in other traits being a realistic estimate. To some this sounds unhelpfully small, but it is of course cumulative and permanent, and delivers the considerable and valuable consequence of steadily reducing the extent of requirement for alternative approaches, such as expensive dietary supplements.

THE GLOBAL METHANE HUB – ONE INITIATIVE, INCORPORATING THE GENETICS APPROACH

It is impossible to cover the range of activities and initiatives that have been conducted or are underway across the globe directed in some way to mitigating climate change – even focussing solely on methane the list would be extensive. But it is worth noting that these include:

- Modelling studies to estimate effectiveness of different technologies (e.g. FAO 2023)
- Economic analyses of different technologies (e.g. Moran *et al.* 2007)
- Policy development and in some cases implementation (e.g. Denmark)
- Development of diet supplements (e.g. Bovaer/3NOP, Asparagopsis)

- Modelling studies investigating genetic approaches (e.g. Cottle *et al.* 2009; de Haas *et al.* 2021; Barwick *et al.* 2019; van der Werf 2023)

Interestingly, despite some political hurdles, Australia and NZ have been active in scientific research around climate change for a considerable time, and in terms of preparedness for signals or triggers of policy and/or market change, are probably as well placed as any countries. That activity over about two decades has included significant government and industry investment (e.g. Meat and Livestock Australia programs, the NZ Greenhouse Gas Consortium, and the Net Zero CRC in Australia).

This background does not imply any widespread agreement on what should or could be done, but does mean that there is a basis of some knowledge particularly relevant to issues of measurement, effects of additives, and early basis for clear thinking on the genetic approach.

Included in this foundation are estimates of emissions from different classes of livestock and different regions (https://www.fao.org/faostat/en/#data/GT). The most significant message from these estimates is that 76% of estimated emissions are from the developing world, and almost 70% are from other than dairy cattle.

One response to this mix of causes for cautious optimism, but against a background of confusion and debate, has been the growth of globally active philanthropy: building funds for potential investment into initiatives targeting climate change mitigation. One example is the Global Methane Hub (https://www.globalmethanehub.org/), whose initiatives include an Accelerator Program, focussed on emissions from enteric fermentation – i.e. ruminants. (NB: the author is a member of the Scientific Oversight Committee for the Accelerator Program).

The Accelerator Program is investing upwards of $US175m over 5 years into:

- Vaccine development,
- Additive development, in particular into understanding their effects,
- Fundamental rumen microbiome research and research infrastructure, and
- Genetic approaches, including interactions with other approaches (e.g. additives), and use of rumen microbiome information in predicting animals' genetic merit.

The genetic component of the Accelerator Program is well underway, with coordination provided by the team at Wageningen University and research lead by Prof. Roel Veerkamp. Progress in the year since the initiative commenced includes:

- Work considering strategy for resource allocation (between species, between regions)
- Projects focussed on collecting methane phenotypes to build genomic reference populations, aiming at c. 100,000 phenotypes collected with corresponding genotypes
- Structured rumen microbiome sampling, to massively expand resources for research into the microbiome, and to augment genomic prediction models
- Design for a global database of phenotypes and associated information, to be accessible for research and genetic evaluations
- Contribution to validation of measurement tools
- Initiating a program of documentation coupled with communication and extension, covering a wide range of topics relevant to breeding for reduced methane emission (this and the previous point are being progressed in close collaboration with the ICAR Feed and Gas Working Group)
- Methods for accounting data, to underpin fair-share principles for access to pooled data
- And extending the program to address current gaps in regional and species coverage

It is anticipated that the program will contribute to development of models for incentivising on-farm change and to government policy development, and to farm- and national-level monitoring (such as by using genetic trends as one measure of change).

An obvious message from the global estimates of emissions is that the major proportion of estimated emissions derive from livestock in the developing world, where infrastructure and systems for genetic improvement is often limited or absent. This fact, in addition to resonating with the focus of much of John Vercoe's research and science leadership, will continue to be front of mind, but it is hoped that encouraging and supporting the development of a global knowledge commons, and shared resources wherever possible, will help address this challenge.

THE CHALLENGE, AND THE RESPONSIBILITY

This brief outline barely scratches the surface of the breadth and rapid growth of activity around methane mitigation, even if we limit our attention to genetic approaches. Climate change is quite literally, the existential challenge of our times. Our community of animal and plant breeders, working together with rumen microbiologists, can play a far more significant role than is currently appreciated, or may ever be acknowledged. We are fortunate that the tools and knowledge available, including in particular genomic selection, offer control, in terms of optimising trait change, and speed of change. Further, the global community of geneticists is characteristically open and amenable to sharing knowledge and resources, including data, and in general is as much motivated by human and planetary good as by financial rewards.

Apart from the direct contribution via implementing effective genetic improvement that reduces methane, our community can make a massive contribution through something less tangible, but equally important: through clear, honest, open communication of what is possible, what is being learned about what is possible, and what precisely are any trade-offs that may need to be evaluated. The theory we work from is one of humanity's finest achievements, and offers clarity that will be vital in the swirling, confusing world of market failures, livelihood fears, trading behaviours etc.

Deploying genetics methods to tackle the climate crisis should be a very significant element of humanity's response, and we have the opportunity to build and deliver a "moonshot" program for our and all time (Mazzucato 2018). Truly, we can walk in the footsteps of John Vercoe, and there can be no greater privilege or opportunity.

ACKNOWLEDGEMENTS

Many colleagues have contributed to these perspectives, but particular thanks to Rich Eckard, Hayden Montgomery, Phil Vercoe, Roel Veerkamp, Birgit Gredler-Grandl and Julius van der Werf.

REFERENCES

Banks R.G. (2022) *Proc. Wrold Cong. Genet. Appl.Livest. Prod* **12**: 1804

Barwick S.A., Henzell A.L., Herd R.M., Walmsley, B.J. and Arthur, P.F. (2019) *Genet. Sel. Evol.* **51**:18.

Cottle D.J., van der Werf J.H.J. and Banks, R.G. (2009) *Proc. Assoc. Advmt. Anim. Breed. Genet.* **18**: 516.

Cunningham E.P. (1979) *Proc. Assoc. Advmt. Anim. Breed. Genet.* **1**: 73.

de Haas Y., Veerkamp R.F., de Jong G. and Aldridge M.N. (2021) *Animal* **15**: 100294.

FAO (2023) https://openknowledge.fao.org/items/b3f21d6d-bd6d-4e66-b8ca-63ce376560b5

IPCC (2022) https://www.ipcc.ch/report/ar6/wg2/

Mazzucato, M (2018) *Ind. Corp. Change* **27**: 803.

Moran D., Barnes A. and McVittie A, (2007) *DEFRA* IF0101

van der Werf J.H.J. (2023) *Proc. Assoc. Advmt. Anim. Breed. Genet.* **25**: 234.

van Breukelen A.E., Aldridge M.N., Veerkamp R.F., Koning L., Sebek L.B. and de Haas Y. (2023) *J. Dairy Sci.* **106**: 4121.

Wittel H.W.J. and Ritter M.W. (1973) *Policy Sci.* **4**: 155.

MEASUREMENT OF FEED INTAKE IN SHEEP USING "OFF THE SHELF" TECHNOLOGIES

P.T. Fitzgerald, R.L. Hergenhan and J.H.J. van der Werf

School of Environmental & Rural Science, University of New England, Armidale, NSW, 2351 Australia

SUMMARY

Feed intake and weight in sheep were measured using technologies designed for the pig industry to determine the suitability and applicability of said technology for sheep. Animal behaviour patterns in relation to proportion of animals accessing the feeder and frequency of visits per day as well as feed intake and weight parameters such as mean, variation, and repeatability are presented. Lessons learned regarding participation rates of animals (over time) and modifications to the system to enhance participation are discussed. Finally estimates of the numbers of days on feed required for valid results on measuring feed intake are presented. The system, with modifications, was found to be a suitable technology for measuring feed intake in ewes and lambs and has the potential to be deployed as a mobile unit to capture feed intake on industry animals.

INTRODUCTION

Feed is a significant cost of production for Australian sheep enterprises (Anderton and Weeks 2023) and is central to productivity and profitability. However, variation in feed intake and in the efficiency at which feed eaten is converted to product (meat, wool etc.) or allocated to bodily functions exists between animals. This efficiency can be expressed in terms of residual feed intake or as feed conversion ratio (Muir *et al.* 2020; Amarilho-Silveira *et al.* 2022; Johnson *et al.* 2022). The current study was undertaken as part of a larger project that aims to phenotype thousands of Australian sheep in order to develop genomically estimated breeding values for methane output and feed intake and correlations between other economically important traits (Fitzgerald *et al.* 2023).

The ability to accurately measure feed intake for sheep is generally hampered by the need to build "custom designed systems" which are generally large fixed structures that involve significant capital investment and require a 42-day trial period (Muir *et al.* 2020; Amarilho-Silveira *et al.* 2022; Johnson *et al.* 2022).

There is an opportunity to have smaller mobile systems that could be rapidly deployed to industry in order to measure smaller cohorts on farm, potentially over shorter periods. The Pig Progeny Tester (PPT) manufactured by FREEDA Solutions was identified as an "off-the-shelf" product that is capable of capturing feed intake and animal liveweights and is widely used in the pig industry. The system (Figure 1) consists of 1) a feed weighing module; 2) an animal weighing platform; 3) a feed hopper that delivers feed via a cup drop mechanism; 4) an electronic control box; 5) an animal weight indicator; 6) a feed weight indicator; and 7) adjustable sidewalls. In simple terms, each time an animal visits the PPT, it is identified via an RFID tag and the system measures the amount of feed eaten and the weight of the animal. A number of PPTs were purchased and trialled to assess their suitability for measuring feed intake for sheep.

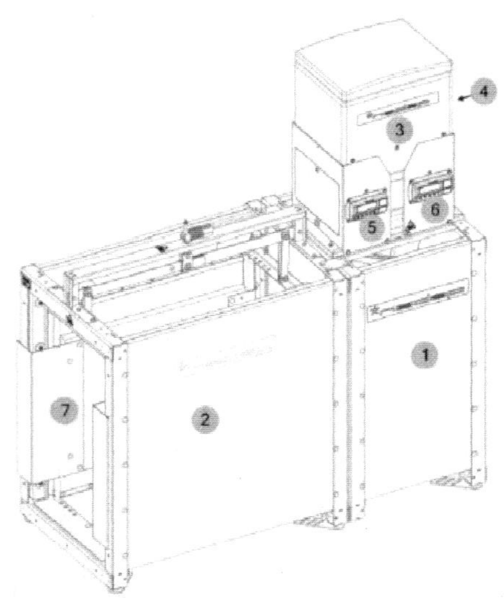

Figure 1. Schematic overview of a PPT

MATERIALS AND METHODS

Data on feed intake using Pig Progeny Testers (PPT, FREEDA Solutions) was collected on three batches of animals in separate trials between June 2024 and January 2025 at Kirby SMART Farm, Armidale, NSW. One group comprised 60 mixed breed lambs and two groups (of size 40 and 80) were Merino ewes. Animals had access to 3 (trial 1 and 2), or 6 (trial 3) PPT units from which they were fed *ad-lib* lucerne pellets (88% dry matter). Trial 1 ran for 6 weeks, while trials 2 and 3 ran for 7-8 weeks. All work was undertaken with the approval of the University of New England Animal Ethics Committee (Approval number ARA 21-086).

At least the first 14 days of each trial were allocated to training to allow animals to adapt and acclimatize to the feed and become familiar with accessing the PPTs. During this period, animals were provided pellets in open troughs in the pens with the same pellets available in the PPTs. Pellets were also placed on the weigh platform to attract animals into the units. The amount of feed and the frequency at which pellets were placed in the pen troughs was gradually reduced over time to encourage animals to access the PPTs. A portion of the Trial 3 animals had an additional training period of 7 days for animals that had not routinely accessed the feeders to maximise participation rates. After the training period, animals sourced all feed directly from the PPTs. Data for every visit that had > 1g feed intake were analysed including liveweights where feed intake events were > 60 seconds. Animals were also weighed in a separate crate pre, post and during the feeding periods to check the accuracy of the system.

Prior to the start of Trial 3, some modifications were made to enhance usage. A grated "foot mat" was placed in front of the entrance to the unit for animals to walk on prior to entry to reduce the amount of mud and faecal matter being walked onto the units during wet periods, which was perceived as reducing the accuracy of liveweight measurements. Non-transparent factory fitted sides were removed from 4 units and replaced with clear polycarbonate sides. These 4 units were placed in pairs side by side. Results presented summarise data from all 3 trials, with focus on the Trial 3 cohort of 80 Merino ewes.

RESULTS AND DISCUSSION

Duration and frequency of visits. Table 1 gives an overview of the number of visits, the duration and the amount of feed eaten per visit. Results are from the raw data, without removing outliers. In general, lambs visited the PPTs more frequently but had many visits of short duration (<30 seconds).

A greater proportion of the lambs accessed the feeder, certainly after accounting for the number of animals per feeder. In general, lambs were more curious and accessed the feeder more frequently relative to ewes. The proportion of animals visiting and the duration of the visits was highest in the third cohort of 80 ewes, and more feed was eaten per visit. This might have been due to the larger number of feeders available in total, and the improved training.

Feed intake parameters during different parts of the test period. Different parts of the feed intake period were evaluated. The first 14 days are considered as the training period, with the number of animals accessing the feeders and visits per animal increasing during this period due to feed being gradually withdrawn from open troughs and only available in the PPTs by Day 14 (Figure 2).

The remaining 5 weeks (days 15-49) were considered as the test period. From the data of the 80-ewe cohort (trial 3) we extracted three feed intake phenotypes from the test period, defined as FI_3wk, FI_4wk, FI_5wk representing the daily feed intake averaged over days 15-35; 15-42; and 15-49, respectively. The average daily feed intake as well as the standard deviation and the average animal weight are summarised in Table 2. A valid measurement was defined as a day where an animal made more than 10 visits as 85% of the time, animals visited at least 10 times per day. The repeatability was defined as the variation between animals over the total variation in a linear model just fitting animal as a random effect.

Table 1. Raw data statistics of the three cohorts of animals measured

Cohort	Lambs	40 Ewes	80 Ewes
Number of visits with feed intake >0g.	131,170	39,531	67,649
Number of animals at start	60	40	80
Number of feeders	3	3	6
Number of animals accessing feeders > once	46	28	70
Number of animals accessing feeders during "test period"	46	24	63
Percentage animals accessing feeders during "test period"	77%	60%	79%
Median number of visits per animal per day	77	26	18
Median duration of visit (sec)	18	40	94
Percentage of visit duration < 30sec	65%	41%	20%
Percentage of visit duration > 5min	1%	6%	8%
Average feed intake(g)/visit	17	47	72
Median of Feed intake (g) per animal per day	1,596	1,860	1,737
Median number of Weight records per animal/day	13	7	12
Median of Weight (kg) per animal per day	37.3	54.0	67.3

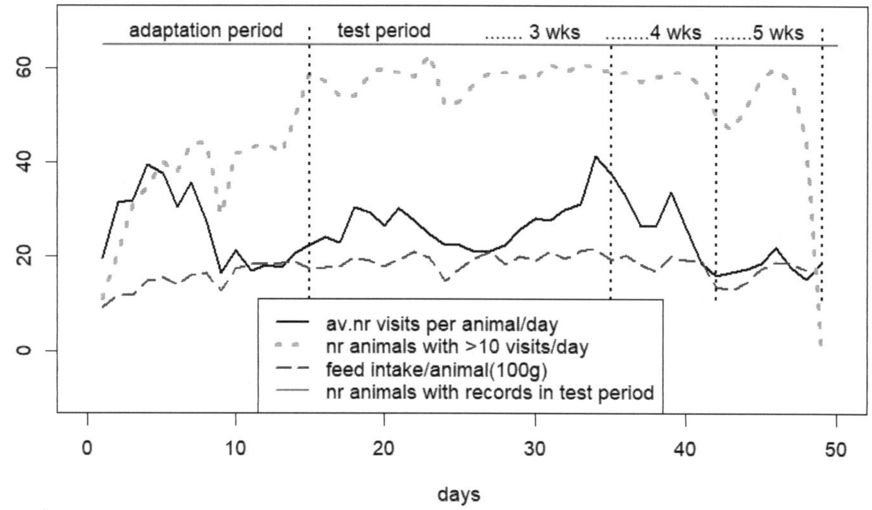

Figure 2. Participation rate, number of visits per day and average feed intake per day over the whole test period for the 80-ewe cohort

Table 2. Mean and standard deviation (Mean ± SD), variance between (Var_Anim) and within animals (Var_Res), repeatability and correlation with final test (r-FI_5wk) for feed intake and weight measurements, and correlation between PPT weight and scale-weight (r scale-PPT).

	Feed Intake (g)			Weight (kg)		
Phenotype	FI_3wk	FI_4wk	FI_5wk	WT_3wk	WT_4wk	WT5wk
Mean ± SD	1972 ± 550	1917± 553	1887 ± 548	68.7 ± 7.4	69.1 ± 7.3	69.7 ±7 .3
Var_Anim	143,133	125,708	118,814	60.77	60.68	59.13
Var_Res	169,448	179,729	181,831	11.31	11.46	11.33
Repeatability	0.47	0.41	0.40	0.84	0.84	0.84
r-FI_5wk	0.98	0.99	1.00	0.992	0.998	1
r scale-PPT				0.94	0.93	0.95

These preliminary results indicate that repeatability of FI is moderate and high for animal weight. The correlation between weight measured by the scale and the PPT units is >0.93 for weeks 3, 4 and

5 indicating that averaged values from the PPT animal weighing system is accurate and reliable. The correlation between mean animal performance for FI_3wk, FI_4wk, and FI_5wk is very high, indicating that it may be possible to shorten the feeding duration period in order to reduce cost and increase the number of animals measured over a year. This result is in agreement with Amarilho-Silveira *et al.* 2022 who found that an RFI model using 35-days FI data compared to 42 days-maintained accuracy. Further work can focus on data editing, for example by discarding deviating values on a within animal basis.

CONCLUSION

The Pig Progeny Testers (PPT) manufactured by FREEDA Solutions, with modifications, was found to be a suitable technology for measuring feed intake and body weight in ewes and lambs. A test period of 35 days may be sufficient for measuring feed intake in sheep. More time is required to further scrutinise the data and correct for additional fixed effects. The system appears to have potential to be deployed by commercial breeders to capture feed intake on industry animals as a mobile unit allowing more animals across industry to be measured compared to fixed facilities

ACKNOWLEDGEMENTS

Without the hard work of Z Brown, H. Speers and staff at Kirby SMART Farm this work would not have been possible. Meat and Livestock Australia contributed financial support for this work.

REFERENCES

Anderton L. and Weeks P. (2023) MLA Market Information Report. Meat & Livestock Australia, North Sydney.

Amarilho-Silveira F., de Barbieri I., Cobuci J.A., Balconi C.M., de Ferreira C.F. and Ciappesoni G. (2022) *Livestock Sci.* **258**: 104889.

Fitzgerald P.T., Clayton E.H., Donaldson A.J., Brown D.J., Oddy V.H. and van der Werf J.H.J. (2023) *Proc. Assoc. Advmt. Anim. Breed. Genet.* **25**: 178.

Johnson P.L., Hickey S., Knowler K., Wing J., Bryson B., Hall M., Jonker A., Janssen P.H., Dodds K.G., McEwan J.C. and Rowe S.J. (2022) *Front. Genet.* **13**: 1.

Muir S.K., Linden N., Kennedy A., Knight M.I., Paganoni B., Kearney G., Thompson A.N. and Behrendt R. (2020) *Small Ruminant Res.* **192**: 106241.

FEED-INTAKE AND METHANE EFFICIENCY FOR WEIGHT AND GROWTH PRE-SLAUGHTER OF MLA RESOURCE FLOCK LAMBS.

B. Paganoni[1], B. Bolt[2], A. Bowden[2], P. Fitzgerald[3], S.F. Walkom[4] and J.H.J. van der Werf[3]

[1] WA Department of Primary Industries and Regional Development, Bunbury, WA, 6230 Australia
[2] WA Department of Primary Industries and Regional Development, Katanning, WA, 6317 Australia
[3] University of New England, Armidale, NSW, 2351 Australia
[4]Animal Genetics Breeding Unit*, University of New England, Armidale, NSW, 2351 Australia

SUMMARY

Improving feed-use efficiency and reducing methane emissions in sheep are critical strategies for enhancing market access, profitability, and the long-term sustainability of the sheep industry. In this study, 758 lambs from Merino, Maternal and Shedding ewes mated artificially to Merino, Maternal, Terminal and Shedding sires were assessed for weight, growth, and methane production prior to slaughter across six kill groups (2 years x 3 ages). Weight and growth accounted for 72-89% of the observed variation in daily feed intake. No additional variation was explained by body composition or methane production, as measured in this study. Residual feed intake exhibited high variability across all breed types (sire means ranged from -0.17 to +0.18 kg/hd/day, SD=0.12), suggesting that genetic selection for feed efficiency and methane reduction is achievable for Terminal, Shedding, and Merino sheep. This research highlights the potential for genetic improvements in reducing methane emissions and enhancing feed efficiency in sheep, offering pathways to a more sustainable and profitable sheep industry.

INTRODUCTION

Efficient use of feed is a critical determinant of productivity and sustainability in livestock systems, particularly in the red meat industry. Understanding factors that influence feed intake and efficiency is essential to develop strategies that enhance growth performance, reduce costs, and mitigate environmental footprints.

Breeding sheep that use less feed to produce the same or more productivity is an attractive solution, however progress is likely to be slow due to the low-moderate heritability (0.1-0.45) of residual feed intake for sheep and trait instability due to feed and environmental variation (Snowder and Van Vleck 2003; Cammack *et al.* 2005; Paganoni *et al.* 2017; Tortereau *et al.* 2020; Ellison *et al.* 2022). Centralised research facilities and protocols that can estimate feed intake traits should play a valuable role in establishing protocols and increasing trait screening and accessibility to all sheep producers through existing services such as Sheep Genetics Australia and established resource flocks such as the Information Nucleus (Fogarty *et al.* 2007). Additionally, established resource flocks allow breeders to take advantage of genomic selection to accelerate genetic improvement exploiting opportunities to increase red meat production while simultaneously reducing methane emissions.

This study investigated variation in feed intake and associated traits including methane production under *ad libitum* feeding pre-slaughter.

---

* A joint venture of NSW Department of Primary Industries and Regional Development and the University of New England

MATERIALS AND METHODS

Lambs from the MLA resource flock at Katanning born in July 2022 (n=929) and 2023 (n=899) were measured for methane and body composition at early post weaning age (5-6 months old), and measured for feed intake (total=736), then methane (n=1,300) and body composition (n=1575) a second time prior to slaughter (7-10 months old). Kill groups 2,3 and 4 of both cohorts entered indoor feedlot facilities between Jan-May prior to slaughter (7-10 months old; Table 1). In each kill group, approximately half had average daily feed intake captured for approximately 28-32 days.

Table 1. Kill group numbers and dates for the 2022-born and 2023-born cohorts of lambs from the MLA resource Flock at Katanning

Kill Grp	2022-born		2023-born	
	No.	Breed	No.	Breed
1	224	Dorper	198	Merino/Maternal/Terminal/Dorper
2	231	Merino/Maternal/Terminal/Dorper	240	Merino/Maternal/Terminal
3	231	Merino/Maternal/Terminal/Dorper	204	Dorper
4	243	Merino/Maternal/Terminal/Dorper	241	Merino/Maternal/Terminal
Total	929	(slaughtered)	883	(slaughtered)

*Slaughter numbers were lower in 2023 due to less ewes mated/lambs weaned

Methane. Post-weaning methane production (CH_4) using Portable Accumulation Chambers (PACs) was captured at early post-weaning (5 months old) in December 2022 (n= 929) and 2023 (n=899). A second CH_4 measurement was completed prior to slaughter for kill groups 2-4 (8-10 months old) for both cohorts. In both years, pasture had senesced, therefore all lambs were being supplemented with grower pellets and oaten hay in 2022, and oaten hay and a 70% oat:30% lupin mix in 2023. Prior to measurement lambs were drafted into groups of 16 and. time off-feed ranged between 1-5 hours. CH_4 was recorded 3 times over 45 minutes using an Eagle 2 gas analyser (*E2*, RKI Instruments, Union City, USA). All non-linear CH_4 accumulation was removed (2%).

Body composition. Muscle and fat were measured twice via ultrasound at the C-site (Robinson *et al.* 1992) using a Sheep Genetics accredited scanner.

Feed intake. All lambs prior to entry into the feed intake shed were adapted to an *ad libitum* pellet diet over 14 days (11MJ,13%CP). The sheep were then stratified by weight (max range of 5kg per pen to minimise bullying) and randomised by sire into up to 20 pens with 10-14 sheep per pen. Animals in all pens were weighed twice weekly. Daily feed intake measurements were recorded automatically for about half of each kill group in the facility (10 reading and 10 non-reading pens; total=758 lambs). Intake data was captured as individual meals. A cleaning algorithm removed meals with blank or not-present EID, where feed intake was 0 or negative and outlier meals (3 x SD from mean meal intake rate – to remove events involving cleaning/vacuuming of the feed trays). After cleaning, the data was summarised into individual daily intake for each sheep. All sheep with >40% coefficient of variation in daily intake were removed as confidence of a stable trait was too low (n=41 or 5% of records).

Residual Feed Intake (RFI). RFI was analysed using a two-step approach. First multiple linear regression was used where average daily intake (ADI) was adjusted by fitting (1) mean metabolic mid weight; MMWT and (2) Average Daily Gain (ADG) as covariates:

(1) MMWT = the predicted mid weight during the feed intake test calculated via linear regression of the twice weekly weights throughout the intake test

(2) ADG = the slope of the linear regression used to calculate MMWT

Each kill group was modelled separately with pen fitted as group. Date of birth, birth type and sex were added and dropped from all models as their effects were insignificant (excepting one model/kill group) after fitting MMWT and ADG. The unexplained variation after fitting the above model was the residual feed intake. In the second step, Residual Maximum Likelihood was used to test feed intake for significant sire effects for each cohort separately. Kill group, MMWT and ADG and sire were fitted as fixed effects (plus all significant two-way interactions) with dam and pen fitted as random effects. All non-significant terms were dropped from the final model(s).

RESULTS AND DISCUSSION

Differences in sire means for RFI were significant for the 2022 cohort (P<0.001; Figure 1). For this cohort 99 sires were represented amongst the intake measures with an average of 3.4 progeny per sire (range 2-7 progeny per sire). For the 2023 cohort there were 101 sires screened for intake with an average of 3.0 progeny per sire (range 2-8 progeny per sire), sire differences were undetectable for this cohort. (Only half of the progeny from each sire is represented at Katanning and only half of each kill group was measured for intake, this limitation meant no common sires were measured across years). More importantly, sires from all breed types were represented at the extreme ends of RFI across both cohorts. Sires means for RFI (using only sires with at least 3 progeny per sire) varied from -170 ± 0.10 g/hd/day (n=10) to +180 ± 0.08 g/hd/day across all kill groups and cohorts Using the average standard deviation in RFI within breed and kill group (0.12) and an estimated heritability in sheep of 0.1 (Cammack *et al.* 2005; Paganoni *et al.* 2017), this indicates potential genetic gains of 0.3% per year may be possible through feed-intake testing under ad libitum feeding conditions pre-slaughter (assuming selection pressure is 1 and a 4-year generation interval).

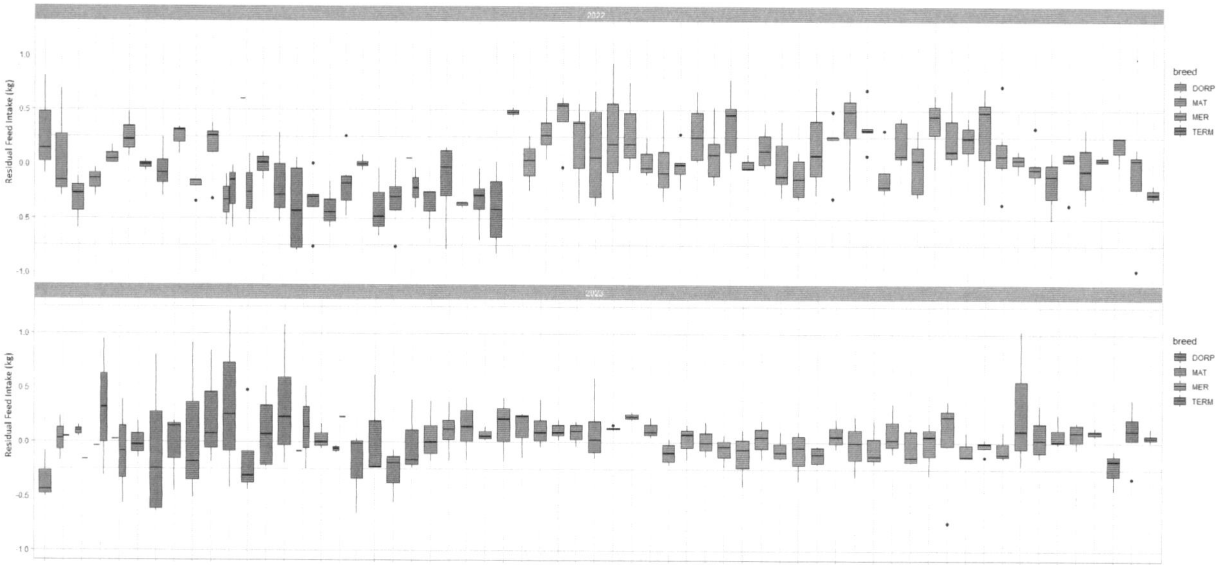

Figure 1. Sire predicted means for residual feed intake (kg) for the 2022-born (upper) and 2023-born (lower) lambs grouped by key breed types (Dorper = orange, Maternal = green, Merino = blue, Terminal = purple) from the MLA resource flock

MMWT varied from 47-58 kg and ADG pre-slaughter from 79-225g/hd/day and ADI varied between 1.4 and 1.9 kg/hd day between kill groups across both cohorts.

MMWT and ADG explained 72-89% of the variation observed in ADI (Table 2), consistent with, or higher, than previous reports for sheep (Paganoni *et al.* 2017). MMWT had the largest and most significant effect on ADI (strongest correlation) across most of the kill groups with consistent increase in intake of 60-70g for every 1kg increase in weight (P<0.05; Table 2). ADG had additional significant positive effects to MMWT for four of the six kill groups measured (P<0.05; Table 2). For every 1g/hd/day increase in ADG there were additional increases in ADI of 1.5-2g per day

($P<0.05$; Table 2). There were no significant effects of birth type or sex on ADI (beyond differences accounted for by MMWT). Date of birth had a small significant additional effect on ADI for one of the six kill groups measured (19g of additional ADI/day older at slaughter, $P<0.05$; Table 2). Body composition as measured by ultrasound scanning of muscle and fat pre-slaughter had no significant effects on ADI, neither did changes in muscle and/or fat between the post-weaning and pre-slaughter measurements. There were also no significant relationships detected between ADI and methane production measured via PACs pre-slaughter or the average of both PACs (pre-slaughter and post-weaning).

Table 2. Estimates of significant terms from the linear regression model (first-step) for residual feed intake for the kill groups 2-4 of the 2022&2023-born lambs from the MLA resource flock at Katanning. The variation explained by the significant model terms appears in brackets (%VE)

Year, grp (%VE)	MMWT	ADG	Date of Birth
2022 2 (89%)	0.11 ± 0.004	n.s.	0.019 ± 0.005
2022 3 (72%)	0.07 ± 0.025	n.s.	n.s.
2022 4 (84%)	0.06 ± 0.022	1.52 ± 0.730	n.s.
2023 2 (89%)	n.s.	1.81 ± 0.840	n.s.
2023 3 (75%)	0.07 ± 0.026	2.03 ± 0.794	n.s.
2023 4 (77%)	0.06 ± 0.018	1.53 ± 0.804	n.s.

CONCLUSION

The findings of this study provide key preliminary insights into variation for feed intake and efficiency traits in Australian sheep. Initial findings indicate that genetic gains in residual feed intake, may be possible for permanent and cumulative efficiency gains in lamb finishing systems. Notably, over 75% of variation in daily feed intake is explained by weight and growth, with methane production and body composition, as measured in this project, contributing minimally. This underscores the central role of weight and growth in driving feed-use efficiency, under *ad libitum* finishing systems. Importantly, sires across all major breed types exhibit extremes in residual feed efficiency, suggesting broad genetic variability that can be leveraged for selection.

ACKNOWLEDGEMENTS

This research was funded by DPIRD and MLA with support from UNE and AGBU.

REFERENCES

Ellison M.J., Cockrum, R.R., Means W.J., Meyer A.M., Ritten J., Austin K.J. and Cammack K.M. (2022) *Small Ruminant Res.* **207**: 1066.

Fogarty N.M., Banks R.G., van der Werf J.H.J., Ball A.J. and Gibson, J.P. (2007) *Proc. Assoc. Advmt. Anim. Breed. Genet.* **17**: 29.

Hopkins D.L., Pirlot K.L., Roberts A.H.K. and Beattie A.S. (1993) *Aus. J. Exp. Agr.* **33**: 707.

Snowder G.D. and Van Vleck, L.D. (2003) *J. Anim. Sci.* **81**: 2704.

Cammack K.M., Leymaster K.A., Jenkins T.G. and Nielsen M.K. (2005) *J. Anim. Sci.* **83**: 777.

Paganoni B., Rose G., Macleay C., Jones C., Brown D.J., Kearney G., Ferguson M. and Thompson A.N. (2017) *J. Anim. Sci.* **95**: 3839.

Tortereau F., Marie-Etancelin C., Weisbecker J.L., Marcon D., Bouvier F., Moreno-Romieux C. and Francois D. (2020) *Animal* **14**: 681.

SELECTING FOR EFFICIENCY IN LONG-FED CATTLE IS IMPORTANT

W.S. Pitchford

Davies Livestock Research Centre, School of Animal and Veterinary Sciences, University of Adelaide, Roseworthy, SA, 5371 Australia

SUMMARY

Genetic parameters for 17 traits were obtained from a leading Wagyu nucleus program. Genetic variation (SD) in daily feed intake was 0.58 kg/d, residual feed intake 0.46 kg/d and after adjustment for carcass traits (weight, muscle, subcutaneous and intramuscular fat) was still 0.37 kg/d. Assuming this variation remains through feedlot finishing, this represents large variation in the cost of production so steers from top 1% bulls joined to average cows could cost $146 less to feed than using average bulls. It is most likely this is associated with variation in visceral fat depots.

INTRODUCTION

The largest cost in livestock production systems is feed. Thus, it makes sense that feed intake is an important trait to be measured if aiming to make genetic gain in profitability. In the 1990's there was a large Australian research effort placed on understanding biological mechanisms of growth through intense measurement of Merino lambs from High vs Low growth selection lines (Herd *et al.* 1993; Oddy *et al.* 1995). They found very few differences in partial efficiencies and concluded that differences in growth were driven by differences in appetite.

Koch (1963) introduced the concept of residual feed intake (RFI) representing variation in intake which is net of that associated with animal size and growth rate during the intake test. Robinson and Oddy (2004) and Barwick *et al.* (2009) demonstrated strong relationships between RFI and fatness in separate datasets. Cows from the beef High and Low-RFI selection and High and Low-Fat EBV lines were run on two properties with two stocking rate treatments to evaluate the impact of RFI selection on maternal productivity. Efficiency results were reported by Hebart *et al.* (2018) and variation in fat by Accioly *et al.* (2018). It was clear differences in RFI were associated with fat. Lines *et al.* (2018) quantified protein metabolism on the beef RFI lines and also concluded that differences in RFI were primarily associated with differences in fatness rather than efficiency.

Pitchford *et al.* (2018) published a summary of results from multiple projects that demonstrated more generally that variation in RFI is associated with variation in appetite rather than maintenance efficiency. They concluded that when appetite is not satisfied, then there is unlikely to be variation in RFI and, therefore, efficiency. They questioned the value of selection for low-RFI in many production systems. The obvious exception is for animals on expensive long-fed regimes. The aim of this analysis is to estimate the amount of genetic variation in residual feed intake that is independent of (conditional on) multiple growth and composition measures in long-fed cattle.

MATERIALS AND METHODS

Genetic parameters were supplied from a leading Wagyu nucleus program's within-herd analysis. The 17 traits were birth weight (BWT, kg), weaning weight (WWT, kg), yearling weight (WT400, kg), ultrasound scanned eye muscle area (UEMA, cm^2), rump fat depth (UP8, mm), rib fat depth (URib, mm), intramuscular fat (UIMF, %), post-weaning feed test daily feed intake (DFI, kg/d), growth rate (FTADG, kg/d), metabolic mid weight (MMWT, $kg^{0.75}$), residual feed intake (RFI, kg/d), 600 day weight (WT600, kg), carcass weight (HSCW, kg), AUSMEAT rump fat depth (AP8), marbling score (AMarb) and two traits measured with the Meat Industry Japan camera being digital marbling percentage (DMP, %) and digital muscle area at the 5th rib (DMA5, cm^2). The genetic standard deviations and correlations after bending to ensure positive definite (using nearPD in R, https://github.com/joehl, following Higham 2002) are presented (Table 1).

The method of calculating conditional genetic covariances was:

$$C_{X,Y} = \begin{bmatrix} \Sigma_{11} & \Sigma_{12} \\ \Sigma_{21} & \Sigma_{22} \end{bmatrix}$$

$$C_{Y|X} = \Sigma_{22} - \Sigma_{21}\Sigma_{11}^{-1}\Sigma_{12}$$

where Σ_{11} is the genetic covariance matrix for Y variables, Σ_{22} is the genetic covariance for the X variables, and Σ_{12} is the genetic covariances between growth, composition and intake traits. Matrix calculations to estimate variation in post-weaning residual feed intake being conditional on other traits was conducted three times: 1) pre-weaning growth (BWT, WWT); 2) yearling (WT400, UEMA, UP8, URib, UIMF); and 3) carcass traits (WT600, HSCW, AP8, AMarb, DMP, DMA5). Each time was cumulative so when conditional on carcass traits, genetic variation was evaluated conditional on both 2) yearling and 1) early growth.

Table 1. Genetic variation (SD) and genetic correlations (x100, trait abbreviations in text)

	BWT	WWT	WT400	UEMA	UP8	URib	UIMF	DFI	FTADG	MMWT	RFI	WT600	HSCW	AP8	AMarb	DMP	DMA5
GenSD	2.2	10.4	19.1	2.1	1.2	0.7	0.4	0.6	0.1	4.3	0.5	23	33	2.9	1.0	5.2	4.6
BWT																	
WWT	84																
WT400	84	94															
UEMA	53	73	70														
UP8	-29	-10	-11	14													
URib	-26	-9	-5	17	83												
UIMF	-10	-1	-10	7	43	36											
DFI	38	48	61	50	4	17	1										
FTADG	59	59	65	70	10	19	13	74									
MMWT	79	96	86	67	-10	-8	5	47	61								
RFI	-15	-18	-14	-4	13	13	11	29	6	-21							
WT600	85	94	99	70	-12	-7	-9	60	65	84	-10						
HSCW	67	67	79	49	-10	0	-7	71	73	62	7	82					
AP8	-10	1	8	13	70	50	10	11	6	-3	16	9	26				
AMarb	-21	-7	1	-3	6	13	32	20	3	0	10	-1	9	-5			
DMP	-22	-10	-11	-9	9	11	43	2	3	-2	4	-12	-1	-13	90		
DMA5	18	23	25	44	-11	-15	6	29	33	24	8	27	32	-2	32	21	

RESULTS AND DISCUSSION

Standard practice in Australia for genetic evaluation of feed efficiency is to conduct a feed test post-weaning and this is becoming increasingly common on commercial properties. During the post-weaning test, the average dry matter intake (DFI) was 7.16 kg/d with a genetic SD of 0.58 kg/d (Table 2). Residual feed intake (RFI) is calculated by regressing feed intake on feed test growth rate (FTADG) and animal size as metabolic mid-weight (MMWT) which is average of start and end test weight raised to power of 0.75. Thus, by definition, there is less variation in RFI (SD 0.46 kg/d) than DFI.

Regressing feed efficiency test traits on birth and weaning weight accounted for 92% of the variation in MMWT but only 38% of the variation in FTADG (Table 2). As RFI is already adjusted for weight it was not surprising that only 3% of the variation was associated with pre-weaning growth. Regressing on yearling growth, muscling, fatness and marbling accounted for an additional 30% of variation in DFI but only 5% of variation in RFI. Regressing on WT600 and carcass traits accounted for 29% of the variation in RFI. Thus, there is both significant variation accounted for (37%) and variation remaining (63%).

Pitchford *et al.* (2018) proposed that measuring RFI in seedstock herds with clients producing grass-fed or short-fed steers was unlikely to be profitable because of the relationship between RFI and appetite leading to increased fatness. However, for long-fed steers where both the cost of feed and fat levels are high it is possibly worthwhile. Even after adjustment for two measures (live ultrasound and carcass) of muscle, sub-cutaneous fat and intramuscular fat, 63% of the variation in RFI remained.

The genetic SD of RFI conditional on growth and carcass traits was 0.37 kg/d (Table 2) for animals with a mean intake of 7.16 kg/d, representing a CV of 5.2% (0.37/7.16). High Content (Purebred & Fullblood) Wagyu are typically fed for 450 days at 12kg as fed (72% DM) costing $450/tonne which equates to $2,430 per head. If the genetic variation in RFI remains from post-weaning to throughout the long finishing period, then 5.2% would represent a genetic SD of $126. With this much variation, the top 1% of bulls would have breeding values $292 above average and steers from breed average cows sired by these bulls would have $146 lower feed costs with the same carcass value.

At 12 MJ ME/kg DM, the total energy intake during finishing would be 46,656 MJ (i.e. 450 days x 12 kg as fed x 72% DM x 12 MJ ME/kg DM) and the genetic SD would be 2,426 MJ (i.e. 5.2%). Oddy *et al.* (2024) reported that 70-80% of metabolisable energy intake above maintenance is lost as heat. Thus, of the 2,426 MJ, there may be just 25% or 606 MJ retained. The null hypothesis implied by Pitchford *et al.* (2018) was that the variation in RFI remaining after adjustment for growth and carcass fatness would primarily be associated with other fat depots.

Table 2. Genetic variation (SD and proportion of variance) in feed test traits conditional on other traits

Scenario	DFI	FTADG	MMWT	RFI
Base	0.58, 100%	0.12, 100%	4.3, 100%	0.46, 100%
1) Pre-weaning	0.51, 77%	0.10, 62%	1.2, 8%	0.45, 97%
2) Yearling	0.40, 47%	0.07, 34%	1.1, 6%	0.44, 92%
3) Carcass	0.27, 22%	0.02, 4%	0.3, 0.4%	0.37, 63%

Yamada *et al.* (2020) reported that the phenotypic SD in visceral fat in Wagyu steers was 8 kg. Given that fat contains 39.3 MJ/kg and has an efficiency of conversion of approximately 70%, 8kg would cost approximately 450 MJ (8 x 39.3 / 0.7) to deposit. This 450 MJ represents 74% of the 606

MJ retained based on the calculations herein. Thus, variation in fat deposition does account for the majority of the residual variation in RFI in Wagyu cattle.

The remaining 26% could be explained by four primary reasons: 1) variation in residual feed intake throughout feeding is greater than on feed test; 2) Yamada's (2020) cattle being fed for less time (not stated in paper) resulting in less variation in fat; 3) other biological mechanisms such as variation in visceral weight which is associated with intake, movement, digestibility and energy lost as methane; and 4) simply variation in the genetic covariances between traits which all have significant standard errors despite being diligently recorded on 2,459 animals.

CONCLUSION

This paper has affirmed the value of measuring RFI in Wagyu herds by demonstrating significant variation independent of carcass value. As expected, it is likely that the bulk of this variation is associated with variation in body fat. Given this is mostly visceral fat, it can be considered a waste which selection for low-RFI would minimise.

ACKNOWLEDGEMENTS

This analysis was made possible due to the availability of data from Joe Grose which had been processed by Maddie Facy and Adam Lloyd, all of 3D Genetics Pty Ltd. (www.3dgenetics.com.au). The conclusions were enhanced from invaluable discussion with Hutton Oddy.

REFERENCES

Accioly J.M., Copping K.J., Deland M.P.B., Hebart M.L., Herd R.M., Lee S.J., Jones F.M., Laurence M., Speijers J., Walmsley B.J. and Pitchford W.S. (2018) *Anim. Prod. Sci.* **58**: 67.

Barwick S.A., Wolcott M.L., Johnston D.J., Burrow H.M. and Sullivan M.T. (2009) *Anim. Prod. Sci.* **49**: 351.

Hebart M.L., Accioly J.M., Copping K.J., Deland M.P.B., Herd R.M., Jones F.M., Laurence M., Lee S.J., Lines D.S., Speijers E.J., Walmsley B.J. and Pitchford W.S. (2018) *Anim. Prod. Sci.* **58**: 80.

Herd R.M., Oddy V.H. and Lee G.J. (1993) *Aust. J. Exp. Agric.* **33**: 699.

Higham, N.J. (2002) *IMA J. Numerical Anal.* **22**: 329.

Koch R.M., Swiger L.A., Chambers D. and Gregory K.E. (1963) *J. Anim. Sci.* **22**: 486.

Lines D.S., Pitchford W.S., Bottema C.D.K., Herd R.M. and Oddy V.H. (2018) *Anim. Prod. Sci.* **58**: 175.

Oddy V.H., Speck P.A., Warren H.M. and Wynn P.C. (1995) *J. Agric. Sci., Cambridge.* **124**: 129.

Oddy V.H., Dougherty J.C.H., Evered M., Clayton E.H. and Oltjen J.W. (2024) *J. Anim. Sci.* **102**: 1.

Pitchford W.S., Lines D.S. and Wilkes M.J. (2018) *Anim. Prod. Sci.* **58**: 1414.

Robinson D.L. and Oddy V.H. (2004) *Livest. Prod. Sci.* **90**: 255.

Yamada T, Kamiya M. and Higuchi M. (2020) *Anim. Sci. J.* **91**: 13449

Proc. Assoc. Advmt. Anim. Breed. Genet. 26: 30-33

GENETIC INSIGHTS INTO METHANE EMISSION TRAITS IN INDOOR-FED GROWING CATTLE

S.B. Crowley[1, 2]. D.C. Purfield[2], S.B. Conroy[3], D.N. Kelly[3], R.D. Evans[3] and D.P. Berry[1]

[1] Department of Animal Bioscience, Animal and Grassland Research and Innovation Centre, Teagasc, Moorepark, Fermoy, Co. Cork, Ireland
[2] Department of Biological Sciences, Munster Technological University, Bishopstown, Co. Cork, Ireland
[3] Irish Cattle Breeding Federation, Link Road, Ballincollig, Co. Cork, Ireland

SUMMARY

The present study aimed to examine the genetic variability of various methane production traits in growing beef cattle and quantify the proportion of genetic variance in daily methane production independent of performance traits such as feed intake, growth rate and body size. Methane emissions were measured using GreenFeed systems from 1,700 crossbred cattle from a commercial feedlot. Performance traits including feed intake, average daily gain, liveweight and carcass weight data were also available. Genetic parameters were estimated using animal linear mixed models. Daily methane production was moderately heritable (0.42 ± 0.09), with a genetic standard deviation of 23.43 g/d. Daily methane production exhibited moderate genetic correlations with feed intake (0.51), ADG (0.39), metabolic liveweight (0.27) and carcass weight (0.42). Genetic adjustment of daily methane production for these traits resulted in a 25% reduction in the genetic standard deviation (from 23.43 g/d to 17.55 g/d), with only 56% of daily methane productions genetic variance remaining independent of the performance traits.

INTRODUCTION

Genetic improvement provides cumulative and permanent benefits in agriculture, making it a powerful tool for sustainable advancements. Methane-related traits, such as daily methane production and residual methane production (RMP), have shown moderate heritability (0.09 to 0.43) and genetic variability (Donoghue *et al.* 2020; Ryan *et al.* 2024), making methane a potentially suitable trait for breeding programs. However, methane is often genetically correlated with key performance traits like feed intake and liveweight (Donoghue *et al.* 2020) which are already part of most national breeding objectives. These correlations raise questions about the additional benefit of directly selecting for methane if these other traits are already under selection. The primary objective of the present study was to quantify the genetic variability in methane related traits, especially those independent of traits already directly considered in most cattle breeding goals.

MATERIALS AND METHODS

Methane and carbon dioxide flux data were collected from 1,700 growing beef cattle at the Irish Cattle Breeding Federation Progeny Test Centre in Tully, Co. Kildare, Ireland, between the years 2018 and 2024. Animals included 170 bulls, 576 heifers, and 954 steers, sourced from commercial farms and representing crossbreeds of 13 breeds: Aberdeen Angus, Aubrac, Belgian Blue, Charolais, Hereford, Holstein Friesian, Jersey, Limousin, Montbéliarde, Norwegian Red, Salers, Shorthorn, and Simmental. Animals were grouped into unisex pens based on sex, breed, and liveweight, with an average pen size of 36 animals.

Pens were equipped with 10 Greenfeed emission monitoring systems (C-Lock Inc., Rapid City, SD) for methane and carbon dioxide flux measurements and 10 Insentec feed stations (Hokofarm Group BV, Marknesse, The Netherlands) for feed intake monitoring. Methane and carbon dioxide emissions were recorded as flux measurements (g/day) using the Greenfeed system. To ensure data accuracy, only animals with at least 30 emission measures, each greater than 3 minutes in duration and at least one emission measure within 10 days of the test end were included. Methane production was calculated as the average daily methane flux across all valid measures.

Animals were fed *ad libitum* total mixed rations (TMR) comprising 13.95% hay, 40.70% water, and 45.35% concentrates, with a dry matter (DM) content of 51% and metabolisable energy (ME) value of 12.1 MJ/kg DM. Feed intake was monitored by the Insentec system, with feed disappearance assumed equal to intake. Energy intake was calculated as the sum of energy consumed from TMR and Greenfeed supplements.

Liveweight was recorded every three weeks. Average daily gain during the test period for each animal was calculated by fitting a linear regression through all liveweight observations of the animal; the same approach was also fitted through serial measures of metabolic liveweight. Animals were slaughtered at a commercial abattoir approximately 148 days after entering the test centre and carcass weight was recorded. Carcass records were available for an additional 671,100 paternal half-siblings of animals with methane measurements. Any animal born to a dam in parity >10 was omitted. Only animals with recorded parentage were included.

Statistical analysis. A series of residual methane traits were estimated using animal linear mixed models in ASReml (Gilmour *et al.* 2009) to reflect methane production adjusted for performance traits of ADG, metabolic liveweight, carcass weight and energy intake. The modelling of daily methane production was progressively built up with different combinations of the performance traits as independent variables. The base model used was:

$$Y = Parity_j + Het + Rec + CG_k + age_i + age*sex + sex*variable_l + sex*variable_m + a_i + e_{ijklm}$$

where Y was methane production; Parity was the fixed effect of dam parity j (1, 2, 3, 4, ≥5); Het was the heterosis covariate and Rec was the recombination loss covariate for animal I; CG was the fixed effect of the contemporary group k; age_i was the fixed effect of age in months at the end of test for animal j; age*sex was the interaction effect of age in months and animal sex (bull, heifer or steer); $variable_l$ was one of liveweight, energy intake, average daily gain, carcass weight; $variable_m$ was of the subset null, metabolic liveweight, energy intake, average daily gain or carcass weight where $variable_m \neq variable_n$; a_i was the additive random effect of the animal j; and e representing the residual variance. The same animal linear mixed model structure was used to estimate variance components for the additional traits (i.e., energy intake, liveweight, ADG, and carcass weight), excluding the sex*variable interaction terms, as these were only applicable when modelling methane production with performance traits as covariates.

RESULTS AND DISCUSSION

The coefficient of genetic variation (CVg) for daily methane production in the present study was 9.27%, comparable to key performance traits such as average daily gain (CVg = 7.23%) and carcass weight (CVg = 9.61%), for which genetic gain has been consistently achieved in beef breeding programs. This similarity suggests that, with adequate selection pressure and accurate genetic evaluations, methane production can also respond to selection.

While direct selection to reduce methane emissions is feasible, beef breeding objectives are multi-trait by design, designed for balanced progress across economically and biologically important traits. The key determinant of the potential additional rate of genetic progress within a breeding program is how much of genetic variability is genetically independent of other traits already being considered in the breeding programme (Rendel and Robertson 1950). The extent of genetic correlations among the goal traits impacts the response to selection. The genetic correlation between energy intake and daily methane production (0.51) indicates that only 75% of the genetic variance in daily methane production is genetically independent of energy intake. Traits like growth rate and carcass weight also make up breeding objectives, all of which were antagonistically correlated with methane production in the present study (Table 2), suggesting that, heavier, faster growing animals with heavier carcasses are, on average, genetically predisposed to produce more methane per day.

The presence of such (genetic) correlations between methane production and performance traits was the motivation in the present study to explore the genetic variability in residual-type traits. Residual traits, when derived via least squares regression, are independent of included regressors for the population in which they were derived. Residual traits have been traditionally calculated at the phenotypic level as a variable in themselves with the variance components of this new trait then calculated. In the present study, phenotypic adjustment of methane production had negligible impact on its heritability (Table 1). This process was undertaken in a single step, with the regressor traits included as covariates in the same model used to simultaneously estimate the variance components. Although all residual methane traits were phenotypically independent of their respective regressor traits by design, this adjustment did not eliminate underlying genetic correlations (Table 2). The three RMP traits adjusted for differences in metabolic liveweight (i.e. RMP_{lw}, RMP_{la}, RMP_{el}) were all negatively genetically correlated with metabolic liveweight (-0.41 to -0.38), highlighting that phenotypic adjustment alone was insufficient to fully remove shared genetic signal. Nonetheless, all RMP traits were strongly genetically correlated with each other (r > 0.88) and with unadjusted methane production (r >0.83), indicating substantial shared genetic architecture.

Table 1. Mean, genetic standard deviation (σ_g), environmental standard deviation (σ_e), and heritability (h^2; standard error in parentheses) for the methane related traits (N=1700), feed intake (N=1700), performance traits (N=1700), and carcass weight (N=45,779)

Trait	Mean	σ_g	σ_e	h^2 (SE)
Methane, g/day	252.70	23.43	27.69	0.42(0.09)
Energy intake, MJ/day	152.60	11.41	8.61	0.63(0.10)
Liveweight, kg	575.70	34.24	25.64	0.74(0.10)
Metabolic liveweight, kg	117.30	5.62	3.65	0.70(0.09)
ADG, kg/day	1.56	0.15	0.18	0.42(0.09)
Carcass weight, kg	321.60	23.26	24.97	0.46(0.03)
RMP_{adg}[1]		21.59	27.70	0.38(0.09)
RMP_{energy}[1]		20.34	25.70	0.38(0.09)
RMP_{cw}[1]		23.28	25.45	0.46(0.09)
RMP_{lw}[1]		23.48	25.69	0.46(0.09)
RMP_{ca}[1]		22.40	25.75	0.43(0.09)
RMP_{la}[1]		22.71	25.99	0.43(0.10)
RPM_{el}[1]		20.76	25.30	0.40(0.10)

[1] RMP_{adg}, methane adjusted for ADG; RMP_{energy}, methane adjusted for energy intake; RMP_{cw} methane adjusted for carcass weight; RMP_{lw}, methane adjusted for liveweight; RMP_{ca}, methane adjusted for carcass weight and ADG; RMP_{la} methane adjusted for liveweight and ADG; RMP_{el} methane adjusted for energy intake and liveweight

Using the genetic correlations estimated in the present study (Table 2 and 3), the genetic variability in daily enteric methane production genetically independent of a combination of other traits was also calculated as

$$\sigma_A^2(1 - R^2) = \sigma_A^2(1 - \boldsymbol{C}'\boldsymbol{V}^{-1}\boldsymbol{C})$$

where σ_A^2 is the additive genetic variance for daily methane production, R^2 is the proportion of variation in daily methane production explained by the performance traits, **C** is the vector of genetic correlations between the daily methane production and the performance traits, and **V** is the matrix of genetic correlations among the performance variables.

When daily methane production was adjusted for the same covariate(s) either phenotypically or genetically, the resulting traits were strongly genetically correlated (>0.97), indicating that both adjustment methods captured nearly identical genetic signals. When daily methane production was genetically adjusted for energy intake, metabolic liveweight, ADG, and carcass weight simultaneously, only 56% of the genetic variance of daily methane production remained independent of these traits. This adjustment reduced the genetic standard deviation by 25%, from 23.43 g/d to 17.55 g/d, with the CVg reducing from 9.27% to 6.94%.

Table 2. Genetic correlations between methane production and the residual traits with energy intake, ADG, metabolic liveweight and carcass weight (SE in parenthesis)

	Energy	ADG	MWT	Carcass
Methane	0.51 (0.09)	0.39 (0.11)	0.27 (0.11)	0.42 (0.09)
RMP_{adg}[1]	0.43 (0.11)	0.06 (0.13)	0.07 (0.13)	0.15 (0.14)
RMP_{lw}[1]	0.26 (0.12)	0.06 (0.14)	-0.38 (0.11)	-0.21 (0.13)
RMP_{cw}[1]	-0.13 (0.15)	0.06 (0.14)	-0.25 (0.12)	-0.16 (0.14)
RMP_{energy}[1]	-0.11 (0.14)	-0.11 (0.14)	-0.07 (0.14)	-0.13 (0.15)
RMP_{ca}[1]	0.42 (0.11)	0.13 (0.14)	-0.28 (0.12)	-0.17 (0.14)
RMP_{la}[1]	0.25 (0.12)	-0.05 (0.15)	-0.38 (0.12)	-0.22 (0.13)
RMP_{el}[1]	-0.12 (0.13)	-0.08 (0.15)	-0.41 (0.12)	-0.21 (0.12)

[1] RMP_{adg}, methane adjusted for ADG; RMP_{energy}, methane adjusted for energy intake; RMP_{cw} methane adjusted for carcass weight; RMP_{lw}, methane adjusted for liveweight; RMP_{ca}, methane adjusted for carcass weight and ADG; RMP_{la} methane adjusted for liveweight and ADG; RMP_{el} methane adjusted for energy intake and liveweight

Table 3. Genetic correlations among energy intake, average daily gain (ADG), metabolic liveweight (MWT), and carcass weight (SE in parenthesis)

Trait	ADG	MWT	Carcass weight
Energy intake	0.63 (0.08)	0.71 (0.06)	0.76 (0.05)
ADG		0.51(0.10)	0.55(0.09)
MWT			0.95(0.01)

CONCLUSION

Genetic adjustment showed that only 56% of the genetic variance in methane production was independent of intake, growth, and carcass traits. This suggests a significant overlap in genetic control, highlighting the importance of evaluating which traits are already under selection. To justify including a methane trait in a breeding program, it is essential to ensure it captures unique genetic variation beyond existing performance traits.

REFERENCES

Donoghue K.A., Bird-Gardiner T., Herd R.M., Hegarty R.S. and Arthur P.F. (2020) *J. Anim. Sci.* **98**: 1.

Gilmour A.R., Gogel B.J., Cullis B.R. and Thompson R. (2009) ASReml User Guide Release 3.0. VSN International Ltd, Hemel Hempsted, UK.

Rendel J.M. and Robertson A. (1950) *J. Genet.* **50**: 1.

Ryan C.V., Pabiou T., Purfield D.C., Berry D.P., Conroy S., Murphy C.P. and Evans R.D. (2024) *J. Anim. Sci.* **102**: 1.

Proc. Assoc. Advmt. Anim. Breed. Genet. 26: 34-37

METHANE EMISSIONS OF LOW AND HIGH METHANE YIELD SELECTION LINE SHEEP FED RYEGRASS-BASED PASTURE ACROSS MULTIPLE PERIODS

A. Jonker[1], S. Hickey[2], J.C. McEwan[3], E. Sandoval[1], S. MacLean[1] and S. Rowe[3]

[1] AgResearch Limited, Grasslands Research Centre, Palmerston North 4442, New Zealand
[2] AgResearch Limited, Ruakura Research Centre, Hamilton 3214, New Zealand
[3] AgResearch Limited, Invermay Agricultural Research Centre, Mosgiel 9092, New Zealand

SUMMARY

There is a lack of information on actual CH_4 emissions from low and high CH_4 selection line sheep-fed pasture *ad libitum* across different periods of the year. The objective of the two experiments was to determine the CH_4 emissions in respiration chambers from low and high CH_4 yield selection line sheep while fed cut pasture *ad libitum* across multiple measurement periods. In both experiments, CH_4 yield was consistently less in low compared to high CH_4 yield selection line sheep being, respectively, 8.9, 8.4 and 9.7 % less in periods 1 to 3 in Exp. 1 and 7.6, 9.6, 7.7 and 8.9 % less in periods 1 to 4 in Exp. 2. These results indicate that the magnitude of CH_4 yield differences between selection lines is consistent across different periods of the year.

INTRODUCTION

Breeding for low CH_4 in sheep is a viable mitigation option (Pinares-Patiño *et al.* 2013) and breeding values (BV's) for low CH_4 emissions are now available for sheep farmers in New Zealand who genotype their sheep and whose sheep are phenotyped for CH_4 using portable accumulation chambers (Jonker *et al.* 2018; Rowe *et al.* 2020). However, the initial phenotyping for the research CH_4 BV's was performed while the sheep were fed lucerne pellets at a fixed feeding level of 2 × maintenance energy requirements (Pinares-Patiño *et al.* 2013), which is different from farm grazing conditions. The CH_4 emissions and CH_4 yield between the two CH_4 selection lines were also found to differ using the SF_6 tracer technique in the field while the sheep were gazing at ryegrass-based pasture (Jonker *et al.* 2017). However, the magnitude of difference between selection line sheep was not the same as on lucerne pellets and there were questions about the accuracy of the DMI estimates. Furthermore, the consistency of the CH_4 yield ranking across repeat periods or seasons with different pasture compositions was not determined. Therefore, there is a lack of robust information on actual 24 h CH_4 emissions from low and high CH_4 selection line sheep-fed pasture across different periods of the year when fed pasture under controlled conditions where both CH_4 emissions and DMI can be accurately determined.

The objective was to determine CH_4 emissions and DMI in respiration chambers from low and high CH_4 yield selection line sheep while fed *ad libitum* cut pasture across multiple measurement periods.

MATERIALS AND METHODS

The two sheep studies were approved by the AgResearch Grasslands Animal Ethics Committee (Palmerston North, New Zealand; AE1364 and 14007).

Animal source. The sheep used are part of the 'Methane yield selection' flock sheep, being bred solely, since 2012, based on having a low or high CH_4 yield (Jonker *et al.* 2018). A cohort of 96 female yearling progeny born in spring 2014 were used in Exp. 1 and a cohort of 72 ram lambs, born in spring 2016 were used in Exp. 2 (same sheep as in Jonker *et al.* 2021). The sheep born in a birth year were raised as one flock at AgResearch Woodlands (Woodlands, New Zealand) and were transported to AgResearch Grasslands (Palmerston North, New Zealand) before the current two experiments in June 2015 and February 2017, respectively.

Measurement periods. The CH_4 measurements were performed in three periods between 7 October 2015 and 12 February 2016 in Exp. 1 and in four periods between 6 March to 13 December 2017 in Exp. 2. In both studies, the sheep were fed *ad libitum* cut ryegrass-based pasture.

Sheep were initially housed in group pens for at least 14 days on cut pasture before being moved into individual metabolic crates for two days followed by two days in 24 respiration chambers as described previously (Pinares-Patiño and Waghorn, 2018; Jonker *et al.* 2021). The sheep were grazing ryegrass-based pasture before and between measurement periods.

The pasture was cut daily at around 1100 h and stored in a chiller at 4°C till feeding at around 1530 h and 0830 h the next day. Representative samples from each daily pasture cut were dried at 105 °C for 24 h in triplicate for dry matter (DM) content determination. Refusals were collected once daily before the afternoon feeding, and an aliquot was dried at 65 °C for 48 h. The DMI was calculated as feed DM offered – feed DM refusals.

Statistical analysis. The DMI and CH_4 emissions were analysed separately for each measurement period in each experiment using a standard least squares model (SAS 2012) with a repeated measures model using ASREML (Gilmour *et al.* 2009), fitting animal as a random effect and selection line, sampling group (24 in a chamber measurement group) and day as fixed effects, and live weight (LW) as a covariate for the three CH_4 emission parameters. The covariate was not used to analyse DMI to enable the expression of *ad libitum* DMI. Significance was declared at $P < 0.05$.

RESULTS AND DISCUSSION

The DMI was on average 6.4% greater ($P>0.05$) across the three periods in Exp. 1 in low compared to high CH_4 yield selection line sheep, while the DMI was similar between selection lines in Exp. 2 (Tables 1 and 2).

Table 1 Dry matter intake (DMI) and methane (CH_4) emissions in low and high CH_4 yield selection sheep fed ad libitum cut pasture across three periods in Experiment 1

Item	DMI (kg/day)	CH_4 (g/day)	CH_4 (g/kg DMI)	CH_4/CO_2 (mol/mol)
Period 1, Oct/Nov 15				
High line	1.44 ± 0.030	34.4 ± 0.48	23.6 ± 0.29	0.070 ± 0.0005
Low line	1.55 ± 0.029	32.2 ± 0.46	21.7 ± 0.28	0.067 ± 0.0005
Difference (%)	-6.6	6.8	8.9	5.3
High – Low	-0.10 ± 0.04	2.19 ± 0.68	1.93 ± 0.42	0.004 ± 0.001
P-value	**0.018**	**0.002**	**<0.0001**	**<0.0001**
Period 2, Nov/Dec 15				
High line	1.56 ± 0.030	35.1 ± 0.47	22.3 ± 0.32	0.069 ± 0.0006
Low line	1.66 ± 0.029	32.9 ± 0.45	20.6 ± 0.31	0.065 ± 0.0006
Difference (%)	-5.8	6.7	8.4	5.4
High – Low	-0.10 ± 0.040	2.19 ± 0.67	1.73 ± 0.46	0.0035 ± 0.0009
P-value	**0.0244**	**0.002**	**0.0003**	**0.0001**
Period 3, Jan/Feb 16				
High line	1.84 ± 0.031	38.4 ± 0.46	20.7 ± 0.24	0.068 ± 0.0005
Low line	1.97 ± 0.030	36.2 ± 0.44	18.9 ± 0.23	0.063 ± 0.0005
Difference (%)	-6.7	6.1	9.7	7.2
High - Low	-0.13 ± 0.04	2.22 ± 0.65	1.83 ± 0.34	0.005 ± 0.001
P-value	**0.003**	**0.001**	**<0.0001**	**<0.0001**

This DMI followed the same trend as LW, which was on average 5.3% greater across the three periods in Exp. 1 in low compared to high CH_4 yield selection line sheep, while LW was similar between selection lines across the four periods in Exp. 2. It is generally expected that a sheep with a larger LW would be able to consume more feed dry matter.

In both experiments, CH_4 yield was consistently less in low compared to high CH_4 yield selection line sheep being, respectively, 8.9, 8.4 and 9.7 % less in periods 1 to 3 in Exp. 1 and 7.6, 9.6, 7.7 and 8.9 % less in periods 1 to 4 in Exp. 2. The repeatability for CH_4 yield was 0.43 in Exp. 1 and 0.32 in Exp. 2 (Jonker *et al.* 2021). To note, when interpreting the results of the two experiments, the difference between CH_4 selection lines was likely bigger in Exp. 1 than in Exp. 2 because the 12 extreme low and 12 extreme high CH_4 rams in Exp. 2 were not part of this study as they were retained in the main flock for the breeding season.

Table 2 Dry matter intake (DMI) and methane (CH_4) emissions in low and high CH_4 yield selection sheep fed ad libitum cut pasture across four periods in Experiment 2

Item	DMI (kg/day)	CH_4 (g/day)	CH_4 (g/kg DMI)	CH_4/CO_2 (mol/mol)
Period 1, March/April 17				
High line	1.23 ± 0.019	23.4 ± 0.27	19.1 ± 0.26	0.065 ± 0.0006
Low line	1.25 ± 0.018	22.0 ± 0.26	17.8 ± 0.25	0.060 ± 0.0006
Difference (%)	-1.3	6.4	7.6	8.5
High – Low	-0.02 ± 0.03	1.40 ± 0.37	1.36 ± 0.36	0.005 ± 0.001
P-value	0.523	**0.0004**	**0.0003**	**<0.0001**
Period 2, June/July 17				
High line	1.29 ± 0.019	26.2 ± 0.44	20.3 ± 0.41	0.064 ± 0.0008
Low line	1.34 ± 0.019	24.6 ± 0.45	18.5 ± 0.41	0.061 ± 0.0008
Difference (%)	-3.7	6.2	9.6	5.7
High – Low	-0.05 ± 0.02	1.52 ± 0.58	1.79 ± 0.54	0.004 ± 0.001
P-value	**0.044**	**0.012**	**0.0016**	**0.0009**
Period 3, Sep 17				
High line	1.53 ± 0.022	36.3 ± 0.52	23.7 ± 0.26	0.070 ± 0.0006
Low line	1.54 ± 0.022	33.6 ± 0.52	22.0 ± 0.26	0.065 ± 0.0006
Difference (%)	-0.7	7.9	7.7	7.8
High – Low	-0.01 ± 0.03	2.66 ± 0.74	1.69 ± 0.36	0.005 ± 0.001
P-value	0.744	**0.0006**	**<0.0001**	**<0.0001**
Period 4, Nov/Dec 17				
High line	1.87 ± 0.028	39.9 ± 0.48	21.5 ± 0.31	0.078 ± 0.0006
Low line	1.86 ± 0.029	36.6 ± 0.50	19.7 ± 0.31	0.071 ± 0.0006
Difference (%)	-0.6	8.9	8.9	10.0
High - Low	0.01 ± 0.04	3.27 ± 0.69	1.75 ± 0.44	0.007 ± 0.0009
P-value	0.782	**<0.0001**	**0.0002**	**<0.0001**

The findings of the current two experiments were similar to the findings of Pinares-Patiño *et al.* (2011ab), who observed 11 and 12 % differences in CH_4 yield between sheep selected to have extremely low or high CH_4 yield measured approximately 8 months apart while fed cut pasture. The mean research BV's for CH_4 yield for the low and high CH_4 yield selection line sheep (on lucerne pellets) were, on average 8.4 and 11.9 % different in 2016 and 2018, respectively (Rowe *et al.* 2019), again being similar differences between selection lines to that observed in the current study. In absolute terms, the g of CH_4 yield difference between low and high CH_4 selection line sheep was 1.7-1.9 g in Exp. 1 and 1.4-1.8 g in Exp. 2, which was also similar to the difference in CH_4 yield selection BV's of 1.4 and 2.0 g in 2016 and 2018, respectively. This was even though the average CH_4 yield was lower at 16.0 g/kg DMI during phenotyping for BV's on lucerne pellets (Rowe *et al.* 2019) than the average CH_4 yield of 20.8 g/kg DMI on fresh pasture in the current two studies. The

CH_4 yield in the current two experiments was in a normal CH_4 yield range for sheep fed fresh pasture only (Swainson *et al.* 2018).

CONCLUSION

The results of the current study indicate that the magnitude of CH_4 yield differences (as % and g difference) between low and high CH_4 yield selection lines sheep were consistent across different periods when fed *ad libitum* cut pasture in both experiments. Therefore, the CH_4 yield BV's generated for NZ sheep can be used to adjust CH_4 emission predictions in farm carbon calculators across all seasons.

ACKNOWLEDGEMENTS

The sheep studies were funded by the Pastoral Greenhouse Gas Research Consortium (PGgRc) and the New Zealand Agricultural Research Centre (NZAGRC).

REFERENCES

Gilmour A.R., Gogel B.J., Cullis B.R. and Thompson R. (2009) ASReml User Guide Release 3.0. VSN Int. Ltd., Hemel Hempstead, UK.

Jonker A., Hickey S., Boma P., Woyimo Woju C., Sandoval E., MacLean S., García Rendón Calzada M., Yu W., Lewis S., Janssen P.H., McEwan J.C. and Rowe S. (2021) *Anim. Prod. Sci.* **61**: 300.

Jonker A., Hickey S., Pinares-Patiño C., McEwan J., Olinga S., Díaz A., Molano G., MacLean S., Sandoval E., Harland R., Birch D., Bryson B., Knowler K. and Rowe S. (2017) *J. Anim. Sci.* **95**: 3905.

Jonker A., Hickey S.M., Rowe S.J., Janssen P.H., Shackell G., Elmes S., Bain W.E., Wing J., Greer G.J., Bryson B., MacLean S., Dodds K.G., Pinares-Patiño C.S., Young E.A., Knowler K., Pickering N.K. and McEwan J.C. (2018) *J. Anim. Sci.* **96**: 3031.

Pinares-Patiño C.S., Ebrahimi S.H., McEwan J.C., Clark H. and Luo D. (2011a) *Proc. N. Z. Soc. Anim. Prod.* **71**: 219.

Pinares-Patiño C.S., Hickey S.M., Young E.A., Dodds K.G., MacLean S., Molano G., Sandoval E., Kjestrup H., Harland R., Hunt C., Pickering N.K. and McEwan J.C. (2013) *Animal* **7**: 316.

Pinares-Patiño C.S., McEwan J.C., Dodds K.G., Cárdenas E.A., Hegarty R.S., Koolaard J.P. and Clark H. (2011b) *Anim. Feed Sci. Technol.* **166–167**: 210.

Pinares-Patiño C.S. and Waghorn G.C. (2018) Technical Manual on Respiration Chamber Design (September 2018). In: Pinares-Patiño C.S., Waghorn G.C. (Eds.). Ministry of Agriculture and Forestry, Wellington, New Zealand.

Rowe S., Hickey S., Johnson P., Bilton T.P., Jonker A., Bain W.E., Veenvliet G., Plie G., Bryson B., Knowler K., Amyes N., Newman S. and McEwan J.C. (2020) *Proc. Assoc. Advmt. Anim. Breed. Genet.* **24**: 15.

Rowe S.J., Hickey S.M., Jonker A., Hess M.K., Janssen P.H., Johnson P., Bryson B., Knowler K., Pinares-Patino C., Bain W.E., Elmes S., Young E., Wing J., Waller E., Pickering N. and McEwan J.C. (2019) *Proc. Assoc. Advmt Anim. Breed. Genet.* **23**: 306.

Swainson N., Muetzel S. and Clark H. (2018). *Anim. Prod. Sci.* **58**: 973.

EFFECT OF WINTER FORAGE BRASSICA GRAZING ON METHANE SELECTION LINE SHEEP

F.L. Booker[1], S.M. Hickey[2], T.P. Bilton[1], I. Hampton[1], B. Bryson[3], M. Ward[3], P.L Johnson[1], J.C. McEwan[1] and S.J. Rowe[1]

[1] AgResearch, Invermay Agricultural Centre, Mosgiel, New Zealand
[2] AgResearch, Ruakura Research Centre, Hamilton, New Zealand
[3] AgResearch, Woodlands Research Farm, Invercargill, New Zealand

SUMMARY

Breeding for lowered methane emissions has been shown to be an effective strategy for reducing methane emissions to meet national and international reduction targets. Other mitigation strategies have been explored such as different feed types, feed additives, and vaccines. The additivity of methane mitigation strategies, however, remains unknown. Divergent methane selection line female sheep under single trait selection for either high or low methane emissions were measured at two time points. They were first measured for methane emissions on rye grass pasture at ~8 months of age followed by a measure of methane emissions after 6 weeks of grazing a forage brassica crop at ~10 months of age. A pilot study was conducted where 52 high line and 52 low line females grazed on a forage brassica which has previously been shown to reduce methane emissions in sheep. Methane measures were collected using portable accumulation chambers (PAC). Three traits were evaluated, methane emissions in grams per day (CH4), carbon dioxide emissions in grams per day (CO2) and a ratio of mols of CH4/(mols CO2 + mols CH4) (CH4ratio). This ratio trait is a predictor of methane yield or grams CH4/ kg DMI. Line was significant for all three traits. Feed type was significant for CH4 and CH4ratio but not for CO2. There was no significant interaction between line and feed type. Feeding the brassica reduced CH4 on average by ~22%. Differences between lines in CH4, CH4ratio and CO2 when grazing grass were ~20%, ~35%, and ~34% respectively and when fed forage brassica differences between lines in CH4, CH4ratio and CO2 were ~23%, ~36% and ~27% respectively.

INTRODUCTION

Breeding for lowered methane emissions has been shown to be an effective strategy for reducing methane emissions (Rowe *et al.* 2019). Other mitigation strategies are being developed. Feed additives and feed types have been shown to reduce methane emissions (Hristov 2023). Brassicas, in particular, have been shown to effectively lower enteric methane emissions in livestock. Brassicas are a common supplementary feed, often fed in Summer or Winter, when feed shortages can occur. Although feeding of brassicas such as forage rape and breeding have both been shown to reduce methane emissions in sheep and cattle (Sun *et al.* 2018), to date, no studies have looked at the additivity or "stacking" of the two strategies. Here we have used divergent methane selection lines (Rowe *et al.* 2019) to explore the effect of brassica feeding on animals already under selection for methane emissions.

MATERIALS AND METHODS

Experimental animals and protocols applied in this study were approved by the AgResearch Ruakura (Hamilton, NZ) Animal Ethics committee (application number 2173). The AgResearch Methane Selection Lines were established in 2009 and have been previously described (Rowe *et al.* 2019). This study used 104 10-month-old females from this flock, born in 2023, with 52 animals from each of the high (H) and low (L) methane emitting lines measured. Methane data was collected using Portable Accumulation Chambers (PAC) which have been previously described (Jonker *et al.* 2020).

Animals had their first methane measure collected while grazing on a ryegrass and clover-based pasture system (Round 1). Following this initial measurement, animals were transitioned to a forage brassica crop (PGG Wrightson's Titan Forage Rape). Animals were transitioned to the crop over 14 days, increasing time on crop by 2 hours every other day for the first week, followed by a 2 hour increase daily for the second week. Following this transition period, the animals continued to graze the crop for a further 4 weeks. At the end of the 6 weeks on brassica, the final PAC measures were collected (Round 2). Three traits were evaluated, methane emissions in grams per day (CH4), carbon dioxide emissions in grams per day (CO2) and a molar ratio of CH4/ (CO2 + CH4) hereafter CH4ratio.

Statistical Analysis. Methane selection line, feed type, and fixed effects and covariates of birth rearing rank (brr) and age of dam (aod) were fitted using a linear mixed model in SAS JMP statistical software. Interactions between line and feed type and between line and liveweight were tested and dropped from the final model. The final model used was:

$$Trait_{ijklm} \sim feed_i + selection\ line_j + aod_k + brr_l + e_{ijklm}$$

where $Trait_{ijkl}$ is the PAC trait (CH_4, CO_2 or $CH_4/(CH_4 + CO_2)$), $feed_j$ is feed type of grass or brassica, *selection line*$_j$ is divergent selection lines for high or low methane emissions per kg dry matter intake, aod_k is age of dam, brr_l is birth rearing rank of four levels (born as a single or twin and raised as a single or twin) and e_{ijklm} is residual error term where $e_{ijklm} \sim N(0,\sigma_e^2\mathbf{I})$.

RESULTS AND DISCUSSION

There were no significant differences in liveweight between the selection lines at both time points. Interactions between selection line and liveweight were tested but were not significant. Therefore, liveweight was not included in the final model. However, average liveweights for the animals, as well as estimated feed offerings are displayed in Table 1.

Table 1. Average feed offering and liveweights

Feed type	Approximate Feed Offering (Kg/DM/Animal/Day)	Average Liveweight (Kg)	
		L	H
Grass	2.38	48.2	51.7
Brassica	4.28	49.1	51.5

Table 2. Average, standard error of mean (SEM) and percentage change (% Δ) of methane, methane yield, and carbon dioxide by feed type and selection line

	Average Methane (g/day)				Average Methane Ratio (mol CH4/Mol CH4 + mol CO2)				Average Carbon Dioxide (g/Day)			
	L	H	SEM	% Δ	L	H	SEM	% Δ	L	H	SEM	% Δ
Grass	11.67	14.58	0.416	19.95	0.029	0.045	0.001	35.5	1153	861	47.00	33.9
Brassica	8.95	11.68	0.416	23.37	0.023	0.036	0.001	36.1	1092	859	47.00	27.1

Table 2 shows proportional differences between lines in CH4, CH4ratio and CO2 were similar regardless of feed type. When fed on pasture, the methane emissions in grams per day of the low methane animals were 19.95% less than the high methane animals. A similar difference between the lines was seen when animals were grazed on brassica (23.37%). Differences between highs and lows when grazing grass were ~35% and ~34% for CH4ratio and CO2 respectively, and ~36% and ~27% when fed the forage brassica. These results reflect previous results shown by Jonker *et al.* (2019)

where seasonal nutritional variation affected the mean of absolute emissions but the proportional differences between the lines stayed consistent.

Line was highly significant for all three traits (Table 3). Feed type was significant for CH4 and CH4ratio but not for CO2. There was no significant interaction between line and feed type.

Table 3. Estimated differences for methane, methane yield and carbon dioxide by Line and feed type

	CH4 (g/day)			CH4ratio			CO2 (g/Day)		
	Δ	P-Value	Mean	Δ	P-Value	Mean	Δ	P-Value	Mean
Line	2.96	1e-10	11.37	0.0147	1e-24	0.1404	262.6	3.7e-7	999.0
Feed	2.81	3e-15	11.37	0.008	1e-14	0.1556	31.35	0.186	999.0

Compared to grazing a rye grass clover pasture, methane emissions in grams per day when sheep were fed on brassica were reduced by 22%. This is less than the 37% reduction reported by Sun *et al.* (2012, 2016) for the same brassica species. Nonetheless, this still shows a significant reduction in methane emissions in sheep fed forage rape.

CONCLUSIONS

In this pilot study, we investigated the effect of feeding brassicas on methane emissions of animals already under selection for methane. Feeding a brassica crop reduced absolute methane emissions (CH4) by ~22%. This was regardless of selection line status and there was no significant difference between selection lines on the effect of brassica. High and low selection lines differed by similar amounts when either fed grass or brassica. The reduction in methane emissions seen when grazing brassicas in this study was in line with the trends reported in the literature, but to a different magnitude. While these are preliminary results, this work suggests that these two mitigation technologies are "stackable". It also suggests that evaluation of other mitigation technologies in these selection lines could be a valuable component of their evaluation prior to release.

ACKNOWLEDGEMENTS

Funding provided through the New Zealand Agricultural Greenhouse Gas Research Centre (NZAGRC) science programme. We would like to acknowledge the hard work of the Woodlands Research Farm team, without whom this work would not be possible.

REFERENCES

Hickey S.M., Bain W.E., Bilton T.P., Greer G.J., Elmes S., Bryson B., Pinares-Patiño C.S., Wing J., Jonker A., Young E.A. and Knowler K. (2022) *Front. Genet.* **13**: 910413.

Hristov A. (2023) *J. Dairy Sci.* **107**: 4129.

Jonker A., Hickey S.M., McEwan J.C. and Waghorn G.C. (2020) In 'Guideline for estimating methane emissions from individual ruminants using: GreenFeed, 'sniffers', hand-held laser detector and portable accumulation chambers', pp. 49-56, editors A. Jonker and G. Waghorn, New Zealand Agricultural Greenhouse Gas Research Centre, New Zealand.

Jonker A., Hickey S., MacLean S., Woju C.W., Calzada M.G.R., Yu W., McEwan J. and Rowe S.J. (2019). Magnitude of difference in methane emissions between sheep bred for low or high methane yield is persistent across four seasons on pasture. 8th International Symposium on Non-CO2 Greenhouse Gases (NCGG8), June 12-14, 2019, Amsterdam, the Netherlands.

Rowe S.J., Hickey S.M., Jonker A., Hess M.K., Janssen P., Johnson T., Bryson B., Knowler K., Pinares-Patino C., Bain W. and Elmes S. (2019) *Proc. Assoc. Advmt. Anim. Breed. Genet.* **23**: 306.

Sun X.Z., Waghorn G.C., Hoskin S.O., Harrison S.J., Muetzel S. and Pacheco D. (2012) *Anim. Feed. Sci. Technol.* **176**: 107.

Sun X.Z., Pacheco D. and Luo D. (2016) *Anim. Prod. Sci.* **56**: 451.

Proc. Assoc. Advmt. Anim. Breed. Genet. 26: 42-45

PORTABLE ACCUMULATION CHAMBERS FOR MEASURING ENTERIC METHANE IN CATTLE

T.P. Bilton[1], S.M. Hickey[2], A. Searle[1], I. Hampton[1], F. Booker[1], C. Rodgers[1], S. Sevier[3], S. Gebbie[3], K.G. Dodds[1], G.M. Jenkins[4], J.A. Archer[4], J.C. McEwan[1] and S.J. Rowe[1]

[1]Invermay Agricultural Centre, AgResearch, Mosgiel, NZ
[2]Ruakura Research Centre, AgResearch, Hamilton, NZ
[3]Lincoln Research Centre, AgResearch, Lincoln, NZ
[4]Beef + Lamb New Zealand, Dunedin, NZ

SUMMARY

Breeding has been shown to be a viable solution for reducing methane emissions in ruminants but requires obtaining on-farm measurements of methane. One technology that has successfully been used on-farm to obtain short-term commercial methane measurements in sheep and goats is portable accumulation chambers (PAC). However, the original PAC units are unsuitable for cattle. In this project, PAC units designed and engineered specifically for cattle were developed and used to measure methane and carbon dioxide emissions from 125 Angus-Hereford (pure or crossbred) heifers across two time-periods. PAC performed as expected with no animal behaviour problems and were shown to be a suitable high-throughput method for the measurement of methane emissions from young beef cattle. The repeatability was found to be 0.35 ± 0.08 for methane emissions (g/day) (CH_4), 0.13 ± 0.09 for carbon dioxide emissions (g/day) (CO_2), and 0.31 ± 0.09 for the molar ratio trait of $CH_4/(CH_4 + CO_2)$. These results suggest that the newly designed cattle PAC units could enable large-scale on-farm methane measurements to be collected for the purpose of ranking animals on their methane emissions. This would facilitate the development of breeding schemes to select for low methane emitting cattle. Future studies should be conducted to investigate the heritability of PAC traits obtained from the cattle chambers.

INTRODUCTION

Concern regarding the impact of greenhouse gases on climate change has increased in recent years and has led to renewed efforts for developing solutions to reduce global emissions in line with international reduction targets set by governments and international organizations. In New Zealand, one third of all greenhouse gas emissions are generated through enteric methane emissions from grazing livestock, with over 70% coming from cattle (MfE 2024). Current targets in New Zealand for biogenic emissions are a 24-47% reduction by the year 2050. Many mitigation technologies, such as feed additives, require daily individual feeding which is untenable in extensive pasture systems. Breeding is currently the only technology commercially available for methane mitigation in ruminants grazing pasture and for the New Zealand sheep industry, methane genomic breeding values are generated as part of routine genetic evaluations (Archer *et al.* 2023). However, breeding for low methane animals requires phenotypic measures of methane and measurements of methane emissions in grazing ruminants is challenging as current technologies (i.e., GreenFeed (C-Lock Inc., Rapid City, SD), SF_6 tracers (Johnson *et al.* 1994)) are low-throughput and expensive for pastoral based systems. In sheep, one technology that has successively been implemented commercially is portable accumulation chambers (PAC) with over 5,000 commercial measures per year routinely collected in New Zealand (Archer *et al.* 2023; Jonker *et al.* 2018). However, the original PAC units were not designed for large animals such as cattle. The objective of this study was to test a newly designed cattle PAC unit to measure methane emissions on a young beef cattle, and whether there is sufficient variation and repeatability in the measures to warrant upscaling to more herds.

MATERIALS AND METHODS

The trial was carried out in accordance with the Animal Welfare Act 2006 and AgResearch Code of Ethical Conduct, and experimental animals and protocols applied in this study were approved by the AgResearch Animal Ethics committee (application number 2008).

Animals. Beef heifers (n=125) bred for the 'Informing New Zealand Beef programme' were transported to AgResearch, Invermay in July 2023, for a period of one month. They were 10 months of age at time of transportation and were a reciprocal cross of Angus (A) and Hereford (H) with four genetic groups represented (AA, AH, HA and HH). The heifers were progeny of cows in their first (dam born in 2020) or second parity (dam born in 2019). There were four management groups that had equal representation of genetic groups but were separated into age of dam and liveweight, where these groups were born 2019 dams with light (L19) and heavy (H19) heifers and born 2020 dams with light (L20) and heavy (H20) heifers (Table 1). An additional 4 animals were originally included in the trial, but were excluded from the analysis due to being identified as steers or failure to be successfully measured using PAC.

PAC measurements. Animals were grazed in management groups on ad-libitum pasture and baleage for 10 days prior to measurement in sealed PAC units designed and engineered specifically for cattle. For the first round of measurements, animals were removed from pasture at 8 am and held in yards with no feed prior to measurement. The animals in each management group had been randomly allocated into measurement 'lots' (n=6 per lot) within sire and breed and were measured on the same day. After one hour off feed, the first lot of six heifers were each loaded into one of the six PACs. Liveweight was taken immediately prior to the animals entering the chambers. At the start of the measurement period, methane, carbon dioxide, and oxygen levels were recorded using a hand-held analyser for each chamber together with ambient temperature and pressure outside the chamber. After 25 minutes an interim gas measure was collected and after 50 minutes a final gas measure was taken from each chamber. After 14 days, a second 'round' of measures were repeated on the same animals that were re-allocated to different lots of 6.

For direct emissions, methane in parts per million (ppm) and percentage of carbon dioxide in the air volume were recorded at 50 min, which were converted to litres, moles and then to grams emitted per measurement period. These were calculated using the equations presented in Rodrigues *et al.* (2024) to produce estimates of grams of methane (CH_4) and grams of carbon dioxide (CO_2) emitted per day. A ratio trait was also derived as an indicator of methane yield (i.e., grams of methane per kg of dry matter intake) and was calculated as the moles of methane per day divided by the total moles ($CH_4/(CH_4 + CO_2)$).

Table 1. Number of animals (mean liveweight[1]) per management group and breed

Breed	Management Group				
	L19	H19	L20	H20	Total
Angus-Angus (AA)	9 (223)	9 (253)	6 (223)	6 (252)	30 (238)
Angus-Hereford (AH)	6 (209)	6 (238)	7 (216)	6 (246)	25 (227)
Hereford-Angus (HA)	8 (240)	8 (271)	6 (218)	7 (245)	29 (245)
Hereford-Hereford (HH)	11 (208)	11 (242)	10 (207)	9 (232)	41 (223)
Total	34 (220)	34 (251)	29 (215)	28 (243)	125 (232)

[1]Liveweight (kg) of the heifers 3-months prior to PAC measurements being collected.

Statistical Analysis. Repeatability of PAC methane traits for cattle, fixed effects and covariates were tested using a linear mixed model in SAS JMP statistical software. The model used to estimate repeatability was:

$$Trait_{ijklm} \sim Round_j + Group_k + Round_j{*}Group_k + Lot_l + Wait\_Time_m + \alpha_i + e_{ijklm} \qquad (M1)$$

where $Trait_{ijkl}$ is the PAC trait (CH_4, CO_2 or $CH_4/(CH_4 + CO_2)$), $Round_j$ is the four-day measurement period (1 or 2), $Group_k$ is the management group, $Round_j{*}Group_k$ is the interaction between $Round_j$ and $Group_k$, Lot_l is the group of 6 animals measured at the same time on a given day, $Wait\_Time_m$ is a covariate of the time (minutes) between the animal being removed from feed and entering the chambers, α_i is the animal random effect where $\alpha_i \sim N(0,\sigma_p^2\mathbf{I})$ and **I** is the identity matrix, and e_{ijklm} is residual error term where $e_{ijklm} \sim N(0,\sigma_e^2\mathbf{I})$. The repeatability is $\sigma_p^2/(\sigma_p^2+\sigma_e^2)$.

RESULTS AND DISCUSSION

The mean liveweight of the heifers at measurement was 250 kg and the mean methane emission was 85.5 g/day across both rounds of PAC (Table 2). In comparison, the average methane emissions for a research sheep flock (Booker *et al.* 2024) in the same season was 12.8 g/day from 323 growing lambs weighing an average of 48.7 kg. Thus, the mean methane emissions per kg of liveweight was 0.342 for cattle, which is similar to sheep (0.263). The coefficient of variation for methane emissions from the heifers was 22.46%, which is in line with values observed in sheep (26.4%), and for the measurement technologies GreenFeed (25.8%) and sulfur hexafluoride tracer (38.6%) when used on cattle (Hristov *et al.* 2016). These results suggest that the cattle PAC units are performing as expected when comparable to PAC data obtained from sheep after adjusting for differences in liveweights.

The mean carbon dioxide percentage was 1.6% for the heifers, which is also in line with values from the same sheep flock which had a mean carbon dioxide percentage of 2.2%. These carbon dioxide percentages were measured after approximately 50 minutes of the animals being in the chambers for both cattle and sheep. The decision to measure cattle for 50 minutes was based on sheep protocols and a pilot study of 30 dairy heifers that showed that a 50-minute measurement period yielded sufficient methane (>800-1000ppm) for animals to express individual differences. The mean chamber temperature was 12.4°C for the heifers, where chambers tended to be cooler than the outside temperature by 1 or 2 degrees on average. Chamber temperatures ranged from 4.7°C to 29.6°C and below 20°C for 93.2% of the lots.

Table 2. Summary statistics[1] for various traits and measurement collected during PAC for the 125 beef cattle at ~10 months of age

Trait	Mean	S.D.	Minimum	Maximum
Liveweight (kg)	250	23.0	194	317
Methane (g/day) (CH_4)	85.5	19.2	43.2	152.2
Carbon dioxide (%)	1.6	0.27	1.1	3.3
Chamber temperature (°C)	12.4	5.6	4.7	29.6

[1]Computed across both PAC rounds.

Table 3 gives the results from fitting the linear mixed model for the three PAC traits. The factors for round and management group were found to be significant (except for CO_2), while PAC lot was slightly significant and wait time was non-significant across all traits. Note that management group is partly confounded with liveweight. The variation in the traits explained by the model was 84.0% for methane emissions and 75.9% for the ratio trait, while there was a poorer fit for carbon dioxide emissions (46.9%). The repeatability estimates were moderate for methane emissions (0.35) and ratio trait (0.31), but the repeatability for carbon dioxide was low (0.13). The repeatability estimate for CH_4 is similar to values obtained in sheep (0.33) but the repeatability estimates for CO_2 and the ratio trait were 0.41 in sheep (Jonker *et al.* 2018) which is higher than the estimates obtained in this study. The lower repeatability for CO_2 could be due to the management groups being partly confounded with liveweight (and hence feed intake). Overall, these results suggest that PAC methane for cattle is repeatable for the methane traits, but further studies should be performed to validate these conclusions.

Table 3. Model fit, explanatory variables and repeatability for methane emissions (CH_4), carbon dioxide (CO_2) and the molar ratio $CH_4/(CH_4 + CO_2)$ traits

	CH_4 (g/day)	CO_2 (g/day)	$CH_4/(CH_4 + CO_2)$
Adjusted r-square	0.840	0.469	0.759
Fixed effects (Prob>*F*)			
Round	0.166	0.759	0.271
Group	<.0001	<.0001	0.0004
Round.Group	<.0001	0.042	<.0001
Lot	0.022	0.061	0.008
Wait_Time	0.703	0.069	0.091
Animal variance (σ_p^2)	43.52	28399	2.62e-05
Error variance (σ_e^2)	79.95	185945	5.85e-05
Repeatability (mean ± s.e.)	0.35 ± 0.08	0.13 ± 0.09	0.31 ± 0.09

CONCLUSION

Portable accumulation chambers (PACs) performed as expected to measure methane emissions in cattle based on previous experience in sheep, with no animal behaviour problems. Furthermore, methane traits computed from the cattle PAC measurements were found to be repeatable. These results are promising and suggest that PAC is a suitable high-throughput method for the measurement of methane emissions from young beef cattle. Additional trials should be undertaken to obtain sufficient measurements to investigate the heritability of PAC traits from the cattle units.

ACKNOWLEDGEMENTS

We acknowledge Beef + Lamb New Zealand, Pāmu farms and the Informing New Zealand Beef programme for funding, trial design and use of their data. We thank Travis Leslie and Bella Taylor who managed and ran the breeding program and collection of measurements on farm.

REFERENCES

Archer J.A., Aspin M.D., Brier D.E., Powdrell S.-J.H., Campbell D., Winkworth-Lawrence C.L., Rowe S.J., McEwan J.C. and Lee M. (2023) *Proc. Assoc. Advmt. Anim. Breed. Genet.* **25**: 186.

Booker F., McEwan J.C. and Rowe S.J. (2024) *Proc. N. Z. Soc. Anim. Prod.* **84**: 4.

Johnson K., Huyler M., Westberg H., Lamb B. and Zimmerman P. (1994) *Environ Sci Technol.* **28**: 359.

Jonker A., Hickey S.M., Rowe S.J., Janssen P.H., Shackell G., Elmes S., Bain W.E., Wing J., Greer G.J., Bryson B., MacLean S., Dodds K.G., Pinares-Patiño C.S., Young E.A., Knowler K., Pickering N.K. and McEwan J.C. (2018) *J. Anim. Sci.* **96**: 3031.

Hristov A.N., Oh J., Giallongo F., Frederick T., Harper M.T., Weeks H., Branco A.F., Price W.J., Moate P.J., Dieghton M.H., Williams S.R.O., Kindermann M. and Duval S. (2016) *J. Dairy Sci.* **99**: 5461.

MfE (2024). New Zealand`s Greenhouse Gas Inventory 1990-2022. In Ref. MFE 1824. Ministry for the Environment, Wellington, New Zealand.

Rodrigues D.P.A., da Silva I.S., Vogel P.G., Machado B.F., Miranda L.B., Kepplin E.J., Dutra E.B., Quinhones L.M., Giovelli R.L., Scheid D.L., da Rosa C.A., Pötter L., Giacomini S.J. and de Azevedo E.B. (2024) *Small Rumin. Res.* **240**: 107364.

Proc. Assoc. Advmt. Anim. Breed. Genet. 26: 46-49

FACTORS AFFECTING METHANE PRODUCTION OF COMPOSITE AND CROSSBRED CATTLE GRAZING TROPICAL AND SUBTROPICAL PASTURES IN NORTHERN AUSTRALIA

C. Whistler[1], K. McCosker[1], C. Warburton[1], D. Johnston[2], T. Grant[3], B. Taylor[3], K. Goodwin[3], M. Dayman[3], N. Scott[4], S. Cullen[4], M.H. Dekkers[4], S. Clark[5] and B.J. Hayes[1]

[1] University of Queensland, QAAFI, Centre for Animal Science, Brisbane, QLD, 4000 Australia
[2] Animal Genetics and Breeding Unit*, University of New England, Armidale, NSW, 2351 Australia
[3] Department of Primary Industries, Queensland Government, QLD, 4000 Australia
[4] University of Queensland, Queensland Animal Sciences Precinct, Gatton, QLD, 4343 Australia
[5] University of New England, Armidale, NSW, 2351 Australia

SUMMARY

Understanding the factors affecting variation in methane (CH_4) production rate among individual animals is an essential step in developing methane phenotypes to enable genomic selection for lower CH_4 emissions. GreenFeed units (C-Lock inc., Rapid City, SD, USA) are an increasingly popular method for recoding methane production of grazing cattle. GreenFeed units take short-term breath measurements (of several minutes duration) when visited by cattle. Our aim was twofold 1) understand the factors associated with variation in CH_4 production records from grazing beef cattle across tropical and sub-tropical cattle grazing regions and 2) given these factors and frequencies of visitation, determine trial length necessary to derive accurate phenotypes for genomic prediction. In total 5 trials were conducted across 3 locations resulting in repeated measurements on 328 mixed sex cattle. Factors including test day, trial location, hour of visit and pipe temperature had a significant effect on CH_4 production. GreenFeed unit contributed a significant proportion of the total variation in methane emissions, across both intraday and daily methane production ($P < 0.05$). Testing the effect of trial length ranging from 5 and 52 days, showed that within-animal variation slightly decreased with repeated observations. However, the addition of cattle not previously recorded increased between animal variation, this slowed post 30 days. The optimal trial length balancing maximising cattle recorded, and repeated observations, was 28 days resulting in 20,440 visits from 324 cattle.

INTRODUCTION

Enteric fermentation from cattle accounts for 17 - 30% of CH_4 produced from human activities (Beauchemin *et al.* 2009). Methane production from cattle has been shown to be moderately heritable (Donoghue *et al.* 2016; Hayes *et al.* 2016), suggesting that selection for lower emitting animals to reduce emissions is possible. Collecting abundant and accurate phenotypic data on CH_4 is the initial phase in implementing genomic selection, however this has been very limited in tropical and sub-tropical environments. To address this, CH_4 production phenotypes have been collected utilising GreenFeed units (GFU) to measure CH_4 emissions of grazing cattle across both tropical and sub-tropical regions. The focus of this paper was to identify factors contributing to individual cattle CH_4 variation in tropical and sub-tropical grazing conditions and optimizing trial design by reducing length of future trials by focusing on reducing within animal variation whilst maximising visitation to GFU by cattle.

---

* A joint venture of NSW Department of Primary Industries and Regional Development and the University of New England

MATERIALS AND METHODS

Background. Five trials were conducted across central and northern Queensland grazing regions between 25/7/2023 and 24/2/2024 to record CH_4 production from growing beef cattle. Trials were conducted in accordance with animal ethics approval from the University of Queensland animal ethics committee, Number; 2022/AE000657. The number of GreenFeed units per trial varied between 4 and 12 with cattle able to freely visit any of the units over the trial period. Units were clustered in groups of 4 towards the extremity of each trial paddock, typically close to water sources. Number of cattle per trial varied depending on available pasture and paddock area with the number of recorded trial animals varying between 36 and 127. Cattle were familiarised with the units over a period of 21 days preceding the initiation of CH_4 recording, this involved more frequent release of attractant and supplementary feeding in proximity to the GFU. Cattle that were observed repeatedly using the GFU were removed before being reintroduced after 5 days to allow other cattle to become accustomed to the units. Trials were conducted over varying periods between 28 and 51 days across the 3 trial sites all with varying pastures, soils and climatic conditions. Cattle were composite and mixed breeds representative of production animals suited to conditions. A full summary of the 5 trials is provided in Table 1.

Table 1. Summary data regarding trial location, length and cattle number, visitation

	T1	T2	T3	T4	T5
Trial location	Brian Pastures Research Station		Goldsborough Station	Spyglass Research Station	
Length (days)	51	45	35	32	28
Recorded cattle	36	40	127	68	62
Paddock area (ha)	44.47	48.05	693.51	115.07	115.07
Pre-trial weight (kg)	369 ± 30	358 ± 28	289 ± 30	197 ± 24	196 ± 19
Av. daily visits per animal (Max-Min)	5 (1-15)	7 (1-27)	4 (1-16)	4 (1-11)	4 (1-11)
Av. total visits per animal (Max-Min)	194 (35-292)	206 (39-355)	72 (6-184)	77 (14-144)	82 (6-135)

Statistical Analysis. Quality control resulted in the exclusion of observations with unknown radio frequency identification (RFID), cattle with less than five visits over each trial, observations with airflow rate below 20L/s and negative CH_4 values were removed. A linear mixed model was fitted to both CH_4 production per individual based on both intraday and averaged daily CH_4 production:

$$y = \mu + b + a + e \quad (1)$$

where y is the resulting vector of estimates for CH_4 production; μ is the mean; **b** is a vector of fixed effects including hour of record, visit duration, pipe temperature all nested within machine, total visitation and initial trial weight nested with in trial; **a** is the random animal effect, ~ N(0, $R\sigma^2$), where $R\sigma^2$ is the between individual variance by test day; **e** is the random residual term. *Bos taurus indicus* (BI) percentage and sex (nested within trial) were excluded as non-significant factors contributing to CH_4 production. When averaging multiple visitations within the period of a day for an individual, all observations within that day were assigned the daily average CH_4 and were than merged with observational data within that day. Observations greater than 4 standard deviations of the mean of the residuals were excluded. Repeatability was calculated as between individual variance over the total variance as per equation 2:

$$Repeatability = \frac{\sigma_a^2}{(\sigma_a^2 + \sigma_w^2)} \quad (2)$$

where σ_a^2 is the between animal variance and σ_w^2 is the within animal variance. Proportion of variation for each factor was determined using the effectsize package (Ben-Shachar *et al.* 2020). Potential to decrease future trials was tested using an expanding window of additional test days between 5 and 52. Optimal trial length aimed to maximise the number of recorded cattle and observations.

RESULTS AND DISCUSSION

Most cattle made repeated visits within a 24hr period, to produce daily CH_4 estimates. Variation between intraday observations and daily averaged values were investigated to determine how the important of factors changed between the two measures, Table 2. High variation between intraday short-term breath measurements was observed which made predicting daily CH_4 production unreliable. Both repeatability (0.11) and the coefficient of determination (0.36) of the intraday breath measurements were lower than daily averaged data. This was due to considerable within animal variation observed in the intraday period.

Table 2 R^2 of model, within animal variance and between animal variance for methane measurements. Percentage of variance explained by factors in model, for factors $P < 0.05$

	Intraday	Daily
Cattle in estimates	328	328
Total observations	28746	28684
R^2 of model	0.36	0.64
Within animal variance	1859.21	515.06
Between animal variance	239.42	276.84
Repeatability (CI)	0.11 (0.09-0.12)	0.31 (0.27-0.35)
Percentage of variance explained		
Hour of day	0.39	0.10
Duration	1.31	0.59
Initial weight	9.60	8.20
Pipe temperature	1.13	0.15
Total visitation	9.69	8.51
Trial	6.29	3.73
Machine ID	0.51	0.20

No relationship was found between CH_4 production and either BI percentage or sex ($P > 0.05$). Total visitation frequency and initial weight had the largest effect on CH_4 production ($P < 0.05$), in both intraday (9.7%, 9.6%) and daily (8.5%, 8.2%) respectively. Trial explained slightly less variation (6.3%, 3.7%). All factors explained less variation using the daily average method than intraday observations. Variance attributed to GFU based on machine ID was higher in the intraday period at 0.5% decreasing to just 0.2%, whilst remaining significant ($P < 0.05$) only a small proportion of variation is observed from different GFU. Inclusion of multiple GFU to ascertain and attribute variation is therefore necessary in trial design to ensure variation between machines is captured.

Between-animal variation was similar for both intraday and daily observations. However, within-animal variation was lower for daily averaged data. Repeatability for intraday data per individual is presented in Figure 1, repeatability was higher as a result. Repeatability trended upwards over the expanding period with high variability observed pre 20 days. Recorded cattle increased rapidly from only 211 at 5-day trial length to 305 after 14 days of recording, successfully recorded cattle increased slowly thereafter reaching 315 at 21 days.

Within-animal variation increased across the expanding trial periods but was highly variable and reflected visitation, cattle returning multiple times a day and for consecutive days had lower variation than cattle with less frequent visitation. Post 28 days only 4 additional cattle were recorded and 8,315 additional observations. Generally, repeatability could be improved by reducing within animal variation by more consistent and earlier visitation. Early adoption and consistent visitation could reduce trial length however ad libitum visitation necessitates longer trials. Dressler *et al.* (2023) reported that a minimum of 40 visits were required to produce a significantly correlated result as with 100 visits, this occurred at 29.5 ± 8.7 days. Our study shows that cattle make ~ 2 visits to GFU per day resulting in a mean of 40 visits falling earlier around day 20 however, recording for at least 28 days is recommended.

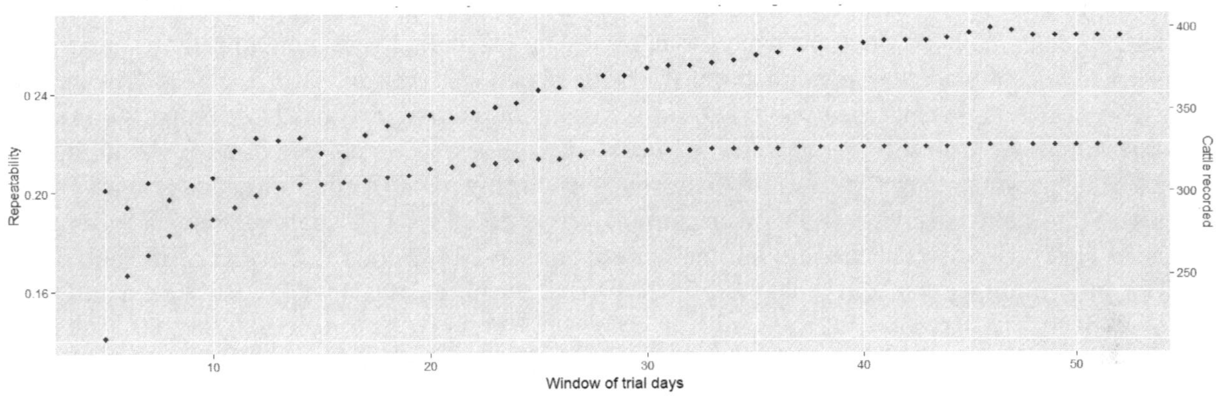

Figure 1 Total cattle recorded across an expanding number of trial days and the effect on repeatability

CONCLUSION

Total visitation and initial weight were significant and influential predictors of individual CH_4 emission estimates for both intraday and daily averaged data. Although duration of visitation, hour of visit and pipe temperature were also statistically significant predictors of CH_4 production, their proportional contributions to the model were comparatively smaller. Inclusion of multiple GFU per trial is suggested to observe between unit variation. A minimum trial length of 20 days is required to reduce variation in repeatability and produce approximately 40 repeated observations per animal. Based on the results from this study a slightly longer trial length of at least 28 days is suggested if the goal is to maximise the number of repeated observations and cattle recorded.

ACKNOWLEDGEMENTS

Funding for this project was generously provided by the Meat and Livestock Australia Donor Company, project number P.P.S.H 1406.

REFERENCES

Beauchemin K.A., McAllister T.A. and McGinn S.M. (2009) *CAB Reviews* **4**: 1.

Ben-Shachar M.S., Lüdecke D. and Makowski D. (2020) *J. Open Source Softw.* **5**: 2815.

Donoghue K., Bird-Gardiner T., Arthur P., Herd R.M. and Hegarty R. (2016) *J. Anim. Sci.* **94**: 1438.

Dressler E.A., Bormann J.M., Weaber R.L. and Rolf M.M. (2023). *J. Anim. Sci.* **101**: 1.

Hayes B., Donoghue K., Reich C., Mason B., Bird-Gardiner T., Herd R. and Arthur P. (2016). *J. Anim. Sci.* **94**: 902.

GENETIC ANALYSIS OF SENSOR-BASED ACTIVITY IN AUSTRALIAN MERINO SHEEP

H. Keshavarzi[1], E. Dehnavi[2], J. McNally[1], J.L. Smith[1] and S. Schmoelzl[1]

[1] CSIRO Agriculture and Food, Armidale, NSW, 2350 Australia
[2] Animal Genetics Breeding Unit*, University of New England, Armidale, NSW 2351 Australia

SUMMARY

The use of on-animal sensor data enables continuous monitoring, providing valuable behavioural observations with the potential to be used for assessing an individual's health and productivity. This study aimed to estimate the genetic parameters of sheep general activity, measured as vector magnitude, which represents the overall magnitude of the movement to provide an objective activity level measurement. Activity records of 1,149 Merino ewes were collected with ActiGraph™ wGT3X-BT® devices over a period of 17 days (after editing 10 non-continuous days used in the analysis). Single trait estimates (STE) were initially considered for this analysis; however, random regression (RR) was also checked, given the patterns observed on raw records over time and days. The STE of heritability (h^2) and repeatability (Rep) in the whole dataset were 0.19 ± 0.06 and 0.56 ± 0.08, respectively. Across 10 days, single estimates of h^2 ranged from 0.15 ± 0.06 to 0.25 ± 0.08 and Rep from 0.45 ± 0.08 to 0.66 ± 0.08, while RR models resulted in lower h^2 (0.11 to 0.23) and Rep (0.22 to 0.34) over these days. Genetic correlations between days (up to day 16) and times (2 hours) were high, which did not indicate the potential of random effects of time and day in genetic parameters for sheep activity. These estimated parameters can be considered for future welfare-related breeding programs, and the amount of repeatability can guide the development of an effective measurement protocol.

INTRODUCTION

Recent research emphasises using sensor technologies to monitor animal better and measure the animals' behaviour, health, and production traits in commercial systems (Almasi *et al.* 2023; Brito *et al.* 2020). Accelerometer data can be used to classify behaviours (Wang *et al.* 2023) or quantified as activity levels (Marchesini *et al.* 2018; Hu *et al.* 2024). These novel traits may be heritable and, once validated, could be incorporated into breeding programs by providing more detailed and consistent information which enables superior genetic comparisons over time (Morris *et al.* 2012) or establishing correlations between behavioural and health traits and improve animal behaviour and welfare. Therefore, this study aimed to estimate the genetic parameters of sheep activity, measured as vector magnitude count (VMC) using linear, repeated measurement and random regression models.

MATERIALS AND METHODS

In this study, the activity data of two contemporary groups of Merino ewes born in 2017 (n=607, recorded in Dec 2022 - Jan 2023) and 2018 (n =542, recorded in Dec 2023 - Jan 2024) were used. The sheep were F1 ewes from the New England Merino Lifetime Productivity flock (MLP), managed by CSIRO at the FD McMaster Laboratory, Chiswick, Uralla, NSW, Australia (Ramsay *et al.* 2019). All experimental procedures conducted on animals were approved by the CSIRO Armidale Animal Ethics Committee (Animal Research Authority no. 21/24 and 22/20).

* A joint venture of NSW Department of Primary Industries and Regional Development and the University of New England

ActiGraph™ wGT3X-BT® devices (Actigraph LLC, Pensacola, FL, USA) were used to collect sheep activity data continuously at 30 Hz. In its default setting, the 3-axis accelerometer within the device measures accelerations for each axis (X, Y, Z) in G value, estimating VMC as:

$$\text{VMC}= \sqrt{axisx^2 + axisy^2 + axisz^2}.$$

For each experiment, the complete data was available for 17 days. On day one, devices were attached to each sheep's neck using a collar. Raw data was downloaded using ActiLife software, which filtered noise and provided summary outputs as hourly epochs. Outliers and VMC less than 100 were removed from the dataset using a histogram or boxplots for values outside the limit at each timepoint and overall. The days when the animal moved between paddocks or through yards were excluded. To ensure data consistency and reduce number of analyses, only days with records from both deployments (in total 10 non-consecutive days) were used for further analysis. From 24-hour records, 4 one-hour time blocks in the afternoon between 1500 and 1900 h were selected for analysis, based on inter-animal variability during this period, combined with an absence of disruptive events such as moving animals between paddocks (in total times=4, day=10, and daytime=40 records per animal). Variations in activity between animals and over time were expected. However, heterogeneity across contemporary groups was addressed with the method described by Brown *et al.* (2005). In total, 45,641 records from 824 fully pedigreed ewes, representing 28 sires were used.

Statistical analysis. A univariate animal model for repeated measures was fitted using ASReml (Gilmour *et al.* 2009) to estimate heritability (h^2) and repeatability (Rep) of activity across the dataset (single estimates using repeated measurement model (RMM)) and intervals for each time (n=4), day (n=10), and daytime (n=40; although there was not any evidence of autocorrelation in this study, it might be a limitation) to assess the necessity of running RR. Given the unknown pattern of this trait across time, day, and daytime, an RR model with polynomial orders (1 to 6) was also applied to activity across day, time and the interaction of day and time. Models were compared using the likelihood ratio test (LRT), Akaike Information Criterion (AIC), and Bayesian Information Criterion (BIC).

Fixed effects included contemporary groups while day, time, and day-time interaction were fitted separately as repeated measurements. Bodyweight (kg) at 4yo, rainfall (mm), temperature humidity index, wind speed (km/h), and solar radiation (W/m^2) were considered as covariates when time and/or day were fitted as repeated measurements in the model. Random effects comprised genetic and permanent environment effects, with genetic group effects assessed. The covariance function was estimated in RR models using $\hat{G} = \Phi K \Phi'$, where $\hat{G}$ is the genetic covariance matrix for breeding values, Φ is the matrix of orthogonal polynomials, and K is the genetic variance coefficients matrix.

RESULTS AND DISCUSSION

The results showed all covariates and fixed factors were significant ($P < 0.001$) except rainfall information. Animal weight and weather conditions significantly influenced animal activity. Genetic groups (assigned by Sheep Genetics) had no significant effect and were excluded.

A single estimate of h^2 and Rep was obtained by fitting daytime as a repeated measure, with h^2 at 0.19 ± 0.06 and Rep at 0.56 ± 0.08 for all days and times in the dataset. Heritability and Rep values varied (with a pattern) across different times (Figure 1a), days (Figure 1d), and daytimes (Figure 1g; Rep estimates were not computed as they were not repeated) ranging from low to moderate.

Polynomial regression was fitted to the fixed effect of time, day and daytime to find the best order in the raw records. The best fit for the records was achieved by using regression models with polynomial orders of 2 for time and 6 for day (Figure 1c, and f, respectively), resulting in R^2 values close to 1 (Figures 1b and e, respectively). Fitting RR with an order of six for daytime (Figure 1i) in raw records showed lower performance ($R^2 = 0.29$; Figure 1h). RR models had lower AIC and BIC values than RMMs, indicating a better fit despite having lower parameter estimates. They also aligned with the raw data pattern. To our knowledge, no study has applied RR models to actual time series data, as previous studies have relied on standardised records based on group means (e.g., time, day).

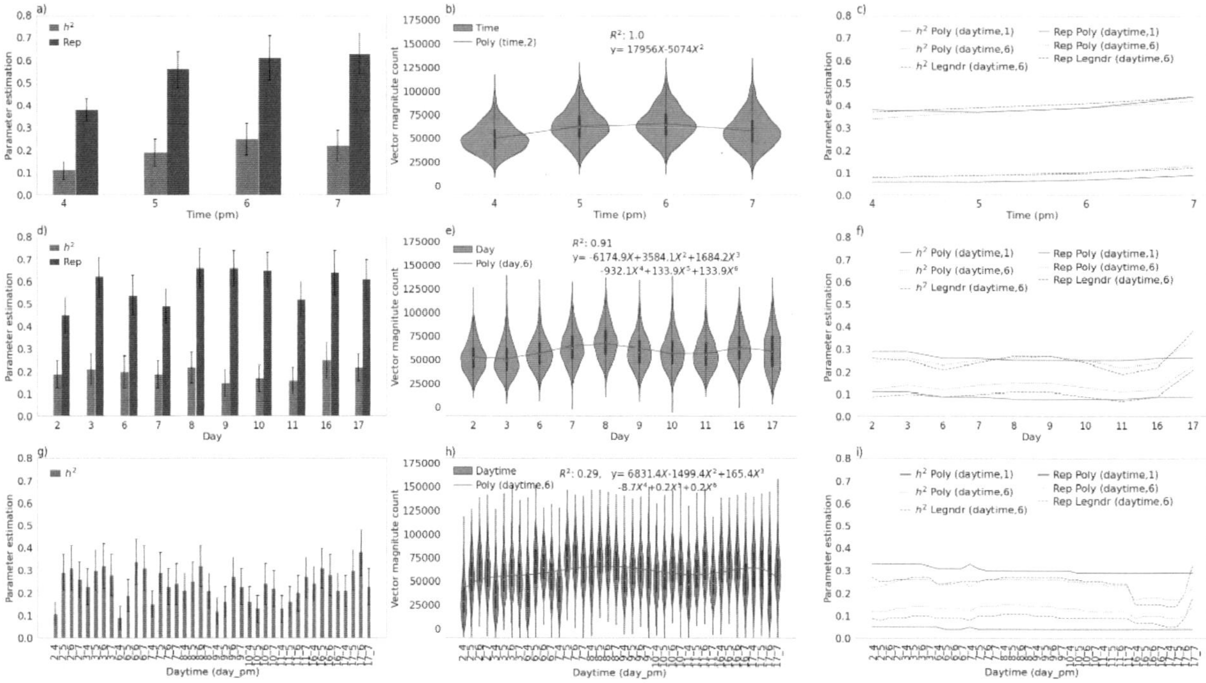

Figure 1. Plot showing parameter estimates (heritability-(h^2) and repeatability (Rep)) and data distribution across different times (a-c), days (d-f), and daytimes (g-i) obtained using various animal models. Note: Models included single estimates at intervals (time, day and daytime), linear random regression (first-order), Polynomial (poly) and Legendre (legendr) with a second order for the time, a sixth order for day and daytime.

Table 1. Genetic correlations across days (upper), and time points (lower diagonal) modelled using sixth-order and second-order polynomials, respectively

.

Day 3	Day 6	Day 7	Day 8	Day 9	Day 10	Day 11	Day 16	Day 17	Days
0.90	0.86	0.84	0.82	0.83	0.86	0.88	0.83	0.54	Day 2
	0.91	0.89	0.90	0.91	0.93	0.91	0.79	0.37	Day 3
		0.99	0.97	0.96	0.95	0.91	0.78	0.34	Day 6
			0.99	0.98	0.96	0.89	0.75	0.26	Day 7
				0.99	0.97	0.90	0.74	0.22	Day 8
					0.99	0.92	0.76	0.25	Day 9
						0.97	0.81	0.37	Day 10
							0.87	0.56	Day 11
								0.68	Day 16

5 PM	6 PM	7 PM	Times
0.90	0.84	0.74	4 PM
	0.97	0.73	5 PM
		0.86	6 PM

* Standard errors ranged from 0.03 to 0.04 for genetic correlations estimated between different times and from 0.02 to 0.04 between different days.

High genetic correlations (r_g) were found between times, and days, except for day 17, which had lower correlations with all other days (0.22–0.68; Table 1). There were also moderate to high r_g between different daytimes except for 78 daytime combinations with r_g lower than 0.6 (not shown). Almasi *et al.* (2023) also represented that there was a low intraclass correlation between measurements longer than one week. Overall, the result of this study indicate that there was no marginal difference in r_g across studied times and days. Therefore, considering the measurements across a broader range of time points and days might provide a more accurate view of actual h^2 and Rep estimated via the RR model compared to a single estimation of RMMs.

Previous studies on sheep activity have shown traits such as Daily Distance Travelled have moderate h^2 (0.36 ± 0.09) and Rep (0.51 ± 0.03) (Johnson *et al.* 2021). Almasi *et al.* (2023) reported that accelerometer-derived traits, such as grazing time, show moderate to high h^2 (0.44 ± 0.23) and Rep (0.70 ± 0.03) that can be useful for ranking animals for feeding activities in selective breeding programs (Almasi *et al.* 2023). Similarly, grazing behaviour traits in semi-extensively reared Boutsko sheep demonstrate significant h^2 and Rep for grazing and speed but low non-significant h^2 for distance (0.05 ± 0.06) (Vouraki *et al.* 2025). These findings highlight the value of wearable sensors for advancing genetic selection in sheep by providing continuous data that enables the development of traits for improved health and production.

CONCLUSION

This study showed moderate heritability and repeatability estimates for sheep activity, suggesting the potential of this novel trait for selection that may offer favourable correlated responses for health and production. Further work is required to determine the relationships between activity and health status, and between activity and economically important production traits. Although a weak pattern was observed in the results when considering RR models, our findings suggest that RMMs suffice for accurately estimating the genetic parameters in this dataset.

ACKNOWLEDGEMENTS

The authors thank Australian Wool Innovation and the Merino Lifetime Productivity (MLP) Project for use of the sheep and data in this study, and CSIRO Chiswick technical staff for conduct of the field work component.

REFERENCES

Almasi F., Stear M.J., Khansefid M., Nguyen H., Desai A., and Pryce J.E. (2023) *Front. Anim. Sci.* **4**: 1154797.

Brown D.J., Atkins K. and Huisman A.E. (2005) *Proc. Assoc. Advmt. Anim. Breed. Genet.* **16**: 84.

Brito L.F., Oliveira H.R., McConn B.R., Schinckel A.P., Arrazola A., Marchant-Forde J.N. and Johnson J.S. (2020) *Front. Genet.* **11**: 793.

Gilmour A.R., Gogel B.J, Cullis B.R. and Thompson R. (2009) ASReml User Guide Release 3.0. VSN International Ltd, Hemel Hempsted, UK.

Johnson P., Cullen N., Hickey S., Knowler K., Bryson B., Hall M. and Pletnyakov P. (2021) *P.N.Z. J. Anim. Sci. Prod.* **81**: 68.

Hu S., Reverter A., Arablouei R., Bishop-Hurley G., McNally J., Alvarenga F. and Ingham A. (2024) *Animals* **14**: 301.

Marchesini G., Mottaran D., Contiero B., Schiavon E., Segato S., Garbin E., Tenti S. and Andrighetto I. (2018) *Vet. J.* **231**: 41.

Morris J.E., Cronin G.M. and Bush R.D. (2012) *Anim. Prod. Sci.* **52**: 665.

Ramsay A., Swan A. and Swain B. (2019) *Proc. Assoc. Advmt. Anim. Breed. Genet.* **23**: 512.

Vouraki S., Papanikolopoulou V., Argyriadou A., Priskas S., Banos G., and Arsenos G (2025) *Applied Anim. Behav. Sci.* **282**: 106473.

Wang L., Arablouei R., Alvarenga F.A.P. and Bishop-Hurley G.J. (2023) *Comput. Electron. Agr.* **206**:107647.

Proc. Assoc. Advmt. Anim. Breed. Genet. 26: 54-57

ASSESSING BEHAVIOURAL RIGIDITY AND ITS GENETIC INFLUENCES IN AUSTRALIAN WAGYU CATTLE

A. Lloyd[1,2], M. Facy[1,3] and J. Grose[1]

[1] 3D Genetics Pty Ltd, Pukawidgi, Bukkulla, NSW, 2360 Australia
[2] Evolved Projects Pty Ltd, Geelong, VIC, 3220 Australia
[3] Davies Livestock Research Centre, University of Adelaide, Roseworthy, SA, 5371 Australia

SUMMARY

Animals that can adapt well to a particular environment and exhibit beneficial behaviours have the potential to exhibit improvements in traits such as productivity and meat quality. Behavioural rigidity (BR) is defined as a lack of flexibility to an individual's approach to life meaning animals that are rigid would adapt poorly to new environments and stressors. In the current study BR was defined as the number of different feed bunks an animal visits per day, demonstrating animals that visited fewer or even exclusive feed bunks were classified as having a higher behavioural rigidity than those who visited a greater variety of feed bunks. Sex, pen, year and average daily gain all had significant correlations with BR in Wagyu cattle. This would indicate that animals that are willing to eat from multiple bunks would have a higher average daily gain and improved adaptivity to a feedlot environment. BR was heritable with an estimate of 0.17 indicating that genetic gain can be made on this trait.

INTRODUCTION

Breeding programs should consist of animals that are well adapted to a particular environment, this includes exhibiting behaviours that are associated with improved production (Adamczyk *et al.* 2013). Animals react differently to environmental stimuli, either adaptively or maladaptively which can be due to either environmental factors (how that animal has been handled in the past) or genetic variation (Haskell *et al.* 2014). These reactions to stimuli can be either positive or negative, with their response affecting metabolism and social interactions within the herd. Some studies have shown that cattle that are highly stressed or temperamental have lower meat quality or reduced weight gain when compared to other cattle in the same herd (Titterington *et al.* 2022).

Behavioural rigidity is known as lack of flexibility in a person's thoughts and approach to life. This behavioural rigidity can cause issues in a person's life as they struggle to adapt to new situations (Schultz and Searleman 2002). In cattle this behavioural rigidity could have a potential to negatively affect productivity. For example, in a feedlot situation, an animal that has high behavioural rigidity may not adapt to the new environment and have lower feed intake with resultant reduced growth. As far as the authors are aware, there has been no published research in this area.

This study aimed to measure behavioural rigidity using feed trial data collected with GrowSafe technology, investigate the effect of year, sex, growth rate and pen and identify the genetic parameters for behavioural rigidity.

MATERIALS AND METHODS

The data used was provided and collected by 3D Genetics, a Wagyu bull breeding program situated in Northern New South Wales. The data included the results of two years of feed trial data collected using the GrowSafe System (Vytelle LLC., Canada) on 3D Genetics' property "Pukawidgi". The feed trials ran for a period of approximately 50 days with adlib feeding of a corn silage based ration. In total there were 1,993 individual animal records included in the analysis, with an average pen size of 75. The average daily gain (ADG) was calculated from the results of the feed trial to give an indication of growth.

Behavioural rigidity (BR) herein was defined as the average number of bunks visited each day with a total of eight bunks in each pen. To calculate the proportion of visits, the number of visits to a specific bunk was divided by the total number of visits by the animals. For the animal to have a one-bunk preference, more than 50% of total visits needed to be to a single-bunk. For an animal to have a two-bunk preference more than 50% of total visits needed to be from one or two bunks.

A general linear model was used to calculate the potential effects of sex, year, pen and growth rate. It can be written as,

$$BR_{ijkl} = \mu + Sex_i + Pen_j + Year_k + Pen_j{:}Year_k + \beta \cdot ADG_{ijkl} + \varepsilon_{ijkl}$$

where BR_{ijkl} is behavioural rigidity, μ is the overall mean, Sex_i is fixed effect of ith level of sex, Pen_j is fixed effect of jth level of pen, $Year_k$ is fixed effect of kth level of year, $Pen_j{:}Year_k$ is the interaction between pen and year, β is the regression coefficient of ADG_{ijkl} and ε_{ijkl} is the residual error term.

The genotype data used included 8,675 animals that were collected over many years. Seven different SNP chips were used including Illumina 777k, Illumina GGPLD V3 30K, Illumina GGPLD V4 30K, Illumina ICB 50K, Illumina GGP 100K and Weatherby's Scientific Versa50K. The common SNPs across all chips was 10 830 which were imputed up to GGPLD30K (21, 791 SNPs) using Minimac4 (Howie *et al.* 2012). All duplicated animals and SNPs with less than 1% minor allele frequency were removed. The heterozygosity fraction (Het) was calculated as the proportion of imputed heterozygote SNPs for each individual animal.

Genetic parameters were estimated through a univariate analysis using ASREML-R (Butler *et al.* 2017). The fixed effects included in the model were sex, Het and contemporary group. The contemporary group was defined as birth group, pen and feed trial date. The genomic relationship matrix was calculated using VanRanden's first method (VanRaden 2008). The variance components and heritability were estimated for behavioural rigidity.

RESULTS AND DISCUSSION

Behavioural rigidity (BR) values ranged from 1.75 to 7.92 with a mean of 6.04 and a standard deviation of 1.12. This demonstrates there was variation in the trait and phenotypically animals seemed to prefer certain bunks. Therefore, in Wagyu cattle there is some behavioural rigidity and in the feedlot environment, individual cattle show a preference for which bunk they will eat from. There was 6.71% of animals that had a one-bunk preference and 31.27% of animals that showed a two-bunk preference. There were some animals who would consume from any bunk (Figure 1a) whereas some animals preferred one or two bunks only (Figure 1b).

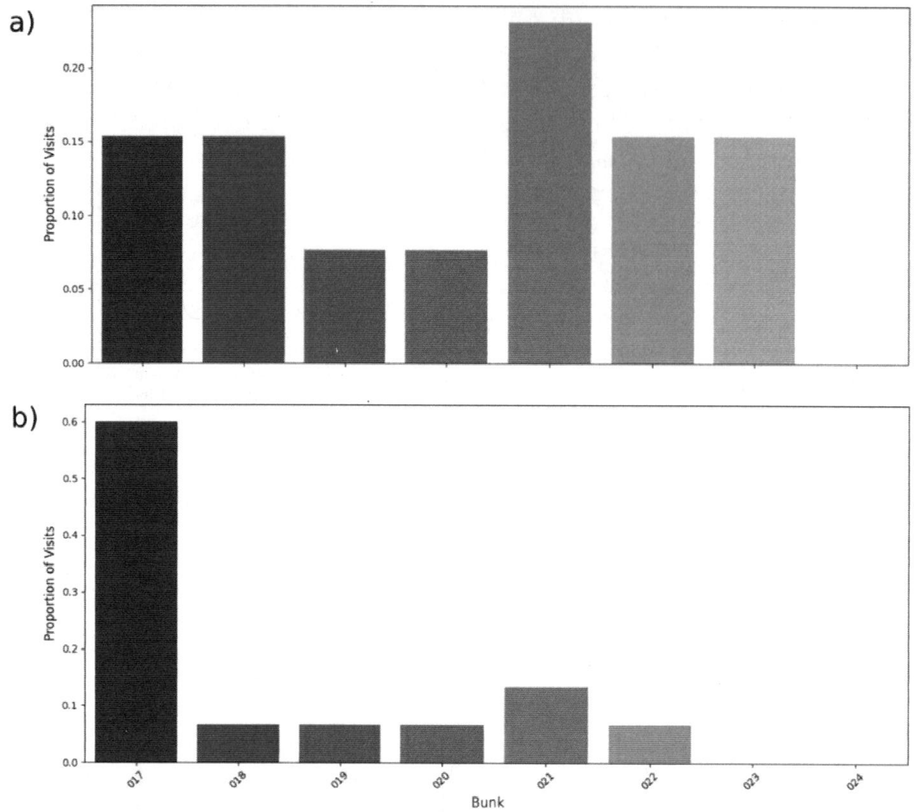

Figure 1. An example of proportion of visit to each feed bunk during a feed trial. a) had no preference of bunk, b) had a preference of majority one bunk only

BR was then included in a linear model to look at the effects of sex, pen, year and average daily gain. It was found that sex, pen, year and average daily gain were significant (Table 1). The sex had a significant impact on BR, with entire males and steers having estimates of -0.16 and -0.94 lower than females (Table 1). Feed pen also had a significant effect on BR, with feed pen 2 (FP2) having the largest effect of 1.10 above feed pen 1 (FP1; Table 1). The significant difference between feed pens would indicate that there is social hierarchy in the pen that determines what bunks animals will eat from. The interaction between pen and year was also significant with FP6:2024 had the largest difference of -1.88 below the baseline of FP1:2023. Average daily gain (ADG) also had a significant association on BR with a regression coefficient of 0.57 (Table 1). This would indicate that animals that have less BR and visit a higher number of bunks would have a higher daily gain. The interaction between ADG and BR could have an influence on the potential growth of cattle in feedlots. If the animal is only eating out of one bunk, that animal may not each as much as other animals who exhibit less behavioural rigidity and are willing to eat from any bunk.

Table 1. Coefficient estimates of linear model of behavioural rigidity with standard errors in brackets. The base values were female for sex, FP1 for pen and 2023 for year

Effect	Estimate	P-value
Mean	4.85 (0.14)	<0.001
Sex – Male	-0.16 (0.08)	0.04
Sex – Steer	-0.94 (0.11)	<0.001
Pen – FP2	1.40 (0.09)	<0.001
Pen – FP6	1.21 (0.09)	<0.001
Pen – FP7	1.21 (0.09)	<0.001
Year – 2024	0.52 (0.12)	<0.001
Average daily gain	0.57 (0.13)	<0.001
Pen – FP2:Year – 2024	-1.05 (0.16)	<0.001
Pen – FP6:Year – 2024	-1.88 (0.16)	<0.001
Pen – FP7:Year – 2024	-1.16 (0.16)	<0.001

The heritability and variance components were then estimated for BR. The heritability was 0.17 with a standard error of 0.04. Other behaviour traits in cattle have been estimated to be low-moderate (Adamczyk *et al.* 2013; Haskell *et al.* 2014; Titterington *et al.* 2022). The variance components were 0.14 for additive variance and 0.68 for residual variance. The heritability was low-moderate, demonstrating that there is a large environmental influence over the trait. One of these environmental effects could be the social hierarchy in the pen, some animals may be at the bottom of the hierarchy and only felt comfortable to eat out of certain bunks. Despite these environmental effects, there is still potential to select against BR and improve animals' ability to adapt to eat from multiple bunks in a feedlot situation.

CONCLUSION

Behavioural rigidity is known as lack of flexibility in someone's approach to life, this can cause issues as it can cause them struggle to adapt to new situations. Behavioural rigidity (BR) was defined as the average bunks visited each day over a feed trial. A general linear model found that sex, pen, year and average daily gain (ADG) were significant for BR. ADG had a significant effect of 0.57, indicating as an animal is less rigid or uses more bunks, they would have a high ADG. This could be important for animals entering the feedlot system, as an animal that shows more behavioural rigidity may exhibit lower growth rates. Further, the genetic parameters were estimated for BR with a heritability of 0.17. BR has a genetic component, and genetic gain could be achieved through selection.

REFERENCES

Adamczyk K., Pokorska J., Makulska J., Earley B. and Mazurek M. (2013) *Livest. Sci.* **154**: 1.

Butler D., Cullis B., Gilmour A.R., Gogel B. and Thompson, R. (2017) 'R Reference Manual Version 4.' In VSN International Ltd, Hemel Hempsted, UK.

Haskell M.J., Simm G. and Turner S.P. (2014) *Front. Genet.* **5**: 368.

Howie B., Fuchsberger C., Stephens M., Marchini J. and Abecasis G.R. (2012) *Nat. Genet.* **44**: 955.

Schultz P.W. and Searleman A. (2002) *Genet. Soc. Gen. Psych. Mono.* **128**: 165.

Titterington F.M., Knox R., Morrison S.J. and Shirali M. (2022) *Animals.* **12**: 2602.

Proc. Assoc. Advmt. Anim. Breed. Genet: 26: 58-61

AN EXAMINATION OF MULTIBREED TEMPERAMENT MEASURES RECORDED IN THE SOUTHERN MULTIBREED RESOURCE POPULATION

B.J. Walmsley[1,2], K.L. Moore[1], R. Woodgate[2], S.A. Clark[3], F.E.M. Haynes[4] and K.A. Donoghue[5]

[1] Animal Genetics Breeding Unit*, University of New England, Armidale, NSW, 2351 Australia
[2] NSW Department of Primary Industries and Regional Development, Livestock Industries Centre, Armidale, NSW, 2351 Australia
[3] School of Environmental and Rural Science, University of New England, Armidale, NSW, 2351 Australia
[4] NSW Department of Primary Industries and Regional Development, 105 Prince St, Orange, NSW, 2800 Australia
[5] NSW Department of Primary Industries and Regional Development, Agricultural Research Centre, Trangie, NSW, 2823 Australia

SUMMARY

Temperament has important impacts on on-farm profitability through its impact on other production traits like reproduction, production and meat quality in addition to its direct impact on the welfare of farm staff. The Southern Multibreed project to date has recorded over 9,800 flight time (FT) and crush score (CS) phenotypes to create a multibreed reference population for cattle temperament. This paper also examines the impact animal handling due to extensive trait recording has on the quality of temperament measures using genetic parameters. The heritabilities for FT were moderate in steers (0.27-0.31) and low to moderate in heifers (0.15-0.25). The heritabilities for CS were moderate for both steers (0.24-0.25) and heifers (0.29-0.36). The phenotypic variance for FT decreased between the weaning and yearling stages in steers but increased in heifers. The phenotypic variance for CS increased between the weaning and yearling stages in steers but decreased in heifers. The genetic correlation between FT measures at weaning and yearling stages was 0.95 in steers and 0.69 in heifers. The correlations between weaning and yearling stages for CS were 0.72 and 0.59 in steers and heifers, respectively. The correlations between steer and heifer FT and CS at the weaning stage were 0.95 and 0.87, respectively. At the yearling stage these correlations were 0.69 and 0.84. These results suggest that the heifers assimilated to the crush environment as time progressed due to extensive phenotypic recording while the steers displayed their inherent behaviour when they were exposed to the stress associated with the new environment when entering the feedlot. These outcomes indicate it is important that animals are in an environment where learnt behaviours don't impact the expression of inherent behaviour when temperament traits are recorded.

INTRODUCTION

Temperament is a trait that has been shown to impact profitability through reproduction and maternal behaviour (Phocas *et al.* 2006), meat quality (Kadel *et al.* 2006), animal and handler welfare (Grandin 1993) and production costs (Burrow 1997). Therefore, improving temperament through selection would help to improve on-farm profitability. To successfully improve profitability, selection tools must provide predictive accuracy for all traits that impact profitability and form the basis of current and future estimated breeding values and selection indexes. The capacity to do this is impacted by the circumstances under which traits are measured. Currently, temperament-related traits are recorded at weaning and included in only a limited number of BREEDPLAN genetic

---

* A joint venture of NSW Department of Primary Industries and Regional Development and the University of New England

evaluations (Docility/Crush Score (CS): Angus/Hereford, Flight Time (FT): Brahman) and only on a within-breed basis (Walkom *et al.* 2018). The Southern Multibreed project (SMB: Walmsley *et al.* 2021) provides a multibreed reference population that can be used to enhance current BREEDPLAN evaluations to better target profitability gains through selection for temperament. This paper provides a brief update on the progress of recording temperament-related traits in the SMB Project and an examination of the impact of animal handling due to extensive trait recording on the quality of temperament measures using genetic parameters.

MATERIALS AND METHODS

Animal Data. Temperament traits were recorded on 1,988 SMB heifers and 1,979 steers from five *Bos taurus* breeds (Angus (n=1,319), Charolais (n=437), Hereford (n=913), Shorthorn (n=563), Wagyu (n=591)) and the Brahman breed (n=144). The SMB animals were bred as part of the SMB project – an extensive reference population collecting many phenotypes from birth to slaughter (Walmsley *et al.* 2021). The animals were born in 2020, 2021 and 2022 (Walmsley *et al.* 2023) across five New South Wales Department of Primary Industries and Regional Development research facilities; Trangie Agricultural Research Centre; Grafton Primary Industries Institute; Tocal Agricultural Centre; Glen Innes Agricultural Research and Advisory Station; and Elizabeth MacArthur Agricultural Institute; Menangle.

All animals had FT and CS recorded at the weaning and yearling stages, while only heifers were recorded post-mating (final). The heifers were only recorded on their birth property while the steers were recorded on their birth property at weaning and at the University of New England (UNE) research feedlot, "Tullimba" (Kingstown) as they entered the feedlot as yearlings. Crush scores were visually assessed when animals were confined in a crush for 5 seconds based on a 5-point scale of agitation; 1=calm/docile, 2=restless, 3=nervous, 4=flighty, or 5=aggressive (Grandin 1993). The scale was modified to include half scores to increase the number of levels. All crush scores were assessed by one assessor. Flight time was measured as the time to break 2 infrared sensors 1.7m apart after exiting the crush into an open yard (Burrow *et al.* 1988). All animals were managed in mixed breed groups. Animals were excluded from the analysis where the sire or dam was unknown or where the animal was born as a twin. Animals with a FT phenotype three standard deviations greater than the overall mean were capped, so their FT phenotype was equal to the mean + 3 standard deviations.

Statistical Analyses. For analysis, FT was multiplied by 100. Genetic parameters and heritabilities were estimated from univariate mixed linear animal models using ASReml (Gilmour *et al.* 2021). The statistical model fitted the contemporary group and breed as fixed class effects and animal age as a fixed covariate. The contemporary group was defined by the herd of birth, year of birth, sex and the calving/management group. The animal was fitted as a random effect, and the dam's permanent environment effect was tested for significance. The categorical nature of CS was ignored because genetic correlation estimates are not affected by the statistical treatment of the categorical trait (Kadarmideen *et al.* 2003). Five generations of pedigree were used, with 359 sires represented by an average half-sib family size of 11.1, ranging from 1 to 52.

RESULTS AND DISCUSSION

The heifers were, on average, 233, 405 and 604 days of age when recorded at the weaning, yearling and final stages, respectively, while the steers were 231 and 493 days of age at the weaning and yearling stages. The mean FT and CS were similar at weaning for steers and heifers while at the yearling stage mean FT was higher and mean CS tended to be lower in heifers. Variance component estimation (Table 1) showed that FT had moderate heritabilities of 0.31 (0.06) and 0.27 (0.06) in steers at the weaning and yearling stages, respectively, and low to moderate heritabilities of 0.25 (0.06), 0.22 (0.05) and 0.15 (0.05) in heifers at the weaning, yearling and final stages. The heritabilities estimated agreed with those found for tropically adapted breeds (Kadel *et al.* 2006). The heritability estimates for crush score in this study tended to be higher than those reported for tropically adapted (0.15; Kadel *et al.* 2006) and Angus (0.21; Walkom *et al.* 2018) cattle. However, higher estimates were reported for Limousin cattle (0.39; Walkom *et al.* 2018). The dam's permanent environmental effect was significant (data not shown) for CS when assessed on steers entering the feedlot at the yearling stage. This was evident in the means and phenotypic variances of the two sexes between stages. The mean and phenotypic variance of FT decreased, and CS increased between the weaning and yearling stage measurements for steers (Table 1). This suggests the steers were more agitated upon entering the feedlot than when they were measured at their birth property. In comparison, FT's mean and phenotypic variances increased, and CS decreased between the weaning, yearling and final stage measurements for heifers. In contrast to the steers, this suggests the heifers became more assimilated rather than agitated to the crush environment on their birth property due to continued handling throughout the project.

Table 1. Summary statistics, phenotypic variance (V_P) and heritability (h^2: standard error) estimates for flight time (FT: seconds*100) and crush score (CS) recorded on steers and heifers at weaning, yearling and final stages

Sex	Trait	Stage	N	Mean	Std	Min	Max	V_P	h^2
Steer	FT	Weaning	1973	139.9	68.5	35.8	446	3492.2	0.31 (0.06)
		Yearling	1930	108.5	51.3	37.7	444	2268.2	0.27 (0.06)
	CS	Weaning	1973	1.4	0.4	1	3	0.14	0.24 (0.06)
		Yearling	1930	1.7	0.6	1	3.5	0.29	0.25 (0.07)
Heifer	FT	Weaning	1982	139.5	70.5	40	446	3719.7	0.25 (0.06)
		Yearling	1974	140.9	81.9	11	444	4747.1	0.22 (0.05)
		Final	1958	179.8	112.1	41.1	605	8609.6	0.15 (0.05)
	CS	Weaning	1982	1.4	0.4	1	3	0.15	0.30 (0.06)
		Yearling	1974	1.4	0.5	1	3	0.18	0.29 (0.06)
		Final	1958	1.2	0.4	1	3	0.11	0.36 (0.06)

The genetic correlation between FT weaning and yearling measures were high for steers (Table 2), suggesting the same FT trait is being measured at weaning and yearling stages. In comparison, the correlations in heifers were lower. The correlations between the final stage measures and those at the weaning and yearling stages varied. The correlation between the weaning and final stages was 0.93 (0.18) for FT and 0.58 (0.12) for CS. In contrast, the yearling and final stages correlations were 0.70 (0.17) for FT but 0.95 (0.07) for CS. When comparing steers and heifers at weaning, the correlations were 0.95 (0.14) for FT and 0.87 (0.14) for CS (Table 2). The high correlations between steers and heifers at the weaning stage suggest the same trait is being measured. At the yearling stage, the correlations were 0.69 (0.14) for FT and 0.84 (0.12) for CS. The reduction in the correlation at the yearling stage between steers and heifers for FT supports that some assimilation may have occurred in the heifers due to continued handling through the crush on the home property while the steers were exposed to a foreign environment.

CONCLUSIONS

The SMB project has measured over 9,800 FT and CS traits in the first three cohorts of the project to establish a multibreed reference population to help make genetic improvements for temperament in Australian cattle. The estimated heritabilities agree with past studies. The genetic correlations were high between measurements at different stages in steers but were less consistent and tended to be lower in heifers. The correlations between steers and heifers were high at the weaning stage but lower at the yearling stage. These results suggest that the SMB heifers tended to assimilate to the crush environment on their birth property due to continued handling, while the steers were better able to display their inherent behaviour due to being exposed to a foreign environment for the yearling measurement taken in the feedlot before entry. These outcomes indicate that it is important that whenever temperament traits are recorded, animals are in an environment where learnt behaviours don't impact the expression of inherent behaviour.

Table 2. Estimated genetic (r_g) and phenotypic (r_p) correlations (standard error) for flight time (seconds*100) and crush score between weaning, yearling and final stages for steers and heifers, and between steers and heifers at weaning and yearling stages

Sex x Stage		Flight time		Crush score	
		r_g	r_p	r_g	r_p
Steer weaning	Steer yearling	0.95 (0.10)	0.33 (0.02)	0.72 (0.13)	0.33 (0.02)
Heifer weaning	Heifer yearling	0.69 (0.13)	0.30 (0.02)	0.59 (0.12)	0.30 (0.02)
Heifer weaning	Heifer final	0.93 (0.18)	0.23 (0.02)	0.58 (0.12)	0.26 (0.02)
Heifer yearling	Heifer final	0.70 (0.17)	0.33 (0.02)	0.95 (0.07)	0.47 (0.02)
Steer weaning	Heifer weaning	0.95 (0.14)	0.26 (0.04)	0.87 (0.14)	0.26 (0.04)
Steer yearling	Heifer yearling	0.69 (0.14)	0.22 (0.05)	0.84 (0.12)	0.27 (0.05)

ACKNOWLEDGEMENTS

The Southern Multi-Breed project is an initiative co-funded by the NSW Department of Primary Industries and Regional Development, the University of New England and the Meat and Livestock Australia Donor Company. The authors thank research station managers and their staff, our skilled technicians, ultrasound scanners, AI, DNA lab, data, project managers and scientists. Individual breeders, participating breed societies and the beef industry are thanked.

REFERENCES

Burrow H.M. (1997) *Anim. Breed. Abstr.* **65**: 477.

Burrow H.M., Seifert G.W. and Corbet N.J. (1988) *Proc. Aust. Soc. Anim. Prod.* **17**: 154.

Gilmour A.R., Gogel B.J., Cullis B.R., Welham S.J. and Thompson R. (2021) 'ASReml SA User Guide Release 4.2' VSN International Ltd, Hemel Hempstead, HP1 1ES, UK.

Grandin T. (1993) *Appl. Anim. Behav. Sci.* **36**: 1.

Kadel M.J., Johnston D.J., Burrow H.M., Graser H-U. and Ferguson D.M. (2006) *Aust. J. Agri. Res.* **57**: 1029.

Kadarmideen H.N., Thompson R., Coffey M.P. and Kossaibati M.A. (2003) *Livest. Prod. Sci.* **81**: 183.

Phocas F., Boivin X., Sapa J., Trillat G., Boissy A. and Le Neindre P. (2006) *Anim. Sci.* **82**: 805.

Walmsley B.J., Donoghue K.A., Johnston D.J., Clark S.A., Siddell J.P., Walkom S.F., Granleese T. and Arthur P.F. (2021) *Proc. Assoc. Advmt. Anim. Breed. Genet.* **24**: 423.

Walkom S.F., Jeyaruban M.G., Tier B. and Johnston D.J. (2018) *Anim. Prod. Sci.* **58**: 213.

Walmsley B.J., Moore, K.L., Walkom, S.A., Clark S.A., Granleese T. and Donoghue K.A. (2023) *Proc. Assoc. Advmt. Anim. Breed. Genet.* **25**: 310.

Proc. Assoc. Advmt. Anim. Breed. Genet: 26: 62-65

GENETIC RISK FACTORS FOR TAIL BITING IN AUSTRALIAN PIGS AS DETERMINED BY BINOMIAL REGRESSION MODELS

D. van Winden[1], R.Y. Galea[2], M.N. Aldridge[1], P.M. Gurman[1] and S. Hermesch[1]

[1] Animal Genetics Breeding Unit*, University of New England, Armidale, NSW, 2351 Australia
[2] AWSC, The University of Melbourne, Parkville, Victoria, 3010 Australia

SUMMARY

Tail biting has adverse welfare implications and is a costly challenge in commercial pig farms. Tail biting incidents can be reduced by breeding animals with reduced susceptibility to becoming a victim of tail biting. This paper aims to quantify genetic risk factors for being a victim of tail biting. Data on 55,373 pigs representing 9 genetic lines were collected. Pigs were housed on 8 commercial farms in Australia under a range of housing and climatic conditions. Genetic risk factors were compared between lines, breeds and sire or dam types. Line, breed, and line type were all significant risk factors for tail biting ($P < 0.0001$). Landrace pigs and sire lines were least affected by tail biting. Female pigs were generally bitten more often ($P = 0.005$) than males, and while there were significant sex-line interactions, these were not observed for breed or line type. Pigs were more likely to receive tail biting if they experienced more days outside thermoneutral temperatures, defined as 15 to 30°C. Some lines were more affected by temperature stress than others. The Large White breed and dam lines showed the least tail biting during temperature stress. To minimise tail biting, pig farmers are encouraged to consider estimating tail biting risk per line, identify sex differences, and/or reducing temperature stress.

INTRODUCTION

Tail biting behaviour occurs regularly on commercial pig farms around the world (Henry *et al.* 2021), resulting in welfare implications ranging from mild marks to serious infections. Pigs with tail lesions or wounds grow slower, require more feed, and are often condemned at the abattoir (Van Staaveren *et al.* 2021). Pig tail biting is a multifactorial problem, therefore it is difficult to both analyse and consider solutions for this behaviour. Identified risk factors include limited feeding space and lack of available enrichment (D'Eath *et al.* 2014); however, adjusting these may not always be feasible for farmers. Additionally, while docking the tail of the animals is a common mitigation strategy, this may cause pain and does not eliminate incidents (Li *et al.* 2017). Alternative strategies to reduce tail biting are needed.

The incidence rate may also be reduced by selectively breeding pigs that display less tail biting (Canario *et al.* 2020). For example, different breeds have shown varying levels of aggressive behaviour (Breuer *et al.* 2003; Sinisalo *et al.* 2012). However, research on genetic aspects of tail biting is sparse, and it is unclear how genetic effects translate to commercial populations. We hypothesize that there are significant genetic risk factors in commercial Australian pigs for being a victim of tail biting, which also involve significant genotype-by-sex and genotype-by-temperature interactions.

MATERIALS AND METHODS

Data. For this research, data on 55,373 pigs were recorded on 8 commercial farms located across Australia from January 2022 (first birth) to December 2023 (last test). Pigs were kept under ranging

---

* A joint venture of NSW Department of Primary Industries and Regional Development and the University of New England

housing and climate conditions. Enrichment was not provided systematically. Animals were considered part of the experiment from birth until date of performance testing or death (whichever occurred first), with an average age of 146 days. Animals that lived for less than 30 days were not used in the analysis. The animals included 9 genetic lines from 3 breeds: Large White, Duroc and Landrace, comprising 5 dam lines and 4 sire lines. To allow the separation of farm and breed effects, each line was either held on multiple farms, or on a farm housing several other lines. Males and females were kept in separate pens, but otherwise handled equally.

Tail biting was defined as a binary trait, using scores of 0 (no lesions) or 1 (moderate to severe lesions). To collect tail lesion data, staff manually went through individual pens and recorded tail lesions per pig weekly from weaning until pigs entered performance testing. Average daily temperature was measured on the exterior of every building using automated weather stations. This was converted into temperature stress for each pig by dividing the number of temperature-stress days by the total number of days with a recorded temperature. Temperature-stress days were defined as days when pigs experienced outside average temperatures outside of the recommended thermoneutral zone of 15 to 30°C (The State of Victoria, 2012).

Statistical analysis. The data were analysed by binomial logistic regression models in R. All models were used to test if recorded tail biting could be predicted by genetic and environmental effects. All models fitted an intercept (i.e. base odds), farm, birth month, and sex as fixed factors, and temperature stress as a covariate. Model 1A additionally fitted genetic line as a fixed factor, and Model 1B fitted both a sex-line interaction, and a temperature-line interaction. Model 2A fitted breed instead of genetic line and 2B the same interactions as for line replaced by breed. Similarly, Model 3A fitted line type instead of genetic line and 3B replaced breed with line type in the interactions.

The robustness of the models was examined using train/test splits. A random test set was generated containing 250 entries with no tail bite and 250 entries with a tail bite. All other entries were used as a training set. This process was repeated 10 times, and the model used an odds-based prediction on each repeat. The model accuracy was defined as the proportion of times where the model correctly predicted the tail biting observation.

RESULTS AND DISCUSSION

Base line effects. Farm and month of birth were significant factors in all models ($P < 0.0001$). Model 1A estimated that different genetic lines had significantly ($P < 0.0001$) different tail biting risks: lines ranged from 0.61 to 3.91 times as likely to be bitten compared to the base animal, which was a female Large White pig in a specific farm born in July 2022. Model 1B (Table 1) included interaction effects. Most sex-line and temperature-line effects were different (both $P < 0.001$) from each other and line estimates were similar to Model 1A.

Breeds and line types. Models 2A and 3A were investigated to estimate differences in tail biting between breeds and line types. Tail biting between breeds and between line types were significantly different ($P < 0.0001$) from each other. Model 2A estimated that compared to Large White pigs, Duroc pigs were 1.25 times as likely to be bitten, and Landrace pigs were 0.83 times as likely to be bitten. This range matches with a previously found 1.38 times difference in tail biting between breeds (Sinisalo *et al.* 2012). According to estimations from Model 3A, dam lines were 1.09 times more likely to be bitten than sire lines.

Model 2B and 3B included interaction effects. There were two discrepancies with previous models: firstly, model 2B (Table 2) found that Duroc pigs were 0.65 times as likely to be bitten than Large White pigs, which was the other way around in model 2A. Secondly, model 3B (Table 3) estimated pigs from dam lines to be 1.569 as likely to be bitten than pigs from sire lines, which was a much larger difference than model 3A.

Table 1. Model 1B: relative odds of being tail bitten on-farm as reported by a binomial regression model (* denotes P < 0.05, ^ denotes base odds)

Factor	Line	As male	Per 1% increased temperature stress
Intercept II: line A, female	1^	0.887	1.002
Line B	0.418*	1.402*	1.005
Line C	0.464	0.387*	1.017
Line D	1.432	0.407*	1.020*
Line E	0.153*	2.832*	1.017*
Line F	3.946*	0.234*	1.010
Line G	0.461*	0.905	1.012*
Line H	0.561	0.911	1.015
Line I	0.893	0.434*	1.014

Table 2. Model 2B: relative odds for pig breeds to be tail bitten on-farm as reported by the binomial regression model (* denotes P < 0.05, ^ denotes base odds)

Factor	Line	As male	Per 1% increased temperature stress
Intercept III: Large White, female	1^	0.945	1.003
Duroc	0.652*	0.776	1.019*
Landrace	0.553*	0.865	1.013*

Table 3. Model 3B: relative odds for pig line types to be tail bitten on-farm as reported by the binomial regression model (* denotes P < 0.05, ^ denotes base odds)

Factor	Line	As male	Per 1% increased temperature stress
Intercept IV: sire line, female	1^	0.868	1.016*
Dam line	1.569*	0.889	1.008*

Sex as a risk factor. According to Model 1A, female pigs were 1.130 times as likely to receive lesions (P = 0.005) than males. However, Model 1B found a significant sex-line interaction for 6 out of 9 lines (Table 1). Line C and F had large differences in tail lesion frequency between sexes (males bitten 0.387 and 0.234 times as often as females, respectively). The interaction was reversed in lines B and E (males bitten 1.402 and 2.832 times as often as females, respectively). Line-sex interactions disappeared in Model 2B and 3B. The differences per lines explain other literature finding sex differences to be ambiguous (Zonderland *et al.* 2010).

Temperature stress as a risk factor. Model 1A estimated pigs that lived more days outside of comfort temperature, experienced 1.009 times more tail biting per percentage of temperature-stress days (P < 0.0001). However, Model 1B found a temperature-line interaction effect, with 3 out of 9 lines showing different effects under temperature stress (Table 1). Pigs from lines A and B were resilient to temperature stress, having slightly increased odds of being tail bitten when more often outside of the thermoneutral zone. Comparatively, pigs from line D were the most affected by temperature stress with an estimate of 1.020 times as likely per percentage of temperature-stress days. Model 2B and 3B estimated differences in temperature sensitivity to persist throughout breed (Table 2) and line type (Table 3). Large White pigs were most resilient to temperature stress (1.003 times more tail biting per increase in temperature stress), and Duroc pigs the least (1.019 times more tail biting). Sire lines were less resilient to temperature stress than dam lines: 1.016 and 1.008 times more tail biting, respectively. Practically, this means that from 0% temperature stress to 50% temperature stress, sire and dam lines get bitten 2.21 and 1.49 times more often respectively. These differences explain the change in order of line estimates between breeds and line types when accounting for interaction effects. Most temperature-stress days were below 15°C (51%). This information can be valuable to farmers designing strategies to reduce risk of tail biting caused by temperature stress.

The robustness of the models was tested. Using the random equal train-test split, models reached accuracies of 0.68-0.70. These values are higher than random chance (0.5), indicating that the estimates of the models were relevant, and the models were reasonably robust.

CONCLUSIONS

According to the models presented, genetic line, breed, and line type had significant differences in tail biting incidence rates. Female pigs were more likely to be bitten than male pigs in most lines, but this sex-line interaction disappeared when grouping lines by breed or line type. Exposure to more temperature-stress days increased tail biting, but individual lines had different sensitivities to temperature stress. Large White pigs and dam lines were the most resistant to temperature stress, and Duroc pigs and sire lines had the most increased tail biting when exposed to temperature stress.

ACKNOWLEDGEMENTS

This project was funded in part by the Australian Government (DISR) CRC-P program. The first author received UNE IPRA and APRIL top-up scholarships. The research could not have been completed without the extensive data collection at Rivalea and Sunpork Group, and their invitation for farm visits to gain extensive insight in farm management.

REFERENCES

Breuer K., Sutcliffe M.E.M., Mercer J.T., Rance K.A., Beattie V.E., Sneddon I.A. and Edwards S.A. (2003) *Appl. Anim. Behav.* **84**: 59.

Canario L., Bijma P., David I., Camerlink I., Martin A., Rauw W.M., Flatres-Grall L., van der Zande L., Turner S.P., Larzul C. and Rydhmer L. (2020) *Front. Genet.* **11**: 611073.

D'Eath R.B., Arnott G., Turner S.P., Jensen T., Lahrmann H.P., Busch M.E., Niemi J.K., Lawrence A.B. and Sandøe P. (2014) *Animal* **8**: 1479.

The State of Victoria (2012) 'Victorian Standards and Guidelines for the Welfare of Pigs' Revision 1, Department of Primary Industries Victoria.

Henry M., Jansen H., del Rocio Amezcua M., O'Sullivan T., Niel L., Shoveller A.K. and Friendship R.M. (2021) *Animals* **11**: 2002.

Li Y.Z., Zhang H.F., Johnston L.J., Martin W., Peterson J.D., and Coetzee J.F. (2017) *J. Anim. Sci.* **95**: 4835.

Sinisalo A., Niemi J.K., Heinonen M. and Valros A (2012) *Livest. Sci.* **143**: 220.

Van Staaveren N., Boyle L.A., Manzanilla E.G., O'Driscoll K., Shalloo L. and Díaz J.A.C. (2021) *Vet. Rec.* **188**: e13.

Wallgren P., Johansson M., Wallgren T., Susic Z., Sigfridson K. and Johansson S. (2024) *Acta. Vet. Scand.* **66**: 2.

Zonderland J.J., Schepers F., Bracke M.B.M., den Hartog L.A., Kemp B. and Spoolder H.A.M. (2011) *Animal* **5**: 767.

GENETIC PARAMETERS OF BEING A TAIL-BITING RECIPIENT IN PIGS

A.A. Reda[1, 2], M.N. Aldridge[1], P.M. Gurman[1] and S. Hermesch[1]

[1] Animal Genetics Breeding Unit*, University of New England, Armidale, NSW, 2351, Australia
[2] School of Veterinary Medicine, Wollo University, Dessie, 1145, Ethiopia

SUMMARY

This study aimed to estimate the heritability of being a tail-biting recipient and to evaluate the genetic associations with performance traits in pigs. Tail bite lesions were recorded weekly from weaning until finisher test (TBPW) and as a single observation at finisher test, with an average age of 147 days (TBFT) on 13,458 pigs from 6 genetic lines between August 2022 and December 2023. Genetic analysis was carried out using linear and binomial animal models. Three models were investigated: fitting the direct genetic effects (M1), fitting common litter (M2), and the line group effects (M3) as additional random effects. The heritability for TBPW was 0.13 ± 0.02 for M1 with a linear model. The heritability estimates dropped to 0.05 ± 0.01 when the litter effect (M2) and line group effects (M3) were fitted. Estimates for the proportion of variance explained by the common litter effects were 0.05 and 0.06, and the line group effects were 0.03 for TBPW. The heritability for TBFT was 0.06 ± 0.01 for M1 with a linear model. Similarly, the heritability for TBFT decreased to 0.03 ± 0.01 for M2 and M3. Heritability estimates for TBPW and TBFT for M1, based on a logit liability scale within a binomial model, were 0.18 ± 0.03 and 0.17 ± 0.04, respectively. Both tail-biting traits had no significant genetic correlations with performance traits, including average daily gain, back fat thickness, and eye muscle depth. Overall, TBPW had a higher estimate of heritability than TBFT. Thus, TBPW should be used in selection strategies of pigs to reduce tail-biting and improve pig welfare.

INTRODUCTION

Socially driven behavioural traits, primarily vulva, ear, and tail-biting, are common in modern pig farming and negatively affect animal welfare, health, and production (Canario *et al.* 2020). Pig breeding companies globally are interested in using genetic selection techniques to lessen the prevalence and effects of harmful behavioural traits, primarily tail-biting. Heritability estimates have been presented for tail-biting recipients recorded at performance testing in finisher pigs in a Tai Zumu line (Canario and Flatres-Grall 2018) and using medication records as a proxy for tail-biting in Large White Australian pigs (Hermesch and Guy 2018). However, the genetic components of tail-biting recipients and their relationship to key breeding goal traits have not been studied in post-weaned pigs. This study aimed to estimate the genetic parameters of alternative traits for recording tail-biting recipients and evaluate their genetic associations with performance traits using data collected post-weaning (TBPW) and at the finisher test (TBFT) at an Australian commercial pig farm.

MATERIALS AND METHODS

Data. Tail-biting and production measurements of post-weaned and finisher pigs were obtained from one piggery in Australia. Tail-biting was recorded between August 2022 and December 2023 on 13,458 tail-docked pigs from six maternal and sire lines of three breeds (Landrace, Large White, and Duroc). Pigs with tail-biting records descended from 1,762 litters, 1,353 dams, and 155 sires.

* A joint venture of the NSW Department of Primary Industries and Regional Development and the University of New England

Pigs were housed in commercial conditions, with males and females housed in separate pens. Tail-biting phenotypes were obtained in growing pigs post weaning. Post-weaning tail-biting (TBPW) was based on weekly observations of pigs from weaning until the end of the performance test at an average age of 147 days. Only the first observation of tail-biting was used for this study. In addition, finisher-test tail-biting (TBFT) was recorded when other performance traits were recorded. Tail-biting was scored based on observed tail lesions as an ordinal categorical variable (0 = no tail lesions, 2 = mild tail lesions, 3 = severe tail lesions, infected tails, and/or tail losses). The categorical scores were converted into a binary trait as tail lesions present (1) or not (0) for the genetic analysis. Performance traits, including average daily gain (ADG) from birth to test, backfat thickness at the P2 site (BFT), and eye muscle depth (EMD), were collected at the end of the test period when pigs reached market weight of 95.2 kg on average.

Statistical analysis. Variance components for tail-biting traits defined as binary traits were estimated using a single-trait animal model with REML assuming a Gaussian distribution. Additionally, the heritabilities of tail-biting traits were calculated based on a binomial distribution employing a logit link function with the residual error variance fixed at $\pi^2/3$ (Gilmour *et al.* 2021). Three models were tested, M1: direct additive genetics effects, M2: common litter effects, and M3: line group effects as additional random effects were compared. Fixed effects of sex (2 levels) and month of birth (16 levels) were fitted for all models, while line (6 levels) was fitted as a fixed effect for M1 and M2. Furthermore, a linear covariate for test weight was fitted for BFT and EMD. All the analyses were conducted using ASReml version 4.2 (Gilmour *et al.* 2021), and the models can be summarised in matrix notation as:

$$\boldsymbol{y} = \boldsymbol{Xb} + \boldsymbol{Z}_1\boldsymbol{a} + \boldsymbol{Z}_2\boldsymbol{c} + \boldsymbol{Z}_3\boldsymbol{g} + \boldsymbol{e},$$

where $\boldsymbol{y}$ is the vector of phenotypes; $\boldsymbol{b}$ is the vector of systematic effects; **a**, **c**, and **g** is the vector of the random additive genetic effects; random common litter effects and random line group effects, respectively, while **e** is the vector of residuals. Individual observations were linked to their respective levels of systematic, random additive genetics, common litter, and random line group effects through incidence matrices **X**, $\mathbf{Z}_1$, $\mathbf{Z}_2$, and $\mathbf{Z}_3$, respectively. Genetic correlations between all traits were estimated using a series of bivariate animal models assuming a Gaussian distribution.

RESULTS AND DISCUSSION

The overall prevalence of tail-biting recipients was 4.42% and 2.72% in TBPW and TBFT, respectively. The weekly observations of TBPW resulted in a higher detection rate of tail-biting compared to the single observation at the end of the growing phase when pigs were performance tested.

Heritability estimates. The heritability of TBPW was 0.13 ± 0.02 for M1 with a linear model (Table 1). When including the litter effect and line genetic effects, heritability estimates decreased to 0.05 ± 0.01. This was expected because the common litter effect accounted for 5 to 6% of the phenotypic variation in TBPW (M2). The line group effect explained 3% of the phenotypic variation in TBPW (M3). The heritability of TBFT was 0.06 ± 0.01 for M1 with a linear model. When the common litter effect and the line group effects were included in models (M2 and M3), heritability estimates for TBFT decreased to 0.03 ± 0.01. This reduction was due to the common litter effect, which accounted for 4% of the phenotypic variation for TBFT, and the line group effect, accounting for 2% of the phenotypic variation in TBFT. Based on binomial models, heritabilities ranged from 0.06 (M2, M3) to 0.18 (M1), and common litter effects varied from 0.11 to 0.13 for both tail-biting traits. Fitting lines either as a fixed or random effect did not significantly affect estimates of additive genetic effects or common litter effects, and genetic lines may be fitted as a fixed or random effect in genetic models. These findings show that the expressions of tail-biting recipient traits were influenced by the pig's additive genetic effect, the common litter effect, and genetic differences between lines.

Line-wise heritabilities were estimated for TBFT using M1 with a linear model. Estimates were highest in the Duroc terminal sire line (0.13 ± 0.05), followed by the Duroc maternal line (0.05 ± 0.02), Landrace maternal line (0.04 ± 0.02), Large White terminal sire line (0.02 ± 0.02), Landrace sire line (0.01 ± 0.02), and the Large White maternal line (0.00 ± 0.00). These estimates indicate the need for line-specific estimates of genetic parameters for tail-biting traits.

The current heritability estimates for TBFT were within the range of heritability estimates found in previous studies using linear models, which ranged from 0.06 to 0.08 (Canario and Flatres-Grall 2018). Moreover, heritabilities for both tail-biting traits from the binomial models were higher than the linear models, which agrees with the findings of Hermesch and Guy (2018). The heritability estimate of TBPW was higher than TBFT, indicating the potential for a response to selection. However, recording TBPW is more time-consuming and costly.

Table 1. Heritability and standard errors (h^2 ± SE), common litter effects (c^2 ± SE), line group effects (g^2 ± SE), and phenotypic variance (σ_p^2) for post-weaning tail biting (TBPW) and finisher-test tail biting (TBFT) traits, backfat thickness (BFT, mm), eye muscle depth (EMD, mm), and average daily gain (ADG, g/day) in recipient pigs

Traits	M	AIC	σ_p^2	h^2 ± SE	c^2 ± SE	g^2 ± SE
TBPW - linear	M1	-30396	0.040	0.13 ± 0.02		
	M2	-30463	0.040	0.05 ± 0.01	0.06 ± 0.01	
	M3	-30488	0.041	0.05 ± 0.01	0.05 ± 0.01	0.03 ± 0.02
TBPW - binomial	M1	61366	4.013	0.18 ± 0.03		
	M2	61077	4.091	0.07 ± 0.04	0.13 ± 0.03	
	M3	60912	4.589	0.06 ± 0.03	0.12 ± 0.03	0.11 ± 0.07
TBFT - linear	M1	-35673	0.026	0.06 ± 0.01		
	M2	-35709	0.026	0.03 ± 0.01	0.04 ± 0.01	
	M3	-35738	0.026	0.03 ± 0.01	0.04 ± 0.01	0.02 ± 0.01
TBFT - binomial	M1	66665	3.979	0.17 ± 0.04		
	M2	66176	4.076	0.06 ± 0.04	0.13 ± 0.04	
	M3	65897	4.569	0.06 ± 0.04	0.11 ± 0.04	0.11 ± 0.07
BFT	M1	27154	3.523	0.54 ± 0.02		
	M2	27059	3.389	0.37 ± 0.03	0.08 ± 0.01	
	M3	27062	3.690	0.34 ± 0.04	0.07 ± 0.01	0.08 ± 0.05
EMD	M1	59249	36.579	0.34 ± 0.02		
	M2	59220	35.800	0.24 ± 0.03	0.05 ± 0.01	
	M3	59241	47.085	0.18 ± 0.03	0.03 ± 0.01	0.24 ± 0.12
ADG	M1	125773	5816.300	0.59 ± 0.03		
	M2	125623	5357.500	0.28 ± 0.04	0.12 ± 0.01	
	M3	125662	5752.000	0.26 ± 0.04	0.11 ± 0.01	0.07 ± 0.04

M: model, M1: σ_a^2, M2: $\sigma_a^2 + \sigma_c^2$, and M3: $\sigma_a^2 + \sigma_c^2 + \sigma_g^2$, AIC: Akaike Information Criterion.

Genetic correlations. There was a very high genetic correlation between TBPW and TBFT (r_g = 0.95 ± 0.02), indicating that the two tail-biting traits are genetically identical. Hence, selection against tail-biting at the end of the finisher test will improve tail-biting at weaning. No genetic correlations were found in the literature between alternative traits describing tail-biting. At the phenotypic level, Hakansson and Hans (2020) found that pigs that received tail-biting earlier or during the post-weaning phases are most likely to get tail bitten in later stages (i.e., finishing phases). Moreover, the genetic correlations between tail-biting traits and performance traits were not significantly different from zero (Table 2). In line with our results, previous studies have reported no significant genetic associations between tail biting and production traits in pigs (Hermesch and

Guy 2018). Together, these results indicate that the current selection for improved productivity does not impact the prevalence of tail-biting recipients. These tail lesion scores could be incorporated into breeding programs using a multiple-trait selection index to mitigate tail biting, thereby improving pig welfare and productivity.

Table 2. Genetic and phenotypic correlations with their standard errors (SE) between both tail-biting traits, post-weaning tail biting (TBPW) and finisher-test tail biting (TBFT) in recipient pigs, and back fat thickness (BFT), eye muscle depth (EMD), and average daily gain (ADG)

Traits	**Genetic correlations ± SE**		**Phenotypic correlations ± SE**	
	TBPW	TBFT	TBPW	TBFT
TBFT	0.95 ± 0.02		0.62 ± 0.01	
BFT	0.02 ± 0.07	0.08 ± 0.08	-0.01 ± 0.01	-0.01 ± 0.01
EMD	0.01 ± 0.06	-0.20 ± 0.09	0.00 ± 0.01	-0.03 ± 0.01
ADG	-0.07 ± 0.06	-0.09 ± 0.08	-0.03 ± 0.01	-0.03 ± 0.01

CONCLUSIONS

Heritability estimates for TBPW and TBFT were low to moderate based on linear and binomial models. Selection lines differed genetically in their susceptibility to receive tail bites, which was also reflected in differences in heritabilities for tail-biting recipients between lines. None of the tail-biting traits showed any antagonistic genetic associations with performance traits that would impede selective breeding against this behaviour. Despite labour and logistical challenges, TBPW had a higher estimate of heritability than TBFT. Therefore, TBPW offers more opportunities for selection to reduce the incidence of tail-biting recipients in pigs.

ACKNOWLEDGEMENTS

This project was partially funded by the Australian Government (DISR) CRC-P program. The leading author is a recipient of the UNE IPRA, DAP, and APRIL top-up scholarships. The authors acknowledge SunPork and its staff for organising a farm visit and providing data for this study.

REFERENCES

Canario L. and Flatres-Grall L. (2018) *Proc. World Cong.Genet. Appl. Livest. Prod.* **11**: 572.

Canario L., Bijma P., David I., Camerlink I., Martin A., Rauw W.M., Flatres-Grall L., Zande Lvd., Turner S.P., Larzul C. and Rydhmer L. (2020) *Front. Genet.* **11**: 611073.

Gilmour A.R., Gogel B.J., Cullis B.R., Welham S.J. and Thompson R. (2021) ASReml User Guide Release 4.2. VSN International Ltd, Hemel Hempstead, HP2 4TP, UK.

Hermesch S. and Guy S.Z.Y. (2018) *Proc. Assoc. Advmt. Anim. Breed. Genet.* **23**: 334.

Hakansson F. and Houe H. (2020) *Prev. Vet. Med.* **184**: 105160.

GENETIC CORRELATIONS BETWEEN PUREBRED AND CROSSBRED PRE- AND POST-WEANING PIGLET MORTALITY

J. Harper[1] and K.L. Bunter[2]

[1] Rivalea Australia, Corowa, NSW, 2646 Australia
[2] Animal Genetics Breeding Unit*, University of New England, Armidale, NSW, 2351 Australia

SUMMARY

Purebred performance can be a poor predictor of future crossbred performance. This study investigated the genetic correlations between purebred and crossbred individual piglet pre- and post-weaning mortality traits. The estimates of the additive genetic correlations between purebred and crossbred performances (r_{pc}) were high (pre-weaning: $r_{pc} = 0.78 \pm 0.097$ and post-weaning: $r_{pc} = 0.94 \pm 0.112$). High correlations ($r_{pc} > 0.75$) demonstrate that survival traits of crossbred pigs will be improved using genetic evaluation and selection based on purebred records.

INTRODUCTION

Mortality traits have low heritability but still have the potential to be genetically improved in purebred populations. The development of successful breeding programs relies on whether purebred performance, recorded in the nucleus tier, accurately predicts outcomes in crossbred performance, the commercial tier (Abell *et al.* 2016). A limiting factor of the typical pyramidal structure of pig breeding systems is that purebred performance can be a poor predictor of future crossbred performance for a range of traits (Ibáñez-Escriche *et al.* 2014).

The genetic correlation between the same trait recorded in purebred vs crossbred populations (r_{pc}) is a key parameter that determines the benefit of recording additional information in crossbred populations (Kramer *et al.* 2021). A high r_{pc}, for example greater than 0.75, indicates that the additive genetic component of the trait was essentially the same in the two populations and the value for selection of recording additional crossbred data may be limited (Kramer *et al.* 2021). Generally, recording systems and performance test records for crossbred animals are insufficient. This limits genetic evaluation and published genetic parameters, particularly for mortality traits. The hypothesis was that the genetic correlation between purebred and crossbred individual piglet mortality traits is high when recorded in the same production environments.

MATERIALS AND METHODS

Data. Purebred and crossbred mortality records were available for this study, from single sire matings producing purebred and crossbred progeny in a common environment, with some overlap of common parentage. The data included individual piglet mortality (0=alive, 1=dead) for pre- (PREw) and post-weaning (POSTw). All piglets were pedigreed from purebred matings within two maternal (genotype A and B) or one terminal (genotype C) selection line, or crossbred, from first-cross progeny (F1 progeny: genotype A × B and B × A) and commercial-cross progeny (F1 × C: F1 sows × C boars). Mortality data was collected as part of a commercial breeding operation located in southern New South Wales, Australia, between January 2011 and December 2019, with a shortened period for crossbred piglets being recorded from August 2015 to December 2019. A total of 477,780 progeny represented 1398 sires, 19,807 biological dams and 19,706 nurse sows, extending over 10

---

* A joint venture of NSW Department of Primary Industries and Regional Development and the University of New England

generations, born in 43,094 litters. Of these 590 sires, 2321 dams and 4627 nurse sows were common to purebred and corresponding crossbred progeny.

The trait phenotypes and animal management have been outlined in Harper *et al.* (2019). Sows and piglets of all genotypes and crosses recorded during the same time-period were managed identically within the same production environment pre- and post-weaning. Purebred piglets were generally fostered within line to nurse sows who had purebred litters. Crossbred piglets were fostered to either purebred sows with first-cross progeny or crossbred sows who had commercial-cross progeny. Piglets in post-weaning, were reared in the same facility with similar health challenges, however purebred piglets were generally penned separately to crossbred progeny.

Each mortality trait was measured on purebred (PurePREw; PurePOSTw) and crossbred (CrossPREw; CrossPOSTw) progeny, defined as different traits for analysis. In this study, purebred animals had their own purebred traits but were given missing values for the corresponding crossbred traits, and vice versa.

Statistical analysis. The nurse sow models identified in Harper *et al.* (2019) formed the basis for the bivariate analyses performed in this study, using linear models in ASReml (Gilmour *et al.* 2021). For PREw, the best model for random effects accounted for direct effects of the animal (piglet), common litter effects of both the nurse sow and biological dam, along with the permanent environment and maternal genetic effect of the nurse sow. Biological dam and nurse sow effects (genetic and permanent environment) were fitted across purebred and crossbred traits to accommodate common sows producing or nursing both purebred and crossbred piglets. For POSTw, the most parsimonious model included only direct animal (piglet) effects along with common litter effects of both the nurse sow and biological dam. The fixed effects fitted for both traits included gender, gestation length group, total born group, birthweight group and birth year quarter. The concatenation of biological dam and nurse sow parity grouping, fostering status of the piglet and farrowing farm described the rearing environment. Additionally, wean age group and weaning farm were fitted for POSTw. Levels for piglet genotype were extended to accommodate crossbred data (purebred traits, contained three levels: A, B and C; and crossbred traits contained three levels: A × B, B × A and F1 × C). Genotype was nested within the purebred or crossbred traits as the latter were only present in the second half of the project time-period. Genetic correlations were then estimated, with both common parents and ancestors providing genetic links between purebred and crossbred individual piglets.

RESULTS AND DISCUSSION

The design of this study provided good cross classification of purebred and crossbred piglet performance for sires and dams. Trait means for PurePREw and CrossPREw were 17.8% (A: 20.6%; B: 15.4%; C: 17.3%) and 16.8% (A × B: 18.2%; B × A: 15.3%; F1 × C: 17.0%), respectively. Suggesting that in this population, raised in a common environment, commercial crossbred piglets with 100% direct and maternal heterosis were no better at surviving than purebred piglets. For PurePOSTw and CrossPOSTw, a proportionally larger difference was seen, 5.59% (A: 5.46%; B: 6.11%; C: 4.67%) and 3.01% (A × B: 3.88%; B × A: 3.15%; F1 × C: 2.43%), respectively. The increased crossbred piglet survival may have been the result of increased direct heterosis relative to purebreds, which is beneficial when piglets are weaned and reared separately from their dams. Phenotypic variances reflected (dis)similarity in incidence, being similar for PurePREw (13.6%) and CrossPREw (13.2%), but approximately halved between PurePOSTw (5.17%) and CrossPOSTw (2.86%). It is important to note that there was no difference in the environment and health status experienced between genotypes as all piglets were reared in the same facility under the same management.

Table 1. Variance component estimates (× 100), resulting ratios of variance components (± standard error) and correlations (± standard error) between random effects for each bivariate model; purebred pre-weaning mortality (PurePREw) and crossbred pre-weaning mortality (CrossPREw); purebred post-weaning mortality (PurePOSTw) and crossbred post-weaning mortality (CrossPOSTw).

			Pre-weaning mortality Models			
			Variance component estimates (x 100)			
	σ^2_a	σ^2_{cnl}	σ^2_{cbl}	σ^2_{mpe}	σ^2_m	σ^2_p
PurePREw	0.24	0.46	0.64	0.12	0.16	13.6
CrossPREw	0.20	0.41	0.34	0.12	0.16	13.2
			Ratios of variance components (± standard error)			
	$h^2 \pm SE$	$cnl^2 \pm SE$	$cbl^2 \pm SE$	$mpe^2 \pm SE$	$m^2 \pm SE$	
PurePREw	0.02 ± 0.002	0.03 ± 0.003	0.05 ± 0.003	0.01 ± 0.001	0.01 ± 0.001	-
CrossPREw	0.02 ± 0.003	0.03 ± 0.005	0.03 ± 0.005	0.01 ± 0.001	0.01 ± 0.001	-
			Correlations (± standard error) between traits within random effects			
	$r_a \pm SE$	$r_{cnl} \pm SE$	$r_{cbl} \pm SE$	$r_{mpe} \pm SE$	$r_m \pm SE$	-
Correlation	0.78 ± 0.097	0.98 ± 0.210	n.e.	n.e.	n.e.	-
			Post-weaning mortality Models			
			Variance component estimates (x 100)			
	σ^2_a	σ^2_{cnl}	σ^2_{cbl}	σ^2_{mpe}	σ^2_m	σ^2_p
PurePOSTw	0.10	0.11	0.16	-	-	5.17
CrossPOSTw	0.03	0.08	0.05	-	-	2.86
			Ratios of variance components (± standard error)			
	$h^2 \pm SE$	$cnl^2 \pm SE$	$cbl^2 \pm SE$	$mpe^2 \pm SE$	$m^2 \pm SE$	
PurePOSTw	0.02 ± 0.002	0.02 ± 0.003	0.03 ± 0.003	-	-	-
CrossPOSTw	0.01 ± 0.002	0.03 ± 0.005	0.02 ± 0.005	-	-	-
			Correlations (± standard error) between traits within random effects			
	$r_a \pm SE$	$r_{cnl} \pm SE$	$r_{cbl} \pm SE$	$r_{mpe} \pm SE$	$r_m \pm SE$	-
Correlation	0.94 ± 0.112	0.77 ± 0.264	n.e.	n.a.	n.a.	-

Abbreviations: σ^2_a = additive genetic variance; σ^2_{cnl} = common nurse litter variance; σ^2_{cbl} = common biological litter variance; σ^2_{mpe} = maternal permanent environmental variance attributed to nurse sow; σ^2_m = maternal genetic variance attributed to nurse sow; σ^2_p = phenotypic variance; h^2 = heritability estimate; cnl^2 = proportion of phenotypic variance attributed to common nurse litter effect; cbl^2 = proportion of phenotypic variance attributed to common biological litter effect; mpe^2 = proportion of phenotypic variance attributed to permanent environmental effect of nurse sow; m^2 = proportion of phenotypic variance attributed to maternal genetic variance of nurse sow; r_a = genetic correlation; r_{cnl} = common nurse litter correlation; r_{cbl} = common birth litter correlation; r_{mpe} = permanent environmental effect of the biological dam correlation; r_m = biological dam genetic correlation; n.a. = not applicable = effect not fitted; n.e. = not estimable; SE = standard error.

The estimates of the additive genetic correlations between purebred and crossbred performances (r_{pc}) within trait were high (PREw: $r_{pc} = 0.78 \pm 0.097$ and POSTw: $r_{pc} = 0.94 \pm 0.112$; Table 1). High correlations ($r_{pc} > 0.75$) indicate that these two traits were largely controlled by the same genes and that obtaining crossbred data may be of limited use to improve accuracy of breeding values for mortality traits, other than through enabling more data. Estimates of the correlation between common nurse litter effects (r_{cnl}) within both PREw and POSTw were also very high ($r_{cnl} = 0.98 \pm$

0.21 and $r_{cnl} = 0.77 \pm 0.26$ respectively) and were not significantly different from one based on the log-likelihood ratio test, indicating that the environment provided by the nurse sow was the same regardless of the piglet genotype, which was expected.

The variance due to nurse sow effects (maternal: $m^2 = 0.01 \pm 0.001$ and permanent environment: $mpe^2 = 0.01 \pm 0.001$) for PREw were similar in magnitude to previous estimates from purebred data only (Harper *et al.* 2019). Correlations were not estimable for the maternal genetic or permanent environmental components of the nurse sow, as convergence failed when these were fitted within trait due to poor data structure. Correlations between common birth litter effects were also not estimable, as birth litters could only be either purebred or crossbred.

Very few studies evaluate purebred and crossbred performance for any trait in the same production environment. For mortality traits, literature tends to only be available for purebred populations; or only on three-way crossbred piglets (F1 sows crossed with terminal sire lines) in commercial production environments (Cecchinato *et al.* 2008, Dufrasne *et al.* 2013, Strange *et al.* 2013); or the combination of purebred and crossbred piglets that are with different environments and management practices, such as a high health nucleus or commercial production farm (Cecchinato *et al.* 2010, Abell *et al.* 2012). All of these scenarios may act to reduce the resemblance in performance between purebred and crossbred progeny, reducing the correlations between these representations of the same trait, making direct comparisons with literature difficult. Overall, r_{pc} here tended to be higher than present in other studies.

CONCLUSIONS

For genetic improvements in the commercial tier of the breeding pyramid to be made, it is important that genetic improvement in the nucleus is reflected by corresponding changes in crossbred performance. The current study demonstrates that survival traits of crossbred pigs will be improved using genetic evaluation and selection based on purebred records due to the high genetic correlations between these traits. Therefore, selection to reduce mortality in purebred populations are expected to deliver downstream improvements for crossbred populations.

REFERENCES

Abell C.E., Fernando R.L., Serenius T.V., Rothschild M.F., Gray K.A. and Stalder K.J. (2016) *J. Anim. Sci. Biotechnol* **7**: 1.

Abell C.E., Stalder K.J. and Holl J. (2012) *Iowa State University Animal Industry Report* **9**: 1.

Cecchinato A., Bonfatti V., Gallo L. and Carnier P. (2008) *J. Anim. Sci.* **86**: 2486.

Cecchinato A., De Los Campos G., Gianola D., Gallo L. and Carnier P. (2010) *J. Anim. Sci.* **88**: 481.

Dufrasne M., Misztal I., Tsuruta S., Holl J., Gray K.A. and Gengler N. (2013) *J. Anim. Sci.* **91**: 5565.

Gilmour A.R., Gogel B.J., Cullis B.R., Welham S.J. and Thompson R. (2021) 'ASReml user guide.' VSN International Ltd, Hemel Hempstead.

Harper J., Bunter K.L. and Hermesch S. (2019) *Proc. Assoc. Advmt. Anim. Breed. Genet.* **23**: 159.

Ibáñez-Escriche N., Varona L., Magallon E. and Noguera J.L. (2014) *Proc. World Cong. Genet. Appl. Livest. Prod.* **10**: 927.

Kramer L.M., Wolc A., Esfandyari H., Thekkoot D.M., Zhang C., Kemp R.A., Plastow G. and Dekkers J.C.M. (2021) *J. Anim. Sci.* **99**: skab270.

Strange T., Ask B. and Nielsen B. (2013) *J. Anim. Sci.* **91**: 1562.

Proc. Assoc. Advmt. Anim. Breed. Genet. 26: 74-77

GENETIC ANALYSIS OF SOW UDDER TREATMENTS

A.M.G. Bunz[1], J. Harper[1] and K.L. Bunter[2]

[1] Rivalea Australia, Corowa, NSW, 2646 Australia
[2] Animal Genetics Breeding Unit*, University of New England, Armidale, NSW, 2351 Australia

SUMMARY

Compromised udder health affects the performance and welfare of the sow and her piglets. Indicator traits for udder health, such as somatic cell count, are difficult to collect and record in sows. In the current study, on-farm udder treatment records were available which were used as an indicator trait for udder health. Genetic parameters were estimated using a threshold model for the udder treatment trait (0 = no udder treatment, 1 = udder treatment). Udder treatments were found to be lowly heritable 0.07 (95% confidence interval between 0.04 and 0.11), and therefore, udder treatment records captured routinely in the herd recording system could potentially be used to genetically improve udder health in sows and reduce sow medication on farms. It was also found that sows who were treated for udder-related conditions weaned half a piglet less per litter compared to sows who were not treated for udder reasons, which equates to $30 AUD / litter and shows the economic importance of improving udder health in sows. Future work could investigate the correlation between udder treatments and other traits, such as sow longevity to further support these findings.

INTRODUCTION

Postpartum disgalactica syndrome (PDS) in sows is a multifactorial disease not only affecting udder health but the reproductive tract, which in turn compromises the performance and welfare of the sow and her piglets (Kemper 2020). PDS is characterised by reduced milk production in the first days after farrowing, leading to increased piglet mortality and lower weaning number (Kemper 2020). One of the factors causing PDS are bacteria, however, to enable ubiquitous bacteria to cause PDS requires non-infectious risk factors to be present, such as prolonged farrowing duration, constipation and high ambient temperature (Kemper 2020). Indicator traits for udder health, such as somatic cell count and milk pH, are limited in sows due to the difficulties in collecting adequate milk (Kemper 2020), making the ability to study udder and milk traits difficult. As an alternative, routinely recorded treatments for udder-related conditions could be used. The aim of this study was to investigate the use of routinely recorded on-farm udder treatments as an indicator trait for udder health and its association with litter survival. It is hypothesised that the trait defined as on-farm recorded udder treatment is heritable and can be used to genetically improve udder health in sows.

MATERIALS AND METHODS

Data. The trait investigated was any medication related to an udder treatment (UT) that sows received during lactation, recorded by staff in the herd recording management system (UT: 0 = no udder treatment, 1 = yes udder treatment). Sows, which required UT treatment were treated with anti-inflammatory drugs, and / or antibiotics. UT is one of multiple treatment reasons recorded on the herd recording management system, with UT having the most frequent reason for sow

* A joint venture of NSW Department of Primary Industries and Regional Development and the University of New England

medication on this farm. The data from one maternal line (Landrace, LD) and one maternal crossbred line (LD × Large White (LW)) were collected from a single farm in southern New South Wales, Australia. Data included 7,518 records from 4,347 sows (daughters of 309 sires) collected between November 2023 and October 2024. Records for UT were not censored by other potential concurrent treatments. Fostering of piglets was generally conducted on the first day of the piglets' life to ensure that all piglets had access to a functional teat and colostrum. Sows were treated for udder-related conditions when staff observed that milk production was reduced (with or without mastitis), and piglets' vitality, weight gain and health appeared compromised. Veterinary diagnosis of udder treatments was not performed, but authorisation for treatment was obtained.

Statistical analysis. Parameter estimates for UT were obtained using an animal threshold model via the Bayesian approach with Gibbs sampling as outlined below:

$$\lambda_i = \boldsymbol{X}\beta + \boldsymbol{Z}a_i + \boldsymbol{Z}pe_i + e_i$$

where λ_i was a vector of the threshold liability value for UT (0 = no udder treatment, 1 = yes udder treatment). $\boldsymbol{X}$ is an incidence matrix for the significant fixed effect (β) included parity group (4 levels: parity 1; parity 2; parity 3 and parity 4, > parity 4) and farrowing year month contemporary group (CG: 11 levels). Preliminary analysis found that line, age at first farrowing, and the number of still-born piglets were not significant (p-value > 0.05) for UT, resulting in their exclusion from the operational model. **Z** represents incidence matrices linking records to random effects, which included the additive genetic effect (a) and the permanent environmental effect (pe) of the sow, accounting for repeated records per sow

The model was implemented using the gibbsf90 + software from the blupf90 suite (Misztal *et al.* 2002). The Gibbs sampler comprised a chain of 100, 000 cycles, with a burn-in of the first 10, 000 iterations and samples stored every 10 cycles. Hence, the posterior means of genetic parameters were estimated from 9, 000 samples. Convergence was evaluated through visual inspection of the trace plot and the Geweke diagnostic (Misztal *et al.* 2002).

The associations between UT, weaning number (WN) and litter survival (LS) after day 1 of age were evaluated. Day 1 litter size was calculated to account for fostering movements and deaths that had occurred very early in lactation, which were not related to the UT trait. LS was estimated as the ratio between the number of piglets alive at weaning and day 1 litter size.

LS was not normally distributed and, therefore, log transformed. LS and WN were analysed using a linear mixed model including parity group, line (2 levels: LD; LD × LW), CG, the permanent environment effect of the sow and UT to characterise the impact of UT on these traits. Further, WN was adjusted for Day 1 Litter Size as a linear covariate. The predicted least square mean for WN or back-log transformed LS were weighted proportional to the frequency of other model factors that were averaged, using R package *emmeans* (Lenth and Lenth 2018).

RESULTS AND DISCUSSION

Phenotypic observations. The prevalence of udder treatment in this study (17%; Table 1), were very similar to those reported in literature (13%; Kemper 2020), with PDS the main reason for udder treatments as described by farm staff. Most udder treatment medications (93%) occurred within the first 7 days after farrowing, as expected (Kemper 2020). Udder treatment was the most frequent reason for treating sows generally and the main reason for administering antibiotics (~50% of the total antibiotic treatment in sows). These results show the importance of improving udder health to also reduce the usage of antibiotics in sows.

In contrast to the current study, Preissler *et al.* (2011) defined the cases of PDS in sows when they showed a rectal temperature above 39.5°C and / or had clinical signs of mastitis. The current study's broad definition of udder treatment can be seen as a limitation from a veterinary perspective. However, diagnosing PDS can be difficult due to the high variation in clinical expression. Consequently, it has been suggested that both the mammary gland appearance and behaviour change in piglets (e.g., weakness) and sows (e.g., limiting access for piglets to teats by lying on the udder) should be used to identify PDS in sows (Kemper 2020).

Further, this study found that the prevalence of UT was highest in parity one sows compared to sows in older parities (23%; Table 1). The reported results of the parity effect on the prevalence of PDS in the literature were inconsistent (Kemper 2020). However, UT treatment in parity one seems not to have a negative effect on staying in the herd until parity two on this farm (P = 0.54).

Table 1. Descriptive statistics of the sow udder treatment trait (UT: 0 = no udder treatment, 1 = yes udder treatment) across parity groups

Trait (units)	Parity	Number of records	Number of Sows	Prevalence (%)
UT (0/1)	1	1,744		23
	2	1,444		19
	3-4	2,256		13
	>4	2,071		15
	overall	7,518	4,347	17

Genetic parameter estimation. The mean heritability of udder treatment was low, h^2: 0.07 (95% confidence interval CI: 0.04 and 0.11; Figure 2). Preissler *et al.* (2012) found slightly higher heritability for PDS (h^2: 0.09), possibly due to a more refined trait definition, as previously mentioned. These results indicate the importance of management strategies to mitigate the incidence of udder treatment. Nevertheless, low heritability for mastitis traits has been found in dairy cattle, and genetic selection for mastitis resistance in the last decade has reduced the incidence of clinical mastitis (Weigel and Shook 2018).

The effect of the permanent environmental effect of the sow was close to zero (pe^2: 0.03; CI: 0.0 and 0.07) using a threshold model, which was also found when fitting a linear model; the permanent environmental effect of the sow was on the boundary to zero. These results indicate that the likelihood of a sow being treated again in subsequent parities is no higher than a sow who has not been treated. This phenomenon seems not to be caused by culling of sows with UT, as UT did not have an effect of parity one sows to stay in herd until parity two, as mentioned above.

Association between UT and LS. Litter survival and weaning number were significantly lower in sows treated for udder-related conditions (Table 2). The difference between litters whose nurse sow was treated for udder-related conditions versus those not treated in weaning number was 0.5 piglets per litter. This equates to $30 AUD / litter and shows the economic importance of improving udder health in sows.

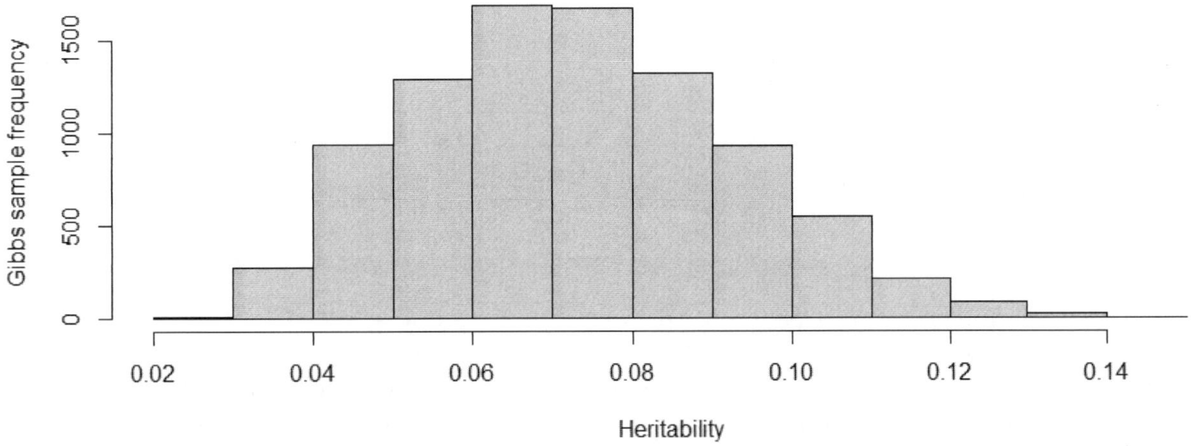

Figure 1. Posterior distribution of heritability for udder treatment in sows

Table 2. Estimated least square means (standard error) for litter survival and weaning number by udder treatment group (0 = no udder treatment, 1 = yes udder treatment)

Traits	Udder Treatment		P value
	0	1	
Number Sows	3,980	1,193	
Litter Survival (proportion)*	0.90 (0.00)	0.87 (0.00)	<0.001
Wean Number (piglets / litter)	11.5 (0.03)	11.0 (0.07)	<0.001

*Back-log transformed from natural logarithms scale

CONCLUSION

Udder treatment records routinely recorded on farms were lowly heritable ($h^2 = 0.07$) and can potentially be used to genetically improve udder health, even without a veterinary diagnosis. However, the low heritability indicates that the environment (including management) plays an important role in UT. Closer monitoring or revising management practices may be required for sows in parity one. Further, the genetic correlation between UT and other traits, such as litter size and sow longevity, should be evaluated.

ACKNOWLEDGEMENTS

Thank you to Charlotte Garner and Arris Ranido for your support and valuable feedback.

REFERENCES

Kemper N. (2020) *J. Anim. Sci.* **98**: 117.

Lenth R. and Lenth M.R. (2018) *Am. Stat.* **34**: 216

Misztal I., Tsuruta S., Strabel T., Auvray B., Druet T. and Lee, D. (2002) *Proc. World Cong. Genet. Appl. Livest. Prod.* **28**: 743.

Preissler R., Hinrichs D., Reiners K., Looft H. and Kemper N. (2012) *J. Anim. Breed. Genet.* **129**: 98.

Weigel K.A. and Shook G.E. (2018) *Vet. Clin. Am-Food A* **34**: 457

Proc. Assoc. Advmt. Anim. Breed. Genet. 26: 78-81

IMPLEMENTATION AND DELIVERY OF THE WAGYU FEEDER CHECK

C.F. Teseling[1], M.B. McDonagh[1] and D.P. Garrick[2,3,4]

[1] Australian Wagyu Association, Armidale, NSW, 2350 Australia
[2] The Helical Company Limited, Rotorua, BOP, New Zealand
[3] Theta Solutions LLC, Lacey, WA, USA
[4] A.L. Rae Centre for Genetics and Breeding, Massey University, New Zealand

SUMMARY

Most Wagyu animals were imported into Australia between 1990 and 1997. Since then, the number of Wagyu and Wagyu-influenced animals in feedlots continued to increase to the point where many high-quality beef export brands are now Wagyu or Wagyu-cross animals. Wagyu and Wagyu-cross animals are typically feedlot fed finished for 350 to 450 days. These extended periods in the feedlot makes it valuable to identify non-performers and to remove them from long-fed programs. The Wagyu Feeder Check (WFC) was developed through partnership with Australian Wagyu Association (AWA), Neogen and CSIRO and is a commercially available DNA test, delivered using a web-based information system developed by AWA and The Helical Company. The system provides producers or feedlot operators with the ability to test feeders and rank them on genetic merit for five feedlot and carcase performance traits as well as the Wagyu Feeder Index which combine the five trait scores into a single value. The WFC system also enables the registration and collection of carcase data of Wagyu and Wagyu-cross animals. This significantly increase the Wagyu reference population for further recalibration of the WFC algorithms as well as genetic analysis of registered Fullblood and Purebred animals.

INTRODUCTION

The Wagyu breed in Australia was established from 221 Fullblood foundation animals mostly exported from Japan between 1990 and 1997. In comparison to other breeds Wagyu is a relative recent introduction to the Australian beef industry. In the past decade the number of Wagyu and Wagyu-cross animals on feed increased very significantly to around 300,000 (Australian Wagyu Association annual survey data, unpublished). In 2018 Single-Step BREEDPLAN, which incorporates the use of genomic information in conjunction with all traditional sources of data in the estimation of breeding values, was implemented for high Wagyu content Fullblood and Purebred animals. This development stimulated high density SNP testing of seedstock animals and the collection of carcase data increased sharply for those animals which were not retained for breeding purposes. The Animal Genetics and Breeding Unit (AGBU) developed a genomic data pipeline which incorporates a range of quality assurance (QA) checks to ensure genotype integrity (Connors *et al.* 2017). However, in many cases these QA checks exclude the genotypes of animals with a genomic Wagyu content lower than 80%. Most F1 to F3 Wagyu crosses therefore do not have genomic information used to determine their genetic merit through EBVs.

Wagyu and Wagyu-cross animals are often feedlot fed for 350 to 450 days. Due to these extended periods in the feedlot, it is very beneficial to identify animals that are likely to perform poorly and eliminate them from long-fed programs. The Wagyu Feeder Check (WFC) was developed to address the need of the Wagyu sector for the feedlot assessment of F1, F2 and F3 Wagyu feeder cattle for economically important traits as well as the AWA developed Wagyu Feeder Index.

DEVELOPMENT OF WAGYU FEEDER CHECK

The WFC prediction is based on a commercially available DNA test which was developed through partnership with AWA, Neogen and CSIRO.

The development of the WFC genomic algorithms (Reverter *et al.* 2023) utilised 8,316 genotypes supplied by Neogen and corresponding carcase phenotypes supplied by the AWA. The carcase data supplied by AWA originated from eight independent commercial supply chains with phenotypes for five economically important traits. This reference population consisted of 2,120 Fullblood Wagyu and 6,196 Wagyu crosses, mainly Wagyu-Angus F1 animals.

The WFC provides a score to reflect the genetic merit of Wagyu content animals for five key traits: average daily gain during feedlot finishing (ADG), hot carcase weight (CWT), carcase eye muscle area (CEMA), AUS-MEAT marbling score (MARB), and carcase subcutaneous fat depth at the rump or P8 site (CP8). The genetic merit scores (Molecular Breeding Values; MBVs) range between 1 and 10 (10 being the highest genetic merit). According to Reverter *et al.* (2023) the genomic prediction accuracies were in the range of 55-65% for all traits except CEMA which were in the 40-50% range. They concluded that the use of the WFC will benefit feedlot operators who finish Wagyu breed content animals and who wish to remove poor performers from their long-fed feeding programs.

THE WAGYU FEEDER INDEX - TRAIT WEIGHTED PROFIT RANKING

Based on the relative impact of the WFC traits on feedlot performance and carcase value, the Wagyu Feeder Index (WFI) was developed by the AWA to simplify sorting animals based on a single value reflecting their genetic merit for feedlot profit.

The WFI calculation is highly aligned to the Wagyu F1 $Index published by the AWA as part of its Wagyu BREEDPLAN reported selection indexes. The relative contributions of each of the WFC traits to the Wagyu Feeder Index is 50% for Marble Score, 16.7% for Carcase Weight, 12.5% for Eye Muscle Area and P8 Fat depth and 8.3% for Average Daily Gain.

WAGYU FEEDER CHECK TESTING AND REPORTING OF RESULTS

The results of the WFC are mainly being used to remove low genetic merit animals from long-fed programs, or to identify animals for which feeding periods can be reduced but still achieve required carcase performance. To ensure timely assignment of animals to the most appropriate feeding program, the reporting of results needs to be automated to minimise human intervention which normally slows down the process. It was also very important to simplify the ability to download and interpret results and to graphically present progeny distributions to assist in the comparison of sires.

The AWA in conjunction with The Helical Company developed a web-based information system to collate the WFC test requests, collect, store, process and report the results (Garrick *et al.* 2025). This process is automated and allows the member to log in and complete the test request, invoices the member for the testing, notifies Neogen of the test request, downloads the test results and genotypes from the Neogen server, loads the results into the Helical database, runs a parent discovery analysis and notifies via email and an online dashboard to the member that the results are available. The parent discovery is a key enabler of the overall value proposition by facilitating the linkage of animals with otherwise unknown parentage to their AWA registered sire so that the commercial sire population can also be managed.

In addition to the tabular display and ability to download the results, each animal's results are also displayed graphically in a radar plot. Figure 1 shows how the WFC scores are graphically displayed for animals with low, average and high marbling potential.

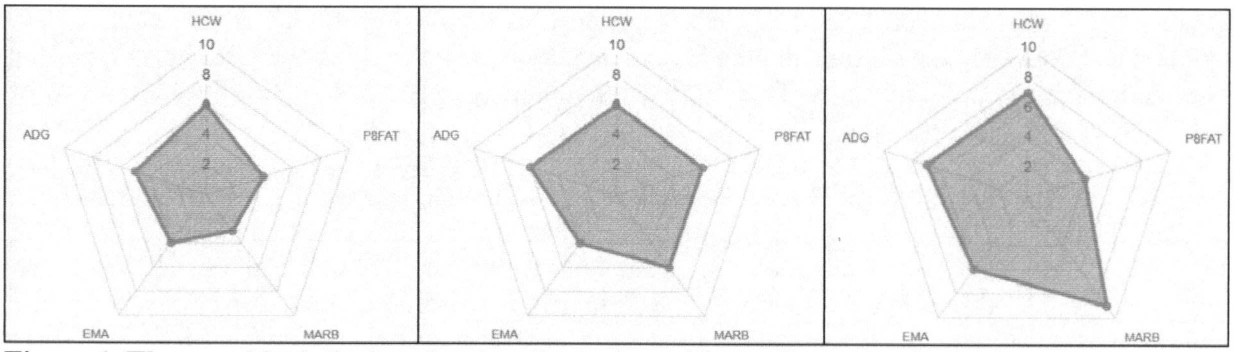

Figure 1. The graphical display showing the trait rankings of an animal with low (left), average (middle) and high (right) marbling WFC scores

The genotypes of WFC tested animals are automatically included in a parent discovery analysis which now includes over 500,000 genotyped animals.

Where sires are identified through the parent verification process, a breeder can evaluate the sires by comparing the WFC score distributions of their respective progeny. Figure 2 shows the proportional progeny trait score distribution graph of two sires and how their trait distributions compare to all animals tested in the specific batch. The average trait ranking distributions of progeny makes it possible to identify and cull a genetically inferior sire well before kill data of progeny is analysed and remove these from impacting future production. Likewise, a breeder may decide to use a high performing bull in an artificial insemination program to increase the number of genetically superior progeny.

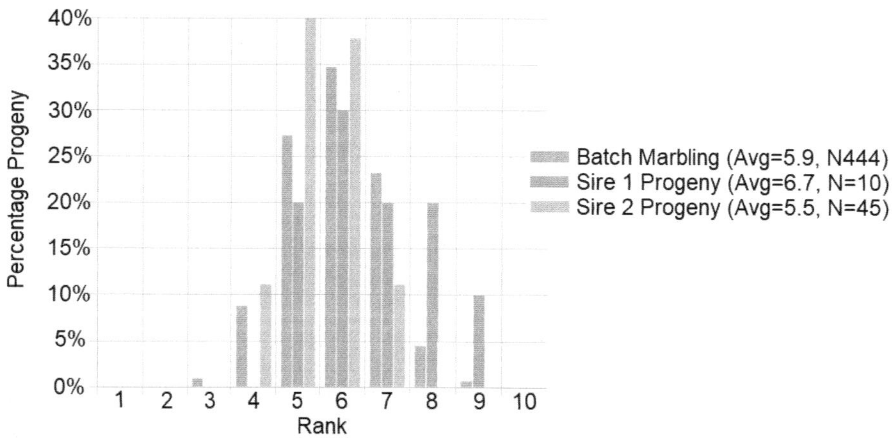

Figure 2. Distribution graph comparing marble trait scores of the progeny of two sires and all animals tested in that batch

USING WAGYU FEEDER CHECK

Development of the WFC delivers a novel approach in optimising management strategies to meet profitability and sustainability targets in the Wagyu industry and is aimed to support supply chain participants who produce export quality Wagyu beef using a feeding period of typically around 350 - 450 days.

The WFC provides the commercial producer or feedlot operator with the ability to test feeders before or at feedlot entry to verify sire parentage and identify low genetic merit carcase performance animals. Animals that are identified as not meeting the performance criteria for a supply chain can be drafted into short-feeding programs or other endpoints, saving on resource use and investment of capital.

The AWA estimates that removing the bottom 10% of crossbred Wagyu feeders to short-fed programs (150 days) would save AUD $22 million in feed costs per year, limiting resource use inefficiencies and reducing financial losses on low performing animals. Removing the bottom 5% of sires and their progeny from future years would save a further $11 million in production losses.

There are a significant number of large northern commercial enterprises which are in the process of introducing Wagyu into their herds to improve fertility and age at first calving. Many of these enterprises are now confronted with the prospect of deciding which of the F1, F2 and F3 females are genetically superior for carcase traits and should be retained as breeding replacements. When these females are WFC tested, the sire discovery and trait scores significantly simplify the process of deciding which females are genetically superior for the feedlot traits.

CARCASE DATA SUBMISSION

The AWA encourages the registration of parent-verified WFC tested animals in the free-of-charge AWA Slaughter Register. A carcase data refund of $5 per WFC tested Slaughter Registered animal is provided when carcase data is submitted and loaded into the AWA database. This has stimulated a significant volume of commercial carcase data submission, with a recent extract for recalibration of the WFC genomic predictions increasing the number of animals in the reference population from 8,000 to close to 18,000.

These additional genotypes and phenotypes constitute a significant increase to the Wagyu reference population and allows the AWA to improve the WFC genomic predictions. It also provides the AWA with large numbers of carcase data and genotypes necessary for improving its genetic estimates of registered cattle within the AWA public genetic evaluation (Wagyu BREEPLAN) for Fullblood and Purebred cattle.

CONCLUSION

The Wagyu Feeder Check is an industry driven tool for the commercial Wagyu sector to determine the genetic merit of F1, F2 and F3 Wagyu animals. The automation of the test request and results delivery pipeline allows for reduced timelines and accuracy due to minimal human intervention. The independent development and delivery of the WFC pipeline demonstrates the maturation of genomic services provision with the Australian beef industry to provide rapid implementation of commercially relevant services.

The cost-free registration of WFC tested animals and rebate for carcase data are innovative ways of stimulating the submission of carcase data for animals which already have a genotype available thereby increasing the Wagyu reference populations. Due to the increase in the number of carcase records available, updated WFC validations are currently under development and promise to deliver even higher levels of accuracy to identify genetically superior animals.

The WFC provides an excellent example of the benefits that can accrue when a Breed Association champions a collective approach to delivering an information system that leverages the activities of a range of stakeholders that span the beef production chain and its supporting industries.

REFERENCES

Connors N., Cook J., Girard C., Tier B., Gore K., Johnston D. and Ferdosi M. (2017) *Proc. Assoc. Advmt. Anim. Breed. Genet.* **22**: 317.

Reverter A., Li Y., Alexandre P.A., Dominik S., Teseling C.F., Van den Heuvel A., Schutt K., McDonagh M. and Porto-Neto L. (2023) *Anim. Prod. Sci.* **64**: AN23246.

Garrick D.P., Garrick D.J. and Howse J.E. (2025) *Proc. Assoc. Advmt. Anim. Breed. Genet.* **26**: *These proceedings*.

COST-BENEFIT OF A GENOMIC REFERENCE POPULATION FOR GOATS IN AUSTRALIA

T. Granleese[1] and S.F. Walkom[2]

[1]NSW Department of Primary Industries and Regional Development, Grafton, NSW, 2460 Australia
[2] Animal Genetics Breeding Unit*, University of New England, Armidale, NSW, 2351 Australia

SUMMARY

There is currently no genomic-based information available to Australian goat meat breeders to make selection decisions. This is due to a lack of performance-recorded goats with genomic information and a lack of genetic evaluation infrastructure. The Measured Goats project is the beginning of a genomic reference population that will help provide the data infrastructure to rectify both challenges. This study reports a cost-benefit analysis of investment of genomic reference populations for goat meat where commercial breeders breeding their own bucks could genomic test their "clean-skin" male selection candidates. The study demonstrates that the Measured Goats project is an excellent start to building a genomic reference population to underpin commercial genomic selection. All testing scenarios return a profit on investment to the industry in years 11-13. The genomic prediction accuracy and hence the response to selection is dependent on effective population size and the number of tested animals in the genomic reference population.

INTRODUCTION

Currently, little is known about the phenotypic and genetic performance of Rangeland goats (captured feral goats) and their crossbred progeny. The Australian Rangeland goat population, however, underpins the Australian goat meat industry which is an export industry valued at $288 million per annum (MLA 2024). Rangeland does will continue to be the basis of the goat meat industry owing to their numbers relative to the limited number of pure or crossbred does.

There is considerable potential for genetic improvement in Rangeland goat herds, although the benefits from research in genetics and genomics are yet to be realised (Kijas 2012; Aldridge and Pitchford 2018). Kijas (2012) reported that the Rangeland goat population was one of the most genetically diverse domesticated species in the world. Aldridge and Pitchford (2018) suggest that performance recording and understanding the genetic parameters of Rangeland goats would help increase breeding goat enterprise's profitability by better understanding the genetic capability of Rangeland goat genetics and hence enable better informed selection decisions.

The Measured Goats project (Granleese *et al.* 2023) is building a genomic reference population to complement the current KIDPLAN database. When a genomic reference population is established, there are possibilities to both enhance "single-step" analysis as well as make genomic predictions for individuals who have no known pedigree and/or performance recording (defined as "clean-skins" in this paper). Genomic prediction of clean-skin individuals could be of particular value to goat breeders who run extensive and rangeland production systems where precise breeding programs are impossible.

Genomic prediction accuracy is determined by factors such as effective population size (Ne), reference population size, and the relationship between reference and target populations. Smaller Ne increases linkage disequilibrium (LD), improving marker-trait associations. Larger genomic

---

* A joint venture of NSW Department of Primary Industries and Regional Development and the University of New England

reference populations enhance accuracy by capturing more genetic variance and reducing sampling error. A close genetic relationship between reference and target populations also boosts accuracy due to shared genetic architecture.

This paper will investigate the impact of the size of genomic reference population targeted for the Australian goat meat industry and the modelled cost-benefit and will focus on prediction for two populations the Rangeland and combined meat breeds (Boer and Kalahari) populations.

MATERIALS AND METHODS

Genomic selection accuracy. Genomic selection accuracy was calculated using Goddard *et al.* (2011) methods with notation from Dekkers (2007) where the correlation between true and predicted breeding values was estimated. This method reflects how well genomic predictions capture genetic variation. Accuracy was estimated using: $r_g = \sqrt{\frac{Nh^2}{Nh^2+Me}}$, where N is the size of the reference population, h^2 is the heritability of the trait, and Me is the effective number of independent chromosome segments. Me is calculated by: $Me = \frac{2NeLk}{ln(NeL)}$ where L is the average chromosome size, k is the number of chromosomes goats have (30), ln is the natural logarithm and the effective population size (Ne) for meat breeds for Boer and Kalahari combined was estimated using pedigree data obtained from Boer Goats Breeder Association of Australia. Given we have no pedigree information and little genomic information on Rangelands animals, it was assumed that this population had little inbreeding and two effective population sizes of 2,500 and 1,000 animals were tested following estimates for other global wild goat populations. The equation to calculate Ne for meat breeds was: $N_e = \frac{1}{2\Delta F}$ where ΔF is the rate of inbreeding per generation (Falconer & Mackay 1996). The proportion of genetic variance captured by markers is defined by $q^2 = \frac{M}{Me+M}$ where M is the number of markers on the genomic test (70,000). The variance ratio of residual and effective QTL variance was calculated as: $\lambda = \frac{Me}{q^2h^2}$. The accuracy of genomic prediction as a predictor of effects captured by markers was calculated by $r\hat{Q} = \sqrt{\frac{T}{T+\lambda}}$ where T is the number of training individuals in the reference population. The accuracy of genomic prediction as a predictor of the true breeding value was calculated as: $r_{GBV} = \sqrt{q^2 r\hat{Q}}$.

Response to selection. Using the accuracies of genomic prediction calculated above, genomic accuracies for each trait (Table 1) adapting the "Kid Survival Plus" (KSV+) index developed by Aldridge and Pitchford (2018) were used in deterministic simulations using truncated selection. Heritabilities of traits within the KSV+ index ranged from 0.1 to 0.5. The genetic standard deviation of the KSV+ index is $11.53. Response to selection was calculated assuming all traits in the KSV+ index were measured in the reference population and only genotyping of male selection candidates were used in selection in clean skin herds. The only variable that changed within scenarios was accuracy of genomic prediction over time (Table 1) for each breeding scenario. This simulates the model where extensive breeders do no phenotyping on their property and genotype only eligible male candidates.

Cost-benefit to industry. The cost-benefit to the industry was calculated as genetic gain multiplied by total breeding does mated annually to genetically improved bucks using the geneflow method (Hill 1974), applying a cumulative discount of expression of 0.07. The starting operating cost of $700,000 to run the reference population in year one was deducted as a cost. Each year the cost increased on the previous year by 3%. The number of commercial does mated to genetically improved bucks began in year 6 starting with 50,000 commercial does mated, then increased by

50,000 annually until capped at 750,000 does in year 20. The cost-benefit was calculated cumulatively.

Testing scenarios. Three scenarios were assessed with straight meat breeds and Rangelands goats with two different Ne sizes of 1000 and 2500.

RESULTS AND DISCUSSION

Benefits to the industry from genetic gain were not realised until year six of the program due to the establishment phase of the genomic reference population where 1,000 or 2,500 individual animals for each breed are recorded in the genomic nucleus each year. We can observe that meat breeds get the highest genomic prediction accuracy (Table 1) due to the lower effective population size which is associated with fewer independent chromosome segments. Goddard *et al.* (2011) emphasizes that M_e depends on the balance between genetic diversity and the recombination potential of the genome. A smaller N_e or shorter genome size reduces M_e, leading to stronger linkage disequilibrium and fewer independent chromosome segments.

Table 1. Genomic prediction accuracies (r_g) for traits with three heritabilities (h^2), the possible genetic gain using KSV+ selection index for cleanskin herds testing male selection candidates only with 2500 and 5000 individuals recorded in the genomic reference population for each breed

Scenario			2500 ref individuals rec				5000 ref individuals rec			
	ΔF	Ne	r_g_h^2 =0.1	r_g_h^2 =0.3	r_g_h^2 =0.5	dG/yr KSV+	r_g_h^2 =0.1	r_g_h^2 =0.3	r_g_h^2 =0.5	dG/yr KSV+
Meat	0.0061	75	0.40	0.60	0.69	$3.68	0.52	0.72	0.80	$4.58
R'land 1,000	0.0020	1000	0.13	0.23	0.29	$1.69	0.19	0.31	0.39	$2.15
R'land 2,500	0.0002	2500	0.07	0.14	0.18	$1.27	0.10	0.19	0.24	$1.47

As the effective population size is much smaller in meat breeds, it allows up to a $4.58 gain per year increase in clean-skin herds when simply genotyping male selection candidates. We observed a larger response to selection as more animals were recorded in the reference population (Table 1). Rangelands Ne 1000 and Ne 2500 had a more modest annual response of up to $2.15 and $1.47, respectively. There are a number of assumptions underpinning Rangeland goat prediction, which may be inaccurate. Firstly, it is assumed Rangeland goat genetics have the same genetic parameters as meat breeds, where in reality, the current KSV+ genetic parameters are only based on meat breeds. The actual difference in breeds and/or combined genetic analysis will not be estimated until the end of the Measured Goats project. The study also assumes that the Ne of the Rangeland goats is somewhere between 1,000 and 2,500 as the Rangeland goat population has no known pedigree information and a small representation of animals genomic tested. Once the project has more Rangeland genomic data on a broader range of the Rangeland goat population, the project will be able to get a better understanding of Ne by calculating the expected decay of LD with distance between markers, which reflects historical recombination and genetic drift.

All breeding scenarios demonstrate that benefits of selection are generated from year 6, and that the industry cost-benefit hits break-even at years 11-13 (Figure 1) Further, the simulation assumes that either meat breeds or Rangeland breeds are run separately. In reality, Rangeland does are usually mated to meat breed bucks where some level of "breeding-up" from Rangeland does to a higher meat content animal is bred over time in the pursuit of higher carcase weights. If this were to happen genomic prediction accuracy for first cross animals would be lower than Rangeland animals given the Ne of first cross animals is higher than Rangelands. But as meat breed content increases over generations, the Ne would then sharply decrease and genomic prediction accuracy and hence response to selection would increase for these clean skin animals.

The modelling in this paper suggests that a better understanding of genetic architecture of the Rangeland goat population will help design and allocate funding towards goat meat genomic resource herds. This modelling demonstrates that the investment in a genomic resource herd is beneficial when assessed from an industry perspective. It also demonstrates that despite low genomic prediction accuracy for Rangeland goat populations, it is still an economically worthwhile exercise to do. Further, it demonstrates that the Measured Goats project is building the initial foundations required to help build a database that allows goat breeders to use genomic selection.

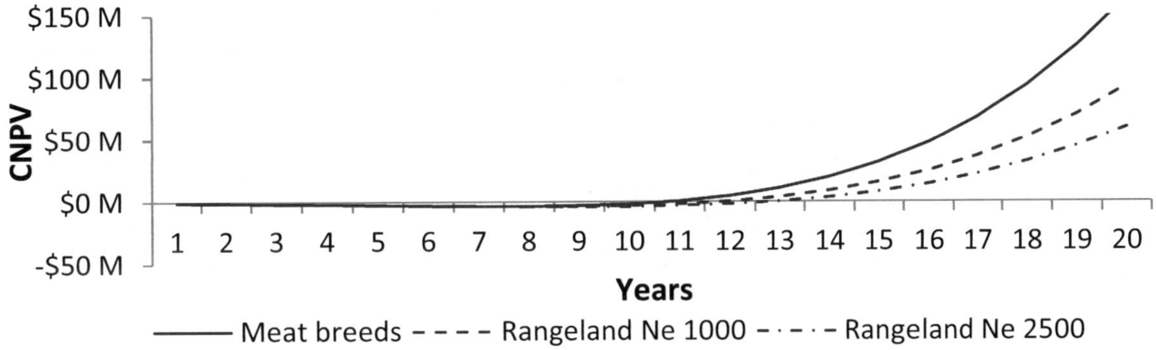

Figure 1. Cumulative net present value (CNPV) for investment of reference population and extra profitability earned by the goat industry under three investment scenarios

CONCLUSION

The Measured Goats in the Rangelands project is forming a valuable resource population for commercial breeders to genomic test "clean skin" animals. It will be important to focus on performance recording Rangeland animals to help build the number of animals given their hypothesised large effective population size. A better understanding of the effective population of Rangelands goats is required to help with reference population design as well as cost-benefit analysis.

ACKNOWLEDGEMENTS

The Measured Goats in the Rangelands project is an initiative co-funded by NSW Department of Primary Industries and Regional Development and Meat and Livestock Australia Donor Company. The authors give thanks to the Boer Goat Breeders Association of Australia (BGBAA) for their pedigree data, as well as Julius van der Werf for his help with calculations of genomic prediction accuracy. The authors would also like to acknowledge the legacy of Andrew Swan in both genetic evaluation and as a person. We miss you Swanny.

REFERENCES

Aldridge M. and Pitchford W. (2018) *B.STU.0295 MLA Final Report.*

Dekkers J.C.M. (2007) *J. of An. Br. and Gen.* **124**: 331.

Falconer D.S., and Mackay T.F.C. (1996) 'Introduction to Quantitative Genetics' 4th edn.

Goddard M.E., Hayes B.J. and Meuwissen T.H.E. (2011) *Anim. Prod. Sci.* **51**: 69.

Granleese T., Mortimer S.I., Atkinson T., Refshauge G., Bird-Gardiner T., Haynes F., Brown D.J., Alexandri R. and Walkom S.F. (2023) *Proc. Assoc. Advmt. Anim. Breed. Genet.* **24**: 230.

Kijas J. (2012) *B.GOA.0072 MLA Final Report*

Proc. Assoc. Advmt. Anim. Breed. Genet. 26: 86-89

GENOMIC MODELS FOR ESTIMATING GENETIC PARAMETERS FOR LIFETIME WOOL TRAITS IN MERINO SHEEP

P.K. Wahinya[1], D.J. Brown[1], B.E. Clarke[2], J.L. Smith[3], T. Bird-Gardiner[4], A.A. Swan[1]

[1] Animal Genetics Breeding Unit*, University of New England, Armidale, NSW, 2351 Australia
[2] Centre for Animal Production and Health, Murdoch University, Murdoch, WA, 6150 Australia
[3] CSIRO, Agriculture and Food, F.D. McMaster Laboratory, Armidale, NSW, 2350 Australia
[4] DPIRD, Agricultural and Biosecurity, Trangie Research Centre, Trangie, NSW, 2823 Australia

SUMMARY

Lifetime production is a key driver of the profitability of Merino flocks. While the genetic control and relationships between young and adult assessments have been evaluated previously, the conclusions have not been validated using data from a progeny test design, which extensively captures lifetime data from diverse sires and ewe bases. This paper estimated genetic correlations between lifetime age stages of greasy fleece weight, fibre diameter and body wrinkle, and between the traits. The results show that assessments for these three traits made at a young age have a moderate correlation (r_g>0.6) with adult performance, however, they are genetically different from adult assessments. These results suggest measurement of adult performance is beneficial. Genetic groups derived from admixture analysis using genotypes provided a good alternative to account for population diversity when the pedigree-derived genetic groups are challenging to define accurately. Further work is required on the other important traits which determine lifetime productivity.

INTRODUCTION

The Australian Merino breed has, over time, been bred and reared under different environments and production systems leading to a significantly admixed population (Swan *et al.* 2016). A genetic evaluation system has been developed over time to drive genetic gain through selection across flocks and Merino bloodlines (MERINOSELECT) (Brown *et al.* 2007). However, since lifetime traits can be difficult and expensive to measure, genetic evaluation has been dominated by early-age preselection measurements. The Merino Lifetime Productivity project was designed to capture lifetime data on the progeny of 134 diverse industry sires for both production and visual traits, allowing a comprehensive analysis of the genetic basis of lifetime production, and current genetic improvement strategies (Ramsay *et al.* 2019). This study evaluated the results of genetic evaluation models a) using pedigree versus genotypes to define genetic relationships, and b) alternative methods of accounting for population diversity using pedigree information versus genomics, for estimating genetic parameters for lifetime wool traits and body wrinkle.

MATERIALS AND METHODS

Data. Greasy fleece weight (GFW), fibre diameter (FD) and body wrinkle (BDWR) records from 5,464 fully pedigreed Merino ewes were used for this analysis. All the ewes were genotyped with a 50k SNP chip and were the progeny of 134 sires and 4,266 dams from the Balmoral, Macquarie, MerinoLink, New England, and Pingelly sites which were involved in the Merino Lifetime Productivity (MLP) project (Ramsay *et al.* 2019). All sites provided two cohorts of ewes with lifetime data up to seven years of age, however, for ease of presentation, this analysis only considered one record for the young stage (P) and four adult measurements up to five years of age (A2 – A5). The two cohorts in each site included were: animals born in 2015 and 2016 at Balmoral,

* A joint venture of NSW Department of Primary Industries and Regional Development and the University of New England

2016 and 2017 at Pingelly and MerinoLink, and 2017 and 2018 at Macquarie and New England sites. Table 1 shows the number of records, means and standard deviation for these traits.

Table 1. Number of records, mean and standard deviations of lifetime trait measurements for greasy fleece weight, fibre diameter and body wrinkle

Trait	Model	Records, means and standard deviation				
		P	A2	A3	A4	A5
Greasy fleece	Records	5414	5283	5156	4898	4587
weight	Mean (sd)	2.89 (0.88)	5.69 (1.77)	5.31 (1.30)	5.29 (1.28)	5.00 (1.24)
Fibre diameter	Records	5441	5294	5169	4918	4638
	Mean (sd)	16.14 (1.43)	17.84 (1.71)	17.75 (1.80)	18.07 (1.94)	18.08 (2.03)
Body wrinkle	Records	5430	4636	4459	4884	4618
	Mean (sd)	2.36 (0.82)	2.39 (0.84)	2.47 (0.91)	2.66 (0.85)	2.56 (0.82)

P – first shearing/post-weaning; A2 – 2 years old; A5 – 5 years old

Statistical analysis. Data were analysed using a linear mixed model using ASReml software (Gilmour *et al.* 2015): $y = \boldsymbol{X}\hat{\beta} + \boldsymbol{Z}_1\boldsymbol{Q}g + \boldsymbol{Z}_1a + \boldsymbol{Z}_2m + e$, where y was a vector of the trait measurements, $\hat{\beta}$ a vector of fixed effects, g a vector of random genetic group effects, a a vector of additive genetic effects within genetic groups for animals in the pedigree, m a vector of maternal permanent environment effects (for post-weaning or first shearing traits only), and e a vector of residual effects. $\boldsymbol{X}, \boldsymbol{Z}_1, \boldsymbol{Q}$ and $\boldsymbol{Z}_2$ were design matrices for their corresponding effects. Fixed effects included contemporary group (67 groups defined by site, year of birth and management group), birth and rearing type (9) within sites and reproduction status (6) within sites for adult stages with age of dam as a covariate. Univariate and bivariate models were used to estimate (co)variances, variance ratios and genetic correlations. Models without genetic groups (NoG), with pedigree (PedG) and genomically (GenG) derived genetic groups were used to evaluate the effect of population structure on genetic parameters. The model PedG included 205 genetic groups derived using pedigree, flock and year of birth information, while GenG included five groups derived from an admixture analysis (Alexandri *et al.* 2025) using genotyped animals in the MERINOSELECT genetic evaluation system (Brown *et al.* 2007). Genetic relationships (60k SNPs) between animals were also independently derived using the pedigree (NRM) and genotypes only (GRM) (Vanraden 2008).

RESULTS AND DISCUSSION

Heritabilities from pedigree models were consistently higher than those estimated from genomic models for all traits (Table 2). This may be due to the GRM capturing relationships by both descent and state, accounting for relationships beyond the NRM. The heritabilities for the first (P) GFW and FD assessments were lower than those at the adult stages, which was also observed by Brown *et al.* (2013). However, the heritabilities estimated with our dataset were slightly higher than those reported by Brown *et al.* (2013) most likely due to the very diverse sires used in the project. The proportion of genetic group variance relative to the genetic variance ranged between 9 – 28% (GFW), 11 – 24% (FD), and 27 – 96% (BDWR), using pedigree models, and 3 – 24% (GFW), 3 – 20% (FD) and 24 – 75% (BDWR) using genomic models. The proportion of genetic group variance increased with age stages for the wool traits and declined with age stages for BDWR. This explains the differences in heritability estimates between the models when diversity was not accounted for. The genetic group effects were lowest or not significant for the first shearing wool traits hence the similar heritabilities across models. Moderate maternal permanent environmental effects were estimated for GFW (0.12) and were low for FD (0.03) and BDWR (0.05).

Table 2. Heritabilities for greasy fleece weight, fibre diameter and body wrinkle from models using pedigree (NRM) vs genomic (GRM) relationships with no genetic groups (NoG) vs pedigree derived (PedG) and genomically derived genetic groups (GenG)

Trait	Genetic groups	NRM					GRM				
		P	A2	A3	A4	A5	P	A2	A3	A4	A5
Greasy fleece weight	NoG	0.41	0.58	0.60	0.54	0.70	0.34	0.54	0.54	0.51	0.54
	PedG	0.37	0.57	0.59	0.52	0.61	0.33	0.54	0.53	0.50	0.52
	GenG	0.41	0.60	0.60	0.54	0.64	0.34	0.54	0.52	0.49	0.52
Fibre diameter	NoG	0.72	0.81	0.85	0.81	0.88	0.61	0.71	0.71	0.68	0.68
	PedG	0.63	0.73	0.71	0.80	0.78	0.59	0.68	0.69	0.65	0.66
	GenG	0.70	0.80	0.76	0.83	0.82	0.61	0.70	0.70	0.66	0.67
Body wrinkle	NoG	0.64	0.58	0.60	0.59	0.56	0.41	0.38	0.33	0.37	0.34
	PedG	0.46	0.43	0.45	0.45	0.38	0.37	0.33	0.27	0.31	0.28
	GenG	0.51	0.47	0.51	0.52	0.49	0.36	0.35	0.31	0.36	0.32

P – post-weaning/first shearing; A2 to A5: 2 to 5 years old; standard errors (se) ranged from 0.02 to 0.07

Genetic correlations between the early measurements and adult performance for GFW, FD and BDWR were positive and moderate to high, ranging from 0.63 to 0.89 (Table 3), similar to Brown *et al.* (2013). These moderate to high genetic correlations suggest that assessments of young animals for these three traits are a good genetic predictor of adult performance. The correlations were highest for young age (P) with the first adult performance, decreasing slightly across the older stages suggesting that the first shearing (P) traits are genetically different from adult assessments, hence the need for adult measurements. The phenotypic correlations for the wool traits (0.42 – 0.69) and BDWR (0.33 – 0.38) were moderate, indicating that these traits are moderately repeatable. The genetic correlations among the adult stages (not shown) were above 0.95, indicating that adult performance across adult stages is effectively the same trait genetically. However, these correlations were not 1, suggesting that some sires could express different production trajectories over time.

Table 3. Genetic and phenotypic correlations between measurements at first shearing (P) and adult stages (2 to 5 years; A2 to A5) for greasy fleece weight, fibre diameter and body wrinkle

Trait	Genetic groups	**Genetic** (se = 0.01 to 0.07)				**Phenotypic** (se = 0.01 to 0.02)			
		A2	A3	A4	A5	A2	A3	A4	A5
Greasy fleece weight (P)	NRM-PedG	0.80	0.76	0.75	0.76	0.55	0.46	0.45	0.44
	NRM-GenG	0.78	0.71	0.70	0.70	0.55	0.45	0.44	0.43
	GRM-PedG	0.76	0.72	0.66	0.67	0.54	0.45	0.44	0.43
	GRM-GenG	0.75	0.69	0.63	0.63	0.54	0.45	0.44	0.42
Fibre diameter (P)	NRM-PedG	0.85	0.79	0.75	0.75	0.69	0.64	0.61	0.58
	NRM-GenG	0.84	0.78	0.74	0.73	0.68	0.63	0.60	0.58
	GRM-PedG	0.89	0.83	0.81	0.78	0.68	0.64	0.61	0.58
	GRM-GenG	0.88	0.83	0.80	0.77	0.69	0.64	0.61	0.58
Body wrinkle (P)	NRM-PedG	0.85	0.80	0.73	0.73	0.38	0.38	0.36	0.34
	NRM-GenG	0.78	0.69	0.66	0.68	0.37	0.37	0.35	0.34
	GRM-PedG	0.80	0.79	0.72	0.73	0.36	0.36	0.34	0.33
	GRM-GenG	0.76	0.70	0.68	0.69	0.36	0.36	0.34	0.33

PedG – pedigree-derived genetic groups; GenG – genomically derived genetic groups; se – standard error

The genetic correlations between post-weaning BDWR and GFW were positive (0.22 to 0.51), and between post-weaning GFW and FD were negative (-0.16 to -0.41) (Table 4). That is, Merino sheep with higher degree of body wrinkle tend to have moderately higher wool production and finer wool (-0.10 to -0.33). Higher wool production is also associated with higher fibre diameter, which is antagonistic to selection for higher fleece value. Index selection has been shown to accommodate this to ensure desired genetic gains for all traits included (Swan *et al.* 2016). Lower phenotypic correlations among these traits were estimated in the same directions as the genetic correlations (not shown), as in the literature (Brown *et al.* 2013). Genetic correlations within the age stages and NRM and GRM models were not significantly different. The slight differences between correlations for early stages, are difficult to explain due to the high standard errors associated with the correlations.

Table 4. Genetic correlations between body wrinkle at post-weaning and greasy fleece weight and fibre diameter (standard errors: 0.04 to 0.14)

Trait (age)	Model	Genetic groups	Greasy fleece weight P	A2	A3	A4	A5	Fibre diameter P	A2	A3	A4	A5
Body wrinkle (P)	NRM	PedG	0.22	0.41	0.24	0.20	0.22	-0.23	-0.25	-0.20	-0.27	-0.27
		GenG	0.30	0.51	0.37	0.35	0.36	-0.24	-0.19	-0.10	-0.19	-0.18
	GRM	PedG	0.33	0.41	0.34	0.27	0.26	-0.33	-0.22	-0.23	-0.21	-0.23
		GenG	0.32	0.44	0.38	0.32	0.32	-0.30	-0.20	-0.18	-0.17	-0.19
Greasy fleece weight (P)	NRM	PedG						0.41	0.34	0.29	0.25	0.29
		GenG						0.26	0.23	0.23	0.19	0.23
	GRM	PedG						0.18	0.28	0.26	0.23	0.22
		GenG						0.16	0.23	0.23	0.20	0.19

CONCLUSION

Greasy fleece weight, fibre diameter, and body wrinkle assessments conducted at a young age are moderate genetic predictors of adult performance, but are genetically different from adult assessments and highlight the importance of obtaining adult measures to accurately select for lifetime productivity. Genomic relationships derived from genotypes accounted for part of the trait variation due to diversity unaccounted for using pedigree relationships. Genetic groups derived from admixture analysis using genotypes provide a good alternative to account for population diversity when the pedigree-derived genetic groups are challenging to accurately define. Further work is required on the other important traits which determine lifetime productivity.

ACKNOWLEDGEMENTS

The authors acknowledge Australian Wool Innovation, Australian Merino Sire Evaluation Association and the Merino Lifetime Productivity project sites for supporting and funding this work and thank all site managers and committees for facilitating data collection. Acknowledgement is also made to the Australian Government that supports research, development and marketing of Australian wool.

REFERENCES

Alexandri P., Gurman P.M. *et al.* (2025) *Proc Assoc Adv Anim Breed Genet.* **25**: *These proceedings.*

Brown D., Huisman A., Swan A., Graser H., Woolaston R., Ball A., Atkins K. and Banks R. (2007) *Proc. Assoc. Advmt. Anim. Breed. Genet.* **17**: 187.

Brown D., Swan A. and Gill J. (2013) *Proc. Proc Assoc Adv Anim Breed Genet.* **20**: 110.

Gilmour A., Gogel B., Cullis B., Welham S. and Thompson R. (2015) 'ASReml user guide release 4.1 structural specification.' VSN International ltd: Hemel Hempstead.

Ramsay A., Swan A. and Swain B. (2019) *Proc. Assoc. Advmt. Anim. Breed. Genet.* **23**: 512.

Swan A.A., Brown D.J. and van der Werf J.H.J. (2016) *Anim. Prod. Sci.* **56**: 87.

Vanraden P.M. (2008) *J Dairy Sci* **91**: 4414.

EVALUATING THE ACCURACY OF SNP-BLUP AND ADMIXTURE FOR BREED COMPOSITION PREDICTION IN LIMITED AND WELL-REPRESENTED CATTLE BREEDS

C.A. Ryan

Teagasc, Moorepark, Co. Cork, Ireland

SUMMARY

The present study provides the first comprehensive comparison of SNP-BLUP and Admixture for predicting breed composition in datasets comprising both underrepresented and well-represented cattle breeds. When analysing 6 breeds with small reference populations, both SNP-BLUP and Admixture demonstrated high accuracy (≥98% correct assignment) with minimal mean absolute difference (0.01) in predicted proportions. However, when the analysis included all 19 breeds (13 well-represented and 6 breeds with limited representation) the accuracy of SNP-BLUP decreased to 82%, while Admixture maintained 99% accuracy. The performance gap was largely resolved by reducing the training population size of well-represented breeds to match the breeds with limited representation (50 animals per breed). This adjustment increased the accuracy of SNP-BLUP to 93%. These results demonstrate that unequal training population sizes negatively impact SNP-BLUP's performance due to its statistical assumptions, while Admixture remained robust regardless of training population balance.

INTRODUCTION

An animal's breed composition is typically assumed to reflect the average of its parents' breed proportions. However, due to chromosomal recombination during gametogenesis, offspring from crossbred parents may deviate from these expectations. Several tools have been developed to estimate breed composition, including Admixture (Alexander *et al.* 2009), and SNP-BLUP (Single Nucleotide Polymorphism Best Linear Unbiased Prediction), which is commonly used in genomic evaluations. These methods employ fundamentally different statistical approaches. Admixture uses a likelihood-based approach to estimate ancestry proportions by modelling allele frequencies across populations. In contrast, SNP-BLUP is based on the infinitesimal model, where SNP effects are assumed to be small, random, and drawn from a shared variance structure. Given the widespread use of SNP-BLUP in genomic evaluations, its applicability to predicting breed composition offers the advantage of utilising existing pipelines, potentially streamlining the process and enhancing computational efficiency compared to standalone software tools. Despite these methodological differences, only two previous studies have compared SNP-BLUP and Admixture for breed composition prediction, and both were limited to specific population structures; Ryan *et al.* (2023) focused on well-represented breeds with large training populations (500 purebred animals per breed) and Struken *et al.* (2017) focused on uniformly small populations (< 60 purebred animals per breed). Therefore, a critical knowledge gap exists regarding the comparative performance of SNP-BLUP and Admixture in datasets containing breeds with limited numbers as well as well-represented breeds. The objective of this study was to evaluate the accuracy of SNP-BLUP for predicting the breed composition of 6 breeds of cattle with limited numbers of purebreds and to compare its performance to that of Admixture. An additional objective was to investigate the impact of including data from 13 well-represented breeds on the accuracy of breed composition predictions for breeds with limited numbers.

MATERIALS AND METHODS

Genotypes for 49,213 SNPs were available post-quality control for a dataset of 19 breeds.

Purebred animals were verified using principal component analysis and unsupervised Admixture analysis (breed composition > 0.9). For the 13 well-represented breeds, the dataset from Ryan *et al.* (2023) was used, consisting of 500 animals per breed in the training population, 3,146 purebred validation animals, and 4,330 crossbred animals. For the additional 6 breeds with limited purebred representation (Piedmontese, Dexters, Montbeliarde, Irish Maol, Jersey, and Romagnola), training population sizes ranged from 22 (Jersey) to 264 (Dexter), with a total of 652 animals, and a purebred validation population of 333 animals. An additional crossbred validation population of 228 animals from these 6 breeds was also included.

Breed composition was estimated using SNP-BLUP in the MIX99 software (MiX99 Development Team, 2022) and Admixture following the methodology outlined in Ryan *et al.* (2023). Analyses were conducted first using only the 6 breeds with limited numbers (K = 6), followed by a combined analysis of all 19 breeds (K= 19). Breed assignment was considered accurate when the predicted proportion for an animal in the purebred validation population was ≥0.90 for a specific breed. The differences in the main breed proportion estimates for crossbred animals predicted using SNP-BLUP and Admixture were compared. To investigate the impact of training population size imbalance, subsets of 200 and 50 animals were randomly sampled from the 13 well-represented breeds to match the smaller training populations of the rare breeds.

RESULTS AND DISCUSSION

When the 6 breeds with limited representation were analysed independently, both SNP-BLUP and Admixture accurately assigned ≥98% of the purebred validation population to their respective breeds (Figure 1). The mean absolute difference between the Admixture and SNP-BLUP breed proportion estimates was minimal (0.011 ± 0.03), consistent with previous studies using balanced training populations consisting of 500 animals per breed (Ryan *et al.* 2023) or uniformly small training populations of < 60 animals per breed (Strucken *et al.* 2017). This demonstrates that both methods perform well when analysing breeds with similar representation levels.

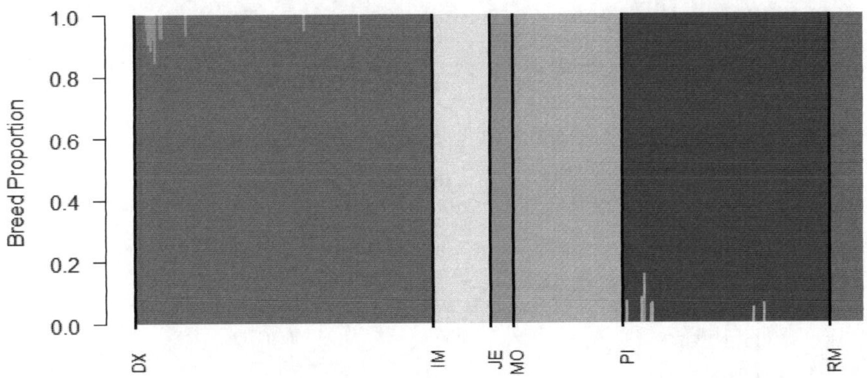

Figure 1. Estimated breed proportions from Admixture for each purebred individual in the validation population for K=6. Each animal is represented by a thin vertical line and each colour represents one inferred population. The breed proportion of each animal is represented by the length of each colour in that animal's vertical bar. Breeds include Dexters (DX), Irish Maol (IM), Jersey (JE), Montbeliarde (MO), Piedmontese (PI), and Romagnola (RM)

However, the mean absolute difference between SNP-BLUP and Admixture breed composition predictions for the purebred validation population increased to 0.077 when 13 additional, larger breeds (from Ryan *et al.* 2023) were included in the analysis (K=19). Admixture maintained a high overall accuracy, correctly assigning 98% of purebred validation animals, while SNP-BLUP correctly assigned only 82% (Table 1). For crossbred animals, the discrepancy between methods was larger, with a mean absolute difference (standard deviation) of 0.12 (±0.15).

The greatest discrepancies occurred for Shorthorns (mean absolute difference = 0.24) and Simmentals (mean absolute difference = 0.50). SNP-BLUP failed to correctly assign any purebred Shorthorn or Simmental animals, instead estimating their average breed proportions as 0.76 and 0.50, respectively. This misclassification was asymmetric: Shorthorns were predominantly assigned as part Shorthorn and part Irish Maol, and Simmentals as part Simmental and part Montbéliarde. This bias likely stems partly from the genetic similarity between the breeds but also from SNP-BLUP's modelling assumptions. As a linear mixed model, SNP-BLUP applies shrinkage to SNP effects, regressing estimates toward a genomic mean that is disproportionately influenced by well-represented breeds due to their larger contribution to allele frequency estimates (Meuwissen *et al.* 2001). In imbalanced datasets, this shrinkage may potentially create a systematic misclassification pattern where animals from well-represented breeds (e.g., Shorthorn, Simmental) are erroneously assigned partial ancestry to breeds with limited representation (e.g., Irish Moiled, Montbéliarde). This may occur through two interacting mechanisms. First, shared haplotype segments between breeds are preferentially attributed to the less-represented breed because their smaller training populations provide weaker constraints on the shrinkage process. Second, breed-distinguishing SNPs for well-represented breeds have their effects disproportionately shrunk toward zero due to the overwhelming influence of numerically dominant breeds on the genomic mean. In contrast, breeds with limited representation likely maintain more stable predictions because their sparse representation minimises their contribution to the global genomic mean, and their unique alleles, being rare in the overall population, are less affected by shrinkage and thus retain stronger predictive value. This potentially explains why Admixture, which models population-specific allele frequencies without shrinkage, maintains high accuracy regardless of training population structure. Admixture successfully assigned 99.6% of Shorthorns and 100% of Simmentals correctly.

Table 1. The percentage of animals within each breed correctly assigned (i.e., estimated breed proportion for a specific breed was predicted to be ≥0.90) to their respective breeds using SNP-BLUP and Admixture when all 19 breeds were included in the analysis

Breed	SNP-BLUP	Admixture
Angus	99.4	100
Aubrac	100	100
Blonde d'Aquitaine	100	99.5
Belgian Blue	99.0	100
Charolais	99.7	100
Dexter	99.2	97.7
Friesian	94.8	98.7
Hereford	99.2	100
Holstein	98.1	74.7
Irish Maol	96.1	100
Jersey	100	100
Limousine	79.9	100
Montbeliarde	92	100
Piedmontese	93.7	97.9
Parthenaise	100	100
Romagnola	86.6	100
Saler	98.6	100
Shorthorn	0	99.5
Simmental	0	100

To test whether training population imbalance was indeed the primary cause of SNP-BLUP's performance decline, the size of the training population for each of the 13 well-represented breeds was systematically reduced. When the training population of the well-represented breeds was reduced to 250 animals per breed, Shorthorns and Simmentals were still assigned as part Irish Maol and Montbéliarde, respectively. However, when training populations were equalised at 50 animals per breed, SNP-BLUP's overall accuracy improved to 93%, confirming that unequal training population sizes, not genetic similarity, were the primary cause of misclassification. Notably, Admixture's accuracy remained stable regardless of training population structure, consistent with its likelihood-based framework.

These findings have important practical implications for genomic analyses. While SNP-BLUP offers computational efficiency and integration with existing genomic evaluation pipelines, the study herein demonstrate that similar training population sizes are essential, the size of which will be limited by the breed with the smallest representation. However, it should be noted that Admixture has its own limitations, including potential sensitivity to input file order (Crum *et al.* 2019) and greater computational demands.

CONCLUSION

The present study provides the first systematic evaluation of breed composition prediction methods in datasets containing both breeds with limited representation and well-represented cattle breeds. While SNP-BLUP and Admixture performed comparably for breeds with small reference populations in isolation (≥98% accuracy), SNP-BLUP's accuracy declined substantially (to 82%) when well-represented breeds were added due to training population imbalance. This performance gap was largely resolved by equalising training population sizes (93% accuracy with 50 animals per breed), demonstrating that SNP-BLUP's limitations stem from statistical assumptions rather than intrinsic genetic relationships. In contrast, Admixture maintained 99% accuracy regardless of dataset structure. These results provide clear guidance for breed prediction methodologies in different scenarios: (1) for balanced datasets or when computational efficiency is paramount, SNP-BLUP performs adequately; (2) for imbalanced datasets where maintaining large reference populations is valuable, Admixture provides more robust results despite greater computational demands.

ACKNOWLEDGEMENTS

This publication has emanated from research supported by the European Commission in the frame of the Horizon 2020 INTAQT project (INnovative Tools for Assessment and Authentication of chicken meat, beef and dairy products' QualiTies, Grant agreement ID: 101000250).

REFERENCES

Alexander D.H., Novembre J. and Lange K. (2009) *Genome Res.* **19**: 1655.
Crum T.E., McHugo G.P., Dover M.J. and MacHugh D.E. (2019) *PLoS ONE* **14**: e0215912.
Meuwissen T.H.E., Hayes B.J. and Goddard M.E. (2001) *Genetics* **157**: 1819.
MiX99 Development Team (2022) MiX99: A software package for solving large mixed model equations. Release I/2022. Natural Resources Institute Finland (Luke).
Ryan C.A., Berry D.P., O'Brien A., Pabiou T. and Purfield D.C. (2023) *Front. Genet.* **14**: 1125678.
Strucken E.M., Al-Mamun H.A., Esquivelzeta-Rabell C., Gondro C., Mwai O.A. and Gibson J.P. (2017) *Genet. Sel. Evol.* **49**: 1.

Proc. Assoc. Advmt. Anim. Breed. Genet. 26: 94-97

IBS VERSUS IBD – NEW INSIGHTS FROM WHOLE GENOME SEQUENCE DATA

C.L. Warburton[1], R. Costilla[2], M.E. Goddard[3], B.J. Hayes[1] and T.H.E. Meuwissen[4]

[1] Queensland Alliance for Agriculture and Food Innovation, University of Queensland, St. Lucia Qld, 4072 Australia
[2] Cawthron Institute, Nelson 7010, New Zealand
[3] Centre for AgriBioscience, Agriculture Victoria, Bundoora, Vic, 3083, Australia
[4] Faculty of Biosciences, Norwegian University of Life Sciences, Ås Norway

SUMMARY

An implicit assumption in most methods for genomic prediction is that identical by state (IBS) genome regions identified by high density (HD), genome wide markers will also be identical by descent (IBD). With the availability of whole genome sequence on many individuals, we can now directly test this assumption. Here we describe a methodology in cattle that tests IBS haplotypes for IBD within and across breeds, and across sub-species *Bos taurus indicus* (*Bos indicus*) and *Bos taurus taurus* (*Bos taurus*). Within breeds, the number of variant alleles in the sequence data that were different for 250kb IBS haplotypes (IBS for SNP on HD array) were very small. Across *Bos taurus* breeds, this number increased slightly, and was consistent with previous estimates of the number of SNP required for multi-taurus breed predictions (360,000 SNP). Notably, in populations consisting of *Bos indicus* and *Bos taurus* breeds of cattle, our results indicate that approximately 1.5 million SNP are required to ensure 100kb IBS haplotypes are also IBD. These findings suggest that for multi-sub species predictions in cattle, a higher density of markers is required than the current HD arrays.

INTRODUCTION

Many authors have assumed that high density SNP arrays would be sufficient for multi-breed and across breed genomic predictions in cattle (de Roos *et al.* 2009; Hayes *et al.* 2019). This follows earlier work that investigated the conservation of linkage disequilibrium (LD) phase of SNP across breeds, which concluded that, for *Bos taurus* breeds, 300,000 SNP would be required for across breed predictions (de Roos *et al.* 2009; Goddard and Hayes 2009, Porto-Neto *et al.* 2014). With the availability of whole genome sequence (WGS), this hypothesis can be tested in more detail. Animals may share common alleles on a chromosome segment for two possible reasons; the chromosome segments are identical by descent (IBD) where the alleles have been inherited from a common ancestor and thus the SNP-QTL allele phase is likely to be conserved; or haplotypes may be identical by state (IBS) where they share alleles by chance, which may result in spurious associations between QTL and SNP allele phase (Powell *et al.* 2010). Haplotypes from different animals that are IBS on the HD SNP array can be compared for how many SNP they differ by in the whole genome sequence, and this gives some information about the likelihood of the haplotype being identical by descent (IBD). Few studies have investigated the density of SNP that is required to estimate IBD relationships between animals (de Roos *et al.* 2009), especially across the two cattle subspecies. The aim of this study is to determine if high density SNP panels are sufficient to identify IBD relationships, and thus maintain LD phase between SNP and QTL alleles, in genetically distinct breeds of cattle originating from two subspecies, *Bos indicus* and *Bos taurus.*

MATERIALS AND METHODS

Whole genome sequence data consisting of approximately 52 million SNP were obtained from the 1000 Bulls Genomes Run 9 dataset (Hayes and Daetwyler 2019). Quality control in sequence data involved removing non-bi-allelic SNP and SNP that had a minor allele count less than 10. High density marker panel data (HD) were obtained by subsetting out 640K SNP in the whole genome sequence that corresponded to the BovineHD SNP chip positions (Matukumalli *et al.* 2011). Phasing was performed upon the HD SNP and the WGS SNP genotypes using Eagle version 2.4.1 (Loh *et al.* 2016). For simplicity, the results presented here are arbitrarily for chromosome 25.

To ensure unbiased comparisons between breeds, we standardised the number of animals per breed by randomly selecting 50 animals per breed. If this is not done, it would be difficult to determine if differences in results were due to different LD patterns or just sampling. The data included 3 purebred *Bos taurus* breeds: Hereford (average WGS read depth 11.52, minimum 1.35, maximum 37.97), Charolais (average WGS read depth 10.75, minimum 5.92, maximum 20.74) and Angus (average WGS read depth 11.96, minimum 2.03, maximum 29.02), 1 *Bos indicus* breed: Brahman (average WGS read depth 9.54, minimum 1.85, maximum 47.62); and 1 stabilised composite breed of *Bos indicus* x *Bos taurus* origins: Droughtmaster (average WGS read depth 9.34, minimum 5.77, maximum 19.45).

Haplotypes were generated for both the WGS data and the HD genotypes for all animals using fixed, non-overlapping haplotype windows of 100kb and 250kb, as previously described in Warburton *et al.* (2023). For each haplotype window size, haplotypes generated with the HD markers were compared across all haplotype loci for each pair of animals and identical haplotypes were considered to be IBS. If a haplotype was identified as IBS using the HD marker panel, the equivalent WGS haplotype loci was compared between the same pair of animals. Identical by descent haplotypes were identified by calculating the proportion of WGS SNP different between a haplotype pair. Proportion of WGS SNP different was calculated by counting the number of SNP different on the WGS haplotype, then dividing this number by the total number of SNP within the WGS haplotype window. A higher proportion of SNP different between haplotype pairs suggests that WGS haplotypes are less likely to be IBD at a potential QTL. For each pair of animals with IBS haplotypes, the breed of the animals was recorded as a breed-pair, for example if both animals were Angus then the breed-pair would be Angus_Angus. The proportion of WGS SNP different were averaged across each of the breed-pair comparisons to investigate probability of IBD across breeds for a given haplotype size. Each haplotype within the 100kb and 250kb window was further allocated to a haplotype bin based on the number of HD SNP within the haplotype; <10 (0 – 9 SNP), 10 (10 – 19 HD SNP), 20 (20 – 29 HD SNP), 30 (30 – 39 HD SNP), 40 (40 – 49 HD SNP) and 50 (50 – 59 HD SNP). The proportion of SNP different was averaged across each of the haplotype bins to calculate the average proportion of SNP different for haplotypes with varying density of HD SNP. The average proportion of SNP different for each haplotype window and bin was calculated for each of the breed-pairs and plotted using ggplot in R (Wickham 2016).

There are approximately 3 billion base pairs in the bovine genome (Zimin *et al.* 2009) and the number of SNP required can be calculated by dividing the size of the genome in base pairs by haplotype length in base pairs (hap_size) to obtain the total number of haplotypes required to span the genome. Once the number of HD SNP required per haplotype to identify and IBD segment (nSNP_hap) is obtained this number can be multiplied by the number of haplotypes required to span the genome, giving the estimated total number of SNP required to identify IBD haplotypes in multi-breed beef populations (see Equation 1).

$$Number\ of\ SNP\ required = \frac{3{,}000{,}000{,}000}{hap\_size} * nSNP\_hap \qquad \text{Equation 1}$$

It is important to consider sequencing error rate, as this will affect both the number and proportion of SNP different per haplotype window. Daetwyler *et al.* (2014) showed that the concordance between HD SNP genotypes and sequence genotypes in the 1000 Bulls data was approximately 99% after BEAGLE correction, and the rate of opposing homozygotes was, on average, 0.6%. Therefore, a threshold of 1% of SNP different between WGS haplotypes would allow for this estimated rate of genotyping error in the 1000 Bulls data. As such, in this study, whole genome sequence haplotypes that have less than or equal to 1% SNP different between animals at a loci will be considered to be IBD for the purposed of calculating the density of SNP required to identify IBD relationships in multi-breed and multi-subspecies populations.

RESULTS AND DISCUSSION

Figure 1 shows the proportion of WGS SNP different averaged across haplotype bin for each haplotype window and breed-pair comparison. There were no 250kb haplotypes with <10 SNP or 20 SNP in a bin, and as such, comparisons between proportion of SNP different are unable to be drawn for this bin. As the number of HD SNP in a haplotype bin increases there is a trend for the proportion of WGS SNP different to decrease across all haplotype windows and breeds, suggesting that the density of SNP within a haplotype will affect the accuracy of IBD detection from IBS haplotypes. Haplotypes with lower SNP density are more likely to be identified as IBS because there are few SNP per haplotype that are being used to identify haplotypes that are IBS. However, by increasing the number of SNP within a haplotype window, it is more likely that haplotypes that are identified as IBS using marker panel haplotypes will be IBD in WGS.

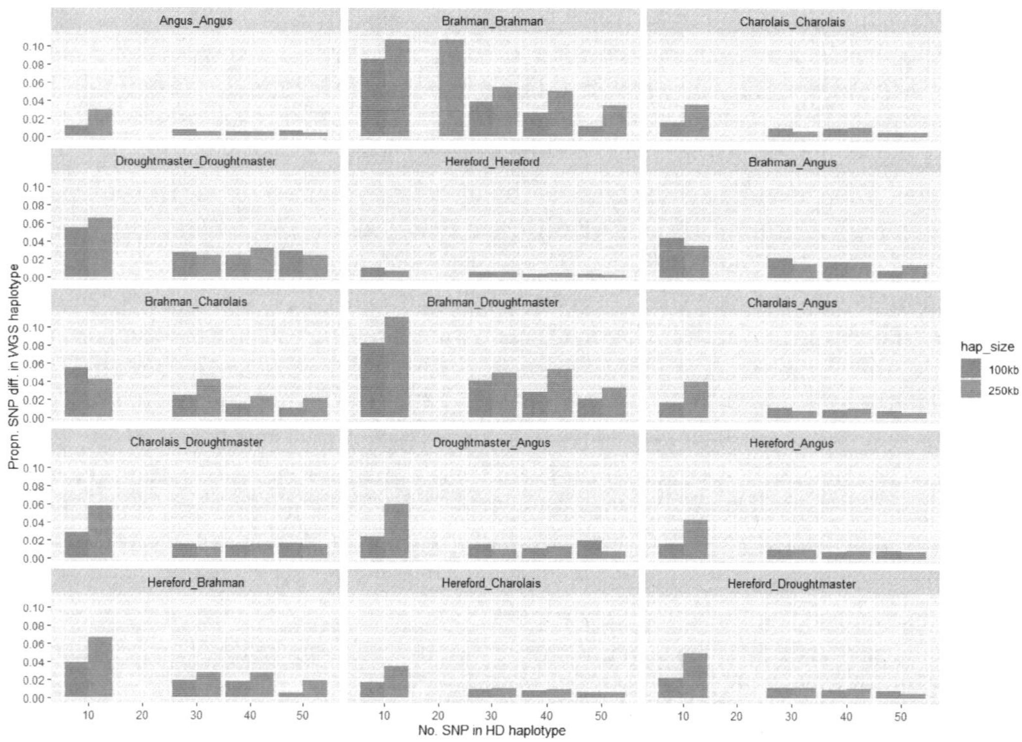

Figure 1. Proportion of SNP different (y-axis) in whole genome sequence haplotype as a function of the number of SNP in the high density marker panel genotype (x-axis) for each breed-pair comparison, across the 100kb and 250kb haplotype windows

The optimal number of SNP per haplotype and haplotype window size can be identified by the haplotypes with a proportion of WGS SNP different that is less than 1% across all breed pairs. In the results presented, the 250kb haplotype window with 30 HD SNP resulted in a 1% or less proportion of WGS SNP different across the multi-breed *Bos taurus* comparisons. Therefore using Equation 1, the estimate of the number of SNP required to identify IBD relationships that are likley to maintain LD phase between SNP and QTL in multi-breed *Bos taurus* populations is 360,000. This finding is corroborated by earlier studies that have demonstrated that 300,000 SNP are required to maintain LD phase between SNP in QTL in *Bos taurus* populations (de Roos *et al.* 2009; Goddard and Hayes 2009). In this study, the 100kb haplotype window with 50 HD SNP resulted in a 1% proportion of WGS SNP different in the multi-subspecies (*Bos indicus* and *Bos taurus*) breed-pair comparisons. Using Equation 1, it is estimated that 1.5 million evenly spaced SNP would be required to achieve haplotypes are IBD in multi-breed and multi-subspecies beef populations. Currently, the Bovine HD marker panel consists of approximately 728,000 SNP, so this estimate is double the number of SNP that are currently available on the HD marker panel.

CONCLUSIONS

In multi-breed and multi-subspecies populations of beef cattle, high density SNP are required to ensure IBS haplotypes are IBD, and therefore carry the same QTL allele. Currently the Bovine HD SNP panel consists of ~728,000 SNP, which was sufficient to identify IBD haplotypes between *Bos taurus* animals. However, higher density of SNP are required when the population consists of *Bos indicus* and *Bos taurus* animals. From our analysis we conclude that approximately 1.5 million evenly spaced SNP are required to identify IBD haplotypes and maintain SNP-QTL LD across these multi-subspecies and multi-breed populations.

ACKNOWLEDGEMENTS

The authors would like to acknowledge and sincerely thank to 1000 Bulls Genomes Project Consortium for the data provided in this manuscript.

REFERENCES

Daetwyler H.D., Capitan A., Pausch H., Stothard P., van Binsbergen R., Brøndum R.F., Liao X., Djari A., Rodriguez S.C., Grohs C., Esquerré D., Bouchez O., Rossignol M-N., Klopp C., Rocha D., Fritz S., Eggen A., Bowman P.J., Coote D., Chamberlain A.J., Anderson C., VanTassell C.P., Hulsegge I., Goddard M.E., Guldbrandtsen B., Lund M.S., Veerkamp R.F., Boichard D.A., Fries R. and Hayes B.J. (2014) *Nat. Genet.* **46**: 858.

Goddard M.E. and Hayes B.J (2009) *Nat. Rev. Genet.* **10**: 381.

Hayes B.J., Corbet N.J., Allen J.M., Laing A.R., Fordyce G., Lyons R., McGowan M.R. and Burns B.M. (2019) *J. Anim. Sci.* **97**: 55.

Hayes B.J. and Daetwyler H.D. (2019). *Annu. Rev. Anim. Biosci.* **7**: 89.

Loh P.R., Danecek P., Palamara P.F., Fuchsberger C., Reshef Y.A., Finucane H.K., Schoenherr S., Forer L., McCarthy S., Abecasis G.R., Durbin R. and Price A.L. (2016) *Nat. Genet.* **48**: 1443.

Matukumalli, L. K., Schroeder S., DeNise S.K, Sonstegard T., Lawley C.T., Georges M., Coppieters W., Gietzen K., Medrano J.F, Rincon G., Lince D., Eggen A., Glaser L., Cam G., and Van Tassel C. (2011) *Illumina Inc.*

Porto-Neto L.R., Kijas J.W. and Reverter A. (2014) *Genet. Sel. Evol.* **46**: 22.

Powell J.E., Visscher P.M. and Goddard M.E. (2010) *Nat. Rev. Genet.* **11**: 800.

de Roos A.P.W., Hayes B.J. and Goddard M.E. (2009). *Genetics.* **183**: 1545.

Warburton C.L., Costilla R., Engle B.N., Moore S.S., Corbet N.J., Fordyce G., McGowan M.R., Burns B.M. and Hayes B.J. (2023). *Heredity.* **131**: 350.

Wickham H. (2016) (Springer-Verlag New York) https://ggplot2.tidyverse.org.

Zimin A.V., Delcher A.L., Florea L., Kelley D.R., Schatz M.C., Puiu D., Hanrahan F., Pertea G., Van Tassell C.P., Sonstegard T.S., Marçais G., Roberts M., Subramanian P., Yorke J.A. and Salzberg S.L. (2009). *Genome Biol.* **10**: R42.

Proc. Assoc. Advmt. Anim. Breed. Genet. 26: 98-101

THE MOLECULAR CATTLE FOR PRECISION BREEDING

R. Xiang, M.E. Goddard, A.J. Chamberlain and J.E. Pryce

Agriculture Victoria, AgriBio, Centre for AgriBiosciences, Bundoora, VIC, 3083 Australia

SUMMARY

Genomic selection has revolutionised livestock breeding. The success of genomic selection is attributed to the polygenic architecture of most economically important traits and the small effective population size in livestock. The latter results in high linkage equilibrium (LD) between SNP markers and causal mutations, which allows genomic selection to work without knowing the causal mutations or biological mechanisms behind phenotypes. However, emerging studies have shown that the knowledge of causal mutations can be used to improve genomic selection. Also, genomic selection cannot directly remove some large-effect mutations with undesired phenotypic effects. With a potential increasing number of gene-edited animal products entering the food chain, the knowledge of causal mutations may warrant future interventions. Causal mutations are challenging to identify. However, with increasingly advanced statistical methods, molecular techniques and the availability of unique or large populations/samples, identifying causal mutations has become possible. For example, causal mutations may have significant effects on transcriptomics, epigenomics and proteomics, before they eventually affect phenotypes. Therefore, fine-mapping the effects of sequence variants on these biological cascades can break LD to inform causal mutations. We propose the concept of Molecular Cattle, describing the effort in developing a deep phenotyped cattle cohort to enhance causal variant discovery which can improve both polygenic genomic selection and monogenic interventions. We propose this concept in this paper and invite collaborations.

INTRODUCTION

The genetic or breeding value of animals can be predicted using genome-wide SNP markers, and this approach is called genomic prediction or selection (Meuwissen *et al.* 2001). Genomic selection has been successful in improving animal breeding. Genomic selection works without the need to know the causal mutations. This is due to the low effective population size in most livestock species where many SNP markers well tag causal variants via LD. However, the predictions performed in such a way have accuracies far away from perfect. Further, as LD structure differs dramatically between populations, genomic prediction accuracy further declines when the training and validation populations are genetically distant. Therefore, if we know the causal mutations and use them in the genomic prediction, the accuracy of genomic prediction is expected to increase and be better maintained across populations.

While livestock genomic research has a long history, few causal mutations underlying quantitative trait loci (QTLs) have been mapped and confirmed. One of the most famous QTL in dairy cattle for milk traits is *DGAT1* which explains more than 30% of genetic variations in milk fat. However, the causal mutations underlying this QTL are still being debated (Grisart *et al.* 2004; Fink *et al.* 2020). Another more recently identified QTL with undesirable effects on milk yield, fertility and survival is on chromosome 18 (e.g. *CTU1* (Xiang *et al.* 2017)). The identification of causal mutations could be used for intervention, such as gene editing, a technique that has been heavily regulated in animals (Solomon 2020). However, in recent years, there has been an increasing amount of research on genetically edited animal products entering the food chain worldwide (Ledesma 2024). Ideally, before the wet-lab experiments, one needs to know which candidate mutations among 10s of millions of sequence variants are to be edited.

Causal mutations affect phenotypes via their biological pathways related to gene expression, regulation, epigenetic modification and protein translation. Therefore, by profiling multi-omics and phenotypes in a large cohort, one can portray the effects of causal mutations. As there are 10s of millions of sequence variants in the animal genome, fine-mapping, i.e., joint analysis of all variants, should be conducted. For example, a Bayesian genome-wide fine-mapping using prior information based on multi-omics data has identified a set of potentially causal mutations in dairy cattle (Xiang *et al.* 2021; Xiang *et al.* 2023). The main challenge of genome-wide fine mapping is the requirement of computing power for ever-growing numbers of animals and sequence variants. Therefore, we propose the concept of Molecular Cattle, the proposal of developing a deep phenotyped dairy cattle cohort that can be used to fine-map causal variants. This choice of dairy cattle is due to our existing access to large datasets and experienced research farms, but this concept can be applied to any cattle breed. This effort will allow us to identify causal mutations that can be used for both improving genomic prediction and genetic intervention.

PROPOSED POPULATION AND ASSAYS

We propose to start to build this cohort using the existing dairy cow population, with a starting sample size of 1,000 (ideally 10,000). This considers good phenotypic records in dairy cattle, statistical power, and sequencing experiments' expenses. We expected this cohort to have conventional phenotypes already, such as milk production, fertility, mid-infrared spectroscopy (MIR) and if possible, methane emissions. All environmental and management exposures should be recorded as precisely as possible. We also expect this cohort to have genotype data which will be imputed to full sequence. We then deep phenotype this cohort with transcriptomics (RNA-seq from blood), epigenomics (DNA methylation from blood), proteomics (mass-spectrum from blood), rumen microbiome (metagenome-sequencing from rumen), metabolomics (LCMS from blood), and clinical biochemistry markers (e.g., Albumin, glucose and etc). As a collaboration effort, we also propose that partners join the effort with all data from multi-omics experiments from collaborators that can be shared openly. Depending on the privacy restrictions of different parties, one can decide to share either raw or summary data. One key hypothesis of our study is that variants with consistent effects on phenotypes across different environmental conditions are more likely to be causal. We propose to conduct a multi-trait Bayesian genome-wide fine-mapping on this dataset to prioritise causative mutation based on BayesR3 (Breen *et al.* 2022), with a few additional functions to be developed to suit this particular dataset. We will use massively parallel reporter assays (MPRA) (Cooper *et al.* 2022) to confirm these statistically finely mapped variants.

DISCUSSION

We use the following graph to illustrate the concept of Molecular Cattle (Figure 1), where we aim to identify causal mutations that can inform both genomic selection and intervention (e.g., gene editing). One key difference of our proposal from previously proposed phenomics is the inclusion of the larger number of molecular phenotypes (hence the molecular cattle). As described above, the effect of variants on different multi-omics data can be fine-mapped using advanced statistical models. The fine mapping result of all variants analysed can be treated as a variant ranking. Such ranking of variants can be used in a different dataset as weights to improve genomic prediction. This task can be done via either GBLUP (Meuwissen *et al.* 2024) or BayesRC (MacLeod *et al.* 2016). One challenge of this analysis is the strong LD between variants that prevents precise mapping of the causal variants. We expect this challenge to be resolved using the data from diverse multi-omics data because these data are independent of LD and can be used to inform biological pathways where causal mutations act. A starting point for intervention could be finely mapped causal variants behind known loci, as summarised in Table 1.

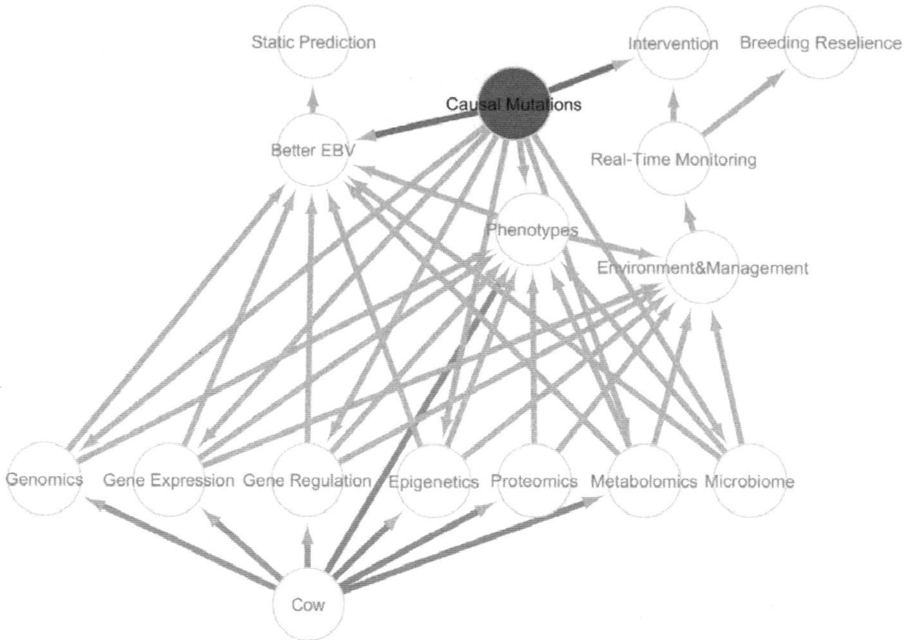

Figure 1. Illustration of the concept of Molecular Cattle. Arrows indicate the direction of data flow. Orange arrows indicate the sequencing experiments; grey arrow indicates the data analysis while the purple arrows indicate the use of causal mutations

Table 1. Examples of some known candidate mutations behind cattle QTLs to be intervened

Loci	Candidate mutation (ARS-UCD1.2)	Phenotypic effects	Molecular phenotype
DGAT1	Chr14:611019-20	Fat	eQTL, sQTL
GHR	Chr20:31888449	Milk production	Conserved across 100 species
CTU1	Chr18:57062518	Survival and fertility	Missense, conserved and regulatory
MGST1	Chr5:93516066	Milk fat yield	eQTL
GC	Chr6:86949653(starting position)	Mastitis	12-kb structure variant

Before taking causal variants to *in vivo* experiments, we expect to validate these regulatory variants using MPRA. MPRAs offer a flexible high-throughput framework to study elements regulating transcription and posttranscriptional events. MPRAs are widely used in humans to verify causal variants identified from GWASs (Cooper *et al.* 2022). This technique is rarely used in animals, likely due to its high costs. However, if wet-lab interventions are considered in the downstream work, it would be necessary to use MPRA to confirm identified causal variants.

Apart from improving genomic selection and providing targets for intervention, the results of this Molecular Cattle can be used to improve animal breeding in several areas that require further development, including real-time management of animals, and quantifying effects of GxE. Because DNA does not change over the lifespan of animals, the breeding values estimated from them represent a static prediction based on the genetic merit of animals. While EBV helps farmers make decisions at an early stage, real-time monitoring of cattle based on informative biomarkers could be more useful when intervention is needed. In this case, the creation of the Molecular Cattle dataset can be used to identify useful biomarkers for real-time monitoring of cattle health (e.g. mastitis onset). One future challenge is the expense of life-time assay of such biomarkers in herds. One potential solution is to take samples from milk cells instead of from blood, as they have shown similar properties in terms of the identification of regulatory variants (Xiang *et al.* 2018). Another solution is to create predicted biomarkers using more easily obtained phenotypes such as MIR.

The lack of high-throughput measurements of environmental exposures, partly due to multiple measurements in non-dairy livestock or on traits that are not automatically recorded, has prevented the proper model of GxE in animal breeding. While the host DNA does not change, how, when and where this DNA information should be expressed depends on environmental factors. Therefore, the multi-omics data, including DNA methylation (Clarke *et al.* 2021) in well-designed experiments, are important for inferring environmental variations that could be useful to inform GxE analysis.

CONCLUSION

We propose the concept of Molecular Cattle as a future research direction, where we could discover causal variants to improve both genomic selection and genetic intervention. Other benefits included better real-time management of herds and a better understanding of GxE factors. We also propose this effort as a national and international collaboration so we can achieve a larger sample size and better breed and environmental diversity for this experiment. We see this process starting from the Oceania regions with the potential to expand across the globe.

REFERENCES

Breen E.J., MacLeod I.M., Ho P.N., Haile-Mariam M., Pryce J.E., Thomas C.D., Daetwyler H.D. and Goddard M.E. (2022) *Comm. Biol.* **5**: 661.

Clarke S., Caulton A., McRae K., Brauning R., Couldrey C. and Dodds K. (2021). *Anim. Front.* **11**: 90.

Cooper Y.A., Teyssier N., Dräger N.M., Guo Q., Davis J.E., Sattler S.M., Yang Z., Patel A., Wu S. and Kosuri S. (2022) *Science* **377**: eabi8654.

Fink T., Lopdell T.J., Tiplady K., Handley R., Johnson T.J., Spelman R.J., Davis S.R., Snell R.G. and Littlejohn M.D. (2020) *BMC Genomics* **21**: 1.

Grisart B., Farnir F., Karim L., Cambisano N., Kim J.-J., Kvasz A., Mni M., Simon P., Frere J.-M. and Coppieters W. (2004) *Proc. Nat. Acad. Sci.* **101**: 2398.

Ledesma A.V. and Van Eenennaam A.L. (2024) Vet. J. **305**: 106142.

MacLeod I.M., Bowman P., Vander Jagt C., Haile-Mariam M., Kemper K., Chamberlain A., Schrooten C., Hayes B.J. and Goddard M. (2016) *BMC Genomics* **17**: 1.

Meuwissen T., Eikje L.S. and Gjuvsland A.B. (2024) *Genet. Sel. Evol.* **56**: 17.

Meuwissen, T.H., Hayes B.J. and Goddard M. (2001) *Genetics* **157**: 1819.

Solomon S.M. (2020) *Nat. Biotechnol.* **38**: 142.

Xiang R., Fang L., Liu S., Macleod I.M., *et al.* (2023) *Cell Genomics* **3**.

Xiang R., Hayes B.J., Vander Jagt C.J., MacLeod I.M., Khansefid M., Bowman P.J., Yuan Z., Prowse-Wilkins C.P., Reich C.M. and Mason B.A. (2018) *BMC Genomics* **19**: 1.

Xiang R., MacLeod I.M., Bolormaa S. and Goddard M.E. (2017) *Sci. Rep-UK* **7**: 9248.

Xiang R., MacLeod I.M., Daetwyler H.D., de Jong G., O'Connor E., Schrooten C., Chamberlain A.J. and Goddard M.E. (2021) *Nat. Comms.* **12**: 1.

RECOVERING OBSCURED GENETIC ARCHITECTURES THROUGH DECONVOLUTION OF GWAS STATISTICS

Z. Loh

School of Environmental & Rural Science, University of New England, Armidale, NSW, 2351 Australia

SUMMARY

The aim for this study was to propose a novel deconvolution technique to improve the recovery of genetic architecture information for a trait that was obscured by strong marker linkage disequilibrium. With a simulated datasets with sample size of 3,500, this deconvolution technique has reduced the departure in distribution between true and estimated marker test statistics by up to 5.82-folds on a trait with heritability of 0.3 and 2,000 quantitative trait loci. Under the same scenario, deconvolution also improved the correlation between true and haplotype significance of association by 24.9%. Therefore, deconvolution could serve as a promising approach in recovery of genetic architecture information that could help in its elucidation in livestock traits.

INTRODUCTION

Strong linkage disequilibrium (LD) between markers has complicated the estimation of genetic architecture of various livestock traits (Lloyd-Jones *et al.* 2019). While marker pruning can be used to obtain markers that were roughly in linkage equilibrium (Chang *et al.* 2015), this method depends on arbitrarily defined parameters such as window sizes and LD thresholds, which could affect the representativeness of retained markers and thus potentially genetic architecture estimation. An alternative approach is the deconvolution of marker effect sizes using a LD matrix. This technique has been used in signal processing in other fields such as radio astronomy (Clark 1980), but studies on its use in genetic architecture parameter estimation remained sparse.

The aim for this study was to propose a novel technique for deconvolution of marker effect size and test statistics from a Genome-Wide Association Study (GWAS). This technique was tested on simulated datasets, with the aim of recovering the underlying effect size distribution and improving the correlation between true and estimated significance of association of haplotype regions.

THE DECONVOLUTION METHOD

The marker effect sizes from a GWAS $\widehat{\boldsymbol{\beta}}$ can be modelled as the result of convolution between the true effect size $\boldsymbol{\beta}$ and LD matrix $\boldsymbol{R}$ with noise $\boldsymbol{e}$ (Cheng *et al.* 2020):

$$\widehat{\boldsymbol{\beta}} = \boldsymbol{R} * \boldsymbol{\beta} + \boldsymbol{e} \qquad [1]$$

This $\boldsymbol{\beta}$ can be estimated (let this estimate be $\widetilde{\boldsymbol{\beta}}$) based on Hogböm's CLEAN algorithm. Originally introduced in radio interferometry, this algorithm iteratively removes the effects of point spread function (PSF) from the raw signals while transferring the maximums onto clean signals (Clark 1980). In this study, $\boldsymbol{R}$, $\widehat{\boldsymbol{\beta}}$ and $\widetilde{\boldsymbol{\beta}}$ represented the PSF, raw and clean signals respectively.

To improve the computational efficiency in cases when number of markers M exceeds sample sizes N, matrix $\boldsymbol{Z}$ (defined such that $\boldsymbol{Z}^T\boldsymbol{Z} = \boldsymbol{R}$) was used in place of $\boldsymbol{R}$. Note that $\boldsymbol{Z}$ is not the Cholesky's factor of $\boldsymbol{R}$; $\boldsymbol{Z}$ is a rectangular matrix its j-th columns defined as follows:

$$\boldsymbol{z}_j = \frac{\boldsymbol{x}_j - 2p_j}{\sqrt{\sum\left(\boldsymbol{x}_j - 2p_j\right)^2}} \qquad [2]$$

where $\boldsymbol{X}$ is the raw genotype array, and p_j is the allele frequency for marker j.

Let $\hat{\boldsymbol{t}}$ be the raw test statistics from GWAS. While other definitions such as Aguilar *et al.* (2019)

can be used with this method, for this study the $\hat{\boldsymbol{t}}$ was defined as $\hat{\boldsymbol{\beta}}$ scaled by its standard deviation s (Gondro 2015). The deconvolution starts by identifying the genome-wide top marker in term of $|\hat{\boldsymbol{t}}|$ (let m be this marker), from which the corresponding marker effect size $\hat{\beta}_m$ and m-th column of $\boldsymbol{Z}$, $\boldsymbol{z_m}$, were obtained. This marker effect size was recorded as the m-th elements of $\tilde{\boldsymbol{\beta}}$. To prevent overcorrection of marker effect sizes which could destabilize the algorithm, the m-th element of $\hat{\boldsymbol{\beta}}$ was "muted" by setting it as "NAN" and no longer used in any subsequent iterations. The effect sizes and test statistics unexplained by $\hat{\beta}_m$, $\hat{\boldsymbol{\beta}}_{(1)}$ and $\hat{\boldsymbol{t}}_{(1)}$, were estimated as follows (Gondro 2015):

$$\hat{\boldsymbol{\beta}}_{(1)} = \hat{\boldsymbol{\beta}} - \hat{\beta}_m \boldsymbol{Z}^T \boldsymbol{z}_m \quad [3]$$

$$\hat{\boldsymbol{t}}_{(1)} = \frac{\hat{\boldsymbol{\beta}}_{(1)}}{s} \quad [4]$$

with the subscript (k) denotes the k-th iteration of this algorithm. The next iteration starts by identifying the top marker in term of $|\hat{\boldsymbol{t}}_{(1)}|$ and its corresponding $\hat{\beta}_m$ and $\boldsymbol{z}_m$. This process iterates until all markers were deconvolved. The deconvolved test statistics $\tilde{\boldsymbol{t}}$ could be obtained by scaling $\tilde{\boldsymbol{\beta}}$ with s as in [4], as empirical simulations suggested that $\tilde{\boldsymbol{\beta}}$ scaled with s better reflects the true QTL effect sizes as it considers the inflation of estimated marker effect sizes in the $\hat{\boldsymbol{\beta}}$.

TESTING THE DECONVOLUTION METHOD

The deconvolution method was tested using Python (version 3.11.5, released 11 September 2023) with simulated genotypes and phenotypes under changing parameter values, with the parameter tested and their associated values provided in Table 1.

Table 1. Parameters and their associated values tested in this experiment

Parameters	Default Value	Alternative Value
Sample size	3500	1500
Number of QTL	2000	500
Genotype Array	$\boldsymbol{X_{2K}}$	$\boldsymbol{X_{10K}}$

Two 60k genotype arrays (denoted as $\boldsymbol{X_{2K}}$ and $\boldsymbol{X_{10K}}$ respectively) were simulated using QMSim (Sargolzaei and Schenkel 2009) with historical population size set at 2,000 and 10,000 respectively, and were gene-dropped for 12,000 and 60,000 generations respectively. 20 chromosomes of 100 cM and mutation rate at 2.5×10^{-5} were used for both arrays, and sample sizes as provided in Table 1.

Within a genotype array $\boldsymbol{X}$, 2000 or 500 markers were designated as QTL and their effect size $\boldsymbol{\beta}$ simulated using gamma distribution $gamma(0.4\,, 1)$. Phenotype $\boldsymbol{y}$ was then calculated as follows:

$$\boldsymbol{y} = \boldsymbol{X\beta} + \boldsymbol{e}$$

where $\boldsymbol{e}$ is the residual component simulated using normal distribution $N\left(0\,, \frac{(1-h^2)var(\boldsymbol{X\beta})}{h^2}\right)$ with the heritability h^2 set at 0.3. The GBLUP-backsolved marker effect sizes $\hat{\boldsymbol{\beta}}$ and test statistics $\hat{\boldsymbol{t}}$ were obtained as described by VanRaden (2008) and Gondro (2015). The expected test statistics with zero LD between QTL and estimation error, $\boldsymbol{t}$, were also calculated from $\boldsymbol{\beta}$ as $t_i = \beta_i\sqrt{(N-2)var(\boldsymbol{x}_i)}/\sqrt{var(\boldsymbol{y}) - \beta_i^2 var(\boldsymbol{x}_i)}$ (Wang and Xu, 2019). Deconvolution was then applied onto $\hat{\boldsymbol{\beta}}$ and $\hat{\boldsymbol{t}}$ to yield the corresponding $\tilde{\boldsymbol{\beta}}$ and $\tilde{\boldsymbol{t}}$. Minor allele frequency filtering of 0.05 was applied on GWAS and the deconvolution algorithm. The deconvolved test statistics $\tilde{\boldsymbol{t}}$ (denoted as DCV) were tested against raw test statistics $\hat{\boldsymbol{t}}$ (denoted as RAW), as well as LD pruning and equal distance pruning methods (denoted as LDT and EQL respectively). For LDT, the markers were pruned as in "indep-pairwise" from PLINK 2.0 (Chang *et al.* 2015) with window size, step size and LD threshold at 50, 10 and 0.4 respectively. For EQL, the fifth of every non-overlapping 10 SNPs windows were chosen.

Two metrics were used to measure the methods' performance: departure in distribution (DiD) and correlation between true and marker haplotype scores (r_h^2). The DiD was measured with Wasserstein's distance between the distribution of $|\boldsymbol{t}|$ and $|\tilde{\boldsymbol{t}}|$ as defined by Vasershtein (1969), with a lower statistic indicates a reduced departure and improved performance. To ensure the retained markers accurately represent the haplotypes' significance of association, the DiD was calculated using the top marker within a marker window, with the marker being used as the window's midpoint. To maintain the genotyping density, markers that were removed during the pruning process were assigned as zero in $\tilde{\boldsymbol{t}}$. For r_h^2, the correlation between the sums of $|\boldsymbol{t}|$ within each haplotype blocks with that from raw marker $|\hat{\boldsymbol{t}}|$ or deconvolved $|\tilde{\boldsymbol{t}}|$ were obtained (Villiers *et al.* 2024). The blocks were defined as the window of markers in the DiD, with both EQL and LDT methods being tested.

This experiment was repeated 40 times for each set of parameter values. Welch's t-test was used to test the significance of differences in performance between methods and parameters, with two tests considered significantly different if the logarithmically transformed p-values $logpval > 2$.

RESULTS AND DISCUSSION

Compared to other methods tested, deconvolution significantly reduces departure in distributions DiD between true and marker test statistics (Table 2); for example, under default parameter values, deconvolution reduces the DiD by 3.22-folds from 0.954 to 0.296, which was also 2.86-folds and 1.15-folds decrease compared to LDT and EQL respectively ($logpval$ up to 114.8 between DCV and RAW) (Figure 1). This differences in performance were even more significant in $\boldsymbol{X_{10K}}$, achieving a 5.82-folds reduction from 0.972 in RAW to 0.167 in DCV ($logpval = 127.7$). Reducing the sample size from 3500 to 1500 also reduces the DiD for DCV from 0.297 to 0.221 ($logpval = 73.6$). This however contrasted with the significantly increased DiD for all other methods ($logpval$ up to 51.5 for RAW and 39.1 for LDT). These results highlighted the negative impacts of LD between markers in obscuring a trait's QTL effect size distribution especially with small sample sizes, and that the effects of LD can be removed using deconvolution.

Table 2. Departure in distribution (DiD) between true and raw (RAW) test statistics, test statistics pruned with LD threshold (LDT) and by 10 markers window (EQL) and from deconvolution (DCV). Superscripts with different letters denote significant differences between methods

Parameters	Values	Method			
		RAW	LDT	EQL	DCV
Sample size (default)	3500	0.954[a]	0.849[b]	0.341[c]	0.296[d]
(alternative)	1500	0.975[a]	0.868[b]	0.356[c]	0.221[d]
Number of QTL	500	0.953[a]	0.849[b]	0.347[c]	0.295[d]
Genotype array	$\boldsymbol{X_{10K}}$	0.972[a]	0.957[b]	0.369[c]	0.167[d]

Deconvolution also significantly increases the correlations of haplotype scores between true and estimated haplotype scores r_h^2 for both LD threshold pruning (LDT) and 10 marker windows (EQL) haplotype block definitions (Table 3); with the exception of default parameter values with EQL where the improvement of DCV from RAW was not deemed statistically significant ($logpval = 1.27$); significant improvements were observed for all other parameter values with both EQL and LDT methods. These improvements were even more significant for $\boldsymbol{X_{10K}}$ and in the oligogenic case of 500 QTL; for example, deconvolution improves the correlation from 0.650 to 0.825 in LDT ($logpval = 16.1$) and from 0.748 to 0.843 in EQL ($logpval = 8.04$) in the case with 500 QTL, a more significant improvement than with 2,000 QTL, suggesting deconvolution improves how well the marker haplotype score in reflecting the true contribution toward the phenotype.

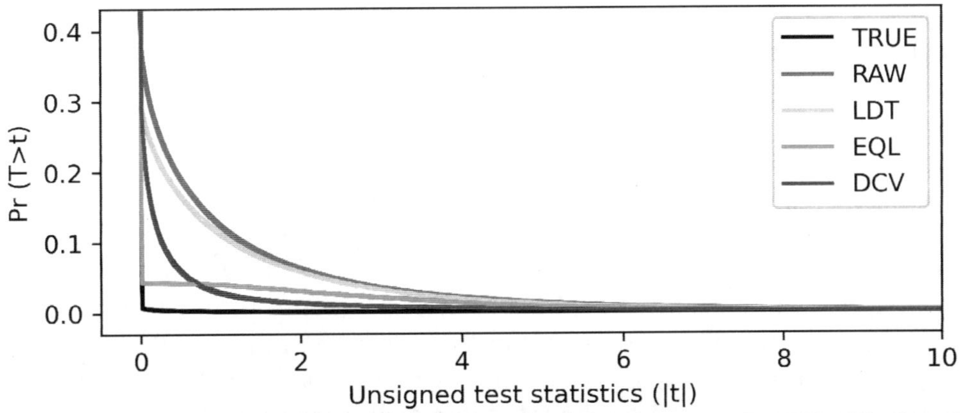

Figure 1. Top 40% tail distribution of true (TRUE), raw (RAW), LD threshold-pruned (LDT), 10 marker window (EQL) and deconvolved (DCV) test statistics under default scenario

Table 3. Correlation in haplotype score for raw (RAW) and deconvolved (DCV) test statistics, r_h^2, for haplotype block by LD threshold or by 10 markers window under changing parameter values. Superscripts with different letters denote significant differences between methods

Parameters	Values	LD threshold pruned (LDT)		10 marker windows (EQL)	
		RAW	DCV	RAW	DCV
Sample size (default)	3500	0.457[a]	0.578[b]	0.571[b]	0.610[b]
(alternative)	1500	0.253[a]	0.351[c]	0.328[b]	0.391[c]
Number of QTL	500	0.650[a]	0.825[c]	0.748[b]	0.843[c]
Genotype array	$\boldsymbol{X}_{10K}$	0.288[a]	0.431[c]	0.332[b]	0.444[c]

CONCLUSION

A novel technique for deconvolution of GWAS statistics that improves the recovery of genetic architecture information was proposed. This technique has successfully reduced the departure in distribution between true and inferred effect size distributions and improved how well the marker haplotype score reflects the true contribution under varying scenarios. It is anticipated this method could be used on a real datasets to elucidate the genetic architectures for various livestock traits.

REFERENCES

Aguilar I., Legarra A., Cardoso F., Masuda Y., Lourenco D. and Misztal I. (2019) *Genet. Sel. Evol.* **51**: 28.
Chang C.C., Chow C.C., Tellier L.C.A.M., Vattikuti S., Purcell S.M. and Lee J.J. (2015) *GigaScience* **4**: 7.
Cheng W., Ramachandran S. and Crawford L. (2020) *PLOS Genet.* **16**: e1008855.
Clark B.G. (1980) *Astron. Astrophys.* **39**: 377.
Gondro C. (2015) 'Primer to analysis of genomic data using R'. Springer, Cham.
Lloyd-Jones L.R., Zeng J., Sidorenko J., Yengo L., Moser G., Kemper K.E., Wang H., Zheng Z., Magi R., Esko T., Metspalu A., Wray N.R., Goddard M.E., Yang J. and Visscher P.M. (2019) *Nat. Commun.* **10**: 5086.
Sargolzaei M. and Schenkel F.S. (2009) *Bioinformatics* **25**: 680.
VanRaden P.M. (2008) *J. Dairy Sci.* **75**: 3136.
Vasershtein L.N. (1969) *Probl. Inf. Transm.* **5**: 64.
Villiers K., Voss-Fels K.P., Dinglasan E., Jacobs B., Hickey L. and Hayes B.J. (2024) *Plant Genome-US* **17**: e20467.
Wang M. and Xu S. (2019) *Heredity* **123**: 287.

SORT & PASTE: A NUMERICAL METHOD TO RENDER ANY RANDOM VARIABLE HERITABLE GIVEN A GENOTYPE

A. Reverter and P.A. Alexandre

CSIRO, Agriculture and Food, QBP, St Lucia, QLD 4067, Australia

SUMMARY

We introduce Sort & Paste, a numerical approach that links an existing genotype file with a randomly sourced variable, thereby allowing genetic control through the variation inherent in the genotype file. This approach is particularly beneficial when there are insufficient records for the genomic analyses of a specific trait of interest, yet the distributional properties of such a trait are known to some degree.

INTRODUCTION

In the context of quantitative genetic and genomic studies, it is often of interest to test competing models for their predictive performance, or to examine the computational efficiency of numerical algorithms. In such instances, the specific characteristics of the phenotype, whether it is derived from real data or simulated, are often less relevant than its association with the genotype information. Additionally, there may be scenarios where the available records for a costly or challenging trait to measure are limited.

We introduce Sort & Paste, a numerical approach that links a pre-existing genotype file with a randomly sampled variable, thereby allowing genetic control through the variation inherent in the genotype file. With similarities with a permutation-based algorithm proposed by Pérez-Enciso *et al.* (2021), Sort & Paste was originally conceived on an *ad-hoc* basis in Alexandre *et al.* (2024) to link non-observed data on semen microbiome abundance with observed data on bull prolificacy.

In the present study, the properties of Sort & Paste are examined using existing genotype data from both single- and multi-breed contexts, as well as differing SNP densities.

MATERIALS AND METHODS

The mechanics of Sort & Paste. The Sort & Paste algorithm proceeds as follows:

1. Let **G** be an $(n{\times}s)$ matrix of genotypes with n rows (for as many animals) and s columns (for as many SNP). Elements in **G** are {0,1,2} for genotypes {AA,AB,BB}.
2. Let **y** be an $(n{\times}1)$ vector of arbitrary values in R^n.
3. Sort step: Sort ***G*** and ***y*** in increasing or decreasing order to obtain ***G_sort*** and ***y_sort*** such that a strong correlation, potentially 1, exists between ***G_sort*** and ***y_sort***. When sorting, note that the rows of ***G*** remain intact.
4. Paste step: Assign the index of the genotype row in ***G_sort*** to the corresponding value in ***y_sort*** with same index.
5. Estimate variance components with GREML or equivalent approaches.

The effect of SNP density and sample size. To investigate the performance of the Sort & Paste algorithm under varying SNP densities and sample sizes, we utilized genotype data from Porto-Neto *et al.* (2023), containing a multi-breed population of 6,063 bulls from six tropically adapted breeds (Brahman, Tropical Composite, Santa Gertrudis, Droughtmaster, Ultra-Black and Belmont Tropical Composite), with genotypes for 680,758 autosomal SNPs. We implemented a 3-by-3 experimental design that incorporated three levels of SNP density: high-density (HD) with all SNPs, medium-density (MD) selecting every tenth SNP, and low-density (LD) choosing every hundredth SNP. Additionally, we examined three sample sizes: the complete dataset (6K), every second record (3K), and every sixth record (1K). Across the nine experimental scenarios, we conducted ten replicates both with and without treatment by the Sort & Paste algorithm. This resulted in a total of 180 runs: 3 SNP densities × 3 sample sizes × 2 treatments × 10 replicates. In each run, the response vector **y** of (n×1) values was drawn from a random normal distribution with a mean of zero and a variance of one. In all cases, the genomic relationship matrix (GRM) was computed using Method 1 of VanRaden (2008) without scaling for breed-specific allele frequencies. Variance components were estimated by GREML using the BLUPF90 family of programs (Misztal *et al.* 2022).

The effect of partial sorting. Next, we aimed to investigate the effect on variance components estimates when the Sort & Paste algorithm is applied to distinct proportions of the data. To this end, we accessed the Angus Australia database to retrieve genotypes for 61,105 autosomal SNPs from 4,873 cattle contributing carcase records in the Angus SteerSELECT genomic tool (Reverter *et al.* 2025). The Sort & Paste algorithm was applied to nine distinct proportions of the data from 0% (i.e. not applied) to 100% (fully applied) by 12.5% increments. Once again, 10 replicates were conducted for each condition, and in all runs a random vector of 4,873 values was drawn from a normal distribution with a mean of zero and a variance of one. Also, variance components were estimated by GREML using the BLUPF90 family of programs (Misztal *et al.* 2022).

RESULTS AND DISCUSSION

Toy example. Figure 1 illustrates the mechanics of Sort & Paste using a toy example where ten animals have genotypes from 10 SNPs. A random dataset across ten records is also available (which in this case have been sampled from a (0,1) uniform distribution. If the genotype and the random data are processed through GREML, the expectation is for a very low estimate of h^2. However, after sorting both the genotype and the dataset, pasting them in this sorted manner the expectation is for very large estimate of h^2.

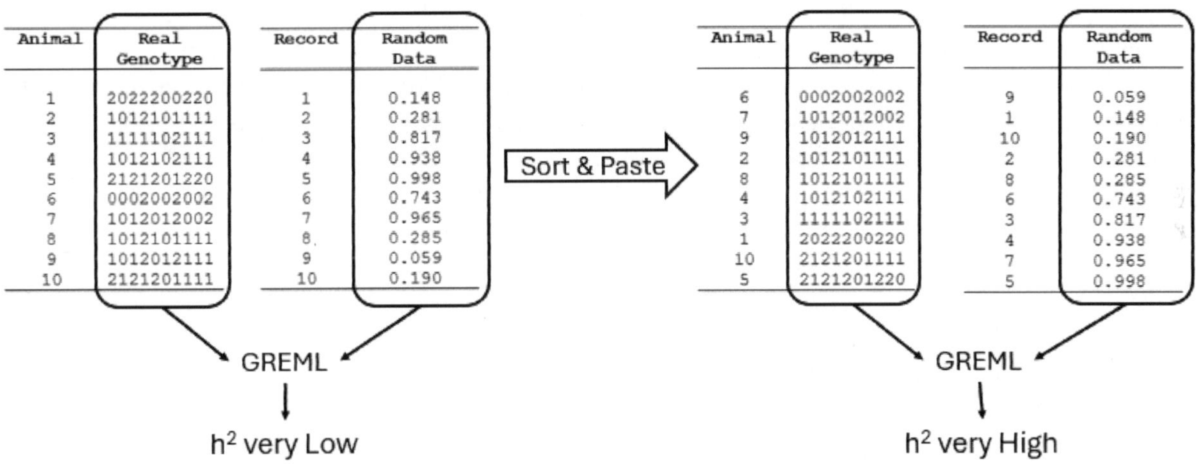

Figure 1. Toy example involving a real 10-SNP genotype across 10 animals is matched with a random data. Before and after Sort & Paste the GREML estimate of h^2 is likely to be very low and very high, respectively

The effect of SNP density and sample size. As expected, when Sort & Paste was not applied, the estimates of phenotypic variance (Vp) and heritability (h^2) were one and zero, respectively, and unaffected (P-value > 0.10) by neither SNP density nor sample size. Across the 90 replicates, the estimate of Vp averaged 0.999 and ranged from 0.951 to 1.039. Similarly, the estimate of h^2 averaged 0.001 and ranged from -8.70e-7 to 0.038.

When Sort & Paste was applied, estimates of h^2 were significantly different from zero (P-value < 0.0001) and affected by SNP density and sample size, such that they were higher with increasing sample size and at low SNP density (Figure 2). Averaged across the ten replicates, the lowest estimate of h^2 (± SD) was 0.397 ± 0.008 found at MD SNP density with 1K sample size, while the highest was 0.862 ± 0.006 found at LD density with 6K sample size.

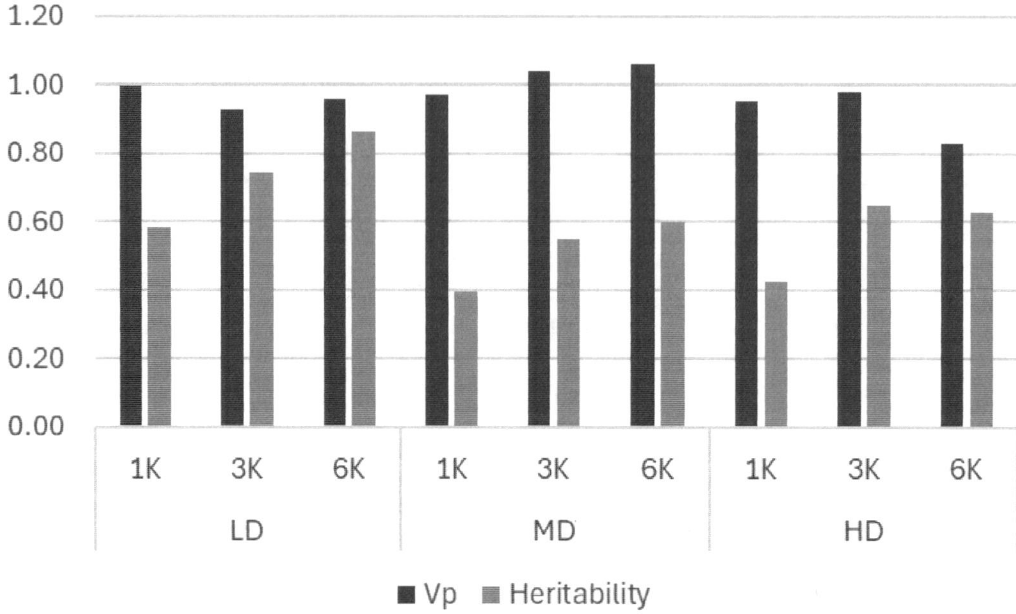

Figure 2. Estimates of phenotypic variance (V_P) and heritability when applying the Sort & Paste algorithm to a multi-breed genotype dataset at either low (LD), medium (MD) or high density (HD) SNP genotypes, and with a sample size of 1K, 3K or 6K animals

The elevated h^2 estimates associated with larger sample sizes can be explained by a stronger link being established between genotype and phenotype by the sorting command with larger samples. Conversely, the lower h^2 estimates observed in MD and HD scenarios was attributed to the genomic diversity inherent in a genotype file from a multi-breed population. This diversity is less pronounced, or inadequately captured with SNP at LD, allowing the Sort & Paste algorithm to yield higher h^2 estimates. Similarly, the variation around the simulated value of 1.0 for the estimates of Vp, particularly for the scenario with HD SNP and 6K animals was attributed to having multiple breeds in the GRM.

The effect of partial sorting. Figure 3 illustrates the increasing pattern of h^2 estimates when the Sort & Paste algorithm was applied to increasing proportions of the Angus genotype dataset. The extreme h^2 estimates of ~1 when the whole data was subjected to Sort & Paste was attributed to the genotype information belonging to a single-breed, highly homogenous population.

Remarkably, the increasing pattern of h^2 estimates showed a non-linear trend. When the Sort & Paste algorithm was applied to 12.5% of the data, the estimated h^2 averaged 0.056 and ranged from 0.034 to 0.075. When applied to 87.5% of the data, the estimated h^2 ranged from 0.702 to 0.727.

CONCLUSION

Sort & Paste is a simple numerical algorithm to render any random variable heritable given a genotype. By simultaneously sorting the genotype and the random variable, a link is created between the two pieces of information which, in turn, makes the variable to be under genetic control. The performance of Sort & Paste can be affected by the variation inherent in the genotype data, including its density and the sample size, as well as the proportion of the data subjected to the Sort & Paste algorithm.

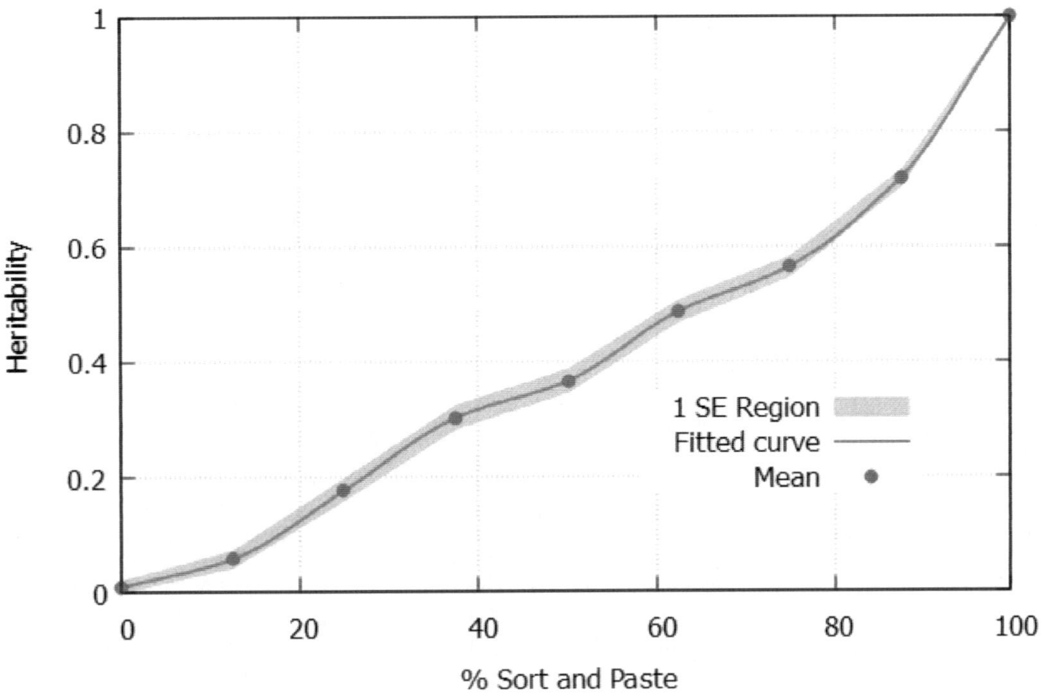

Figure 3. Estimates of heritability applying the Sort & Paste algorithm to various proportions of the Angus genotype dataset

Sort & Paste is particularly relevant in the presence of genotype data when competing models are being tested for their predictive ability and/or computational efficiency. It is also of value when the available records for a costly or challenging trait to measure are limited, but its distributional properties, including mean, variance and heritability are known. Further research is needed to prove the value of Sort & Paste in research focussed on the correlated responses to selection on traits that have not yet been observed.

ACKNOWLEDGEMENTS

The authors acknowledge the contribution of Angus Australia for kindly providing the genotype dataset used in this study. Silvia Rodríguez-Ramilo and Núria Mach contributed through early discussions that gave rise to the initial ideation of the Sort & Paste algorithm. CSIRO acknowledges the Traditional Owners of the lands that we live and work on across Australia and pays its respects to Elders past and present.

REFERENCES

Alexandre P.A., Rodríguez-Ramilo S.T., Mach N. and Reverter A. (2024) *J. Anim. Breed. Genet.* **00**: 1.

Misztal I., Tsuruta S., Lourenco D., Masuda Y., Aguilar I., Legarra A. and Vitezica Z. (2022) Manual for BLUPF90 family of programs. Athens: University of Georgia.

Pérez-Enciso M., Zingaretti L.M., Ramayo-Caldas Y. and de los Campos G. (2021) *Genet. Sel. Evol.* **53**: 65.

Porto-Neto L.R., Alexandre P.A., Hudson N.J., Bertram J., McWilliam S.M., Tan A.W.L., Fortes M.R.S., McGowan M.R., Hayes B.J. and Reverter A. (2023) *PLoS O*ne **18**: e0279398.

Reverter A., Duff C., Samaraweera M., Alexandre P.A., Hine B.C., Ingham A.B., Mowbray L.A. and Porto-Neto L.R. (2025) *Proc. Assoc. Advmt. Anim. Breed. Genet.* **26**: *These proceedings.*

VanRaden P.M. (2008) *J. Dairy. Sci.* **91**: 4414.

ASSESSING THE EFFECTIVENESS OF LOW-COVERAGE NANOPORE SEQUENCING IN GENOMIC SELECTION: A COMPARISON BETWEEN CLAIR3 AND QUILT PIPELINES

L.T. Nguyen[1], H. Lamb[1], J. Copley[1,2], B.J. Hayes[1] and E.M. Ross[1]

[1] Centre for Animal Science, Queensland Alliance for Food and Agriculture Innovation, The University of Queensland, Brisbane, QLD, 4072 Australia
[2] Copley Pastoral, Anduramba, QLD, 4355 Australia

SUMMARY

QUILT and Clair3 bioinformatic tools have been used in studies for genomic prediction using low-coverage sequencing from Oxford Nanopore Technologies (ONT). Both tools have demonstrated their potential in achieving accurate predictions of genomic breeding values (GBVs). However, different coverage has been employed (e.g. 0.1X in Lamb et al. (2023) and 2X in González-Recio *et al.* (2023). To assess the efficacy of these two bioinformatics pipelines for GBV calculation from low-coverage ONT sequencing, we sequenced 16 animals at 0.1X coverage using the latest Q20 and kit V14 chemistry, as used in the study by González-Recio *et al.* (2023). These 16 animals were also genotyped using 35K SNP array, allowing for benchmarking of the two tools. Our results showed that QUILT outperformed Clair3 at very low coverage (0.1X). The correlation between QUILT-derived GBVs and 35K imputed-to-700K GBVs was slightly higher than that observed between Clair3-derived GBVs and 35K imputed-to-700K GBVs, indicating QUILT's superior performance in ultra-low-coverage genomic prediction.

INTRODUCTION

Third-generation sequencing technologies, such as Oxford Nanopore Technologies (ONT), have demonstrated their capacity for genomic selection using information from low sequencing depth. Genotyping-by-sequencing using low-coverage sequence data paired with imputation is becoming an increasingly popular SNP genotyping method for genomic prediction, especially as sequencing costs continue to decrease and portable long-read sequencers from ONT enable on-site sequencing. Lamb *et al.* (2023) showed that at a coverage of 0.1X, the QUILT imputation pipeline, when using an imputation reference panel with a minor allele frequency (MAF) greater than 0.1, achieves comparable accuracy to low-density SNP arrays, as benchmarked against the high-density (HD) 700K SNP array. Additionally, González-Recio *et al.* (2023) demonstrated that when using Clair3 software for genotype calling with ONT sequencing depth as low as 2X and the latest Q20 and kit V14 chemistry, the accuracy of GBVs ranged from 0.79 to 0.99, depending on the trait. Although Clair3 has been widely used and recommended for SNP calling from long-read data, these pipelines have not yet been directly benchmarked against each other to determine which is more accurate for calculating GBVs. Here, we assessed the efficacy of two bioinformatics pipelines for GBV calculation from low-coverage ONT data (0.1X): one using QUILT and the other using Clair3. The accuracy of low pass sequencing using ONT for genomic prediction was compared against SNP genotyping arrays.

MATERIALS AND METHODS

Animal ethics. Animal ethics approval was obtained from the University of Queensland ethics board (Animal Ethics Approval Number 2021/AE000541).

Animal information. A total of 16 Brahman beef cattle, consisting of 8 cows and their 8 calves, were used in this study. All animals were female.

DNA extraction. DNA was extracted from approximately 30mg ear punch of 16 Brahman samples using the Puregene Gentra DNA Tissue kit (QIAGEN, Australia) according to the manufacturer's instructions. DNA concentrations were measured using the Qubit dsDNA Broad Range assay kit (Thermo Fisher Scientific). The purity of the extracted DNA was determined with the NanoDropND 1000 (v.3.5.2, Thermo Fisher Scientific).

Nanopore sequencing at 0.1 X coverage. Extracted DNA samples were processed using the native barcoding kit 24 V14 (SQK-NBD114.24, Oxford Nanopore Technology, Oxford, UK) following the manufacturer's instructions, with adjustments as detailed in Hayes *et al.* (2021). Sequencing was performed on a PromethION (v10.4.1) flow cell in a PromethION P24 device, with sequencing halted once all samples reached approximately 0.1X coverage. The basecalling of the sequence data was performed using super accuracy mode (Guppy version 6.4.2).

Genotype calling and imputation. Following sequencing, the generated read files (fastq) for each sample underwent alignment to the *Bos taurus* ARS-UCD2.0 reference genome (Rosen *et al.* 2020) using minimap2 (Li 2018) with -x map-ont option. The resulting mapped files (bam) were then employed in two approaches: 1) genotype and imputation through QUILT version 1.0.1 (Davies et al. 2021), and 2) genotype via Clair3 version 0.1.10 (Zheng *et al.* 2021) with subsequent imputation using Beagle version 5.2 (Browning *et al.* 2018). A SNP panel of 15,000,408 million SNPs and 1,208 animals comprised of breeds found in northern Australia was used as the reference for imputing the missing genotypes.

SNP genotyping using 35K SNP array. These 16 animals were also genotyped on the 35K GGP TropBeef SNP array (Neogen, Lansing, MI) and their genotypes were subsequently imputed up to the bovine HD array density using FImpute (Sargolzaei *et al.* 2014) and a panel of 3,140 cattle genotyped on the bovine HD SNP array (Illumina, San Diego, CA).

Genomic breeding values calculation using SNP-BLUP method. Genomic breeding values for four traits: body weight (BW), hip height (HH), body condition score (BCS), and corpus luteum (CL) score were calculated using 641,163 SNP loci, with SNP effects from Hayes *et al.* (2019), as described in Lamb *et al.* (2021) using $\hat{a} = \mathbf{M}\hat{g}$. Here, the imputed genotypes using QUILT were contained in an n by m matrix **M**, where n represents the number of animals and m represents the number of SNP loci, while $\hat{g}$ was a vector of predicted SNP effects with length m. The SNP effects used to calculate the GBVs were derived from 28,000 animals with genotype and phenotypes as described in Hayes *et al.* (2023) and Lamb *et al.* (2021). The estimated GBVs from 35K array genotypes and Clair3 genotypes were calculated using the same equation as the imputed QUILT genotypes, with **M** replaced by the imputed Clair3 genotypes and 35K SNP array genotypes, respectively.

Correlation Analysis. The accuracy of the sequence based GBVs was determined by calculating the Pearsons correlation between the GBVs generated by either Clair3 or QUILT and the GBVs generated from the SNP array genotypes.

RESULTS AND DISCUSSION

This study confirms the potential of genotyping-by-sequencing at ultra-low coverage for genomic prediction of four economically important traits including BW, BCS, HH and CL600 in cattle. Despite the limitation of a small sample size (n=16), our results demonstrated strong correlations between GBVs derived from QUILT and those from the 35K SNP array, ranging from 0.86 to 0.96 (Figure 1), consistent with the findings of Lamb *et al.* (2021, 2023) and Davies *et al.* (2021).

Regarding the imputation of Clair3 output, genotypes could not be successfully imputed using GLIMPSE at low coverage sequence. Therefore, Beagle v5.2 was paired with Clair3 genotypes for imputation. The correlation between GBVs calculated from Clair3 genotypes and 35K SNP array was within an acceptable range, varying from 0.83 to 0.92, except for the CL600 trait, where the correlation was only 0.21 (Figure 1). The lowest accuracy trait for QUILT was also CL600 (r=0.86). Notably, the correlation between CL600 GBV from Clair3 and the 35K imputed to the 700K SNP array was not significant (P = 0.42), whereas all other correlations were highly significant.

Since puberty is a complex trait influenced by many genes with small effects, low-coverage sequence data may lead to the loss of key SNPs associated with pubertal traits. Additionally, as demonstrated by Lamb *et al.* (2023), Beagle at coverage below 1X does not perform as well as QUILT and GLIMPSE, potentially resulting in inaccurate genotype assignments and a weakened correlation.

The low correlation observed when using Clair3 for CL600 could be attributed to the challenges associated with low-coverage sequencing handled by Clair3. Notably, the study by González-Recio *et al.* (2023) also used Clair3, but at a sequencing depth above 2X, where the accuracy of GBVs ranged from 0.79 to 0.99. Unlike QUILT, Clair3 does not perform phased haplotype-based imputation or leverage long-range linkage disequilibrium information to improve accuracy. This limitation reduces Clair3's effectiveness when handling low-coverage sequencing data, particularly for complex traits like CL600. In summary, QUILT demonstrated robust performance across all traits.

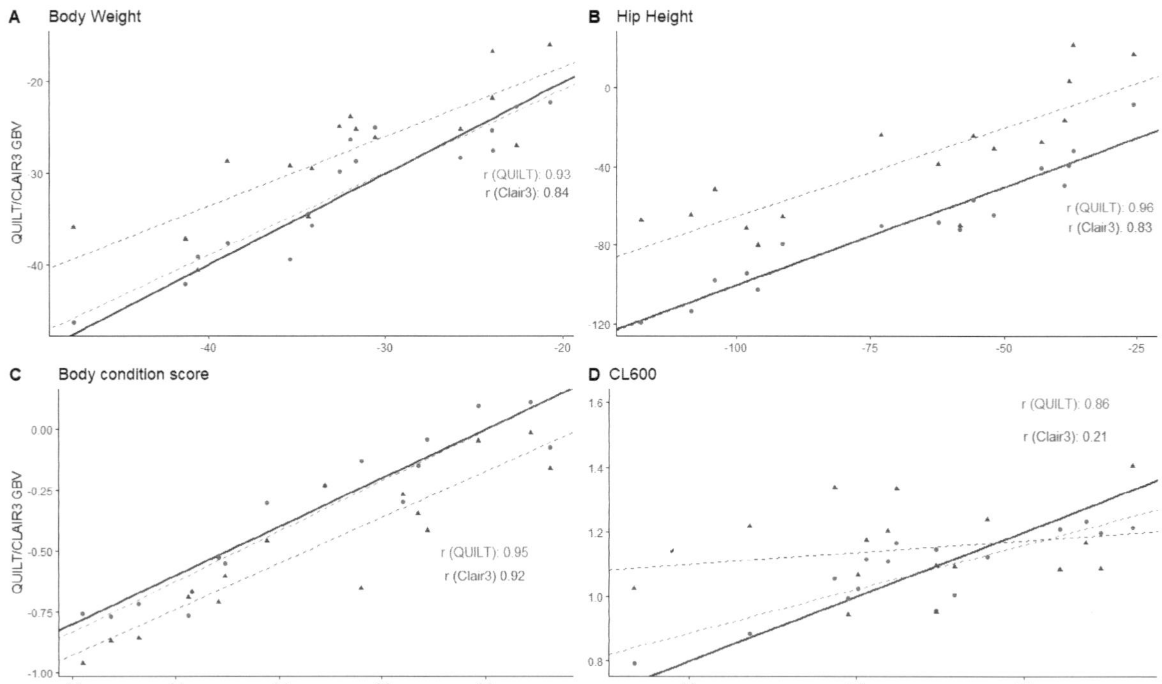

Figure 1. Correlation between QUILT GBV/Clair3 GBV and 35K SNP array GBV for four traits. Each dot in the figure represents a sample, with the dot color indicating the imputation approach: red signifies the use of QUILT, while blue denotes the use of Clair3

In the subsequent analysis, rankings of animals for the four tested traits based on QUILT-derived GBVs and Clair3-derived GBVs were compared to GBVs calculated from 35k SNP array imputed to 700K (Figure 2). QUILT demonstrated superior performance in ranking animals when compared to 35K SNP array for the body condition score trait. For the other three traits, QUILT outperformed Clair3 in the re-ranking of animals. As expected from the correlation results, Clair3-derived GBVs for CL600 showed the greatest re-ranking.

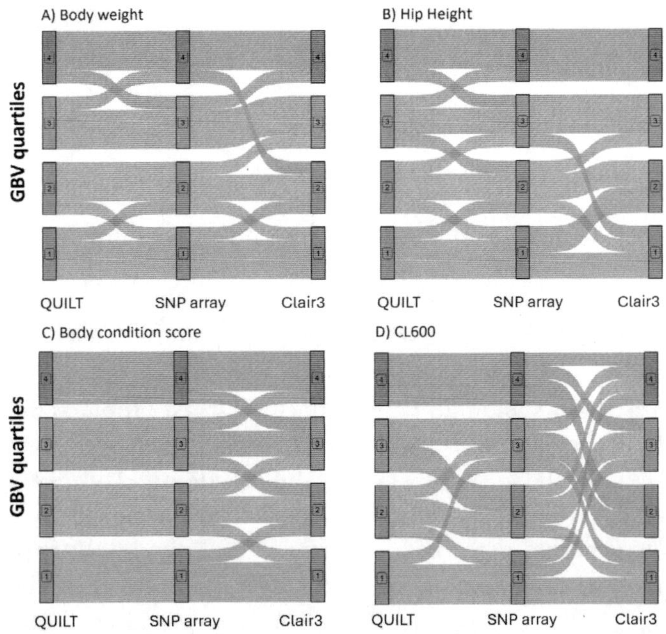

Figure 2. Change in GBV quartile rankings between GBV derived from QUILT and Clair3 compared to 35K SNP array for the four tested traits

CONCLUSION

In this study, we demonstrated the capability of achieving accurate genomic predictions using ONT sequence data, even at remarkably low sequencing coverage (0.1X), by applying QUILT, Clair3, and Beagle v5.2 for genotyping and imputation. Our results highlight that genotyping-by-sequencing at 0.1X coverage with QUILT outperforms the accuracy achieved through imputation from Clair3 and Beagle v5.2. Additionally, Clair3 at low coverage can still produce reasonable GBVs, particularly for less complex traits.

ACKNOWLEDGEMENTS

The authors gratefully acknowledge Mr and Mrs Tom and Marie Copley for access to their cattle for sampling. This project was funded by the Meat and Livestock Australia (project P.PSH.1401). The authors declare no potential conflicts of interest.

REFERENCES

Browning B.L. and Browning S.R. (2016) *Am. J. Hum. Genet.* **98**: 116.
Davies R.W., Kucka M., Su D., *et al.* (2021) *Nat Genet.* **53**: 1104.
González-Recio O., López-Catalina A., Peiró-Pastor R., *et al.* (2023) *J. Anim. Sci. Biotechnol.* **14**: 98.
Hayes B.J., Copley J. Dodd E., *et al.* (2023) *Genet Sel Evol.* **55**: 71.
Hayes B.J., Nguyen L.T., Forutan M., *et al.* (2021) *Front. Genet.* **12**: 760450.
Lamb H.J., Hayes B.J., Randhawa I.A.S., *et al.* (2021) *Plos One.* **16**: e0261274.
Lamb H.J, Nguyen L.T, Copley J.P., *et al.* (2023) *BMC Biol.* **21**: 286.
Li H. (2018) *Bioinformatics.* **34**: 3094.
Rosen B.D., Bickhart D.M., Schnabel R.D., *et al.* (2020) *Gigascience.* **9**: 3.
Sargolzaei M., Chesnais J.P. and Schenkel F.S. (2014) *BMC Genomics* **15**: 478.
Zheng Z., Li, S., Su, J., *et al.* (2022) *Brief. Commun.* **2**: 797.

Proc. Assoc. Advmt. Anim. Breed. Genet. 26: 114-117

ESTIMATING THERMAL TOLERANCE IN SELECTIVELY-BRED NEW ZEALAND GREENSHELL MUSSELS (*PERNA CANALICULUS*)

M.R. Scholtens[1], R. Roberts[2], N.J. Delorme[1], N. King[1], N.L.C. Ragg[1], R. Costilla[1] and J.A.Ericson[1]

[1] Cawthron Institute, Nelson, New Zealand
[2] Shellfish Production and Technology New Zealand SPATnz) Ltd., Nelson, New Zealand

SUMMARY

New Zealand's aquaculture industry is under growing pressure from recurring marine heatwaves, that threaten the productivity and sustainability of marine farms, emphasizing the need for targeted genetic solutions. This study focuses on the Greenshell™ mussel (*Perna canaliculus*) (GSM), a key aquaculture species in New Zealand (NZ) that has been selectively bred since 2002. This well-established family-based breeding programme provides an opportunity to breed for thermal tolerance and improve survival and growth during periods of elevated temperature for this farmed species. We defined thermal tolerance as the number of days to death (DTD) at 26°C. This study explores the appropriate model for estimating the heritability for thermal tolerance in GSM. A total of 850 mussels produced in 2021 and 2022 were evaluated, consisting of 55 families from 49 sires and 43 dams. The estimated heritability was moderate (0.48±0.10), suggesting that if thermal tolerance is included in a selection index, there is adequate genetic variation for thermal tolerance in this mussel to allow genetic improvement in this trait.

INTRODUCTION

The GSM is NZ's most valuable aquaculture export (www.aquaculture.org.nz). Although selective breeding programs and hatchery production methods are well established for this species (Camara and Symonds *et al.* 2014) most of the industry relies on wild caught seed 'spat' (Jeffs *et al.* 2018). However, with increasing sea water temperatures, this wild spat supply is becoming scarce (Wu *et al.* 2024) and the need for more consistent and sustainable sources, including hatchery-reared spat, is growing. Seawater temperatures between 16 – 19°C are ideal for the GSM (Venter *et al.* 2023) while Delorme *et al.* (2024) reported that sea surface temperatures are surpassing 24°C in many areas of the North Island. Selective breeding has the potential to develop mussels that are more resilient to increasing sea water temperatures.

Survival during heat stress is becoming an economically important trait in shellfish, with genetic parameters being published for oysters (Jiang *et al.* 2023; Chi *et al.* 2024) and abalone (Liu *et al.* 2022). For GSM, recent studies have showed that thermal tolerance is influenced by genetics (Ericson *et al.* 2023) and the interaction between genetics and age (Delorme *et al.* 2024), but there are no reports on the heritability of survival during heat stress in GSM. Therefore, this study aims to explore the genetic potential of selecting for thermal tolerance in the GSM.

MATERIALS AND METHODS

Data. We evaluated a total of 850 mussels for thermal tolerance: 55 families with 13-16 individuals/family. These mussels are part of a commercial breeding program with 22 years of selection and come from the 2021 and 2022 cohorts. The 55 families consist of 54 full-sib families, some of which are also half-sib families (n=14), selected for commercial production traits.

Trial design. On 12th February 2024, we collected adult mussels from the farms in the Marlborough Sounds, NZ and transported to the Cawthron Aquaculture Park (CAP). We cleaned all mussels of any biofouling, engraved them with an ID and divided them (2 mussels/family, 114 mussels/tank) into eight 100 L tanks of flowing seawater (2.5 - 4 L min^{-1} $tank^{-1}$). We maintained the temperature at 20°C for the first eight days, then increased by 1°C per day until reaching 26°C, which exceeds the chronic thermal maximum for this species (Delorme *et al.* 2024). The mussels were not fed during the experiment. We kept the mussels at 26°C and checked individual survival twice daily (dead = shell gaping).

Statistical analysis. We defined thermal tolerance as the number of days to death (DTD) at 26°C. By day 29 only 1 mussel remained, which we censored for the analysis. We performed all statistics using R 4.3.3 software. We obtained descriptive statistics using the 'psych' package and performed the genetic models using the Restricted Maximum Likelihood procedure in ASReml-R v4 (Gilmour *et al.* 2009). We estimated the variance components and genetic parameters for DTD and shell length using a bivariate animal which is represented as;

$$\begin{bmatrix} \mathbf{y}_i \\ \mathbf{y}_j \end{bmatrix} = \begin{bmatrix} \mathbf{X}_i & 0 \\ 0 & \mathbf{X}_j \end{bmatrix} \begin{bmatrix} \mathbf{b}_i \\ \mathbf{b}_j \end{bmatrix} + \begin{bmatrix} \mathbf{Z}_i & 0 \\ 0 & \mathbf{Z}_j \end{bmatrix} \begin{bmatrix} \boldsymbol{\alpha}_i \\ \boldsymbol{\alpha}_j \end{bmatrix} + \begin{bmatrix} \mathbf{W}_i \\ 0 \end{bmatrix} \boldsymbol{c} + \begin{bmatrix} \mathbf{e}_i \\ \mathbf{e}_j \end{bmatrix}$$

where for traits i (length) and j (DTD), $\begin{bmatrix} \boldsymbol{y}_i \\ \boldsymbol{y}_j \end{bmatrix}$ is a vector of phenotypes, $\begin{bmatrix} \mathrm{b}_i \\ \mathrm{b}_j \end{bmatrix}$ is a vector of fixed effects of cohort and tank, $\begin{bmatrix} \boldsymbol{\alpha}_i \\ \boldsymbol{\alpha}_j \end{bmatrix}$ is a vector of random animal genetic effects, **c** is a vector of the common environmental (CE) effects for trait i and $\begin{bmatrix} \boldsymbol{e}_i \\ \boldsymbol{e}_j \end{bmatrix}$ is a vector of random residuals. The **X**, **Z** and **W** are design matrices for the corresponding fixed and random effects for traits i and j. It was assumed that genetic effects were distributed as $\sim N\left(\begin{bmatrix} 0 \\ 0 \end{bmatrix}, A \otimes \begin{bmatrix} \sigma_{a_i}^2 & \sigma_{a_{ij}} \\ \sigma_{a_{ji}} & \sigma_{a_j}^2 \end{bmatrix}\right)$, where $\begin{bmatrix} \sigma_{a_i}^2 & \sigma_{a_{ij}} \\ \sigma_{a_{ji}} & \sigma_{a_j}^2 \end{bmatrix}$ is the additive genetic variance-covariance structure, the CE effects were distributed as $\sim \mathrm{N}\left(\begin{bmatrix} 0 \\ 0 \end{bmatrix}, I \otimes \begin{bmatrix} \sigma_{c_i}^2 & 0 \\ 0 & 0 \end{bmatrix}\right)$, where $\begin{bmatrix} \sigma_{c_i}^2 & 0 \\ 0 & 0 \end{bmatrix}$ is the CE variance-covariance structure and the residual effects were distributed as $\sim \mathrm{N}\left(\begin{bmatrix} 0 \\ 0 \end{bmatrix}, I \otimes \begin{bmatrix} \sigma_{e_i}^2 & \sigma_{e_{ij}} \\ \sigma_{e_{ji}} & \sigma_{e_j}^2 \end{bmatrix}\right)$, where $\begin{bmatrix} \sigma_{e_i}^2 & \sigma_{e_{ij}} \\ \sigma_{e_{ji}} & \sigma_{e_j}^2 \end{bmatrix}$ is the residual variance-covariance structure, where **A** is the additive genetic relationship matrix and **I** are identity matrices. To determine inclusion in the models, the significance of fixed and CE effects were tested using Wald-F statistics, likelihood ratio test (LRT test) and the Bayesian information criterion (BIC) (Lynch and Walsh 1998). Narrow-sense heritability for each trait was calculated as $h_i^2 = \frac{\sigma_{\alpha_i}^2}{\sigma_{p_i}^2}$, where $\sigma_{\alpha_i}^2$ are the additive genetic variance and $\sigma_{p_i}^2$ is the total phenotypic variance which was $(\sigma_{\alpha_i}^2 + \sigma_{c_i}^2 + \sigma_{e_i}^2)$ for length and $(\sigma_{\alpha_i}^2 + \sigma_{e_i}^2)$ for DTD. The effect of the CE (c_i^2) was calculated as $c_i^2 = \frac{\sigma_{c_i}^2}{\sigma_{p_i}^2}$ for length. The Z-score was used to assess whether the h^2 was significantly different from zero.

RESULTS AND DISCUSSION

Descriptive statistics for DTD and shell length are in Table 1. The average survival of GSM when exposed to water temperatures of 26°C was 15±4 days and the average mussel shell length was 84.5±15 mm.

In a commercial breeding program, juvenile mussels are reared in the same full-sib family tank until they are mature enough to be placed on single-family rope droppers in the sea, where they complete their growth through to harvest size (~1-2 years). If statistical models don't account for this process (fit the CE effect), h^2 are likely to be overestimated (Alcapan *et al.* 2007). However, when the CE is included as a random effect along with the individual animal, depending on the data structure, this can cause confounding effects and lead to model convergence issues, or underestimate the genetic component of the trait. Results from the LRT test showed that the CE had a significant effect on shell length but not on DTD. This variation captured by fitting CE in the model is evident for length with the c^2 of 0.20±0.15.

Table 1. Descriptive statistics of days to death (DTD) and length (mm) of Greenshell™ mussels in NZ

Trait	N	Mean	SD[1]	Min	Max	CV[2]
Length	850	84.46	15.14	51	166	18
DTD	850	15.38	3.99	6	29	26

[1]SD = standard deviation, [2]CV = coefficient of variation.

The estimated h^2 for DTD was 0.48±0.10 which is significant and at the upper range of h^2 estimates for physiological resilience to stress in other shellfish/aquatic species. For example, thermal tolerance h^2 have been estimated for *C. gigas* (0.16-0.20±0.03-0.04 (Chi *et al.* 2024)), hybrid oysters (*Crassostrea gigas* ♀ x *C. angulata* ♂) which ranged from 0.19-0.27±0.03-0.05 (Jiang *et al.* 2023), from 0.35-0.42±0.03-0.06 for abalone (Liu *et al.* 2022), and from 0.41-0.48±0.06-0.07 in rainbow trout (Perry *et al.* 2005). Although common environmental effects were not significant for DTD, the h^2 observed in this study may be partially the result of overestimation of additive genetic variance via common rearing of full-sib families.

The r_g between DTD and shell length was moderate and negative (-0.49±0.27) suggesting that faster growing mussels may be less resilient to thermal stress. However, this relationship might be confounded by size because faster growing mussels were larger at the time of testing. This observation aligns with findings from Delorme *et al.* (2024) where larger mussels from the same family were less tolerant to heat stress, indicating a potential size-related vulnerability to thermal conditions. Other studies have found the r_g between thermal tolerance and growth traits tend to be low or weak in oysters, ranging from -0.28 to +0.24 in *C. gigas* (Chi *et al.* 2024), and 0.016 in hybrid oysters (Jiang *et al.* 2023). Similarly, low to no correlations between thermal tolerance and growth traits have been observed in finfish. For example, Perry *et al.* (2005) reported genetic correlation of upper thermal tolerance and body weight in rainbow trout ranging from -0.28 to +0.04.

Table 2. Estimates of additive genetic variance (σ_α^2), common full-sib variance (σ_c^2), residual variance (σ_e^2), heritability (h^2), common full-sib environmental effect (c^2), genetic (r_g) and phenotypic (r_p) correlations for days to death (DTD) and shell length of Greenshell™ mussels in New Zealand

Trait	Model	σ_α^2	σ_c^2	σ_e^2	h^2 ± SE[1]	± SE[1]	r_p ±SE[1]	r_g ±SE[1]
Length	Full	.78	6.91	31.18	0.41 ± 0.34	!0 ± 0.15	0.19 ± 0.16	-0.49 ± 0.27
DTD	Full	.87	-	7.43	0.48* ± 0.10	-		

[1]SE = standard error. *Statistically significant

CONCLUSIONS

The moderate heritability of DTD suggests that there is adequate genetic variation to achieve genetic improvement for thermal tolerance in GSM in NZ. The common full-sib rearing environment does not significantly influence thermal tolerance but there does seem to be a moderate negative correlation between thermal tolerance and shell length, which is influenced by the common full-sib environment. The estimated genetic parameters in this study provide evidence for the inclusion of DTD into a selection index.

ACKNOWLEDGEMENTS

The authors would like to thank SPATnz for collecting and providing the mussels for the trial and the Cawthron scientists and technicians for conducting the trial. This work was funded by the New Zealand Ministry of Business, Innovation and Employment under the Cawthron Shellfish Aquaculture Research Platform (Contract CAWX1801).

REFERENCES

Alcapán A.C., Nespolo R.F. and Toro J.E. (2007) *Aquac. Res.* **38**: 313.

Aquaculture New Zealand. (2024) Industry — Aquaculture New Zealand. www.aquaculture.org.nz.

Chi Y., Li Q., Xu C., Liu W. and Liu H. (2024) *J. Appl. Genet.* **65**: 155.

Camara M.D. and Symonds J.E. (2014) *N. Z. J. Mar. Freshw. Res.* **48**: 466.

Delorme N.J., King N., Cervantes-Loreto A., South P.M., Baettig C.G., Zamora L.N., Knight B.R., Ericson J.A., Smith K.F. and Ragg N.L.C. (2024) *Sci. Rep.* **14**: 19130.

Ericson J.A., Laroche O., Biessy L., Delorme N.J., Pochon X., Thomson-Laing J., Ragg N.L.C., Smith K.F. (2023) *Front. Physiol.* **14**: 1265879.

Jeffs A.G., Delorme N.J., Stanley J., Zamora L.N. and Sim-Smith C. (2018) *Aquac.* **488**: 30.

Gilmour A.R., Gogel B., Cullis B., Thompson R. and Butler D. (2009) ASReml User Guide Release 3.0. VSN International Ltd, Hemel Hempstead, UK.

Jiang G., Xu C. and Li Q. (2023) *Aquat. Living Resour.* **36**: 30.

Liu J., Peng W., Yu F., Shen Y., Yu W., Lu Y., Lin W., Zhou M., Huang Z., Luo X., You W. and Ke C. (2022) *Evol. Appl.* **15**: 992.

Lynch M. and Walsh B. (1998) ‘Genetics and analysis of quantitative traits’ 1st ed. Oxford University Press USA.

Perry G.M., Martyniuk C.M., Ferguson M.M. and Danzmann R.G. (2005) *Aquac.* **250**: 120.

Wu W., Chaput R., Lundquist, C.J., Montaño Orozco M.M. and Jeffs A.G. (2024) *Aquac.* **578:** 740025.

Venter L., Alfaro A.C., Ragg N.L.C., Delorme N.J. and Ericson J.A. (2023) *J. Therm. Biol.* **117**: 103702.

Proc. Assoc. Advmt. Anim. Breed. Genet. 26: 118-121

GENOTYPE-BY-ENVIRONMENT INTERACTIONS IN YELLOWTAIL KINGFISH: IMPLICATIONS FOR GENETIC IMPROVEMENT

M. Camara[1], A. Setiawan[1], J.W.M. Bastiaansen[2] and N.K. Jacob[2,3]

[1] National Institute of Water and Atmospheric Research (NIWA), Northland Aquaculture Centre, 101 Sime Road, Ruakākā 0116, New Zealand
[2] Animal Breeding and Genomics Group, Wageningen University, P.O. Box 3386700 AH Wageningen, The Netherlands
[3] Department of Animal Breeding and Domestic Animal Genetics, University of Goettingen, Albrecht-Thaer-Weg 3, 37075 Goettingen, Germany

SUMMARY

New Zealand's National Institute of Water and Atmospheric Research (NIWA) supports a genomics-based genetic improvement programme for *Seriola lalandi* (Yellowtail Kingfish or Haku in te reo Māori). Supplying genetically elite larvae, fry, and broodstock to diverse production systems, requires an understanding of potential genotype-by-environment (GxE) interactions and influence the configuration and design of the programme. We raised genetically related groups of progeny in tanks configured for either flow-through (FT) or recirculation (RAS) and estimated the between-environment genetic correlations for harvest length (HL) and harvest weight (HW). The genetic correlation estimates were modest (0.46 – 0.57) suggesting that GxE is quite strong and should be taken into consideration in the programme design.

INTRODUCTION

New Zealand's National Institute of Water and Atmospheric Research (NIWA) has recently initiated a genomics-based genetic improvement programme to improve production traits in *Seriola lalandi* (Yellowtail Kingfish or Haku in te reo Māori). The programme aims to supply genetically elite larvae, fry, and broodstock to the rapidly growing, global kingfish industry, which currently uses several growout systems. Consequently, the configuration and design of the programme must consider the potential for genotype-by-environment (GxE) interactions. If different genotypes are favoured in different systems, we could either select for generalist genotypes, potentially at the expense of performance in specific systems, or for specialist genotypes for specific systems.

GxE interactions are quantified using multi-trait genetic analyses with the same phenotype in different environments/systems treated as different traits (Falconer and Mackay 1996). High favourable between-environment genetic correlations indicate similar genetic architectures and responses to selection. High unfavourable genetic correlations imply that "winners" in one environment are "losers" in others because many genetic variants have opposite effects in different environments. Near zero genetic correlations suggest that different genes control variation in the same trait in the different environments.

In this paper, we report the between-environment genetic correlations between harvest weight (HW) and harvest length (HL) in a research-scale high density recirculating aquaculture system (RAS) and, because there is no sea cage production in New Zealand and biosecurity considerations prohibited exporting live juveniles, an identical system configured for single-pass flow-through (FT) and lower density as an admittedly rough proxy for alternative systems such as sea cages.

MATERIALS AND METHODS

Because it is impractical to strip spawn Haku for designed mating plans, we used group spawning in six large spawning tanks to produce our experimental fish. Between 15 October and 19 November 2020, we collected fertilised eggs from 13 spawning events and pooled them into five larger temporal groups for larval rearing in RAS. We reared the later-spawning groups at a slightly higher temperature for a shorter time to equalise sizes at the tagging stage (~165g after an average of 187 and 160 days of growth), after which we combined and distributed them randomly but unevenly into the RAS and FT systems to mimic commercial densities. The FT system received mechanically filtered and UV sterilized seawater from Bream Bay NZ. The RAS used mechanical drum filtration, moving bed biofiltration, a protein skimmer and UV treatment. RAS tanks had a 24h light photoperiod and FT an ambient photoperiod. We automatically fed the RAS tanks 10 times over 24 hours and FT tanks hourly during daylight hours. We split the RAS fish into two tanks approximately halfway through growout to maintain commercially relevant densities, but not the FT fish. We fed both systems a ration based on NIWA's size and temperature-dependent growth model (unpublished). Figure 1 shows the between systems and temporal variation in temperature and dissolved oxygen.

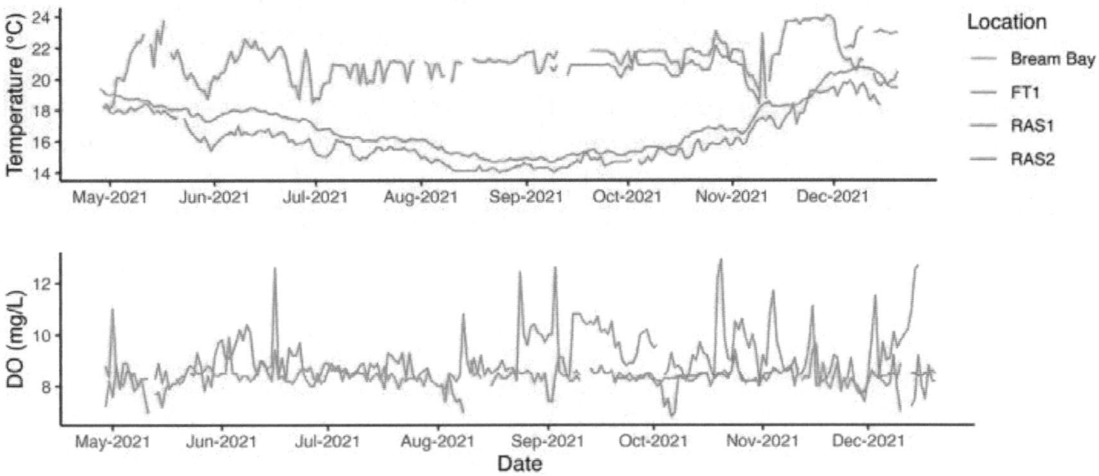

Figure 1. Temperature and DO (dissolved oxygen) variation in Bream Bay and the experimental systems

We measured harvest weight (HW) and harvest length (HL) on all experimental fish after 378 days (RAS, n=1820, mean HW ~1.9kg) and 412 days (FT, n = 306, mean HW ~1.4kg), and collected a 3mm fin clip stored in 100% ethanol. We sent these and parental fin clip samples to AgResearch (Lincoln, New Zealand) for genotyping using genotyping-by-sequencing (GBS, Elshire *et al.* 2011), parentage assignment, and genomic relationship estimation using R scripts developed by AgResearch (Dodds *et al.* 2015, Dodds *et al.* 2019). Based on the parentage assignments, individuals were also assigned to their spawning/hatching events (HatchGroup). Downstream data processing and analyses used R software (R Core Team 2024).

Statistical analysis. We fitted a four-trait animal model treating the same trait measured in different environments treated as separate traits using ASReml-R (Butler 2023) in the following model:

$$y_{i,j,k,l} = u_i + \text{HatchGroup}_j + \text{Tanks(Env)}_{i,k} + r_l + e_{i,j,k,l}$$

where:$y_{i,j,k,l}$ is a vector of phenotypic values of the four traits (HW-RAS, HL-RAS, HW-FT, and HL-FT); u_i is a vector containing the mean for the four traits (i=1-4); HatchGroup is a fixed effect to account for differences between the five batches of experimental fish produced on different dates (j = 1-5); $\text{Tanks(Env)}_{i,k}$ is the fixed effect for the combination of tanks a fish passed through, nested within each environment (k = 1-2 for FT, 3-6 for RAS); r_l is random additive genetic effect of the l-th individual; $e_{i,j,k,l}$ is a vector of random residuals for each individual l. The between-environment residual covariances are assumed to be zero because individual fish only have data from one of the two environments.

The heritability for each trait was estimated as the additive genetic variance ($\sigma^2{}_a$) divided by the phenotypic variance ($\sigma^2{}_p$), $h^2 = \frac{\sigma^2{}_a}{\sigma^2{}_p}$ and the between-environment genetic correlations as the genetic correlation $r_{a,RjFj}$ of the same trait j in different systems as $r_{a,RjFj} = \frac{cov(\sigma^2{}_{a,Rj},\sigma^2{}_{a,Fj})}{\sigma_{a,Rj}\sigma_{a,Fj}}$.

RESULTS AND DISCUSSION

Table 1 presents the estimated (co)variance components. Table 2 presents the heritabilities, genetic, and phenotypic correlations between the four traits. These estimates indicate that all four traits are sufficiently genetically variable and heritable for improvement. As well, the between-environment genetic correlations that represent the magnitude of GxE are all moderate (0.46 to 0.61), indicating that there is substantial re-ranking of the parents between the two environments.

Table 1. Additive genetic variance (σ_a^2), residual variance (σ_E^2), and phenotypic variance (σ_P^2 for both traits in both environments estimated with the G from the full GBS dataset. Standard errors are shown in parenthesis, HL = Harvest Length, HW = Harvest Weight, RAS = Recirculating Aquaculture System, FT = Flow Through System

TRAIT	ENVIRONMENT	σ_a^2	σ_E^2	σ_P^2
HL	RAS	3.96 (0.40)	3.63 (0.20)	7.59 (0.33)
HW	RAS	0.066 (0.006)	0.047 (0.003)	0.11 (0.005)
HL	FT	2.54 (0.62)	1.60 (0.39)	4.14 (0.39)
HW	FT	0.034 (0.007)	0.015 (0.004)	0.05 (0.005

As the two production systems were smaller-scale analogues of commercial farms, more research is required to rigorously extend these analyses to real-world production systems and to identify the specific environmental factors that drive them using more complex models (Sae-Lim *et al.* 2014). However, the differences between the two environments in the current study were most likely smaller than those between commercial RAS and alternative production systems such as sea cages. As a "rule of thumb" breeders consider between-environment genetic correlations < 0.8 as potentially problematic as they indicate substantial re-ranking of selection candidates (Robertson 1959) and therefore different selection decisions.

Table 2. Genetic correlations (se), below shaded diagonal; phenotypic correlations (se), above shaded diagonal; and heritabilities (se, shaded diagonal for both traits in both environments. n.e. = non- estimable phenotypic correlations due to fish being in different environments. HL = Harvest Length, HW = Harvest Weight, RAS = Recirculating Aquaculture System, FT = flow through system

	HL-RAS	HL-FT	HW-RAS	HW-FT
HL-RAS	0.52 (0.03)	n.e.	0.90 (0.01)	n.e.
HL-FT	0.57 (0.10)	0.61 (0.11)	n.e.	0.88 (0.01)
HW-RAS	0.92 (0.01)	0.46 (0.10)	0.59 (0.03)	n.e.
HW-FT	0.54 (0.10)	0.88 (0.04)	0.54 (0.10)	0.70 (0.10)

More quantitatively these estimated genetic correlations imply that if selection was undertaken exclusively in one of these production systems, approximately half to 60% of the genetic gain generated would transfer to the other system. Consequently, these patterns may have important implications for the design of the genetic improvement programme.

The most obvious option to address this complication is to establish largely separate breeding programmes to maximise the rate of genetic gain in specific production systems. However, that approach dramatically increases the costs of the programme(s) and/or constrain the utility of the genetic gains produced. An alternative is to select for stable performance across production systems (i.e. generalists) that perform well in a range of conditions. This "norm of reaction" approach (Sae-Lim *et al.* 2014) would likely result in slower genetic gain in all systems but is simpler and less costly than setting up separate breeding programmes for each production system and would produce "across-the-board" benefits for all producers. A third option is to estimate breeding values in multiple environments in a single programme and offer producers custom spawns between parents with high breeding values in their systems. This would likely require maintaining a larger population of candidates to provide a wider range of options.

CONCLUSION

GxE interactions between research-scale RAS and alternative production systems represented by FT are strong enough to complicate the design and implementation of genetic improvement programmes producing genetic gain to a range of Haku production systems. More research is required to extend these conclusions to real world systems and identify the causal environmental parameters driving these interactions.

ACKNOWLEDGEMENTS

We want to thank the technical staff of NIWA's Northland Aquaculture Centre at Bream Bay for their tireless and diligent work to conduct the grow out trials, particularly during the challenging period of the Covid 19 pandemic, which made their lives much harder.

REFERENCES

Butler D.G., Cullis B.R., Gilmour A.R., Gogel B.G. and Thompson R. (2023) ASReml-R Reference Manual Version 4.2. VSN International Ltd., Hemel Hempstead, HP2 4TP, UK.

Dodds K.G., McEwan J.C., Brauning R., R.M. Anderson R.M., van Stijn T.C., Kristjánsson T. and Clarke S.M. (2015) *BMC Genomics* **16**: 1047.

Dodds K.G., McEwan J.C., Brauning R., van Stijn T.C., Rowe S.J., McEwan K.M. and Clarke S.M. (2019) *G3-Genes Genom. Genet.* **9**: 3239.

Elshire R.J., Glaubitz J.C., Sun Q., Poland J.A. and Kawamoto K. (2011) *PLoS ONE* **6**: 19379.

Falconer D.S. and Mackay T.F.C. (1996) An Introduction to Quantitative Genetics. Fourth edition. Addison Wesley Longman Limited, Essex, England.

R Core Team (2024) R: A Language and Environment for Statistical Computing. R Foundation for Statistical Computing, Vienna, Austria.

Robertson A. (1959) *Biometrics* **15**: 469.

Sae-Lim P., Komen H., Kause A. and Mulder H.A. (2014) *Genet. Sel. Evol.* **46**: 16.

Proc. Assoc. Advmt. Anim. Breed. Genet. 26: 122-125

MARKER-BASED FAMILY ASSIGNMENT WITHOUT PARENT GENOTYPES

K.G. Dodds[1], J.E. Symonds[2], S.J. Rowe[1] and J.C. McEwan[1]

[1] AgResearch, Invermay Agricultural Centre, Mosgiel, New Zealand
[2] Cawthron Institute, Nelson, New Zealand

SUMMARY

We investigate how to assign individuals to families (and therefore, implicitly, to parents) when parent genotypes are not available. In the first case, a set of progeny were assigned to putative full-sib families using an algorithm based on clustering and expected relatedness values. A plausible set of full-sib families was found, but the method is subject to some arbitrary thresholds. In the second case, individuals were putatively assigned to their (paternal) grandparents based on relatedness or the frequency of opposing homozygotes in the pair and these were compared to the true pedigree. The assignments were good when they were restricted to the set of paternal grandparent mate pairs and/or when dam genotypes were available to infer sire alleles, which could then be matched to the paternal grandparents using methods designed for low-depth sequencing-based genotypes.

INTRODUCTION

Sometimes breeders or researchers want to infer family relationships using genetic marker information, but do not have genotypes or DNA samples from the putative parents. Two such use cases are to 1) assign a set of progeny to full- or half-sib families, or 2) assign individuals to their grandparents. In the absence of parents, exclusion methods, such as relying on opposing homozygotes, are not available to unequivocally exclude some potential relationships. Modern technologies like SNP chips or sequencing-based assays allow inference based on relatedness estimates (Moore *et al.* 2019) and/or on the distribution of genotype combinations in pairs of individuals (VanRaden *et al.* 2013). These approaches are applied to two real data sets to assign individuals to families using genomic data.

MATERIALS AND METHODS

Example 1: full-sib families. For this research, a set of 5,216 Chinook salmon were obtained from a breeding company for undertaking a research project (Symonds *et al.* 2025). The company advised that this population comprised 133 full-sib families, however, the family membership was unknown, and the parental DNA was not available. The progeny were genotyped using genotyping-by-sequencing (GBS) and QC filters applied as described in Scholtens *et al.* (2023). A genomic relationship matrix (GRM), **G**, was calculated following the method of Dodds *et al.* (2015) and then rescaled by dividing each row and each column by the square root of the diagonal element. This was to allow focus on the relatedness derived from parents alone and not additional common ancestry (creating inbreeding).

Fish were assigned to full-sib families as follows:

1) k-means clustering was applied using a distance matrix calculated as the Euclidean distance between pairs of rows in **G**, with 120 clusters and using the *kmeans* function in base R software with 50 random starting sets (*nstart*=50 in the *kmeans* function) but default settings otherwise.

2) split families further cutting hierarchical cluster dendrogram (on sqrt(**G**)-min(**G**)) of each family at 1.6

3) split families with mean relatedness < 0.2 into singleton families

4) join families related at 0.4

5) split families with evidence of two groups, based on the top join of a hierarchical cluster and where the resulting two groups have mean relatedness < 0.35, or a mean squares ratio (original group

size -2) times squared difference between groups divided by the sum of squared differences within groups > 4.

Example 2: finding grandparents. A set of 263 sheep (the 'progeny') from an experimental flock (Rowe *et al.* 2019), born in 2023, along with their parents and grandparents were genotyped using various SNP chips containing at least 16,000 SNPs. The 12,682 autosomal SNPs that were common across all these SNP chips, had less than 30% missingness and were not monomorphic were used to construct a GRM using the imputation-free method described by Dodds *et al.* (2015). The 2023-born cohort and their parents previously had their parents assigned based on their SNP chip genotypes. Only those progeny whose parents were not also in the set of grandparents were included for this study. These progeny had 10 sires with unique parentage and 145 dams, which had 24 sires and 118 dams. In total, there were 31 grandsires and 126 granddams.

Paternal grandparents (PGPs) were assigned using different methods and levels of associated information ('scenarios', Table 1). In all cases all true PGPs were present, but scenarios without an 'a' in the label (Table 1) also included the maternal grandparents. Grandparents were putatively assigned to the progeny either as the highest related ('relatedness method') or the progeny-grandparent pairings with the lowest opposing homozygote rate ('EMM method'). EMM refers to the excess mismatch rate compared to the expected rate for the given level of sequencing depth for methods such as GBS; the expected mismatch rate with actual genotypes (e.g. chip-based) is zero. For the '1' scenarios (Table 1), only the grandparents were available. For scenarios labelled as '2' where the dams were present, the additional criteria of grandparent-dam relatedness or grandparent-dam-progeny mismatch rates were examined to help determine the PGPs. If the best match failed to meet the criteria, the 2nd best match was then considered.

Inferred sire alleles were used in the '3' scenarios (Table 1). These alleles are inferred from progeny and dam genotypes. For example, if the progeny is AB and the dam is AA the sire must have a B allele. The inferred alleles were then treated in the same way as a GBS genotype with depth of 1 or 0 corresponding to when an allele could be inferred or not, respectively, allowing a search for the sire's parents. Assignments were made using the parentage functionality within the KGD software (Dodds *et al.* 2019).

RESULTS AND DISCUSSION

Example 1: full-sib families. There were 148 families found (compared to 133 recorded families), ranging in size from 1 (11 families) to 51. However, 14 of the assigned families had five or fewer members while the remaining 134 families had at least 17 members. It is possible that a few fish from outside these families may have escaped into this group and that two of the assigned families are actually a single family.

The large family sizes in this study have likely assisted the clustering algorithm to define these groups effectively. The likely presence of relatedness between families (such as different families being half-sibs) has made the process of assigning families more difficult as the closest related half-sibs may have similar relatedness to the least related full-sibs, especially if there is some relatedness between the non-common parent for the half-sibs.

The method is not guaranteed to produce the same groupings if rerun (as the k-means method uses randomised starting sets). However, in this case, a repeat run of the method did produce the same results. Several arbitrary criteria have been used in this process; the initial number of families to use depends on the expected number of families, but it is not clear if the other thresholds will need to be adjusted for other situations. For the current example it appeared more effective to initially assign fewer than the expected number of families and then subdivide them if needed.

The proposed method could also be adapted (using different thresholds) for finding half-sib families. This is relevant in livestock populations when only the progeny generation is genotyped, and dams have only one or a few progeny each. Assigning with a target relatedness of 0.25 (for non-inbred half-sibs) may present more challenges than the full-sib case, but it should still be possible for reasonably sized families.

Example 2: finding grandparents. A summary of the PGP assignment is shown in Table 1. In this study the best matches were always 'assigned', which is sensible if the true relatives are present in the data (as known in this case). In practice, assignments depend on the measures passing a certain threshold (e.g. for relatedness and/or excess mismatch rate). The KGD software reports only the top two matches; however, it might be possible to find suitable matches (Scenarios 2, 2a) in lower ranked matches when additional criteria (to exclude the maternal relatives) are used.

Table 1. Paternal grandparent assignment rates

Scenario*	Match method	Additional criteria†	PGS % correct	PGD % correct	PGS & PGD % correct	% unassigned
1	relatedness		80	44	30	0
	EMM		80	44	30	0
1m	relatedness		94	94	93	0
	EMM		94	93	93	0
1am	relatedness		100	100	100	0
	EMM		100	100	100	0
2	relatedness		41	44	11	0
		DamRel	83	85	59	26
		TrioMM	87	88	69	32
	EMM		41	54	17	0
		DamRel	73	90	58	22
		TrioMM	88	93	76	30
2a	relatedness		76	65	49	0
		DamRel	97	83	79	11
		TrioMM	100	89	89	14
	EMM		78	75	57	0
		DamRel	94	90	82	8
		TrioMM	98	95	92	11
3	relatedness		98	97	95	0
	EMM		97	98	95	0
3a	relatedness		100	99	99	0
	EMM		99	99	98	0
3m	relatedness		100	100	100	0
	EMM		100	100	100	0
3am	relatedness		100	100	100	0
	EMM		100	100	100	0

* Scenario labels with 'a' include only PGP in the test set, otherwise all GPs; scenario labels with 'm' use known grandparent mating pair information. In scenarios 1, 1m, assignments were considered correct if the PGP was among the best two matches (allows the maternal grandparent to be the best match). In scenarios 2(a), the dam genotypes were available, so the 2nd best female match is assigned as PGD (best match is the dam). In scenarios 3(a/m) the matching is to the inferred sire alleles.

† DamRel criterion is that an assigned paternal grandparent must have a relatedness less than 0.4 with the dam. TrioMM criterion is that the progeny, dam and proposed grandparent should have a parent-progeny mismatch rate less than 0.1.

PGS is paternal grandsire and PGD is paternal grandam.

Although haplotypes were inferred for the sires, no SNP positional information was subsequently used, i.e., the fact that the sire alleles were together in the same gamete was disregarded.

Additionally, there was no attempt to collate sire haplotypes from different progeny, as at that stage it was not known which progeny have the same sire. It might be possible to find paternal half-sibships through clustering methods and then collate the sire haplotypes. The corresponding 'read depths' would be the number of offspring with inferred sire alleles for that SNP. The use of haplotypes (VanRaden *et al.* 2013) and/or the number of shared identity-by-descent segments (Jewett *et al.* 2021) could provide a more powerful assignment tool, as might be required in more extensive searches.

The approaches to making assignments (relatedness or EMM) performed similarly with neither being uniformly better. In practice combining these approaches is likely lead to better results. Moreover, it is also possible that half-sib clustering could aid the assignments, however this approach was not investigated.

CONCLUSIONS

Using genetic data to assign family relationships is simpler for parent-progeny than for sibships or more distant relationships. The assignment process can be helped with the use of ancillary information such as known mating pairs or expected number of family groups. The proposed method for assigning full-sib families placed most of the fish into almost the expected number of groups, but required a set of arbitrary thresholds. Finding paternal grandparents is difficult when dams are also related as in the sheep dataset. If the dams are genotyped (and assigned), a useful approach is to infer the sire haplotype and then apply a parentage assignment to that haplotype, using methods designed for low-depth sequence-based genotyping. The proposed methods are a useful addition to the tools available for parentage testing.

ACKNOWLEDGEMENTS

We thank the GenomNZ (AgResearch) laboratory for genotyping the animals in both examples. Sheep were funded by the New Zealand Agricultural Greenhouse Gas Centre (NZAGRC) and the Pastoral Greenhouse gas research consortium (PGgRc). The salmon research was supported by the NZ Government Ministry for Business Innovation and Employment (contract no. CAWX2304).

REFERENCES

Dodds K.G., McEwan J.C., Brauning R., Anderson R.A., Van Stijn T.C., Kristjánsson T. and Clarke S.M. (2015) *BMC Genomics* **16**: 1047.

Dodds K.G., McEwan J.C., Brauning R., Van Stijn T.C., Rowe S.J., McEwan K.M. and Clarke S.M. (2019) *G3: Genes Genom. Genet.* **9**: 3239.

Jewett E.M., McManus K.F., Freyman W.A. and Auton A. (2021) *Am. J. Hum. Genet.* **108**:.2052.

Moore K.L., Vilela C., Kaseja K., Mrode R. and Coffey M. (2019) *J. Anim Sci.* **97**: 35.

Rowe S.J., Hickey S.M., Jonker A., Hess M.K., Janssen P.H., Johnson T., Bryson B., Knowler K., Pinares-Patiño C., Bain W., Elmes S., Young E., Wing J., Waller E., Pickering N. and McEwan J.C. (2019) *Proc. Assoc. Advmt Anim. Breed.Genet.***23**: 306.

Scholtens M., Dodds K., Walker S., Clarke S., Tate M., Slattery T., Preece M., Arratia L. and Symonds J. (2023) *Aquaculture* **563**: 738936.

Symonds J.E., Scholtens M.R., Dodds K.G., Costilla R., Clarke S.M., Arinez J., Kenny N., Walker S.P. and Waddington Z. (2025) *Proc. Assoc. Advmt Anim. Breed. Genet.* **26**: *These proceedings.*

VanRaden P.M., Cooper T.A., Wiggans G.R., O'Connell J.R. and Bacheller L.R. (2013) *J. Dairy Sci.* **96**:1874.

APPLYING MACHINE LEARNING TO UNLOCK GENOMIC INSIGHTS IN SNAPPER BREEDING

J. Blommaert[1] and M. Wellenreuther[1,2]

[1] The Institute of Plant and Food Research, Nelson, New Zealand
[2] School of Biological Sciences, University of Auckland, Auckland, New Zealand

SUMMARY

The Australasian snapper (*Chrysophrys auratus*) is a promising candidate for aquaculture. In the snapper breeding programme at Plant and Food Research, identifying genetic variants associated with key economic traits is essential for optimising selective breeding. Using machine learning for phenotyping and to analyse genomic data, our work focuses on detecting single nucleotide polymorphisms (SNPs) that influence growth traits within the breeding programme. Our approach and resulting knowledge has the potential to accelerate breeding programmes and provide more efficient ways to improve broodstock quality and sustainability in aquaculture.

INTRODUCTION

Traditionally, genomic-informed breeding programmes have focused on using approaches such as GWAS to identify single nucleotide polymorphisms (SNPs) as markers for trait-associated genetic variation (Dekkers 2012). While these approaches have been invaluable for advancing selective breeding, relying on these approaches can miss the more complex genetic architecture of traits such as growth (Chafai *et al.* 2023). Machine learning approaches accounting for interactions between variants have shown promise in other breeding programmes, despite the problem of high-dimensionality of genomic data (Feldner-Busztin *et al.* 2023). This kind of approach can facilitate the development of more robust and resilient aquaculture populations (Gill *et al.* 2022).

To diversify and build resilience in the aquaculture sector in New Zealand, the Australasian snapper (*Chrysophrys auratus*) has been the focus of a two decade long selective breeding program (Samuels *et al.* 2024). This breeding program has demonstrated substantial advancements in growth rates, survival, and feed conversion ratios, marking significant progress toward optimising snapper for aquaculture (Moran *et al.* 2023; Samuels *et al.* 2024).

Here, we apply computer vision and machine learning techniques to unlock genomic insights in snapper breeding. By cataloguing and integrating SNP variant data with extensive phenotypic data from 1,011 snapper in the F_4 cohort of a long-term selective breeding programme, we conducted genome-wide association studies (GWAS) to identify variants associated with growth traits. We also evaluated whether we could also identify genetic variants for genomic prediction, employing XGBoost as a predictive model. Our findings provide a valuable framework for enhancing genomic selection in aquaculture and offer insights into the potential applications of machine learning in selective breeding programmes, with implications for improving resilience and sustainability in aquaculture and beyond.

MATERIALS AND METHODS

Genotype and phenotype data. At 3 months post-hatch, fish were manually measured for weight and fork length and imaged. The images were used to measure a further 13 phenotypes via an in-house computer vision pipeline (Figure 1a). Genetic data was generated using a SNP chip designed for use in snapper (Montanari et al. 2023). Overall, 1,011 F_4 fish were included. After quality filtering, 11,006 SNPs were retained for analyses. For both the genotype and phenotype datasets, PCAs were performed using prcomp in R. PC1 of the phenotype data was also included as a trait in the GWAS and genomic prediction. To infer family relationships, kmeans clustering was performed on a genetic relatedness matrix for all fish.

GWAS and genomic prediction. FarmCPU was used for GWAS within rMVP. For genomic prediction, we used XGBoost trained on 80% of the fish, using the relatedness cluster assignments in the stratification step. We included the first two genetic PCs as covariates in the models. ML analyses were applied to three phenotypes- condition factor, weight, and the distance from the top lip to the eye. We included condition factor since it is a derived trait which considers weight and length. We included weight because it is a common target for breeding programmes, and distance between eye and top lip since this yielded the highest number of significant SNPs from the GWAS. Each model was tuned using tidymodels and Root Mean Squared Error (rmse) and R^2 were recorded for both training and test data sets.

RESULTS AND DISCUSSION

Across all fish included, weight ranged from 7.46g to 45.19g, while fork length ranged from 74.47mm to 123.03mm. Across the height measurements, the biggest variation was at the 75% of the length of the fish point, with 24.25mm difference between the minimum and maximum. Overall, traits were highly correlated with each other (Figure 1b), correlation coefficients ranged from 0.27 (eye width vs. distance between the caudal peduncle and pectoral joint) to 1 (fork length vs distances between each lip and the tail fork, and distance between top lip and tail fork vs. distance between bottom lip and tail fork). For the phenotypes, 96.96% of the variance was explained by PC1 of the phenotypes. For this reason, PC1 was used also included as predictive trait for the GWAS and XGBoost model.

Overall, 24 SNPs were identified by the GWAS (e.g. Figure 2) as being involved in the 15 phenotypes and phenotype PC1. Of these, 8 were shared between at least two traits, and 16 were unique to the trait they were identified in. Significant SNPs were across 14 chromosomes in the snapper genome and found to overlap with genes potentially involved in growth via metabolic pathways and even appetite signalling.

The machine learning models were found to be moderately accurate on the training data, but less so on the testing data (Table 1), suggesting that the high dimensionality of the SNP data is leading to overfitting. This was not addressed by reducing the number of input SNPs by random selection or by ranking SNPs by GWAS p-value (data not shown). Together, this suggests that the dataset is prone to overfitting and would benefit from the addition of additional fish samples.

Table 1. Residual mean squared error and R^2 statistics for XGBoost models on the test and train datasets for three selected traits

Trait	rmse train	R^2 train	rmse test	R^2 test
Condition factor	0.1	0.671	0.132	0.241
Weight	3.94	0.71	5.97	0.146
Top lip to eye (mm)	0.221	0.975	0.965	0.0675

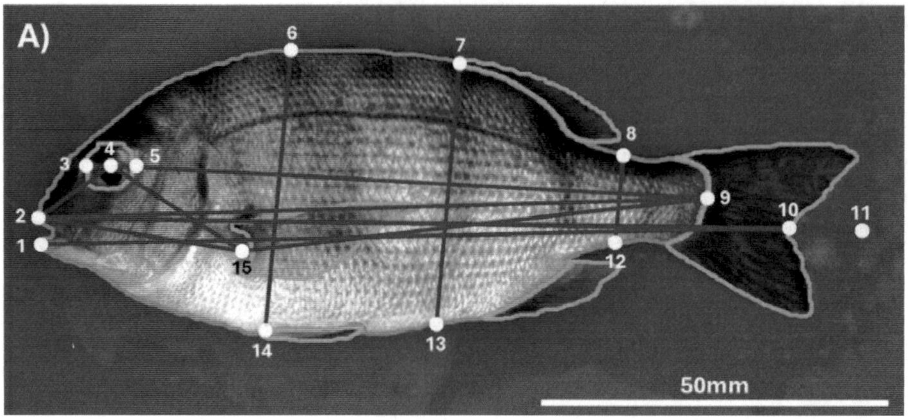

b)

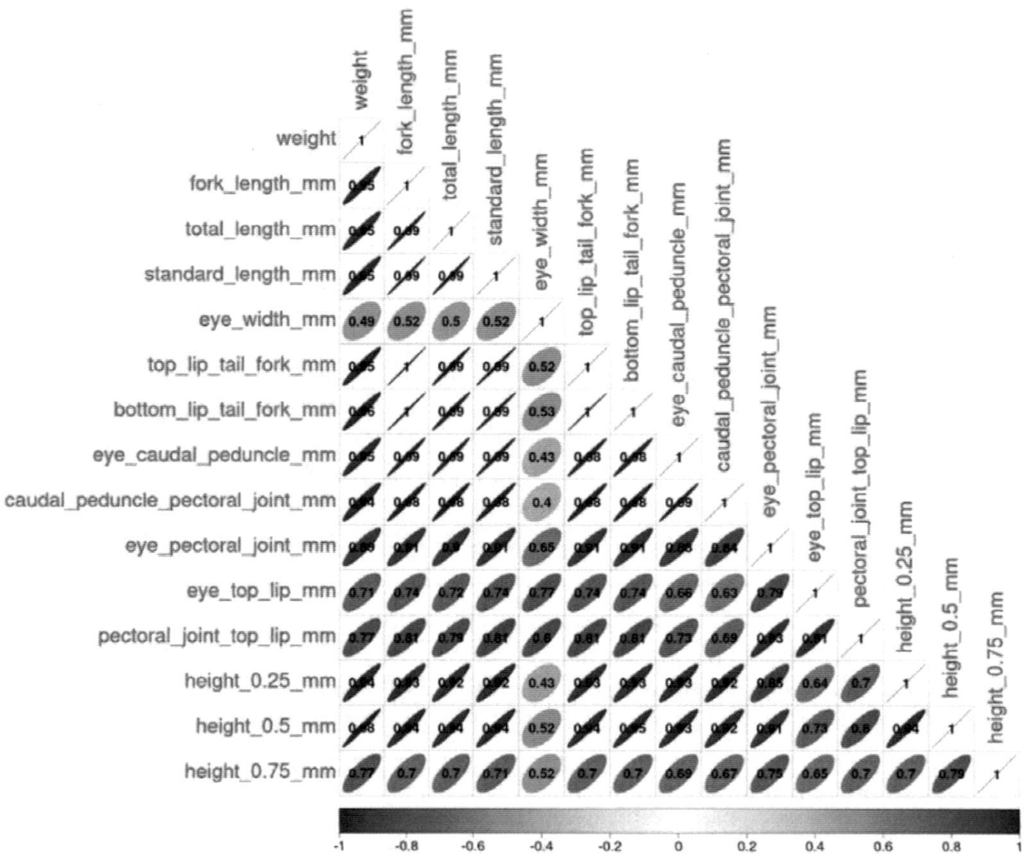

Figure 1. A) An output of the computer vision phenotyping pipeline showing the contours of fish body parts (orange) and landmarks (white points) used for each of the 13 measurements gathered by this pipeline (purple). The labelled points indicate landmarks used for measurements from the computer vision pipeline. Landmarks are 1- bottom lip, 2- top lip, 3 and 5- the left and right edges of the eye respectively, 4- centre of the eye, 6 and 14- top and bottom of the fish at 25% of its total length respectively, 7 and 13- top and bottom of the fish at 50% of its total length respectively, 8 and 12- top and bottom of the fish at 75% of its total length respectively, 9- peduncle, 10- tail fork, 11- total length end point. B) Correlation matrix of all 15 measured traits in this study. The colour and shape of each ellipse represents the R^2 value for that correlation (written in the corresponding box for each correlation). R^2 values were only included where p-values were < 0.05.

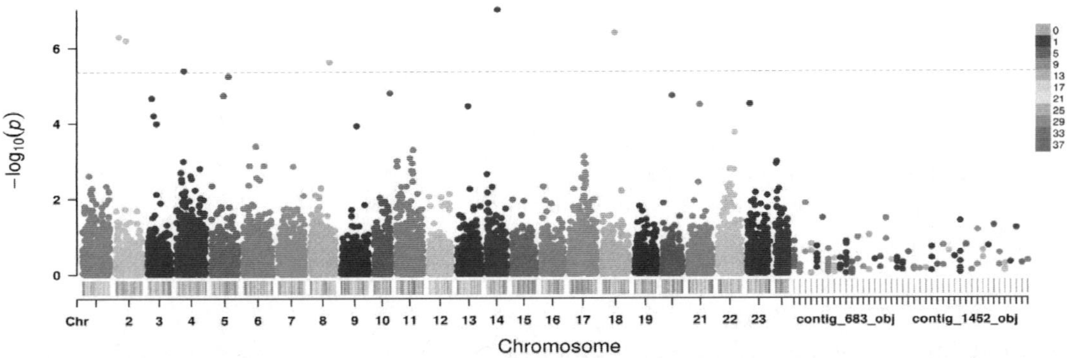

Figure 2. Manhattan plot showing SNPs significantly associated with weight in snapper. The heatmap along the bottom of the figure shows SNP density across the genome.

CONCLUSION

Including traditional genomic selection methods in our breeding program has already seen large gains in growth phenotypes, and while high dimensionality still hampers the machine learning approach for genomic prediction, applying more sophisticated statistical methods would allow us to consider the genetic, and eventually environmental interactions in complex traits. Future work will investigate similar analyses on other traits, such as internal organ measurements and look to increase the size of the data set used for the machine learning approach.

ACKNOWLEDGEMENTS

We thank the staff at Plant & Food Research for their contributions to breeding and rearing the snapper populations in this programme. This project was supported by the MBIE Endeavour Fund (C11X1603) under the 'Accelerated Breeding for Enhanced Seafood Production' programme and a Technology Development Fund, both awarded to MW. We also extend our thanks to Philipp Bayer from the Minderoo Foundation.

REFERENCES

Chafai N., Hayah I., Houaga I. and Badaoui B. (2023) *Front. Genet.* **14**: 1150596.

Dekkers J.C.M. (2012) *Curr. Genom.* **13**: 207.

Feldner-Busztin D., Nisantzis P.F., Edmunds S.J., Boza G., Racimo F., Gopalakrishnan S., Limborg M.T., Lahti L. and de Polavieja G.G. (2023). *Bioinformatics* **39**: btad021.

Gill M., Anderson R., Hu H., Bennamoun M., Petereit J., Valliyodan B., Nguyen H.T., Batley J., Bayer P.E. and Edwards D. (2022) *BMC Plant Biol.* **22**: 180.

Montanari S., Deng C., Koot E., *et al.* (2023). *G3-Genes Genom. Genet.* **13**: jkad170.

Moran, D., Schleyken J., Flammensbeck C., Fantham W., Ashton D. and Wellenreuther M. (2023) *Aquaculture* **563**: 738970.

Samuels G., Hegarty L., Fantham W., Ashton D., Blommaert J., Wylie M.J., Moran D. and Wellenreuther M. (2024) *Aquaculture* **586**: 740782.

Proc. Assoc. Advmt. Anim. Breed. Genet. 26: 130-133

FAST-TRACKING FARMED KING SALMON CLIMATE CHANGE ADAPTATION

J.E. Symonds[1], M.R. Scholtens[1], K.G Dodds[2], R. Costilla[1], S.M. Clarke[2], J. Arinez[3], N.J. Kenny[3], S.P. Walker[1] and Z. Waddington[4]

[1] Cawthron Institute, Nelson, New Zealand
[2] AgResearch, Invermay Agricultural Centre, Mosgiel 9053, New Zealand
[3] Biochemistry Department, University of Otago, Dunedin 9016, New Zealand
[4] The New Zealand King Salmon Co. Ltd, Picton 7220, New Zealand

SUMMARY

The thermal tolerance of pedigree Chinook salmon (*Oncorhynchus tshawytscha)* was tested in two tank-based challenges: (a) High temperature (HT) at 23.5 °C and (b) Temperature and Oxygen (TO) at 21.0 °C with low dissolved oxygen (DO), using commercial breeding programme fish. Fish were measured at the start and end of both challenges and survival was monitored. Survival and days to death in both the HT and TO treatments were moderately heritable. Low genetic correlations between these traits at the different challenge conditions indicate they are separate traits. The results demonstrate the significant potential to breed for improved thermal tolerance in all-female farmed Chinook salmon.

INTRODUCTION

Climate change is altering Aotearoa New Zealand's marine environment at an unprecedented rate with major changes severely impacting farming and natural systems. Farmed Chinook salmon are negatively impacted when temperatures exceed 17.0-18.0 °C for prolonged periods and marine heatwaves have resulted in reduced survival. An important climate resilience trait is thermal tolerance, and we explored the potential to improve this trait in pedigree Chinook salmon by investigating survival, time to death and growth at two temperatures: (a) 23.5 °C (HT) and (b) 21.0 °C with lower DO (TO). We report genetic parameters for these experimental challenges.

MATERIALS AND METHODS

A total of 2,609 individually tagged all-female pedigree Chinook salmon from the 2023-year class were obtained from a commercial salmon breeding programme. The tagged salmon were transferred to Cawthron's Finfish Research Centre (FRC) in June 2024, 13 months post-fertilisation, and acclimated to seawater. A timeline of the trial is shown in Figure 1. Weight (g) (WT) prior to the challenge tests (WT1) and after exposure to HT (WT2) and TO (WT3) were measured. Specific growth rates (SGR) (percent change in weight per unit of time) from pre- to post-challenge under the HT (SGRHT) and TO treatments (SGRTO) were also calculated. Further details of the rearing systems, husbandry and phenotypes can be found in Elvy *et al.* (2022) and Scholtens *et al.* (2023).

For the challenges 2,609 fish were randomly distributed across 9 x 8000 L tanks, 289-290 fish per tank. Four tanks of fish were challenged at 23.5 °C (DO 6.0 to 8.0 mg/L) and 5 tanks were challenged at 21.0 °C with lower DO (3.5-4.5 mg/L). In all tanks, the temperature was increased by 0.5 °C per day from 15.0 °C until the target temperature was reached. Lower DO levels were achieved by reducing the amount of oxygen added to the tanks over 7 days. Mortalities and time of death were recorded regularly. The end-point for HT was 60 % mortality. For TO the end-point was after 73 to 77 days under the challenge conditions.

Statistical analysis. Thermal tolerance was evaluated based on survival and growth. Survival traits were defined in two different ways - binary and continuous. Binary test survival (BS) after exposure to either HT (BSHT) or TO (BSTO) was coded as 1 for survivors and 0 for dead. The continuous survival trait was measured as the number of days to death (DTD) at HT (DTDHT) and

TO (DTDTO). To accommodate censoring of survivors in the challenges, DTD traits were transformed after giving survivors a DTD beyond the maximum in the mortalities (Gianola and Norton 1981). All analyses were run using R 4.3.3. Descriptive statistics were obtained using the 'psych' package. Specific growth rate was calculated as $(\ln(Wt_y) - \ln(Wt_x))/\text{days}) \times 100$ where x and y represent pre- and post-challenge weights, respectively (Scholtens *et al.* 2023).

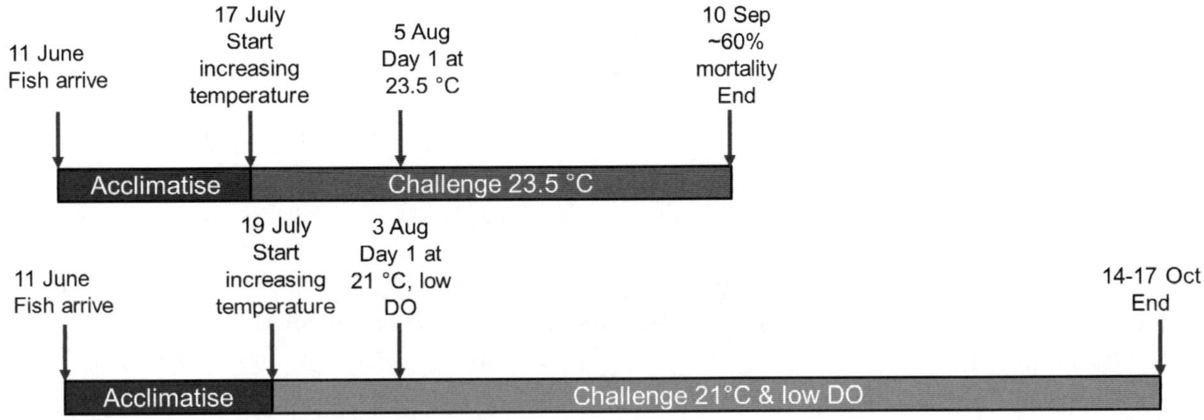

Figure 1. Timeline of the two challenge treatments

Genetic parameter estimation. Estimates of variance and covariance components for WT, SGR, DTD, and BS were obtained using the Restricted Maximum Likelihood (RML) procedure in ASReml version 3 (Gilmour *et al.* 2009). Continuous traits (WT, SGR and DTD) were analysed fitting a mixed linear animal model (LM); $y_{ij} = \mu + f_i + a_j + e_{ij}$, where y_{ij} is the phenotype observed; μ is the overall mean; f_i is the fixed effect of tank (tank history); a_j is the additive genetic effect for the jth individual and e_{ij} was random residual. The additive genetic effect was assumed to be $a_i \sim N(0, \mathbf{G}\sigma_a^2)$, and residuals assumed to be $e_{ij} \sim N(0, \mathbf{I}\,\sigma_e^2)$, where **G** is the genomic relationship matrix, calculated using the GBS data while taking read depth into account (the KGD method) (Dodds et al., 2015); and **I** is an identity matrix. The phenotypic variance was the sum of genetic (σ_a^2) and residual (σ_e^2) variances. For the binary traits (BSHT and BSTO), a generalized linear mixed model using a binomial distribution with a logit function was used. Narrow-sense heritability (h^2) was estimated for each trait as the proportion of genetic variance with respect to the total variance, except for BSHT and BSTO, where the residual variance was fixed at $\pi^2/3$.

Genetic and phenotypic correlations were estimated using bivariate animal models with the same components as the univariates, it was assumed that genetic effects were distributed as $\sim N\left(\begin{bmatrix}0\\0\end{bmatrix}, G\otimes\begin{bmatrix}\sigma_{a_i}^2 & \sigma_{a_{ij}}\\ \sigma_{a_{ji}} & \sigma_{a_j}^2\end{bmatrix}\right)$, where $\begin{bmatrix}\sigma_{a_i}^2 & \sigma_{a_{ij}}\\ \sigma_{a_{ji}} & \sigma_{a_j}^2\end{bmatrix}$ is the additive genetic variance-covariance structure and the residual effects were distributed as $\sim N\left(\begin{bmatrix}0\\0\end{bmatrix}, I\otimes\begin{bmatrix}\sigma_{e_i}^2 & \sigma_{e_{ij}}\\ \sigma_{e_{ji}} & \sigma_{e_j}^2\end{bmatrix}\right)$, where $\begin{bmatrix}\sigma_{e_i}^2 & \sigma_{e_{ij}}\\ \sigma_{e_{ji}} & \sigma_{e_j}^2\end{bmatrix}$ is the residual variance-covariance structure.

RESULTS AND DISCUSSION

Fish phenotypic results. Descriptive statistics for growth and survival traits at multiple timepoints are presented in Table 1. As the temperature increased above 19.0 °C the feed response in all tanks decreased and most fish stopped feeding at 23.5 °C. This is reflected in the negative mean SGR over the HT challenge period. The TO fish had a mean SGR of 0.43. Mean SGR under optimal trial conditions (17.0 °C) was 0.84 for fish of similar size in a different trial (Scholtens *et al.*, 2023).

Table 1. Descriptive statistics of growth and survival traits in Chinook salmon

Trait	N	Mean	SD	Min	Max
WT1	2,609	468.87	119.10	194.20	882.70
WT2	414	431.36	151.14	166.00	823.00
WT3	1,250	787.49	287.61	172.60	1760.00
SGRHT	414	-0.02	0.29	-0.83	0.67
SGRTO	1,273	0.43	0.30	-0.42	1.08
DTDHT	718	21.28	7.75	2.0	37.0
DTDTO	145	59.17	11.97	19.0	75.0
BSHT	1,134	0.37	0.48	0	1
BSTO	1,419	0.90	0.30	0	1

WT = weight (g); SGR = SGR between weight assessments pre- (WT1) and post-challenge (WT2 or WT3); DTD = days to death for mortalities only and BS = binary survival.

Challenge results. Mortalities occurred once the temperature reached 23.5 °C and HT was terminated at close to 60% cumulative mortality after 37 days. The TO cumulative mortality reached only 10.18 % after 73-77 days. The heritabilities (h^2) for the temperature tolerance traits BS and DTD at both temperatures were moderate (Table 2) but not as high as in some reports (Benfey *et al.* 2024; Perry *et al.* 2005).

Table 2. Estimates of variance components for growth and survival traits in Chinook salmon

Trait	$h^2 \pm$ SE	$\sigma_e^2 \pm$ SE	$\sigma_a^2 \pm$ SE
WT1	0.37±0.03	8270.21±295.90	4824.24±459.88
WT2	0.43±0.08	13619.97±1691.73	10408.32±2525.65
WT3	0.34±0.04	53988.63±2884.67	27557.13±4003.27
SGRHT	0.37±0.09	0.01±0.01	0.01±0.01
SGRTO	0.21±0.04	0.01±0.01	0.01±0.20
DTDHT	0.29±0.06	0.70±0.06	0.28±0.07
DTDTO	0.38±0.20	0.59±0.18	0.36±0.20
BSHT	0.23±0.04	3.29	0.96±0.20
BSTO	0.25±0.05	3.29	1.08±0.30

SE = standard error; h^2 = heritability; σ_e^2= error/residual variance; σ_a^2= additive genetic variance; WT = weight; SGR = SGR between weight assessments pre and post-challenge; DTD = days to death and BS = binary survival.

Weight and SGR heritabilities were moderate to high in both challenges. Pre-challenge weight (WT1) was positively correlated with survival in TO but negatively correlated with survival in HT (Table 3). The genetic correlations between the survival traits in HT and TO were all low and negative, indicating they are different traits. Although WT1 was negatively correlated with survival at 23.5 °C, the SGR during this challenge was positively correlated with survival (0.254±0.150). Similarly, the SGR during the 21.0 °C challenge was also positively correlated with DTDTO and BSTO (0.512±0.134 and 0.576±0.131) respectively, indicating that animals with good growth genetics also tended to survive longer. Further bivariate analysis will consider latent threshold models with the binary traits.

Table 3. Estimates of genetic (below diagonal) and phenotypic (above diagonal) correlations ± (standard errors) for growth and survival traits in Chinook salmon

	WT1	WT2	WT3	SGRHT	SGRTO	DTDHT	DTDTO	BSHT	BSTO
WT1		**0.916** (0.012)	0.775 (0.013)	0.505 (0.055)	0.314 (0.032)	-0.308 (0.037)	0.269 (0.030)	-0.274 (0.036)	0.288 (0.030)
WT2	**0.909** (0.027)			0.788 (0.041)		0.533 (0.069)		0.432 (0.079)	
WT3	**0.900** (0.022)	0.688 (0.120)			0.797 (0.014)		0.330 (0.043)		0.200 (0.047)
SGRHT	0.421 (0.107)	0.764 (0.077)	0.355 (0.143)			0.218 (0.089)		0.205 (0.089)	
SGRTO	0.490 (0.081)	0.600 (0.150)	**0.806** (0.043)	0.492 (0.170)			0.066 (0.046)		0.097 (0.046)
DTDHT	-0.165 (0.082)	0.490 (0.110)	-0.155 (0.105)	0.254 (0.150)	-0.171 (0.121)				
DTDTO	0.488 (0.092)	0.464 (0.163)	0.588 (0.098)	0.192 (0.187)	0.512 (0.134)	-0.168 (0.128)		0.778 (0.016)	**0.974** (0.002)
BSHT	-0.279 (0.089)	0.167 (0.152)	-0.250 (0.112)	0.062 (0.172)	-0.236 (0.129)	**0.957** (0.021)	-0.181 (0.142)		
BSTO	0.549 (0.090)	0.525 (0.164)	0.582 (0.107)	0.256 (0.189)	0.576 (0.131)	-0.217 (0.130)	**0.988** (0.006)	-0.239 (0.144)	

Key:
>0.8
0.60-7.9
0.40-5.9

CONCLUSION

The results indicate that selecting for improved thermal tolerance in Chinook salmon is possible but the design of the tank-based challenge has to be carefully considered as survival and SGR at 21.0 °C with low DO and at 23.5 °C are different traits.

ACKNOWLEDGEMENTS

This work was funded by the NZ Government Ministry for Business Innovation and Employment (contract CAWX2304). Thanks to the GenomNZ lab for genotyping the salmon. Fish manipulations were approved by the Nelson Marlborough Institute of Technology Animal Ethics Committee (AEC-2024-CAW-02). Thanks to the FRC and NZKS staff for their valuable assistance.

REFERENCES

Benfey T.J., Gonen S., Bartlett C.B. and Garber A.F. (2024) *Aquac. Rep.* **37**: 102249.

Dodds K.G., McEwan J.C., Brauning R., Anderson R.A., Van Stijn T.C., Kristjánsson T. and Clarke S.M. (2015) *BMC Genomics* **16**: 1047.

Elvy J.E., Symonds J.E., Hilton Z., Walker S.P., Tremblay L.A., Casanovas P. and Herbert N.A. (2022) *Aquaculture* **554**: 738184.

Gianola D. and Norton H.W. (1981) *Genet.* **99**: 357.

Gilmour A.R., Gogel B., Cullis B., Thompson R. and Butler D. (2009) ASReml User Guide. Release 3.0. VSN International Ltd, Hemel Hempstead, UK.

Gjedrem T. (1983) *Aquaculture* **33**: 51.

Perry G.M.L., Martyniuk C.M., Ferguson M.M. and Danzmann R.G. (2005) *Aquaculture* **250**: 120.

Scholtens M., Dodds K.G., Clarke S.M., Walker S.P., Tate M.L., Miller R., Slattery, T. and Symonds J.E. (2022) *Proc. World Cong. Genet. Appl. Livest. Prod.* **11**: 44.

Scholtens M., Dodds K., Walker S., Clarke S., Tate M., Slattery T., Preece M., Arratia L. and Symonds J. (2023) *Aquaculture* **563**: 738936.

Proc. Assoc. Advmt. Anim. Breed. Genet. 26: 134-137

ETHANE AND CARBON DIOXIDE EMISSIONS IN AUSTRALIAN ANGUS – PAST AND PRESENT

M.D. Madsen[1], C.J. Duff[2], M. Samaraweera[2], L.A. Mowbray[2], T. Granleese[3], R. Hergenhan[1], T. Bird-Gardiner[3], P.F. Arthur[3] and S.A. Clark[1]

[1] Department of Animal Science, School of Environmental and Rural Science, University of New England, Armidale, NSW, 2351 Australia
[2] Angus Australia, 86 Glen Innes Road, Armidale, NSW, 2350 Australia
[3] Agriculture and Biosecurity, Department of Primary Industries and Regional Development, Grafton, NSW, 2460 Australia

SUMMARY

Methane production (MP) and carbon dioxide production (CP) records have been collected on Australian Angus cattle in both past and current projects. The historic records consisted of animals measured in respiration chamber (n=887 animals) and using Greenfeed Emission Monitor (GEM, n=318 animals), while recent records were collected using GEM units (n=892 animals). The aims of this study were to estimate variance components and genetic relationships between emission traits based on historic respiration chamber records, current GEM records, and combined current and historic GEM records. The heritabilities of MP were similar in the respiration chamber (0.39, SE 0.10), current GEM (0.38, SE 0.11), and combined current and historic GEM (0.36, SE 0.08) datasets. The heritability of CP was higher in the respiration chamber dataset (0.49-0.50, SE 0.26) than the current GEM (0.38, SE 0.10) or the combined current and historic GEM (0.33, SE 0.08) datasets. Low to moderate positive genetic correlations were observed between the respiration chamber records and either the current or combined current-historic GEM datasets. The large SEs were likely due to the relatively small data sizes, particularly for the respiration chamber CP records (263 records). The results indicated that including both historic and current records may be beneficial for the estimation of genetic parameters of emissions in Australian Angus cattle.

INTRODUCTION

Beef is considered a high-quality protein source, but also has a considerable environmental impact, particularly in the form of methane (CH_4) emissions. Consequently, research into increasing the sustainability of Australian beef production has been, and continues to be, a priority for the industry. Genetic selection and breeding of cattle that produce less CH_4 is a method offering a cumulative and permanent reduction in emissions from the beef industry.

Multiple projects have collected CH_4 and carbon dioxide (CO_2) emissions data on Australian Angus cattle over the last two decades, including the National Livestock Methane Program (NLMP) and the Low Methane Beef (LMB) project. Thus, records are available on both current and historic animals, and the potential value of including all available data in the genetic evaluation of emissions can be explored. The aims of the current study were to estimate variance components of emission traits based on current and historic records, and to examine the genetic correlations between emission traits based on records measured using different methods.

MATERIAL AND METHODS

Low methane beef dataset (CurGEM). The LMB records were collected in 2022-2024 on 892 genotyped Angus steers, comprising: 477 Angus Sire Benchmarking project (ASBP) and 415 Southern MultiBreed project (SMB) steers. All LMB data were collected using Greenfeed Emission Monitor (GEM) units (C-Lock Inc, USA) at the "Tullimba" research feedlot, near Armidale, NSW, Australia. The GEM units collect spot-samples of CH_4 and CO_2 when an animal visits the machine.

Only animals with a minimum of five CH_4 and CO_2 spot-samples of a minimum recording duration of 2 minutes were used. The spot-samples were averaged over the trial period to calculate CH_4 production (MP, g/day) and CO_2 production (CP, g/day) for each steer.

National livestock methane program datasets (RESP and HisGEM). The NLMP project produced two different emission datasets; one containing emission records measured using open-circuit respiration chambers (RESP), and one comprised of emission records obtained using GEM units (HisGEM). Detailed descriptions of cattle populations, management and recording procedures are provided in Herd *et al.* (2014) for RESP and Arthur *et al.* (2017) for HisGEM.

The RESP included MP on 887 Angus animals (517 bulls and 370 heifers), and 263 animals also had a CP record. Initial analysis showed genetic correlations of ~1.00 between bulls and heifers for both MP and CP and the traits were therefore treated as one trait for both bulls and heifers.

The HisGEM contained GEM emission records for 318 Angus steers collected at "Tullimba" feedlot. The dataset only included animals with a minimum of 30 spot-samples with a minimum recording duration of 3 minutes. The spot-samples were averaged over the trial period to calculate MP and CP.

Combined dataset (ComGEM). Analysing the records from the HisGEM dataset alone was not possible, due to the small data size. A dataset combining the HisGEM and CurGEM records were created (ComGEM), and it contained MP and CP records on 1,210 steers. Descriptive statistics for the four datasets can be found in Table 1. All animals were genotyped.

Table 1. Descriptive statistics for methane and carbon dioxide production (g/day) in the respiration chamber dataset (RESP) and the current (CurGEM), historic (HisGEM), and combined current and historic Greenfeed Emission Monitor (ComGEM) datasets

Trait	Statistic	RESP	CurGEM	HisGEM	ComGEM
Methane production	Records	887	892	318	1,210
	Mean (SD)	138 (24)	179 (34)	203 (31)	185 (35)
	Range	84-251	62-317	126-308	62-317
Carbon dioxide production	Records	263	892	318	1,210
	Mean (SD)	3,155 (643)	9,900 (1,350)	8,933 (1,214)	9,646 (1,382)
	Range	2,169-4,973	5,433-14,928	6,045-12,833	5,433-14,928

Statistical analysis. The data were analysed within trait using two bivariate genomic best linear unbiased prediction models, one model with RESP and CurGEM records and one model with RESP and ComGEM records. The models followed:

$\begin{bmatrix}\mathbf{y}_{RESP}\\ \mathbf{y}_j\end{bmatrix} = \begin{bmatrix}\mathbf{X}_{RESP} & 0\\ 0 & \mathbf{X}_j\end{bmatrix}\begin{bmatrix}\mathbf{b}_{RESP}\\ \mathbf{b}_j\end{bmatrix} + \begin{bmatrix}\mathbf{Z}_{RESP} & 0\\ 0 & \mathbf{Z}_j\end{bmatrix}\begin{bmatrix}\mathbf{a}_{RESP}\\ \mathbf{a}_j\end{bmatrix} + \begin{bmatrix}\mathbf{e}_{RESP}\\ \mathbf{e}_j\end{bmatrix}$, where the subscript RESP referred to the RESP datasets and subscript j referred to the CurGEM or ComGEM datasets. **y**, **b**, **a** and **e** were vectors containing the phenotypes (MP or CP), fixed effects, random genetic effects and random residuals for each dataset. Contemporary group and age were used as fixed effects for both traits in all datasets. Age of dam was included for both traits in the RESP dataset. The random genetic and residual effects were assumed multivariate normally distributed following $\begin{bmatrix}\mathbf{a}_{RESP}\\ \mathbf{a}_j\end{bmatrix} = MVN\left(\begin{bmatrix}0\\0\end{bmatrix}, \begin{bmatrix}\sigma^2_{a_{RESP}} & cov\\ cov & \sigma^2_{a_j}\end{bmatrix} \otimes G\right)$ and $\begin{bmatrix}\mathbf{e}_{RESP}\\ \mathbf{e}_j\end{bmatrix} = MVN\left(\begin{bmatrix}0\\0\end{bmatrix}, \begin{bmatrix}\sigma^2_{e_{RESP}} & 0\\ 0 & \sigma^2_{\epsilon_j}\end{bmatrix} \otimes \begin{bmatrix}\mathbf{I} & 0\\ 0 & \mathbf{W}_j\end{bmatrix}\right)$, respectively. σ^2_a and cov were the additive genetic variance within dataset and the covariance between the datasets, respectively. **G** was the genomic relationship matrix (VanRaden 2008) consisting of 2,103 animals. $\sigma^2_{e_{RESP}}$ was the residual variance of the RESP dataset. $\sigma^2_{\epsilon_j}$ was a scalar.

I was an identity matrix. $\mathbf{W}_j$ was a diagonal matrix used to weight the residuals based on the number of spot-samples used to calculate the trial average phenotype ($w_{ii} = 1 - \frac{1}{ns_i}$, where ns_i was the number of spot-samples for animal *i*). The residual variances in the CurGEM and ComGEM datasets were then calculated as $\frac{\sigma^2_{\epsilon_j}}{W_j}$. All analysis was completed using the DMUAI packet of DMU (Madsen and Jensen, 2013).

RESULTS AND DISCUSSION

The heritabilities of MP and CP were moderate for RESP, CurGEM, and ComGEM, with all three datasets showing considerable additive genetic variance for both traits (Table 2). The heritability of MP (0.39) and CP (0.49-0.50) in the RESP dataset differed from the previously reported heritabilities of 0.27-0.31 for MP (Donoghue *et al.* 2016; Donoghue *et al.* 2020; Hayes *et al.* 2016) and 0.53 for CP (Donoghue *et al.* 2020) which were estimated using the complete NLMP respiration chamber dataset. This differences in heritabilities were likely due to genotypes only being available for a subset of the NLMP historic respiration chamber dataset for this study (887 out of 1,096 animals).

The SEs of the estimated variance components were lower (both numerically and when expressed as a ratio of the estimate) for the ComGEM dataset compared to the CurGEM dataset. The lower SEs indicate that combining the CurGEM with the HisGEM improved the accuracy of estimation of variance components compared to using only the CurGEM dataset.

The genetic correlations between the RESP and either CurGEM or ComGEM dataset were moderately positive for MP and low to moderately positive for CP (Table 2) indicating that MP and CP collected via respiration chambers and GreenFeed units are different traits. The observed moderate genetic correlations between datasets were expected due to the differences in recording methods and environments. The SEs were large for all genetic correlations. However, the SEs of the genetic correlations between the RESP and ComGEM datasets were reduced compared to the SEs of the genetic correlations between the RESP and CurGEM datasets.

It was not surprising that more variation was observed in the GEM based datasets than the RESP dataset. For example, the variances of MP were at least 3.8 times larger in the GEM based datasets relative to the RESP dataset. For CP, the variances were at least 14.7 times larger in the GEM based datasets than the RESP dataset. The short duration spot-sampling done by GEM units is expected to contain more variability, even after averaging across the trial, than the continuous 24 to 48-hour measurement of emissions obtained from respiration chambers. In addition, feeding method and environmental differences may have contributed to the observed differences in estimated variance components. The RESP records were collected in single animal chambers, with temperature control, restricted feed intake and limited option for physical activity. In contrast, the GEM records were collected in open air pens with *ad lib* bunk feeding. Given the moderate genetic correlation between RESP and GEM, these datasets will be analysed separately for MP and CP in the future. The difference in data size likely also contributed to the differences between estimates from the RESP and the GEM based datasets. The number of CP records was less than a third in the RESP dataset relative to the GEM based datasets. This resulted in SEs-to-estimate ratios for CP that were 2 to 4 times larger in the RESP compared to the other two datasets. It also resulted in larger SEs of the genetic correlations between datasets for CP compared to the SEs of the genetic correlations of MP.

While this study indicates that inclusion of historic records could be beneficial for evaluation of emission traits, it did not examine the prediction ability of the different datasets. Further research should be conducted to assess the impacts of including the historic datasets on predictability of breeding values.

Table 2. Number of records and estimates and standard errors in parenthesis for additive genetic variance (σ_a^2), residual variance (σ_e^2) and heritability (h^2) of methane production and carbon dioxide production for the respiration chamber records (RESP), current Greenfeed Emission Monitor records (CurGEM) and combined historic and current Greenfeed Emission Monitor records (ComGEM), as well as the genetic correlation (r_g) between the RESP and either CurGEM or ComGEM datasets

Analysis	RESP and CurGEM		RESP and ComGEM	
	RESP	**CurGEM**	**RESP**	**ComGEM**
Methane production				
Records	887	892	887	1,210
σ_a^2	78 (16)	339 (78)	77 (16)	300 (59)
σ_e^2	120 (12)	551 (61)	120 (12)	542 (47)
h^2	0.39 (0.10)	0.38 (0.11)	0.39 (0.10)	0.36 (0.08)
r_g	0.55 (0.26)		0.39 (0.23)	
Carbon dioxide production				
Records	263	892	263	1,210
σ_a^2	19,806 (8,013)	365,755 (83,530)	20,155 (7,997)	295,414 (62,055)
σ_e^2	20,490 (5,588)	597,627 (65,735)	20,275 (5,548)	611,833 (51,464)
h^2	0.49 (0.26)	0.38 (0.10)	0.50 (0.26)	0.33 (0.08)
r_g	0.19 (0.40)		0.38 (0.38)	

CONCLUSION

The results of the current study suggest utilising all available data may be beneficial when evaluating emissions in Australian Angus cattle. However, further research is needed, particularly regarding the impact of including the historic records on the predictive ability of the models.

ACKNOWLEDGEMENT

The authors would like to thank the NLMP for providing historic emission records and the SMB and ASBP for providing animals for the LMB project. We would also like to thank all staff at Angus Australia and Rangers Valley Feedlot involved in the ASBP project, as well as the staff at Tullimba feedlot and New South Wales Department of Primary Industries involved in the NLMP, LMB and SMB projects.

REFERENCES

Arthur P.F., Barchia I.M., Weber C., Bird-Gardiner T., Donoghue K.A., Herd R.M. and Hegarty R.S. (2017) *J. Anim. Sci.* **95**: 645.

Donoghue K.A., Bird-Gardiner T., Arthur P.F., Herd R.M. andHegarty R.S. (2016) *J. Anim. Sci.* **94**: 1438.

Donoghue K.A., Bird-Gardiner T., Herd R.M., Hegarty R.S., and Arthur P.F. (2020) *J. Anim. Sci.* **98**: 1.

Hayes B., Donoghue K.A., Reich C., Mason B., Bird-Gardiner T., Herd R.M. and Arthur P.F.(2016) *J. Anim. Sci.* **94**: 902.

Herd R.M., Arthur P.F., Donoghue K.A., Bird S., Bird-Gardiner T. and Hegarty R.S. (2014) *J. Anim. Sci.* **92**: 5267.

Madsen P. and Jensen J. (2013) *Aarhus Univeristy* **1**: 1.

VanRaden P.M. (2008) *J. Dairy Sci.* **91**: 4414.

Proc. Assoc. Advmt. Anim. Breed. Genet. 26: 138-141

VALIDATION OF GENOMIC PREDICTION FOR METHANE EMISSION IN AUSTRALIAN MERINO SHEEP

M.N. Aldridge[1], S.I. Mortimer[2], D.J. Brown[1], P.T. Fitzgerald[3], E.H. Clayton[2], A. Donaldson[2], B. Paganoni[4], A. Thompson[5], V.H. Oddy[2,3] and J.H. van der Werf[3]

[1] Animal Genetics Breeding Unit*, University of New England, Armidale, NSW, 2351 Australia
[2] NSW Department of Primary Industries and Regional Development, Armidale NSW, 2351 Australia
[3] School of Environmental and Rural Science, University of New England, Armidale, NSW, 2351 Australia
[4] WA Department of Primary Industries and Regional Development, Bunbury, WA, 6230 Australia
[5] Centre for Animal Production and Health, Murdoch University, Murdoch, WA, 6150 Australia

SUMMARY

Methane emissions are being recorded in Merino sheep across Australia at a large scale. These records will form a reference population that will be used to estimate genomically predicted breeding values for methane for the sheep industry. This paper aims to determine the predictive ability of the current data to estimate methane rate (ml/min) EBVs using single step GBLUP. Phenotypic data was collected in research and commercial seedstock flocks using portable accumulation chambers. Accuracy of prediction was tested through cross validation, using individual flock data as independent test sets. Accuracy ranged from 0.02 and 0.42 for individual animal EBVs. The large range is likely due to variation in linkage with the reference population, environmental conditions prior to measurement, and some inconsistency of measuring protocol. Further work is needed to improve recording strategies and the size and diversity of the reference population.

INTRODUCTION

Selecting sheep that produce less methane (CH_4) is an achievable goal. High and low sheep selection lines for CH_4 had an average of 10 to 12% difference after 10 years of selection (Rowe *et al.* 2019). The complexity and cost of measuring CH_4 makes it a trait that can benefit from genomic prediction. Current projects are designed to collect enough phenotypes for a reference population to provide the industry with genomically predicted CH_4 breeding values. Additionally, there are several past projects that measured CH_4 on large numbers of animals (Robinson *et al.* 2014; Goopy *et al.* 2016; Paganoni *et al.* 2017; Wahinya et al. 2022; Sepulveda *et al.* 2022). Merino sheep are the most abundant breed in Australia are and consequently the largest CH_4 source in the sheep industry, substantiating a larger representation of Merinos in both recent and historic data. The primary aim of this work was to validate the ability of Merino data recorded in the current Australian projects to predict CH_4 breeding values of industry seedstock flocks. The secondary aim was to determine if the data from earlier projects provides valuable information for improving prediction accuracy.

MATERIALS AND METHODS

All CH_4 rate (ml/min) measurements were recorded using the portable accumulation chamber (PAC) methodology (Goopy *et al.* 2011). The data was divided into data collected during projects

---

* A joint venture of NSW Department of Primary Industries and Regional Development and the University of New England

from 2021 onwards (recent), and projects from between 2009 and the start of 2021 (historic). The data was split as recording protocols and research flock structures have changed over time. The protocol for recent animals included 5 to 7 experimental runs per day, with 12 animals per run. Before each run the animals to be measured were taken off feed 1 hour before entering the PAC. Ewes were then measured for 40 minutes and lambs for 50 minutes. The historic data had varying numbers of runs per day, number of animals per run, animal ages, and the maximum time in the PAC was up to 1 hour. While CH_4 rate was recorded on multiple breeds, this analysis was limited to only include purebred Australian Merino sheep. The recent data includes 5,628 genotyped animals with CH_4 rate measured, from across 11 flocks which includes both industry and research flocks. The historic data includes 3,616 genotyped animals with CH_4 rate recorded from 7 sites, 3 of which were also used in the recent projects. This historic data provides potential linkage between the two data sources. The flocks were spread across Australia, but were predominately from NSW and WA. The historic data was sourced from both the Information Nucleus Flock and Sheep Genomics Flock, while the recent data was recorded on ewes from the AWI Merino Lifetime Productivity (MLP) project, industry ewes from seedstock breeders, and lambs from the MLA Resource Flock (RF). Due to heterogeneity of the data, CH_4 rate was standardised by flock mean and flock standard deviation.

Univariate animal models via restricted maximum likelihood (REML) were used to estimate all variance components, using WOMBAT (Meyer 2007). The model can be summarised as:

$$\boldsymbol{y} = \boldsymbol{X\beta} + \boldsymbol{ZQg} + \boldsymbol{Za} + \boldsymbol{e}$$

where $\boldsymbol{y}$ is a vector of trait observations for CH_4 rate (CH_4 ml/min), $\boldsymbol{\beta}$ is a vector of fixed effects including Date x Run x Flock, Birth type x Rear type, Sex, Age, Feed Level (FL) x Flock, **g** is a vector of random genetic group effects based on flock of origin, $\boldsymbol{a} \sim (\mathbf{0}, \mathbf{H}\sigma_a^2)$ is vector of additive genetic effects (within genetic groups), where **H** is a combined **A** (pedigree) and **G** (genomic) relationship matrix, and **e** is a vector of residual effects. The matrices **X**, **Z**, and **Q** are design matrices for the corresponding model effects. There were differences in available feed between flocks, and differences were observed in feeding behaviour of individuals within a flock. Feed level is an estimate of feed intake relative to estimated maintenance during the 24 hours before measuring CH_4. This was obtained from measurements of CH_4, CO_2 and O_2 fluxes in the PAC, calibrated against data where feed intake was known (V.H. Oddy *pers. comm.*). Analyses of the recent data included both the full model and ignored the Feeding Level effect; a combined analysis of recent and historic data did not fit Feeding Level due to lack of data to allow its calculation.

For validation of prediction accuracy, single step genomic best linear unbiased prediction EBVs of all animals in the recent data were estimated (Legarra *et al.* 2014), using a training dataset with all phenotypes but those from one test flock removed, and comparing these EBVs with phenotypes of the test flock animals. This procedure was repeated by using each flock as a test set, and repeated again with one test of 512 random animals (mean number of animals removed per test flock). These phenotypes were regressed on EBVs to estimate prediction bias with an expected slope of 1. The accuracy was estimated as $COR(Predicted\ phenotype,\ EBV)/\sqrt{h^2}$. To estimate linkage, the mean of the top 10 genomic relationships was calculated between animals in the test set and the reference population (animals in the training set that are both genotyped and phenotyped). The mean of the top 10 genomic relationships between the test set and each flock in the training set was also calculated. A final analysis was conducted to determine the predictive ability of historic data for recent animals. Whereby, all the historic data was treated as the training set and all of the recent data as the test set, and then repeated with the reciprocal training and test set.

RESULTS AND DISCUSSION

The heritability for CH_4 rate using recent data was higher when using the full model (0.19), than when the FL x Flock effect was not fitted (0.17). Fitting FL x Flock mostly decreased the residual variation (0.35 down to 0.32) while additive genetic variance (0.07) and genetic group variances

(<0.01) were unchanged. This change in residual variance shows the importance of measuring CO_2 and O_2 to adjust for different feeding levels before recording CH_4. Alternatively, if more feed intake data becomes available, we recommend further investigation of correcting CH_4 rate measurements for time since last feeding. Using the combined recent and historic data resulted in a lower heritability (0.16), which could be due to the changes in recording practices and the changes to population structure over time.

For some flocks, CH_4 rate EBVs were predicted with high accuracies (Table 1). The validation results were highly variable between test subsets, which can partially be explained by the linkage between flocks (Figure 1). The mean of the top 10 relationships between test flocks and the reference population ranged between 0.35 and 0.40. While the range was small, there was a tendency for higher accuracies to have higher linkage with the reference population. Ewe flocks with higher relationships with MLP or other industry flocks, had higher accuracies than flocks that had lower relationships with other ewe flocks and higher relationships with the lamb flocks. Further investigation is required to determine if CH_4 rate measured on ewes and lambs should be treated as the same genetic trait. Future recording strategies should consider recording CH_4 rate on industry lambs and targeting industry flocks that can help improve linkage between industry flocks. While there were 3 flocks that were poorly predicted with low accuracies (0.02 to 0.06), the majority of flocks had reasonable accuracies (0.14 to 0.42) that would improve genetic selection. As more data becomes available and as the measuring protocol is further refined to reduce environmental errors, it is expected that accuracies will further improve.

The bias was different to the expectation of 1 for most flocks (Table 1). The variation in bias across the different test sets demonstrates that the EBVs were either over- or under-dispersed relative to the variation in progeny performance observed in each flock. These results could potentially be explained by differences in genetic diversity for each flock, environmental differences (especially diet) and measurement error. This needs further investigation, but the analysis would be limited by the number of animals measured per flock in the available data.

The historic data as a training set was a better predictor of recent animals (accuracy of 0.12), than recent data as a training set was for historic animals (accuracy of 0.04). This is beneficial as recent animals are the selection candidates. This analysis should be repeated for other breeds or with a multi-breed evaluation, in which case the historic data could be useful. While the recent data in this analysis only used purebred Merinos, there are over 3,000 animals recorded that represent other breeds and Merino crosses, additional recording of animals that are not Merino are planned for 2025.

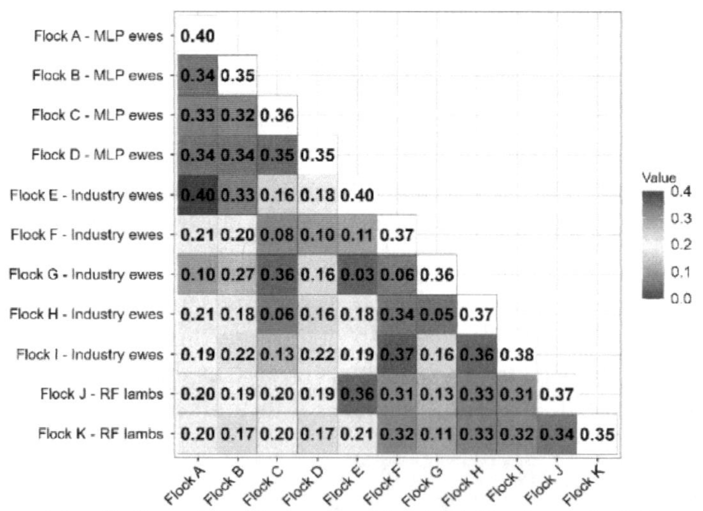

Figure 1. Mean of the top 10 relationships between the test flock and either the full reference population (diagonal) or other individual flocks in the reference population (below diagonal)

Table 1. Validation results from regression of pre-corrected CH_4 rate (ml/min) on EBVs

Test subset	Source of animals in flock	Sires	Animals	Bias	Acc.
Random	Reference pop.	246	512	0.91 ± 0.38	0.41 ± 0.10
Flock A	MLP ewes	29	824	0.60 ± 0.23	0.38 ± 0.08
Flock B	MLP ewes	30	502	0.98 ± 0.29	0.22 ± 0.10
Flock C	MLP ewes	28	575	0.23 ± 0.51	0.18 ± 0.10
Flock D	MLP ewes	25	501	0.30 ± 0.28	0.02 ± 010
Flock E	Industry ewes	46	478	1.43 ± 0.44	0.42 ± 0.10
Flock F	Industry ewes	35	382	1.75 ± 0.60	0.33 ± 0.12
Flock G	Industry ewes	31	484	0.77 ± 0.28	0.22 ± 0.10
Flock H	Industry ewes	45	442	0.59 ± 0.47	0.19 ± 0.11
Flock I	Industry ewes	41	494	0.53 ± 0.47	0.06 ± 0.10
Flock J	RF lambs	109	479	0.19 ± 0.28	0.14 ± 0.11
Flock K	RF lambs	74	467	-0.05 ± 0.41	0.02 ± 0.11

CONCLUSION

Methane traits are expensive to collect, biologically complex, and environmental factors affect the measurements. The validation in this paper indicates a reasonable prediction can be made and used for genetic improvement of Australian Merinos. This prediction depends on linkage to the reference population, which has implications for future recording strategies. The historical data has some value for improving prediction accuracy of recent animals. Further investigation is needed for using both ewe and lamb measurements and for a multi-breed analysis.

ACKNOWLEDGEMENTS

Supported in part by Meat and Livestock Australia (MLA), P.PSH.2011 and L.GEN.2204. Ethics approval; University of New England's Animal Ethics Committee (ARA21-086). With support for the historic data provided by WA-DPIRD, The Australian Federal Government, Murdoch University, MLA, and Australian Wool Innovation. We also acknowledge all the people involved with the data collection.

REFERENCES

Goopy J.P., Woodgate R., Donaldson A., Robinson D.L. and Hegarty R.S. (2011) *Anim. Feed Sci. Technol.* **166**: 219.

Goopy J.P., Robinson D.L., Woodgate R.T., Donaldson A.J., Oddy V.H., Vercoe P.E. and Hegarty R.S. (2016) *Anim. Prod. Sci.* **56**: 116.

Meyer K. (2007) *J. Zhejiang Univ. Sci. B.* **8**: 815

Legarra A., Christensen O.F., Aguilar I. and Misztal I. (2014) *Livest. Sci.* **166**: 54.

Paganoni B., Rose G., Macleay C., Jones C., Brown D.J., Kearney G., Ferguson M. and Thompson A.N. (2017) *J. Anim. Sci.* **95**: 3839.

Robinson D.L., Goopy J.P., Hegarty R.S., Oddy V.H., Thompson A.N., Toovey A.F., Macleay C.A., Briegal J.R., Woodgate R.T., Donaldson A.J. and Vercoe P.E. (2014) *J. Anim. Sci.* **92**: 4349.

Rowe S., Hickey S., Jonker A., Hess M., Janssen P., Johnson T., Bryson B., Knowler K., Pinares-Patino C. and Bain W. (2019) *Proc. Assoc. Advmt. Anim. Breed. Genet.* **23**: 306.

Sepulveda B.J., Muir S.K., Bolormaa S., Knight M.I., Behrendt R., MacLeod I.M., Pryce J.E. and Daetwyler H.D. (2022) *Front. Genet.* **13**: 883250.

Wahinya P.K., Oddy V.H., Dominik S., Brown D.J., Macleay C.A., Paganoni B., Thompson A.N., Donaldson A.J., Austin K., Cameron M. and van der Werf J.H.J. (2022) *Anim. Prod. Sci.* **62**: 818.

Proc. Assoc. Advmt. Anim. Breed. Genet. 26: 142-145

BUILDING A REFERENCE POPULATION TO ENABLE GENOMIC SELECTION FOR METHANE RELATED TRAITS IN BEEF CATTLE

S. Clark[1], M. Madsen[1], R. Hergenhan[1], C. Duff[2] and T. Granleese[3]

[1] School of Environmental and Rural Science, University of New England, Armidale, NSW, 2350 Australia
[2] Angus Australia, 86 Glen Innes Road, Armidale, NSW, 2350 Australia
[3] NSW Department of Primary Industries and Regional Development, Beef Industry Centre, Armidale, NSW, 2350 Australia

SUMMARY

Breeding for cattle that produce lower methane emissions, whilst maintaining or improving productivity has the potential to be a permanent and cumulative strategy for the minimisation of greenhouse gas emissions. Emissions traits in beef cattle are heritable and previous studies have shown that breeding for lower methane emissions is possible. However, a large number of animals measured for emissions traits, simultaneously with production traits, is required to underpin EBVs to allow for effective selection. This paper discusses key measurement technologies, concepts underpinning reference population design and summarises the current reference population for temperate beef cattle in Australia.

INTRODUCTION

Enteric methane emissions from ruminants, such as cattle, contributes up to 80% of agricultural greenhouse gas (GHG) emissions. Hence, reducing methane production is essential in any GHG emissions reduction strategy. An added benefit of this is that selection for reduced methane also provides opportunities to improve feed efficiency as 6-10% of potential energy is lost from the animal as methane. Previous studies have shown that traits related to methane emissions in beef cattle are heritable; that breeding for lower methane emissions is possible and that average emissions may be reduced by ~0.5-2.5% pa (with an average across studies of ~1% pa) (Barwick *et al.* 2019; Fennessy *et al.* 2018). Some opportunities for the management of methane emissions are available by improving productivity, especially reproductive performance and days to slaughter. However, a large volume of animals measured for greenhouse gas emission traits, simultaneously with production traits, is required to underpin EBVs to allow for balanced selection. Previous studies have summarised the technologies available for methane measurements. This study will expand on these with a perspective on measurement technologies for individual animal recording for the construction of a reference population to underpin selection tools. Measurement tools, which animals to sample and ways to value methane to enable balanced selection will be discussed.

METHODS FOR RECORDING METHANE

For animal breeding purposes the overall goal is to decrease overall lifetime methane production. This section will discuss alternative ways to record methane and their ability to estimate lifetime methane. Methane has been measured with a variety of tools for the past ~100 years, with tools ranging from closed circuit respiration chambers to sniffer-based sensors. Initial studies have focused on mean daily methane outputs for primarily animal science studies that have attempted to quantify the differences between treatments in controlled experiments.

Closed circuit respiration chambers. Often viewed as the gold standard for methane measurement, respiration chambers (RC) measure methane over a relatively short period of time (24-48 hrs) very accurately. The major challenge for this method is the short duration (24 hrs), ability to get large numbers of animals recorded and the cost of the system. Generally, chamber-based systems are unlikely to be the key measurement technologies for large scale reference data generation. The largest study of beef cattle recorded in chambers is Donaghue *et al.* (2016) where 1,046 Angus animals were cycled through RC and heritability and GEBV estimated. Heritabilities were generally moderate (~0.3) and GEBV accuracies of 0.5-0.6 achieved within the population (Hayes *et al.* 2016).

Modified SF6. An alternative to respiratory chambers is the modified sulphur hexafluoride (SF6) tracer method. It also gives an accurate measure of total methane released over a short period of time. Previous studies have shown that SF6 has a high correlation with chamber-based measures (0.83, Manzanilla-Pech *et al.* 2021). The challenges for SF6 are that the method is costly and labour intensive per animal recorded and is often not used for large scale animal recording (100's not 1000's of animals recorded). The SF6 tracer is also a potent greenhouse gas and collection of permeation tubes post recording is challenging for processors. One of the largest examples of using the SF6 method for reference population data recording is 463 cows measured repeatedly using this system at the Department of Economics Development, Jobs, Precincts and Regions, Ellinbank (National Centre for Dairy Research and Development) from 2013 to 2017 (Richardson *et al.* 2021). Heritability from this data was reported to be 0.2 (SE 0.11) (Manzanilla-Pech *et al.* 2021). Whilst this is a start, much larger numbers of animals are required, it is unlikely that the use of SF6 will allow for reference populations with thousands of animals to be recorded in beef cattle.

GreenFeed. In the past decade one of the most common methods for recording methane on beef cattle is using GreenFeed Emissions Monitors. GreenFeed is an automated feeder system that monitors CH_4 and CO_2 fluxes from breath and eructation gas from cattle that was created as an alternative to SF6 (Hammond *et al.* 2016). These machines use multiple short methane samples (2-3mins) captured over a testing period. A major challenge for this type of system is that animal access is voluntary so careful management is required to achieve high rates of animal visitation over a trial period. In intensive situations units can be placed at a ratio of 1 unit to 40 animals and in more extensive situations 1:20 is more suitable. This equates to approximately 240 animals per unit annually if recording is for 28 days. Whilst this is considerably more than SF6, GreenFeed units are expensive and still require some labour to manage the units over a trial period. Recording of methane with GreenFeed is still expensive with more cost-effective alternatives required.

Methane sensors (sniffers). Methane concentration (ppm) can be measured with standalone methane sensors. Examples of these have been described previously with two common sniffers being: the nondispersive infrared CH_4 sensor (Guardian NG, Edinburgh Instruments Ltd.) and the portable Fourier transform infrared Gasmet DX-4000 (Gasmet, Gasmet Technologies Oy). The correlation between respiration chamber CH_4 emissions and sniffers has been reported to be high under controlled conditions (0.77, Manzanilla-Pech *et al.* 2021). However, a challenge for sniffer systems is to validate their accuracy in extensive, outdoor conditions. Another limitation for sniffer systems is that methane concentration is measured (ppm) and a conversion is required to understand the volume of methane the animal has emitted. Sniffer-based recording protocols have been designed for recording dairy cattle; however, the extensive nature of beef cattle production has limited their use in recording programs.

Portable (or semi portable) Accumulation Chambers (PAC). These have been used to spot sample animals and larger, less portable, versions have also been proposed and tested for use in beef cattle in New Zealand. Animals are confined to a sealed chamber for a given time and the amount of methane is recorded. The volume of methane produced is then a function of initial methane, final methane and the volume of the chamber (Goopy *et al.* 2016). One of the challenges for the PAC system, and other spot sampling methods, is the adjustment for the time since the animal had eaten. Methane follows a diurnal pattern as the animal grazes and ruminates and this needs to be considered to improve precision of methane measurements from spot sampling methods. The advantages for sniffer and PAC methods are that they enable high volume recording of large cohorts of animals, whilst they may not be suitable for absolute methane (in g/day) they may provide useful tools to measure animals as part of the reference population for emissions traits.

REFERENCE POPULATION DESIGN

A reference population is simply animals recorded for a trait, that have known relationships to selection candidates (either derived by pedigree or genomics) and known management information. The creation of a reference population for methane has 3 key principles: 1) build accuracy of selection, 2) understand correlations between methane and other important traits and 3) provide tools to breeders to select for improved methane in a large part of the population (diversity).

Previous studies have shown that genomic prediction accuracy is impacted by the relationship within and to the reference population (Pszczola *et al.* 2012). When the reference population is small the key is to record animals closely related to selection candidates, as the reference population grows the baseline accuracy of prediction increases and relationship to the reference becomes less important (Kamprasert *et al.* 2024). In beef cattle, the diversity of breeds used for production make it challenging to create reference datasets for all breeds.

To ensure balanced selection occurs, animals with key traits recorded on them should be prioritised for methane recording so that correlations between existing traits and new traits like methane can be estimated. This also enables the benefits of multi-trait selection to be realised, especially allowing for the impact of indirect selection for other traits like reproduction, productivity, welfare and feed intake on methane outputs of the breeding program to be estimated. If multiple methods are used to measure methane careful consideration is required as to whether alternative ways to measure methane require separation as different traits. Other traits like microbiome profiles and volatile fatty acid content from rumen fluid or saliva have also been suggested as proxies for methane outputs. A further complication for methane is that animals can be recorded in the feedlot or at pasture and this may further require additional trait definitions.

CURRENT REFERENCE POPULATION FOR TEMPERATE BEEF CATTLE

In Australia, a reference population for methane traits is currently being generated using GreenFeed emissions monitoring units to record animals. In the construction of this reference population there has been a focus on recording close relatives of key sires by utilising two key populations in temperate Australia, the Southern Multi-Breed project (6,000 steers and heifers) (Walmsley *et al.* 2021) and the Angus Sire Benchmarking project (2,000 steers) (Parnell *et al.* 2019). Overlaying on existing projects has meant that other key traits like birth, growth, carcase and reproductive performance have also been recorded on the same animals. Table 1 illustrates the animals that have been recorded to date. These animals were from breeds that were chosen based on their importance to the beef herd in temperate Australia (Walmsley *et al.* 2021).

Table 1. Number of records per individual and per breed in the reference population

Breed	Feedlot (steers)		Pasture (heifers)	
	Individuals	Records	Individuals	Records
Angus	1,048	65,591	509	33,751
Charolais	155	8,333	129	7,595
Hereford	329	20,754	281	15,022
Shorthorn	181	7,360	133	9,239
Wagyu	166	6,976	113	5,882
Brahman	25	291	43	698
BrahmanX	87	1,958	141	4,636
Total	1,991	111,263	1,349	76,823

The mean performance for animals at pasture and in the feedlot has been similar, as shown by Table 2. Slightly higher means and standard errors observed in the feedlot, which is expected given the steers are generally heavier than heifers at point of recording.

Table 2. Means for methane and carbon dioxide for animals in the reference population (SE)

Gas (g/day)	Feedlot	Pasture	Both
CH4	168 (77)	166 (57)	167 (69)
CO^2	9711 (2479)	7038 (1737)	8617 (2567)

CONCLUSION- TRAIT DEFINITIONS AND VALUING METHANE

The final piece of the puzzle for including methane in the breeding program is including it in a selection index with other key traits. Methane can be minimised at the breeding program level by either per head or per kg product (van der Werf 2023). Each is essentially a breeding objective; a further question is often asked when considering methane; should methane output be optimised per unit of feed intake (methane yield) or per unit of production (methane intensity)? The main impact this has on the development of the reference population for methane is regarding the other traits that are to be recorded. If methane output per kg of feed is used, then it is essential for daily feed intake to be recorded on animals which is an added challenge for recording methane yield on pasture.

ACKNOWLEDGEMENTS

The reference population recording is funded by the NSW department of Primary Industries and regional development, Angus Australia, UNE and MLA donor company (P.PSH.2012).

REFERENCES

Barwick S.A., Henzell A.L., Herd R.M., Walmsley B.J. and Arthur P.F. (2019) *Genet. Sel. Evol.* **51**: 18.

Donoghue K.A., Bird-Gardiner T., Arthur P.F., Herd R.M. and Hegarty R.F. (2016) *J. Anim. Sci.* **94**: 213.

Fennessy P.F., Byrne T.J., Proctor L.E. and Amer P.R. (2019) *Anim. Prod. Sci.* **59**: 1598.

Goopy J.P., Robinson D.L., Woodgate R.T., Donaldson A.J., Oddy V.H., Vercoe P.E. and Hegarty R.S. (2016) *Anim. Prod. Sci.* **56**: 116.

Hammond K. J., Waghorn G. C. and Hegarty R. S. (2016) *Anim. Prod. Sci.* **56**: 181.

Hayes B.J., Donoghue K.A., Reich C.M., Mason B.A., Bird-Gardiner T., Herd R.M. and Arthur P.F. (2016) *J. Anim. Sci.* **94**: 902.

Kamprasert N., Aliloo H., van der Werf J.H.J., Duff C.J. and Clark S.A. (2024) *J. Anim. Breed. Genet. In press.*

Manzanilla-Pech C.I.V., Løvendahl P., Mansan Gordo D., Difford G.F., Pryce J.E., Schenkel F., Wegmann S., Miglior F., Chud T.C., Moate P.J., Williams S.R.O., Richardson C.M., Stothard P. and Lassen J. (2021) *J. Dairy Sci.* **104**: 8983

Parnell P.F., Duff C.J., Byrne A.I. and Butcher N.M. (2019) *Proc. Assoc. Advmt. Anim. Breed. Genet.* **23**: 492.

Pszczola M., Strabel T., Mulder H.A. and Calus, M.P.L. (2012) *J. Dairy Sci.* **95**: 5412.

Richardson C.M., Nguyen T.T.T., Abdelsayed M., Moate P.J., Williams S.R.O., Chud T.C.S., Schenkel F.S., Goddard M.E., van den Berg I., Cocks B.G., Marett L.C., Wales W.J. and Pryce J.E. (2021) *J. Dairy Sci.* **104**: 539.

Walmsley B.J., Donoghue K.A., Johnston D.J., Clark S.A., Siddell J.P., Walkom S.F., Granleese T. and Arthur P.F. (2021) *Proc. Assoc. Advmt. Anim. Breed. Genet.* **24**: 423.

Van der Werf J.H.J. (2023) *Proc. Assoc. Advmt. Anim. Breed. Genet.* **23**: 492.

Proc. Assoc. Advmt. Anim. Breed. Genet. 26: 146-149

REPEATABILITY AND HERITABILITY OF DAIRY COW METHANE CONCENTRATION USING SNIFFER SENSORS

F. Almasi[1], S.R.O. Williams[2], C.J. Vander Jagt[1], L.C. Marett[2,3], J.L. Jacobs[2,3] and J.E. Pryce[1,3]

[1] Agriculture Victoria Research, Agribio, Bundoora, VIC, 3083 Australia
[2] Agriculture Victoria Research, Ellinbank, VIC, 3821, Australia
[3] La Trobe University, Bundoora, VIC, 3083, Australia

SUMMARY

Large-scale phenotyping of methane emissions in dairy cows is necessary for genetic selection, yet gold-standard methods such as respiration chambers and the SF_6 technique are expensive and impractical for broader implementation. In this study, 50 Arcoflex (Arcoflex International Pty Ltd) sniffer sensors were installed one per bay in a herringbone dairy facility, with one per day. Methane concentrations (100 to 5,000 ppm) were recorded every 10 seconds between March and November 2024. Data were gathered from 532 genotyped cows during both morning and afternoon milkings, with an average of 227 visits per cow. Methane concentration measurements were summarised with two methods: (1) an arithmetic mean of all observations per visit, and (2) a "peak" average, defined as data points that exceeded their immediate neighbours. Univariate linear mixed models in ASReml were then applied to estimate heritability (h^2) and repeatability (r). Overall, arithmetic means per visit averaged 445 ppm, whereas peak-based visits averaged 639 ppm. The highest heritability (0.12±0.02) was estimated for peak-based measurements at the per-visit level, while the highest repeatability (0.38±0.01) occurred when peak data were aggregated weekly. While these findings highlight the potential of sniffer sensors for large-scale methane phenotyping, more data from additional herds and farm settings are needed to further validate this approach.

INTRODUCTION

Methane is a potent greenhouse gas, and its accurate measurement is essential for developing effective mitigation strategies to address climate change. Notably, enteric fermentation in ruminants, contributes approximately of 6% of the entire global greenhouse gas emissions (Beauchemin *et al.* 2020), highlighting the importance of targeting this source for emission reductions. While gold standard methods such as respiration chambers and the SF_6 tracer technique provide high-quality data (Deighton *et al.* 2014), their high costs and labour-intensive nature limit their feasibility for large-scale application which is required for reliable genetic selection. Alternatively, sniffer sensors installed in milking parlours offer a cost-effective, though spot-sampling, approach to monitor methane emissions under practical farm conditions. This study aims to evaluate the repeatability and heritability of sniffer-based measurements under various scenarios to identify the most reliable method for large-scale methane monitoring in herringbone Australian dairy operations.

MATERIALS AND METHODS

Sensor Setup and Data Collection. Fifty Arcoflex sensors (Arcoflex International Pty Ltd) were installed in a herringbone dairy, with one per feed bin (Ellinbank SmartFarm, Victoria, Australia). The heated electrode sensors continuously recorded methane concentrations in parts per million (ppm) at ten-second intervals while cows were being milked. The sensors function within a measurement range of 100 to 5,000 ppm. Data were extracted using Arcoflex Data Exporter V1.7.1.0 from March 1 to November 30, 2024.

Each side of the milking parlour (left and right) had 25 bays with individual feed bins; once milking concluded on a side, the bins were raised, allowing all cows in those 25 bays to exit simultaneously. This setup ensured a clear pattern of entry and exit that facilitated consistent data capture throughout the study period.

For genomic estimation, we utilised the XST-74K genotype set. This set comprises approximately 74,000 single nucleotide polymorphisms (SNPs) and is a combination of AVR's (Agriculture Victoria Research) custom DairyBio XT-50K chip and the standard Illumina Bovine 50K chip. The integration of these two chips enhances the density and coverage of genomic data, providing a more robust foundation for prediction models (Van Den Berg *et al.* 2024).

Matching Methane Records to Individual Cows. Because the Arcoflex sensors were not linked to cow radio-frequency identification (RFID), a unique system was developed to associate methane concentration (MeC) readings with individual cows. The milking system captured each cow's National ID and milking order, as animals entered and exited each bay. These data were used alongside information from the milking parlour's gate system, which recorded the presence or absence of cows in the bays. By aligning the sequence in which cows entered the bays (entry order) with gate status (confirming occupancy), a unique identifier was generated for each cow during each milking session. This identifier provided a reliable link between the milk production data and the methane concentration measurements once the timestamps from both datasets were synchronised. Cross-checking these matched records ensured that methane concentration readings were correctly attributed to the right cow.

Data Cleaning. Prior to analysis, any cows receiving a methane mitigating feed additive(s) were excluded from the dataset. Records were also removed if the cow's days in milk (DIM) exceeded 450. Sensor records of 0 ppm were discarded because these often occur when an animal's head position is too far away from the sensor or when ambient methane concentrations are too low to be detected. Cows were milked twice daily, in the morning and afternoon, and recorded visits that were shorter than 2.5 minutes or longer than 25 minutes were also excluded (due to uncertainty of correct identified cow). The remaining visits had a minimum duration of 2.6 minutes, a maximum of 25 minutes, and a mean of 14.2 minutes (median 13.8 minutes). Although the sensors can measure up to 5,000 ppm, preliminary analyses indicated that values exceeding 3,000 ppm were rare and significantly skewed the distribution, therefore values above 3,000 ppm were removed.

Overall, 127,669 valid records remained (79% of total observations), with individual cows having a minimum of 1 visit, a maximum of 452 visits, and an average of 227 visits (median 226), over an average of 138 days. After all exclusions, data from 532 cows from 74 sires remained in the final dataset.

Statistical analysis. To determine the most repeatable time frame for MeC, 1,260 scenarios were created to examine repeatability estimates and compare Akaike Information Criterion (AIC) and Bayesian Information Criterion (BIC) values. Each scenario varied in two ways: (i) window length of recording (from a minimum of 100 seconds to a maximum of 500 seconds) and (ii) start time of recording (ranging from 10 seconds to 1500 seconds after milking began). The response variable was the average MeC recorded per visit. A linear mixed-effects model was then specified, with milking session (morning vs. afternoon), bay number, and polynomial for both cow age (months) and days in milk (DIM) as fixed effects, and individual cows fitted as a random effect to account for repeated measurements within cows. Model estimation employed restricted maximum likelihood (REML) using the nlme package (Pinheiro *et al.* 2024) in RStudio (R Core Team 2023).

Based on the highest repeatability and lowest AIC and BIC values, the optimal interval was identified as second 90 to second 410, spanning 320 seconds. All subsequent analyses therefore focused on measurements collected within this time window, and data outside of it were excluded. Two averaging methods were then defined for further investigation: (1) calculating the simple arithmetic mean of all data points within the selected interval, and (2) identifying a "peak" within each visit (an observation exceeding its immediate neighbours) and computing the average of these.

Heritability and repeatability for each methane trait (average and peak average) were estimated using univariate linear mixed models in ASReml 4.2.1 (Gilmour *et al.* 2021). The general model is described by:

$$\mathbf{y} = \mathbf{Xb} + \mathbf{Zu} + \mathbf{Wc} + \mathbf{e}, \qquad (1)$$

where **y** is the vector of the response variable (either the average MeC or the average of MeC peaks). The vector **b** contains the fixed effects, which include the overall mean, bay number, milking session, and polynomial for both cow age (months) and DIM. In this framework, **u** represents the random additive genetic effects for each animal ($\mathbf{u} \sim N(0, \mathbf{A}\sigma_a^2)$) where A is the relationship matrix derived from the genotypes of 532 animals. The vector **c** denotes the random permanent environmental effects ($\mathbf{c} \sim N(0, \mathbf{I}\sigma_c^2)$) and **e** is the random residual effect ($\mathbf{e} \sim N(0, \mathbf{I}\sigma_e^2)$). Matrices X, Z, and W link observations to the corresponding fixed or random effects.

When data were summarised on a daily or weekly basis, the variables for bay number and milking session were removed from the model. This adjustment was made because the daily or weekly aggregated data no longer allowed for meaningful differentiation by bay or milking session.

Narrow-sense heritability (h^2) was calculated as the ratio of the additive genetic variance (σ_e^2) to the total phenotypic variance ($\sigma_a^2+\sigma_c^2+\sigma_e^2$). Repeatability was estimated by dividing the sum of the additive genetic (σ_e^2) and permanent environmental (σ_e^2) variances by the total phenotypic variance.

RESULTS AND DISCUSSION

Average-based measures recorded mean values of around 445 ppm, with SD ranging from 122 to 217 ppm (Table 1). By contrast, peak-based measures had higher means, approximately 639 ppm, and displayed larger SD (202 to 364 ppm), indicating greater variability in maximum emissions. In a study, Van Breukelen *et al.* (2022) reported a mean of 367 ppm, which is considerably lower than the 444 ppm observed here. This discrepancy may reflect differences in sensor technology and farm practices (e.g., automated milking systems versus herringbone). Additionally, Van Breukelen *et al.* (2022) attributed some of their lower measurements to sensor drift toward zero which was not observed in our data as the Arcoflex sensors do not detect values below 100 ppm.

The heritability estimates ranged from 0.03 to 0.12, depending on how visits were defined, with the highest value (0.12) achieved when peaks were averaged within each visit and the model included bay, milking session (morning vs. afternoon), DIM, and cow age as fixed effects (Table 1). This aligns relatively well with the 0.13 heritability reported by Van Breukelen *et al.* (2022) for mean MeC per visit. However, whereas Van Breukelen *et al.* (2022) observed a considerable increase in heritability (up to 0.32) when data were aggregated weekly, our analysis showed a decrease to 0.05. One likely explanation is the differing recording systems of automated milking systems (AMS) often rely on a single sensor for the entire herd, making the effect of "bay" irrelevant, whereas our herringbone setup employs multiple sensors, one per bay. Although 532 cows were phenotyped in this study, this sample size may not fully capture the genetic variation present in the broader population, suggesting future work should encompass more animals and potentially multiple herds to strengthen parameter estimates. While the bay effect was significant ($P < 0.05$), we chose not to pre-adjust for it prior to aggregating daily or weekly measurements. Instead, we accounted for bay effects directly in the model, which was feasible only at the per-visit level. Furthermore, in a separate analysis (unpublished), we observed that bays located closer to the shed entrance, where they were more exposed to wind, had lower repeatability than bays situated farther inside the facility. Consequently, pooling observations over time may have introduced additional, unaccounted-for variability at the bay level, lowering our heritability estimates for aggregated traits.

Table 1. Mean, variance components, repeatability, and heritability of MeC. V_A is additive genetic variance, V_{PE} is permanent environment variance, and V_P is phenotypic variance

Trait	Mean±SD	V_A	V_{PE}	V_P	h^2±SE	r±SE
Average	445±217	3484	3770	7254	0.09±0.02	0.19±0.01
Average of peaks	639±364	13442	9437	89213	0.12±0.02	0.2±0.01
Daily Average	444±188	1671	4180	35617	0.05±0.01	0.17±0.00
Daily Average of Peak	637±314	6480	10639	98669	0.07±0.01	0.17±0.01
Weekly Average	445±122	347	4652	14376	0.03±0.02	0.35±0.01
Weekly Average of Peaks	638±202	1888	12582	38501	0.05±0.02	0.38±0.01

The repeatability estimates in this study increased when methane measurements were aggregated over longer periods, reflecting how weekly averages can reduce day-to-day fluctuations in the data. Although the weekly repeatability values we obtained were lower than some previously published findings, the general pattern aligns with the results of Van Breukelen *et al.* (2022), who observed a repeatability of 0.30 per visit and 0.68 for weekly means. The rise in repeatability with temporal aggregation highlights the benefit of smoothing out short-term noise that can arise from factors such as cow movement or sensor proximity. However, in our herringbone system, the effect of bay-specific variability could no longer be accounted for once measurements were collapsed into weekly averages, potentially limiting the full advantage of data aggregation. In future work, we plan to correct for bay effects before aggregation to minimise unaccounted variability and improve the accuracy of weekly methane estimates.

CONCLUSION

This study revealed higher heritability and repeatability for peak-based measurements than for arithmetic means, suggesting that peaks may better capture genetic variation in methane output. From a practical standpoint, it remains to be established whether peak- or average-based approaches align more closely with total daily methane production and thus offer greater potential for mitigation through genetic selection. Overall, while these findings indicate that sniffer sensors can supply practical and sufficiently robust data, additional records from more cows and different herds are needed to further validate and refine methane measurement approaches in dairy operations.

ACKNOWLEDGEMENTS

I thank Majid Khansefid for his valuable advice on the statistical analyses and Iona MacLeod for providing the genotypic data used in this study.

REFERENCES

Beauchemin K.A., Ungerfeld E.M., Eckard R.J. and Wang M., (2020) *Animal* **14**: s2.

Deighton M.H., Williams S.R.O., Hannah, M.C., *et al.*, (2014) *Anim. Feed Sci. Tech.* **197**: 47.

Gilmour A.R., Gogel B.J., Cullis B.R., Welham S.J. and Thompson R. (2021) ASReml User Guide Release 4.2 Functional Specification. VSN International Ltd, Hemel Hempstead, HP2 4TP, UK.

Pinheiro J., Bates D. and R Core Team, (2024) nlme: Linear and Nonlinear Mixed Effects Models.

R Core Team (2023) n.d. R: A language and environment for statistical computing. R Foundation for Statistical Computing, Vienna, Austria.

Van Breukelen, A.E., Aldridge M.A., Veerkamp R.F. and De Haas Y. (2022) *J. Dairy Sci.* **105**: 4256.

Van Den Berg I., Nguyen T.V. Nguyen T.T.T., Pryce J.E., Nieuwhof G.J., MacLeod I.M. (2024) *J. Dairy Sci.* **107**: 9591

Proc. Assoc. Advmt. Anim. Breed. Genet. 26: 150-153

METHANE INDEX EXPLORER: OPTIMISING A BREEDING VALUE FORMAT FOR SIMULTANEOUS INCLUSION OF ENTERIC METHANE EMISSIONS IN BREEDING SCHEMES AND NATIONAL INVENTORIES

P.C. Ariyarathne[1], S.M. Hickey[2], T.P. Bilton[1], P. Johnson[1], N.K. Pickering[3], S.J. Rowe[1] and J.C. McEwan[1]

[1] AgResearch Ltd, Mosgiel, New Zealand
[2] AgResearch Ltd, Hamilton, New Zealand
[3] Focus Genetics Ltd, Napier, New Zealand

SUMMARY

This study developed a method to incorporate portable accumulation chamber methane (PACCH4) breeding values (BVs) into ram selection criteria. The PACCH4BVs of 1349 rams born in 2021 and 2022 were estimated different genotypes inclusions levels and evaluated using a flock-specific selection index. The selection response under nine economic weights (from NZ$0-NZ$200/tonne GWP100 CO2e) on PACCH4 showed that progress depended on selection weight, genotype inclusion level, and between-year variability. A selection weight of between NZ$50 and $100/tonne can reduce methane emissions while not dramatically affecting progress in production traits. However, animals should be PAC measured and genotyped to achieve up to 1% annual reduction in PACCH4BVs. This software is suggested for use as a breeder aid. We also recommend that in the future PACCH4BVs are expressed as percentages to aid end-user and inventory calculators' understanding and interoperability.

INTRODUCTION

Enteric methane (CH_4) emissions present a significant challenge for ruminant livestock production. It has been shown that enteric CH_4 emission is a heritable trait, and breeding is an effective mitigation strategy (Rowe *et al.* 2019). The impacts of breeding for low CH_4 and how to incorporate the trait into a breeding scheme are less well understood. Although most genetic correlations with other traits appear neutral (Rowe *et al.* 2022), an important factor is its implicit relationship with feed intake and live weight: both important traits in national genetic selection programmes. National inventories account for greenhouse gas emissions using standardised formulae published by the Intergovernmental Panel for Climate Change describing ratios of methane emissions to energy intake for given stock classes. Currently, there is an opportunity to co-design national inventories and breeding schemes to ensure that the data from each is interchangeable and independent. A breeding value (BV) is an estimate of the relative genetic merit of an animal. It is used to select parents of future generations. BVs for distinctive characteristics are often combined using economic weights (EW) into a profit index. As breeding is a long-term commitment spanning generations, careful thought must be given before finalizing the format. The primary aim here was to develop a methodology including BV that is fit for purpose for national evaluations, government inventory, and allows breeders to optimise genetic progress for their own flocks and indices in a simple graphical format.

MATERIALS AND METHODS

Estimating indices. The genomic BV (gBV) and indices for the rams born in 2021 and 2022 in the AgResearch Woodlands genetic research flock (Flock 2638) were extracted from Sheep Improvement Ltd (SIL, www.sil.co.nz) database. Results from ram labs born in 2021 and 2022 were analysed separately to assess within flock selection response consistency. The gBVs extracted on 22-Nov-2024 (Job no. 2427567) are called version 5 animal evaluation (AEV5), whereas gBVs extracted on 9-Dec-2024 (Job no. 2434492) are called version 6 (AEV6). AEV5 and AEV6 differ in the analytical method and the number of genotypes used. A total of 744 and 605 animals from the 2021 and 2022 cohorts were analysed, with 384 and 324 recorded for PACCH4. Genotype inclusion varied by evaluation: AEV5 had minimal genotyping in 2021 (0.1%, n=1) but increased to 50.2% (n=304) in 2022. In contrast, AEV6 had high genotype inclusion, reaching 99.3% (n=739) in 2021 and 99.7% (n=603) in 2022. The index utilised for the evaluation of flock 2638 contained the sub-indices of New Zealand maternal worth (NZMW: weaning weight, weaning weight maternal, ewe live weight, carcass weight, number of lambs born, lamb survival, survival maternal) with faecal egg count (FEC: FEC1, FEC2, adult FEC), dag scores (DAGS: lamb DAGS, adult DAGS), meat (MEAT: hindquarter lean yield, loin lean yield, shoulder lean yield, carcass weight yield), and portable accumulation chamber methane (PACCH4). This index is referred to as the NZMW-plus index from here onwards, and the traits' EW (cents) are presented as SIL Technical notes (2022). The code utilized for the analysis and SIL Technical notes indicating EW for the traits are publicly available at https://github.com/AgResearch/MIE.The gBVs for PACCH4BV (expressed as unit g/day per 5 month lamb) were divided by scaled mean PAC emissions and multiplied by a hundred to convert it to a percentage (PACCH4BV%). This latter value is easily understood by breeders and allows transparent changes in BV and on-farm inventory calculator methodology. The conversion of PACCH4BV% to a dollar index value is called here the dual-purpose CH4 (DPCH4) subindex. Nine EW (GWP100 ($/tonne)) for CH4gBV% were assessed (from $0 to $200 in $25 steps) to determine its impact on CH4gBV% and production. A DPCH4 for each EW was calculated. A $100 per tonne of GWP100 CO2e equated to an index value of -681 cents for PACCH4BV and -51 cents for PACCH4BV%. The NZMW-plus+DPCH4 index for each animal and each EW was calculated as the sum of the NZMW-plus and DPCH4 indices.

Potential progress of PACCH4BV%. The linear regression (r^2) between the NZMW-plus index and NZMW-plus+DPCH4 index for each PACCH4BV% EW($/tonne) was estimated, multiplied by a hundred, and was considered as the percentage potential progress in the NZMW-plus index. Similarly, the r^2 between DPCH4 and NZMW-plus+DPCH4 for each PACCH4BV% EW($/tonne) was estimated, and this was considered as the percentage potential progress in CH_4. The reduction of PACCH4BV% per year in each NZMW-plus index was calculated as follows.

PACCH4BV% reduction/per year = (potential progress in CH_4 (%) / 100) x Std Dev. PACCH4BV% x (i/L)

The intensity of selection (i) was estimated considering a selection of 2% of males (i = 2.421) and 30% of females (i = 1.159) and the generation interval (L) of 3.5 years for females and 1.3 years for males, therefore, i/L of 0.75.

RESULTS AND DISCUSSION

Figure 1 shows the potential progress in the NZMW-plus index and PACCH4BV% under each EW for CH_4 for AEV5 and AEV6 for both years. The selection responses were consistent over years. There was a clear difference in the gains achieved between evaluations. Notably, there was little gain in PACCH4BV% in 2022 born rams until $50/tonne in both AEV5 and AEV6. AEV6 achieved a greater PACCH4BV% progress compared to the AEV5 under all the EW for CH_4 and higher gains were achieved above $100/tonne but with a higher trade-off in the production index gains (NZMW-plus). For example, approximately 30% and 70% gains in PACCH4BV% and production can be achieved, respectively, at $150 EW in born 2021 animals through AEV6.

The per-year PACCH4BV% reductions under each EW are presented in Figure 2. The trends indicate an average 1.00% and 0.52% reduction when an EW of $100/tonne for years 2021 and 2022, respectively. A maximum of up to a 2.6% per-year PACCH4BV% reduction can be achieved in the AEV6. A favourable observation is that even with zero or low index weights on PACCH4BV%, the NZMWplus index used here did not result in a predicted increase in

PACCH4BV% per animal emissions, implying a change in either feed efficiency or a decrease in methane per unit of dry matter eaten, or both. However, further work is required to investigate this aspect.

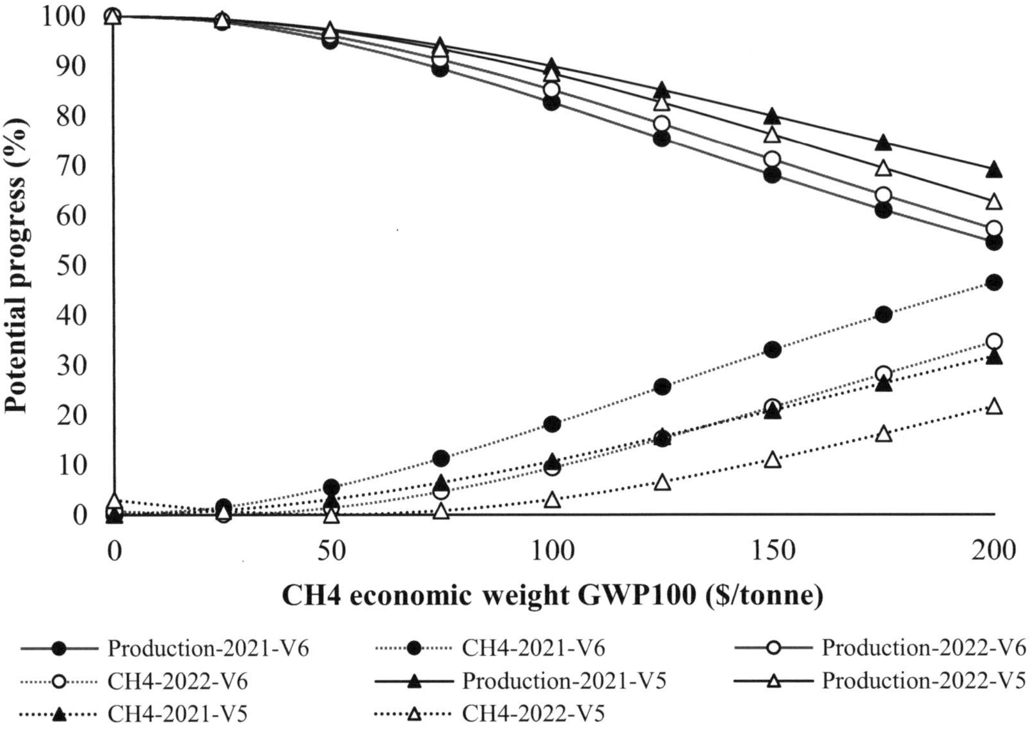

Figure 1. Percentage progress of the methane (PACCH4) breeding values and NZMW+FEC+DAGS+MEAT index (NZMW-plus) in the year 2021 and 2022

The differences in achieved progress in AEV5 vs AEV6 for the same rams can be attributed to the variations in the relationships among gBVs of traits between evaluations and accuracy of estimated gBVs. The gBVs estimated in AEV6 include a higher proportion of genotyped animals (>99%) and, therefore, should possess improved accuracy of estimated gBVs compared to AEV5. The higher genetic gains of PACCH4BV% per year achieved through the PACCH4BVs estimated in AEV6 were closer to what was observed in practice (Booker *et al.* 2024). They also observed that current industry progress in NZMW-plus is only about 50% of its potential so it is possible to increase the NZMW-plus at a faster rate than industry while also decreasing PACCH4BV%. This suggests that the introduction of methane into selection indices should be accompanied by detailed advice on how to also increase overall flock genetic gain. One caveat of this approach is that the EW of the production traits assume all changes in flock methane emissions are captured by PACCH4BV, which does not fully account for increases in lamb survival and lambs weaned.

In this analysis, the gBVs of rams are generated by evaluating the flock data. Therefore, the results obtained are more applicable to the ram breeding flock concerned. This is because genotyping, trait recording, and culling strategies selection indices vary widely, and the process can be implemented when the required genetic parameter estimates between the various subindices are either not available or poorly estimated. Further, the results generated by this method might differ from what is obtained from the selection index theory, given that the latter does not account for the flock-specific genetic parameters and variations in the breed composition of the flock.

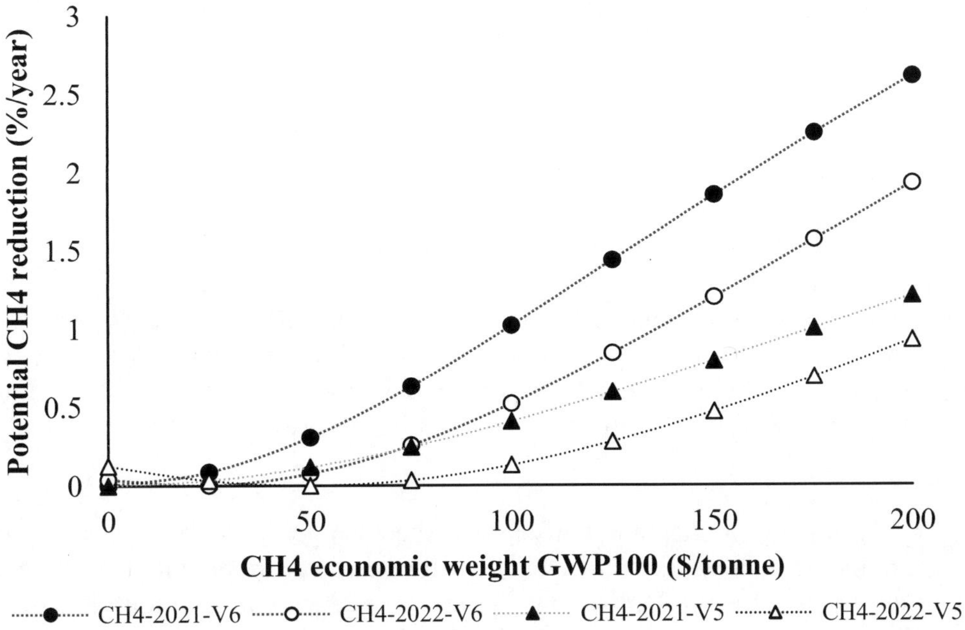

Figure 2. Per year reduction of methane (CH_4) breeding value percentage showing the effect of additional genotypes used in V6 vs V5 for born 2021 and 2022 animals

CONCLUSION

CH_4 mitigation is possible through selective breeding for PACCH4BV%. Greater reductions in CH_4 coupled with a reduction in gain in production traits, therefore, the EW should be cautiously selected. The superior gains achieved by incorporating higher levels of genotypic data for BV estimations suggest that animal evaluations should be based on genotype data if rapid change is desired. Given the stability of the results over the two years evaluated, the PACCH4BV% can be recommended as a desirable form of the PACCH4 trait to include in the selection objectives for mitigating CH_4 emissions.

ACKNOWLEDGEMENTS

Thanks to the Woodlands and Invermay staff who manage and record flock 2638, to the breeders who are now recording methane traits, to B+LNZ Genetics for the breeding value estimates, and to Focus Genetics who funded and encouraged the development of the methane index explorer.

REFERENCES

Booker F., McEwan J.C. and Rowe S.J. (2024) *NZ J. Anim. Sci. Prod.* **84**: 4.

Sheep industry Breeding objectives (2022) Available at https://www.sil.co.nz/files/1661310874929.pdf (accessed 10 December 2024).

Rowe S.J., Hickey S.M., Jonker A., Hess M.K., Janssen P., Johnson T., Bryson B., Knowler K., Pinares-Patino C., Bain W., Elmes S., Young E., Wing J., Waller E., Pickering N. and McEwan J.C. (2019) *Proc. Assoc. Advmt. Anim. Breed. Genet.* **23**:306.

Rowe S.J., Hickey S.M., Bain W.E., Greer G.J., Johnson P.L., Elmes S., Pinares-Patiño C.S., Young E.A., Dodds K.G., Knowler K., Pickering N.K., Jonker A. and McEwan J.C. (2022) *Front. Genet.* **13**: 911355.

INCORPORATION OF DIRECT METHANE TRAITS IN NATIONAL BREEDING OBJECTIVES

F. Hely, D. Kenny, N. Howes and P. Amer

AbacusBio Limited, PO Box 5585, Dunedin 9058, New Zealand

SUMMARY

This study applied selection index theory to model potential enteric methane emissions reductions within the dairy, beef, and sheep industries of New Zealand and Ireland under various trait recording, evaluation, and selection scenarios. Reductions in gross methane are projected to range between 0.5 to 3% of total current industry emissions by 2040, doubling to 1 to 6% by 2050. These could be further increased through more effective breeding programs, improved recording methods, and stronger selection emphasis on methane traits, despite opportunity costs.

INTRODUCTION

Genetic selection for reduced enteric methane emissions is increasingly being recognised as an attractive mitigation opportunity to address the significant contribution that ruminant livestock make to the Green House Gas emissions of New Zealand and Ireland. However, claims about the speed and magnitude of potential mitigation impacts vary widely, and the likely costs of implementation are expected to be high.

The objective of this project was to develop robust modelling tools to predict industry impacts of GHG reductions through genetic selection of novel methane and conventional traits. This study represented the first round of results within the joint Ireland and New Zealand research initiative targeting efficient ruminant breeding programs to reduce enteric methane production (NZAGRC 2022).

MATERIALS AND METHODS

Selection index theory. This work employed selection index theory (Van Vleck 1976; Dekkers 2007) to model the long-term effects of incorporating methane and feed intake traits into breeding programs. The focus was on understanding how these traits would influence responses to selection in conventional traits, feed intake, and methane emissions across the dairy, beef and sheep industries of New Zealand and Ireland. The study distinguished between two groups: species with an existing index framework for incorporating methane traits (New Zealand sheep, and both Irish sheep and cattle) and species without such a framework (New Zealand dairy and beef cattle). For the latter group, assumptions about future selection index development, including the uncorrelated nature of methane traits with key production traits, were made. The study's methodology involved defining methane traits, gathering genetic parameters, estimating economic values based on carbon price scenarios, and using selection index modelling to assess trait responses. These responses were then converted into industry-level methane reductions and opportunity costs.

Key to the methodology was the accurate definition and inclusion of methane traits, such as methane yield (g CH_4/kg DMI) and residual methane (methane adjusted for feed intake).

Trait definitions. The breeding objective traits modelled included methane yield (g CH_4/kg DMI), methane production (g CH_4/animal/day), and residual feed intake (kg DM/day adjusted for liveweight and production). The recorded traits for the NZ scenarios were methane yield in bulls and methane adjusted for liveweight and production in cows and heifers (dairy), methane adjusted for liveweight (sheep) and residual methane (adjusted for feed intake) in steers (beef). Genetic parameters were sourced from existing evaluations where available, with additional estimates for methane traits drawn from relevant literature. Genetic parameters for methane yield traits in New Zealand dairy cattle were derived from the studies reported by Manzanilla-Pech *et al.* (2021) and

Richardson *et al.* (2021). For beef cattle, the parameters were based on Donoghue *et al.* (2016), and for an adjusted methane trait in sheep Jopson (personal communication). In Ireland, the genetic parameters for methane production traits expressed as methane per animal per day were sourced from Ryan *et al.* (2024) for beef cattle, Berry (personal communication) for dairy cattle, and McHugh *et al.* (2022) for sheep.

Economic Weight Calculations. Most economic weights were sourced from existing industry indexes. The economic weights for methane traits were based on carbon price scenarios and the global warming potential (GWP) of methane. The economic weight for methane yield for a given animal class (EW_{MY}^{Class}) was calculated as:

$$EW_{MY}^{Class} = \frac{CP \times GWP_{100}^{Methane} \times DGE_{Class} \times TDMI_{Class}}{1{,}000{,}000\ g/tonne}$$

where CP is the carbon price in \$ or € per tonne CO2e, $GWP_{100}^{Methane}$ is the 100 year GWP coefficient for methane, DGE_{Class} is the discounted genetic expression (Berry *et al.* 2006) effect for the given animal class and $TDMI_{Class}$ is the total dry matter intake in kg for the animal class over the relevant period (e.g. annual for cow traits, or weaning to first calving for heifers). The economic weight for RFI for a given animal class (EW_{RFI}^{Class}) was calculated as:

$$EW_{RFI}^{Class} = DGE_{Class} \times C_{Class}^{DMI} \times DOF_{Class}$$

where C_{Class}^{DMI} is the cost per kg of dry matter for the given animal class and DOF_{Class} is the number of days on feed for the given animal class. The economic weight for methane production for a given animal class (EW_{MP}^{Class}) was calculated as:

$$EW_{MP}^{Class} = \frac{CP \times GWP_{100}^{Methane} \times DGE_{Class} \times DOF_{Class}}{1{,}000{,}000\ g/tonne}$$

where the other inputs are as described in previous equations.

Selection Index Modelling. Selection index models predicted changes in traits under various selection scenarios, using assumptions about heritability, correlations, and genetic relationships. The models generated expected trait unit responses, allowing comparison across different species and industries using a custom-developed R Shiny app. This app facilitated the modification of key parameters such as heritabilities, correlations, and selection proportions, while integrating genomic and pedigree data.

Post-processing. Post-processing of the selection index model outputs involved converting trait responses into industry impacts, using population numbers, GWP methane coefficients, and assumed carbon pricing. The outputs included national emissions baselines, economic responses, relative methane reductions, and opportunity costs. The opportunity cost was defined as the difference between the sum of the base model economic responses (i.e. response in the current index) and new model economic responses in non-methane traits, divided by the annual CO2 reduction.

As part of the post-processing, emissions intensity responses for most breeding programmes were also estimated. These estimates were based on previous work (Amer *et al.* 2018; Quinton *et al.* 2018; Zhang *et al.* 2019), whereby conventional traits were assigned emissions intensity coefficients which defined the marginal change in emissions per unit of farm output expected with each change in conventional traits. The contribution of genetic gains in methane per unit of feed and feed intake to marginal changes in emissions intensity was derived specifically for this study. The information to complete this analysis for Irish sheep was not available.

RESULTS AND DISCUSSION

The modelling of methane emissions reductions for the dairy, beef, and sheep industries of New Zealand and Ireland predicted potential reductions in gross methane emissions ranging from 0.5% to 3% by 2040 and doubling by 2050 (Table 1). As the Irish scenarios used methane production as the breeding objective trait, a status quo change in gross methane was estimated based on the correlated response from changes in existing index traits, alongside the status quo changes expected in emissions intensity.

Table 1. Summary of 2040 and 2050 impacts of incorporating methane traits within sheep, beef, and dairy breeding objectives for both New Zealand and Ireland.[1]

Industry by trait	Carbon price ($/€)	Change in farm profit index response (%)	Opportunity cost/tonne CO2e	Reduction annual methane (%)		Change in emissions intensity (%)	
				2040	2050	2040	2050
NZ Sheep							
Status quo						-3.40	-6.80
CH4 Indirect	$300	-8.7	$143.97	1.39	2.79	-4.24	-8.48
Direct CH4	$300	-15.7	$136.06	2.66	5.33	-5.00	-10.0
NZ Dairy							
Status quo						-5.39	-10.77
CH4 yield in bulls	$350	-4.4	$155.59	1.21	2.43	-6.35	-12.70
Residual CH4 females	$350	-4.9	$223.15	2.71	5.42	-7.81	-15.62
NZ Beef							
Status quo						-0.49	-0.98
Residual CH4 steers	$200	-2.7	$19.48	0.65	1.29	-1.58	-3.17
Residual CH4 & RFI steers	$200	36.4	-$385.70	0.43	0.87	-2.13	-4.25
Ireland Sheep							
Status quo	€0	0	€0	0.02	0.04	N/A	
CH4 production	€300	-6.1	€75.21	0.62	1.24	N/A	
Ireland Dairy							
Status quo	€0	0	€ 0.00	0.47	0.94	-3.42	-6.84
CH4 production	€300	-2.4	€ 54.75	1.80	3.61	-4.32	-8.64
Ireland Beef							
Status quo	€0	0	€0	-0.11	-0.22	-1.85	-3.70
CH4 production	€300	-47.53	€277.54	2.48	4.96	-6.03	-12.06

[1]Multiple scenarios are presented to contrast different breeding strategies and economic assumptions. Results shown are the expected impacts relative to the current selection index.

These estimates considered the balance between genetic gains in conventional traits and methane-specific traits and reflected realistic assumptions about the accuracy of methane measurements and genetic correlations. Larger reductions may be achievable with more accurate proxy measurements, accelerated breeding programs, and incentivized selection emphasis on methane traits.

Despite the efficiency of current breeding programs at reducing emissions intensity, improvements in emissions reductions through recording and selection for methane and feed efficiency traits were expected to be modest. Opportunity costs for methane trait selection were estimated to range from $/€ 19 to 277 per tonne of CO2e with relatively modest shifts in selection emphasis, using carbon price scenarios that were considered to be at the high end. Some of the scenarios where feed intake was part of the new methane initiative showed negative opportunity costs (i.e., benefits) for farmers. This was because the new opportunities to make economic gains from feed efficiency improvements offset the sacrifice in selection emphasis given to feed efficiency.

The differing approaches to defining methane traits between New Zealand and Ireland highlight significant considerations in genetic selection strategies. New Zealand's preferred method of adjusting methane breeding values for feed intake, or a proxy for it, is preferred because reducing methane by selection for lower feed intake per animal will not reduce national emissions as long as pastoral farmers are highly motivated to fully use their available forage resources. Including feed intake recording along with methane could create beneficial outcomes in New Zealand's beef and dairy industries, though it demands research investment and government support, especially to justify a stronger weighting on the methane trait. The Irish industries currently estimate methane production breeding values as a standalone trait for a subset of animals, with expected improvements in accuracy and dispersion as more records become available. New Zealand's focus on recording feed intake and methane output in young bulls stands out globally, with most other countries measuring housed dairy cows. Results indicate that adding methane trait recording in young heifers and cows in a genomic reference population could add value, but greater adoption of semen from young genomic sires would be necessary for realization of these benefits.

Key knowledge gaps and areas for further research for future updates include the impact of adjusting methane breeding values for feed intake, the genetic variation in methane emitted at given feed intake levels, and potential antagonisms with digestive efficiency and animal resilience traits. These areas require further research to inform future updates to the model and assumptions.

CONCLUSION

This study provides initial projections for methane emissions reductions by 2040 and 2050 in the dairy, beef, and sheep industries of New Zealand and Ireland. It highlights the potential for genetic selection to contribute to GHG reductions, while also identifying opportunities for further improvements and the importance of addressing knowledge gaps.

ACKNOWLEDGEMENTS

Funded by the New Zealand Government to support the objectives of the Global Research Alliance on Agricultural Greenhouse Gases.

REFERENCES

Amer P.R., Hely F.S., Quinton C.D. and Cromie A.R. (2018) *Animal* **12**: 5.

Berry D.P., Madalena F.E, Cromie A.R. and Amer P.R. (2006) *Livest. Prod. Sci.* **99:** 159.

Dekkers J.C.M. (2007) *J. Anim. Breed. Genet.* **124**: 331.

Donoghue K.A., Bird-Gardiner T., Arthur P.F., Herd R.M. and Hegarty R.F. (2016) *J. Anim. Sci.* **94**: 1438.

Manzanilla-Pech C.I.V., Løvendahl P., Gordo D.M., Difford G.F., Pryce J.E., Schenkel F., Wegmann S., Miglior F., Chud T.C., Moate P.J. and Williams S.R.O. (2021) *J. Dairy Sci.* **104**: 8983.

McHugh N., O'Connor E., Berry D.P., Dunne E., Farrell L., Rafter P., McDermott K. and McGovern F.M. (2022) *Proc. World Cong. Genet. Appl. Livest. Prod.* **12**: 2948.

NZAGRC 2022: Roadmap for efficient and effective ruminant breeding programs to reduce enteric methane production (agreement number 2022-NZIRL-ROADMAP-ABB).

Quinton C.D., Hely F.S., Amer P.R., Byrne T.J. and Cromie A.R. (2018). *Animal* **12**: 889.

Richardson C.M., Nguyen T.T.T., Abdelsayed M., Moate P.J., Williams S.R.O., Chud T.C.S., Schenkel F.S., Goddard M.E., van den Berg I., Cocks B.G. and Marett L.C. (2021) *J. Dairy Sci.* **104**: 539.

Ryan C.V., Pabiou T., Purfield D.C., Berry D.P., Conroy S., Murphy C.P. and Evans R.D. (2024) *J. Anim. Sci.* **102**: skae034.

Van Vleck L.D. (1976) *Biometrics* **32**: 173.

Zhang X., Amer P.R., Jenkins G.M., Sise J.A., Santos B. and Quinton C. (2019) *J. Dairy Sci.* **102**: 11153.

Proc. Assoc. Advmt. Anim. Breed. Genet. 26: 158-161

THE GREEN COW PROJECT: DEVELOPING A METHANE EFFICIENCY GENOMIC BREEDING VALUE FOR THE AUSTRALIAN DAIRY INDUSTRY

C.J. Vander Jagt[1], F. Almasi[1], B.J. Sepulveda[1], L.C. Marett[1,5], S.R.O. Williams[1], R. Xiang[1], S. Bolormaa[1], M. Shaffer[2], J. Penry[3], C. Hill[3], D. Gibson[4], J.L. Jacobs[1,5] and J.E. Pryce[1,5]

[1] Agriculture Victoria Research, Victoria, Australia
[2] DataGene Ltd, Bundoora Victoria, Australia
[3] Dairy Australia, South Bank Victoria, Australia
[4] Gardiner Foundation, South Bank Victoria, Australia
[5] La Trobe University, Bundoora Victoria, Australia

SUMMARY

Methane (CH_4) represents a critical target for near-term climate mitigation and the Australian dairy industry is under increasing pressure to further reduce its emissions footprint, particularly enteric methane emissions (EME). While management practices and feed additives show promise in reducing EME, the temporary nature of additives and public concerns surrounding their use highlight the need for more sustainable and permanent solutions. The Green Cow project, led by Agriculture Victoria Research (AVR), aims to address this challenge by developing a multi-trait Methane Efficiency Australian Breeding Value (ME-ABV) for the dairy industry. By leveraging multiple selection criteria (proxy traits) in a multi-trait model, the ME-ABV will initially be integrated into the Australian dairy Sustainability Index (SI) before being released as a standalone breeding value. This approach provides a practical and publicly acceptable solution to reducing EME in the dairy sector while simultaneously improving feed efficiency and maintaining productivity.

INTRODUCTION

Methane (CH_4) is a potent greenhouse gas with a warming potential 80 times greater than CO_2 over 20 years, making its reduction crucial for near-term climate mitigation (Black *et al.* 2021). The Australian dairy industry has one of the world's best emissions footprints, averaging 0.93 kg CO_2e/kg Fat and Protein Corrected Milk (Dairy Australia 2023), it faces growing pressure to further cut gross EME (g/day) and EME-intensity (g/kg milk) for environmental and social reasons.

Various strategies can reduce gross EME and EME-intensity, including improved management practices and feed additives. While feed additives have been shown to reduce EME, their effect is temporary (Black *et al.* 2021). Recent controversy surrounding the use of 3-NOP in the dairy and beef livestock industries (Dalton 2024) highlight a significant challenge: public perception and acceptance of feed additives. The most effective and broadly supported long-term solution is likely through genetics, as genetic gains are permanent and cumulative, with EME reductions improving progressively with each generation. Moreover, genetic approaches could complement other strategies, offering the potential for a stacked approach to maximise impact. Genomic selection is a cost-effective solution to address this challenge, requiring only EME measurements (or proxy measures) from genotyped cows to develop a genomic breeding value.

In 2022 DataGene, the organisation responsible for providing breeding values to the Australian dairy industry, introduced the SI (Nguyen *et al.* 2023, Richardson *et al.* 2021), which places additional emphasis on green traits such as feed efficiency, longevity, and milk production. This index is projected to reduce EME-intensity by 5 to 10% by 2050 (Richardson *et al.* 2022). However, incorporating EME into the breeding objective could achieve even greater reductions, with an estimated up to 30% reduction in EME-intensity by 2050 (Richardson *et al.* 2022). The Green Cow project, led by AVR, aims to address this challenge by developing a Methane Efficiency Australian Breeding Value (ME-ABV) for the Australian dairy industry. The Green Cow project is a 4-year project that commenced in December 2023. Co-developed with the Victorian Dairy Innovation Agreement (VDIA, a collaboration between Gardiner Foundation, Dairy Australia and AVR) and DataGene, the project will provide farmers with a powerful tool to reduce emissions through genomic selection. Central to this work are key national and international collaborations including VDIA's DairyBio and DairyFeedbase programs, the ZNE-Ag CRC and the GMH, an international initiative dedicated to reducing methane emissions. This paper details these efforts and the project plan, and highlights progress made so far.

MATERIALS AND METHODS

The Green Cow project will take a multi-trait approach to develop the ME-ABV, using EME proxy-data from microbiome profiling (ruminal and faecal), milk mid-infrared spectrometry (MIR) and CH_4 sniffers. Validation of these EME proxies against the at-pasture baseline EME measuring approach, SF_6 tracer technique, is ongoing. A multi-trait approach will provide a more accurate and robust prediction by leveraging trait correlations and capturing different aspects of EME biology. Additionally, using multiple phenotyping methods reduces the risk of systematic errors that might arise from relying solely on a single approach. Since genotype-by-environment interactions are expected to significantly influence trait expression, utilising a diverse range of phenotyping techniques allows us to better capture these interactions, leading to more robust genomic predictions of EME across varying environmental conditions. This approach also ensures that breeding programs remain resilient to potential failures or inaccuracies from any single data source.

The current prioritised order for phenotyping CH_4 is as follows:

1) **Rumen microbiome:** recent studies suggest that the ruminal microbiome is heritable and strongly correlated with EME (González-Recio *et al.* 2023). Additionally, it offers the potential to link EME data from other countries for improved prediction (Sepulveda 2024), as all EME traits primarily originate from the rumen microbiota. However, there are challenges such as recruiting farmers to allow ruminal sampling, ensuring farms have appropriate cattle restraints, and securing enough trained personnel to carry out the sampling. Furthermore, the project may benefit from potential access to collaborative international microbiome data through the GMH.
2) **MIR:** milk composition measured by MIR light absorption can be used to predict traits such as EME (Rojas de Oliveira *et al.* 2024). It is cost-effective, non-invasive and suitable for large-scale phenotyping and already used for reproductive outcomes (Ho and Pryce 2020, DataGene 2022) and the Canadian Methane Efficiency breeding value (Rojas de Oliveira *et al.* 2024). If validated in Australia, it could maintain the ME-ABV cost-effectively. As MIR analysis is already routinely conducted at herd test centres, integrating it for methane phenotyping would minimise additional costs and logistical challenges while enabling large-scale, ongoing data collection to support the ME-ABV's long-term sustainability. However, reliability may be affected by the diversity of Australian dairy systems due to genotype-by-environment interactions.
3) **CH_4 sniffers:** these sensors measure CH_4 concentrations in the air near cows, offering a non-invasive, cost-effective method for large-scale EME phenotyping. ArcoFlex CH_4 sniffers have been installed in feed bins at AVR's Ellinbank SmartFarm, measuring CH_4 ppm every 10 seconds while cows are in the milking stall. If validated, these sensors could enable scalable EME phenotyping, leveraging the existing network of farms where these devices are already installed and operational. Other CH_4 sniffers in Australia, including La Trobe University's devices, may be used in the Green Cow project. Their approach estimates CH_4 from ruminal gas by calculating oxygen absence as a dilution factor.
4) **Faecal microbiome:** the faecal microbiome has shown potential as a useful proxy for predicting daily methane emissions in beef cattle (Manafiazar *et al.* 2021). If the faecal

microbiome proves to be predictive of EME in dairy cows, it might not be as accurate as the ruminal microbiome but could allow for easier phenotyping, albeit with the need for a much larger sample size.

The Green Cow project will target specific types of farms to optimise participation and outcomes. The priority is research farms, as most herd members are already genotyped, reducing costs and saving time. These farms are more likely to have staff available for animal handling and can better tolerate production disruptions. Additionally, they typically possess appropriate cattle restraints, which are essential for ruminal sampling. The second focus is Ginfo farms (Ginfo is Australia's national reference herd for genetic information), where most animals are genotyped and detailed animal records are maintained. These farms have a strong history of research participation and established relationships with AVR, ensuring smooth collaboration. A third target is farms that participated in the VDIA's DairyBio 10K bloods project (VDIA 2025). These herds are genotyped, have comprehensive records, and farmers are willing to engage in research. If the ZNE-Ag CRC project, which aims to determine whether blood volatile fatty acid profiles can predict CH_4 emissions, proves successful, these farms could contribute an additional 10,000 phenotypes. Lastly, if the ArcoFlex CH_4 sniffers validate against SF_6, the network of ArcoFlex farms could provide ~6,000 EME phenotypes from diverse environments, with the potential for more over time. If validated, ArcoFlex farms will become the top priority for phenotyping.

As more phenotypes are collected, they will be incorporated into EME genomic prediction models to refine and develop iterations of the ME-ABV. This ABV will be introduced in phases, starting with inclusion in the SI. Once higher reliability is achieved, it will be released as a standalone ME-ABV. By 2030, the goal is to integrate the ME-ABV into the main industry index, the Balanced Performance Index. Reliability will improve as more animals are phenotyped, either through direct EME measurements or proxies. Achieving the typical reliability of a standalone ABV (≥30%) requires data from 10,000 to 15,000 animals. However taking a multi-trait model approach, which combines multiple interconnected measurements, may reduce the number of phenotypes needed to reach this level of reliability (Song *et al.* 2020). Given the high cost of phenotyping, a stepwise approach is essential. Introducing the ME-ABV through the SI first, combined with the use of multi-trait modelling, ensures the fastest and most impactful route for delivering benefits to the industry.

RESULTS AND DISCUSSION

Progress to date. Since commencing December 2023, the Green Cow project has progressed significantly. Focus has been on analysing existing EME-related data from experiments conducted at the Ellinbank SmartFarm, integrating it with ongoing data collection. This has enhanced the reliability of the EME genomic breeding value (based on SF_6) developed through the VDIA's DairyBio program, raising it from below 0.10 to around 0.15. Recent incorporation of MIR and microbiome phenotypes has further improved the model's reliability to ~0.24, potentially increasing its resilience to genotype-by-environment interactions. Despite these improvements, results remain inconsistent across different analysis scenarios, highlighting significant data noise and underscoring the need for more phenotypic data to reduce variability and stabilise the genomic prediction model. The model requires testing in diverse environments and farming systems beyond Ellinbank.

Efforts to validate the ArcoFlex CH_4 sniffers at the Ellinbank SmartFarm have advanced, with 50 devices operational since February 2024. The team has matched sniffer measurements to individual cows and estimated the heritability and repeatability of the data. Work to determine the correlation of the sniffer data to SF_6 data is still underway. While initial results highlight the sensors' potential for predicting EME, further validation using additional SF_6 data is required.

Early analysis of 25 animals revealed that faecal metagenome features had phenotypic correlations of up to 0.40 ± 0.18 with EME production (g/d) and up to 0.62 ± 0.12 with features from the ruminal metagenome (Sepulveda *et al.* 2025). While not as strong as the associations observed with ruminal features (genetic correlations between 0.59-0.93, González-Recio *et al.* 2023), it remains highly promising. Notably, DNA extraction and sequencing from ~2,000 additional ruminal and faecal samples, is underway (to conclude in 2025). This expanded dataset will enhance the reliability and stability of the genomic prediction model(s). The team has also developed scalable protocols for faecal microbiome sampling on commercial farms, offering a non-invasive and efficient method for phenotyping. While activities beyond Ellinbank are yet to occur, these preparations position the project for rapid expansion in the near future.

CONCLUSION

Designing a reference population for EME is challenging. The Australian dairy industry is addressing this through a multi-trait approach to developing the ME-ABV, aiming to deliver more accurate and robust predictions by capturing various aspects of EME biology. Close to 1,000 SF_6 CH_4 records from AVR's Ellinbank SmartFarm support validation of other selection criteria. Higher-throughput phenotypes, such as MIR, show promise as selection criteria but require external validation beyond Ellinbank. The ruminal microbiome is a validated proxy trait and may capture genotype-by-environment interactions, a key challenge in developing EME breeding values. Scaling ruminal microbiome data for reliable breeding values is difficult. Early research into the faecal microbiome and CH_4 sniffers is also encouraging, and, if validated, these methods could provide a substantial number of phenotypes. Initial development of the ME-ABV has shown that reliability levels fluctuate depending on the analysis method used, highlighting significant data noise and emphasising the need for more phenotypic data. Plans are underway to expand phenotyping efforts beyond Ellinbank in 2025, which should strengthen these models and reduce data variability. In addition, global and domestic collaborations through initiatives such as the GMH and ZNE-ag CRC will improve access to additional data and support the development of EME breeding values.

REFERENCES

Black J.L., Davison T.M. and Box I. (2021) *Animals* (Basel) **11**: 951.

Dairy Australia (2023) Available at: https://www.dairy.com.au/sustainability/sustainability-framework-reports-and-scorecards

Dalton A. (2024) Available at: https://www.theage.com.au/national/nsw/panic-over-additive-in-cattle-feed-sparks-furore-over-milk-and-meat-20241204-p5kvr1.html.

DataGene (2022) Available at: https://www.datagene.com.au/wp-content/uploads/2023/11/Fact-sheet-35-MIR-Conception-Tool.pdf.

González-Recio O., Martínez-Álvaro M., Tiezzi F., *et al.* (2023) *Livest. Sci.* **269**: 105171.

Ho P.N. and Pryce J.E. (2020) *J. Dairy Sci.* **103**: 11535.

Manafiazar, G., Fitzsimmons C., Zhou M., *et al.* (2021) *Animal* **15**: 100304.

Nguyen T.T.T., Richardson C.M., Post M., *et al.* (2023) *Anim. Prod. Sci.* **63**: 1126.

Richardson, C.M., Amer P.R., Quinton C., *et al.* (2022) *J. Dairy Sci.* **105**:4272.

Richardson C.M., Sunduimijid B., Amer P., *et al.* (2021) *Anim. Prod. Sci.* **61**: 1781.

Sepulveda, B.J. (2024) Doctor of Philosophy thesis. La Trobe University, Victoria, Australia.

Rojas de Oliveira H., Sweett H., Narayana S., *et al.* (2024) *JDS Commun.* **5**: 756.

Song, H., Zhang Q. and Ding X (2020) *J. Anim. Sci. Biotechno.* **11**: 88.

Victorian Dairy Innovation Agreement, VDIA (2025) Available at: https://dairybio.com.au

Proc. Assoc. Advmt. Anim. Breed. Genet. 26: 162-165

INDUSTRY PERSPECTIVES ON DAIRY CALF TRAITS

M.M. Axford [1,2,3], M. Khansefid[1,2] and J.E. Pryce [1,2]

[1] School of Applied Systems Biology, La Trobe University, Bundoora, VIC, 3083 Australia
[2] Agriculture Victoria, AgriBio, Centre for AgriBioscience, Bundoora, VIC, 3083 Australia
[3] DataGene Ltd, Bundoora, VIC, 3083 Australia

SUMMARY

Breeding tools are under development in Australia to improve dairy calf health. A survey revealed that dairy farmers value calf traits similarly to well-known traits such as survival, mastitis, and conformation. Results suggest that calf traits should be presented in genetic evaluations so that the favourable direction is a higher value and as individual traits rather than a general calf index. Findings from this study will assist industry to optimise the release of this new information.

INTRODUCTION

Calf morbidity increases heifer rearing costs, directly through increased labour and health costs and indirectly through poorer lifetime performance. In addition, improving heifer rearing improves two sustainability metrics, outlined by the Australian Dairy Sustainability Framework (Dairy Australia 2023). Firstly, reducing disease, deaths and reliance on antimicrobials to treat disease, contributes to the goal of providing best care for animals. Secondly, rearing less animals to achieve the same level of production reduces the amount of methane emitted per kg milk solids produced.

Previous studies investigating the genetic variation in calf health traits suggest that it is feasible to breed for fewer stillbirths and lower pre-weaning mortality in Australian dairy herds (Axford *et al.* 2024) and further work is underway to measure genetic variation for additional calf traits. However, producing new Estimated Breeding Values (EBV) is only the first step to using genetic selection to influence calf performance. Herd managers need to be aware of and understand genetic tools and consider any barriers to using the tools in their herd before they can decide to implement (Monks *et al.* 2021). Using an industry survey, we aimed to analyse the perspectives of farmers and service providers on improving calf health traits. The purpose of the survey was to provide details on the language associated with potential traits and their perceived importance to improve the ease with which adoption can take place following implementation by DataGene, Australia's dairy genetic evaluation centre.

MATERIALS AND METHODS

An on-line survey was conducted between 23 October 2023 to 10 June 2024 using SurveyMonkey (SurveyMonkey Inc. 2024). The survey was distributed through traditional and social media to dairy farmers, calf growers (those who raise calves but do not milk cows) and industry stakeholders. Farmers and calf rearers were separated from industry stakeholders as they have an economic interest in a farm business. Respondents were asked about their business activity, region, role, number of calves reared, breed, calving system, calf record keeping, trait preferences as well as how they would like to see calf traits presented for genetic evaluation. Trait preferences were gathered using two question types with the first being a scale from 'lowest emphasis' to 'highest emphasis' and the second asking respondents if they would trade off protein kg/cow/year for an improvement in calf traits, similar to the method applied by Axford *et al.* (2021).

RESULTS AND DISCUSSION

A total of 109 responses were received of which 66% of respondents were dairy farmers and calf rearers located in all dairying regions of Australia, 14% provided technical support to farmers and

13% were from bull breeding companies and resellers. For simplicity, calf rearers responses were grouped with those from farmers for the remainder of this report. Most respondents were farm owners (66%). Of these respondents, 44% reared between 51-150 calves per year and 31% reared between 151-300 calves per year. Holstein-Friesian (58%) and Jersey (19%) were most frequently reported to be the 'main breed'. While a small number of farmers (3%) reared mainly beef cross calves, half of respondents reared some beef cross calves.

All farmers reported that calf records were kept but there were multiple methods for recording this data as illustrated in Figure 1. Calving ease and calf fate was the most common information to record and it was usually recorded electronically (73%). It was more common for calf illness records to be kept on paper rather than electronically (52% compared to 37%). From a survey of Canadian farmers, Edwards *et al.* (2024) reported that 56% reported illness was captured using computer software or an app. In the same study, the authors concluded that farmers were more likely to record calf illness and treatments if records were collected with a computer software system, recorded in closer proximity to the calves and collected by a nonfamily employee (Edwards *et al.* 2024). It seems that the recording of calf data lacks the same systematic approach that is currently applied to the milking herd with multiple sources of data required to attain a complete picture of the fate and health of calves. Future research and benchmarking activities would benefit from improvements in calf record keeping systems being made available to farmers.

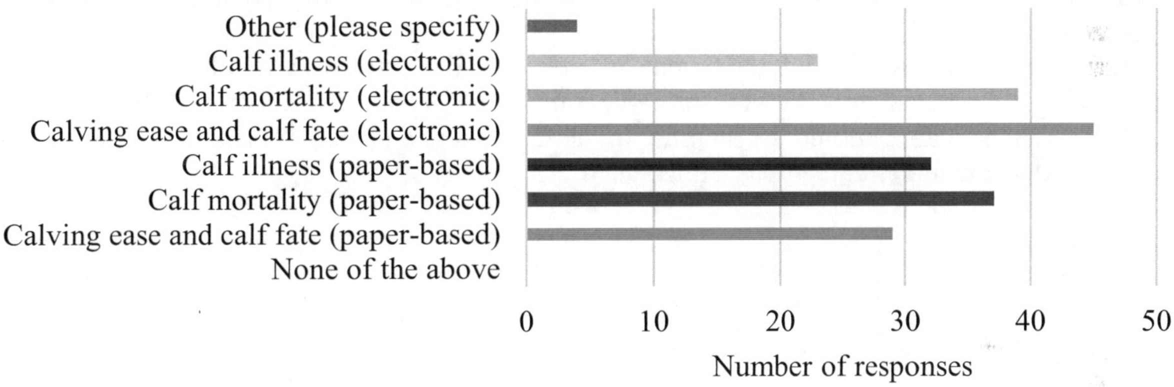

Figure 1. Record keeping practices of Australian dairy farmers and calf growers

Calving ease is known to have a significant effect on the rate of stillborn calves. To understand the extent of calving difficulties, farmers were asked to estimate the proportion of first-time heifers that require assistance. The responses ranged from 0-33 % with a mean of 6.5% (SD 6.2) which is consistent with the percent of calvings with dystocia (5.8%) reported from Australia's centralised data repository by Axford *et al.* (2024). Amongst farmers with Holstein as the main breed, the mean percent of calvings with dystocia was 6.4% (SD 5.7).

The most popular trait names were Stillbirth ABV (55%), Calf Survival ABV (43%) and Calf Health ABV (36%). There was close alignment between farmers and service providers in their choice of breeding value name. As expected, respondents preferred that calf traits were presented so that higher ABV values reflect healthier calves (88%). Interestingly, respondents preferred that traits were presented separately, rather than as components of an index which may be related to a desire for transparency, especially when traits are new.

National genetic evaluation services routinely report calving ease, gestation length and stillbirth breeding values. However, calf health traits are less common in genetic evaluations and their economic value is harder to define compared to production traits. The survey asked questions to help understand industry perspectives on the value of various calf traits and the results are shown in

Figure 2. On a preference scale of 1-5 where 5 was most important, the mean score for all respondents ranged between 3.5 (SD 1.1) for heifer survival from weaning to first calving and 3.8 (SD 1.1) for calf health. These results were significantly lower than scores for some existing breeding values such as protein yield, but higher than scores for gestation length, feed saved and heat tolerance. Between calf traits, there was no significant difference in the mean score for calf health, calf survival at birth (stillbirth), calf survival to weaning (pre-weaning mortality), calving ease and heifer survival from weaning to first calving. Calf trait scores were similar to those for survival, mastitis and type traits that are included in Australia's national breeding indices.

A similar survey with a larger number of respondents (n=254) and many overlapping traits was conducted in Australia in 2019. That survey included calving ease but did not include the recently conceived calf traits. Interestingly, the order of trait preferences in 2019 was similar to that observed in the current study with fertility, protein and fat ranking highly whereas feed saved, gestation length and heat tolerance had lower median scores. The similarity in the ranking of traits across the two studies suggests that the current study is broadly reflective of farmer views.

Farmers with Jersey cattle (n=10) put significantly less emphasis on calving ease (2.7, SD 1.2) compared to farmers with Holstein cattle (4.07, SD 0.8). The mean score for calf health and calf survival at birth for farmers with Jersey cattle was ranked second and third, next to protein, and these traits were scored significantly higher than calving ease. This is understandable considering the prevalence of dystocia in Jerseys (3.6%) is lower than Holsteins (10.8%) in calvings with first parity dams (Axford *et al.* 2024).

Farmers generally placed a higher emphasis on calf traits compared to advisers. While not statistically significant, advisers ranked survival, somatic cell count and mastitis higher than calf traits and this observation could be helpful to better understand contextual differences when developing extension material.

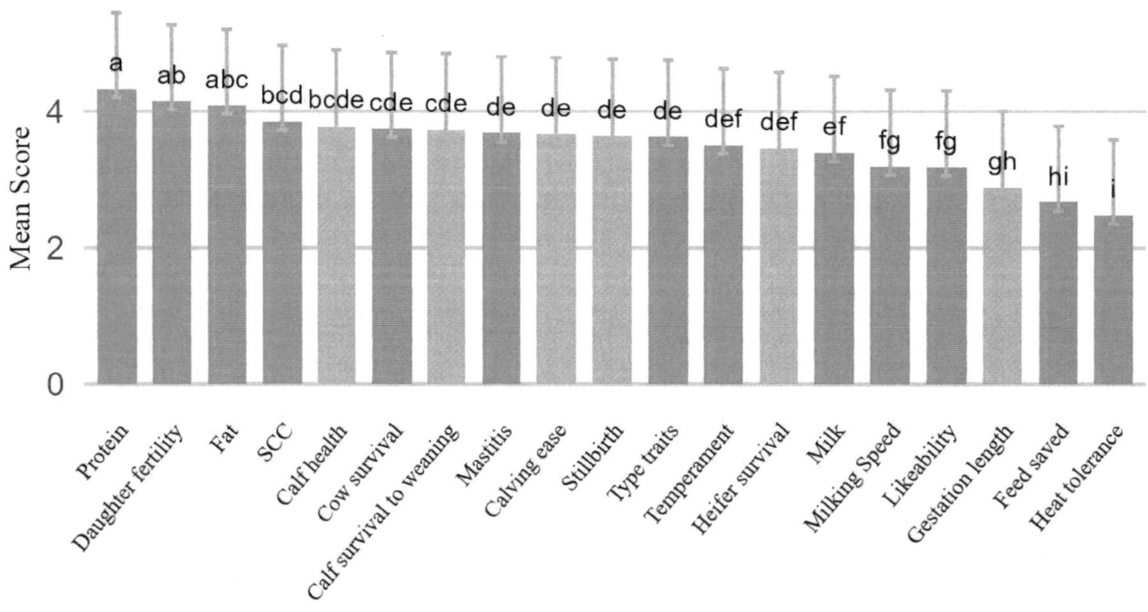

Figure 2. Weighted mean scores (bars) and standard error (whiskers) for trait preferences for all respondents (5 is most important, 1 is least important). Letters identify traits with mean scores that are significantly different ($P<0.05$). Yellow bars denote calf traits and blue bars are cow traits included in the survey

An alternative approach to valuing calf traits is by comparing that trait to a trait with a known value, such as protein kg. In this survey, farmers were asked how much protein they would trade to achieve an improvement in a calf trait. This question was asked specifically of farmers as they are most closely associated with dairy farm businesses. There was a wide margin of responses with a median trade-off of 6.5-10 kg protein to achieve a reduction in calf health issue from 10% to 5%. For example, in the case of Stillbirth, farmers were willing to trade 10 kg protein to lower the number of stillbirths from a hypothetical 10% to 5%.

Surveys are a cost effective and accessible mechanism for gathering feedback from farmers and their advisers. Over time, the insights gathered from similar surveys have provided reliable feedback to guide decision making related to the delivery of breeding values and indexes (Martin-Collado *et al.* 2015; Byrne *et al.* 2016). Surveys raise awareness and can act as an engagement tool in the lead up to delivery of new EBV. Surveys have limitations, such as the potential bias towards internet users as discussed by Martin-Collado *et al.* (2017) or enthusiasm for the subject matter. It is difficult to tell the impact of this bias without having purposeful discussions with farmers that do not participate in surveys because they are worried about scams, are not interested, do not have time and other reasons. The authors agree with findings by Martin-Collado *et al.* (2017) that industry co-ordination to avoid overwhelming farmers with too many surveys is likely a useful strategy to maintain and improve response rates.

CONCLUSIONS

Based on this research, we recommend that industry apply Stillbirth, Calf Survival and Calf Health as the names of potential EBV and that traits should be presented in such a way that higher values are more favourable. Further, EBV should be presented as separate traits rather than as indexes that group traits together, where possible.

Calf traits were highly valued by survey respondents and were scored similarly for importance to traits such as survival and mastitis traits that are already included in Australia's breeding indexes.

ACKNOWLEDGEMENTS

The authors thank survey respondents. We also thank DataGene, Dairy Australia, GippsDairy, Holstein Australia, Jersey Australia, Genetics Australia and MaxCare for distributing the survey.

REFERENCES

Axford M., Santos B., Stachowicz K., *et al.* (2021) *Anim. Prod. Sci.* **61**: 1940.

Axford M.M., Khansefid M., Haile-Mariam M., *et al.* (2024) *J. Dairy Sci.* **107**: 6994.

Byrne T.J., Santos B.F.S., Amer P.R., *et al.* (2016) *J. Dairy Sci.* **99**: 8146.

Dairy Australia (2023) Australian Dairy Sustainability Framework, Dairy Australia, Melbourne.

Edwards K.Y., LeBlanc S.J., DeVries T.J., *et al.* (2024) *JDS communications* **5**: 42.

Martin-Collado D., Byrne T.J., Amer P.R., *et al.* (2015) *J. Dairy Sci.* **98**: 4148.

Martin-Collado D., Hely F., Byrne T.J., *et al.* (2017) *Animal (Cambridge, England)* **11**: 318.

Monks L.C., Scott B.A., Bullen S.L., *et al.* (2021) *Proc. Assoc. Advmt Anim. Breed. Genet.* **24**: 242.

SurveyMonkey Inc. (2024) Available at www.surveymonkey.com.

Proc. Assoc. Advmt. Anim. Breed. Genet. 26: 166-169

CORRELATION OF COW MORTALITY AND CULLING (SALE) RATE WITH SELECTED TYPE TRAITS IN AUSTRALIAN HOLSTEIN AND JERSEY CATTLE

M. Haile-Mariam[1,2], M.E. Goddard[1,4], M. Axford[1,2,3], J. Newton[1,2] and M. Khansefid[1,2]

[1] Agriculture Victoria Research, AgriBio, Centre for AgriBioscience, Bundoora, VIC, 3083 Australia
[2] School of Applied Systems Biology, La Trobe University, Bundoora, VIC, 3083 Australia
[3] DataGene Ltd, 5 Ring Road, Bundoora, VIC, 3083 Australia.
[4] Faculty of Science, University of Melbourne, Parkville, VIC, 3010 Australia

SUMMARY

Over the last 30 years, cow mortality rate (MR) on dairy farms has increased but the proportion of cows that are sold (i.e. culled rate, CR) has remained relatively unchanged. Cow MR on farms is more costly and of greater animal welfare concern than the removal of cows for other reasons. This study assessed the relationship of cow MR and CR with type traits in Holstein (H) and Jersey (J) cows to better understand cow MR and CR. In H and J, the genetic correlations of MR with udder texture (0.38) and angularity (0.57) were positive and high. In H, the genetic correlation between MR and body condition score (BCS) was also moderate. This suggests increased MR is related to negative energy balance (poor BCS, more angularity). On the other hand, J cows with high scores for angularity, udder texture and udder depth which also produce more milk are less likely to be culled (low CR). In H, udder depth is the only trait associated with CR due to associated with udder health. Overall, this study showed that the value of some type traits for predicting cow MR and CR is different.

INTRODUCTION

Selection for longevity has resulted in positive genetic trend for the trait (VanRaden *et al.* 2016). On the other hand, cow mortality rates (MR) have increased from below 1% in 1990 to over 5% in 2015 in some countries (Compton *et al.* 2017; Thomsen 2023; VanRaden *et al.* 2016), though this could partly be due to improved recording. In Australia during the same period, cow MR increased from below 1% to about 4% (Haile-Mariam *et al.* 2023). The impact of increased cow MR on profitability is greater than that of culling because farmers incur additional disposal costs instead of income from the sale cows. Cow deaths are more costly because they usually happen early in lactation (Thomsen 2023). Cow MR is also of greater animal welfare concern than culling cows. In Australia, the correlations of some economic traits with cow MR were genetically different from those with culling rate (CR). For example, cow MR has unfavourable genetic correlation with milk yield (MY) but favourable correlation with fertility whereas CR had favourable correlations with MY traits. Furthermore, the analyses showed (Haile-Mariam *et al.* 2023) that the relationship between MY and CR was weaker in Holstein (H) than in Jersey (J). The high genetic correlation between MY traits and MR could be related to an increased incidence of metabolic diseases associated with negative energy balance (Schmidtmann *et al.* 2023). Among type traits, traits such as body condition score (BCS) and angularity may be associated with energy balance and resilience (Bengtsson *et al.* 2022; Schmidtmann *et al.* 2023). Therefore, the objective of this study was to assess if the associations of MR with some type traits are different from those with CR. Knowledge of the relationship of MR and CR with type traits can be used to develop approaches to improve health, resilience, and fertility of dairy cattle.

MATERIALS AND METHODS

The data for this study, described by Haile-Mariam *et al.* (2023), comes from Australian dairy herds that record cow termination dates, reasons, and pedigree information. Terminations were categorised as either death or sale. The study includes H and J cows born between January 1997 and 2020 and calved at least once between 1999 and 2023. As the recording of termination data is voluntary in Australia, we selected cows from herd-years with more than 30 calving events and total exit rates between 1.5% and 75% (Haile-Mariam *et al.* 2023). This resulted in removing 47% of herd-years and 18% of the calvings. For defining MR, the cows removed from the herd due to death during the lactation were coded 1 and those that remained in the herd or removed due to culling were coded 0. Similarly, for defining CR, the cows that were removed from the herd for all reasons except for death were coded 1 and the rest including those that died on the farm were coded 0. All cows were coded 0 at the end of each lactation before they are culled or died.

To estimate the correlations of MR and CR with type traits the two datasets were merged. The type traits selected are previously known to be related to survival, fertility, and mastitis resistance (Haile-Mariam *et al.* 2014) in Australian dairy cattle. The type data used for this study included 436,896 cows that had linear type scores on a scale of 1 to 9 and BCS on a scale of 1 to 8 (Haile-Mariam *et al.* 2014). The mean days in milk at classification were 138 and 130 days in H and J cows and the age at classification was 31 in H and 30 months in J cows, respectively. Table 1 shows the number of cows and the mean with the standard deviation of the traits. After the edits, only 25% of the H cows and 12% of the J cows had data on type traits like Angularity but only 23% of the H and less than 3% of the J cows had BCS data. To improve the accuracy of genetic correlation estimates of MR and CR with type traits, some cows with type data but without termination data were included in these analyses. Genetic and residual correlations between MR or CR and type traits were estimated using bivariate models by analysing MR and CR with one type trait (Table 1) at a time. When analysing MR and CR, the fixed effects fitted were: dry-off of year of lactation or termination year of cow and dry-off of month of lactation or termination month of cow, age at calving within parity, and herd-year-season of calving. Age at calving was fitted as a covariate (linear and quadratic). The fixed effects fitted when analysing the type traits were age (in months) and days at classification and age at calving (in months) fitted as covariates and month of calving and herd-classifier-round fitted as classes. When estimating the correlation of MR and CR with BCS, body depth (BD) was also fitted as a covariate. This was done after a preliminary analysis showed that adjusted BCS was more strongly correlated (-0.39) with fertility (calving interval) than the unadjusted (-0.28) suggesting that cows with deep body were given a higher BCS than others. The random effect fitted was the sires of the cows with the numerator relationship between the sires.

Table 1. Description of the data and the mean and standard deviation (SD) for important traits for Holstein (no. of lactations, 4791600) and Jersey (no. of lactations. 752959) cows

Trait	Holstein		Jersey	
	No. cows	Mean (SD)	No. cows	Mean (SD)
Mortality rate (%)	1460179	3.56(18.54)	216623	3.10(17.33)
Culling rate (%)	1460179	17.35(37.87)	216623	16.44(37.06)
Angularity	436227	5.85(1.10)	65677	6.48(0.91)
Body condition score	387157	3.98(0.69)	7221	3.98(0.74)
Body depth	436083	6.03(1.23)	65677	6.17(0.89)
Udder depth	436210	5.54(1.38)	65692	4.72(1.10)
Udder Texture	436226	6.12(1.19)	65692	6.43(0.95)

RESULTS AND DISCUSSION

The focus of this study was the genetic and residual correlations of MR and CR with selected type traits (Table 2) and builds on our earlier report of heritability for MR and CR (Haile-Mariam *et al.* (2023). In both breeds, the likelihood of death is higher for cows that are more angular and with higher scores for udder texture (Table 2). On the other hand, the possibility of culling was lower for cows with higher angularity or higher udder texture scores in J because high MY is an important reason for keeping cows to continue production (Haile-Mariam *et al.* 2025). In the current data, the genetic correlation between angularity and MY was higher in J (0.64±0.04) than in H (0.27±0.02). The genetic correlation of BCS with MR was favourable in H and the correlation became stronger (from -0.37 to -0.45) when BCS was adjusted for BD (Table 2). In J the relationship between BCS (unadjusted) and MR was low (Table 2) and was only -0.07 when BCS was not adjusted for BD. However, this may be partly because the number of J cows with BCS data was limited. On the other hand, BCS had negligible association with CR. The genetic correlations of BD with both MR and CR are similar in H but not in J (Table 2). Overall, in J, the genetic correlations of MR with type traits are different from those with CR which agrees with the negative correlation (-0.31) of MR with CR (Haile-Mariam *et al.* 2023).

Table 2. Genetic and residual correlations of mortality and culling with type traits including body condition score (BCS) in Holstein and Jersey cattle with standard errors (±)

Traits	Holstein			
	Mortality		Culling	
	Genetic	Residual	Genetic	Residual
Angularity	0.57±0.03	0.03±0.00	0.26±0.03	-0.04±0.00
BCS adjusted for BD	-0.45±0.04	-0.03±0.00	-0.08±0.03	-0.01±0.00
BCS unadjusted for BD	-0.37±0.04	-0.03±0.00	-0.08±0.03	-0.01±0.00
Body depth (BD)	0.35±0.03	0.01±0.00	0.38±0.02	-0.01±0.00
Udder depth	-0.12±0.03	-0.01±0.00	-0.34±0.02	-0.02±0.00
Udder texture	0.43±0.03	0.03±0.00	0.07±0.03	-0.04±0.00
	Jersey			
Angularity	0.44±0.08	0.03±0.01	-0.38±0.06	-0.10±0.01
BCS adjusted for BD	-0.13±0.19	0.03±0.03	0.13±0.16	-0.03±0.03
BCS unadjusted BD	-0.07±0.19	0.03±0.03	0.09±0.16	-0.04±0.03
Body depth (BD)	0.34±0.08	0.01±0.01	-0.07±0.06	-0.06±0.01
Udder depth	-0.20±0.09	0.00±0.01	-0.46±0.05	-0.04±0.01
Udder texture	0.38±0.08	0.01±0.01	-0.57±0.06	-0.09±0.01

The residual correlations between MR and type traits are small, but in the same directions as the genetic correlations for all the type traits considered in this study in H. Phenotypically for every unit increase in angularity and udder texture score in H, MR increased by 0.09±0.03% and 0.12±0.02%, respectively. In J, MR increased by 0.11±0.09% with angularity which was not statistically significant. On the other hand, in H for every unit increase in udder texture, angularity, and udder depth, CR (%) was reduced by -0.73±0.05, -0.59±0.06, -0.41±0.05, respectively. In J cows, the corresponding values were -1.55±0.19, -2.28±0.19, and -0.94±0.16. The reduction in CR in both breeds with increasing scores for angularity and udder texture suggests that farmers are reluctant to cull cows with high type scores and want to retain cows with high MY (Haile-Mariam *et al.* 2023). Unlike angularity and udder texture, the correlations and the effect of udder depth on CR are more the outcome of cows with shallow udder (i.e., high score) having better udder health than cows with deep udder, despite their poor MY (Abdelsayed *et al.* 2017). However, it is worth noting that all the phenotypic relationships observed in the current study could be biased because cows that are removed from the herd early before type scoring and data of all cows in herds that do not participate in type classification could not be included.

The unfavourable genetic correlation of MR with angularity and udder texture and its favourable correlation with BCS in the current study, plus the associations between cow MR and high early MY (0.36 to 0.41) and poor fertility (0.54 to 0.68) in our previous study (Haile-Mariam *et al.* 2023) shows these type traits can serve as predictors of poor fertility (Haile-Mariam *et al.* 2014) and poor health that may lead to an increased risk of death. That cow MR data and the data of cows that exit the herd before the end of the lactation have similar genetic correlations with these type traits which was shown in Haile-Mariam *et al.* (2025) means by selective use of culling and MR data it is possible to develop genetic evaluations for health and resilience traits. For example, the genetic correlation estimates in Haile-Mariam *et al.* (2025) show that the likelihood of cow exit, especially up to 180 days, was higher for thin (poor BCS) and more angular cows. This implies that cows with high potential for MY, poor fertility, poor BCS, and high scores for angularity are more likely to exit early due to metabolic stress or die on the farm. Here the link between cow MR and early culling before the end of the lactation is that in both cases most cows exit the herd early in lactation. This agrees with the literature (Lassen *et al.* 2003; Schmidtmann *et al.* 2023) that showed moderate genetic correlations between some health and type traits.

CONCLUSIONS

MR has moderate to high genetic correlation with angularity and udder texture in both H and J suggesting that type these traits can be used as predictors of MR. In H, the genetic correlation between MR and BCS was also moderate suggesting these traits can serve as an indicator of negative energy balance which may lead to increased cow MR and an increased likelihood of exiting before the end of the lactation. On the other hand, type traits associated with high MY (angularity, udder texture) result in a decreased likelihood of CR, particularly in J cows. Overall, this study demonstrates that the usefulness some of type traits for predicting cow MR and CR is different.

ACKNOWLEDGEMENTS

This study was part of the DairyBio research program (funded by Agriculture Victoria, Dairy Australia, and the Gardiner Foundation, Melbourne, Australia). Data used for this study are obtained from DataGene. We thank farmers and breed organisations for supplying the data.

REFERENCES

Abdelsayed M., Haile-Mariam M. and Pryce J.E. (2017) *J. Dairy Sci.* **100**:9643.

Bengtsson C., Thomasen J.R., Kargo M., *et al.* (2022) *J. Dairy Sci.* **105**: 7588.

Compton C.W.R., Heuer C., Thomsen P.T., *et al.* (2017) *J. Dairy Sci.* **100**:1.

Gilmour A. R., Gogel B. J., Cullis B. R. and Thompson R. (2015) ASReml User Guide Release 4.0. VSN International Ltd., Hemel Hempstead, UK.

Haile-Mariam M., Gonzalez-Recio O. and Pryce J.E. (2014). *J. Dairy Sci.* **97**: 3173.

Haile-Mariam M., Khansefid M., Axford M., *et al.* (2023) *J. Dairy Sci.* **106**: 7880.

Haile-Mariam M., Goddard M.E., Newton J., *et al.* (2025) *J. Dairy Sci. In Press.*

Lassen J., Hansen M., Sørensen M.K., *et al.* (2003) *J. Dairy Sci.* **86**: 3730.

Schmidtmann C., Segelke D., Bennewitz J., *et al.* (2023). *J Dairy Sci.* **106**: 421.

Thomsen P.T. (2023) *Vet Rec.* **193**: 179.

VanRaden P.M., Wright J.R., Tooker M.E. and Norman H.D. (2016) *Interbull Bulletin No. **50**.* Puerto Varas, Chile, October 24 - 28, 2016.

UPDATING THE GENETIC PREDICTION MODEL FOR COW SURVIVAL IN AUSTRALIA

M. Khansefid[1,2], J.E. Pryce[1,2], J.E. Newton[1,2], M. Axford[1,2,3], G. Nieuwhof[3], M.E. Goddard[1,4] and M. Haile-Mariam[1,2]

[1] Agriculture Victoria Research, AgriBio, Centre for AgriBioscience, 5 Ring Road, Bundoora, VIC, 3083 Australia
[2] School of Applied Systems Biology, La Trobe University, Bundoora, VIC, 3083 Australia
[3] DataGene Ltd, 5 Ring Road, Bundoora, VIC, 3083 Australia
[4] Faculty of Veterinary and Agricultural Sciences, The University of Melbourne, Parkville, VIC, 3010 Australia

SUMMARY

Cow survival, defined as the ability of the cow to live from first calving to culling or death, is the second most important trait in Australia's main dairy economic selection index, the Balanced Performance Index (BPI), following protein yield. Defining this trait as success or failure in surviving across lactations is suitable for countries with predominantly seasonal, pasture-based dairy systems, where exact culling dates are often unknown. In this paper, we explored the impact of the new bi- and multi-variate prediction models on the estimated breeding values (EBVs) for survival and the correlation between the new survival EBVs and the EBVs for important traits and selection indices. The correlation between the survival EBVs of the old and updated models was high (~0.90). However, new survival EBVs showed a weaker correlation with milk traits compared to old survival. Despite this, the correlation between milk traits (combined in the Australian Selection Index, ASI) and new survival EBVs remained stronger in Jerseys (0.49) compared to Holsteins (0.29). Furthermore, the correlations between the new survival and somatic cell count EBVs increased in Holstein (from 0.55 to 0.58) and Jersey (from 0.09 to 0.17). Similarly, the correlations between the new survival and fertility EBVs also increased substantially in Holsteins (from 0.38 to 0.51) compared to Jerseys (from 0.11 to 0.18). Hence, the new survival EBVs strongly correlated with milk traits in Jerseys and with fertility in Holsteins. In addition to the direct contribution of survival to BPI, the increased correlations of the new survival EBVs with fertility and somatic cell count could be beneficial for reducing mastitis and improving fertility.

INTRODUCTION

Improving cow survival, defined either as productive life or survival across lactations, brings economic, animal welfare, and environmental benefits to dairy farming (Hu *et al.* 2021). Cow survival has a high relative weight in BPI (~8%) and Sustainability Index, SI (~13%). The SI has attracted more attention recently due to greater emphasis on the traits that contribute to reducing emissions intensity, such as production, survival and feed saved, which helps dairy farmers fast-track genetic gain for reduced emissions intensity (Richardson *et al.* 2022). Selecting the cow survival trait can also improve resilience, fertility and disease resistance (Barkema *et al.* 2015).

According to our published results, the reliability of EBVs for survival could be improved by: 1) defining cow survival across lactations rather than across years that allows us to take into account the effect of parity, age at calving, month of calving and herd–year–season of calving in the model, 2) partitioning survival to early survival (survival from the first to the second lactation) and late survival (survival from the second to subsequent lactations) and, consider survival as two different but genetically correlated traits in primiparous and multiparous cows (we indirectly incorporate the culling reasons into genetic evaluations of survival because culling reasons vary between the first and subsequent lactations), and 3) including likeability, pin set and udder depth as predictor traits in a multi-trait linear model for genetic evaluation (Khansefid *et al.* 2021; Khansefid *et al.* 2023).

In this paper, we explore the impact of the new prediction models on the survival EBVs and their EBVs correlations with the important traits and selection indices.

MATERIALS AND METHODS

Data on calving events, milk records, termination dates, workability scores and classification were received from DataGene and Holstein Australia. Although the raw data used in this study was the same as Khansefid *et al.* (2023), we excluded cows born before 1988 due to the limited and low quality of historical survival data. The variance components in a multi-trait prediction model were estimated in Holstein and Jersey breeds using ASReml 4.2 (Gilmour *et al.* 2022) as follows:

$$\begin{pmatrix} \mathbf{y}_1 \\ \mathbf{y}_2 \\ \vdots \\ \mathbf{y}_5 \end{pmatrix} = \begin{pmatrix} \mathbf{X}_1 & \mathbf{0} & \cdots & \mathbf{0} \\ \mathbf{0} & \mathbf{X}_2 & & \mathbf{0} \\ \vdots & & \ddots & \vdots \\ \mathbf{0} & & \cdots & \mathbf{X}_5 \end{pmatrix} \begin{pmatrix} \mathbf{b}_1 \\ \mathbf{b}_2 \\ \vdots \\ \mathbf{b}_5 \end{pmatrix} + \begin{pmatrix} \mathbf{Z}_1 & \mathbf{0} & \cdots & \mathbf{0} \\ \mathbf{0} & \mathbf{Z}_2 & & \mathbf{0} \\ \vdots & & \ddots & \vdots \\ \mathbf{0} & & \cdots & \mathbf{Z}_5 \end{pmatrix} \begin{pmatrix} \mathbf{u}_1 \\ \mathbf{u}_2 \\ \vdots \\ \mathbf{u}_5 \end{pmatrix} + \begin{pmatrix} \mathbf{e}_1 \\ \mathbf{e}_2 \\ \vdots \\ \mathbf{e}_5 \end{pmatrix}$$

where, $\mathbf{y}_1$ and $\mathbf{y}_2$ contain the early and late survival records, $\mathbf{y}_3$, $\mathbf{y}_4$ and $\mathbf{y}_5$ contain predictor traits: likeability, pin set and udder depth scores; $\mathbf{b}_1$ to $\mathbf{b}_5$ contain the fixed effects (early survival: age at calving, month of calving and herd–year–season of calving; late survival: same to early survival and parity; likeability: age, calving month and herd–year–season of first calving; pin set and udder depth: age at calving, calving month, days in milk at classification date and herd–year–classifier); $\mathbf{u}_1$ to $\mathbf{u}_5$ and $\mathbf{e}_1$ to $\mathbf{e}_5$ are vectors of EBVs and residual effects which are driven from $\sim$ N(0, $\begin{pmatrix} \mathbf{A}\sigma^2_{u\,1} & \mathbf{A}\sigma_{u\,1,2} & \cdots & \mathbf{A}\sigma_{u\,1,5} \\ \mathbf{A}\sigma_{u\,2,1} & \mathbf{A}\sigma^2_{u\,2} & & \\ \vdots & & \ddots & \vdots \\ \mathbf{A}\sigma_{u\,5,1} & & \cdots & \mathbf{A}\sigma^2_{u\,5} \end{pmatrix}$) and $\sim$ N(0, $\begin{pmatrix} \mathbf{I}\sigma^2_{e\,1} & \mathbf{0} & \cdots & \mathbf{I}\sigma_{e\,1,5} \\ \mathbf{0} & \mathbf{I}\sigma^2_{e\,2} & & \\ \vdots & & \ddots & \vdots \\ \mathbf{I}\sigma_{e\,5,1} & & \cdots & \mathbf{I}\sigma^2_{e\,5} \end{pmatrix}$), respectively; $\mathbf{X}_1$ to $\mathbf{X}_5$, and $\mathbf{Z}_1$ to $\mathbf{Z}_5$ are design matrices relating phenotypes to the corresponding fixed and random effects.

Similar to the current evaluation model for survival in Australia, all pure and crossbred cows were included in a single genetic evaluation model without accounting for breed effects. We ran the model using Mix99 (Lidauer *et al.* 2013) with just early and late survival (bivariate model) and also including the predictor traits (multivariate model). The variance components estimated using the Holstein breed data were used to estimate EBVs in the model as the data was dominated by Holstein cows and the variance components in Holstein and Jersey breeds were very similar. The EBVs for early and late survival were combined (0.3 × early survival + 0.7 × late survival) to produce one EBV for overall survival. To assess the impact of using the new method for calculating EBVs for cow survival, we calculated the correlation between the new survival EBVs and indices as well as selected traits in Holstein and Jersey bulls. We used the recently published EBVs from the DataGene website (241203_NasisAll.201.csv; accessed December 2024). The Holstein (n = 7,854) and Jersey (n = 1,558) sires with BPI reliability $\geq$ 40 were selected. All selected bulls had more than 50 daughters with survival scores, and almost all had reliable EBVs for the traits in BPI.

RESULTS AND DISCUSSION

The estimated heritabilities (h^2), genetic and phenotypic correlations between traits in the multi-trait model are shown in Table 1. The correlation between early and late survival, and their correlation with predictor traits were close in Jersey and Holstein breeds, making it feasible to run a single evaluation for all pure and crossbred animals. The standard errors (s.e.) of the estimates associated with Holstein were smaller than those in the Jersey breed due to the larger sample size in Holstein. The genetic correlation between early and late survival was high. In Holstein, the traits most strongly correlated with early and late survival were likeability and udder depth, respectively.

Table 1. The estimated heritabilities (diagonal), as well as genetic (lower triangular) and phenotypic (upper triangular) correlations (r) ± s.e. between traits in Holsteins and Jerseys

$r_{genetic}$ / h^2 / $r_{phenotypic}$		Early Survival	Late Survival	Likeability	Pin Set	Udder Depth
Holstein	Early Survival	**0.035±0.001**	0.007±0.001	0.168±0.001	0.019±0.001	0.019±0.001
	Late Survival	0.785±0.026	**0.036±0.001**	0.076±0.001	0.025±0.001	0.043±0.001
	Likeability	0.529±0.043	0.250±0.058	**0.143±0.004**	0.002±0.001	-0.011±0.001
	Pin Set	0.154±0.057	0.141±0.049	-0.005±0.025	**0.338±0.008**	-0.044±0.001
	Udder Depth	0.190±0.126	0.364±0.143	-0.043±0.061	-0.037±0.046	**0.385 ± 0.009**
Jersey	Early Survival	**0.047±0.004**	0.008±0.001	0.226±0.002	0.012±0.002	0.021±0.005
	Late Survival	0.774±0.042	**0.040±0.002**	0.118±0.001	0.012±0.002	0.030±0.003
	Likeability	0.748±0.093	0.480±0.094	**0.226±0.014**	-0.021±0.001	-0.021±0.002
	Pin Set	0.209±0.125	0.204±0.099	-0.098±0.062	**0.265±0.020**	-0.026±0.002
	Udder Depth	0.037±0.449	0.356±0.280	0.032±0.218	-0.007±0.084	**0.272±0.019**

The correlation between survival EBVs and selection indices, and important traits EBVs in Holstein and Jersey sires are shown in Figures 1 and 2, respectively. The updated survival definition affected the correlation between survival and the SI due to the larger contribution of survival in SI compared to BPI. Compared to the old model, the correlation between ASI and survival EBVs was reduced, but it was still stronger in Jersey (0.49) than in Holstein (0.29). The correlations between new survival EBVs and fertility, somatic cell count, and udder depth were increased in Holstein and Jersey. Although the EBVs in bi- and multivariate models were similar, the predictor traits are expected to improve the reliability of survival EBVs for bulls, particularly with a limited number of daughters (Khansefid *et al.* 2021). Negative correlations between survival and feed saved, as well as heat tolerance in Jersey, may be due to complex trait interactions influencing cow survival.

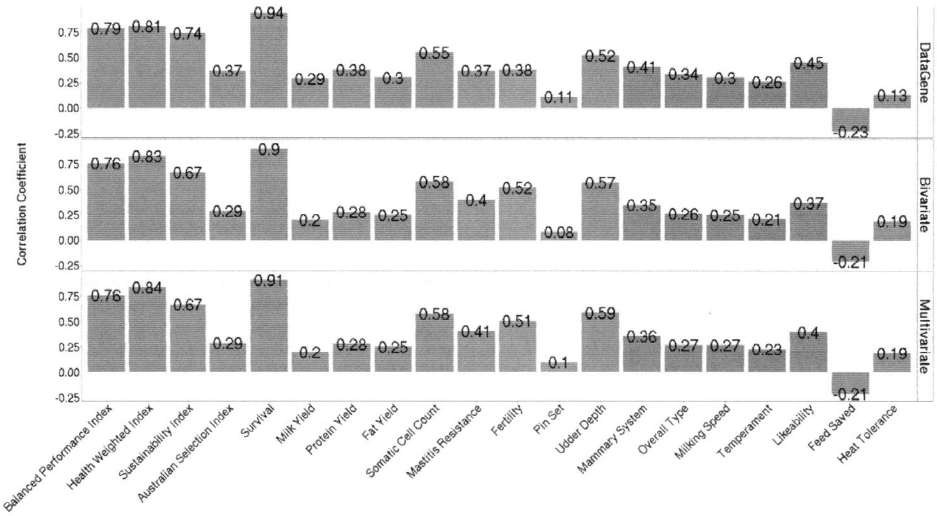

Figure 1. The correlation between current DataGene as well as new survival (bi- and multivariate models) EBVs and selection indices and important traits EBVs in Holstein sires

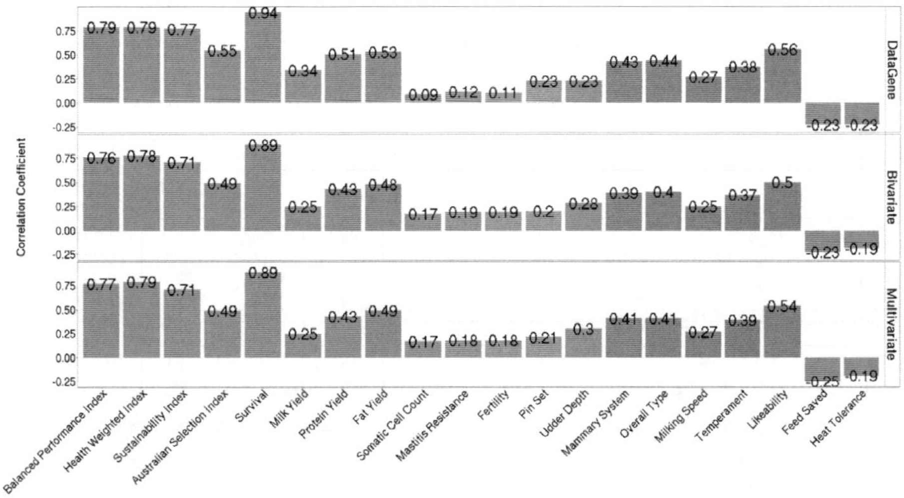

Figure 2. The correlation between current DataGene as well as new survival (bi- and multi-variate models) EBVs and selection indices and important traits EBVs in Jersey sires

CONCLUSION

Compared to current survival EBVs, the new EBVs show a weaker correlation with ASI and milk traits in both Holstein and Jersey breeds. However, the correlation between new survival and somatic cell count, as well as fertility increased. Therefore, in addition to the direct contribution of survival to BPI, selecting a new definition of survival could be beneficial for reducing mastitis, and improving fertility, and potentially resilience and health traits.

ACKNOWLEDGEMENTS

We appreciate financial support from DairyBio, a joint venture of Dairy Australia, The Gardiner Foundation, and Agriculture Victoria (Melbourne, Australia). The authors would like to acknowledge DataGene and Holstein Australia for providing the data for this study.

REFERENCES

Barkema H.W., von Keyserlingk M.A.G., Kastelic J.P., Lam T.J.G.M., Luby C., Roy J.P., LeBlanc S.J., Keefe G.P. and Kelton D.F. (2015) *J. Dairy Sci.* **98**: 7426.

Gilmour A.R., Gogel B.J., Cullis B.R., Welham S.J. and Thompson R. (2022) *VSN International Ltd, Hemel Hempstead, HP1 1ES, UK.* www.vsni.co.uk

Hu H., Mu T., Ma Y., Wang X. and Ma Y. (2021) *Front. Genet.* **12**: 695543.

Khansefid M., Haile-Mariam M. and Pryce J.E. (2021) *Anim. Prod. Sci.* **61**: 1828.

Khansefid M., Pryce J.E., Shahinfar S., Axford M., Goddard M.E. and Haile-Mariam M. (2023) *Anim. Prod. Sci.* **63**: 1031.

Lidauer M., Matilainen K., Mäntysaari E., Pitkänen T., Taskinen M., Strandén I. (2013) General program for solving large mixed model equations with preconditioned conjugate gradient method. Release VII/2013 beta

Richardson C.M., Amer P.R., Quinton C., Crowley J., Hely F.S., van den Berg I. and Pryce J.E. (2022) *J. Dairy Sci.* **105**: 4272.

Proc. Assoc. Advmt. Anim. Breed. Genet. 26: 174-177

SELECTION FOR FIRST PARITY LINEAR-TYPE TRAITS IN THE NEW ZEALAND DAIRY CATTLE POPULATION

M.A. Nilforooshan, R.G. Sherlock and B.L. Harris

Livestock Improvement Corporation, Hamilton 3240, New Zealand

SUMMARY

Unlike production traits, linear-type traits are subjective and expensive to measure. Recording, genetic evaluation and selection for these traits is generally done on the first parity records (2 years of age). It is, however, important to know whether this selection is effective in improving later-in-life performances. This can be assessed by estimating the genetic correlation between early-in-life and later-in-life evaluations of the same trait. Data were extracted from the animals with later-in-life (parities 2–5) phenotypes, and genetic correlations were estimated between the same trait early- and later in life. The estimated genetic correlations of greater than 0.92 indicated that the benefits of early-in-life selection for these traits were retained later in life.

INTRODUCTION

Linear-type traits are among the most economically important traits in the dairy cattle industry. These traits cover and evaluate various aspects of the cow's body conformation, including udder shape and attachment, body frame, and leg structure. Linear-type traits are genetically correlated with production traits (Lin *et al.* 1987), udder health (Rogers *et al.* 1991), lameness (Boettcher *et al.* 1998), and survival (Williams *et al.* 2022). Since these traits are frequently scored in the first parity and genetically correlated with survival in older parities, Williams *et al.* (2022) reported these traits were suitable early predictors of survival for later parity cows. Currently, 15 linear-type traits are routinely evaluated in the New Zealand dairy cattle population (Harris 2022).

It is important to know whether early-in-life (EIL) evaluations are good indicators of later-in-life (LIL) performances for linear-type traits and whether animals selected for EIL performances will perform well LIL. In that case, EIL selection will be effective over an animal's productive life, and the superiority will be passed on to the next generations. This concept can be reformulated to determine whether EIL and LIL observations on the same trait are genetically similar or how strong the genetic × age interaction is for linear-type traits.

In the present study, the effectiveness of EIL (first parity) selection for linear-type traits on the LIL genetic merits was investigated by estimating the genetic correlations between EIL and LIL of the same traits and on the next generation's performance by estimating the heritabilities, in the NZ dairy cattle populations.

MATERIALS AND METHODS

Data. Late (after the first parity or 2 years of age) records on 12 linear type traits were recorded on 7,796 cows, once per cow. Six of these traits were udder conformation traits. Trait names and abbreviations are presented in Table 1. Recording inspections occurred in November-December 2023 and February-March 2024. Contemporary groups were defined as Herd-Inspector-Season. However, since each herd was recorded once and there was no inspector overlap between herds, this definition was equivalent to "Herd". A single record from parity 6 and 67 records from small contemporary groups with less than 5 observations were discarded. All the 7,728 remaining cows in 93 contemporary groups had a single LIL phenotype for all the linear type traits. Table 1 presents the descriptive statistics of the LIL linear type scores. The full definition of the traits can be found in DairyNZ 2020.

Table 1. Trait names and abbreviations, and descriptive statistics of the 7,728 late-in-life (parities 2–5) scores of linear-type traits. The maximum of all the traits is 9

Trait name	Trait abbr.	Min.	Median	Mean	SD
Capacity	CAP	3	7	6.69	0.99
Dairy Conformation	DC	2	7	6.48	1.16
Front Teat Placement	FT	1	5	4.81	0.68
Fore Udder Attachment	FU	1	6	5.65	1.25
Rump Angle	RA	1	4	3.95	0.80
Rear Leg Set	RLS	3	6	6.00	0.61
Rear Teat Placement	RT	2	6	6.49	1.16
Rear Udder Attachment	RU	1	6	5.96	1.12
Rump Width	RW	2	6	5.53	1.08
Stature	ST	2	6	5.75	1.23
Udder Overall Score	UO	1	6	5.80	1.20
Udder Support	US	1	6	6.32	1.14

The definition of EIL contemporary groups (age-'calving season'-herd) differed from that of LIL contemporary groups (herd). For variance components estimation, a pedigree subset including animals with various degrees of relatedness (closely related to unrelated) with the 7,728 animals was extracted from the whole pedigree. Contemporary groups corresponding to EIL records containing any of the 7,728 animals with LIL records, their dams, full-sisters, half-sisters, and progeny were identified. Four generations of parents were extracted from 474,560 animals in those contemporary groups, resulting in a pedigree of 1,369,575 animals and 474,560 EIL animals with phenotype in 7,902 EIL contemporary groups. There were 7,618 animals with phenotypes for both EIL and LIL. Descriptive statistics of the EIL phenotypes are presented in Table 2.

Table 2. Descriptive statistics of the 474,560 early-in-life (parity 1) scores of linear type traits. The minimum and maximum of all the traits are 1 and 9

Trait abbr.*	Median	Mean	SD
CAP	6	6.26	1.03
DC	6	6.31	1.1
FT	5	4.58	0.69
FU	6	5.76	1.12
RA	4	4.44	0.77
RLS	6	6.15	0.63
RT	6	6.16	1.03
RU	6	5.86	1.09
RW	6	5.79	0.98
ST	5	5.50	1.31
UO	6	5.72	1.13
US	6	6.01	1.09

*The full trait names are given in Table 1.

Model. A bivariate model was fitted to EIL (parity 1) and LIL (parities 2–5) of each linear-type trait. Due to the limited number of LIL phenotypes, parities 2 to 5 were combined into a single trait, where a single phenotype was available per phenotyped animal. Since EIL and LIL phenotypes were recorded at different ages, the two were considered residually independent (i.e., residual covariance equal to 0). Both traits (EIL and LIL of each linear-type trait) included random effects of direct additive genetic and residuals. The fixed effects were contemporary groups and covariates for genetic groups, age at calving (AAC), AAC^2, inbreeding coefficient, breed proportions, and heteroses. The EIL model also included Legendre Polynomials of order 4 covariates for days-in-milk. The variance components (genetic covariances and residual variances) were estimated using the AI-REML procedure of the APEX software (GENOMIC COMPUTING PTY LIMITED 2023).

RESULTS AND DISCUSSION

The genetic covariances and residual variances were estimated from the EIL and LIL bivariate analysis of each linear-type trait. Table 3 presents the heritability and genetic correlation estimates from each analysis.

Table 3. The heritability (h^2) and genetic correlation (r_g) estimates between the early-in-life (EIL, parity 1) and later-in-life (LIL, parities 2–5) variates of each linear type trait, in bivariate analyses

Trait abbr.*	h^2_{EIL}	h^2_{LIL}	r_g
CAP	0.270	0.375	0.957
DC	0.204	0.414	0.937
FT	0.267	0.350	0.925
FU	0.245	0.427	0.935
RA	0.331	0.541	0.969
RLS	0.096	0.101	0.995
RT	0.370	0.502	0.969
RU	0.254	0.469	0.935
RW	0.244	0.397	0.929
ST	0.492	0.658	0.974
UO	0.244	0.443	0.935
US	0.257	0.391	0.927

*The full trait names are given in Table 1.

The estimated genetic correlations ranged from 0.925 (FT) to 0.995 (RLS). These high genetic correlations indicate low genetic × age interactions and that EIL linear type traits genetically remain the same trait over time. As such, EIL evaluations are suitable indicators of the corresponding LIL linear-type trait, and cows selected for EIL linear-type traits genetically tend to perform better later in life. Aside from the RLS low heritability, heritabilities were moderate to high, showing great potential for selection on these traits and large gain transferred to the next generation. The LIL heritabilities were higher than those for EIL. The LIL heritabilities are expected to be lower than the EIL heritabilities due to EIL selection. However, higher LIL heritabilities might be due to the low number of LIL records, low selection intensity on the EIL linear type traits, and less important environmental effects in LIL than in EIL for the linear type traits.

There were 2,561, 2,249, 1,705, and 1,213 phenotypes in parity 2 to 5, respectively. With 33.1% of LIL phenotypes belonging to parity 2, there was a risk that the high estimated genetic correlations between EIL and LIL could be due to the short age gap between parity 1 (EIL) and parity 2. In another attempt, the age gap between EIL and LIL was increased by discarding the second parity phenotypes and the variance components estimation analyses were repeated. The results are presented in Table 4. Discarding the second parity, LIL phenotypes caused decreases in the heritabilities of both EIL and LIL traits except for CAP-LIL and RLS-LIL. The estimated genetic correlations between EIL and LIL decreased for most of the traits. Those that did not decrease had a negligible increase. The estimated genetic correlations remained greater than 0.9, indicating that the conclusions made based on the previous results remained valid.

Table 4. The heritability (h^2) and genetic correlation (r_g) estimates between the early-in-life (EIL, parity 1) and later-in-life (LIL, parities 3–5) variates of each linear type trait, in bivariate analyses

Trait abbr.*	h^2_{EIL}	h^2_{LIL}	r_g
CAP	0.267	0.393	0.914
DC	0.199	0.358	0.938
FT	0.264	0.327	0.927
FU	0.241	0.404	0.937
RA	0.324	0.541	0.961
RLS	0.095	0.116	0.977
RT	0.366	0.486	0.966
RU	0.249	0.446	0.935
RW	0.241	0.371	0.945
ST	0.490	0.657	0.975
UO	0.237	0.411	0.902
US	0.253	0.360	0.915

*The full trait names are given in Table 1.

CONCLUSION

Linear-type traits are subjective and expensive to measure. These traits are usually recorded and evaluated in the first parity when the cow is 2 years of age (EIL). Selection is then performed on the EIL evaluations. Whether the selected cows will be genetically superior and perform well later in life is of interest. The high estimated genetic correlations between EIL and LIL of the same linear-type trait confirmed that selection decisions made on EIL evaluations improve LIL performances, and EIL is a good indicator of LIL for linear-type traits. The moderate to high estimated heritabilities showed that selecting these traits will improve the EIL and LIL of the next generation.

REFERENCES

Boettcher P.J., Dekkers J.C., Warnick L.D. and Wells S.J. (1998) *J. Dairy Sci.* **81**: 1148.

DairyNZ (2020) https://www.dairynz.co.nz/media/xsmce10a/top-booklet-2020.pdf (Accessed 10/10/2024).

Genomic Computing Pty Limited (2023) APEX – Linear model suite. https://ghpc.ai/ (Accessed 10/10/2024).

Harris B.L. (2022) *JDS Commun.* **3**: 152.

Lin C.Y., Lee A.J., McAllister A.J., Batra T.R., Roy G.L., Veseley A.J., Wauthy J.M. and Winter K.A. (1987) *J. Dairy Sci.* **70**: 2385.

Rogers G.W., Hargrove G.L., Lawlor T.J. and Ebersole J.L. (1991) *J. Dairy Sci.* **74**: 1087.

Williams M., Sleator R.D., Murphy C.P., McCarthy J. and Berry D.P. (2022) *J. Dairy Sci.* **105**: 7550.

Proc. Assoc. Advmt. Anim. Breed. Genet. 26: 178-181

EVALUATING GENETIC AND PHENOTYPIC RELATIONSHIPS OF SIX TRAITS OTHER THAN PRODUCTION (TOP) IN DAIRY CATTLE PRIOR TO THEIR ADOPTION IN NEW ZEALAND

R.C. Handcock[1,2], D.P. Garrick[2,3] and M.A. Stephen[1,2]

[1] DairyNZ Ltd., Hamilton, Waikato, New Zealand
[2] AL Rae Centre for Genetics and Breeding, Massey University, Waikato, New Zealand
[3] The Helical Company Limited, Rotorua, New Zealand

SUMMARY

This study examined genetic and phenotypic correlations between six internationally recognised Traits Other than Production (TOP) with 12 existing TOP traits routinely scored in New Zealand dairy cattle. Scored on 1,386 first lactation Holstein-Friesian heifers, the aim was to assess whether evaluating these traits could provide additional information for breeding decisions. Of the six new traits, Body Depth (BD) and Udder Depth (UD) showed the greatest genetic variance and heritability, while Chest Width (CW) and Rear Legs Rear View (RLRV) had the lowest. All new traits were moderately to strongly correlated positively (>0.65; Central Ligament [CL], UD, CW and BD) or negatively (<-0.65; RLRV and Foot Angle [FA]) with at least one existing TOP trait. These findings suggest limited potential for capturing additional genetic information beyond what is already available, although some new traits may still serve as useful predictors due to their higher heritability.

INTRODUCTION

The routine scoring of Traits Other than Production (TOP) supports genetic evaluations in New Zealand's (NZ) dairy sector (DairyNZ 2024). Internationally recognised traits, including Central Ligament (CL), Udder Depth (UD), Chest Width (CW), Body Depth (BD), Rear Legs Rear View (RLRV), and Foot Angle (FA) (Brotherstone 1994; Interbull 2024; Williams *et al.* 2022), are not routinely scored in NZ. Incorporating these traits could improve the evaluation of conformation traits and enhance alignment with international genetic evaluation systems, facilitating data exchange with Interbull. This study evaluates the genetic and phenotypic correlations between these novel and established traits and assesses their potential value for inclusion in the TOP system.

MATERIALS AND METHODS

Individual TOP scores (for the traits described in Table 1) for 1,386 first lactation, Holstein-Friesian heifers from 15 herds were collected between 15/03/21 and 01/04/21 by trained inspectors. All traits were scored on a 1 to 9 linear scale. The 12 existing traits are described in DairyNZ (2023) and the six new traits are described in ICAR (2015). The Ruakura Animal Ethics Committee (Hamilton, NZ) approved this study and all manipulations (AE application: 14448).

Cattle were genotyped using Weatherbys (www.WeatherbysScientific.com) Illumina Versa 50K bovine panel and the resultant SNP data were stored in a Helical database for processing and extraction (Garrick *et al.* 2025). Any missing SNPs were imputed using findhap software (Van Raden *et al.* 2011). Only mapped SNPs assigned to chromosomes 1 through 29 and with a minor allele frequency >1% were included in the analysis (n=46,766).

Data were analysed in Julia 1.7.0 using the Julia for Whole-genome Analyses Software (JWAS v1.1.2) package (Cheng *et al.* 2018). Variance parameters were estimated using Markov-chain Monte Carlo (MCMC) approaches and inference was based on MCMC chains of 75,000 samples after a burn-in of 25,000 samples, retaining every 10th sample. Co(variance) components were estimated by fitting the below model equation pairwise:

$$Y = Xb + M\alpha + e$$

where: Y was the trait of interest (Table 1); X is the incidence matrix for fixed herd-visit day contemporary group effects (b); M is the matrix of additive covariates for 46,766 loci with effects (α) assumed independently and identically normally distributed with variance σ_{α}^2; and e is the residual effects independently and identically normally distributed with variance σ_e^2. Inference was based on the posterior means of the MCMC samples, their posterior variance, and their credibility intervals obtained directly from the samples. The upper and lower 90% credibility intervals were calculated by taking the 95th percentile of the MCMC samples as the upper bound and the 5th percentile as the lower bound.

RESULTS AND DISCUSSION

Heritability estimates for the udder traits were low to moderate and ranged from 0.16 for Fore Udder Attachment up to 0.34 for UD (Table 1). Heritability estimates were low for the two leg conformation traits (RLRV and Rear Legs Set [RLS]) and were low to moderate for the remaining body traits (0.13 to 0.39; Table 1). Estimates for both new and existing traits were aligned with values reported internationally (Brotherstone 1994; Berry *et al.* 2004; Williams *et al.* 2022).

Table 1. Descriptive statistics, genetic variances and heritability estimates of 12 existing and six new Traits Other Than Production (TOP) type traits. Genetic variances and heritability estimates are presented as posterior means with lower and upper 90% credibility intervals

Trait Name	Abbr.	Status	Mean ± SD	Genetic Variance	Heritability
Udder Traits					
Central Ligament	CL	New	6.09 ± 0.98	0.16 (0.10,0.23)	0.19 (0.12,0.27)
Fore Udder Attachment	FU	Existing	5.39 ± 1.02	0.15 (0.09,0.20)	0.16 (0.10,0.23)
Front Teat Placement	FT	Existing	4.82 ± 0.68	0.10 (0.07,0.14)	0.24 (0.16,0.32)
Rear Teat Placement	RT	Existing	7.00 ± 0.94	0.23 (0.16,0.30)	0.28 (0.20,0.36)
Rear Udder Attachment	RU	Existing	5.50 ± 0.88	0.09 (0.06,0.13)	0.15 (0.09,0.21)
Udder Depth	UD	New	6.19 ± 1.11	0.34 (0.26,0.43)	0.34 (0.26,0.42)
Udder Overall	UO	Existing	5.40 ± 0.88	0.12 (0.08,0.16)	0.18 (0.12,0.24)
Udder Support	US	Existing	5.97 ± 1.00	0.22 (0.14,0.29)	0.30 (0.20,0.40)
Body Traits					
Body Depth	BD	New	6.18 ± 1.04	0.40 (0.30,0.51)	0.39 (0.30,0.49)
Capacity	CA	Existing	6.21 ± 0.86	0.23 (0.17,0.30)	0.32 (0.24,0.41)
Chest Width	CW	New	5.57 ± 1.08	0.08 (0.04,0.12)	0.13 (0.07,0.19)
Dairy Conformation	DC	Existing	6.09 ± 0.96	0.19 (0.12,0.27)	0.22 (0.14,0.31)
Foot Angle	FA	New	5.36 ± 1.00	0.21 (0.14,0.29)	0.23 (0.15,0.30)
Rear Legs Rear View	RLRV	New	5.21 ± 0.90	0.04 (0.01,0.10)	0.06 (0.01,0.13)
Rear Legs Set	RLS	Existing	6.27 ± 0.59	0.03 (0.02,0.05)	0.10 (0.05,0.16)
Rump Angle	RA	Existing	4.14 ± 0.80	0.14 (0.09,0.19)	0.22 (0.15,0.31)
Rump Width	RW	Existing	5.43 ± 0.86	0.12 (0.07,0.18)	0.18 (0.11,0.27)
Stature	STAT	Existing	5.71 ± 0.85	0.18 (0.14,0.23)	0.38 (0.29,0.46)

The genetic correlations among the new and existing udder traits were higher than the phenotypic correlations (Table 2). Both the new udder traits (CL and UD) had at least one genetic correlation with an existing udder trait that was moderate to highly positive (Table 2). The genetic and phenotypic correlations between CL and UD was moderate at 0.31 (0.07, 0.52) and 0.35 (0.31, 0.39). In NZ, the existing Udder Support (US) trait is scored as a combination between CL and UD; unsurprisingly, the genetic correlation between US and UD was high at 0.79 and moderately high at 0.63 between US and CL (Table 2). The genetic correlations between CL and both front and rear teat placement traits were also high and positive (0.70 and 0.74; respectively); similar to Berry *et al.* (2004) who reported that a stronger central ligament was associated with closer teat placement.

Table 2. Posterior means (and 90% credibility intervals) of the genetic and phenotypic correlations between new and existing Traits Other Than Production (TOP) udder type traits. Trait abbreviations are defined in Table 1

Trait	US	FU	RU	FT	RT	UO
Genetic Correlations						
CL	0.63	0.31	0.31	0.70	0.74	0.35
	(0.44,0.77)	(-0.01,0.54)	(0.00,0.53)	(0.54,0.83)	(0.60,0.85)	(0.06,0.56)
UD	0.79	0.62	0.52	0.36	0.19	0.71
	(0.71,0.87)	(0.44,0.77)	(0.30,0.69)	(0.13,0.55)	(-0.04,0.39)	(0.58,0.82)
Phenotypic Correlations						
CL	0.42	0.20	0.29	0.26	0.37	0.34
	(0.38,0.46)	(0.15,0.24)	(0.25,0.33)	(0.21,0.30)	(0.33,0.41)	(0.30,0.38)
UD	0.45	0.29	0.34	0.11	0.11	0.40
	(0.41,0.48)	(0.25,0.33)	(0.29,0.37)	(0.07,0.16)	(0.06,0.15)	(0.36,0.43)

Table 3. Posterior means (and 90% credibility intervals) of the genetic and phenotypic correlations between new and existing Traits Other Than Production (TOP) body type traits. Trait abbreviations are defined in Table 1

Trait	STAT	CA	RA	RW	RLS	DC
Genetic Correlations						
CW	0.05	0.66	0.26	-0.02	-0.27	0.63
	(-0.25,0.29)	(0.48,0.80)	(-0.06,0.50)	(-0.35,0.27)	(-0.58,0.06)	(0.40,0.79)
BD	0.02	0.87	0.41	0.27	-0.10	0.87
	(-0.17,0.19)	(0.79,0.93)	(0.16,0.65)	(-0.02,0.56)	(-0.39,0.19)	(0.78,0.93)
RLRV	0.00	0.13	-0.24	-0.11	-0.65	0.12
	(-0.50,0.40)	(-0.43,0.54)	(-0.65,0.14)	(-0.63,0.31)	(-0.85,-0.33)	(-0.48,0.55)
FA	0.45	0.09	-0.32	-0.02	-0.89	0.19
	(0.22,0.68)	(-0.16,0.31)	(-0.54,-0.11)	(-0.31,0.26)	(-0.95,-0.80)	(-0.10,0.48)
Phenotypic Correlations						
CW	0.06	0.30	0.06	0.03	0.03	0.25
	(0.02,0.11)	(0.26,0.34)	(0.01,0.10)	(-0.02,0.07)	(-0.02,0.07)	(0.20,0.29)
BD	0.07	0.57	0.09	0.06	0.09	0.49
	(0.02,0.11)	(0.54,0.60)	(0.05,0.14)	(0.01,0.10)	(0.04,0.13)	(0.45,0.52)
RLRV	-0.02	0.11	0.02	-0.03	-0.01	0.14
	(-0.07,0.02)	(0.06,0.15)	(-0.03,0.06)	(-0.08,0.02)	(-0.06,0.03)	(0.09,0.18)
FA	0.12	0.07	0.00	0.01	-0.24	0.12
	(0.07,0.16)	(0.02,0.11)	(-0.04,0.05)	(-0.04,0.05)	(-0.29,-0.20)	(0.07,0.16)

Table 4. Posterior means (and 90% credibility intervals) of the genetic (below diagonal) and phenotypic (above diagonal) correlations between four new Traits Other Than Production (TOP) body type traits. Trait abbreviations are defined in Table 1

Trait	CW	BD	RLRV	FA
CW	1	0.29 (0.25,0.34)	0.18 (0.13,0.22)	0.07 (0.03,0.12)
BD	0.43 (0.08,0.69)	1	0.07 (0.02,0.11)	0.00 (-0.05,0.04)
RLRV	0.32 (-0.27,0.71)	0.16 (-0.13,0.48)	1	0.15 (0.10,0.19)
FA	0.17 (-0.14,0.47)	0.09 (-0.10,0.30)	0.18 (-0.21,0.63)	1

The existing body Capacity (CA) trait is scored as a combination between CW and BD, and expectedly, the genetic correlations between CA and BD or CW were high at 0.87 or moderately high at 0.66, respectively (Table 3). All four new body traits had at least one genetic correlation with an existing trait that was moderate to highly negative (<-0.65; RLRV and FA) or moderate to highly positive (>0.65; CW and BD). Brotherstone (1994) and Berry *et al.* (2004) also reported that FA and RLS had strong negative genetic correlations. Genetic correlations among the four new body traits ranged between 0.09 and 0.43 (Table 4), and were strongest between CW and BD.

The strong genetic correlations observed between the new traits and existing traits highlight their close genetic relationships, suggesting that these new traits may not capture novel genetic information. However, new traits that have strong genetic correlations with existing traits but more genetic variance or a higher heritability could provide value as a predictor trait. For example, while FA exhibited a strong negative genetic correlation with the existing trait RLS, its greater heritability (0.23 vs. 0.10) makes it a promising candidate as a predictor trait. Similarly, BD had strong genetic correlations (0.87) with both CA and DC, with a similar or greater heritability (0.39 vs. 0.32 and 0.22, respectively).

CONCLUSION

This study revealed strong genetic correlations between new body or udder traits and at least one existing TOP trait, indicating that the new traits may largely reflect genetic information already captured by existing traits. The body trait FA demonstrated a higher heritability compared with its correlated existing trait RLS, highlighting its potential utility as a predictor trait. While further research with larger datasets reflecting NZ's mixed-breed herds is needed to confirm these findings, this study provides a foundation for informed decision-making on the inclusion of new body and udder conformation traits into routine genetic evaluations.

ACKNOWLEDGEMENTS

The study was funded by NZ dairy farmers through DairyNZ Inc. and by the NZ Ministry of Business, Innovation and Employment (DRCX1302). Additional in-kind support was received from LIC, CRV, NZAEL, and Holstein Friesian NZ. The authors gratefully acknowledge the contributions of the DairyNZ Ltd. science and technical teams, participating vet clinics, herd owners and graziers, and use of NeSI facilities (www.nesi.org.nz).

REFERENCES

Berry D.P., Buckley F., Dillon, P., *et al.* (2004) *Ir. J. Agr. Food Res.* **43**: 161.

Brotherstone (1994) *Anim. Sci.* **59**: 183.

Cheng H., Fernando R.L. and Garrick D.J. (2018) *Proc. World Cong. Genet. Appl. Livest. Prod.* **11**: 859.

DairyNZ (2023) Retrieved from https://www.dairynz.co.nz/media/h4xd5xke/top-booklet-2023.pdf [accessed Nov 2024].

DairyNZ (2024) Retrieved from https://www.dairynz.co.nz/animal/breeding-decisions/breeding-values/ [accessed Nov 2024].

Garrick D.P., Garrick, D.J. and Howse J.E. (2025) *Proc. Assoc. Advmt. Anim. Breed. Genet.* **26**: *These proceedings*.

ICAR (2015) Retrieved from https://www.icar.org/wp-content/uploads/2015/08/Conformation-Recording-CR-WG.pdf [accessed Nov 2024].

Interbull (2024) Retrieved from https://interbull.org/static/web/conf2408r.pdf [accessed Nov 2024].

VanRaden P.M., O'Connell J.R., Wiggans G.R. andWeigel K.A. (2011) *Gen. Sel. Evol.* **43**: 1.

Williams M., Sleator R.D., Murphy C.P., *et al.* (2022) *J. Dairy Sci.* **105**: 7550.

Proc. Assoc. Advmt. Anim. Breed. Genet. 26: 182-184

DUAL-PURPOSE NORWEGIAN RED CATTLE FOR SUSTAINABLE DAIRY AND BEEF PRODUCTION IN PASTURE-BASED SYSTEMS

C. McKimmie[1], H.A. Najafabadi[1], E. Noble-Campbell[2], J. Hoffmann[3] and J. Jenko[3]

[1] Genetics Department, Samen NZ, Hamilton 3300, New Zealand
[2] Faculty of Agriculture and Life Sciences, Lincoln University, Lincoln 7647, New Zealand
[3] Geno SA, Storhamargata 44, N-2317 Hamar, Norway

SUMMARY

The Norwegian Red (NR) breed has been developed as a dual-purpose dairy and beef cattle breed, offering opportunities to improve sustainability in New Zealand's pasture-based livestock systems. This study evaluates NR cattle's milk production and beef performance in comparison to existing New Zealand dairy breeds. Our results indicate that NR-sired cows maintain milk solid production comparable to New Zealand Holstein-Friesians while demonstrating superior fertility and persistent lactations. Additionally, surplus NR-sired calves achieve significantly higher carcass weights and greater economic returns than pure Holstein-Friesian (HF) calves in the same rearing system, presenting a viable alternative to add value to surplus dairy-origin calves (reducing reliance on low-value bobby calves) without compromising dairy output. The findings highlight NR cattle's potential to enhance farm profitability and environmental sustainability through better dairy-beef integration.

INTRODUCTION

New Zealand's dairy industry faces growing pressure to improve sustainability while maintaining profitability. One key challenge is the surplus of dairy-origin calves with limited beef value. This contributes to ethical concerns and resource inefficiency (Beukes *et al.* 2009). The Norwegian Red (NR) breed has been selectively bred for both milk and beef traits (Aass and Vangen 1998), making it a potential solution to improve calf utilization without compromising dairy production. Historically beef traits accounted for approximately 10% of the total merit index for the NR and today they represent 8% of the index (Geno 2023). NR cattle exhibit high fertility, strong health traits, and favorable feed efficiency, aligning well with New Zealand's pasture-based systems. This study examines NR cattle's dual-purpose potential in New Zealand dairy herds by comparing their milk production, reproductive performance, and beef value against traditional dairy breeds.

MATERIALS AND METHODS

Animals and data collection. In New Zealand, Performance data were collected from 1,621 dairy cows across two age groups (2-year-olds and 3-year-olds) representing three breed groups: Viking Red (n = 482, 435), NR (n = 50, 21), and New Zealand Holstein-Friesian (n = 284, 349). The NR-sired cows were managed alongside NZ Holstein-Friesian herdmates under the same pasture-based conditions. Milk production traits recorded included fat production (kg), milk solids (kg), protein production (kg), lactation days, and total lactation solids. In Norway, beef carcass data from slaughter records were obtained for 160 pure Holstein and 106 crossbred (NR × Holstein) cattle (all dams were Norwegian Red cows). These animals were finished for beef production, and their growth and carcass traits (carcass weight, price per unit weight, total carcass value, and fat score) were recorded.

Statistical analysis. Data were analyzed using R (version 4.3.0). Welch's ANOVA and the Games-Howell post-hoc test assessed breed differences in milk production traits. Carcass performance was compared using t-tests. Effect sizes were calculated using partial eta-squared (ηp^2) to determine the practical significance of observed differences.

RESULTS AND DISCUSSION

Milk production performance.NR cows demonstrated no significant difference (P > 0.05) in total milk solids production (kg/day) compared to Holstein-Friesians, supporting their suitability for New Zealand dairy systems. NR cows had significantly longer lactations (P < 0.001). While lactation length is not a direct measure of fertility, in seasonal systems it can reflect earlier conception. This aligns with previous studies reporting shorter days open and higher conception rates for NR cows compared to Holsteins (Walsh *et al.* 2008; Ferris *et al.* 2014).

Table 1. Means (±SD) of milk production traits by breed and age group

Trait	2-year-olds			3-year-olds		
	Viking	Norwegian	NZ Friesian	Viking	Norwegian	NZ Friesian
Fat (kg/day)	0.72±0.15	0.67±0.13	0.66±0.14	0.78±0.16	0.84±0.17	0.79±0.15
Milk Solids (kg/day)	1.33±0.26	1.22±0.20	1.31±0.27	1.44±0.30	1.46±0.24	1.44±0.28
Protein (kg/day)	0.55±0.09[a]	0.54±0.08[ab]	0.53±0.09[b]	0.61±0.12	0.62±0.09	0.60±0.10
Lactation Days	236.9±26.8[a]	242.9±23.0[a]	227.3±41.1[b]	231.7±35.5[b]	268.8±16.3[a]	225.3±46.3[b]
Lactation Solids (kg)	292.2±72.4	274.3±68.0	299.2±98.9	315.8±72.4[b]	396.7±68.0[a]	323.8±98.9[ab]

Different superscripts a,b within a row indicate significant difference at P < 0.05

Beef performance. In the Norwegian beef dataset, NR-sired calves demonstrated better carcass characteristics compared to Holstein-Friesians, including higher carcass weights (807 lbs vs. 770 lbs, P < 0.05) and higher carcass value ($1,302 vs. $1,204, P < 0.05). Improved muscling and dressing percentages make NR crosses more marketable for beef production, presenting an opportunity to integrate dairy-beef systems effectively. Similar benefits of NR genetics for beef performance have been reported (McNamee *et al.* 2015).

Table 2. Carcass performance of Holstein vs. NR×Holstein crosses

Trait	Holstein	NR X Holstein crosses
Number of cows	160	106
Carcass weight (lb)	770	807*
Carcass price per pound ($/lb)	1.57	1.63*
Carcass value ($)	1204.07	1302.26
Carcass fat (based on grades ranked 1 to 4)	2.43	2.51

* P<0.05 for difference between NR and NR×Holstein.

Environmental and Economic Impact. Adopting NR genetics presents a unique opportunity to reduce the environmental impact of dairy-beef production by optimizing resource utilization. Integrating surplus dairy calves into beef production can lower greenhouse gas emissions (van Selm *et al.* 2021; Yang *et al.* 2020). Although feed efficiency and antibiotic use were not measured in this study, previous research has shown that NR-sired cattle can achieve similar or greater milk solids yield with comparable or lower feed intake than Holsteins, indicating favourable feed efficiency (McClearn *et al.* 2022; Dong *et al.* 2015). Additionally, the Norwegian Red breed has been selected for mastitis resistance and other health traits for decades, contributing to lower disease incidence and reduced reliance on antibiotics (Begley *et al.* 2009; Rajala-Schultz *et al.* 2021). These attributes align with New Zealand's national targets to reduce antimicrobial use in livestock (NZVA 2024).

Future Prospects. The widespread adoption of NR cattle could significantly improve the integration of dairy and beef systems in New Zealand. Further research should focus on long-term genetic evaluations to optimize NR crossbreeding strategies and maximize profitability. Additionally, economic modeling could be performed to assess the financial viability of NR genetics at the farm level, considering both dairy and beef revenue streams.

CONCLUSION

NR cattle could offer a dual-purpose solution for New Zealand dairy herds, combining competitive milk production with enhanced beef value. While this study did not directly measure fertility or health traits, longer lactations observed in NR cows reflects timely conception in the seasonal systems, Improved fertility, longer lactations, and high carcass performance suggest that NR genetics can enhance dairy-beef integration while addressing sustainability concerns. Together with evidence from other studies, these results suggest that NR genetics could support more integrated and sustainable dairy-beef systems. However, the relatively small NR sample size in this study should be considered, and further research is needed to confirm these trends under New Zealand conditions. Widespread adoption of NR genetics in New Zealand could lead to improved farm profitability, reduced environmental impact, and enhanced animal welfare.

ACKNOWLEDGEMENTS

The authors thank Norway's national dairy herd database for providing the beef data used in this study. We clarify that the dairy performance data were collected from cows raised on commercial farms in New Zealand,

REFERENCES

Aass L. and Vangen O. (1998) *Acta Agr. Scand.* **48**: 65.

Beukes P., Gregorini P., Romera A., Levy G. and Waghorn G. (2009) *Agr. Ecosyst. Environ.* **136**: 358.

Begley N., Patton J., Pierce K.M. and Buckley F. (2009) *J. Dairy Sci.* **92**: 749.

Dong L.F., Yan T., Ferris C.P., McDowell D.A. and McCoy M.A. (2015) *J. Dairy Sci.* **98**: 1136.

Ferris C.P., Patterson D.C., Gordon F.J., Watson S. and Kilpatrick D.J. (2014). *J. Dairy Sci.* **97**: 5206.

Geno (2023). https://www.geno.no/fagstoff-og-hjelpemidler/avlsprogram-for-norsk-rodt-fe/avlsmalet-for-nrf/

McClearn B., Delaby L., Gilliland T.J., Guy C., Coughlan F., McCarthy B. and Buckley F. (2022) *J. Dairy Sci.* **105**: 242.

McNamee A., Keane M., Kenny D., Moloney A., Buckley F. and Riordan E. (2015) *Livest. Sci.* **173**: 95.

NZVA (2024) New Zealand Veterinary Association. "Antimicrobial Resistance Strategy." https://nzva.org.nz/clinicalresources/general/amr/ (accessed 4 March 2024).

Rajala-Schultz P.J., Nødtvedt A., Halasa T. and Persson Waller K. (2021) *Front. Vet. Sci.* **8**: 623998.

van Selm B., De Boer I.J., Ledgard S.F. and van MiddelaarC.E. (2021) *Agr. Sys.* **186**: 102936.

Walsh S., Buckley F., Pierce K.M., Byrne N., Patton J. and Dillon P. (2008). *J. Dairy Sci.* **91**: 4401.

Yang W., Rennie G., Ledgard S., Mercer G. and Lucci G. (2020) *Agr. Sys.* **178**: 102747.

Proc. Assoc. Advmt. Anim. Breed. Genet. 26: 185-188

AGRICULTURE INSTITUTE AUSTRALIA'S STUDENT COMPENDIUM IS BUILDING A SUSTAINABLE RESEARCH, DEVELOPMENT AND ADOPTION WORKFORCE

B. Hancock[1] and T. Manning[2]

[1] Agriculture Institute Australia, Nuriootpa, SA, 5355 Australia
[2] School of Animal and Veterinary Sciences, University of Adelaide, Roseworthy, SA, 5371 Australia

SUMMARY

The Student Compendium connects students to primary industries, and primary industry to students. It is a one-stop-shop that presents students with the breadth of opportunities within primary industries and natural resource management. The Compendium results from a collaboration between the Ag Institute SA Division (AIA-SA), a not-for-profit professional association for agriculture and natural resource management students and professionals, and the Adelaide University Agricultural Students Association (AUASA).

Within the 459 entries in the 2025 edition, over 20 Organisations from the dairy, beef, sheep, pork, chicken and alpaca industries have offered opportunities in animal breeding and genetics.

Annually, interns are engaged to collect, collate, communicate, and gain many new skills and confidence by directly performing the tasks to compile the compendium. Their experience is also shared with their fellow student cohort. The intern experience assists them to consolidate their entry point into the workforce or further study, and with the real winner being the industry as the interns are workforce-ready and emerging as future leaders.

INTRODUCTION

Agricultural industries continue their strong desire to gain attention and attract the next generation of youth into life-long careers in agriculture (Hancock *et al.* 2022). The Student Compendium is a unique, innovative, and high-quality resource for students and industry alike. It acts as a conversation starter to call everyone to action on the *building human capacity* pillars of the South Australian Red Meat and Wool, Dairy, Grains, and Horticulture Industry Blueprints (Livestock SA 2024).

In 2021, AIA-SA and AUASA identified that an increasing number of students were keen to enter agriculture, especially during and after COVID-19 when attention was bought to food security and supply chains. However, students didn't know where to get connected to the industry to identify career opportunities. There was good general information on career pathways available through internet resources such as careerharvest.com.au, but the missing link was the opportunities that organisations were offering at the local level, and who to connect with. On approaching industry, they were equally frustrated in being unable to make connections with students across all levels of study, from high school to vocational and tertiary levels. Over the past 5 years, the Student Compendium (Davis *et al.* 2025) has become the connection tool.

To use South Australia as an example, if the industry is to successfully service the needs of the SA agriculture workforce pipeline, it must bring the great career opportunities in agriculture to the attention of students and parents who haven't ever thought about agriculture. The Student Compendium is well connected to the 70 high schools in SA that teach Agricultural Science. However, the majority of these schools are regionally located and have small numbers of students studying the subject. To increase the reach of career opportunities in agriculture, the industry must endeavor to extend its reach beyond this cohort, and appeal to students who may not be studying agriculture at these and the other 180 high schools in the state. South Australia doesn't need every school to teach agricultural science. However, there is a need to immerse examples of agriculture into the curriculum of every subject. The Student Compendium is a valuable aid in this process.

MATERIALS AND METHODS

Over the 5 years from 2021 to 2025, 22 internships were offered to students through the AIA-SA & AUASA. Internships are offered via an annual expression of interest application process to engage agricultural, animal sciences, and agribusiness undergraduate students as a Project Officer Intern, a Communications & Socials Intern, and 3 to 4 Compendium Interns (Hancock 2022; Blight, 2024). The Compendium Interns have undertaken the majority of the collection of new content and sponsorship for the compendium. The Internships are performed over the second half of semester two and during the summer break ensuring the compendium is ready and printed for orientation week at the commencement of the academic year. The internships require a commitment of 80 to 120 hours, completed at the student's own negotiated pace, and they fit it in and around study, exams, other industry work experience, and their social lives.

Organisations are approached by the interns to summarise student opportunities into 130 words or less, to fit within one of the following sections; i) high school and vocational training, ii) internships and work experience, iii) graduate and trainee programs, iv) undergraduate scholarships, v) postgraduate scholarships and studies, vi) early-mid career professional development awards, vii) competitions and travel, viii) conferences, ix) industry and professional associations and networks. Organisations are invited to contribute a simple, low cost, high-impact sponsorship of logo or advert.

Each year, the Student Compendium is collected and collated from October to December, compiled in January, printed in February, and launched in March where student and industry participants are asked to rate the value of the Compendium to students on a scale of zero to five. Five hundred hard copies are printed and distributed, and an electronic PDF version is posted on the AIA, AUASA, and major funding partner's websites, as well as the University's internship and work experience portals.

Opportunities within animal breeding and genetics come from the dairy, beef, sheep, pork, chicken, and alpaca sectors. They include a diverse array of offerings from Career Harvest, industry vacation placements, feedlots, processors, SunPork, Auctions Plus, Elders, Nutrien, Angus Australia, Hereford Australia, SA Livestock Consultants, postgraduate studies, conferences, and professional associations (e.g. Association for Advancement of Animal Breeding and Genetics).

The Communications and Social Plan (Hancock 2024) was developed by the 2023 Student Compendium Working Group. Each year, the interns are trained, briefed, and engaged in the key foundations of the Communications Plan and explore new communications technologies. They also develop the cover design that is subtly distinguishable from the previous edition. During the year the Communication & Social intern develops AIA's socials and posts the sponsor's student opportunities under the brand of Compendium Corner (AIA 2024).

Following training in media and photography by the Stock Journal, Ag Communicators, and Environmental & Science Media, interns co-author conference and journal papers, develop conference posters, attend professional and industry conferences, write media releases and produce videos. They apply for appropriate awards and competitions.

In a new initiative in 2024, the Student Compendium High Schools Special Edition was sent to all 250 High Schools in SA, along with five hard copies, a PDF copy and a QR code to access the document online. At each school, key influencers were shown the value of the compendium and its intended usage. They included the Principal, the Ag Science and Science teachers, SACE coordinators, career pathways advisors and counsellors, and the school and community library. The high schools were encouraged to promote the compendium in school newsletters and encourage local agriculture professional to attend career nights and expos, and to bring it to the attention of parents and guardians as they continue to be major influencers of students' career entry points and pathways (Graham 2021).

RESULTS AND DISCUSSION

The Student Compendium is rated very highly by industry and users for value and usefulness. Over the 5 years from 2021 to 2025, the Student Compendium has rated from 4.5 in 2021 up to 5.0 in 2025 on a scale of 0 to 5 for value to students by both students and professionals. A survey of 2nd and 3rd year ag science students has indicated that students were seeking out 5 to 10 opportunities from the compendium, with some engaging in more than 10. Learnings and experiences of the Compendium Interns frequently form the content or base of the Stock Journal's Undergraduate Insight Opinion Column.

A number of videos were scripted, filmed, edited and produced in 2024. The major one was a new video highlighting the intern experience in a *behind the scenes* feature of the Ag Institute Australia (AIA) Student Compendium program, and captured the journey of the 2024 interns and the practical insights gained. It delved into the hands-on nature of the internship and showcased how agricultural students applied their knowledge and skills in real-world settings. The video highlighted the dynamic and interactive environment that the internship provided, and how it offered students the chance to connect with industry professionals and develop essential skills for their future careers. Individually, it helped the interns build a network as a key stepping stone to career development, expanded a professional network that connected with industry leaders, contributed to securing a dream job as a graduate, expanded agricultural knowledge through hands-on experiences in university and on farms in the United Kingdom. For all of the interns, the program built confidence and important skills in networking and professional communication. The video was filmed across various locations, capturing interviews with these students, showcasing the significant impact the internship had on their career readiness and industry connections (Adams 2024).

Other videos developed during the program have included promotional videos for students, professional membership of the Grasslands Society of Southern Australia, a preview for the Barossa Improved Grazing Group (BIGG) Annual Conference, and The Primary Industries Education Foundation of Australia (PIEFA) Blog Series: student insights highlighting tertiary agriculture studies and career expos.

A number of interns have received notable awards during their time as interns, including James Easter, Student of the Year award, 2022 Kondinin / ABC Rural (Chambers 2023), Grace Maloney, Highly Commended award, 2023 Rural Media & Communications (SA/NT) ,Tayla Blight, Junior LambEx Coles Trailblazer 2024, (Blight 2024), and Emily Adams received two awards, Winner and Highly Commended, Rural Media & Communications SA/NT 2024.

The skills and confidence gained from the Student Compendium experience resulted in interns gaining other internships and conference bursaries such as the Barossa Improved Grazing Group (BIGG) Conference, Overseas Study Experience, Grasslands Conference: Joyce Gillepsie and AW Howard Memorial Trust, BIGG Communication Intern – Drought Preparedness, AgriFutures Horizon Scholarship, Crawford Fund Conference, LambEx'24, Women Together Learning (WoTL), 2024 SA Livestock Advisor Update and the AIA-SA Newsletter Editor Internship.

A literature legacy continues to build and provides interns with valuable experience and skills in co-authoring of papers, and presenting at conferences such Australian Association of Animal Sciences, Primary Industries Education Foundation Australia, Australasian Pacific Extension Network, National Association of Agriculture Educators, and Grasslands Society of Southern Australia.

Following the South Australian lead, there are now 2nd editions of the compendium developed in WA and NSW, and a 1st edition in Queensland/Northern Territory. On the back of these successes, the SA branch of AIA is now facilitating the development of a national compendium.

Building connections and relationships between students and industry professionals and their organisations is achieved by attending University events such as orientation days, AgTivate Your Career, and Growing the Future. AUASA also offers a mentoring program for 3rd year students called *Plant the Seed of Your Career*, and AIA has developed its own mentoring program. Both mentoring programs have seen a noticeable increase in the demand for grazing livestock mentors.

Invitations continue to be extended to high schools, residential boarding colleges, and regional field days and career expos to participate in the program, where there is increasing engagement with non-farm-based students, and students who haven't previously thought about a career in agriculture.

CONCLUSION

The 22 Student Compendium Interns have gained quality skills and confidence and it has set them on their journey to becoming emerging leaders in agriculture, and champions to their student peers to build the research, development, and adoption workforce of the future. The compendium provides over 450 opportunities for high school, vocational and tertiary students to connect with primary industries and further build the workforce. It is available in all 250 high schools in SA and is a key resource for university students. Primary industries organisations value the compendium as a way to connect with students, provide support, guidance, and mentoring, and in-turn, build the workforce.

ACKNOWLEDGEMENTS

The authors wish to acknowledge the financial contributions made by the 6 major funding partners, including the SA Grains Industry Trust Fund , YITPI Foundation, SA Sheep Industry Fund, SA Cattle Industry Fund, Primary Industries Education Foundation of Australia, the Show Society Foundation, the 2024 Student Compendium High Schools Special Edition sponsors, the Show Society Foundation, Citrus SA, Lucindale Lions - SE Field Days, Mannum Lions, Stirling Rotary - Hills Small Acreage Field Days, YP Field Days, and Loudounbrae Farming. We acknowledge the hard work of 2021-25 Student Compendium Working Groups for their concerted drive to grow, develop and increase the outreach of the Compendium, and Miriam Yip (Environmental Science & Media), Belinda Cay (AgCommunicators), and Elizabeth Anderson (Stock Journal) for respective assistance with video, communications and media.

REFERENCES

Adams E., Hancock B. and Davis C. (2024) *J. Ag Inst. Aust.* **35**: 27.

AIA (2024) 'Compendium Corner'. https://www.facebook.com/AgInstituteAustralia [retrieved on 18th December 2024]

Blight T. and Hancock B. (2024) *J. Ag Inst. Aust.* **35**: 20.

Chambers E., Easter J. and Hancock B. (2023) *Proc. Pri. Ind. Ed. Found. Aust.* **4**: 7.

Davis M., Davis C., Hancock B., Manning., Thomas R., Bird R., Norman J., Lokan S. and Gifford S. (2025) 'Student Compendium' 5th ed. Agriculture Institute Australia.

Graham S. (2021) *J. Int. Inn. Sci. Math.* **29**: 11.

Hancock B. and Maloney G. (2024) *J. Rur. Ext. Inn. Sys.* **20**: 47.

Hancock B., Davis C. and Blight T. (2024) *J. Ag Inst. Aust.* **35**: 24.

Hancock B. and Davis C. (2022) *J. Ag Inst. Aust.* **33**: 27.

Hancock B., Peters E., Easter J., Manouge S. and Chambers E. (2022) *Proc. Aust. Assoc. Anim. Sc.* **34**: cxlix.

Livestock SA (2024) SA Red Meat and Wool Blueprint 2030. https://livestocksa.org.au/industry-development/industry-blueprint [Retrieved on 18th December 2024].

MULTI-TRAIT GENOMIC MODELS WITH TRAIT AND MARKER SPECIFIC WEIGHTS CAN SUBSTANTIALLY INCREASE THE ACCURACY OF GENOMIC BREEDING VALUES

T. Meuwissen[1] and V. Boerner[2]

[1] Faculty of Life Sciences, Norwegian University of Life Sciences, 1432 Ås, Norway
[2] GHPC Consulting and Service Pty. Ltd., Armidale, NSW, 2350 Australia

SUMMARY

The GWABLUP (Genome-Wide Association based Best Linear Unbiased Prediction) approach used GWA analysis results to differentially weigh the SNPs in genomic prediction, and was found to improve the reliabilities of genomic predictions. However, the proposed multi-trait GWABLUP method assumed that the SNP weights were the same across the traits. Here we extended and validated the multi-trait GWABLUP method towards using trait-specific SNP weights. In a 3-trait dairy data set, multi-trait GWAS estimates of SNP effects and their standard errors were translated into trait-specific likelihood ratios for the SNPs having trait effects and posterior probabilities using the GWABLUP approach. This produced trait-specific prior (co)variance matrices for each SNP, which were applied in a SNP-BLUP model for genomic predictions, implemented in the APEX linear model suite. In a validation population, the trait-specific SNP weights resulted in more reliable predictions for all three traits. Especially, for somatic cell count, which was hardly related to the other traits, the use of the same weights across all traits was harming genomic predictions. The use of trait-specific SNP weights overcame this problem. In multi-trait GWABLUP analyses of ~30,000 reference population cows, trait-specific SNP weights resulted in up to 13% more reliable genomic predictions than unweighted SNP-BLUP, and improved genomic predictions for all three studied traits.

INTRODUCTION

A limitation of GBLUP and SNP-BLUP is their assumption that all SNPs contribute equally to the total genetic variance. Bayesian variable selection methods allocate more variance/weight to the most important SNPs but are complex and computationally demanding. Recently, GWABLUP was proposed which uses deterministic weights based on GWAS (Genome Wide Association Study) results (Meuwissen *et al.* 2024). Since the SNP weights are (pre)determined by the GWAS signals, GWABLUP is based either on a weighted SNP-BLUP or on a weighted G-matrix in GBLUP, which may both be extended to single step methods (ssGWABLUP). Also a multi-trait extension of GWABLUP was proposed assuming that SNP weights are equal across the traits (Meuwissen *et al.* 2024). The latter assumed that all traits are affected by the same QTL (Quantitative Trait Loci). But generally, different traits will be affected by different QTL, and the use of the same SNP weights across the traits is suboptimal. Our aim is here to extend the multi-trait GWABUP method to using SNP weights that are trait-specific and to compare the results to using equal weights across the traits. The methods are compared in the same dairy cattle data set as Meuwissen *et al.* (2024), and using the APEX linear models suite (Boerner 2024), which implements multi-trait SNP-BLUP with different (co)variance matrices per SNP and thus allows for different weights per SNP and per trait.

MATERIALS AND METHODS

The 3-trait dairy data set of Meuwissen *et al.* (2024) included the yield deviations (YD) of milk and protein yield and somatic cell count (SCC) and their reliability on 32,201 Norwegian Red cows, and was kindly provided by Geno SA (www.geno.no). Estimates of heritabilities, genetic and environmental correlations of the traits are depicted in Table 1.

Table 1. Heritabilities (diagonals), genetic correlations (below diagonals), and residual correlations (above diagonals) of the dairy traits

	Milk	Prot	SCC
Milk	0.26	0.96	-0.17
Protein	0.85	0.20	0.16
SCC	0.10	0.10	0.16

In Meuwissen *et al.* (2024), a canonically transformation of the 3 traits was performed (Ducrocq *et al.* 1993) which resulted in 3 genetically and environmentally independent canonical traits with standardised environmental variances of 1 and genetic variances of 0.16, 0.29 and 1.44, respectively. The data also included imputed HD genotypes on N_{snps}= 617,739 SNPs for all 32,201 cows (see Meuwissen *et al.* (2024) for details). Uniform SNP weights across the traits were obtained from the combined log-likelihood ratios of the GWAS of the 3 canonical traits as described in Meuwissen *et al.* (2024). The GWAS of the three canonical traits resulted also in estimates of their SNP effects and standard errors for each of the canonical traits. To obtain trait-specific SNP weights, these canonical trait SNP effects were back-transformed to original trait SNP effects together with their standard errors. Following Meuwissen *et al.* (2024), GWABLUP log-likelihood ratios per SNP and trait were calculated as $LR_{tj} = 0.5\,\widehat{b_{tj}^2}/se_{tj}^2$, LR_{tj} where is the log-likelihood ratio, $\widehat{b_{tj}}$ is the GWAS estimate of the effect, and se_{tj} is the standard error of SNP j for trait t. The model averaging of MCMC Bayesian genomic prediction schemes is to some extend mimicked by calculating the moving average of the likelihood ratios, i.e. by averaging the likelihood ratios of the current SNP and two adjacent SNPs to the left and two adjacent SNPs to the right, resulting in $\overline{LR_{tj}}$. Following Meuwissen *et al.* (2024), posterior probabilities that the SNPs have non-zero effects, which serve as trait-specific weights for the SNPs, were calculated for trait t and SNP j as $\pi\, e^{\overline{LR_{tj}}}/[\pi e^{\overline{LR_{tj}}} + (1-\pi)]$ where the prior probability of non-zero SNP effects is $\pi = 0.001$.

The model for the multi-trait SNP-BLUP is $y = X\mu + Zb + e$ where y is a vector of yield deviations for the 3 traits (ordered by traits within cows); $X = 1_{n_a} \otimes I_3$ is a design matrix linking the records to their trait means (n_a is the number of animals); μ is a vector of estimates of trait means for the 3 traits; $Z = M \otimes I_3$ with M being a $(n_a \times N_{snps})$ matrix of centred allele counts of the cows; b is a vector of $(3 \times N_{snps})$ multi-trait SNP effects (ordered by traits within SNPs). The prior distribution of the residuals was $e \sim N(0, W \otimes R)$, where R is the residual (co)variance matrix of the traits (Table 1) and W is a $(n_a \times n_a)$ diagonal matrix with the inverses of the weights of the yield deviations on the diagonal. For the analysis the multi-trait SNP-BLUP method of Liu *et al.* (2014) is used, where in our data all animals are genotyped and pedigree relationships are not used and thus set to an identity matrix. The (co)variances of the breeding values and SNP effects (traits within cows and SNPs) is modelled by $var(u) = Z\Theta Z' + \varepsilon I_{n_a}, var(b) = \Theta, cov(u,b) = Z\Theta$, where $u = Zb$ is a vector of multi-trait breeding values; ε is a small number (ε=0.01 was used here) added to regularise the matrix and making $var([u,b])$ non-singular; and Θ is a block-diagonal matrix of 3×3 blocks $\Theta = \sum_{\oplus} V_{SNP_j}$ where $\oplus$ denotes the direct sum. V_{SNP_j} is the 3×3 SNP specific (co)variance matrix across 3 traits, i.e. the SNP effects are a priori assumed unrelated with prior distributions $b_j \sim N(0, V_{SNP_j})$, where b_j is a 3×1 vector of effects of SNP j. Let $V_{SNP} = G/\sum_{k=1}^{N_{snps}} 2\,p_k(1-p_k)$ where G is the genetic (co)variance matrix of traits and p_k is the allele-frequency of SNP k. Further let P be an $N_{snps} \times 3$ matrix of posterior probability of SNPs obtained the for 3 canonical traits (Meuwissen *et al.* 2024) with $P_{t,j} = \pi\, e^{\overline{LR_{tj}}}/[\pi e^{\overline{LR_{tj}}} + (1-\pi)]$, and h_t be a vector of 1 of length 3 and h_s be a vector of 1 of length N_{snp}. In case of regular multi-trait unweighted SNP-BLUP (SNP$_{unw}$-BLUP) $V_{SNP_{unw}} = I_{N_{snps}} \otimes V_{SNP}$, in case of equal weights across the traits (SNP$_{eqw}$-BLUP) $V_{SNP_j} = ((Ph_t)/((h_s/N_{snp})'P(h_t/3))) \otimes V_{SNP}$, and in case of SNP and trait-specific weights (SNP$_{tsw}$-BLUP), $V_{SNP_j} = \Omega(I_{N_{snps}} \otimes V_{SNP})\Omega, \Omega = I_{N_{snps}\times 3}\sqrt{\overrightarrow{(P/((h_s/N_{snp})'P(h_t/3)))'}}$. An efficient iterative double-preconditioned conjugate gradient algorithm (Vandenplas *et al.* 2019) was used to solve these equations as implemented in the APEX linear models suite (Boerner 2024). This analysis of the SNP$_{unw}$-BLUP, SNP$_{eqw}$-BLUP and SNP$_{tsw}$-BLUP models yielded multi-trait estimates of SNP effects and of animal genetic effects.

The records of cows born in 2018 (1988 cows) were used for validation, and their YDs were masked from the above data analyses. The remaining 30,213 cows were used for the training of the models, i.e. they were used for the GWAS analyses and the estimation of SNP effects. These SNP effects were used to obtain multi-trait breeding value estimates (EBVs) of the validation cows. The squared correlations between the EBVs of the 1988 validation cows and their YDs were used as an indicator of the reliabilities of the EBVs.

RESULTS AND DISCUSSION

Table 2 shows the reliabilities of the multi-trait genomic predictions measured as the squared correlations between GEBV and YDs for milk- and protein yields and SCC of the 1988 validation cows. The YD reliabilities for milk, protein and SCC are 0.409, 0.326, and 0.246, respectively (Meuwissen *et al.* 2024). When expressed relative to the YD reliabilities the reliabilities of SNP$_{unw}$-BLUP are 0.49 (=0.199/0.409), 0.54, and 0.68, for milk, protein and SCC, respectively.

The use of SNP weights in SNP$_{eqw}$-BLUP and SNP$_{tsw}$-BLUP models significantly improved genomic prediction reliabilities for milk and protein yields by 11-13%. SNP$_{tsw}$-BLUP obtained the highest reliability for all three traits, albeit the improvement for SCC was minor and not statistically significant.

Table 2. Reliabilities of genomic predictions measured as the squared correlations between GEBVs($\widehat{g_v}$) and yield deviations(y_v) of 1988 validation cows

SNP-BLUP Model	$cor(y_v, \hat{g}_v)^2$ *,**		
	Milk	Protein	SCC
SNP$_{unw}$-BLUP***	0.199^a	0.178^a	0.168^a
SNP$_{eqw}$-BLUP***	0.223^b	0.197^b	0.160^a
SNP$_{tsw}$-BLUP***	0.226^b	0.201^b	0.169^a

*Standard errors of $cor(y_v, \hat{g}_v)^2$ are between 0.006 and 0.008.

**Different letters in the superscripts denote statistically significant differences (P < 0.05)

***Subscripts unw, teqw, and tsw mean unweighted, equal weights across the traits, and unequal weights across the traits, respectively.

Table 3 shows inflation biases of the multi-trait predictions measured as the regression coefficients of the yield deviations on the GEBVs for the 1988 validation cows. The inflation bias was only significant for the SCC analysis without SNP weights, where there was a deflation bias. The analyses that used SNP weights yielded virtually unbiased genomic predictions.

Table 3. Inflation bias of genomic predictions measured as the regression coefficient of yield deviations(y_v) on GEBVs($\widehat{g_v}$) for 1988 validation cows

SNP-BLUP Model	$b_{y_v,\hat{g}_v}$ *		
	Milk	Protein	SCC
SNPunw-BLUP**	1.08	1.07	1.19
SNPteqw-BLUP**	1.01	1.00	1.06
SNPtsw-BLUP**	0.99	0.98	1.01

*Standard errors of $b_{y_v,\hat{g}_v}$ are between 0.04 and 0.06.

**Subscripts unw, teqw, and tsw mean unweighted, equal weights across the traits, and unequal weights across the traits, respectively.

It seems natural to combine the GWAS signals across the traits by a multi-trait GWAS (Meuwissen *et al.* 2024), which makes optimal use of the data. But multi-trait GWAS analyses are computationally rather complicated, and simpler single-trait GWAS based approaches may be preferred. For SNP_{tsw}-BLUP models, single-trait GWAS analyses may be directly used to provide the trait-specific SNP weights. If the single-trait SNP analyses are not very powerful (do not result in clear genome-wide significant QTL signals), the use a multi-trait GWAS analysis may be worthwhile. The trait-specific SNP weights applied in SNP_{tsw}-BLUP adjust the prior variances of the traits on a per SNP basis, but not the correlations between the traits. A more flexible model would also estimate correlations of SNP effects on a per SNP basis. Gebreyesus *et al.* (2017) estimated correlations for groups of SNPs and obtained improved prediction reliabilities for milk composition traits. More research will be needed to investigate whether SNP specific correlations would increase the reliabilities of the genomic predictions.

CONCLUSION

The model with trait-specific SNP weights yielded EBVs with the highest reliability for all three traits analysed. For SCC, the model with identical SNP weights reduced the reliability of the EBV compared to unweighted SNP-BLUP. This problem was remedied by the use of trait-specific SNP weights. The multi-trait GWABLUP models yielded up to 13% more reliable EBV compared to unweighted multi-trait SNP-BLUP.

ACKNOWLEDGEMENTS

The authors are grateful to GENO SA (www.geno.no) for providing the data, to the Norwegian Research Council for funding of project number 309611.

REFERENCES

Boerner V. (2024) www.ghpc.ai

Ducrocq V. and Besbes B. J. (1993) *Anim. Breed. Genet.* **110**: 81.

Gebreyesus G., Lund M.S., Buitenhuis B., Bovenhuis H., Poulsen N.A. and Janss L.G. (2017) *Genet. Sel. Evol.* **49**: 1.

Liu Z., Goddard M.E., Reinhardt F. and Reents R. (2014) *J. Dairy. Sci.* **97**: 5833.

Meuwissen T., Eikje L.S. and Gjuvsland A.B. (2024) *Genet. Sel. Evol.* **56**: 1.

Vandenplas J., Calus M.P.L., Eding H. and Vuik C. (2019) *Genet. Sel. Evol.* **51**: 1.

Proc. Assoc. Advmt. Anim. Breed. Genet. 26: 193-196

INTEGRATING WHOLE GENOME SEQUENCE VARIANTS SELECTED FROM LARGE-SCALE ASSOCIATION ANALYSIS INTO THE SINGLE-STEP MARKER MODEL IN NEW ZEALAND DAIRY CATTLE

Y. Wang, M.A. Nilforooshan, R.G. Sherlock, A.M. Winkelman and B.L. Harris

Research and Development, Livestock Improvement Corporation, Hamilton 3240, New Zealand

SUMMARY

Genome-wide association studies (GWAS) of whole-genome sequence data are used to identify causative variants or single nucleotide polymorphisms (SNPs) closely linked to causative variants that are associated with traits of interest. SNP associated with causative variants are often not included in the standard SNP chips used for routine genomic evaluation, e.g., the Illumina Bovine SNP50 chip. Several studies have shown the potential of including pre-selected sequence variants (SEQ) in genomic prediction. However, studies integrating such SNPs using a single-step genomic prediction model remain limited. A large-scale genome-wide association study of 42 routinely evaluated traits was conducted for New Zealand dairy cattle, and 4431 sequence variants were selected using an iterative GWAS approach. The current study evaluated the performance of genomic prediction by combining pre-selected sequence variants with the standard Illumina 50k SNP panel used in the NZ dairy industry in a single-step marker model. Various levels of increased ratio of prediction accuracy, decreased dispersion, and increased bias were observed in focal cows and bulls for live weight. A clear advantage of including these sequence variants in the model was not observed.

INTRODUCTION

Genomic prediction using single nucleotide polymorphism data relies on the linkage disequilibrium (LD) between SNPs and causative variants. In the dairy industry, genomic predictions of breeding values have primarily used genotypes from the Illumina Bovine 50k SNP panel. However, standard Illumina 50k SNP arrays may not capture causal variants effectively because they consist of randomly selected markers chosen mainly for their polymorphism across breeds. Rare causal variants may not be in strong LD with the SNPs on these arrays and would have little, if any, impact on genomic predictions. In contrast, whole-genome sequence data is expected to capture many rare causal variants. Pre-selected sequence variants, either causal or closely linked to causal variants, are advantageous for genomic prediction and have been reported in several studies (VanRaden *et al.* 2017; Moghaddar *et al.* 2019). This study aimed to assess the impact of incorporating pre-selected sequence variants into the single-step genomic prediction of live weight in the New Zealand dairy population.

MATERIALS AND METHODS

Data. The population included Holstein-Friesian (HF), Jersey (JER), European Red Dairy (RDC) purebred, and Holstein-Friesian Jersey (HFJ) crossbred animals. Animals that didn't fit the above-mentioned groups were categorised as "other" (OTH). Table 1 shows the number of animals with live weight phenotypes and genotypes in each breed category.

Sequence variants selection. The sequence variants were selected from large-scale genome-wide association studies in New Zealand dairy cattle, which included 42 routinely evaluated traits (Wang, unpublished results). A total of 292,667 animals genotyped on various genotype panels were imputed to the whole genome sequence level using Beagle 5.4 (B.L. Browning 2018). After imputation, variants with a minor allele frequency lower than 0.5% or dosage R-squared lower than

0.9 were removed. A total of 16,453,913 variants were retained for the GWAS analysis. In this study, we conducted an iterative GWAS analysis in which the most significant variants identified in the current iteration were then added as covariates in the next iteration. Such iterations continued until no variants that passed the p-value threshold (5×10^{-8}) remained. Using this approach a list of 4799 variants that were significantly associated with the phenotype was generated. Multiple variants selected from different traits have the same locations. Thus, 4431 unique positions were added to the current genomic prediction model.

Table 1. Number of animals with live weight phenotype and genotype by breed included in the model

Breed	Number of animals with phenotype	Number of animals with genotype
Holstein-Friesian	466,513	71,244
Jersey	208,411	41,150
Holstein-Friesian Jersey cross	638,445	157,618
European Red	10,255	757
Other	86,302	15,172
Total	1,409,926	285,941

Statistical analysis. The single-step marker model is described in Harris (2022):

$$y = Xb + Z_g M_g m + Z_n u_n + Za + Zp + e$$

where **y** is the phenotype vector; $\mathbf{X}$, $\mathbf{Z_g}$, $\mathbf{Z_n}$, and $\mathbf{Z}$ are the incidence matrices; **b** is the vector of fixed effects, including hybrid vigor, genetic groups, and breed covariates; $\mathbf{M_g}$ is the SNP marker matrix, where g refers to genotyped animals; **m** is the vector of SNP marker effects with each column centered to have a mean of zero; $\mathbf{u_n}$ is the vector of genomic breeding values for non-genotyped animals and n refers to non-genotyped animals; **a** is the vector of polygenic effects; **p** is the vector of permanent effects and **e** is the random residual effect.

In this study, we used two sets of SNPs for genomic prediction: (i) filtered Illumina 50k and (ii) filtered Illumina 50k and 4431 pre-selected sequence variants. The Linear Regression (LR) validation method (Legarra and Reverter 2018) was used to compare the prediction performance before and after including the pre-selected sequence variants in the model. The impact of the two marker sets on the genomic predictions was assessed using "*whole*" and "*partial*" datasets. The whole dataset included all available pedigree, genotype, and phenotype information, resulting in genomic evaluations that we will refer to as u_w. In the partial data set, phenotypes obtained after 2020-05-31 were removed, and the resulting evaluations will be referred as u_p). The whole dataset contained 1,995,603 live weight records from 1,409,927 animals, whereas the partial dataset contained 1,676,349 records from 1,193,138 animals.

LR validation statistics were obtained for two focal populations - validation bulls and validation cows. The focal population of validation bulls included genotyped bulls with at least 20 daughters and records in the whole but not the partial data. The number of daughters per sire ranges from 20 to 559, with a mean of 97.08. The focal population of validation cows included genotyped cows with at least one record in the whole, but no record in the partial data. The number of focal bulls and cows in each breed category can be found in Table 2. Due to their poor population representation, animals in the RDC and OTH groups were not considered when assessing the genomic prediction performance. For both focal populations, prediction performance was evaluated by bias, dispersion, and ratio of accuracies calculated from GEBVs obtained from the whole and partial dataset described in the LR method. The bias $\hat{\Delta}_p$ was calculated as the difference between the mean u_p and u_w. The expected value of the bias is zero (i.e. unbiased). The dispersion $\hat{b}_p$ was calculated as the regression coefficient of u_w on u_p, which was expected to be one when there is no over- or under-dispersion. Values greater than one indicate under-dispersion, and values less than one indicate over-dispersion. The correlation between u_w and u_p is an estimator of the ratio of prediction accuracies ($\rho_{w,p}$).

Table 2. Number of live weight records for the whole and partial dataset, and the number of validation animals per breed

Breed	Number of records (whole)	Number of records (partial)	Focal	
			Bulls	Cows
Holstein-Friesian	693,439	626,823	327	8,675
Jersey	298,783	262,074	176	5,968
Holstein-Friesian Jersey cross	877,050	680,383	307	37,123
European Red	12,627	11,557	-	-
Other	113,703	95,511	-	-
Total	1,995,602	1,676,348	810	51,766

RESULTS AND DISCUSSION

Ratio of accuracy. Overall, the ratio of accuracy increased to varied degrees in both focal bull and cow populations when pre-selected sequence variants were added to the model. JER animals had a relatively lower ratio of accuracy compared to HF and HFJ. The ratio of accuracy increased from 0.769 to 0.791 for bulls and from 0.844 to 0.854 for cows. After including SEQ variants, the ratio of accuracy increased from 0.818 to 0.830 in HOL bulls and 0.854 to 0.861 in HOL cows. For HFJ animals, the ratio of accuracy increased from 0.826 to 0.846 in bulls and 0.892 to 0.899 in cows, the highest among all breed groups. Standard errors were larger for bulls than for cows using both marker sets (Table 3).

Table 3. The ratio of the prediction accuracies ($\rho_{w,p}$) for the live weight from validation bulls and cows. Standard errors within brackets

Breed	Focal Bulls		Focal Cows	
	Filtered 50k	Filtered 50k+SEQ	Filtered 50k	Filtered 50k+SEQ
Holstein-Friesian	0.818 (0.032)	0.830 (0.031)	0.854 (0.006)	0.861 (0.005)
Jersey	0.769 (0.048)	0.791 (0.046)	0.844 (0.007)	0.854 (0.007)
Holstein-Friesian Jersey cross	0.826 (0.032)	0.846 (0.031)	0.892 (0.002)	0.899 (0.002)

Dispersion. In general, adding the sequence variants resulted in inconsistent changes in the dispersion across the breed and focal groups (Table 4). Among all breed groups of focal bulls, HFJ had the most unbiased prediction; also, after including SEQ variants, bias decreased from 1.040 to 1.030. HF had a slight overdispersion with a 0.6% increase in bias after adding SEQ variants. JER had a higher overdispersion level, with 0.866 before and 0.868 after SEQ variants were included. In contrast, focal cows showed a much smaller dispersion than bulls, which was close to 1 across all breeds. Adding sequence variants in the model only resulted in around a 1% change in all breed groups.

Table 4. Dispersion ($\hat{b_p}$) for the live weight from validation bulls and cows. Standard errors within brackets

Breed	Focal Bulls		Focal Cows	
	Filtered 50k	Filtered 50k+SEQ	Filtered 50k	Filtered 50k+SEQ
Holstein-Friesian	0.927 (0.036)	0.921 (0.034)	1.028 (0.007)	1.020 (0.006)
Jersey	0.866 (0.055)	0.868 (0.051)	0.990 (0.008)	0.979 (0.008)
Holstein-Friesian Jersey cross	1.040 (0.041)	1.030 (0.037)	1.061 (0.003)	1.053 (0.003)

Bias. Overall, bias was slightly increased after including sequence variants in all scenarios aside of HF cows (Table 5). However, in general, the increased level was negligible compared to the standard deviation of genetic variance in both bulls (40kg in the "*whole*" and 37kg in the "*partial*" evaluation) and cows (37kg in the "*whole*" and 32kg in the "*partial*" evaluations). Among all breeds, the highest bias was observed in JER, where the bias increased from 5.20 to 5.26 for bulls after adding sequence variants in the model. For cows, the value increased from 4.04 to 4.27. For all the other breed groups, the bias was all under 3.

Table 5. Bias ($\hat{\Delta}_p$) for the live weight from validation bulls and cows. Standard errors between brackets

Breed	Focal Bulls		Focal Cows	
	Filtered 50k	Filtered 50k+SEQ	Filtered 50k	Filtered 50k+SEQ
Holstein-Friesian	1.800 (0.886)	1.860 (0.860)	0.407 (0.163)	0.352 (0.161)
Jersey	5.200 (1.010)	5.260 (0.965)	4.044 (0.155)	4.275 (0.153)
Holstein-Friesian Jersey cross	1.970 (0.920))	2.340 (0.872)	1.191 (0.076)	1.195 (0.075)

CONCLUSIONS

Enhancing the standard SNP chip with pre-selected sequence variants for genomic prediction of live weight was assessed using changes in the ratio of the accuracies, bias, and dispersion of the resulting GEBVs. The changes differed across breeds within the two focus populations but were generally small. Overall, the improvements in the estimates were minor, and no clear benefit of the sequence variants was found.

ACKNOWLEDGMENTS

The NZ Ministry of Primary Industries financially supported this study through the SFF Futures Program: Resilient Dairy–Innovative Breeding for a Sustainable Dairy Future (grant number: PGP06-17006). The authors wish to acknowledge the use of New Zealand eScience Infrastructure (NeSI) high-performance computing facilities as part of this research.

REFERENCES

Browning B.L., Zhou Y. and Browning S.R. (2018) *Am. J. Hum. Genet.* **103**: 338.

Harris B.L. (2022) *JDS communications* **3**: 152.

Legarra A. and Reverter A. (2018) *Genet. Sel. Evol.* **50**: 53.

Moghaddar N., Khansefid M., van Der Werf J.H.J., Bolormaa S., Duijvesteijn N., Clark S.A., Swan A.A., Daetwyler H.D. and MacLeod I.M. (2019) *Genet. Sel. Evol.* **51**: 72.

VanRaden P.M., Tooker M.E., O'Connell J.R., Cole J.B. and Bickhart D.M. (2017) *Genet. Sel. Evol.* **49**: 32.

Proc. Assoc. Advmt. Anim. Breed. Genet. 26: 197-200

A LINKAGE DISEQUILIBRIUM METHOD TO REPOSITION SINGLE NUCLEOTIDE POLYMORPHISM AND IMPROVE GENOTYPE IMPUTATION ACCURACY

Y. Fazel, K. Moore, S. Miller and M.H. Ferdosi

Animal Genetics Breeding Unit*, University of New England, Armidale, NSW, 2351 Australia

SUMMARY

Incorrect positioning of single nucleotide polymorphisms (SNP) can affect imputation accuracy and decrease the accuracy of genomic prediction. This study aimed to develop a method to identify the most likely genomic position of the misplaced SNPs which have low imputation accuracy by fitting a Spline curve using linkage disequilibrium (LD) information. The accuracy of the method was validated by correctly identifying the masked position of 2,560 out of 45,918 SNP with a 100% correlation between the original and estimated positions. Candidate SNPs with low imputation accuracy (< 0.5) were assumed to be incorrectly positioned on the genome assembly. The pair-wise LD between these SNPs and other SNPs on the genome was used to fit a Spline curve. The Spline peak was considered the most likely position for the candidate SNPs. This LD-based method assigned the new position for 92% of the SNPs with low imputation accuracy and improved the mean imputation accuracy of these repositioned SNPs from 0.21 to 0.97 in Australian Brahman cattle.

INTRODUCTION

Genotype imputation can provide high-resolution genomic information that can be used to accelerate the rate of genetic gain (Piccoli *et al.* 2014). For accurate genotype imputation, a genome assembly providing the genomic positions of the SNPs is required, and for some SNPs, the reported position on the assembly may be inaccurate (Yadav *et al.* 2021). However, structural variants, genotyping errors, chip density, size of the reference population, minor allele frequency (MAF), the magnitude of LD between SNP, the number and size of the shared haplotypes between individuals in the reference and target population, genetic structure and history of the population, and SNP distribution on the genome can all have an effect on genotype imputation accuracy (Bohmanova *et al.* 2010; Lashmar *et al.* 2019; Chen *et al.* 2021). The LD information has been used to identify the correct position of the SNP in the past. Although a few studies have discussed the relationship between the LD and SNP position (Miller *et al.* 2006; Khatkar *et al.* 2010; Yadav *et al.* 2021), identifying and correcting the incorrect position of SNP on the genome is still a challenge. This study aimed to develop a new method to identify the correct position of SNPs with low imputation accuracy (< 0.5).

MATERIALS AND METHODS

Genomic data and quality control. A total number of 50,060 Australian Brahman animals, genotyped with chip densities ranging from 7k to 800k, were used to evaluate SNP imputation accuracy. Quality Control (QC) before imputation removed SNP with minor allele frequency (MAF) less than 0.05 and with a call rate of less than 0.4 across all animals. Similarly, the animals with a call rate of less than 0.2 across all SNPs in the combined data set have also been eliminated during the QC filtration (Connors *et al.* 2017). After QC, 45,973 animals and 45,918 SNPs were left for further analysis.

---

* A joint venture of NSW Department of Primary Industries and Regional Development and the University of New England

Estimation of imputation accuracy. A genotype imputation scenario from 30k to 50k with 10,000 animals in the reference population was considered the optimum scenario to evaluate the SNP imputation accuracy (Ferdosi *et al.* 2021). The population was divided randomly into 10,000 reference animals with the remaining set as the target population. Then, in the target population, SNPs were randomly masked to the lower density and imputed to the original density. The process of random division of the population and masking of the SNPs was repeated 20 times to ensure all SNPs were involved in the imputation process. FImpute version 3 (Sargolzaei *et al.* 2014) with default parameters without using the pedigree information was used for the imputation, and Pearson's correlation between observed and imputed genotypes was considered as the metric for imputation accuracy. SNPs with mean imputation accuracy above 0.99 and a standard deviation (SD) of less than 0.001 across 20 iterations were assumed to be in their correct position, in order to use them to validate the Spline method. These SNPs are hereby referred to as validation SNPs. The SNPs with a mean imputation accuracy lower than 0.5 and SD lower than 0.1 in all iterations (hereafter referred to as candidate SNPs) were identified as SNPs potentially misplaced in the genome assembly for further LD analysis and repositioning.

LD calculation. After identifying low imputation accuracy SNPs, a pair-wise LD calculation was conducted using Plink version 1.9 (Purcell *et al.* 2007) between the candidate SNPs and all other SNPs across the genome.

The Spline model. For each candidate SNP and all other SNP on each chromosome, a piece-wise cubic polynomial Spline model, specified by four coefficients, was used to capture the nonlinear relationship between LD values and positions on the genome using the Splines package, specifically the natural cubic splines (ns) function in R (Ihaka and Gentleman 1996). LD values at each genomic position were used to solve for the Spline equation coefficients. The solved polynomial equations were used to fit the Spline curve across positions on the genome. The Spline model equation is as follows;

$$X = \sum_{i=1}^{k} \beta_i \phi_i(t)$$

where X is the LD between the candidate SNP and other SNPs on the chromosome, t is the position of the SNP on the genome in base pair, β_i are coefficients estimated by the Spline model, ϕ_i is the Spline function in the interval $[t_i, t_{i+1}]$, and k is the number of intervals (the region between two consecutive knots). The coefficients of the Spline equation in each interval were calculated using the least squares fitting approach to minimise the total error between observed LD values and estimated LD values by the Spline model. The number of data points between each two consecutive knots depended on the SNP distribution across the chromosome.

Flexibility control. The flexibility (fitness) of the Spline model was controlled by defining the degree of fitness (df), which quantifies the number of intervals and the smoothness of the fitted Spline curve for each candidate SNP in each chromosome. The number of intervals also determines the model's fitness based on the number and distribution of the data points across the chromosome. Four random degrees of model fitness, based on the number of intervals (10, 20, 50, and 100), were used to find the optimum degree of fitness for the Spline model.

Peak detection. After fitting the Spline curve to identify the position of interest on the genome, the first derivative of the Spline curve equation was calculated to identify the position of the peak of the curve. Peaks and valleys are the points where the slope of the curve is zero. The highest peak across all chromosomes was considered the estimated position for each SNP candidate.

Validation of the Spline model. To test the accuracy of the Spline model in identifying the SNP position using LD information, the positions of 2,560 validation SNPs were masked and re-estimated using the Spline method. The correlation between the original and estimated positions of the validation SNPs was considered as the model accuracy.

Repositioning of the SNP with low imputation accuracy. After repositioning of the candidate SNP, a further 20 rounds of imputation were conducted to assess the imputation accuracy of the repositioned SNPs in their new positions.

RESULTS AND DISCUSSION

Figure 1 shows the accuracy of the Spline model in estimating new positions for the candidate SNPs using different degrees of fitness. The original and estimated positions of the validation SNPs had the strongest correlation when the number of intervals was the highest.

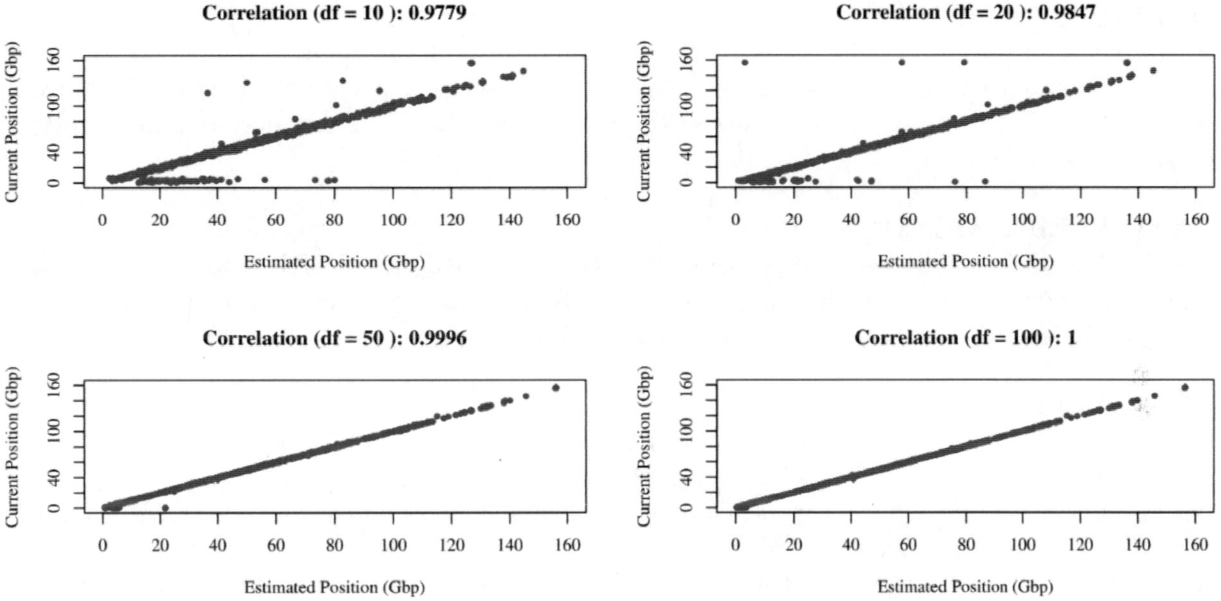

Figure 1. Validation of the Spline model using different degrees of fitness in estimating new positions using pair-wise LD information of the validation SNPs in Brahman

The Spline model was able to identify new positions for 34 out of 37 SNPs with low imputation accuracy. Figure 2 shows the imputation accuracy for these SNPs before and after repositioning.

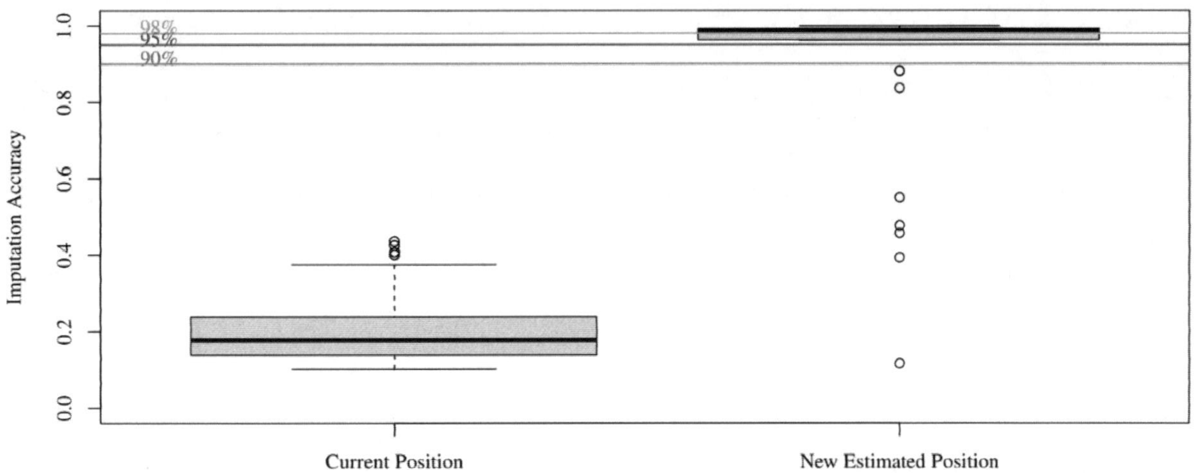

Figure 2. Imputation accuracy before and after SNP repositioning with Spline model

The Spline method was able to accurately identify the most likely position of 92% of the SNPs with low imputation accuracy on the genome using pair-wise LD information and improved the imputation accuracy of these SNPs. We observed that the incorrect position of the SNPs on the genome can contribute to lower imputation accuracy. However, the SNPs, whose new positions were assigned at the end of the chromosomes, still showed low imputation accuracy. Despite the high performance of the Spline method in identifying the correct positions, sensitivity to very low LD values and difficulties in estimating the positions of the SNPs that belong at the ends of chromosomes, can be considered as its limitations.

CONCLUSION

The incorrect position of the SNPs on the genome affects the genotype imputation accuracy. Fitting a Spline curve using pair-wise LD information can help to accurately identify the correct position of the SNPs and using newly assigned positions for the SNPs can improve the genotype imputation accuracy.

ACKNOWLEDGEMENTS

We acknowledge the Agricultural Business Research Institute (ABRI) and the Australian Brahman Breeders Association for data access. The first author was financially supported by the Animal Genetics and Breeding Unit (AGBU) scholarship.

REFERENCES

Bohmanova J., Sargolzaei M. and Schenkel F.S. (2010) *BMC Genomics* **11**: 1.

Chen L., Pryce J.E., Hayes B.J. and Daetwyler H.D. (2021) *Animals* **11**: 541.

Connors N., Cook J., Girard C., Tier B., Gore K., Johnston D. and Ferdosi M. (2017) *Proc. Assoc. Advmt. Anim. Breed. Genet* **22**: 421.

Ferdosi M., Connors N. and Khansefid M. (2021) *Proc. Assoc. Advmt. Anim. Breed. Genet* **23**: 118.

Ihaka R. and Gentleman R. (1996) *J. Comput. Graph. Stat.* **5**: 299.

Khatkar M.S., Hobbs M., Neuditschko M., Sölkner J., Nicholas F.W. and Raadsma H.W. (2010) *BMC Bioinformatics* **11**: 1.

Lashmar S., Muchadeyi F. and Visser C. (2019) *S. Afr. J. Anim. Sci.* **49**: 262.

Miller S., Hayes B., Goddard M. (2006) *Proc. World Cong. Genet. Appl. Livest. Prod.* **8**: 1.

Piccoli M.L., Braccini J., Cardoso F.F., Sargolzaei M., Larmer S.G. and Schenkel F.S. (2014) *BMC Genetics* **15**: 1.

Purcell S., Neale B., Todd-Brown K., Thomas L., Ferreira M.A., Bender D., Maller J., Sklar P., De Bakker P.I. and Daly M.J. (2007) *Am. J. Hum. Genet.* **81**: 559.

Sargolzaei M., Chesnais J.P. and Schenkel F.S. (2014) *BMC Genomics* **15**: 1.

Yadav S., Ross E.M., Aitken K.S., Hickey L.T., Powell O., Wei X., Voss-Fels K.P. and Hayes B.J. (2021) *BMC Genomics* **22**: 773.

MULTI-BREED GENOMIC ESTIMATED BREEDING VALUES PREDICT LIVEWEIGHT PRODUCTION IN NORTH AUSTRALIAN BEEF CATTLE

B.J. Hayes[1], G. Fordyce[1], E. Dodd[1], S. Speight[2] and K. McCosker[1]

[1] University of Queensland, Queensland Alliance for Agriculture and Food Innovation, Centre for Animal Science, Brisbane, QLD, 4000 Australia
[2] Black Box Co Mareeba, QLD, 4880 Australia

SUMMARY

Lifetime productivity of cows is a major driver of profitability of beef enterprises. One of the key components of lifetime productivity is liveweight production, the sum of the weight of cow's calves at weaning and its own annual liveweight change. Here multibreed genomic estimated breeding values (GEBV) for weight at 600 days, body condition score, heifer puberty, pregnant or not 4 months after calving (P4M), temperament, tick score, buffalo fly lesion score, teat score and udder score are assessed for their association with liveweight production that was recorded in two independent populations. GEBV for weight, heifer puberty and P4M were significant predictors of liveweight production in Brahmans. In tropical composites GEBV for weight, P4M, temperament and teat score were significant predictors of liveweight production. Results suggest selecting for GEBV for these traits will improve liveweight production.

INTRODUCTION

Lifetime productivity of beef cows is a key trait driving profitability of beef enterprises, however this trait has only rarely been measured due to the expense and long time frame required. Many predictors of lifetime productivity have been suggested, including heifer age at puberty (Johnston *et al.* 2013) and longevity (Martinez *et al.* 2005). Liveweight production, which combines weight of calves weaned and cow weight change, has also been suggested (Fordyce *et al.* 2023). Note that liveweight production is an imperfect measure of lifetime productivity, for example it does not account for feed eaten. Multi-breed genomic estimated breeding values for northern Australian beef cattle have been developed utilising a large reference population of commercial cattle from 60 properties across northern Australia (Hayes *et al.* 2023). The GEBV include weight at approximately 600 days, body condition score, heifer puberty (from an ultrasound scan assessing the presence or absence of a corpus luteum at approximately 600 days), pregnant or not 4 months after calving, temperament, tick score, buffalo fly lesion score, teat score and udder score. To assess if using these GEBV might be useful in selecting for improved lifetime productivity, we regressed GEBV simultaneously on liveweight production in a population independent of the reference.

MATERIALS AND METHODS

Animals, phenotypes and genotypes. The reference set for calculating GEBV was described in Hayes *et al.* (2023). Recently, 6,000 additional records have been added to that reference. Briefly, the reference set includes 34,292 heifers from sixty beef cattle herds, including crossbred and purebreds from at least 14 breeds, i.e. Angus, Belmont Red, Brahman, Charolais, Droughtmaster, Hereford, Limousin, Murray Grey, Santa Gertrudis, Shorthorn, Wagyu, Boran, Senepol and Tuli. Traits measured on the heifers included live weight, hip height and body condition score (BCS) at an average of 600 days, and a heifer puberty trait. The heifer puberty trait was cycling or not cycling by an average of 600 days assessed by the presence or absence of corpus luteum using ovary scanning. Tick scores (0-5), buffalo fly lesion scores (1-5), temperament (docility) scores (1-5) and teat and udder scores (1-5) were also recorded on some animals (Table 1) following BREEDPLAN guidelines for these traits (Table 1). Pregnant or not 4 months after calving (McCosker *et al.* 2023) was derived from foetal aging. All heifers were genotyped with the 35k or 50k TropBeef SNP array by Neogen, Australasia, or the Weatherbys 50k array. SNPs were removed if more than 10% of the genotypes were missing for that SNP. If an individual genotype had a GC score less than 0.6, it was set to missing and recovered by imputation. Genotypes were imputed up to 709,768 SNPs (bovine high-density (HD) array) using the findhap software (VanRaden *et al.* 2013) and a panel of 4,506 cattle from relevant breeds and crossbreds that were genotyped with the Bovine HD array. The accuracy of imputation was at least 93% for all breeds/crossbreds (Hayes *et al.* 2023).

Statistical models. To predict genomic estimated breeding values (GEBV), the model was $\mathbf{y}=\mathbf{1}_n\mu+\mathbf{Xf}+\mathbf{Zu}+\varepsilon$, where **y** is the vector of trait records; μ is the population mean, and $\mathbf{1}_n$ is a vector of 1s; **f** is the vector of fixed effects including contemporary group, the herd+birth year+paddock year of recording (2015, 2016, 2017, 2018, 2019, and 2020), a linear covariate of the average marker heterozygosity (to capture heterosis effects); X is a design matrix that relates fixed effects to records; **u** is the vector of random genetic effects $\sim N(0,\mathbf{G}\sigma_g^2)$ with **G** being the genomic relationship matrix between all the heifers and σ_g^2 the genetic variance captured by the SNPs, with **G** constructed according to Yang et al. (2010); **Z** is a design matrix that relates records to animals; **ε** is a vector of random deviations $\sim N(0,\mathbf{I}\sigma_e^2)$, with **I** an identity matrix and σ_e^2 the error variance. There were at least 25 animals in each contemporary group. Variance components and heritabilities were estimated with GCTA (Yang et al. 2010). Note this is the NOBREED model of Hayes *et al.* (2023).

Effect of GEBV on liveweight production. GEBV for all traits were predicted for a completely independent dataset; the Beef CRC groups of 894 and 1,088 Brahman and Tropical composite cattle (Johnstone *et al.* 2013). The 50k genotypes of these animals were imputed to 709,768 SNP using a large reference set, then multiplied by the SNP prediction equations from backsolving the system of equations above (Hayes *et al.* (2023). The accuracy of the GEBV were assessed by correlating the GEBV with phenotypes for these cattle for the most closely related trait they were measured for and dividing by the square root of heritability for these traits. Liveweight production for each cow was calculated following Fordyce *et al.* (2023) as the sum of the weight of its calf at weaning and its own annual liveweight change. The GEBV for all traits were regressed on liveweight production simultaneously. As producers receiving GBV also receive which quintiles their animals are in (based on all animals in the reference set), for heifer puberty and P4mM we also plotted actual age at puberty and percentage pregnant in the validation sets for the group of animals in each predicted quintile. Quintiles are always 1 =worst 20% for a trait to 5 = best 20% for a trait.

RESULTS AND DISCUSSION

Heritability of fertility traits was moderate (heifer puberty) to low (P4M; Table 1). Heritability estimates were fairly consistent with previous estimates (Johnston *et al.* 2003; Johnston *et al.* 2009; Johnston *et al.* 2013; Porto-Neto *et al.* 2104; Bunter and Johnston 2013).

Table 1. Number of records, heritabilities and standard errors for GEBV traits

Trait	Number	Heritability
Weight	31,673	0.33±0.01
Hip height	30,903	0.44±0.01
Body condition score	32,104	0.23±0.01
Heifer Puberty	34,292	0.22±0.01
Pregnant 4 months after calving	14,419	0.10±0.01
Temperament	5,650	0.31±0.03
Tick score	5,307	0.32±0.03
Buffalo fly lesion score	13,203	0.11±0.02
Teat score	4,044	0.34±0.04
Udder score	4,044	0.25±0.03

In the validation set, there was a good association between heifer puberty GEBV quintile and age at puberty. Animals in the 5th quintile reached puberty approximately eighty days before the 1st quintile, Figure 1. For P4M, 30% more animals in quintile 5 were pregnant than in the 1st quintile.

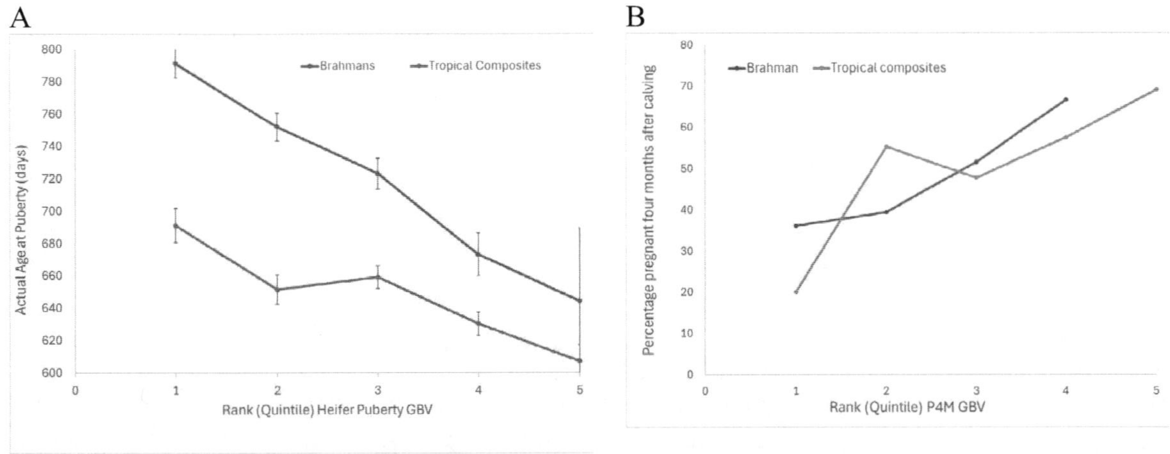

Figure 1. A. Average age at puberty for animals in the validation set by their GEBV quintile for heifer puberty (5=20% most likely to be pubertal at 600 days, 1=20% least likely to be pubertal at 600 days). B. Percentage of animals pregnant for animals by GEBV quintile (5=20% of animals most likely to be pregnant within 4 months of calving, 1 =20% of animals lease likely to be pregnant within 4 months of calving.

Accuracies of GEBV were moderate for most traits (Table 2). The lowest accuracy was for tick score in Brahman. This was because there was very limited variation for this trait in the validation set. The standard errors on the correlations are 0.03 so accuracy for all traits, except tick score in Brahmans, is significantly greater than zero. The accuracies are encouraging given there is likely to be considerable genetic distance between the reference set and validation, given the time that has elapsed since the Beef CRC recording was conducted (20 years), and the difference in traits definition between GEBV reference and CRC validation.

Table 2. Accuracy of GEBV in the Brahman and Tropical Composite Beef CRC validation set

GEBV trait	Beef CRC trait	Accuracy	
		Brahman	Trop Comp
600 day weight	600 day weight	0.24	0.32
Hip height	Hip height	0.57	0.53
Body condition score	Body condition score	0.15	0.36
Heifer puberty	Age at corpus luteum	0.39	0.25
P4M	Post partum anoestrus interval	0.11	0.35
Docility	Flight time	0.30	0.24
Tick score	Tick score	-0.07	0.17
Fly lesion score	Fly lesion score	0.16	0.50
Teat score	Back teat score calving	0.14	0.13
	Front teat score calving	0.12	0.25
Udder score	Udder score calving	0.28	0.29

The GEBVs that had a significant effect on liveweight production were for Brahmans, weight, heifer puberty and P4M, and weight, P4M, temperament and teat score GEBV for tropical composites.

Table 3. Regression of GEBV on liveweight production in Brahman and Tropical composite cattle. Significance of effects, effect estimates (for significant effects)

Trait	P-values		Estimates (kg)	
	Tropical composites	Brahmans	Tropical composites	Brahmans
Weight	**<0.001**	**0.04**	0.5	0.1
Height	0.76	0.07		
BCS	0.14	0.25		
Heifer puberty	0.99	**<0.001**		27.2
P4M	**0.02**	**0.01**	31.9	80.5
Temperament	**0.04**	0.08	-12.7	
Tick	0.30	0.88		
Fly	0.77	0.55		
Teat	**0.04**	0.35	-34.2	
Udder	0.77	0.79		

Effects were mostly in the expected direction, for example better P4M (closer to one) was associated with increased liveweight production in both breed types, and larger (potentially bottle) teats were associated with decreased liveweight production in tropical composites. It should be again noted that these observed regression coefficients where for a population a few decades old, it would be useful to repeat this exercise in a population of more modern cattle.

CONCLUSION

GEBV for weight, heifer puberty and P4M are significant predictors of liveweight production in Brahmans. In tropical composites GEBV for weight, P4M, temperament and teat score are significant predictors of liveweight production. Variation in heifer puberty in tropical composites is much less than in Brahmans, which may explain why this trait is not significant for tropical composites. These results suggest selecting for GEBV for these traits will improve liveweight production. With some (considerable) extension (for example inclusion of the effect of heterozygosity estimated from the model, some measure of feed intake, economics of production), this index could also predict future profit potential of females (eg Dunne *et al.* 2020).

ACKNOWLEDGEMENTS

Funding for this project was generously provided by the Meat and Livestock Australia donor company, project numbers P.P.S.H 1420 and P.PSH.P0833. We would also like to gratefully acknowledge Brian Burns and Nick Corbet, as well as the Beef CRC.

REFERENCES

Bunter K.L.. and Johnston D.J. (2013) *Anim. Prod. Sci.* **54**: 50.
Dunne F.L., Berry D.P., Kelleher M.M., *et al.* (2020) *J. Anim. Sci.* **98**: skaa334.
Fordyce G., Barnes T.S., McGowan M.R., *et al.* (2023) *Anim. Prod. Sci.* **63**: 395.
Hayes B.J., Copley J., Dodd, E., *et al.* (2023). *Genet. Sel. Evol.* **55**: 71.
Johnston D.J., Reverter A., Burrow H.M., *et al.* (2003) *Aust. J. Agri. Res.* **54**: 107.
Johnston D.J., Corbet N.J., Barwick S.A., *et al.* (2013). *Anim. Prod. Sci.* **54**: 74.
Johnston D.J., Barwick S.A., Corbet N.J., *et al.* (2009) *Anim. Prod. Sci.* **49**: 399.
Martinez G.E., Koch R.M., Cundiff L.V., *et al.* (2004) *J. Anim. Sci.* **82**: 1912.
Porto-Neto L.R., Reverter A., Prayaga K.C., *et al.* (2014) *PLoS ONE* **9**: e113284.
VanRaden P.M., Null D.J., Sargolzaei M., *et al.* Cole J.B., *et al.* (2013) *J. Dairy Sci.* **96**: 668.
Yang J., Lee S.H., Goddard M.E., *et al.* (2011). *Am. J. Hum. Genet.* **88**: 76.

ESTIMATES OF ACCURACY FOR GENOMIC-ONLY BREEDING VALUES AND FLOCK PROFILES IN AUSTRALIAN MERINO SHEEP

L. Li, P.M. Gurman, A.A. Swan and D.J. Brown

Animal Genetics Breeding Unit*, University of New England, Armidale, NSW, 2351 Australia

SUMMARY

As uptake of genomic-only estimated breeding values (GEBVs) and Flock Profiles – defined as the average GEBV from a random sample of approximately 20 animals from a commercial flock – expands within the Australian sheep industry, the need for accurate and reliable results becomes increasingly critical. Currently, GEBVs are presented without accuracy estimates, making it difficult to judge the genomic connectedness of the genotype to the reference population. This study aims to develop and implement a robust methodology for estimating the accuracy of GEBVs and Flock Profiles. Building on existing genetic evaluation accuracy frameworks, we propose a new method that can be integrated into routine evaluations, providing users with confidence in the reliability of their results.

INTRODUCTION

Single-step genomic BLUP (SS-GBLUP) combines pedigree, phenotypic, and genomic information and is routinely used in genetic evaluations for the Australian sheep and beef cattle industries (Brown *et al.* 2018; Johnston *et al.* 2018). One of the key advantages of genomic selection is its ability to calculate genomic-only estimated breeding values (GEBVs) for individuals without phenotypes and pedigree information. This application has been implemented in the Australian Merino sheep industry by back-solving SNP effects from SS-GBLUP EBVs, and then calculating GEBVs for animals with only genotype data (Swan *et al.* 2018). Currently, the resulting GEBVs for the genotype-only animals are then averaged to establish a benchmark for commercial flocks, referred to as Flock Profiles, using a random sample of approximately 20 animals from a flock.

However, this process does not estimate the associated accuracy for these predictions. Some flocks are genetically more different from the genotypes of the reference population in MERINOSELECT, which is expected to result in GEBVs of lower accuracy. It is crucial for the interpretation of GEBVs that predictions are sufficiently accurate and, like ASBVs, are only reported if they reach reporting thresholds for accuracy. Additionally, having an accuracy estimate attached to Flock Profiles and individual animal GEBVs is vital as these products are further adopted. This study aimed to derive a methodology for approximating the prediction error variance (PEV) and accuracy for both GEBVs and Flock Profiles. Initially, the methods were validated using a small dataset by comparing the approximated PEV with true PEV. Subsequently, these methodologies were applied to recent data from the Merino sheep industry to assess the accuracy of both GEBVs and Flock Profiles for a subset of traits.

MATERIALS AND METHODS

Derivation of Accuracy for GEBVs. Single-trait SNP-BLUP models can approximate accuracies for the genomic component of the single-step BLUP model (Li *et al.* 2023). In short, consider a single trait SNP-BLUP model with a mixed model equation (MME) $\begin{bmatrix} \mathbf{X'X} & \mathbf{X'ZW} \\ \boldsymbol{W'}\mathbf{Z'X} & \boldsymbol{W'Z'}\mathbf{ZW} + \mathbf{I}\lambda \end{bmatrix} \begin{bmatrix} \boldsymbol{b} \\ \boldsymbol{s} \end{bmatrix} = \begin{bmatrix} \boldsymbol{X'y} \\ \boldsymbol{W'Z'y} \end{bmatrix}$, where **y** is the vector of observations, **b** is the vector of

* A joint venture of NSW Department of Primary Industries and Regional Development and the University of New England

fixed effects, $\mathbf{s}$ is the vector of estimated SNP effects, $\mathbf{X}$ and $\mathbf{Z}$ are incidence matrices which map observations to fixed effects and individual animals' breeding values, respectively, $\mathbf{W}$ is the animals by markers matrix of centred and scaled marker genotypes and $\lambda = \sigma_e^2/\sigma_s^2$, i.e. the ratio of the residual variance and the marker variance. The prediction error covariance (PEC) for SNP effect, $\boldsymbol{C}^{22}$, can be obtained by inverting the left-hand side of MME, which by block matrix inversion rules can be written as $\mathbf{C}^{22} = [\boldsymbol{W}'\mathbf{Z}'(\mathbf{I}-\mathbf{P})\mathbf{Z}\boldsymbol{W} + \mathbf{I}\boldsymbol{\lambda}]^{-1}$, where $\mathbf{P} = \mathbf{X}(\mathbf{X}'\mathbf{X})^{-1}\mathbf{X}'$.

The PEC for GEBVs for all animals can therefore be calculated as $\boldsymbol{PEC}_{\boldsymbol{GEBVs}} = \boldsymbol{W}\boldsymbol{C}^{22}\boldsymbol{W}'\sigma_e^2$. In practice, only the diagonal elements of the PEC are required for individual animal calculations. After the derivation of PEC, the GEBV accuracy of individual animal can be calculated as $\mathbf{Acc} = \sqrt{\mathbf{1} - \frac{\mathbf{diag}(\boldsymbol{PEC}_{\boldsymbol{GEBVs}})}{diag(\boldsymbol{WW}')\sigma_a^2}}$, where $diag(\boldsymbol{WW}')$ are the diagonal elements of the genomic relationship matrix and σ_a^2 is the additive genetic variance. It is worth noting that the PEV for GEBVs calculated by this algorithm used SNP information only. We consider these PEV as approximations for the PEV of EBV of these genotype-only animals if they are derived in the routine SS-GBLUP analysis directly. This model utilizes an H matrix with a lambda value to integrate genomic and pedigree relationship matrices to derive EBVs (Brown *et al.* 2018).

Derivation of PEV and Accuracy for Flock Profiles. The above formula for an individual animal's accuracy can be extended to describe the accuracy for a group of animals. Suppose the flock profile contains genotypes from n individual animals, with the centred and scaled matrix $\boldsymbol{W}_{FP}$ obtained as the subset of $\boldsymbol{W}$ for the animals in the profile. The PEC for the animals is $PEC_{FP} = \boldsymbol{W}_{FP}\boldsymbol{C}^{22}\boldsymbol{W}'_{FP}\sigma_e^2$. If we define a vector $\boldsymbol{v} = \frac{1}{n}\mathbf{1}'$ as an n-dimensional column vector with each weight as 1/n, then the accuracy for the flock profile is $Acc_{FP} = \sqrt{1 - \frac{\boldsymbol{v}'\boldsymbol{PEC}_{\boldsymbol{FP}}\mathbf{v}}{\boldsymbol{v}'(\boldsymbol{W}_{FP}\boldsymbol{W}'_{FP})\mathbf{v}\sigma_a^2}}$. These calculations have been incorporated into the current snpEPN genomic accuracy software (Li *et al.* 2023).

Validation. A small test dataset from the MERINOSELECT analysis was extracted to validate the methodology as a subset of flocks and genotypes. The validation aimed to confirm that our PEV estimates are close to true values and achieve higher accuracy for Merino genotypes versus lower accuracy for non-Merino genotypes. This dataset comprised 106,000 animals, with 38,492 having genotypes for 61,146 SNPs. Approximately half of the genotyped animals were extracted from Merino flocks designated as the reference population, while the remaining genotypes were added from non-Merino flocks, which were expected to be genetically highly different from these reference genotypes for testing purposes. Three traits with different reference sizes were investigated: weaning weight (WWT, 15,279), yearling greasy fleece weight (YGFW, 8,379), and adult staple length (ASL, 1,186). Genetic parameters used in this validation were extracted from the MERINOSELECT with heritability of 0.29, 0.34 and 0.34 for WWT, YGFW and ASL, respectively. In this validation, the PEV of GEBVs from snpEPN were first compared with the PEV based on SS-GBLUP model (SS-PEV) with assumption that these genotype-only animals were included in the routine analysis. SS-PEV were calculated directly from the inverse of mixed model equations using an H relationship matrix with a lambda value of 0.95 to combine genomic and pedigree relationship matrices. Then, the accuracy of GEBVs was calculated and compared by breeds to demonstrate the influence of reference breeds on predictive accuracy.

Example data. The described methods were also applied to a MERINOSELECT analysis extracted from the October 1, 2024 run, including 6,359,892 animals, of which 523,845 were genotyped for 61,260 SNPs. Additionally, 78,852 genotypes were back-solved, including 27,670 genotypes from 996 test flocks. Traits investigated included birth weight (BWT), weaning weight (WWT), yearling weight (YWT), intramuscular fat (IMF), yearling eye muscle depth (YEMD), yearling greasy fleece weight (YGFW), and yearling fleece diameter (YFD). Heritability values used for accuracy approximation, along with the count of animals having both genotypes and phenotypes

(Nref) ranging from 7,357 for IMF to 396,307 for WWT for each trait, are detailed in Table 1.

Table 1. Heritability (h^2) and number of animals with both genotypes and phenotypes (Nref) for live weight, carcass and fleece traits extracted from the MERINOSELECT analysis

Trait	BWT	WWT	YWT	IMF	YEMD	YGFW	YFD
h^2	0.22	0.34	0.41	0.52	0.31	0.34	0.58
Nref	29,775	396,307	273,915	7,357	165,088	197,280	212,179

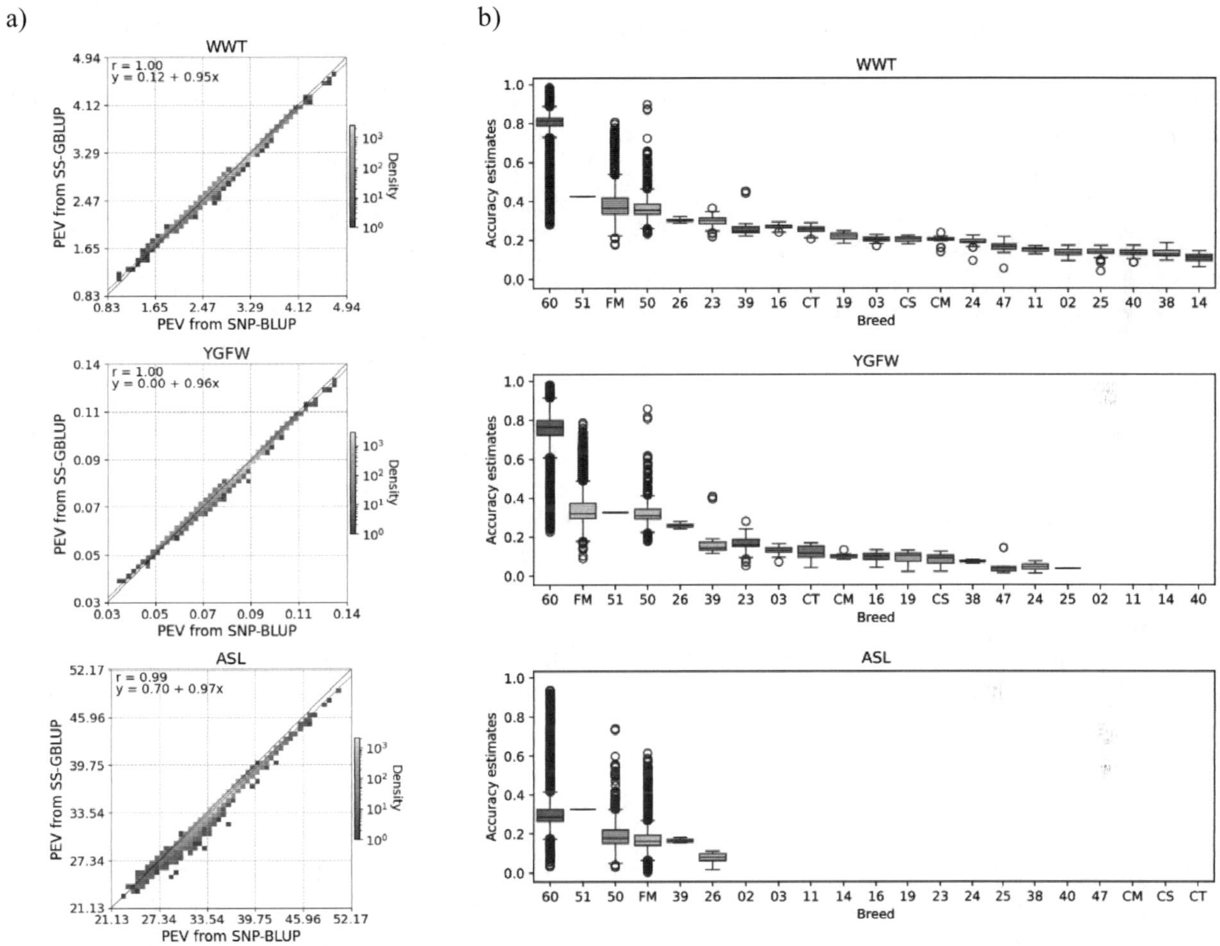

Figure 1. a) PEV from SNP-BLUP versus PEV from SS-GBLUP; b) Accuracy estimates for all animals by breeds for weaning weight (WWT), yearling greasy fleece weight (YGFW), and adult staple length (ASL). Note: Corresponding breed names for the codes listed in b) can be found at: https://www.sheepgenetics.org.au/resources/breeders-guide/animal-identification/.

RESULTS AND DISCUSSION

Validation results. The PEV of GEBVs for genotyped animals estimated from SNP-BLUP was very close to the PEV based on SS-GBLUP model, with correlations near 1 for all traits (Fig. 1a), indicating that the algorithms implemented were functioning as expected. The accuracy of EBVs for all animals across traits by breeds are shown in Fig. 1b. The Merino breed (60) consistently displayed higher accuracy for all traits, with a wider spread indicating variability within the breed. Other breeds less related to the reference flocks generally presented much lower accuracies, especially for YGFW and ASL, where most breeds had accuracy less than 0.2 (YGFW) and close to zero (ASL).

This may be due to the reference animals predominantly being from Merino breeds, with the reference size reducing from 15,279 for WWT to 1,186 for ASL.

Example data results. Figure 2 shows two boxplots illustrating the accuracy of GEBVs and Flock Profiles across all traits. The GEBVs show a wide range of accuracies, with BWT and IMF showing much lower accuracy (~0.4-0.5) and greater variability. In contrast, other traits display higher accuracies (>0.8) with narrower spreads, indicating more accurate and consistent predictions. These results correspond to the size of the reference population, as detailed in Table 1. Not surprisingly, the accuracy of Flock Profiles (Figure 2b) presents a pattern like GEBVs but with generally higher and more uniform accuracies. These high accuracy levels for both GEBVs of genotype-only animals and Flock Profiles are highly promising for the implementation of genomic selection in the Australia Merino sheep industry. However, these high accuracy estimates require further validation. This could be achieved, for instance, by comparing the accuracy of testing flocks with the linkage metrices between genomic reference populations and testing flocks.

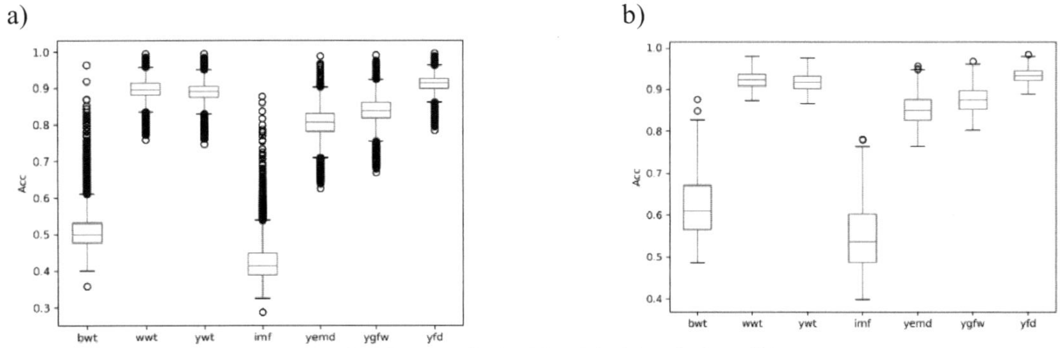

Figure 2. Accuracy by traits for a) GEBVs; b) Flock Profiles

CONCLUSIONS

A methodology for estimating the accuracy of GEBVs and Flock Profiles has been developed and incorporated into the current snpEPN program. This methodology was validated using a small dataset showing high correlations between PEV derived from SNP-BLUP and PEV from SS-GBLUP. Very high accuracy (>0.8) was observed for both GEBVs and Flock Profiles within a real Merino sheep dataset for traits with substantial-sized reference populations, such as WWT, YWT, YEMD, YGFW and YFD. However, lower accuracy (0.4-0.5) was noted for traits with smaller reference populations, such as BWT and IMF. By providing accuracy estimates for GEBVs and Flock Profiles, the Merino sheep industry can benefit from a more transparent and reliable genetic evaluation system. This methodology can be implemented in other sheep breeds and beef cattle.

ACKNOWLEDGEMENTS

This research was funded by Meat and Livestock Australia project L.GEN.2501. The authors acknowledge the contribution of data from Australian sheep breeders.

REFERENCES

Brown D.J., Swan A.A., Boerner V., Li L., Gurman P.M., McMillan A.J., van der Werf J.H.J., Chandler H.R., Tier B. and Banks R.G. (2018) *Proc. World Cong. Genet. Appl. Livest. Prod.* **11**: 460.

Johnston D.J., Ferdosi M.H., Connors N.K., Boerner V., Cook J., Girard C.J., Swan A.A. and Tier B. (2018) *Proc. World Cong. Genet. Appl. Livest. Prod.* **11**: 269.

Li L., Gurman P., Swan A.A. and Tier B. (2023) *Anim. Prod. Sci.* **63**: 1086.

Swan A.A., Gurman P.M., Boerner V., Brown D.J., Clark S., Gore K., Granleese, T. and van der Werf J.H.J. (2018) *Proc. World Cong. Genet. Appl. Livest. Prod.* **11**: 1.

HELICAL: A FULL-STACK AGILE PLATFORM ENABLING DATA FLYWHEELS AND DATA SHARING TO EMPOWER MODERN BREEDING PROGRAMS

D.P. Garrick[1,2,3], D.J. Garrick[1,2] and J.E. Howse[1]

[1] The Helical Company Limited, Rotorua, New Zealand
[2] Theta Solutions LLC, Lacey WA, USA
[3] A.L. Rae Centre for Genetics and Breeding, Massey University, New Zealand

SUMMARY

Helical is a comprehensive and fully integrated but modular informatics software-as-a-service (SaaS) platform that can be deployed as a complete breeding program information and decision support system. Built-in modules can standalone such as the genomic (SNP) database. Data import and extraction is highly flexible to cater for integration of any module with existing legacy systems. Licensees are typically research or breeding organisations who deliver and/or ingest data from stakeholders at an a) administrative level, b) user level, and/or c) a public level. The platform's architecture provides highly configurable but secure access controls for independent but integrated portals which are separately catered to each audience. At an organisation level, the system delivers full control of genomic data while maintaining freedom to work with multiple genotyping providers and service partners. This paper describes the platform's technical implementation and demonstrates its practical impact through commercial case studies, highlighting how organisations can continuously extract value from their genomic investment through sophisticated analytical tools. The platform's multi-tiered modular access facilitates powerful data sharing arrangements between stakeholders such as breed societies and commercial producers, while maintaining data sovereignty for each party. The result is a future-proofed system that enables organisations to maximise utility from their genomic data.

INTRODUCTION

Animal breeding programs face many complex challenges in today's technological landscape, including managing increasingly sophisticated genomic datasets, exchanging data with competing genomic labs and other service providers and linking a variety of data between stud and commercial operations. Advancement of DNA testing technologies has led to an explosion in genomic data volumes, yet most organisations remain constrained by old-fashioned inflexible systems that create data silos and lack the agility to incorporate new technologies, tackle emerging opportunities, or explore new service offerings without substantial manual intervention or costly investment.

The Helical platform addresses these challenges through a full-stack architecture that can be deployed either as a comprehensive information system or in a modular fashion such as a standalone genomic database that can be integrated with existing pedigree and performance systems. This flexible deployment model enables organisations to maintain part or all of their current registry systems while gaining sophisticated genomic capabilities, including the freedom to work with multiple genotyping providers, perform automated parentage analyses, and manage their DNA data through intuitive interfaces. For organisations seeking a complete solution, the platform can provide an end-to-end system encompassing registry, performance recording, genomic data management, analytical functionality, sales catalogues, and evaluation results publication.

The platform enables breeding organisations to precisely control data access at every level while maintaining workflow flexibility. The backend services power complex analyses through sophisticated algorithms and data management systems, while the frontend provides intuitive interfaces for non-technical users to perform complex genetic analyses. This integrated approach ensures that organisations can maintain complete control over their data and analyses while benefiting from modern software architecture and tools, facilitating powerful feedback loops between various stakeholders to enhance genetic improvement programs.

MATERIALS AND METHODS

The Helical platform is a species-agnostic full-stack architecture consisting of frontend web interfaces, backend services with databases, and a comprehensive suite of command-line tools. Some modules and tools are catered specifically for diploid species but can be extended as needed for other organisms. The backend module manages databases and analytical procedures through an Application Programming Interface (API), which the frontend and command line tools utilise for data operations and job submissions (Figure 1).

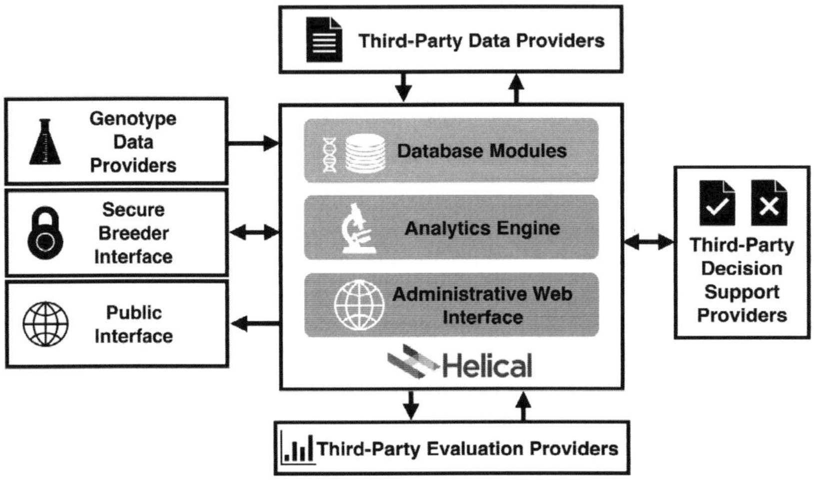

Figure 1. Overview of the Helical platform architecture

The platform is utilised by organisations across many livestock and plant species, with flexible configurations enabling custom terminology to cater for specific industries and use cases.

Frontend interfaces. The platform implements four distinct web portals:

1. Administrative interface for staff to manage system configuration, users and permissions, data templates for import and export, entity management, and analytical workflows including DNA test management and parentage verification or discovery, amongst others.
2. Entity/stakeholder interface for members to manage their animals, view and export DNA and evaluation results, invoices, submit data, and to manage their accounts.
3. Public interface for anonymous users to search animals and estimated breeding values and view published sales catalogues.
4. Technical interface for geneticists, data scientists, or other technical users to access documentation on command line tools and interact with datasets, results, and tools.

Secure backend systems. A web server exposes an API and core database modules including:

- Configuration: business logic and system settings.
- Pedigree and performance: animal data fields.
- DNA: storage/retrieval of sample information and SNP data from various panels.
- Entity management: stakeholder and user data.
- Analysis: various automated and semi-automated processing of pedigree, performance, and genotype data including parent verification and discovery, rigorous genotype sample quality controls, and data export functions.

Optional built-in modules. These extend functionality for:

1) Registry operations (registrations, transfers) including custom business logic and rules.

2) Genetic test results and diagnostics including storing test results for genetic conditions.
3) Performance data for any pre-defined and configured traits.
4) Evaluation results, e.g. estimated breeding values, progeny differences (EBVs/EPDs), accuracies or reliabilities and indexes from one or more evaluation providers.
5) Microsatellite data storage and analysis.

Command line module. The command line tools provide a suite of utilities for technical analyses, data management, and custom pipeline creation. This includes built-in core capabilities for in-house genetic/genomic analyses including single-step evaluations and variance component estimation (heritability estimation), parentage verification and discovery, genotype imputation, DNA data interrogation and analysis, and high-performance linear algebra tools. Helical is routinely used in breeding company scale datasets as well as the largest national and international implementations encompassing tens of millions of animals. The command-line module integrates with the platform's backend API to enable trivial import and export of data stored in Helical database instances. This allows seamless incorporation of any external software tools or pipelines, including enrichment of data from third party systems. Technical users can leverage these capabilities to implement custom analysis pipelines while maintaining version control and reproducibility.

Integrations. The system routinely integrates with third parties:

1) To provide sample IDs and source genotype calls from Neogen, Weatherbys Scientific, Zoetis, and other genotype providers for DNA testing with demonstrated capability to store >1,000,000 genotyped samples on densities up to 700K.
2) To supply pedigree, performance and/or genotype data for national and international evaluations including IGS[1], NZAEL[2], PACE[3], LCC[4], and BREEDPLAN.
3) To display evaluation results and sale catalogues.
4) To produce invoices on Xero software for accrued charges.
5) For mate selection such as provided by MateSel (Kinghorn 2011).

The system is wholly packaged in a single cross-platform executable available under paid license. Because the analytical procedures relating to genomic evaluations exist alongside the backend and frontend modules, there exists a wide scope to surface advanced analyses into the web interfaces, allowing an unlimited amount of extensibility to provide advanced technical functionality to users through secure and accessible web applications.

Any data can be bulk imported or exported in user-customised text formats, e.g. comma separated values. This includes derived data fields generated on the fly via user-defined JavaScript functions. Both the administrative and breeder portal can facilitate such data exchange.

RESULTS AND DISCUSSION

The Australian Wagyu Association's (AWA) implementation of Helical demonstrates the platform's ability to enable sophisticated genomic operations while facilitating valuable data sharing between stakeholders. AWA utilizes Helical as their "One-stop DNA shop," leveraging its ability to integrate with multiple genotyping providers (Neogen and Zoetis) across their international membership base spanning Australia, New Zealand, USA, UK, Canada, and China. The platform manages all aspects of DNA testing from submission through to results delivery, including automated parentage discovery, genetic condition reporting, and trait predictions.

---

[1]International Genetic Solutions run by American Simmental Association
[2]New Zealand Animal Evaluation Limited, a DairyNZ subsidiary running the NZ dairy evaluation
[3]Pan-American Cattle Evaluation run by the American Hereford Association
[4] Leachman Cattle Company $Profit™ evaluation

AWA's implementation also showcases how Helical's genomic capabilities can drive commercial value through products like the Wagyu Feeder Check (Reverter *et al.* 2023), which ranks animals on five feedlot growth and carcass traits. Commercial producers use these predictions for culling and feeding lot decisions for fullblood and F1 crossbred animals. Helical's automated chip-agnostic parent discovery capabilities create a critical link between commercial animals and AWA registered sires. Producers are credited by AWA for submitting carcass data, which enhances prediction accuracy for the entire membership. This data sharing cooperation enables a powerful data flywheel where commercial producers benefit from valuable decision support tools while the AWA breeding program gains improved carcass trait predictions for registered sires.

AWA utilises the platform's genomic analysis tools to compute measures of inbreeding and diversity, including inbreeding from traditional pedigree coefficients, runs of homozygosity (ROH), or diversity assessed by genomic relationships. These analyses provide members with sophisticated insights to aid breeding decisions, demonstrating how maintaining sovereignty over genomic data allows organisations to implement new tools that create ongoing value from their historical genotype investment. The platform's robust quality control pipelines and flexible analysis framework ensure that genomic data collected over time, even representing different chip densities, manifests, genotyping technologies or service providers, can be leveraged for increasingly sophisticated applications as new methodologies and DNA testing technologies emerge.

The success of the AWA implementation, along with Helical's adoption by numerous organisations across beef and dairy cattle, sheep, dogs, horses, and plants globally, demonstrates how the platform's flexible and agile architecture can be adapted to diverse breeding contexts while maintaining robust genomic capabilities and data sovereignty.

CONCLUSION

The Helical platform is a modern full-stack architecture that provides a future-proofed foundation for breeding program data. It can be deployed as a comprehensive breeding database system or as a standalone genomic database integrating with existing systems. Combined with robust data sharing frameworks, sophisticated genomic analysis capabilities, intuitive interfaces for DNA data management, robust backend services for real-time analytics, and extensive command-line tools, the platform provides organisations complete and agile ownership of their data while maintaining freedom to work with multiple service providers, ensuring they can always access the most competitive pricing and service offerings.

The platform breaks down traditional data silos for the benefit of all stakeholders. Organisations gain not only the freedom to choose and switch between genotyping providers, but also retain permanent access to their historical genomic data and sophisticated analysis tools. This data sovereignty enables them to continue extracting new value from their genomic investment through time, whether through automated parentage discovery, implementation of new trait predictions, or creation of novel data sharing arrangements that benefit both seedstock and commercial operations.

As breeding programs continue to evolve, the ability to independently manage genomic data while maintaining competitive vendor relationships or adopt new technologies will become increasingly critical. This independence ensures organisations can continue to leverage their accumulated genomic data in new ways as technology advances, maximising the long-term return on their breeding program data investment.

REFERENCES

Kinghorn B.P. (2011) *Gen. Sel. Evol.* **43**: 1.

Reverter A., Li Y., Alexandre P.A., Dominik S., *et al.* (2023) *Anim. Prod. Sci.* **64**: AN232446.

Proc. Assoc. Advmt. Anim. Breed. Genet. 25: 213-216

ALIGNMENT OF PEDIGREE AND GENOMIC INFORMATION IN A COMBINED TERMINAL SIRE AND MATERNAL SHEEP ANALYSIS

P.M. Gurman, P. Alexandri, D.J. Brown, K.L. Bunter, S. de las Heras-Saldana, K.P. Gore, L. Li, A.J. McMillan, S.F. Walkom and A.A. Swan

Animal Genetics Breeding Unit*, University of New England, Armidale, NSW, 2351 Australia

SUMMARY

Recent efforts to combine the Sheep Genetics LAMBPLAN maternal and terminal sire analyses presented the opportunity to review the genomic model used in single-step genomic BLUP. This paper uses forward validation to examine the suitability of different genomic models, including metafounders, a genomic relationship matrix (GRM) constructed from a single set of allele frequencies and a breed-adjusted GRM using allele frequencies for each identified subpopulation. The metafounder model yielded slightly higher stability values and was slightly less biased than the other genomic models. Metafounders also largely resulted in higher within-breed stabilities. These results support the use of a metafounder model for the combined LAMBPLAN analysis.

INTRODUCTION

The Australian sheep industry utilises diverse sheep breeds for their wool production, maternal abilities, along with meat and wool shedding properties. The genetic analysis for these sheep, LAMBPLAN, currently splits these animals into two groups, maternal dual purpose sheep breeds (primarily based on their mothering abilities) and terminal sire breeds (based mostly on their carcase qualities, Walkom *et al.* 2025), although the rise of composite sheep means breeds are often included in both analyses (McMillan *et al.* 2023). Subpopulations in these analyses have large genetic differences (Alexandri *et al.* 2025), with some subpopulations performing selection within their own population to maintain purity, while others are characterised by a diverse selection across multiple breeds, either to create a new stabilised subpopulation, or to target specific traits.

Single-step analyses incorporating multiple breeds or distinct subpopulations can be challenging due to the misalignment between the pedigree and genomic information. This is due to the numerator relationship matrix (NRM) coefficients ignoring breed differences, which are included in the genomic relationship matrix. This study examines approaches for the alignment of relationship matrices in a combined terminal-maternal sheep dataset with large breed structures present and aims to determine the model with the best predictive ability. The first of these, currently implemented in LAMBPLAN (Gurman *et al.* 2019), removes breed structure from the genomic relationship matrix, to allow alignment of the genomic relationship matrix and the pedigree relationship matrix (Makgahlela *et al.* 2013; Gurman *et al.* 2019). Metafounders (Legarra *et al.* 2015) provides a method to align the pedigree information to the genomics by adding population structure to the NRM.

MATERIALS AND METHODS

The data used here are described by Walkom *et al.* (2025) and the fixed effects component is described by de las Heras-Saldana *et al.* (2025). The pedigree included 4,839,523 animals, with 246,093 genotyped for 62k imputed SNPs, and 23 traits, with 3,658,685 animals being recorded for at least one trait. Multiple BLUP analyses compared different methods for aligning the genomic and pedigree relationship matrices. The first model was a pedigree-only analysis with genetic groups. The genetic groups covered 35 breed groups and up to 3-time windows. The second model was a

---

* A joint venture of NSW Department of Primary Industries and Regional Development and the University of New England

single-step model with a single set of scalar adjustments (Christensen 2012) based on the GRM constructed by Yang *et al.* (2010). The third method was to adjust the GRM for each animal's expected allele frequency given the estimated breed proportions from 27 subpopulation allele frequencies (Makgahlela *et al.* 2013; Gurman *et al.* 2019; Alexandri *et al.* 2025). This third method is currently used in the LAMBPLAN evaluations. The fourth and fifth models were based on transferring the genetic group definitions to metafounders. The matrix of relatedness between and within breeds, Γ, was estimated in two steps. Metafounders covering multiple time groups were merged and Γ was estimated using the back pedigree of the genotyped animals (Legarra *et al.* 2024a). This matrix was converted to F_{st} values (Legarra *et al.* 2024b). The collapsed Γ matrix was then expanded to cover time-based groups (Legarra *et al.* 2024a) based on pedigree inbreeding coefficients. When groups had insufficient genotypes (as observed by low Γ values during estimation of the merged Γ) the values in Γ for the base group were assumed for these metafounders. Metafounder analyses were performed as both pedigree only and genomic models assuming allele frequencies of 0.5 for all SNPs. All genomic analyses were performed using ssGTBLUP, (Mäntysaari *et al.* 2017) and the AGBU commercial solver.

Forward cross validation evaluated the most performant model. Animals born during or after 2021 were used as the validation (removing data for 716,842 animals, or 19.6% of animals with one or more phenotypes), with their phenotypes removed from the part analyses. Pairwise BLUP analyses were performed with (w) and without (p) the validation phenotypes, with the subsequent EBVs for the validation animals labelled as $\boldsymbol{u}_w$ and $\boldsymbol{u}_p$, respectively. Pairwise analyses were performed for each BLUP analysis described in the previous paragraph. The metrics of Legarra *et al.* (2018) were used to evaluate each model. We define here stability as $cor(\boldsymbol{u}_p, \boldsymbol{u}_w)$, which gives a relative indication of the model accuracy. Dispersion was defined as $\frac{cov(\boldsymbol{u}_p,\boldsymbol{u}_w)}{var(\boldsymbol{u}_p)}$, with values less than one indicating overdispersion and bias was defined as $\frac{\overline{u_p}-\overline{u_w}}{\sqrt{v_a}}$ where v_a is the genetic variance.

RESULTS AND DISCUSSION

A metric of genetic similarity, F_{st}, calculated from Γ with the collapsed time groups, is presented in Figure 1, grouped based on the breed types (maternal, terminal, wool, shedding and other). F_{st} values between some subpopulations were large, with the Border Leicester showing genetic differences from several other subpopulations. Differences were also observed between East Friesian and other groups but is likely due to low representation of East Friesians (see Figure 1 for Q matrix column sums for genotyped animals, considering crossbred contributions). Similar F_{st} ranges have been observed between some beef breeds (Manzari *et al.* 2024).

A summary of the cross-validation results across all subpopulations is presented in Table 1, and the stabilities for selected subpopulations are presented in Figure 2. The highest stabilities were observed for the genomic metafounder model. The genomic metafounder model also resulted in the least dispersion error and the least bias compared to the other genomic models. The stabilities within subpopulations show that for most subpopulations, the metafounder model was slightly more accurate than the other genomic models. Adding metafounders to the pedigree model also resulted in better predictive ability (Table 1), even with the pedigree model including genetic groups based on the same mapping of animals to groups/metafounders. One downside of the metafounders model was the higher iteration count and convergence time compared to the breed-adjusted GRM, but this reduced performance could be accepted given the gain in the predictive ability of EBVs.

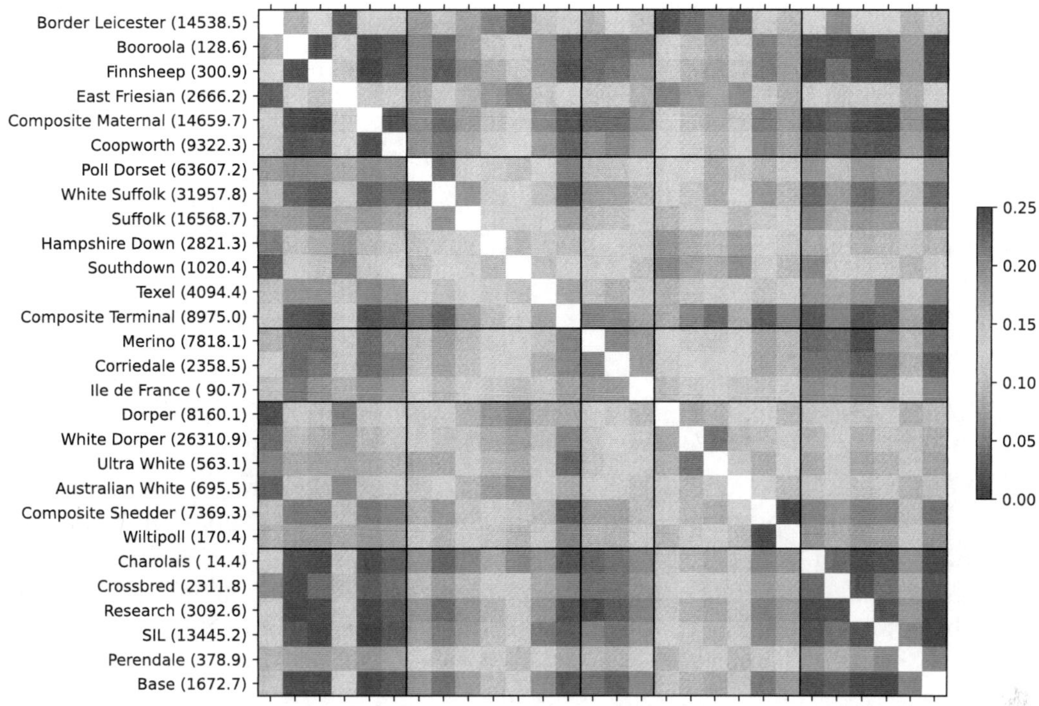

Figure 1. F_{st} value for each subpopulation calculated from the estimated gamma matrix. Black boxes group the subpopulations based maternal, terminal sire, wool, shedding and other subpopulations. Numbers next to each breed are the sums of the Q matrix columns for genotyped animals, considering contributions from crossbreds

Based on this research, metafounders have recently been incorporated into the routine BLUP evaluation pipeline for the combined LAMBPLAN analysis. This improvement results in changes to the routine pipeline including new parameters that need routine updates. For instance, the question of how often the gamma matrix should be re-estimated is unknown, but likely has similarities to how often the allele frequencies need to be re-estimated. Implementing metafounders will also require changes to the accuracy approximation algorithm. Bermann *et al.* (2023) has presented the differences in terms of true accuracies, with approximations requiring further research. Finally, the interpretation of relationships between animals diverges greatly from traditional interpretations (due to metafounder relationships) and changes to diagnostic information will need to be considered. While these are challenges to overcome, changing to a metafounder model will result in more accurate breeding values, and will allow for better across population selection decisions.

Table 1. Cross validation results and iteration counts and solving time for each model

	Stability	Dispersion	Bias	Iterations	Solve Time (hh:mm)
Pedigree	0.758	0.945	0.006	2,294	2:12
Pedigree with metafounders	0.776	0.961	0.020	3,143	3:29
Genomic with metafounders	0.851	0.944	0.025	3,684	7:05
Yang GRM	0.839	0.943	0.064	4,340	8:20
Breed Adjusted GRM	0.824	0.935	0.072	2,809	5:02

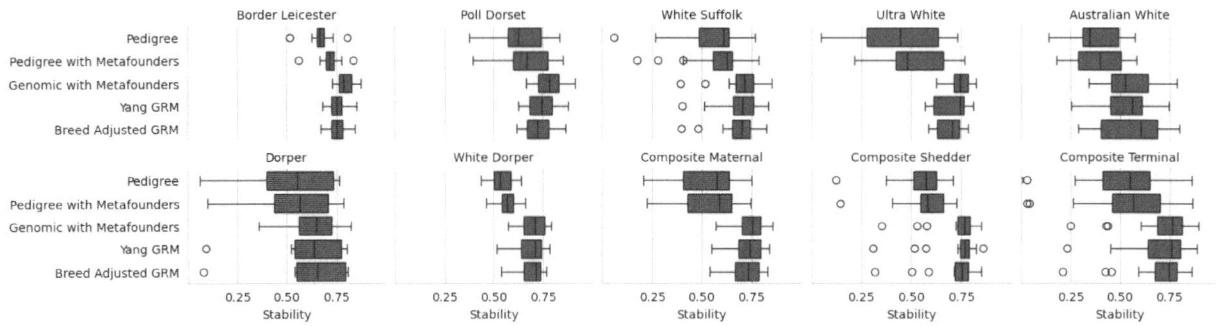

Figure 2. Cross validation stabilities as analysed within subpopulations

CONCLUSION

Metafounders improved predictive ability compared to the other genomic models considered. This work will power a future Combined LAMBPLAN analyses to provide breeders with more accurate breeding values and better across-breed predictions. Further work is required to fully document the benefits and to understand the ramifications of this change in a commercial context.

ACKNOWLEDGEMENTS

This research was funded by Meat and Livestock Australia project L.GEN.2204. The authors acknowledge the teams behind the Sheep CRC Information Nucleus and MLA Resource flocks as well as individual industry flock owners involved in flock management and data collection.

REFERENCES

Alexandri P., Brown D.J., Bunter K.L., De las Heras-Saldana S., Gore K.P., Gurman P.M., Li L., McMillan A.J., Walkom S.F. and Swan A.A. (2025) *Proc. Assoc. Advmt. Anim. Breed. Genet.* **26**: *These proceedings*.

Bermann M., Aguilar I., Lourenco D., Misztal I. and Legarra A. (2023) *Genet. Sel. Evol.* **55**: 6.

Christensen O.F. (2012) *Genet. Sel. Evol.* **44**: 37.

Gurman P.M., Bunter K.L., Boerner V., Swan A.A. and Brown D.J. (2019) *Proc. Assoc. Advmt. Anim. Breed. Genet.* **23**: 254.

De las Heras-Saldana S., Alexandri P., Brown D.J., Bunter K.L., Gore K.P., Gurman P.M., Li L., McMillan A.J., Walkom S.F. and Swan A.A. (2025) *Proc. Assoc. Advmt. Anim. Breed. Genet.* **26**: *These proceedings*.

Legarra A., Bermann M., Mei Q. and Christensen O.F. (2024a) *Genet. Sel. Evol.* **56**: 1.

Legarra A., Bermann M., Mei Q. and Christensen O.F. (2024b) *Genet. Sel. Evol.* **56**: 34.

Legarra A., Christensen O.F., Vitezica Z.G., Aguilar I. and Misztal I. (2015) *Genetics* **200**: 455.

Legarra A. and Reverter A. (2018) *Genet. Sel. Evol.* **50**: 53.

Makgahlela M.L., Strandén I., Nielsen U.S., Sillanpää M.J. and Mäntysaari E.A. (2013) *J. Dairy Sci.* **96**: 5364.

Mäntysaari E.A., Evans R.D. and Strandén I. (2017) *J. Anim. Sci.* **95**: 4728.

Manzari Z., Connors N.K., van der Werf J.H.J., Johnston D.J. and Ferdosi M.H. (2024). *Int Plant Anim. Geno. Conf.* **P007**

McMillan A.J., Walkom S.F. and Brown D.J. (2023) *Proc. Assoc. Advmt. Anim. Breed. Genet.* **25**: 31.

Walkom S.F., Alexandri P., Brown D.J., Bunter K.L., De las Heras-Saldana S., Gore K.P., Gurman P.M., Li L., McMillan A.J. and Swan A.A. (2025) *Proc. Assoc. Advmt. Anim. Breed. Genet.* **26**: *These proceedings*.

Proc. Assoc. Advmt. Anim. Breed. Genet. 26: 217-220

IMPLEMENTING GENOMIC OUTCOMES FOR SIMULATED PROGENY TO SUPPORT MATING DECISIONS

D.P. Garrick[1,2,3], J.E. Howse[1] and C.F. Teseling[4]

[1] The Helical Company Limited, Rotorua, New Zealand
[2] Theta Solutions LLC, Lacey WA, USA
[3] A.L. Rae Centre for Genetics and Breeding, Massey University, New Zealand
[4] Australian Wagyu Association, Armidale, NSW, 2350 Australia

SUMMARY
This study presents results of modelling probable progeny genotypes based on parental genomic data to support mate selection and allocation. We construct genomic profiles for "digital progeny" to evaluate possible mating outcomes, and characterise these in terms of predicted genomic breeding values and inbreeding coefficients via runs of homozygosity (ROH) measures. Results demonstrate the utility of this approach for informing mate selection and allocation decisions which balance genetic gain with inbreeding and maintenance of diversity.

INTRODUCTION
The widespread adoption of genomic testing in cattle breeding has transformed selection decisions by providing detailed genetic information that surpasses traditional pedigree-based relationships and parent averages. While conventional approaches rely on average genetic relationships, genomic data allows breeders to analyse specific chromosomal segments and inheritance patterns, enabling more precise management of inbreeding and genetic diversity while maximizing genetic gain.

In vertically integrated industries, individual purebred lines are maintained to optimise genetic gain for specific traits. The commercial output is a product of outcrossing these lines together to create 2, 3, or 4 way crosses. This approach is common in chickens, turkeys, pigs and maize. In cattle, outcrossing different breeds, such as Angus and Wagyu, is commonly used to leverage the hybrid vigour and exploit complementary genetic gain such breeds have in economically important traits, such as growth and carcass traits. Within a single breed (or line) it is important to drive genetic gain while limiting inbreeding depression. This is more difficult in populations that have undergone a bottleneck such as Wagyu, which globally can be traced back to 220 animals which were exported out of Japan in the 1990s, and which have been subsequently closed to export.

One strategy for improving genetic diversity and reducing inbreeding involves outcrossing unrelated individuals. In the context of a purebred population, one could consider family lines analogously to separate breeds. In a purely pedigree-based approach, family lines are tracked simply by "walking" up the family tree. Algorithms for computing inbreeding and relatedness via pedigree are widely used and can provide an "averaged" picture of such metrics for potential mating outcomes. In practice, the existence of haplotypes and recombination events that break associations of haplotypes make genomic data a more precise measure of descendant outcomes.

Recent studies have demonstrated the value of analysing SNP data for runs of homozygosity (ROH) within the Australian Wagyu population (Garrick *et al.* 2025a), to elucidate inbreeding and diversity at a genomic level. That study showed that genomic inbreeding estimates were ~10% higher than pedigree estimates on average. It also demonstrated that for a given pedigree inbreeding coefficient there exists a wide distribution of realised genomic inbreeding outcomes. Building on this foundation, we demonstrate results of simulating probable genomic outcomes in progeny, enabling characterisation of next generation inbreeding and estimated breeding values for the support of mate selection and allocation decisions.

MATERIALS AND METHODS

The comprehensive Australian Wagyu Association (AWA) Helical database was used in this study. The database contains over 591,500 SNP genotype samples from various genotype chips and service providers. Additionally, it includes the recorded pedigree of over 677,000 animals, encompassing registered Japanese Black, Akaushi, Composite, and commercial animals. A subset of data representing animals genotyped on the GGP100K platform with call rate > 0.95 and MAF>0.15 were utilized for this study. Utilising only the 29 autosomes and after removing duplicated SNPs and those without known positions, 63,894 SNPs remained. "Digital progeny" were simulated for a number of bulls and cows who were both genotyped on the GGP100K.

Probable genomic outcomes for digital progeny were generated using a variation of the meiosis simulation procedure described by Villiers *et al* (2022) independently implemented within the Helical software package (Garrick *et al.* 2025b). Specifically, independent gametes are generated from each parent using a count-location method (Karlin and Liberman 1979) as follows:

1. Determine the number of crossovers by sampling from a Poisson distribution with $\lambda = L$ where L is the length of the chromosome in Morgans.
2. Generate the positions for each crossover by randomly sampling from a uniform distribution across the length of the chromosome.
3. Randomly sample 0 or 1 to designate the first segment as paternal or maternal in origin.
4. Create the entire gamete for the outcome by copying the sire or dam gamete that was chosen from the start of the chromosome until the end, swapping to the alternate gamete at each crossover point, and repeating for each autosome.
5. The recombination rate can be provided as a user input but was set for all simulations to 1.23 cM/Mb (centiMorgans / Megabase) based on estimated recombination rates for Angus and Limousin (Weng *et al.* 2014).

Coefficients for pedigree inbreeding and genomic inbreeding via runs of homozygosity (ROH) were measured on potential parents and their digital progeny as described in Garrick *et al.* (2025a) by assessing regions of 30 or more consecutive homozygous SNPs allowing for 1 missing and 1 heterozygous SNP within a run. Genomic inbreeding was computed as the total proportion of the genome existing within ROH divided by the length of the genome, e.g. 2.9 Gigabase.

Direct genomic values (DGVs) were computed as $DGV = M\alpha$ where M is a genotype matrix and each row represents an individual genotype with columns consisting of -1, 0, or 1 values representing homozygous, heterozygous, or opposite homozygous states at each locus, and α is a column vector of marker effects assumed for the purposes of this study to be a value of 1 at every location, but can be trivially substituted with estimated marker effects for any trait undergoing genomic evaluation.

RESULTS AND DISCUSSION

Figure 1 displays the distribution of ROH inbreeding for 1000 simulated digital progeny out of a sire and dam with a live genotyped offspring. Vertical lines display the pedigree inbreeding coefficient and the actual offspring ROH estimate using its genotype, showing the realised genomic inbreeding in that offspring sits in the upper end of the simulated distribution.

Figure 2 displays the distribution of ROH in the simulated scenario where the offspring from Figure 1 was bred with her sire. The resultant distribution of ROH F shifts right because of the high relatedness between her and her sire, with a ~9% increase in the mean ROH estimate across the 1,000 digital progeny.

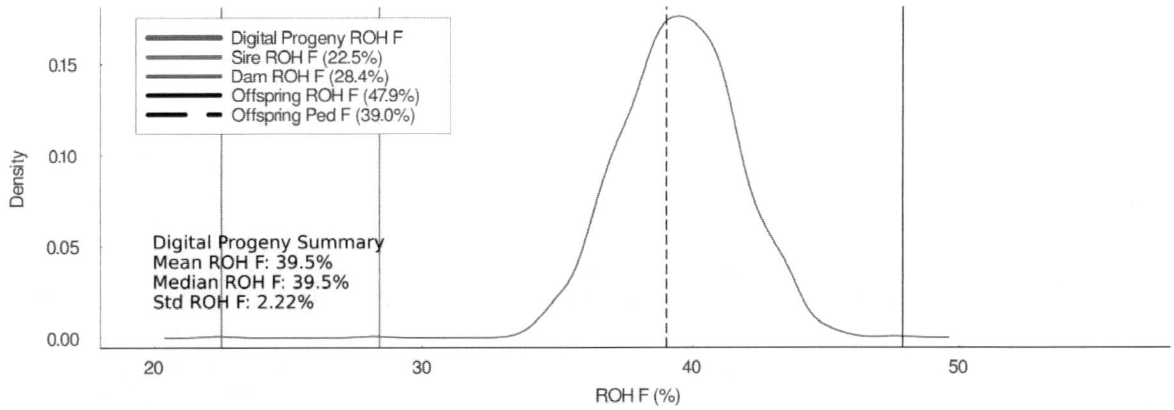

Figure 1. Density distribution of genomic inbreeding via ROH % for 1,000 digital progeny compared to an actual genotyped offspring and her pedigree inbreeding estimate

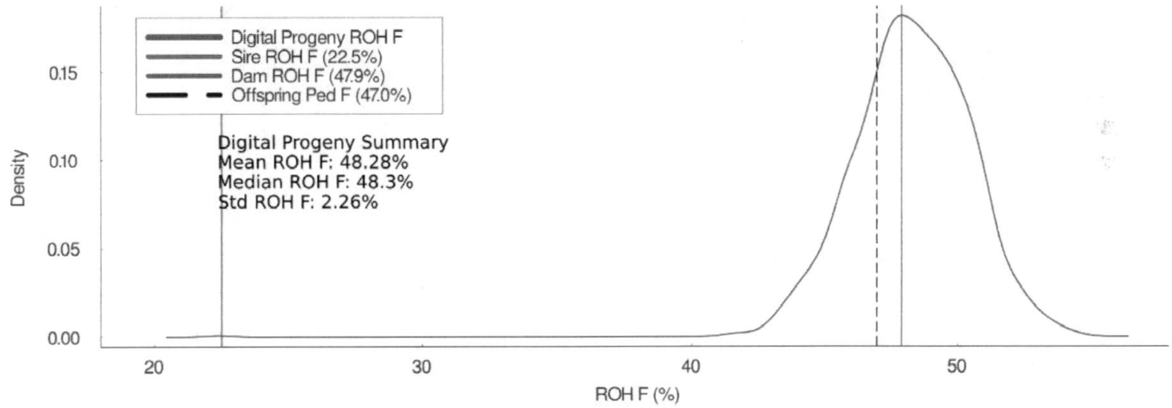

Figure 2. Density distribution of genomic inbreeding via ROH % for 1,000 digital progeny out of a theoretical mating between the genotyped offspring from Figure 1 and her sire

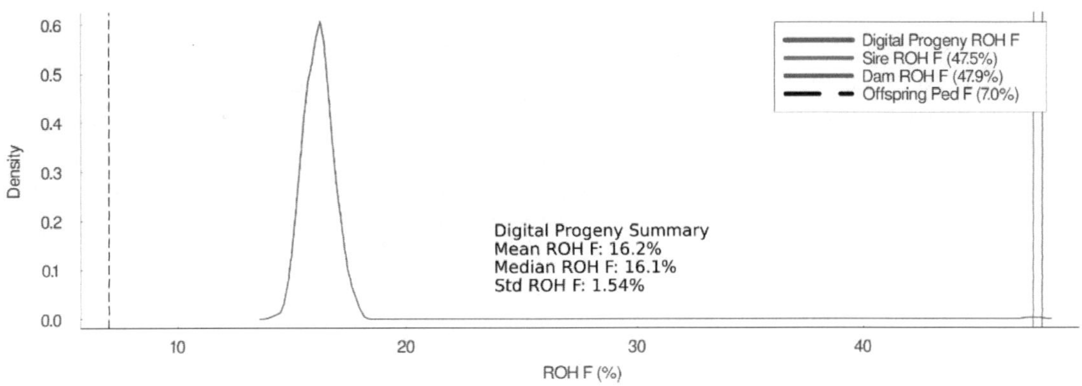

Figure 3. Density distribution of genomic inbreeding via ROH % for 1,000 digital progeny out of a theoretical mating between the genotyped offspring from Figure 1 and a relatively unrelated but highly inbred sire

Figure 3 displays the distribution of ROH for 1,000 digital progeny when the offspring from Figure 1 is mated to a relatively unrelated bull but who also has high estimated inbreeding coefficients. Despite this sire and dam each having genomic inbreeding estimates of ~47%, the estimated pedigree inbreeding coefficient for their offspring is 7% since the sire and dam have pedigree relatedness of 0.14. The distribution of outcomes of their digital progeny show a mean genomic inbreeding coefficient of only 16%. The distribution is also fairly narrow with a standard deviation of 1.54%, however there exists a long tail of outcomes all the way to ~47% genomic inbreeding depending on the exact realisation of meiosis (recombination) that might occur in the simulated progeny. The ROH-based approach captures actual realized genomic inbreeding more accurately than pedigree estimates, which are limited by the depth of recorded ancestry; further discussion of parameter sensitivity for ROH is addressed in Garrick et al. (2025a).

Figure 4 displays a distribution of DGVs for the digital progeny out of the same simulated mating used for Figure 3. The mean DGV of the digital progeny is -883 which is reasonably aligned with the parent average estimate of -807, however the distribution shows a wide variety of possible outcomes.

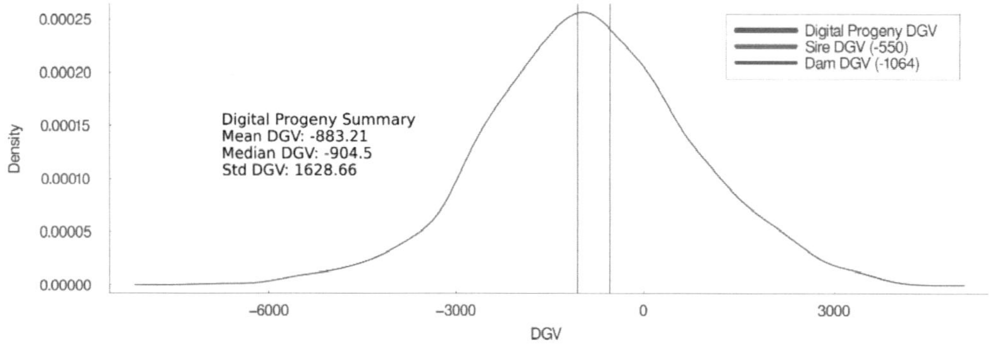

Figure 4. Density distribution of DGVs for 1,000 digital progeny out of a theoretical mating between the genotyped offspring from Figure 1 and a relatively unrelated but highly inbred sire

CONCLUSION

Genetic improvement programs traditionally rely on parent averages and pedigree-based relationships to predict progeny outcomes, but the actual genetic makeup of individual offspring can vary substantially due to Mendelian sampling. This study demonstrates that simulating digital progeny allows breeders to better understand the distribution of possible genomic outcomes for specific matings, both in terms of inbreeding coefficients and estimated breeding values. The considerable variation in possible outcomes highlights why selection can continue to be effective across generations - some progeny will receive a more favourable combination of parental alleles than others. This approach to modelling potential mating outcomes provides a valuable tool for making more informed breeding decisions that balance genetic gain with the management of inbreeding and maintenance of genetic diversity.

REFERENCES

Garrick D.P., Garrick D.J. and Howse J.E. (2025b) *Proc. Assoc. Advmt. Anim. Breed. Genet.* **26**: *These proceedings.*

Garrick D.P., Garrick D.J. and Teseling C.F. (2025a) *Proc. Assoc. Advmt. Anim. Breed. Genet.* **26**: *These proceedings.*

Karlin S and Liberman U. (1979) *Adv. Appl. Probab.* **11**: 479.

Villiers K., Dinglasan E., Hayes B.J. and Voss-Fels K.P. (2022) *G3-Genes Genom. Genet.* **12**: jkac216.

Weng, Z.Q., Saatchi, M., Schnabel, R.D. *et al.* (2014) *Genet. Sel. Evol.* **46**: 34.

ASSOCIATION BETWEEN RUMEN AND FAECAL MICROBIOME AND ENTERIC METHANE EMISSIONS IN DAIRY CATTLE

B.J. Sepulveda[1], C.J. Vander Jagt[1], A.J. Chamberlain[1,3], J. Wang[1], L.C. Marett[2], S.R.O. Williams[2], J.L. Jacobs[2,3] and J.E. Pryce[1,3]

[1] Agriculture Victoria Research, Bundoora, VIC, 3083Australia
[2] Agriculture Victoria Research, Ellinbank, VIC, 3821 Australia
[3] La Trobe University, Bundoora VIC, 3083 Australia

SUMMARY

Ruminant livestock primarily contribute to global warming through enteric methane emissions, which is mainly produced by the rumen microbiota. The rumen microbiota could potentially be used as an indicator of enteric methane emissions. There is growing interest in evaluating the faecal microbiome as an indicator of enteric methane, as it is an easier-to-record alternative to the rumen microbiome. Here, we present the phenotypic correlations between methane production and the principal components from 60 rumen and 60 faecal microbiomes (120 samples in total) of 25 Holstein lactating dairy cows. Methane production exhibited phenotypic correlations of up to $|0.37| \pm 0.20$ (mean 0.14) with the principal components of the rumen microbiome, and up to $|0.35| \pm 0.18$ (mean 0.16) with the principal components of the faecal microbiome. The principal components of the faecal microbiome showed phenotypic correlations of up to $|0.62| \pm 0.12$ (mean 0.22) with the principal components of the ruminal microbiome. These results require validation in larger populations, and genetic correlations need to be determined before implementation in commercial conditions.

INTRODUCTION

Enteric methane emissions (EME) from ruminants contribute to global warming. Methane emission reductions are expected to make a significant contribution to mitigating global warming, particularly in the short term (Forster *et al.* 2021). Enteric methane is the product of a complex interaction between ruminal microorganisms during the fermentation of feed (Gonzalez-Recio *et al.* 2023). Recent studies have identified features of the ruminal metagenome that could potentially be used as indicators of EME in breeding programs aimed at reducing these emissions (Gonzalez-Recio *et al.* 2023). However, recording EME and collecting ruminal samples on commercial farms is logistically challenging and cost-prohibitive on a large scale.

Faecal samples are easier and cheaper to collect than ruminal samples, making the faecal microbiome an attractive alternative to evaluate as an indicator of both EME and the ruminal microbiome. Earlier studies reported no association between ruminal and faecal microbiome features (Ross *et al.* 2012). However, technological advances in long-read sequencing over the past decade have enabled more detailed characterisation of microbiomes. This technology was recently used to calculate principal components (PCs) of ruminal microbiome feature matrices, which were associated with EME (Gonzalez-Recio *et al.* 2023).

The aim of this study was to investigate the phenotypic correlations between EME, and the PCs of matrices constructed with taxonomic and functional features from ruminal and faecal microbiomes obtained with long-read sequencing technology in dairy cattle.

MATERIALS AND METHODS

Twenty-five Holstein lactating cows located at the Ellinbank SmartFarm (Ellinbank, Victoria, Australia) were phenotyped for dry matter intake (DMI) and EME recorded as daily enteric methane production measured as grams per day (MeP; g/d). The animals were assessed in 2 experiments between 2022 and 2023. Experiment 1 included 20 cows, experiment 2 included 10 cows, with 5 cows participating in both experiments. The cows in experiment 1 on average produced 32 ± 2.6 kg (mean ± standard deviation) energy-corrected milk (ECM), at 4.3 ± 1.97 parities, were 74 ± 21.1 days in milk (DIM), 584 ± 43.9 kg body weight and had a total dry matter intake of 18.5 ± 1.55 kg/d. The cows in experiment 2 had an average ECM yield of 23 ± 1.8 kg (mean ± standard deviation), 4.0 ± 1.56 parities, were 217 ± 15.7 DIM, 631 ± 40.5 kg body weight and had a total DMI of 22.4 ± 2.72 kg/d.

The cows had continuous access to feed, water, and a loafing area for rest. The cows were outside except for twice-daily milking at approximately 07:00 h and 15:30 h. All cows were offered 6.2 kg DM/d of grain and *ad libitum* vetch (*Vicia sativa* L.) hay. Individual cow DMI was measured using feed bins equipped with load cells. Methane production was obtained using the modified sulphur hexafluoride (SF_6) tracer method (Deighton *et al.* 2014). Methane production was measured over consecutive 5 days, with emissions transformed to MeP. The average ± standard deviation (SD) of MeP was 534 ± 66.5 g/d.

Two ruminal and faecal samples were collected from each cow during the EME recording periods. Ruminal fluid samples were collected via an oesophageal probe placed into the rumen via the mouth (Moate *et al.* 2014). Faecal samples were collected opportunistically from a fresh pat over days 1 to 2 of the methane measurement period, with any missing samples collected via rectal tickling on day 3 of the methane measurement period. Rumen samples were allowed to drain freely through a cheesecloth layer, separating rumen solids from the ruminal fluid, and then frozen at -80°C. Collected faeces were subsampled (~30 g) then frozen at -80°C.

The microbial genomic DNA from the ruminal and faecal samples was extracted using the ZymoBIOMICS DNA Microprep Kit (Zymo Research) with a reduced volume of binding buffer. Sequencing libraries were prepared using the Native Barcoding Kit 96 V14 (SQK-NBD114.96, Oxford Nanopore Technologies) and sequenced on the PromethION 24/2 (Oxford Nanopore Technologies) according to the manufacturer's instructions. Basecalling was conducted using Dorado software version 0.5.3 with the module FAST. Reads with quality score less than 7 or length less than 250 base pairs were removed. Taxonomic and functional annotation was conducted with the SqueezeMeta software (Tamames and Puente-Sánchez 2019) with version 1.5.2 to obtain the number of reads assigned to taxonomic genera, KEGG Orthology groups (KOs) and Clusters of Orthologous Genes (COGs). The number of reads assigned to the same feature in the two samples from the same animal-experiment combination was summed.

For the ruminal and faecal samples separately, the metagenome features (genera, KOs and COGs) that were not present in at least 90% of the animals were removed, and the remaining missing values were imputed to non-zero values with the geometric Bayesian-multiplicative method from the cmultRepl function of the zCompositions R package (Palarea-Albaladejo and Martín-Fernández 2015). Relative abundance matrices were constructed with the proportion of the reads assigned to each feature relative to the sum of reads assigned to all features of the same type (genera, KOs or COGs) within the same animal-experiment combination. These relative abundance matrices were then transformed with the centered log-ratio transformation with the unweighted option of the CLR function from the easyCODA R package (Greenacre 2018) to account for the compositional nature of the dataset. Then, the first five PCs were calculated with the function prcomp of the R package stats (R Core Team 2022) with the options centre and scale. The phenotypic correlations (r_p) between MeP and the PCs were estimated with bivariate linear models with the structure $\mathbf{y} = \mathbf{X}\boldsymbol{\beta} + \mathbf{e}$ using ASReml-R version 3.0, where $\mathbf{y}$ was a vector of MeP and PCs or two PCs; $\boldsymbol{\beta}$ is a vector of fixed effects; $\mathbf{e}$ is a vector of random residuals with an assumed distribution $N(0, \mathbf{I}\sigma_e^2)$; and $\mathbf{X}$ is an incidence matrix. The effects of the experiment (2 levels), DMI, DIM, parity, ECM, and body weight were evaluated for each combination of two traits, and significant effects were included as fixed effects in the respective model for that combination.

RESULTS AND DISCUSSION

The first five principal components derived from the metagenomes explained between 54 and 69% of the matrices' variance. MeP exhibited phenotypic correlations between zero of and |0.37| ± 0.20 with the PCs of the rumen microbiome (Table 1). The phenotypic correlations between MeP and the PCs from the faecal metagenome ranged from zero to |0.35| ± 0.18. Twenty-one out of 75 (28%) of the phenotypic correlations calculated between the PCs from the rumen and faeces had a magnitude of at least 0.30. Using the KOs, the PCs from faeces showed phenotypic correlations ranging from |0.22| ± 0.18 to |0.62| ± 0.12 with the first PC from the rumen, which accounted for 46% of the variation in the rumen.

Table 1. Phenotypic correlations (r_p) ± standard error (SE), and corresponding *p*-value between the first five principal components (PC1 to PC5) calculated from the content of features from the faecal and ruminal metagenome and methane production (MeP). FE: Fixed effects with significant effect fitted in the prediction model. KO: KEGG Orthology groups. COG: Clusters of Orthologous Genes. Variance explained by each principal component is shown in parenthesis

Features	Faeces	Rumen	FE	r_p ± SE	*p*-value
Genera	PC1 (16%)	PC3 (6%)	-	-0.52 ± 0.16	1.2×10^{-3}
Genera	PC2 (13%)	PC2 (15%)	parity,ECM	-0.48 ± 0.15	1.4×10^{-3}
Genera	PC4 (7%)	PC3 (6%)	experiment,DMI, parity, ECM	-0.43 ± 0.17	1.1×10^{-2}
Genera	PC5 (6%)	PC1 (29%)	-	-0.40 ± 0.18	2.6×10^{-2}
Genera	PC5 (6%)	PC2 (15%)	experiment,DMI, parity, ECM	-0.47 ± 0.16	3.3×10^{-3}
Genera	PC5 (6%)	PC5 (4%)	parity	-0.37 ± 0.17	3.0×10^{-2}
KO	PC2 (11%)	PC3 (6%)	experiment,ECM	-0.35 ± 0.17	4.0×10^{-2}
KO	PC3 (7%)	PC1 (46%)	experiment,DMI,ECM	-0.35 ± 0.17	4.0×10^{-2}
KO	PC3 (7%)	PC2 (9%)	DMI, parity	-0.46 ± 0.16	4.0×10^{-3}
KO	PC5 (4%)	PC1 (46%)	experiment,parity	-0.62 ± 0.12	2.4×10^{-7}
COG	PC2 (11%)	PC3 (7%)	DMI,parity,ECM	-0.37 ± 0.18	4.0×10^{-2}
COG	PC4 (5%)	PC3 (7%)	DMI,parity	-0.39 ± 0.17	2.2×10^{-2}
COG	PC5 (5%)	PC1 (29%)	experiment	-0.34 ± 0.17	4.6×10^{-2}
Features	**Faeces**	**Final trait**	**FE**	**r_p ± SE**	***p*-value**
Genera	PC5 (6%)	MeP	experiment	0.35 ± 0.18	5.2×10^{-2}
Features	**Rumen**	**Final trait**	**FE**	**r_p ± SE**	***p*-value**
Genera	PC4 (5%)	MeP	-	0.37 ± 0.20	6.4×10^{-2}

This study investigated the phenotypic correlations between EME and the PCs from taxonomic and functional features of the ruminal and faecal microbiomes in dairy cattle. Some PCs from the faeces exhibited phenotypic correlations with MeP up to |0.35| and other PCs up to |0.62| with the first principal component of the ruminal KOs content, which explained 46% of the variation in the ruminal microbiota. Using a short-read sequencing technology, a previous study found no stronger correlation between the ruminal and faecal microbiomes of the same cow compared to correlations between different cows (Ross *et al.* 2012), which could indicate no association between ruminal and faecal samples. A possible reason for this difference could be the long-read sequencing technology used here, which has shown unprecedented advances over the past decade in taxonomic and functional characterisation of metagenomes (Nature Methods 2023).

While the laboratory and computational costs of processing the faecal microbiota are approximately the same as those for processing the ruminal microbiota, faecal sampling is easier with less animal welfare implications and is more cost-effective than collecting ruminal samples.

This suggests that faecal microbiota could be a suitable proxy to estimate EME and infer the ruminal microbiota if the correlations can be improved. The correlations estimated in this study still require validation in larger populations in varying environments and production systems. Furthermore, a biological interpretation of the association between the features with larger weights in the PCs and EME requires further investigation. Additional data is also needed to estimate the genetic correlations between these traits. Determining the genetic correlations between EME and the ruminal and faecal microbiota in Australian dairy cattle is essential to assess the viability of using these microbiomes in breeding programs to reduce EME.

CONCLUSIONS

The faecal microbiome showed promise as a proxy for estimating both enteric methane emissions and the ruminal microbiome. However, the results are limited by the small number of animals analysed and therefore, it is essential to validate the results in larger and more diverse populations. Additionally, it is necessary to explore the biological basis of the statistical associations between the principal components of the microbiomes and enteric methane emissions. Furthermore, estimating the heritability of microbial features and their genetic correlations with enteric methane emissions is necessary to establish these features as reliable genetic indicators in breeding programs designed to reduce emissions.

REFERENCES

Deighton M.H., Williams S.R.O., Hannah M.C., Eckard R.J., Boland T.M., Wales W.J. and Moate P.J. (2014) *Anim. Feed. Sci. Technol.* **197**: 47.

Forster P., Storelvmo T., Armour K., Collins W., Dufresne J.-L., Frame D., Lunt D.J., Mauritsen T., Palmer M.D., Watanabe M., Wild M., and Zhang H. (2021) In 'Climate Change 2021. C. Intergovernmental Panel on Climate', pp. 923-1054, editors V., P. Zhai, A. Pirani, S.L. Connors, C. Péan, S. Berger, N. Caud, Y. Chen, L. Goldfarb, M.I. Gomis, M. Huang, K. Leitzell, E. Lonnoy, J.B.R. Matthews, T.K. Maycock, T. Waterfield, O. Yelekçi, R. Yu, and B. Zhou, Cambridge University Press, Cambridge.

González-Recio O., Martínez-Álvaro M., Tiezzi F., Saborío-Montero A., Maltecca C. and Roehe R. (2023) *Livest. Sci.* **269**: 105171.

Greenacre M. (2018) 'Compositional Data Analysis in Practice' 1st ed. Chapman and Hall/CRC.Palarea-Albaladejo J. and Martín-Fernández J. A. (2015) *Chemometr. Intell. Lab. Syst.* **143**: 85.

Moate P.J., Williams S.R.O., Torok V.A., Hannah M.C., Ribaux B.E., Tavendale M.H., Eckard R.J., Jacobs J.L., Auldist M.J. and Wales W.J. (2014) *J. Dairy. Sci.* **97**: 5073.

Nature Methods (2023) *Nat. Methods* **20**: 1.

R Core Team (2022) R Foundation for Statistical Computing, Vienna, Austria.

Ross E.M., Moate P.J., Bath C.R., Davidson S.E., Sawbridge T.I., Guthridge K.M., Cocks B.G. and Hayes B. J. (2012) *BMC Genet.* **13**: 53.

Tamames J. and Puente-Sánchez F. (2019) *Front. Microbiol.* **9**: 3349.

METAGENOMIC PREDICTIONS FOR ENTERIC METHANE EMISSIONS IN SHEEP USING LONG-READ SEQUENCING OF RUMEN FLUID SAMPLES

Y. Li[1], L.T. Nguyen[1], C.T. Ong[1], S. Yadav[1], M. Aldridge[2,3], P. Fitzgerald[2], J. van der Werf [2] and E. M. Ross[1]

[1] The University of Queensland, Queensland Alliance for Agriculture and Food Innovation, Centre for Animal Science, St Lucia, QLD, 4072 Australia

[2] University of New England, School of Environmental and Rural Science, Armidale, NSW, 2350 Australia

[3] Animal Genetics Breeding Unit*, University of New England, Armidale, NSW, 2350 Australia

SUMMARY

Metagenomic data were generated using Oxford Nanopore Technologies (ONT) platform for 396 sheep rumen fluid samples. Taxonomic and gene functional abundance matrices were constructed to develop predictive models for estimating methane emission phenotypes based on methane concentration from portable accumulation chambers (PAC). The greatest microbiability (m^2; proportion of phenotypic variation explained by the microbiome relationship matrices) was generated using Clusters of Orthologous Groups (COGs) functional annotations of the reads ($m^2 = 0.92$). Correlations between the predicted phenotypes and the PAC phenotypes in a 5-fold cross-validation ranged from 0.48 to 0.51. These results indicate that metagenomic predictions could be implemented as a proxy methane phenotype in sheep and could be used to increase the feasible reference population size in genomic predictions for methane using multi-trait models.

INTRODUCTION

Methane emission from ruminant livestock is one of the main contributors to greenhouse gas production from the agricultural sector. Enteric methane emission levels can be predicted from the variation in rumen microbiota composition and metabolic activities (Ross and Hayes 2022). Recent advances in long-read sequencing technology have enhanced recognition of genetic elements, taxonomic identification and functional annotation in metagenomics (Kim *et al.* 2024). We applied long-read metagenomic sequencing of rumen samples to identify which microbial taxonomic or functional classification approach best captures methane phenotypic variation and compared prediction accuracies across these different classification methods.

MATERIALS AND METHODS

Collection of methane emission data and associated parameters. Methane emissions were measured using portable accumulation chambers (PAC) for 50 minutes per animal in 396 lambs aged 5.75 ± 0.12 months. The lambs were sourced from the 2023 lambing season of MLA resource flock at Kirby (Armidale, NSW). Methane concentrations were measured in parts per million (ppm) using an Eagle2 Gas Monitor (RKI Instruments). The recorded gas concentrations were corrected using CH_4 ppm × standard temperature and pressure. Associated fixed effects were measured simultaneously with CH_4 phenotype measurements, including body weight (kg), sampling date, and time off feed, which accounts for key potential confounding effects.

Metagenomic library preparation, long-read sequencing and basecalling. For metagenomic sequencing, libraries of 396 samples were prepared using the Oxford Nanopore Native Barcoding 11Kit (SQK-NBD114.96). Following end preparation and clean-up, DNA fragments were ligated

* A joint venture of NSW Department of Primary Industries and Regional Development and the University of New England

with unique barcodes and sequencing adapters, then pooled for sequencing on the PromethION platform (ONT). Dorado basecaller (v0.7.0) with the "Super Accuracy" model v4.3.0 generated a total of 4TB of long-read sequencing data from the raw dataset. Reads with an average Phred quality score below 10 and length < 200 bp were removed from the dataset.

Taxonomy and functional identification. SqueezeMeta v1.6.3 (Tamames and Puente-Sánchez 2018), was used to assign individual reads to taxa at the levels of Phylum to Species, as well as classify the function of the genes included in each read using both the EggNOG (Huerta-Cepas et al., 2019) and KEGG (Kanehisa and Goto 2000) databases. Within SqueezeMeta, Prodigal v2.6.3 was used for protein-coding gene prediction, DIAMOND v2.1 for protein alignment against NCBI non-redundant protein databases, and EggNOG (v5.0) and KEGG (v110) databases for metabolic pathway reconstruction. Taxa assigned to the kingdoms of *Bacteria*, *Archaea*, and *Eukarya* were retained for microbially abundant matrices, with an average of 65.7% of sequence counts not assigned to these taxonomic classifications being removed. Eight distinct abundance matrices were constructed for microbial features composition: six for taxonomic levels (*Phylum* to *Species*) and two for functional annotations marked as KEGG and COGs (Clusters of Orthologous Groups).

Statistical model definition. Low-abundance microbial taxa and pathways (count < 10) were removed. Counts were normalised to relative abundances (percentages) per sample, summing to 100%. Both methane emissions (ppm) and microbial abundances were scaled to z-scores for modeling. ASReml-R 4.2 (Butler and Cullis 2023) was used to fit a BLUP-based mixed linear model:

$$y = Xb + Zu + e$$

where, y was the vector of phenotypic values (CH_4 emissions). X and Z were design matrices linking observations to fixed and random effects, respectively. The fixed effects, b included linear cofactors such as weight and time off feed prior to entering the PAC. The microbiome random effects u were assumed to follow a multivariate normal distribution as $u \sim N(0, G\sigma_m^2)$. The G was a variance-covariance (microbial relationship) matrix derived from the standardised taxonomic or functional abundance matrix *M*,

$$G = MM'/n$$

where M was the dimensions $i \times n$ standardised taxonomic or functional abundance matrix, n was the number of features (taxa or functional pathways) and i was the number of samples (Ross and Hayes 2022). This matrix represents pairwise covariances among samples based on their microbial abundance profiles. The variance component σ_m^2 associated with the G quantifies the proportion of total variation explained by composition of microbiome features. The microbiome variance component σ_m^2 (V_m), estimated to be using Residual Maximum Likelihood (REML), represented the proportion of CH_4 emission variation attributed to the microbiome abundance matrix. The residuals were assumed to be normally distributed with variance σ_e^2. Environmental variance σ_e^2 (V_e), obtained from the BLUP model, represented the random error effects after accounting for weight, batch-level and microbial abundance effects. The proportion of the phenotypic variance explained by the microbial component (microbiability m^2), was calculated as the ratio of linear microbiome variance to total variance: $m^2 = V_m / (V_m + V_e)$, indicating the predictive ability of microbial characteristics by using abundances matrixes for CH_4 emissions.

Three distinct cross-validation strategies were performed to assess predictive microbial ability. These included 2-fold and 5-fold random partitioning with 20 iterations each, and a leave-one-batch-out approach testing generalisability across different sampling conditions. For the latter, one validation batch from 42 total batches (structured across 6 days with 7 different time off feed intervals) served as the testing set. The accuracy (r) was calculated as the average Pearson's correlation between predicted and observed methane phenotypes across all folds: $r = \frac{1}{k}\sum cor(y_i, \hat{y}_i)$. y_i represented the vector of methane phenotypes, and $\hat{y}_i$ were the corresponding predicted phenotype value. The factor k represented the number of folds in cross-validation. To account for microbiability (m^2), the corrected accuracy was also reported as $\frac{r}{\sqrt{m^2}}$. All accuracy values were reported as a means with standard errors.

RESULTS AND DISCUSSION

The CH_4 concentrations, which were corrected for standard temperature and pressure, ranged from 123.21 to 1415.87 ppm. The mean (±SD) concentration was 647.69 (±238.67) ppm, with a median of 624.92 ppm. The methane yield coefficient of variation between animals was 36.85%.

Filtering removed 27.27% (±5.73%) of low-quality reads. The number of sequencing reads was 1.68×10^6 ($\pm1.08\times10^6$), with an average read length of 642.79 (±221.63) bp, a read length N50 of 1013.74 (±1012.71) bp, and an average sequencing accuracy of 98.06% (±0.25%).

The taxonomic and functional abundance matrices derived from long-read metagenomic analysis were quantified at multiple classification levels, with the feature counts summarised in Table 1. Notably, all taxonomic levels from long-read metagenomic explained more of the variance than short-read approaches in other studies (Xie *et al.* 2021). The microbiability (m^2) (Table 1) increased from phylum (0.27) to genus/species level (0.65), with stable standard errors (± 0.2) as the number of features (i.e. columns in the count matrix) increased. Functional abundance matrices had higher m^2 values compared to taxonomic abundance matrices. The microbiability for methane emission in this study showed more variance explained than previous approaches. Earlier Bayesian modeling studies reported lower microbial effects (comparable to m^2): average of 0.07 based on OTUs (Operational Taxonomic Units) taxonomic groups (Zhang *et al.* 2020), and <0.33 based on taxonomic abundance from long-read metagenomic data (Marcos *et al.* 2024).

Table 1. Microbiability (m^2) and variance components of microbiome data

Level	n	m^2	$V_m \pm$ SD	$V_e \pm$ SD	MSE
Phylum	242	0.27	0.13 ± 0.04	0.35 ± 0.03	0.73
Class	655	0.45	0.23 ± 0.05	0.28 ± 0.03	0.60
Order	1174	0.51	0.25 ± 0.06	0.24 ± 0.04	0.54
Family	2008	0.55	0.26 ± 0.06	0.21 ± 0.04	0.52
Genus	3042	0.65	0.30 ± 0.06	0.16 ± 0.04	0.45
Species	3042	0.65	0.30 ± 0.06	0.16 ± 0.04	0.45
KEGG	9591	0.80	0.48 ± 0.10	0.12 ± 0.04	0.39
COGs	22873	0.92	0.49 ± 0.08	0.04 ± 0.05	0.34

MSE (Mean Squared Error) $=\frac{1}{k}\sum(\hat{y}_i - y_i)^2$; SD: Standard Deviation.

Cross-validation showed robustness of corrected accuracy at 2-fold (range: 0.45-0.51) and 5-fold (range: 0.48-0.51) across all taxonomic and functional levels (Table 2). A previous study in sheep using covariance matrices based on metabolomic matrices and rumen microbial OTUs matrices in sheep revealed average accuracy (r) below 0.4 (Ross *et al.* 2020), although their sample size was much smaller. However, the leave-one-batch-out validation accuracy was lower (range: 0.18-0.33), potentially attributed to either large variation in the fixed effects (weight and time off feed) with unknown environmental conditions or the reduced test sample size which introduced unknown predictive errors. After correcting accuracies by microbiability (m^2), leave-one-batch-out (42-Fold) accuracy ranged from 0.32 to 0.39. The correlation between microbiome-predicted methane and PAC methane was comparable to the correlation between repeated PAC measurements, which ranged from 0.51 to 0.90 in sheep across 7 days (O'Connor *et al.* 2021).

Table 2. Accuracy estimation of cross-validation across taxonomic and functional levels

	2-Fold		5-Fold		42-Fold	
	r	r/m	r	r/m	r	r/m
Phylum	0.42 ± 0.05	0.49 ± 0.06	0.44 ± 0.03	0.49 ± 0.03	0.18 ± 0.06	0.34 ± 0.11
Class	0.41 ± 0.05	0.46 ± 0.04	0.43 ± 0.03	0.48 ± 0.03	0.23 ± 0.05	0.35 ± 0.08
Order	0.44 ± 0.04	0.51 ± 0.03	0.45 ± 0.05	0.51 ± 0.05	0.27 ± 0.06	0.39 ± 0.08
Family	0.40 ± 0.01	0.45 ± 0.05	0.44 ± 0.03	0.50 ± 0.03	0.26 ± 0.06	0.35 ± 0.08
Genus	0.41 ± 0.00	0.46 ± 0.02	0.44 ± 0.04	0.49 ± 0.04	0.26 ± 0.06	0.32 ± 0.08
Species	0.43 ± 0.01	0.49 ± 0.02	0.45 ± 0.04	0.51 ± 0.05	0.26 ± 0.06	0.32 ± 0.08
KEGG	0.42 ± 0.04	0.47 ± 0.04	0.45 ± 0.03	0.51 ± 0.04	0.32 ± 0.05	0.36 ± 0.06
COGs	0.42 ± 0.00	0.48 ± 0.04	0.42 ± 0.02	0.48 ± 0.03	0.33 ± 0.06	0.35 ± 0.06

r: Accuracy; r/m: Corrected Accuracy; All values are mean ± standard error.

CONCLUSIONS

Long-read metagenomic data analysis and appropriate model selection improved the prediction accuracy of methane emission phenotypes in this study. Variation explained by the microbiome relationship matrix was highest for the COGs functional categorisation. Higher prediction accuracy was achieved by including within cohort methane measurements in the reference set.

ACKNOWLEDGEMENTS

This research is funded by Meat and Livestock Australia and MLA Donor Company (P.PSH. 2010 and PSH.2011). The author acknowledges support through the Australian Government and The University of Queensland Research Training Program (RTP) Scholarship. The author gratefully acknowledges Collins Asiamah for essential DNA extraction work and thank Daniel Brown and Ed Clayton for their contribution to this project scope.

REFERENCES

Butler D.G., Cullis B.R., Gilmour A.R., Gogel B.G. and Thompson R. (2023) In 'ASReml-R Reference Manual Version 4.2' pp. 30-105, VSN International Ltd., HP2 4TP, UK.

Huerta-Cepas J., Szklarczyk D., Heller D., *et.al.* (2019) *Nucleic Acids Res.* **47**: D309.

Kanehisa M., and Goto S. (2000). *Nucleic Acids Res.* **28:** 27.

Kim C., Pongpanich M. and Porntaveetus T. (2024) *J. Transl. Med.* **22**: 111.

Marcos C.N., Carro M.D., Gutiérrez-Rivas M., *et al.* (2024) *J. Dairy. Sci.* **107**: 7064.

O'Connor E., McGovern F.M., Byrne D.T., *et al.* (2021) *J. Anim. Sci.* **99**: skab132.

Ross E.M. and Hayes B.J. (2022) *Front. Genet.* **13**: 865765.

Ross E.M., Hayes B.J., Tucker D., *et al.* (2020) *J. Anim. Sci.* **98**: 10.

Tamames J. and Puente-Sánchez F. (2018) *Front. Microbiol.* **9**: 3349.

Xie F., Jin W., Si H., *et al.* (2021) *Microbiome* **9**: 137.

Zhang Q., Difford G., Sahana G., *et al.* (2020) *ISME J.* **14**: 2019.

Proc. Assoc. Advmt. Anim. Breed. Genet. 26: 229-232

PROXIES IN PRACTICE: IMPLEMENTING RUMEN METAGENOMIC PROFILES FOR GENOMIC SELECTION OF METHANE EMISSIONS

T.P. Bilton[1], P.L. Johnson[1], H. Henry[1], B.J. Perry[1], F. Booker[1], J. Peers-Adams[1], M. Ward[2], B. Bryson[2], S.M. Clarke[1], J.C. McEwan[1], L.W. Coleman[3], N.K. Pickering[3], K.G. Dodds[1] and S.J. Rowe[1]

[1] AgResearch, Invermay Agricultural Centre, Mosgiel, New Zealand
[2] AgResearch, Woodlands Research Farm, Invercargill, New Zealand
[3] Focus Genetics, Napier, New Zealand

SUMMARY

Selective breeding for low-methane ruminants is a viable solution for reducing agricultural greenhouse emissions, but requires phenotypic measures of methane. Methods for obtaining on-farm methane emission measurements are available but are expensive and laborious, whereas proxy traits may provide a more efficient and cheaper alternative. A rumen metagenomic profile (RMP) has recently been shown to be one suitable proxy for methane, as it is both predictive of, and genetically correlated with, methane-related traits. We describe a two-step approach for using RMP as a proxy trait to generate genomic breeding values of methane related traits from portable accumulation chambers (PAC). We illustrate the utility of this approach on a commercially reared sheep flock. Our sequence-based, low-cost approach is extremely robust and could be extended to other ruminants (i.e., cattle) and would have greater utility for large ruminants where measuring methane emissions is more difficult compared to sheep.

INTRODUCTION

The global livestock industry is actively developing mitigation tools to reduce methane emissions from ruminants to meet international agricultural greenhouse gas reduction targets. One mitigation tool that is currently available for industry is selective breeding but requires on-farm measures of methane, but obtaining direct on-farm measurements of methane emissions is expensive and laborious. One measurement technology that has successively been used commercially on-farm is portable accumulation chambers (PAC) with over 5,000 commercial measures being collected per year in sheep across New Zealand (Archer *et al.* 2023). However, infrastructure constraints and logistic challenges are limiting access for breeders and farmers to measure their flocks using PAC. Proxy traits can provide an inexpensive alternate that enable a larger number of phenotypic methane measures to be collected. One proxy that has previously been shown to be predictive of and genetically correlated with PAC methane traits is RMP (Bilton *et al.* 2024). This study builds on Bilton *et al.* (2024) by presenting an approach for incorporating RMP information into the generation of PAC methane breeding values that can be used in a selection index for breeding.

MATERIALS AND METHODS

Experimental animals and protocols applied in this study were approved by the AgResearch Ruakura (Hamilton, NZ) Animal Ethics committees (application number 2383).

Animals and phenotypes. This study used 13,748 animals born between 2019 and 2023 from a commercial New Zealand sheep flock where all animals had been successfully genotyped (Table 1) and performance traits recorded. Methane emissions of a subset of these animals (1364 across all birth years) were measured using PAC at approximately 7-9 months of age following the standard protocol for PAC (Jonker *et al.* 2020). PAC data and additional animal information and measurements were downloaded from the Sheep Improvement Limited database (Newman *et al.* 2000). The PAC methane trait used in this study was methane emissions in grams per day (CH_4).

Table 1. Number of animals and phenotypes by year of birth

Phenotype(s)	Birth Year					Total
	2019	2020	2021	2022	2023	
PAC only	163[1]	2	236	252	233	886
PAC + RMP		250			228	478
RMP only					747	747
Neither	3118	2649	2657	2032	1181	11,637
Total	3281	2901	2893	2284	2389	13,748

[1]Born 2019 animals had two rounds of PAC measurements, where measurements were taken two weeks apart

Rumen sampling & metagenomic profiles. Rumen samples were collected from 975 born 2023 rams at approximately 6 months of age via stomach intubation across two days of sampling. Animals were removed from feed in morning and afternoon batches, where each batch of animals was sampled over a five-hour period by two teams of technical staff when the animals were between 1 hour and 6 hours off feed. Rumen samples were preserved using the TNx2 solution (Budel *et al.* 2022) and sequenced using restriction enzyme-reduced representation sequencing (Hess *et al.* 2020) with *Pst*I across multiple libraries. Libraries were sequenced with single end reads (1x100) on an Illumina NovaSeq6000 generating 26 Gb of data per 368 sample library. The reference-free pipeline developed by Hess *et al.* (2020) was used to generate a count matrix of tags (unique raw sequences trimmed to 65 bp) and a cohort adjustment based on year of birth, time off feed group (hour intervals) and sample date was applied. An additional 250 animals (born 2020) also had RMPs that formed part of the dataset of the Bilton *et al.* (2024) study. A microbial relationship matrix (MRM) was constructed using the 4585 animals from the Bilton *et al.* (2024) study (that also had PAC and genotype data) plus the 975 rams born 2023 from this study using the approach described in Hess *et al.* (2020).

Animal genotyping. All animals were genotyped using a variety of nested SNP chips, where 13,232 common SNPs across all chips were retained for analysis. The software KGD (Dodds *et al.* 2015) was used to construct a genomic relationship matrix (GRM) based on VanRaden method 1, where each pairwise relationship was computed using only SNPs with non-missing data.

Statistical analysis. Predictions of methane from RMP for the 975 born 2023 rams were performed using a linear mixed model approach. The model used was of the form:

$$y_{ijkl} = \mu + cg_j + aod_k + brr_l + bdev_i + m_i + e_{ijkl} \qquad (1)$$

where μ is the overall mean, cg_j is the jth contemporary group based on the combination of flock, birth year and sex, aod_k is the effect of the kth age of dam (2, 3, 4+), brr_l is the effect of the lth birth/rear rank group (1/1+, 2/2, 2+/1, 3/2, 3+/3+), $bdev_i$ is the birth deviation from the flock and birth year mean, y_{ijkl} denotes the trait (CH_4), $\mathbf{m} \sim N(\mathbf{0},\sigma_m^2\mathbf{M})$, $\mathbf{m} = (m_1,\ldots,m_n)'$, $\mathbf{e} \sim N(\mathbf{0},\sigma_e^2\mathbf{I})$, $\mathbf{e} = (e_{1jkl},\ldots,e_{njkl})'$, **M** denotes the MRM, and **I** is the identity matrix. The microbial values, m_i, are referred to as the "RMP methane trait" as it provides an estimate of y_{ijkl} after adjusting for fixed effects. Model (1) was fitted on a training set consisting of all 4,585 animals from the Bilton *et al.* (2024) study and predictions of the microbial values ($\widehat{m}_i$) were made on the 975 born 2023 rams with RMP profiles.

To incorporate RMP information into the breeding values for the PAC methane trait, a bivariate model of the form:

$$\mathbf{y}^* = \mathbf{Xb} + \mathbf{Zu} + \mathbf{e} \qquad (2)$$

was used, where $\mathbf{y}^* = (\widehat{\mathbf{m}},\mathbf{y})'$, $\mathbf{X} = ((\mathbf{X}_1,\mathbf{0})',(\mathbf{0},\mathbf{X}_2)')$, $\mathbf{b} = (\mathbf{b}_1,\mathbf{b}_2)'$, $\mathbf{Z} = ((\mathbf{Z}_1,\mathbf{0})',(\mathbf{0},\mathbf{Z}_2)')$, $\mathbf{u} = (\mathbf{u}_1,\mathbf{u}_2)'$, $\mathbf{e} = (\mathbf{e}_1,\mathbf{e}_2)'$, $\mathbf{X}_1$, $\mathbf{X}_2$ and $\mathbf{Z}_1$, $\mathbf{Z}_2$ are the incidence matrices for the fixed and random effects respectively, $\mathbf{b}_1$, $\mathbf{b}_2$ and $\mathbf{u}_1$, $\mathbf{u}_2$ are vector of fixed and random effects respectively, and $\mathbf{e}_1$, $\mathbf{e}_2$ are residual error terms. The model assumes that $\mathbf{u} \sim MVN(\mathbf{0}, \mathbf{G}\otimes\mathbf{B})$, where **G** denotes the GRM and $\mathbf{B} = ((\sigma_{u1}^2,\sigma_{u12})',$

$(\sigma_{u12},\sigma_{u2}^2)')$, and $\mathbf{e} \sim \text{MVN}(\mathbf{0}, \mathbf{I}\otimes\mathbf{C})$ where $\mathbf{C} = ((\sigma_{e1}^2,\sigma_{e12})',(\sigma_{e12},\sigma_{e2}^2)')$. Model (2) was fitted in ASReml 4.2 (Gilmour *et al.* 2015) using the qualifier "!BLUP 3" with prespecified variance and covariance values of $\sigma_{u1}^2 = 0.243$, $\sigma_{u2}^2 = 0.469$, $\sigma_{e12} = 0.242$, $\sigma_{e1}^2 = 0.417$, $\sigma_{e2}^2 = 1.802$, $\sigma_{e12} = 0.252$, which were obtained from the forward prediction analysis in Bilton *et al.* (2024). In addition, the fixed effects cg_j, aod_k, brr_l, and $bdev_i$ were included in $\mathbf{X}_2$ whereas $\mathbf{X}_1$ only contained the overall mean. The bivariate model was fitted using all 13,748 animals from the flock (Table 1) under four scenarios of differing levels of phenotypic information. These were (a) PAC data from born 2019 to 2022 (PAC (-b23)), (b) scenario (a) plus the RMP methane trait (PAC (-b23) + RMP), (c) all available PAC data (PAC all), and (d) all available PAC data and the RMP methane trait (PAC (all) + RMP). The genomic estimated breeding values (gEBVs) for the PAC methane trait ($\hat{\mathbf{u}}_2$) were extracted and examined across the four different scenarios.

RESULTS AND DISCUSSION

The pairwise comparison of the gEBV values for the 975 rams is shown in Figure 1a across the four scenarios with different phenotype information available for the born 2023 rams. The correlation between gEBVs with no phenotype measures (PAC (-b23)) with the gEBVs including the PAC measures (PAC (All)) for the born 2023 rams was 0.7, which was larger than the gEBV correlation of 0.63 between using only RMP (PAC (-b23) + RMP) with only PAC (All). However, gEBVs generated using only RMP (PAC (-b23) + RMP) or only PAC (PAC All) for the born 2023 rams had a similar correlation (0.87-0.89) to the gEBVs generated using all the phenotypes (PAC (All) + RMP). Here, we are assuming PAC (All) + RMP provides the best gEBVs as it has the most phenotype information included.

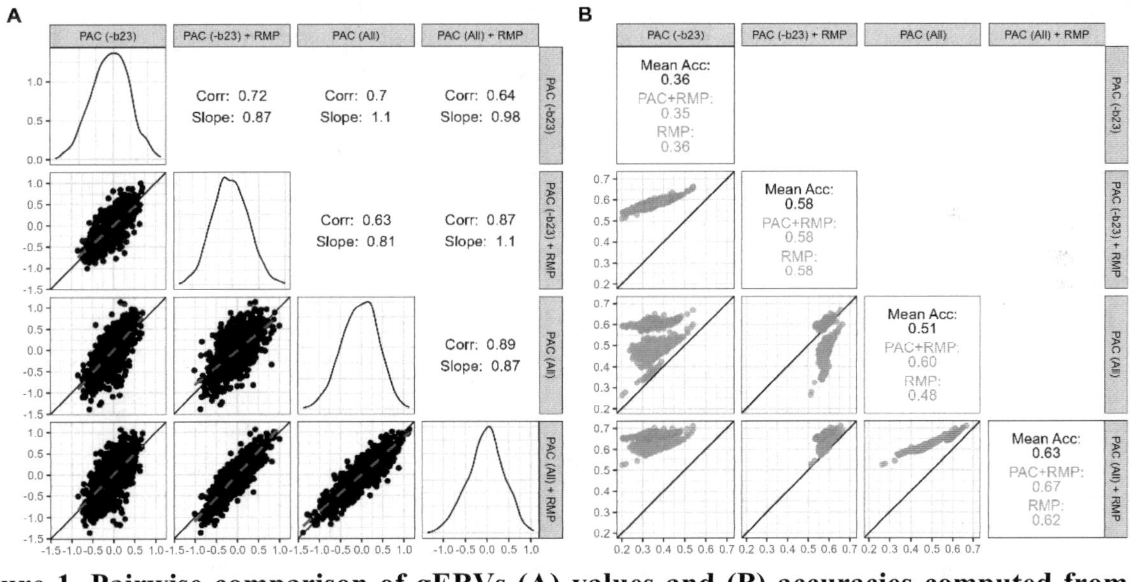

Figure 1. Pairwise comparison of gEBVs (A) values and (B) accuracies computed from the bivariate model using ASReml for the 975 born 2023 rams under four different scenarios of phenotype information available. The four scenarios are PAC data from born 2019 to 2022 (PAC (-b23)), PAC data from born 2019 to 2022 plus the RMP methane trait (PAC (-b23) + RMP), all available PAC data (PAC all), and all available PAC data and the RMP methane trait (PAC (all) + RMP). For plot A, the black solid line is the line of identity, the blue dashed line is the regression line, and the upper diagonals contain the correlation (Corr) and regression slope (Slope). For plot B, points are coloured based on if a ram only had a RMP (blue) or both a PAC and RMP measure (red), and the diagonals provide the mean accuracies.

The gEBV accuracies computed from ASReml are given in Figure 1B and show that using RMP information without PAC measures improves the mean gEBV accuracies from 0.36 (PAC (-b23)) and from 0.48 (PAC (All)) to 0.58 (PAC (-b23) + RMP) for the born 2023 rams with no PAC measures but were slightly lower than the mean accuracy of 0.62 when PAC measures were included (PAC (All) + RMP).

These results indicate that RMP profiles would greatly improve gEBV accuracy for animals without PAC measures but only modest improvement for animals with a PAC phenotype. Hence, RMP could be used as an alternative method for obtaining PAC measures, enabling larger numbers of animals to be phenotyped across the industry. Nevertheless, a sufficient number of animals would need both a RMP and PAC measure for training and calibrating the prediction models.

CONCLUSION

In this study, we have provided a strategy for incorporating RMP information into the generation of gEBVs for PAC methane traits based on using a two-step bivariate modelling approach. RMP could provide farmers with another option for generating phenotypic methane measures to breed for low methane animals that is lower cost (approximately $60NZD per animal) compared to PAC ($100NZD per animal). This would facilitate obtaining phenotypic measures of methane emissions for large number of animals and potentially enable faster genetic progress for methane traits.

ACKNOWLEDGEMENTS

Funding provided through the New Zealand Agricultural Greenhouse Gas Research Centre (NZAGRC) science programme, and the Ministry for Business, Innovation and Employment (MBIE) via the "Mapping the New Zealand Ruminotype landscape" (C10X1807) program. We would like to acknowledge Focus Genetics for access to genotype and phenotype data and the support of the team at Pāmu Freestone for their support in data collection, with special thanks to Tim Smith and Isabel Schmahl.

REFERENCES

Archer M.D., Drier D.E., Powdrell S.-J.H., Campbell D., Winkworth-Lawrence C.L., Rowe S.J., McEwan J.C. and Lee M. (2023) *Proc. Assoc. Advmt. Anim. Breed. Genet.* **25**: 186.

Bilton T.P., Alemu S.W., Dodds K.G., Henry H., Hess M.K., Jordan R., Booker F., Hickey S.M., Amyes N., Knowler K., Sandoval E., Peers-Adams J., van Stijn T.C., Baird H., Watson T., Bain W., Veenvliet B., Pile G., Bryson B., Clarke S.M., Johnson P.L., McEwan J.C. and Rowe S.J. (2024) *Submitted.*

Budel J.C.C, Hess M.K., Bilton T.P., Henry H., Dodds, K.G., Janssen P.H., McEwan J.C. and Rowe S.J. (2022) *Anim. Microbiome* **4**: 39.

Dodds K.G., McEwan J.C., Brauning R., Anderson R.M., van Stijn T.C., Kristjánsson T. and Clarke S.M. (2015) *BMC Genomics* **16**: 1047.

Gilmour A.R., Gogel B.J, Cullis B.R. and Thompson R. (2015) ASReml User Guide Release 4.2 structural specification. VSN International Ltd, Hemel Hempsted, UK.

Hess M.K., Rowe S.J., Van Stijn T.C., Henry H.M., Hickey S.M., Brauning R., McCulloch A.F., Hess A.S., Kirk M.R., Kumar S., Pinares-Patiño C., Kittelmann S., Wood G.R., Janssen P.H. and McEwan J.C. (2020) *PLoS One* **15**: e0219882.

Jonker A., Hickey S.M., McEwan J.C. and Waghorn G.C. (2020). In 'Guideline for estimating methane emissions from individual ruminants using: GreenFeed, 'sniffers', hand-held laser detector and portable accumulation chambers', pp. 49-56, editors A. Jonker and G. Waghorn, New Zealand Agricultural Greenhouse Gas Research Centre, New Zealand

Newman S.-A., Dodds K.G., Clarke J.N., Garrick D.J. and McEwan J.C. (2000) *Proc. N. Z. Soc. Anim. Prod.* **60**: 195.

Proc. Assoc. Advmt. Anim. Breed. Genet. 26: 233-236

TISSUE-SPECIFIC RNA EXPRESSION DATA ENHANCES GENOMIC PREDICTION OF METHANE EMISSIONS IN SHEEP

S.W. Alemu, T.P. Bilton, B.J. Perry, H.M. Henry, K.M. McRae, K.G. Dodds, J.C. McEwan, S.M. Clarke and S.J. Rowe

AgResearch, Invermay Agricultural Centre, Private Bag 50034, Mosgiel 9053, New Zealand

SUMMARY

Reducing methane emissions in sheep through selective breeding requires accurate genomic prediction methods. This study investigated whether incorporating RNA tissue expression data could enhance prediction accuracy for methane emissions. Analysis was conducted on 1,366 sheep, including RNA sequencing data from three tissues (longissimus muscle, rumen wall, and rumen papillae) in a subset of 48 animals. To address the high dimensionality of RNA expression data (>20,000 genes), we employed principal component analysis (PCA) and selected components based on their correlation with methane emissions. These selected components were integrated as intermediate traits in a Neural Network Genomic Best Linear Unbiased Prediction (NNGBLUP) model, evaluating up to nine principal components (PCs) sequentially for each tissue. While the traditional GBLUP approach demonstrated robust performance within our flock sheep population, the integration of tissue-specific RNA expression data through selected PCs provided modest improvements in prediction accuracy. These findings suggest that RNA tissue expression data can complement existing genomic prediction methods, despite constraints from limited sample size and genetic diversity. Our results highlight the potential of incorporating molecular phenotypes in breeding programs aimed at reducing methane emissions in sheep.

INTRODUCTION

Addressing methane emissions from sheep is key for enhancing agricultural sustainability, especially in regions where sheep farming plays a significant economic role. Current genomic prediction methods for methane emissions rely primarily on genetic markers, but these approaches may not fully capture the biological complexity of methane production. Our study focuses specifically on the potential of RNA tissue expression data to enhance prediction accuracy for methane emissions in sheep. Previous research has demonstrated the value of genomic prediction methods using rumen microbiome information (Hess *et al.* 2023), but the role of tissue-specific gene expression in methane production remains poorly understood. This study aims to evaluate whether RNA expression profiles from three tissues - rumen wall, rumen papillae, and longissimus muscle - can improve the accuracy of genomic prediction for methane emission beyond traditional genomic approaches. While rumen wall and papillae tissues are directly involved in digestive processes and methane production, the inclusion of longissimus muscle allows us to explore diverse biological information that might enhance genomic prediction accuracy. By incorporating RNA expression data from these tissues into prediction models, we aim to develop more accurate genomic prediction strategies for methane emissions in sheep breeding programs.

MATERIALS AND METHOD

The animal experiments conducted adhered strictly to the guidelines of the 1999 New Zealand Animal Welfare Act and the AgResearch Code of Ethical Conduct. The trials were approved by the AgResearch Animal Ethics Committee (approval number 15601).

Genotype. All animals were genotyped using a 15K SNP chip, which directly focused on the available genetic markers without any imputation to higher densities. Filtering was performed such that only autosomal SNPs were retained, and SNPs with a minor allele frequency greater than 0.01 and a call rate of 70% were included. This resulted in a total of 13,916 SNPs, with a mean call rate of 0.96. A genomic relationship matrix (GRM) was constructed using the KGD software (Dodds *et al.* 2015). For each pair of individuals, only SNPs with non-missing genotypes were used to compute the genomic relationship, assuming missingness at random.

RNA-seq. Tissue samples (rumen wall, rumen papillae, and longissimus muscle) were collected from 48 males born in 2021. Females were excluded to maintain the breeding population, ensuring herd sustainability. Total RNA from the snap-frozen tissue was extracted using the NucleoSpin RNA extraction kit (Macherey-Nagel), according to the manufacturer's guidelines, following the protocol for RNA purification from cultured cells and tissue. For the longissimus sample, the protocol for difficult-to-lyse tissues was followed. Sequencing libraries were made with the Illumina Stranded Total RNA with Ribo-Zero Plus and sequenced on the Illumina Novaseq 6000. Analysis began with a filtering step, performed independently for each tissue type, where genes with zero counts in more than 10% of samples were removed. The filtered data were transformed using a centred log-ratio transformation using the proper package and standardized by scaling to zero mean and unit variance. Principal component analysis (PCA) was then conducted on each individual tissue.

Prediction Set-Up. The dataset was divided into training and validation sets to test the models' predictive power. The training set included data from 2019 to 2021 (804 animals), while data from 2022 and 2023 (562 animals) served as the validation set. A subset of 48 animals, born in 2021 but recorded in 2022, was enriched with RNA data for analysis.

Table 1. Distribution of experimental sheep by birth year, gender, and RNA sequencing

Year of birth	Males	Females	Total animals	RNA sequenced
2019	130	145	275	-
2020	126	135	261	-
2021	128	140	268	48(males)
2022	124	136	260	-
2023	142	160	302	-
Total	650	716	1,366	48

Statistical analysis. The models included fixed effects of birth and rearing rank, dam age (1, 2, 3+ years), and contemporary group (flock-birth year combination). Birthday deviation from the flock-year mean was used as a covariate to adjust phenotypes. Enteric methane emissions were measured using portable accumulation chambers (PAC). Animals were off feed for 1–4 hours after 3 days of ad-lib grazing on pasture with sufficient dry matter. Using the adjusted methane phenotype, this study employed two genomic prediction approaches. The first method was Genomic Best Linear Unbiased Prediction (GBLUP), which served as our baseline model. The second approach was Neural Network GBLUP (NN-GBLUP), implemented within a Bayesian framework which extended traditional GBLUP by incorporating RNA expression data through a neural network structure (Zhoa *et al.* 2022). This model processed information through three layers: an input layer containing genomic data, an intermediate layer representing RNA expression information through principal components (PCs), and an output layer predicting methane emissions. The analysis involved 60,000 chains with a burn-in phase of 10,000, sampled every 10 steps and implemented in JWAS software (Cheng *et al.* 2018).

To identify the most informative RNA expression patterns, we first performed principal component analysis and selected PCs based on their correlation with methane emissions in the subset of 48 animals with RNA data. We systematically evaluated a series of models incorporating different numbers of principal components (PCs). For each tissue, we created nine distinct models by incrementally including PCs that showed the strongest correlations with methane emissions. Starting with the two most strongly correlated PCs, we sequentially added components to create models with 3 to 10 PCs. The selected PC based on correlation for the three tissues is shown in Table 2.

Table 2. Selected principal components by tissue type based on correlation with methane emissions

Tissue	Selected PC based on correlation with methane
Longissimus	44, 10, 43, 2, 24, 37, 1, 25, 3, 12
Rumen Papillae	35, 8, 24, 30, 25, 16, 41, 45, 32, 16
Rumen wall	30, 10, 40, 37, 42, 43, 23, 19, 39, 33

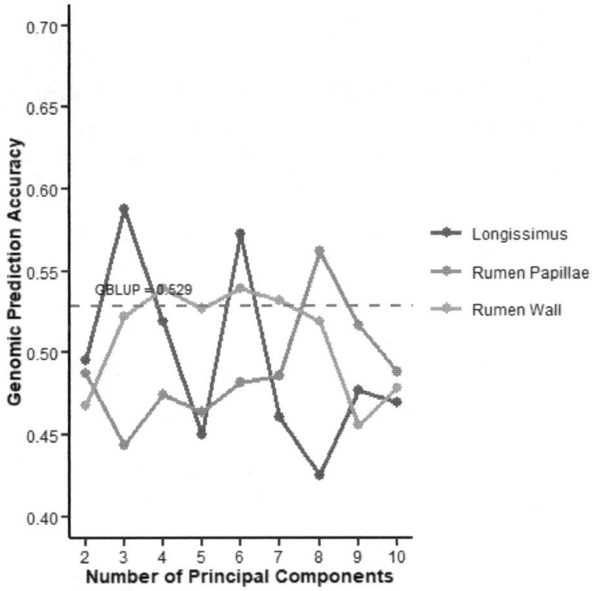

Figure 1. Accuracy of methane emission prediction using tissue-specific principal components

RESULTS AND DISCUSSION

The genomic prediction accuracy for the baseline GBLUP was 0.529 (Figure 1), providing a reference point for evaluating the incorporation of RNA tissue expression data. For rumen wall tissue, while the addition of some PCs resulted in prediction accuracies lower than the GBLUP baseline, specific combinations showed improved performance. The highest prediction accuracy of 0.539 was achieved using PCs: PC1, PC2, PC10, PC24, PC37, and PC43. These specific combinations are detailed in Table 2, where principal components were selected based on their correlation with methane emissions in the subset of 48 animals with RNA data.

In the case of rumen papillae tissue, the integration of 8 PCs resulted in a prediction accuracy of 0.562, representing a substantial improvement over the GBLUP baseline. This improvement suggests that rumen papillae expression patterns may capture important biological signals related to methane production.

Longissimus tissue demonstrated the highest prediction accuracy of 0.588 with 3 PCs, surpassing the GBLUP baseline, suggesting insights into methane production mechanisms despite its indirect role in ruminal processes. While specific studies on muscle gene expression's direct correlation with methane emissions are limited, microbial gene abundance in the gastrointestinal tract has been shown to influence metabolic processes and traits like methane production (Roehe *et al.*2016). This underscores the potential for muscle tissue gene expression patterns to indirectly impact methane emissions through metabolic pathways.

These results are particularly noteworthy given the constraints of the study design. The data consisted of animals from a single flock, and RNA tissue information was available for only 48 individuals. Under these limitations, the fact that the traditional GBLUP model maintained comparable performance highlights its robustness in scenarios with high genetic relationships within the study population. However, the observed improvements in prediction accuracy when incorporating RNA tissue expression data, albeit modest, suggest potential value in integrating molecular phenotypes for enhancing genomic prediction.

The tissue-specific variations in prediction accuracy improvements indicate that different tissues may capture distinct aspects of the biological mechanisms underlying methane production. Future studies with larger RNA-sequenced populations and diverse genetic backgrounds could further validate and potentially amplify these improvements in genomic prediction accuracy.

CONCLUSION

In conclusion, our study demonstrates a proof-of-concept that integrating RNA tissue expression data can modestly improve genomic prediction accuracy for methane emissions in sheep. While our results show promising improvements, it is essential to acknowledge that these findings are preliminary given the limited sample size of 48 individuals with RNA data and the constraints of a single-flock study design. Further validation across larger and more diverse populations is necessary to solidify and potentially amplify these results.

ACKNOWLEDGEMENTS

The authors would like to acknowledge the AgResearch Animal Genomics Research laboratory and field team farm for the RNA extractions and sequencing, and detailed tissue collection, respectively.

REFERENCES

Cheng H, Rohan F.R. and Garrick D (2018) *Proc. World Congr. Genet. Appl. Livest. Prod.* **11**: 859.

Dodds K.G., McEwan J.C., Brauning R., Anderson R.M., van Stijn T.C., Kristjansson T. and Clarke S.M. (2015) *BMC Genomics* **16**: 1047.

Hess M.K., Zetouni L., Hess A.S., Budel J., Dodds K.G., Henry H.M., Brauning R., McCulloch A.F., Hickey S.M., Johnson P.L., Elmes S., Wing J., Bryson B., Knowler K., Hyndman D., Baird H., McRae K.M., Jonker A., Janssen P.H., McEwan J.C. and Rowe S.J (2023) *Genet. Sel. Evol.* **55**: 53.

Roehe R., Dewhurst R.J., Duthie C.A., Rooke J.A., McKain N., Ross D.W., Hyslop J.J., Waterhouse A., Freeman T.C., Watson M., Wallace R.J. and Yu P. (2016) *PLoS Genet.* **12**: e1005846.

Zhao T., Zeng J. and Cheng H. (2022) *Genetics.* **221**: iyac034.

Proc. Assoc. Advmt. Anim. Breed. Genet. 26: 237-240

USE OF MID-INFRARED SPECTROSCOPY DATA FOR IMPROVING ACCURACY OF GENOMIC PREDICTION OF METHANE EMISSIONS

S. Bolormaa[1], P.N. Ho[1], I.M. MacLeod[1,2], M. Haile-Mariam[1,2], L.C. Marett[3], S.R.O. Williams[3], J.L. Jacobs[2,3,4], C.J. Vander Jagt[1], M.E. Goddard[1,4], R. Xiang[1], and J.E. Pryce[1,2]

[1] Agriculture Victoria Research, Agribio, Bundoora, VIC, 3083 Australia
[2] School of Applied Systems Biology, La Trobe University, Bundoora, VIC, 3083 Australia
[3] Agriculture Victoria Research, Ellinbank Centre, Ellinbank, VIC, 3821 Australia
[4] School of Agriculture and Food, University of Melbourne, Parkville, VIC, 3010 Australia

SUMMARY

Worldwide, methane emissions (Me) from ruminants are driving a measurable impact on global warming. Breeding cattle with lower methane output is one option that could contribute to reducing total methane emissions. The aim of this paper was to explore the use of milk mid-infrared (MIR) data for improving the accuracy of Me genomic estimated breeding values (GEBV) for 436 Holstein cows with methane phenotypes and 50K SNP genotypes. Out of these 436 cows with methane phenotypes, 200 had MIR records. We selected the most informative MIR spectra using hierarchical clustering analysis, and subsequently used this information to predict Me phenotypes. The use of informative MIR spectra increased the prediction accuracy of methane phenotypes by up to 8% compared with using the entire 537 spectra and helped to improve the accuracy of GEBVs for methane emissions. However, this finding is based on a small cross-validation study and needs to be confirmed through a larger study.

INTRODUCTION

Methane is a short-lived, but potent greenhouse gas contributing substantially to global warming. Over 70% of agricultural greenhouse gases is from enteric methane emissions (Me) produced by ruminants. In cattle, methane is a heritable trait, ranging from 0.12 to 0.45 (Manzanilla-Pech *et al.* 2021). Thus, genomic selection for reduced methane is potentially a more reliable, cost-effective, cumulative and permanent tool for methane mitigation than some other strategies. In Australia, the Sustainability Index developed using breeding values of traits associated with methane was released by DataGene in 2021. While developing breeding values using direct measures of methane is the best strategy (Richardson *et al.* 2022), enteric methane is an expensive and labour-intensive trait to collect (e.g. SF6 or respiration chambers). Proxy methods such as mid-infrared spectroscopy (MIR) measured from milk samples potentially offers a cost-effective solution to estimate methane at scale. The objective of this study was to explore the potential of MIR for predicting methane, particularly if specific MIR wavenumbers (spectra) among the total of 537 spectra might be more informative than others, and to test their subsequent use to improve the accuracy of genomic breeding values (GEBV) for methane.

MATERIALS AND METHODS

Phenotype and genotype data. Direct methane production records (g/d per animal) were collected from 436 Australian lactating dairy cows as described in Pryce *et al.* (2015). Methane production primarily reflects animal feed intake and consequently milk production in Holsteins (Fresco *et al.* 2023), so we estimated the residual level of methane produced by each animal after accounting for their energy corrected milk production (ECM, Pryce *et al.* 2015) and refer to this as RMe, where RMe = methane – (mean + fixed effects + ECM) and fixed effects are described in Eq 1 below. We also calculated Methane intensity phenotypes (MeI, g/kg of ECM) calculated as methane production divided by ECM. For 200 of the 436 cows, MIR data from milk samples were processed following Ho *et al.* (2019). After processing, a total of 537 spectral wavenumbers between 928 and 1596 as well as between 1693 and 3025 cm^{-1} were retained for further analysis. The 50k SNP array genotypes used in this study included 41,276 SNP that passed quality control and were a subset of high-density genotypes used in Bolormaa *et al.* (2022). The phenotypes of all traits including raw MIR spectra were adjusted for fixed effects using Eq 1:

$$\text{trait} = \mu + \text{HYS} + \text{batch} + \text{parity} + \text{DIM} + \text{pol(age,-2)} \qquad \text{(Eq 1)},$$

where μ is the overall mean of phenotypes of each trait across the population, HYS relates to herd, year and season of calving, batch is 16 cohort groups from experiments. DIM is days in milk at the beginning of the experiment as a covariate, and parity (1, 2, 3, and 4+) were fitted as fixed effects, and poly(age,-2) is age of cows at calving fitted as a second-order orthogonal polynomial function.

MIR analysis. First, GEBV were calculated for each of the 537 MIR spectra using a linear regression model with the 50k SNP genomic relationship matrix using ASReml program (Gilmour *et al.* 2009). To identify the most related groups of MIR spectra, we applied the Hierarchical Clustering method using the Ward's criterion based on the multi-dimension variance (R program, R Core Team 2021). The clustering analysis was performed using R (R Core Team 2021) based on the pair-wise correlations of these GEBV. Number of clusters to be retained was determined by the Elbow method in K-means Clustering, and 11 specific spectra groups (grp) were identified at the Elbow point. Then, using partial least squares regression (PLS) analysis with 10-fold cross-validation, we tested how accurately predict the methane phenotypes using each group (or combined groups) of raw MIR spectra (mirMeI(grp)), compared to the phenotypes using entire 537 spectra (mirMeI). Four of our groups provided the R^2 of prediction of > 0.1 (range: 0.11 to 0.32, Table1), and were selected for the next part of the analysis. Table 1 shows the number of MIR spectra in each selected group. The R^2 using entire 537 spectra for mirMeI was lower compared with using grp5 or combinations with other (non-overlapping) spectra groups.

Table 1. Number of spectra and squared correlation (R^2) for MIR clustering groups

MIR clustering group	No. spectra^	R^{2*}
grp2	7	0.11
grp5	26	0.32
grp7	13	0.18
grp8	66	0.14
grp2+5+7[&]	46	0.31
grp2+5+7+8[&]	104	0.31
mirMeI[J]	537	0.24

[&]represent spectra from clustering group 'grp2, grp5, grp7, and/or grp8' together; [J]MIR-predicted methane using full 537 spectra; ^number of spectra identified in each clustering group; $^*R^2$ is the coefficient of determination obtained from PLS regression analysis.

Genomic prediction. Next, univariate and bivariate genomic REML (GREML) analyses were performed using ASReml (Gilmour *et al.* 2009) to calculate genomic heritability (h^2) and predict Me GEBV. The genomic relationship matrix used in GREML analysis was built based on the 50k SNP genotypes using the method of Yang et al. (2010). In bivariate analyses, RMe together with MIR predicted MeI from specific MIR groups (mirMeI(grp)) were analysed by treating them as correlated traits. Genomic predictions were validated in the 200 MIR cows using a leave-one (cow)-out approach. The prediction accuracy for RMe was calculated as the correlation between GEBV and the RMe, divided by the square root of the heritability (h^2). The accuracies using bivariate models were compared to the accuracy of the univariate model. We also considered a scenario where the direct MIR spectra were included in a bivariate model. For this purpose, using those spectra allocated to each clustered group in Table 1, we performed principal component (PC) analyses, and then the 1st PC for each group was used as the second trait (MIR PC) in the bivariate model.

RESULTS AND DISCUSSION

Mean (SD) of RMe was 47.6 (11.13) g/d. Genomic heritability estimate using 436 cows in univariate analysis was 0.24 for RMe, which is within the reported range in literature (Richardson *et al.* 2021; Manzanilla-Pech *et al.* 2021). The MIR PC traits provided a range of h^2 estimates (0.15-0.51, Figure 2a), but with large standard errors (range: 0.173-0.215) due to the small dataset. The heritability for MeI predicted using all 537 MIR spectra (mirMeI) was high (0.71), probably indicating that mirMeI additionally captures the variance explained by milk production.

a)

trait	No.Cows	h^2 (s.e.)	RMe	mirMel	PC(grp2)	PC(grp5)	PC(grp7)
RMe	436	0.24 (0.109)					
mirMel	200	0.71 (0.186)	0.47				
PC(grp2)	200	0.15 (0.189)	0.40	0.41			
PC(grp5)	200	0.20 (0.202)	0.34	0.31	0.87		
PC(grp7)	200	0.25 (0.205)	0.35	0.31	0.85	1.00	
PC(grp8)	200	0.51 (0.215)	0.32	0.37	0.52	0.69	0.73

b)

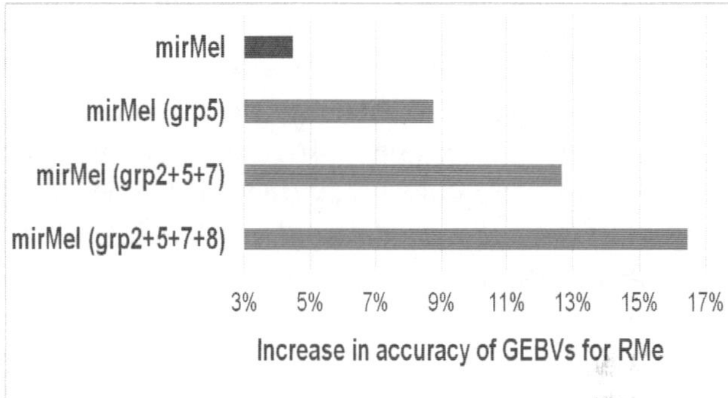

Figure 2. a) Heritability estimates and correlations of GEBV between directly measured methane and MIR predicted methane (mirMeI) or MIR principal component (PC(grp2 to grp8)) traits; b) Increase in accuracies of GEBV for residual methane (RMe)* in bi-variate analyses

*blue bar represents using phenotypes for predicted methane intensity (mirMeI) based on entire 537 MIR spectra and green for MeI predicted based on MIR spectra specific to the clustering group(s): 'grp(2+5+7)' or 'grp(2+5+7+8)' represent spectra from grp2, grp5, grp7, and/or grp8 together.

Since our primary objective is to use more informative spectra groups that may help improve the accuracy of GEBV for methane traits, we are only interested in MIR spectra groups (MIR PC traits) that have the apparent correlations of GEBV ($r > 0.2$) with GEBV for methane traits. All four MIR PC traits shown in Figure 2a exhibited moderate relationships with directly measured and MIR predicted methane traits (ranging from 0.32 to 0.40, Figure 2a). We found that the spectra of these 4 MIR PC traits fall within the infrared spectral regions driven largely by C-O, C=O, C-H, and C-N, which are particles of or bonds to fat (volatile and fatty acids) and protein. This was further confirmed on finding that these MIR PC traits exhibited moderately strong correlations with milk fat percentages (FAT%, r = 0.44-0.53), perhaps indicating that MIR PC traits in these specific groups capture some of the genetic variance of FAT%. Manzanilla-Pech et al (2021) reported moderate genetic correlation (0.45) between direct methane and ECM. As expected, strong correlation of GEBV (r=0.5) was observed between mirMeI and FAT%, whereas the correlation of FAT% with RMe was lower (r=0.26).

To compare accuracy of the GEBV from the univariate GBLUP for RMe (independent from ECM) with the accuracies from the bivariate GBLUP analysis, in addition to RMe, we used either MIR predicted MeI traits (with R^2 of $\geq$0.30 from PLS) or individual MIR PC traits as the second trait. The accuracy of GEBV for RMe was 0.41 in our univariate model. The accuracy of GEBV for RMe was improved 4% by fitting all 537 MIR spectra predicted methane phenotypes (mirMeI) as a second trait in the bivariate model (Figure 2b). An even greater increase in the genomic accuracies for RMe was observed using only the MIR spectra groups rather than using all 537 spectra as a second correlated trait (e.g. up to 16% using 'grp2+5+7+8', Figure 2b). When we compared use of phenotypes based either on same number of spectra generated from PLS or PC approaches, genomic accuracy using the MIR grp5 predicted MeI from PLS was higher (+5%) compared with using the MIR PC phenotypes (results not shown). This shows an advantage of predicted methane phenotypes with acceptable R^2 of ~0.30 from PLS over the PC1 phenotypes from PC analysis. The MIR PC1 phenotypes from each spectra group also helped to increase the accuracy of directly measured RMe due to its correlation with methane. However, it is worth testing if its increase in genomic accuracy of RMe is greater than FAT% with a larger methane dataset. The findings in this study generally showed the potential of the use of MIR as a proxy for improving the genomic accuracy of RMe, but a larger dataset is required to better evaluate this.

CONCLUSION

Identifying informative MIR spectra through hierarchical clustering analysis helped increase the prediction accuracy of methane phenotypes by 8% and subsequently improved the accuracy of GEBV for methane by up to 16%. The phenotypes predicted using PLS approach provided greater genomic prediction accuracy than PC generated phenotypes. However, this study should be interpreted with caution due to the small size of reference and validation sets used in this study and requires confirmation using a larger population.

ACKNOWLEDGEMNTS

We thank DairyBio, jointly funded by Dairy Australia (Melbourne, Australia), The Gardiner Foundation (Melbourne, Australia) and Agriculture Victoria (Melbourne, Australia) for funding this project. We acknowledge use of data from the Agriculture Victoria Ellinbank SmartFarm.

REFERENCES

Bolormaa, S., MacLeod I.M., Khansefid M., Marett L.C., Wales W.J., Miglior F., Baes C.F., Schenkel F.S., Connor E.E., Manzanilla-Pech C.I.V., Stothard P., Herman E., Nieuwhof G.J., Goddard M.E. and Pryce J.E. (2022) *Gen. Sel. Evol.* **541**: 60.

Fresco, S., Boichard D., Fritz S., Lefebvre R., Barbey S., Gaborit M. and Martin P. (2023) *J. Dairy Sci.* **106**: 4147.

Gilmour, A.R., Gogel B.J., Cullis B.R. and Thompson R. (2009) ASReml User Guide Release 3.0. VSN Hemel Hempstead, UK.

Ho, P. N., Bonfatti V., Luke T.D.W. and Pryce J.E. (2019) *J. Dairy Sci.* **102**: 10460.

Manzanilla-Pech, C.I.V., Løvendahl P., Mansan Gordo D., Difford G.F., Pryce J.E., Schenkel F., Wegmann S., Miglior F., Chud T.C., Moate P.J., Williams S.R.O., Richardson C.M., Stothard P. and Lassen J. (2021) *J. Dairy Sci.* **104**: 8983.

Pryce, J.E., Gonzalez-Recio O., Nieuwhof G., Wales W.J., Coffey M.P., Hayes B.J. and Goddard M. E. (2015) *J. Dairy Sci.* **98**: 7340.

Richardson, C.M., Amer P.R., Quinton C., Crowley J., Hely F.S., van den Berg I. and Pryce J. E. (2022) *J Dairy Sci.* **105**: 4272.

Richardson, C.M., Nguyen T.T.T., Abdelsayed M., Moate P. J., Williams S.R.O., Chud T. C. S., Schenkel F.S., Goddard M.E., van den Berg I., Cocks B.G., Marett L.C., Wales W.J. and Pryce J.E. (2021) *J. Dairy Sci.* **104**: 539.

Yang J., Benyamin B., McEvoy N.P., Gordon S., Henders A.K., Nyholt D.R., Madden P.A., Heath A.C., Martin N.G., Montgomery G.W., Goddard M.E. and Visscher P.M. (2010) *Nat. Genet.* **42**: 565.

Proc. Assoc. Advmt. Anim. Breed. Genet. 26: 241-244

EXPLORING ADAPTIVE SEQUENCING FOR EFFICIENT PROFILING OF METHANOGENIC MICROBIOMES IN CATTLE ORAL SAMPLES AS A PROXY FOR METHANE PREDICTION AND MITIGATION STUDIES

Z. Chen[1], J. Silaban[2], E.M. Ross[1] and C.T. Ong[1]

[1] Queensland Alliance for Agriculture and Food Innovation, University of Queensland, St. Lucia, QLD, 4072 Australia
[2] School of Chemistry and Molecular Biosciences, University of Queensland, St. Lucia, QLD, 4072 Australia

SUMMARY

Oxford Nanopore Technologies (ONT) adaptive sequencing enables real-time sequence enrichment or depletion during the sequencing run. However, the efficiency of different adaptive modes on microbiome analysis has not been well characterised. This study sequenced DNA from bovine saliva samples with two adaptive sequencing protocols (depletion and enrichment) and one regular protocol (adaptive function disabled). The depletion protocol with a bovine reference file efficiently reduced the bovine base number in the sequencing output. The enrichment protocol, despite using a methanogen reference file, resulted in the lowest archaeal species diversity. Greater base numbers of targeted methanogens were observed in the depleted samples compared to both the regular and enrichment protocols. The depletion protocol showed no notable effect on the relative abundances of selected archaeal genera. This study demonstrated the high efficiency of ONT depletion sequencing in minimising host contamination of saliva microbiome data, while effectively retaining the original microbial profile.

INTRODUCTION

The process of rumen fluid collection is invasive for animals. Because of the rumination process, oral swabbing has been considered a non-invasive alternative method for rumen microbial profiling. However, it also raises concerns regarding host genome contamination in the samples. Chemical and enzymatic host DNA depletion protocols have been widely used, but they also introduce biases to microbiome analysis (Ganda *et al.* 2021). In recent years, Oxford Nanopore Technologies (ONT) has introduced adaptive sequencing for the enrichment or depletion of sequences of interest and has shown potential applications in reducing host contamination (Ong *et al.* 2022). However, the efficiencies of different ONT adaptive strategies on microbiome characterisation require further evaluation. Therefore, this study hypothesised that ONT adaptive sequencing would significantly reduce host contamination. In the resulting data, we also tested the hypothesis that the relative abundances of archaeal genera would be affected by adaptive sampling.

MATERIALS AND METHODS

Sample collection. The bovine saliva samples of this study were collected under Animal Ethics Number AE000438. The animals used in this study were bulls housed at one of the commercial properties in Queensland, Australia. The Super•SAL™ device (Oasis Diagnostics® Corporation, USA) was used for bovine saliva collection following the manufacturer's instructions with minor modifications. Generally, the sponge was applied to gently swab the inner oral lining of a single bull until the indicator of the device turned fully red. Afterwards, the sponge was cut with sterile scissors and transferred into a 10 ml screw cap tube with 2 ml sterile 7:3 1xPBS-glycerol buffer, followed by ten inversions for mixing. After collection, samples were promptly placed on ice and transferred to -80°C for long-term storage. Samples from three animals (animals 7219, 7233, and 7234) were selected for testing.

DNA extraction and sequencing. A total of three biological replicates (animals 7219, 7233, and 7234) were used for DNA extraction in this study. The saliva was first squeezed from the sponge into a 10 ml screw cap tube by syringe. The squeezed liquid was centrifuged at 14,000 rpm for 5 min at 4°C, followed by removing the supernatant. The pellets were used for DNA extraction by QIAamp PowerFecal Pro DNA Kit (QIAGEN, Germany) according to the manufacturer's instructions. The DNA library was prepared by ONT Native Barcoding Kit 96 V14 (ONT, UK) following the manufacturer's protocol and subsequently loaded on a FLO-MIN114 (R10.4.1) Flow Cell. The sequencing was performed on the PromethION 2 Solo (ONT, UK). For samples sequenced with depletion protocol, a bovine reference file with *ARS-UCD1.2 Bos taurus* (GCA_002263795.2) and Brahman genomes (Ross 2019) was uploaded. Samples sequenced with enrichment mode utilised a reference file with 12 methanogen genomes. In addition, the adaptive function was disabled for samples under the regular protocol. Each adaptive strategy included three biological replicates, each with three technical replicates. The sequencing was terminated after all samples reached 300 Mb.

Bioinformatics and analysis. Sequencing data were base called using ONT Dorado v0.6.2 under the super accurate model with the adapter trimming option enabled. Reads shorter than 100 bp were removed using Nanofilt v.2.8.0 (De Coster *et al.* 2018). The quality-controlled data were mapped to the bovine reference file with ARS-UCD1.2 *Bos taurus* (GCA_002263795.2) and Brahman genomes (Ross 2019) using minimap2 v2.28 (Li 2018). The host sequences were removed using SAMtools v1.13 (Danecek *et al.* 2021), followed by alignment to the reference file with 12 specific methanogen genomes. Taxonomic classification of all non-host data was performed by Kraken2 v 2.1.2 (Wood *et al.* 2019) using the host-removed reads as input under a customised database. The NCBI RefSeq complete genomes of bacteria, fungi, archaea, and protozoa were used for database construction. The calculation of the Shannon index (alpha diversity) was performed by Vegan v2.6-8 and phyloseq v1.48.0 (McMurdie and Holmes 2013). The host sequence removal and methanogen capture efficiency, read length N50, Shannon index, and microbial relative abundance between the two adaptive protocols were compared using the regular protocol as a benchmark. The effects (*P*-values) of adaptive modes, technical replicates, and biological replicates were estimated using Fitting Linear Models.

RESULTS AND DISCUSSION

Sequencing produced 35 Gb of data in total. The filtering process removed 110 Mb (0.31%) from reads under 100 bp. The ONT depletion strategy significantly reduced the amount of host sequence data ($P < 0.01$; Figure 1A) compared to the enrichment. This indicated that the depletion protocol was a more effective alternative in sequencing host-contaminated microbiome samples, consistent with studies of other host-associated microbiomes (Ong *et al.* 2022). Depletion modes demonstrated greater N50 values compared to the enrichment method ($P < 0.001$; Figure 1B), indicating that the enrichment method rejected a substantial proportion of reads during sequencing, which could lead to the distortion of the microbial composition profile (Marquet *et al.* 2022).

The Shannon index of reads assigned to Archaea by Kracken2 was compared. The species diversity of the enrichment protocol was significantly less than the depletion mode ($P < 0.001$; Figure 1C). This indicates that the enrichment method does capture most of the 12 selected methanogen reads, a large proportion of other archaeal reads were off-target, which could be largely disadvantageous in microbiome studies where which species are present is largely unknown.

Host-removed sequences were aligned to the reference file with the 12 selected methanogen genomes. Although the enrichment function was enabled, these samples showed significantly lower percentages of methanogen bases than the depletion protocol ($P < 0.01$; Figure 1A). However, there was no notable difference in the relative abundances for the chosen archaeal genera between the two adaptive protocols in the taxonomic classification analysis ($P > 0.05$; Figure 1D). On the other hand, the enrichment showed notably greater relative abundances of several bacterial genera, such as *Clostridium*, a trend not observed in the depletion protocol ($P < 0.001$), which indicated that the enrichment sequencing might not be a suitable protocol for comprehensive microbial profiling. The distorted bacterial composition in the enrichment may result from the false-negative rejection and high short-read proportions (Martin *et al.* 2022). In addition, these findings also indicate that the depletion strategy can preserve the original microbial distribution in the sample (Marquet *et al.* 2022), which enables the effective microbial profile characterisation with a much smaller dataset.

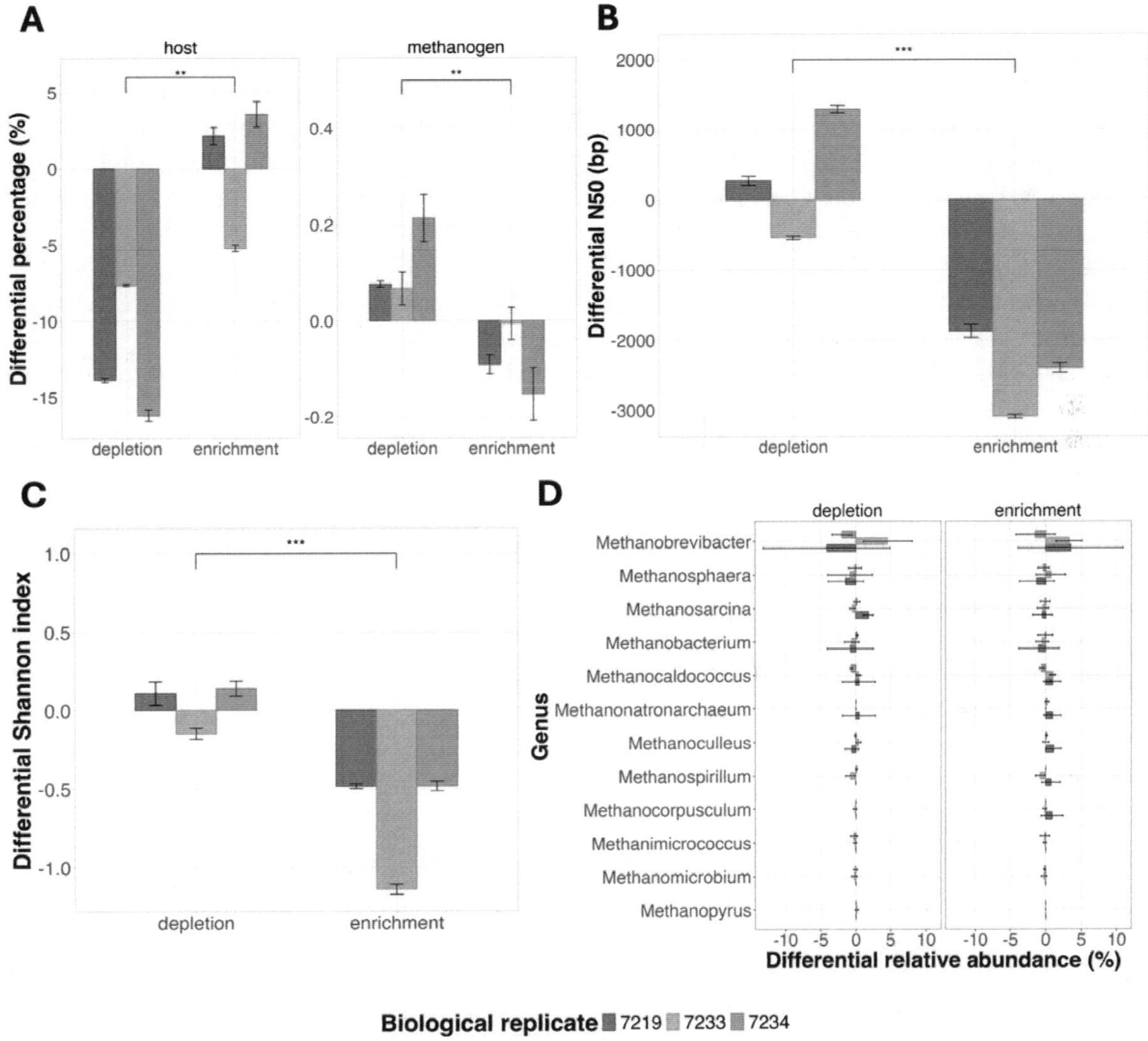

Figure 1. A) Differential proportions of bases aligned to the host genomes and selected methanogen sequences, B) differential read length N50, C) differential Shannon index (alpha diversity) of the archaeal subset, and D) differential relative abundance of selected methanogen genera between the depletion and enrichment protocols. P < 0.05: *; P < 0.01: **; P < 0.001: ***.

The host sequence removal and methanogen capture efficiency, read length N50, Shannon index, and microbial relative abundance between the two adaptive protocols were compared using the regular protocol as a benchmark. Technical replicates of each biological replicate were used to calculate the mean and standard deviation. All values are displayed as differences between the adaptive sequencing results, and the regular protocol results from the same biological sample averaged across the three technical replicates.

CONCLUSION

The ONT depletion protocol efficiently reduced the host-derived bases in the sequencing dataset and revealed a higher archaeal diversity compared to the enrichment strategy. The adaptive sequencing protocols did not affect the selected archaeal genus profile. These findings indicated that the ONT depletion function may reduce the high sequencing depth requirement in microbiome analysis. Therefore, ONT depletion sequencing is an efficient and cost-effective alternative to minimise host contamination in saliva microbiome data while still retaining the original microbial profile.

ACKNOWLEDGEMENTS

We appreciate the valuable support from our colleagues at the Centre for Animal Science within the Queensland Alliance for Agriculture and Food Innovation. We acknowledge the Low Emission Saliva Test for Ruminants project (LESTR), founded by Meat & Livestock Australia, grant number P.PSH.2010. We also thank the Graduate School Scholarships from the University of Queensland.

REFERENCES

Danecek P., Bonfield J.K., Liddle J., Marshall J., Ohan V., Pollard M.O., Whitwham A., Keane T., Mccarthy S.A., Davies R.M. and Li H. (2021) *Gigascience.* **10**: giab008.

De Coster W., D'hert S., Schultz D.T., Cruts M. and Van Broeckhoven C. (2018) *Bioinformatics* **34**: 2666.

Ganda E., Beck K.L., Haiminen N., Silverman J.D., Kawas B., Cronk B.D., Anderson R.R., Goodman L.B. and Wiedmann M. (2021) *Msystems.* **6**: e0061921.

Li H. (2018) *Bioinformatics* **34**: 3094.

Marquet M., Zöllkau J., Pastuschek J., Viehweger A., Schleussner E., Makarewicz O., Pletz M.W., Ehricht R. and Brandt C. (2022) *Sci Rep.* **12**: 4000.

Martin S., Heavens D., Lan Y.X., Horsfield S., Clark M.D. and Leggett R.M. (2022) *Genome Biol.* **23**: 11.

Mcmurdie P.J. and Holmes S. (2013) *Plos One* **8**: e61217.

Ong C.T., Ross E.M., Boe-Hansen G.B., Turni C., Hayes B. and Tabor A.E. (2022) *J Anim Sci.* **100**: skab344.

Ross E. (2019) Final Report P.PSH.0868, Meat and Livestock Australia Limited Locked Bag 1961 North Sydney, Australia.

Wood D.E., Lu J. and Langmead B. (2019) *Genome Biol.* **20**: 257.

Proc. Assoc. Advmt. Anim. Breed. Genet. 26: 245-248

EVALUATING ORAL SWABBING AS A PRACTICAL ALTERNATIVE TO RUMEN SAMPLING FOR LARGE-SCALE METHANE EMISSION STUDIES IN GRAZING ANIMALS

C.T. Ong[1], Z. Chen[1], A.S. Cavallaro[1], M.R.S. Fortes[1,2], M. Dekker[2], A. Boulton[2], B. Kniwesky[2] and E.M. Ross[1]

[1] The University of Queensland, QAAFI, Brisbane, QLD, 4072 Australia
[2] The University of Queensland, SCMB, Brisbane, QLD, 4072 Australia

SUMMARY

The oral microbiome has been proposed as a proxy for the rumen microbiome. This study aimed to evaluate this approach in grazing cattle. Results from 18 cows revealed significant differences in the diversity and composition of oral and rumen microbiomes. However, methanogen profiles were highly similar between the two sample types. These findings indicate that oral samples can effectively capture key aspects of the rumen microbiome for methane studies due to their relationship to rumination behaviour. This study highlights a practical approach for using the analysis of the oral microbiome analysis as a proxy for rumen microbiome in methane emission research in grazing animals.

INTRODUCTION

The rumen microbiome has been explored from various angles in recent years, to evaluate its involvement in ruminant production, disease resistance and environment sustainability. Nonetheless, rumen content collection poses a significant challenge to leveraging the potential of rumen microbiome data. Common methods, such as ruminal fistulation, stomach tubing, or rumenocentesis, are invasive and contrary to animal welfare principles. Moreover, these methods are labour-intensive, increasing the time cattle must remain restrained in a crush and elevating their stress levels. As a result, rumen collection is impractical and unsuitable for large-scale deployment or commercial farm settings. To address this, our study aims to evaluate the potential of using the oral microbiome as a proxy for the rumen microbiome in grazing cattle.

MATERIALS AND METHODS

Samples were collected from 18 non-pregnant Brahman cows grazing at the University of Queensland's (UQ) Gatton pasture. Oral and rumen microbiome samples were taken at three timepoints: Winter (August 2023), Spring (November 2023), and Summer (January 2024). The same cattle were sampled, for both oral and rumen, on each occasion within 5 minutes of entering the crush. All collections occurred between 10 a.m. and 12 p.m. Animal procedures were approved under Animal Ethics Approval AE000438 by the UQ Animal Ethics Committee. In total, 54 oral and 54 rumen samples were collected.

Each sample was vortexed for 15 seconds and centrifuged at 12,000 x *g* to pellet the contents. DNA was extracted using the QIAGEN QIAamp PowerFecal Pro DNA Kit (QIAGEN, Germany) following the manufacturer's instructions. To obtain microbiome signatures, oral and rumen samples were prepared using the ONT Native Barcoding SQK-NBD114.96 kit (ONT, UK). The ligated DNA library was loaded onto FLO-PRO114M PromethION flowcells and sequenced using the ONT PromethION 2 Solo sequencer with MinKNOW software version 24.06.10. Sequencing continued until each sample reached 1 million reads. The raw data was rebase-called using Dorado v0.6.2.

Base-called reads were assessed using NanoPlot 1.3.0. To remove bovine reads, they were mapped to the ARS-UCD1.2 *Bos taurus* genome (GCA_002263795.2) and Brahman genome (Ross *et al.* 2022) using Minimap2 v2.17, and unmapped (non-host) reads were retained. Chopper v0.9.0

trimmed the first 15 bases of non-host reads and selected those longer than 100 bp with a quality score above 7. Each sample was then subsampled to 1.5 million reads using seqtk v1.4. Taxonomic and functional assignments were conducted using SqueezeMeta v1.6.5, performing Diamond BlastX against the nr, COG, and KEGG databases. Non-microbial sequences, such as viruses, animal, and plant, were excluded. Bioinformatic analysis and visualisation were carried out in R Studio using the *phyloseq* v1.42, *vegan* v2.6.6.1, *DeSeq2* v1.38.3, and *ggplot2* v3.5.1 packages.

RESULTS AND DISCUSSION

To determine the potential of using oral microbiome as a proxy for rumen microbiome in grazing cattle, both samples were collected from 18 empty Brahman cows across three seasons. The sequencing results revealed the high host contamination in oral samples, which constituted an average of 81.28% cattle data in the raw sequencing data (Table 1). This observation was consistent with the high host contamination observed in oral samples, including those from other organisms (Kumar *et al.* 2023).

Table 1. The statistics of sequencing data after each step of data processing

	Rumen				Oral			
	Raw	Nonhost	Chopper	Subset	Raw	Nonhost	Chopper	Subset
Mean Read N50	3,888.39	3,873.32	2,032.34	2,032.32	1,670.22	1,977.41	2,426.60	2,426.72
Mean Quality	15.92	15.92	16.85	16.85	13.18	10.24	12.91	12.91
Mean Reads # (K)	2,241.23	2,218.24	1,736.63	1,362.21	2,285.38	439.86	330.60	321.03
Mean Total Bases (Mb)	3,063.40	3,021.46	1,859.18	1,497.82	1,851.19	382.85	338.28	326.95

The alpha diversity analysis indicated that species richness was generally greater in rumen samples, though this difference was not significant in the winter 2023 samples (Figure 1A). The variations in significance levels were attributed to differences in grazing times during different seasons and subsequently the rumination period, which influenced the capture of rumen microbiota in the oral cavity, despite sample collection taking place within similar timeframes. Beta diversity, visualised through principal component analysis (PCoA), revealed distinct clustering between oral and rumen samples across different seasons (Figure 1B).

Permutational multivariate analyses of variance (PERMANOVA) were conducted to evaluate how season, sample type, and individual animal contribute to the beta diversity differences between oral and rumen samples (Table 2). These analyses confirmed significant differences ($P < 0.05$) in overall microbiome composition between oral and rumen samples at multiple taxonomic levels, including Phylum, Class, Order, Family, Genus and Species (result not shown). The differences may be attributed to the different physiological functions of the oral cavity and rumen. Additionally, the rumen is a specialised, nutrient-dense anaerobic environment that fosters a diverse yet distinct community of symbiotic microorganisms essential for the rumination process. The oral cavity, by contrast, is a more dynamic environment due to its anatomical position. Constant fluctuations in pH, nutrient availability, and moisture levels restrict the establishment of a diverse microbial community. However, when the PERMANOVA analysis focused specifically on the phylum *Euryarchaeota*, which contains mainly the methanogens, the compositional differences between oral and rumen samples were not significant ($P = 0.19$) (Table 2). Visualising the relative abundances of the top 5 genera within *Euryarchaeota* further highlighted the similarities in methanogen composition between oral and rumen microbiomes (Figure 2). These findings suggested that, despite the overall differences in microbial communities, oral samples captured the methanogen profiles during rumination. This study showed that the methanogen profiles were consistent between oral and rumen samples, supporting the potential for using oral samples as a practical proxy for rumen sampling in methane-related studies.

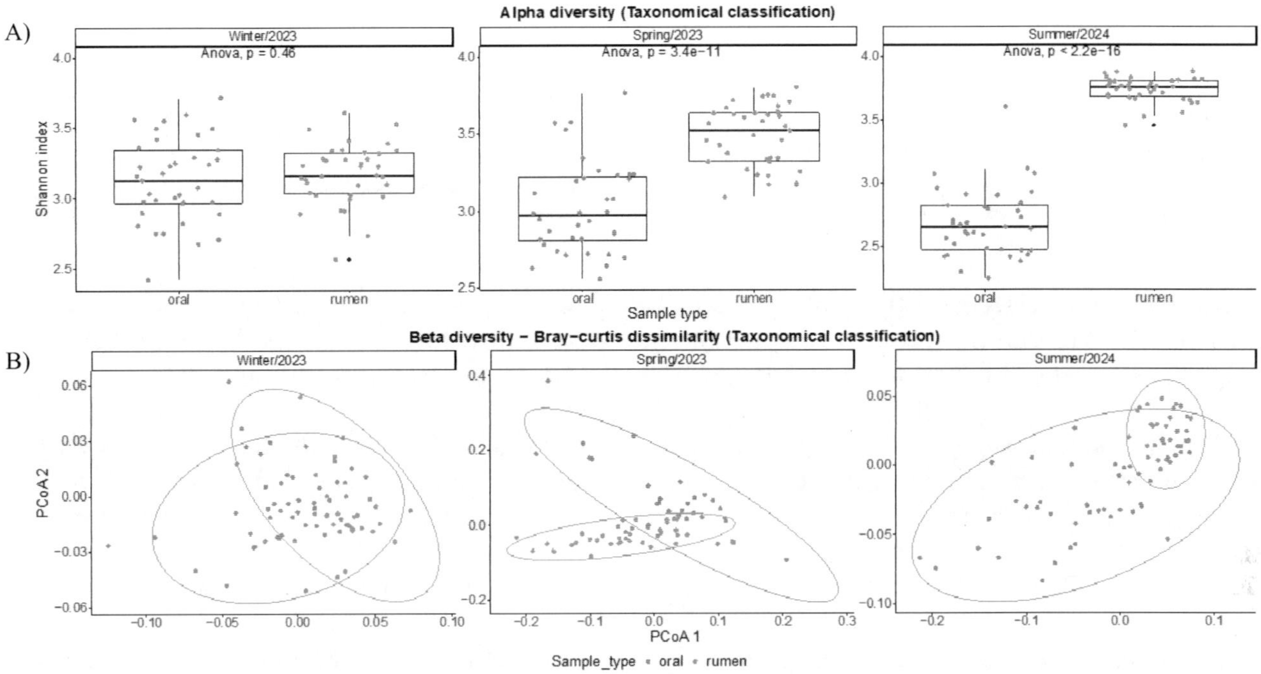

Figure 1. A) Alpha diversity and (B) beta diversity of the oral (red) and rumen (blue) microbiomes collected from 18 empty cows across 3 different timepoints, winter/2023, spring/2023 and summer/2024 in Queensland Australia

Table 2. F statistic, coefficient of determination (R^2) and *P*-values for season, sample type and animal from PERMANOVA analysis testing the differences in beta diversity between oral and rumen samples

Subset	Variable	R^2	F statistic	P-value	Significance
Whole	Season	0.40388	79.3041	0.000999	***
	Sample type	0.0823	32.4109	0.000999	***
	Animal	0.05779	0.5973	0.995005	
Phylum:Euryarchaeota	Season	0.02692	2.8846	0.01099	*
	Sample type	0.00751	1.6085	0.19181	
	Animal	0.13027	0.7346	0.96803	

This study examined the methanogen profile to demonstrate the potential of using the oral microbiome for methane emission research. However, the relative abundances of the top 5 genera in the *Euryarchaeota* subset were low, likely due to the limited sequencing coverage in this study. To address this limitation, future studies should sequence microbiome samples at greater depth to enhance coverage. Additionally, methanogenesis may be taxonomically more diverse than previously assumed (Wu *et al.* 2024). Recent research indicates an advancement in the functional meta-omics and the benefit of its integration in the methane emission studies (Zhao *et al.* 2024). Therefore, future investigations should incorporate accurate methane emission measurements and explore both taxonomic and functional subsets of the metagenome to identify key microbiome components significantly associated with the methane emission phenotype.

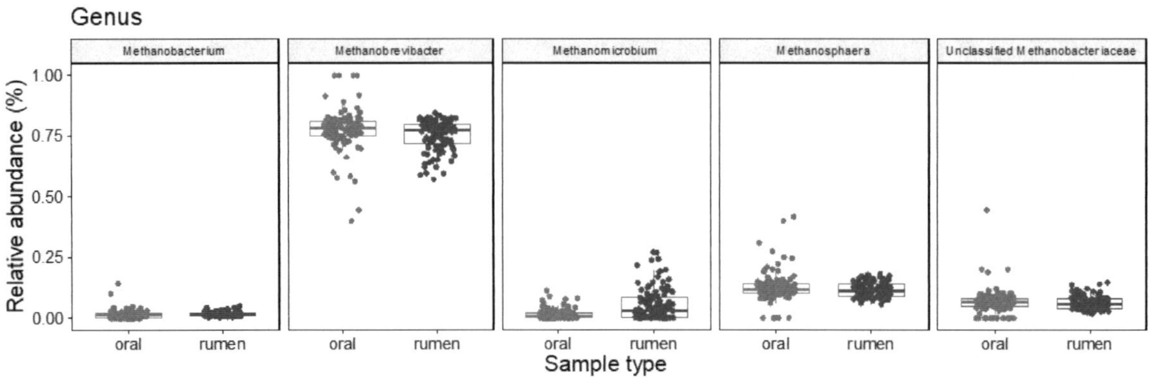

Figure 2. The relative abundances of the top 5 genera in Euryarcheota subset of the microbiome samples collected from oral (red) and rumen (blue) samples collected from 18 empty cows across 3 consecutive seasons, winter/2023, spring/2023 and summer/2024

Previous studies suggested the high similarity between oral and rumen samples when sampling was conducted at a certain timepoint under controlled conditions (Young *et al.* 2020). In this study, significant differences in the diversity and composition of microbiomes from oral and rumen samples collected from grazing animals across different seasons highlight the challenges of using oral samples as a proxy for rumen microbiomes in grazing animals. However, the study proposes that a strategic data analysis approach, such as focusing on methanogen profiles, could overcome these limitations for studying methane emissions. This approach may also be applied to other studies aiming to replace rumen sampling with oral samples, though establishing the most effective analysis strategy requires extensive background research.

CONCLUSION

While oral samples have been proposed as a proxy for rumen microbiomes, this study highlights the challenges of applying this approach to grazing animals. Significant differences in diversity and composition between oral and rumen microbiomes were observed in this study. However, the notable similarity in methanogen profiles suggests that, with strategic data analysis and further research, oral microbiome sampling holds potential as a practical alternative for rumen microbiome studies in methane emission research.

ACKNOWLEDGEMENTS

This work was made possible by the dedication of the LESTR group, UQ QASP staff, and funding from Meat and Livestock Australia (project P.PSH.2010)

REFERENCES

Kumar A., Skrahina V., Atta J., Boettcher V., Hanig N., Rolfs A., Oprea G. and Ameziane N (2023) *Front. Genet.* **14**: 1081424

Ross E.M., Nguyen L.T., Lamb H.J., Moore S.S. and Hayes B.J. (2022) *bioRxiv*, **2022.02.09**.479458.

Wu K., Zhou L., Tahon G., Liu L., Li J., Zhang J., Zheng F., Deng C., Han W., Bai L., Fu L., Dong X., Zhang C., Ettema T.J.G., Sousa D.Z. and Cheng L. (2024) *Nature* **632**: 1124.

Young J., Skarlupka J.H., Cox M.S., Resende R.T., Fischer A., Kalscheur K.F., McClure J.C., Cole J.B., Suen G. and Bickhart D.M. (2020) *Appl. Environ. Microb.* **86**: e00861.

Zhao Y., Tan J., Fang L. and Jiang L. (2024) *Sci. Total Environ.* **951**: 175732.

Proc. Assoc. Advmt. Anim. Breed. Genet. 26: 249-251

EFFECTS OF ADDING A NEW FERTILITY TRAIT TO THE ANGUS AUSTRALIA HEIFERSELECT INDEX

C.D. Quinton[1], S. Harburg[1], L.E. Proctor[1], P.R. Amer[1], C.J. Duff[2] and A.M. Samaraweera[2]

[1] AbacusBio Limited, Dunedin 9016, New Zealand
[2] Angus Australia, Armidale, NSW, 2350, Australia

SUMMARY

The Angus Australia HeiferSELECT index is designed to aid selection of replacement heifers in commercial beef herds. This index contains cow-calf, feedlot and carcase traits. However, when this index was developed in 2020, fertility evaluations were not available. Recently, Angus Australia in collaboration with CSIRO, developed a new fertility evaluation "Heifer Conception" defined as number of weeks pregnant at test following first mating (PREG). In this study we calculated an economic value for PREG ($/wk PREG) by modelling economic impacts from its effects on cow survival, and on calves' birth dates. Adding PREG to the index substantially changed animal rankings. Top heifers selected according to the index with PREG added had better average fertility than those selected according to the index excluding fertility, with only minor or negligible sacrifice in growth and carcase traits superiority.

INTRODUCTION

The Angus Australia HeiferSELECT is a genomic toolkit designed to aid beef producers selecting commercial Angus replacement females. This genomic selection tool includes GEBVs, developed in collaboration with the Commonwealth Scientific and Industrial Research Organisation (CSIRO), for birth weight (BW), weaning weight (WW), yearling weight (YW), mature cow weight (MCW), milk, average daily gain (ADG), daily feed intake (DFI), carcase weight (CWT), eye muscle area (EMA), MSA marbling (MARB), rib fat (RIB), ossification and ImmuneDEX (Angus Australia 2023).

The HeiferSELECT index is designed to apply most of these GEBVs in multi-trait selection and identify commercial heifers with genetic potential for maternal traits they, and future daughters, will express as well as beef traits expressed by their calves. This index contains maternal (BW, WW, milk, MCW) and terminal (post-wean growth PWG, feedlot growth FG, DFI, RIB and MARB) traits (see Quinton *et al.* 2021 for trait details including calculation of PWG and FG from WW, YW and CWT). When this index was developed in 2020 GEBVs for fertility were not yet available.

Recently a new fertility trait "Heifer Conception" defined as number of weeks pregnant at test following first mating, based on foetal aging (PREG; Alexandre *et al.* 2023) was developed and added to the HeiferSELECT toolkit. The objectives of this study are 1) to calculate an economic value for PREG; and 2) to assess the effects of adding PREG to the HeiferSELECT index on animal rankings and selected trait genotypes.

MATERIALS AND METHODS

Fertility Economic Value. The economic value (EV) of PREG ($/wk PREG) is built from three component linear EVs: the EV of the effect of fertility on cow survival, and the EVs of effects of fertility on birth dates of each of heifers' and cows' calves. All component EVs are calculated as the change in profitability ($) per animal, per week change in PREG, independent of changes in other index traits.

The EV of the effect of fertility on cow survival within a producing cow herd of fixed size (EV_{cs}) was modelled from a direct comparison of two herd scenarios: *i.* a BASE herd with population average survival for cows ages 2 to 11 years, and *ii*. a BASE+improved fertility herd where survival in all age classes (except final) was assumed to improve due to less fertility-related culling. The BASE+improved fertility herd has longer average lifespan, lower replacement rate and corresponding lower replacement costs, lower cull cow revenue, and slightly higher profit from weaned calf weight (young cows have smaller calves). This results in a net increase in profit from improved fertility. Note that this model assumes a herd with fixed number of producing cows, and therefore improved survival reduces replacement rate but does not change the number of calves produced and sold. The relationship of fertility with probability of survival is based on Johnston and Bunter (1996) and assumes calving success and survival have a 1:1 relationship; i.e., that non-producing cows are culled. The value is scaled for heifer PREG GEBV based on the relationship of cow days to calving phenotype (approximated with deregressed EBVs) with heifer PREG GEBV, estimated from bull data.

The EV of the effect of fertility on heifers' calves birth date (EV_{hcb}) assumes that in a herd with fixed initial mating date and weaning date, heifers that conceive 1 week earlier have calves with 1 additional week of growth at weaning and require 1 week of additional lactation feed. The difference in calf profit from an additional week of growth at weaning is calculated from the bio-economic model's calf growth model and the model WW EV which incorporates economic effects on market animal carcase revenue and feed costs for both short/mid fed and long fed production systems (Quinton *et al.* 2021). The cost of 1 week of additional cow lactation feed energy is calculated from the model's cow energy model. The overall component EV is calculated from the average calf profit difference minus lactation costs in short/mid fed and long fed production systems, weighted by the industry proportions of these systems.

The EV of the effect of fertility on mature cows' calves birth date (EV_{ccb}) follows the same logic as for heifers' calves but the effects on calf profitability and lactation feed costs were scaled to account for the less-than-unity genetic correlation between heifer fertility and mature cow fertility as per EV_{cs}.

The overall PREG index economic weight was calculated as $b_{PREG} = (EV_{cs} \times DGE_{cs}) + (EV_{hcb} \times DGE_{hcb}) + (EV_{ccb} \times DGE_{ccb})$, where DGE_{cs}, DGE_{hcb} and DGE_{ccb} are the respective component's coefficients of discounted genetic expression in heifers and their offspring over 20 years.

Effects of adding fertility to the index. The structure of the HeiferSELECT index including PREG (I_{HSPREG}) is as follows, where b are linear index economic weights and $f(GEBV)$ represent non-linear economic functions, to calculate an index value in units \$/heifer at selection:

$$I_{HSPREG} = f(GEBV_{BW}) + (b_{WW}GEBV_{WW}) + f(GEBV_{Milk}) + (b_{MCW}GEBV_{MCW}) + (b_{PWG}GEBV_{PWG}) + (b_{FG}GEBV_{FG}) + (b_{DFI}GEBV_{DFI}) + (b_{RIB}GEBV_{RIB}) + f(GEBV_{MSA}) + (b_{PREG}GEBV_{PREG})$$

The effects of adding PREG to the index were assessed using a set of heifer GEBVs that represent the range of genotypes in the population (Table 1). Index values were calculated for each animal according I_{HSPREG} and according to the index excluding fertility (I_{HS}, calculated with the above equation excluding the PREG term). Mean GEBVs were calculated and compared for all heifers in the set and for the top 20% of heifers according to I_{HS} and I_{HSPREG}. Means analysis included the traits previously listed plus heifer age at first calving (AFC; Alexandre *et al.* 2023).

RESULTS AND DISCUSSION

The Pearson correlation between I_{HS} and I_{HSPREG} was 0.80; therefore, adding PREG to the HeiferSELECT index had substantial effect on animal rankings. This was caused by moderate negative correlations between PREG and major index traits WW and PWG (correlation of $GEBV_{PREG}$ with $GEBV_{WW}$ = -0.13, with $GEBV_{PWG}$ = -0.11). PREG had favourable correlation with

MCW (correlation of $GEBV_{PREG}$ with $GEBV_{MCW}$ = -0.14) but approximately neutral correlations with other index traits.

Mean GEBVs of the top 20% of heifers selected according to I_{HS} and I_{HSPREG} are shown in Table 1. Overall, adding PREG to the HeiferSELECT index selected heifers with better fertility, with minor or negligible sacrifice in other traits. Top heifers according to I_{HSPREG} had significantly better PREG than top heifers according to I_{HS} (paired t-test P<0.0001). WW and YW means were slightly lower with I_{HSPREG} (P=0.0633 and 0.0551 respectively), but the indexes did not differ significantly for other traits means (P>0.1).

Table 1. GEBV means, SD, minimum and maximum values for all heifers in data set and top 20% selected according to the HeiferSELECT Index excluding and including fertility

	All heifers (N=921)			**HSIndex Top 20%**			**HSIndex+fert Top 20%**		
GEBV, unit	**mean**	sd	*min, max*	**mean**	sd	*min, max*	**mean**	sd	*min, max*
BW, kg	**-0.8**	1.3	*-4.6, 3.4*	**-0.7**	1.4	*-4.2, 3.4*	**-0.7**	1.4	*-4.2, 3.4*
WW, kg	**-1.2**	6.0	*-21.9, 18.6*	**1.9**	6.2	*-12.2, 18.6*	**0.7**	5.9	*-14.4, 13.3*
MCW, kg	**4.2**	12.8	*-39.0, 44.7*	**7.8**	14.1	*-26.4, 44.7*	**6.8**	13.6	*-29.7, 44.7*
MILK, kg	**2.5**	2.3	*-5.0, 8.8*	**3.3**	2.0	*-0.9, 8.3*	**3.1**	2.1	*-1.6, 8.3*
YW, kg	**2.4**	7.2	*-22.4, 25.4*	**6.4**	7.3	*-10.6, 25.4*	**5.0**	7.0	*-11.4, 18.6*
PWG, kg	**3.6**	2.0	*-2.0, 9.6*	**4.5**	1.9	*-0.7, 9.6*	**4.3**	1.9	*-0.7, 9.6*
FG, kg	**-9.0**	16.8	*-59.7, 39.9*	**0.9**	14.7	*-40.5, 39.9*	**0.2**	15.7	*-42.9, 39.9*
ADG, kg/d	**0.1**	0.0	*-0.1, 0.2*	**0.0**	0.0	*0.0, 0.2*	**0.0**	0.0	*0.0, 0.2*
DFI, kg/d	**0.2**	0.4	*-0.9, 1.3*	**0.1**	0.4	*-0.8, 1.2*	**0.2**	0.4	*-0.8, 1.2*
EMA, cm^2	**-0.4**	2.5	*-8.3, 8.2*	**0.3**	2.8	*-5.1, 8.2*	**0.3**	2.8	*-6.6, 8.2*
MARB, score	**46.7**	30.1	*-43.7, 153.8*	**68.4**	25.5	*11.5, 153.8*	**64.2**	26.2	*11.0, 153.8*
RIB, mm	**0.0**	1.1	*-3.3, 3.3*	**0.0**	1.1	*-2.7, 3.3*	**0.0**	1.1	*-3.1, 3.3*
CWT, kg	**-3.6**	9.7	*-30.1, 27.2*	**4.0**	8.0	*-16.1, 26.4*	**2.8**	9.1	*-19.5, 27.2*
AFC, d	**0.8**	8.0	*-24.2, 29.2*	**0.2**	8.2	*-23.2, 20.1*	**-0.5**	8.0	*-23.2, 20.1*
PREG, wk	**-0.4**	1.0	*-3.9, 3.1*	**-0.5**	1.1	*-3.9, 2.5*	**0.4**	0.9	*-2.4, 3.1*

CONCLUSION

The HeiferSELECT genomic tools offer cost-effective evaluation of commercial Angus heifers for cow-calf, feedlot, carcase and resilience traits. Adding the new Heifer Conception fertility trait to the HeiferSELECT index should aid producers in multi-trait selection decisions by identifying commercial heifers with optimal genetic profile for fertility and maternal traits combined with growth and carcase quality traits to be expressed in their market progeny.

ACKNOWLEDGEMENTS

The Authors wish to acknowledge CSIRO for developing the Angus HeiferSELECT prediction models. We also thanks Angus Australia members and personnel for collecting the data used to underpin the genomic tools.

REFERENCES

Alexandre P.A., Porto-Neto L.R., Hine B.C., Samaraweera A.M., Byrne A.I., Duff C.J. and Reverter A. (2023) *Proc. Assoc. Advmt. Anim. Breed. Genet.* **25**: 67.

Angus Australia (2023) https://www.angusaustralia.com.au/content/uploads/2023/08/HS-TechDoc_A4_Aug23-3.pdf. Accessed April 11, 2025.

Johnston D.J. and Bunter K.L. (1996) *Livest. Prod. Sci.* **45**: 13.

Quinton C.D., Archer J.A., Amer P.R., Harburg S., Petersen G., Byrne A., Duff C. and Parnell P. (2021) *Proc. Assoc. Advmt. Anim. Breed. Genet.* **24**: 308.

Proc. Assoc. Advmt. Anim. Breed. Genet. 26: 252-255

THE INFLUENCE OF SPERMATOZOAL MOTILITY ON CONCEPTION AND PREGANCY STATUS FOLLOWING ARTIFICIAL INSEMINATION IN SHEEP

M.J. Hodge[1, 2, 3], S.J. Rindfleish[2], L. Li[4], C.P. Stephen[1], S.D. Pant[1] and S. de las Heras-Saldana[4]

[1] Charles Sturt University, Wagga Wagga, NSW, 2678 Australia
[2] Apiam Genetics Services, Dubbo, NSW, 2830 Australia
[3] Sheep Genetics, Meat & Livestock Australia, Armidale, NSW, 2351 Australia
[4] Animal Genetics Breeding Unit*, University of New England, Armidale, NSW, 2351 Australia

SUMMARY

Reproductive efficiency is a key driver of productivity and profitability in many sheep enterprises. Both ewe and ram genetics influence reproductive efficiency, along with environmental influences. However, past research has heavily focused on the influence of the ewe. Reproductive failure from the paternal side has a significant influence, as a single ram may inseminate hundreds of ewes during their lifetime. Hence, the objective of this study was to investigate the influence of ram spermatozoal motility on conception outcomes and foetal scan count following artificial insemination (AI). Spermatozoal motility was found to significantly influence conception outcomes following logistic regression and approached significance for foetal scan count following ordinal regression. Therefore, spermatozoal motility could play a key role in improving reproductive efficiency in sheep.

INTRODUCTION

Host genetics from the ram and ewe both impact factors which influence reproductive efficiency, including conception outcomes and foetal scan count. Comparatively, relatively little emphasis has been placed on investigating the ram's influence on reproductive efficiency despite rams having the potential to inseminate hundreds of ewes annually, and many more over their lifetime, having a substantially larger impact on the flock. The objective of this study was to characterise the influence of spermatozoal motility on both conception outcomes (binary variable) and foetal scan count (ordinal variable).

MATERIALS AND METHODS

Historic AI data was provided by a commercial artificial breeding company which included progressive spermatozoal motility via visual microscopic assessment of ejaculates used in on-farm AI. A retrospective dataset for pregnancy scan results (scanning to identify the number of foetuses) and pedigree was also supplied by Sheep Genetics (Meat & Livestock Australia) for all ewes inseminated. Ejaculates with at least 40% motility were used in AI. Data quality control steps included the removal of AI records where the ejaculate's motility was not assessed, the type of semen (fresh, straw- or pellet-frozen) was not noted, and small number of records for ewe and sire breed AI combinations. A total of 2,608 insemination records from 2,289 ewes across 4 breeds and 90 rams from 5 breeds, spanning 11 AI sites across 5 years were contained in the final dataset.

Statistical analysis. Fixed effects including the AI site (n=11), type of semen used for insemination (n=3), ewe age (n=8), and spermatozoal motility were included in the final model following significance when fit individually in preliminary linear regression via ASReml (version

* A joint venture of NSW Department of Primary Industries and Regional Development and the University of New England

4.2) (Gilmour *et al.* 2021). As breed was confounded with AI site, it was not fitted in the final model. Logistic regression was performed using conception outcomes (binary trait) including fixed effects of the AI site and type of semen used for insemination, as well as covariates like spermatozoal motility and ewe age, while the inseminating ram was fit as a random polygenic effect. The full model used for logistic mixed-model regression analysis in ASReml was as follows;

$$Logit(n) = Wb + Za + e$$

where $Logit(n) = ln\frac{N(Y=1)}{1-N(Y=1)}$; $N(Y = 1)$ is the probability of conception success as the outcome Y following on-farm AI; b is the vector of fixed effects; a is the random polygenic effect; W and Z are the incidence matrix relating to the fixed effects and random polygenic effects, respectively. Log odds of conception at each level of spermatozoal motility was estimated using the predict statement. The heritability of conception outcomes was estimated via ASReml.

An ordinal regression was used for foetal scan count (ordinal variable = 0, 1, 2, 3) via ASReml. Fixed effects included AI site and type of semen used for insemination with covariates (spermatozoal motility and age of the ewe) and a random polygenic effect fitted in the final model. A multinominal error distribution was assumed.

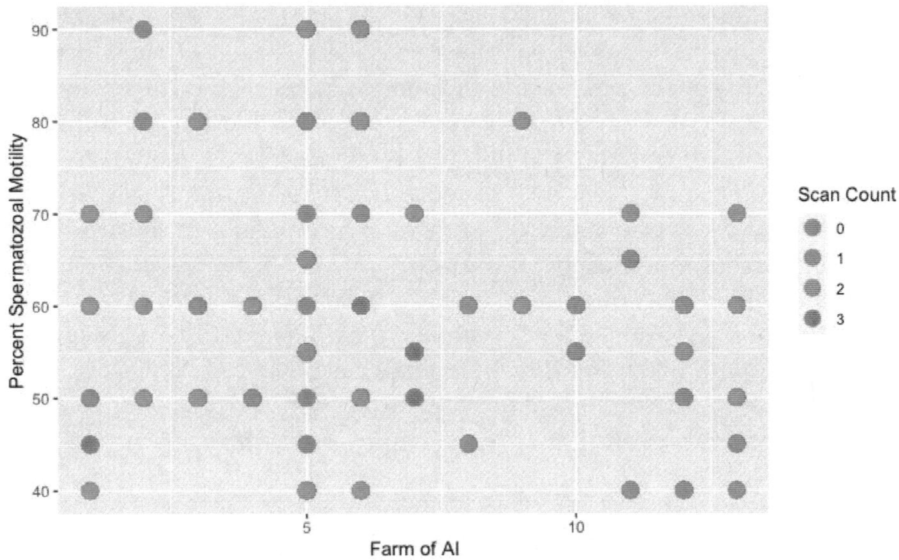

Figure 1. Range in spermatozoal motility for each AI site and the corresponding foetal scan count result following on-farm AI

RESULTS AND DISCUSSION

AI site had a significant influence on conception outcomes (P <0.001) and foetal scan count (Table 1). Spermatozoal motility also significantly influenced conception outcomes (P = 0.024) but not foetal scan count (P = 0.079). The age of the ewe had a significant impact on foetal scan count (P <0.001) but not conception outcome. The type of semen used for insemination was not significant for either conception outcome or foetal scan count. Results from past studies (Spanner *et al.* 2024; David *et al.* 2015; Wierzbowski and Kareta 1993; and Morris *et al.* 2001) align with the results of the present study showing that spermatozoal motility has an influence on conception outcomes.

The heritability for conception outcomes following on-farm AI was 0.27 (±0.06). This was higher than previous heritability estimates for ewe conception outcomes 0.06 (Peñagaricano *et al.* 2012), 0.05 - 0.23 (Bunter *et al.* 2016; Bunter *et al.* 2021). Differences in the number of animals inseminated and time period may have contributed to the higher heritability in this study. The breed of both the ewes and rams was confounded with the site of AI, hence, the heritability of the present study could have potentially overestimated the additive genetic variance, leading to the higher heritability estimate. As such, it would be beneficial to conduct further research on a larger dataset to evaluate the influence of both breed and AI site on conception outcomes following on-farm AI.

Table 2. Fitted effects and their significance for logistic and ordinal regression with conception outcomes and foetal scan count, respectively

Fixed effect	Effect type	P-value
	Conception outcome (Logistic regression)	
AI site	Fixed (n = 11)	<0.001
Spermatozoal motility	Covariate	0.024
Age of ewe	Covariate	0.477
Type of semen used for insemination	Fixed (n = 3)	0.419
	Foetal scan count (Ordinal regression)	
AI site	Fixed (n = 11)	<0.001
Spermatozoal motility	Covariate	0.079
Age of ewe	Covariate	<0.001
Type of semen used for insemination	Fixed (n = 3)	0.407

As spermatozoal motility increased, so too did the log odds of a positive conception outcome following on-farm AI (Table 2). While further research and validation is required using a larger dataset, these results indicate the potential to improve AI conception outcomes by using sperm with higher percent motility. A recent study by Spanner *et al.* (2024) reported spermatozoal motility assessed via CASA influences the odds of pregnancy following on-farm AI, amongst other parameters like spermatozoal morphology and ewe uterine tone. However, Spanner *et al.* (2024) only used Merino sheep, hence, further validation on a larger dataset include multiple breeds is required. It is important to note that ejaculates with less than 40% motility were not used for AI, due to industry practice of discarding low motility ejaculates (Van Metre *et al.* 2012).

Table 2. Log odds ratio for each level of spermatozoal following logistic regression for conception outcomes

Spermatozoal motility (%)	Lod odds (±SE)
40	1.534 (±0.557)
45	1.719 (±0.497)
50	1.903 (±0.442)
55	2.087 (±0.397)
60	2.272 (±0.365)
65	2.456 (±0.348)
70	2.640 (±0.349)
80	3.009 (±0.403)
90	3.378 (±0.504)

Spermatozoal motility has been reported to significantly influence lambing percentages (David *et al.* 2015; Wierzbowski and Kareta 1993) and *in vitro* fertilisation (Morris *et al.* 2001). However, these studies used a different phenotype to determine reproductive efficiency (i.e. lambing rate, *in vitro* fertilisation (IVF), semen assessment (i.e. gross motility) and method of insemination (i.e. cervical insemination, IVF). The retrospective data available for the present study was limited to foetal scanning data, hence litter size was not able to be directly assessed. Furthermore, other factors, like parity and ewe nutrition (Kelly *et al.* 1992) as well as the weather conditions at lambing (Masters *et al.* 2023) have a large influence on conception, fecundity and lamb survival. It is likely these factors have a cumulatively larger impact on overall reproductive efficiency than spermatozoal motility.

The present study did not assess the influence of spermatozoal morphological abnormalities on conception outcomes, as this data was not available. While morphological abnormalities are not routinely assessed as part of industry protocols, the inability to account for the potential influence of morphological abnormalities on conception outcomes is a limitation of the present study. Porcine studies have shown that morphological abnormalities in spermatozoa can affect reproductive efficiency, whereas spermatozoal motility has not been shown to have an impact (McPherson *et al.* 2014). Spanner *et al.* (2024) suggest spermatozoal morphology also has a significant influence on pregnancy in sheep following on-farm AI.

CONCLUSION

Ram spermatozoal motility potentially influences conception outcomes, along with the AI site. Further, the farm of AI and age of the ewe significantly influence foetal scan count. While further research is recommended to determine the exact extent of impact that ram semen has on reproductive efficiency, including investigations across multiple breeds as well as cross breeds at the same AI site, performing AI with a higher spermatozoal motility may ultimately lead to increased reproductive efficiency.

ACKNOWLEDGEMENTS

Thank you to both Sheep Genetics (Meat & Livestock Australia) and Apiam Animal Health for their in-kind contributions of data for this research, as well as the Australian Government Research Training Program for scholarship funding.

REFERENCES

Bunter K.L., Swan A.A., Purvis I.W. and Brown D. (2016) *Anim. Prod. Sci.* **56**: 679.

Bunter K.L., Swan A.A., Gurman, P.M. and Brown D.J. (2020) *Anim. Prod. Sci.* **61**: 333.

David I., Kohnke P., Lagriffoul G., Praud O., Plouarboué F., Degond P. and Druart X. (2015) *Anim. Reprod. Sci.* **161**: 75.

Gilmour A.R., Gogel B.J., Cullis B.R., Welham S.J. and Thompson R. (2021) ASReml User Guide Release 4.2, V.I. Ltd, Editor. 2021, Hemel Hempstead: UK. HP2 4TP.

Kelly R.W., Speijers E.J., Ralph I.G. and Newnham J.P. (1992) *Aust. J. Agric. Res.* **43**: 339.

Masters, D.G., Blache, D., Lockwood, A.L., Maloney, S.K., Norman, H.C., Refshauge, G. and Hancock, S.N. (2023) *Anim. Prod. Sci.* **63**: 623

McPherson F.J., Nielsen S.G. and Chenoweth P.J. (2014) *Anim. Reprod. Sci.* **151**: 28.

Morris L.H., Johnson W.H., Leibo S.P. and Buckrell B.C. (2001) *Reprod. Fert. Dev.* **13**: 193.

Peñagaricano F., Weigel K.A. and Khatib H. (2012) *Anim. Genet.* **43**: 65.

Spanner, E.A., de Graaf, S.P. and Rickard, J.P. (2024) *Sci. Rep.* **14**: 27556.

Van Metre, D.C., Rao, S., Kimberling, C.V. and Morley, P.S. (2012) *Prev. Vet. Med.* **105**: 118.

Wierzbowski S. and Kareta W. (1993) *Theriogen.* **40**: 205.

GENETIC PARAMETERS FOR CAUSE OF DEATH TRAITS IN LAMBS AND INDICATORS TRAITS LINKED TO NEONATAL SURVIVAL

M. Hebart, R. McEwin and F.D. Brien

University of Adelaide, Roseworthy, SA, 5371 Australia

SUMMARY

Improving lamb survival has implications for production, economics and welfare. Direct selection for lamb survival is difficult given its low heritability. Selection based on cause of death could provide an alternative or supplementary strategy for improving lamb survival. This study estimated heritabilities for cause of death traits determined from necropsy, and genetic correlations between these and indicators of neonatal lamb survival. Our results support using lambing ease as an indicator trait to improve lamb survival. Although birth weight was positively genetically correlated with Dystocia A, it was negatively correlated with the starvation and mismothering and starvation, mismothering and exposure complexes, complicating its use as an indicator trait. However, the moderate genetic correlations between observed birth vigour and the starvation and mismothering complex (0.53) and between thorax circumference and Dystocia B (0.58) and a composite dystocia trait (0.47) indicate potential for indirect selection for improved lamb survival.

INTRODUCTION

Reproductive efficiency of the Australian sheep flock has remained unchanged for the past five decades. Hinch and Brien (2014) reported that on average 20-30% of lambs die before weaning, with 74% of those dying within three days of birth. Lamb survival is a complex trait that is influenced by many environmental and management factors, and heritability estimates of lamb survival are very low (Brien *et al.* 2010; Safari *et al.* 2005). McFarlane (1965) suggested that an alternative to selecting directly to improve lamb survival is to select against the cause of death, as determined by necropsy. Brown *et al.* (2014) estimated heritabilities for different causes of death and suggested little advantage to using these over direct selection for lamb survival. However, understanding the genetic relationships between different causes of death and other production and indicator traits will help understand and improve selection decisions for lamb survival. The aim of this study was to determine the genetic relationships between the cause of death traits and a suite of neonatal lamb metrics and behaviour characteristics that are potentially indicators of survival.

MATERIALS AND METHODS

Data from 27,240 lambs born between 2007 – 2011 from the Sheep CRC Information Nucleus Flock (IN) were used in this study. Detailed descriptions of the sites, design and management of the IN was provided by Geenty *et al.* (2014). Briefly, an average of 3,832 Merino ewes and 796 crossbred ewes were annually joined by AI to key industry sires at 8 sites across southern Australia. Lifetimewool guidelines (ewe body condition score of 3.0 at joining, mid-pregnancy, lambing and weaning; Young *et al.* 2011) were followed at all sites to manage the flocks. The average lamb survival was 79.8% (lambs from Merino ewes 79% and crossbred 85%; Refshauge *et al.* 2016).

Necropsies were performed on 3,198 neonatal lambs that died (Refshauge *et al.* 2016), and one of ten death categories were assigned; Dystocia A (DysA: oedema present, n = 282); Dystocia B (DysB: no oedema, significant cranial and central nervous system haemorrhage but no metabolised fat, n = 660); Dystocia C (DysC: as per DysB but with metabolised fat, n = 577), starvation mismothering complex (SM; no significant cranial and CNS haemorrhage, n = 796); Predation (n=214); deaths *in utero* pre-birth and in premature lambs (DIUPB, n = 328); Exposure (n=172); Infection (n=18); Misadventure (n = 21); and Undiagnosed (n = 130). Composite traits of Dystocia

(Dys; Dystocia A, B or C, n = 2315); and Starvation mismothering exposure (SME; n = 968) were also analysed. A pedigree was constructed with founder animals assigned a genetic group (Swan *et al.* 2016) to account for differences in breed, selection strategy, selection lines and flocks. The survival indicator traits, recorded at lamb tagging, included birth weight (BWT), metacarpal bone length (ML), crown-rump length (CRL), thorax circumference (THO), and rectal temperature (RT), and visually assessed lamb vigour (OBV), birth coat score (BCS), and lambing ease score (LE). Five timed behaviour traits were also analysed; time taken after release for the lamb to bleat (BLT), stand (STD), contact the ewe (CONT), contact the udder (UDD) and follow the ewe (FOLL). Data collection protocols for the IN are described by Brien *et al.* (2010). Descriptive statistics for the indicator traits can be found in McEwin *et al.* (2025).

Statistical analysis. Genetic correlations between the cause of death and timed behaviour traits and other survival indicator traits were estimated using the restricted maximum likelihood (REML) method within an animal model in ASREML-R 4.2 (Butler *et al.* 2023). All traits including necropsy traits were analysed as continuous variables. The main effects for the necropsy data included in the model were contemporary group (year x site; 41 levels), sire type (Maternal, Merino, Terminal), dam breed (Merino, Crossbred), type of birth (single, twin or multiple), age of dam (1-8 years), sex (male or female), and day of birth (nested within contemporary group). The random effects included in the model were direct additive genetic, maternal (comprising both maternal genetic and environmental effects), and genetic groups.

RESULTS AND DISCUSSION

The estimates of phenotypic variance were heritability estimates for the cause of death traits were generally low, ranging from 0.006 to 0.027 (Table 1), which were slightly lower than those of Brown *et al.* (2014), who included BWT and BWT^2 in their statistical model. In contrast, we did not fit BWT (or BWT^2) as covariates in analyses as the genetic relationship between BWT and cause of death traits were under investigation. The composite traits of Dys and SME had higher heritabilities than their components but were not higher than those for lamb survival itself (Brien *et al.* 2010).

Table 1. Estimates of phenotypic variance (V_P), direct heritability (h^2), combined maternal additive and environmental variance (mpe^2), and genetic group variance (gg^2)

Trait[a]	V_P	h^2	mpe^2	gg2
DysA	0.009	0.010 ± 0.006	0.007 ± 0.009	0.001 ± 0.002
DysB	0.022	0.008 ± 0.006	0.006 ± 0.007	0.007 ± 0.007
DysC	0.020	0.008 ± 0.006	0.022 ± 0.009	0.020 ± 0.011
Dys	0.049	0.014 ± 0.006	0.014 ± 0.008	0.038 ± 0.016
SM	0.026	0.006 ± 0.005	0.006 ± 0.009	0.001 ± 0.003
SME	0.031	0.014 ± 0.006	0.014 ± 0.011	0.000 ± 0.002
DIUPB	0.010	0.027 ± 0.007		0.008 ± 0.006

[a]DysA (Dystocia A: oedema present); DysB (Dystocia B: no oedema, significant cranial and central nervous system haemorrhage but no metabolised fat); DysC (Dystocia C: as per DysB but with metabolised fat), SM (starvation mismothering complex - no significant cranial and CNS haemorrhage); Dys (composite traits of Dystocia A, B or C); SME (Starvation mismothering exposure); and DIUPB (deaths *in utero* pre-birth and in premature lambs).

The cause of death traits DysA, DysB and Dys were positively genetically correlated with LE (Table 2). These estimates were slightly higher than those reported by Brown *et al.* (2014) and indicates that selection for improved LE may reduce cases of dystocia. SM and SME were not genetically correlated with LE. RT was negatively genetically related with DysA, DysB, Dys, SM, SME and DIUBP (range -0.18 to -0.54), indicating that selection for higher RT may genetically reduce losses generally from all these causes of death. BWT was positively genetically correlated with DysA (0.46) and Dys (0.23), but negatively genetically correlated with SM and SME (-0.32 and -0.28 respectively). This is consistent with the curvilinear relationship observed between BWT and lamb survival where lambs with lower BWT are more likely to die from SME while those with high BWT are more likely to die from dystocia.

Table 2. Estimates of genetic correlations ± s.e. between lamb survival indicator traits and cause of death traits

Trait[a]	DysA	DysB	DysC	Dys	SM	SME	DIUPB
LE	0.50 ±0.15	0.23 ±0.19	0.01 ±0.19	0.30 ±0.13	0.03 ±0.22	0.07 ±0.17	0.35 ±0.12
BWT	0.46 ±0.13	0.08 ±0.16	0.06 ±0.15	0.23 ±0.11	-0.32 ± 0.17	0.28 ±0.12	-0.06 ±0.10
RT	-0.35 ±0.17	-0.34 ±0.20	n.e.[b]	-0.18 ±0.16	-0.54 ±0.25	-0.42 ±0.18	-0.54 ±0.11
THO	0.23 ±0.21	0.58 ±0.22	0.37 ±0.22	0.47 ±0.16	0.07 ±0.01	0.01 ±0.18	-0.19 ±0.16
CRL	0.31 ±0.13	0.07 ±0.16	0.03 ±0.15	0.14 ±0.11	-0.12 ±0.18	-0.07 ±0.13	-0.01 ±0.10
ML	0.37 ±0.19	-0.16 ±0.20	0.16 ±0.20	0.09 ±0.15	-0.24 ±0.23	-0.13 ±0.17	-0.01 ±0.15
OBV	-0.11 ±0.15	0.29 ±0.17	0.13 ±0.16	0.15 ±0.12	0.53 ±0.22	-0.07 ±0.13	0.17 ±0.11
BCS	-0.01 ±0.12	-0.27 ±0.15	0.01 ±0.13	-0.12 ±0.10	-0.37 ±0.18	-0.29 ±0.12	0.14 ±0.09
BLT	-0.06 ±0.23	0.09 ±0.26	0.06 ±0.26	0.06 ±0.20	-0.20 ±0.30	-0.12 ±0.22	n.e.
STD	0.06 ±0.19	0.22 ±0.24	0.08 ±0.24	0.06 ±0.19	0.39 ±0.29	0.24 ±0.21	n.e.
CONT	-0.03 ±0.21	0.14 ±0.22	0.14 ±0.22	-0.01 ±0.17	0.21 ±0.25	0.11 ±0.19	n.e.
UDD	-0.25 ±0.36	-0.05 ±0.36	0.47 ±0.37	0.07 ±0.17	-0.27 ± 0.41	-0.20 ±0.30	n.e.
FOLL	-0.25 ±0.25	0.14 ±0.26	0.12 ±0.27	0.03 ±0.20	0.75 ±0.33	0.56 ±0.22	n.e.

[a]LE: lambing ease score; BWT: birth weight (kg); RT: rectal temperature (^{0}C); THO: thorax circumference (cm), CRL: crown-rump length (cm); ML: metacarpal bone length (cm); OBV: assessed lamb vigour, BCS: birth coat score; timed behaviour traits were recorded from time taken after release for the lamb to bleat (BLT), stand (STD), contact the ewe (CONT), contact the udder (UDD) and follow the ewe (FOLL). The cause of death traits are defined in the footnote to Table 1 and in the Materials and Methods
[b]n.e. not estimable

Moderate correlations between THO and DysB and Dys (and 0.37 with DysC) suggest that lambs with a large thorax circumference are more likely to die from dystocia causes, particularly from birth injury. However, THO appears unrelated to deaths caused by starvation, mismothering and exposure. Positive correlations between CRL and DysA, and between ML and DysA indicate that lambs with longer crown-rumps and metacarpals are more prone to die of classic dystocia. A moderate correlation (0.53) between OBV and SM indicates that more vigorous lambs are less likely to die from SM although this trend was not observed in the composite trait SME. The correlations between OBV and the dystocia traits were generally low indicating selection for OBV will not reduce deaths caused by birthing difficulties. High positive correlations between FOLL and SM and SME and moderate positive correlations between STD and SM and SME indicate that lambs that took longer to follow and stand are more likely to die from SM and SME. The correlations between the other timed behaviours and SM/SME, and dystocia traits were generally low.

CONCLUSION

Heritabilities for all cause of death traits were low which suggests little advantage in using these over selecting directly for lamb survival. Despite this, understanding the genetic relationships between causes of death and other survival indicator traits may allow more accurate selection decisions for improving lamb survival. Firstly, using LE in selection to improve lamb survival is supported by our results. Secondly, selection for OBV could reduce death from SM and thirdly, selection for reduced THO could reduce the incidence of all categories of dystocia.

ACKNOWLEDGEMENTS

The IN and associated research programs were supported by the Australian Government's CRC Program, MLA and AWI Ltd. Resources were provided by 5 partner organisations, NSW Department of Primary Industries and Regional Development (NSWDPIRD), the University of New England, the Victorian Department of Primary Industries, the South Australian Research and Development Institute, and the WA Department of Primary Industries and Regional Development. Supporting science and technical staff of these partner organisations are sincerely thanked. We acknowledge the key role of Gordon Refshauge of NSWDPIRD in coordinating the necropsy work and for data cleaning. We also acknowledge Daniel Brown of AGBU for providing data from the IN program. Finally, this study is part of the MDC project 'Fit to Lamb' P.PSH.1399, led by Murdoch University, administered by MLA, with funding provided by the Australian government, Murdoch University, the University of Adelaide, AWI Ltd, Animal Health Australia and CSIRO.

REFERENCES

Brien F.D., Hebart M.L., Smith D.H., Edwards J.E. Hocking, Greeff J.C., Hart K. W., Refshauge G., Bird-Gardiner T.L., Gaunt G., Behrendt R., Robertson M.W., Hinch G.N., Geenty K.G. and van der Werf J.H.J. (2010) *Anim. Prod. Sci.* **50**: 1017.

Brown D.J., Jones R.M. and Hinch G.N. (2014) *Anim. Prod. Sci.* **54**: 736.

Butler D.G., Cullis B.R., Gilmour A.R. Gogel, B.G. and Thompson, R. (2023) 'ASReml- R Reference Manual Version 4.2.' VSN International Ltd., UK.

Geenty G.N., Brien F.D., Hinch, G.N., Dobos R.C., Refshauge G., McCaskill M., Ball A.J., Behrendt R., Gore K.P., Savage D.B., Harden S., Hocking-Edwards J.E., Hart K. and van der Werf J.H.J. (2014) *Anim. Prod. Sci.* **54**: 715.

Hinch G.N. and Brien F. (2014) *Anim. Prod. Sci.* **54**: 656.

McEwin R.A. Hebart M.L. and Brien F.D. (2025). *Proc. Assoc. Advmt. Anim. Breed. Genet.* **26**: *These proceedings.*

McFarlane D. (1965) *New Zealand Vet. J.* **13**: 116.

Refshauge G., Brien F.D., Hinch G.N. and van de Ven R. (2016). *Anim. Prod. Sci.* **56**: 726.

Safari E., Fogarty N.M. and Gilmour A.R. (2005) *Livest. Prod. Sci.* **92**: 271.

Swan A.A., Brown D.J. and van der Werf J.H.J. (2016) *Anim. Prod. Sci.* **56**: 87.

Young J.M., Thompson A.N., Curnow M. and Oldham C.M. (2011) *Anim. Prod. Sci.* **51**: 821.

Proc. Assoc. Advmt. Anim. Breed. Genet. 26: 260-263

A REVISITING OF THE GENETICS OF LAMB SURVIVAL AND RELATED TRAITS, AS PART OF THE FIT TO LAMB PROJECT

R.A. McEwin, M.L. Hebart and F.D. Brien

Davies Livestock Research Centre, School of Animal and Veterinary Sciences, University of Adelaide, Roseworthy, SA, 5371 Australia

SUMMARY

Measurable genetic gains in lamb survival have been reported in research and industry flocks, despite low heritability estimates. The study reported herein utilised Sheep CRC Information Nucleus Flock data (2007 to 2011) and MLA Resource Flock data (2012-2023) to complete an updated univariate analysis for lamb survival and its indicator traits. Direct genetic, maternal genetic, maternal permanent environmental, common litter and genetic groups variances were estimated for 18 traits using an animal model. Direct heritability estimates were low for all lamb survival traits (0.01) and generally low for indicator traits except for birth weight (0.12), metacarpal length (0.29), crown-rump length (0.20), thorax circumference (0.12) and birth coat score (0.39). Maternal heritabilities were low and equal to or greater than direct heritabilities for lamb survival traits and were generally low for indicator traits. The proportion of phenotypic variance accounted for by permanent environmental effects or genetic groups effects was often larger than direct genetic effects for lamb survival traits and varied across indicator traits. Common litter effects accounted for the largest proportion of variation across all traits, ranging from 11% to 76%.

In summary, lamb survival traits are lowly heritable with equal proportions of variance explained by direct and maternal effects. Some indicator traits appear more heritable and may be useful in selection programs as potentially correlated traits.

INTRODUCTION

Poor lamb survival is a major contributor to sheep reproductive inefficiency in Australia (Hinch and Brien 2014) and internationally, with only 70-85% of lambs born surviving to weaning (Dwyer *et al.* 2016). The prospects for achieving genetic improvement of lamb survival within a breed have generally been viewed as limited, based on very low estimated heritability (Everett-Hincks *et al.* 2005; Safari *et al.* 2005a; Safari *et al.* 2005b; Everett-Hincks and Cullen 2009). Despite this, measurable genetic gains in lamb survival have been reported in research flocks (Cloete *et al.* 2009) and in New Zealand (SIL 2024) and Irish industry flocks (N. McHugh, *pers. comm.*).

For genetic evaluation purposes and for designing effective breeding strategies, it is essential to have reliable estimates for genetic parameters. In an earlier paper, Brien *et al.* (2010) provided phenotypic and genetic parameters for lamb survival and related traits, based on data collected from 2007 to 2009 in the Sheep CRC's Information Nucleus (IN) flock. In this paper, we update and add to the earlier estimates, in relation to results from univariate analyses only, using IN data collected from 2007 to 2011 and data from the MLA Resource Flock (MLA RF) collected from 2012 to 2023 including data from up to an additional 12,352 lamb records.

MATERIALS AND METHODS

The Sheep CRC's Information Nucleus (IN) flock consisted of genetically linked flocks run at eight sites in Australia, containing Merino and crossbred ewes that were artificially inseminated annually to Merino, maternal and terminal breed rams. The MLA RF began as the IN, but has been run at two research sites only, with management following a similar protocol to the IN, except that pedigrees are verified by DNA analysis.

IN and MLA RF data on 27,603 lambs was utilised to estimate genetic parameters for lamb survival at birth, 3 days of age, 7 days of age and weaning (LS0, LS3,LS7 and LSW respectively), as well as indicator traits; birth weight (BWT), metacarpal length (ML), crown-rump length (CRL), thorax circumference (THO), rectal temperature (RT), observed birth vigour (OBV), birth coat score (BCS), maternal behaviour score (MBS; currently defined as a trait of the lamb however will be analysed as a trait of the dam at a later date) and lambing ease score (LE). Five timed behaviour traits were also analysed; time taken after release for the lamb to bleat (BLT), stand (STD), contact the ewe (CONT), contact the udder (UDD) and follow the ewe (FOLL). Further details on the design, data collection and management of the IN is reported by Brien *et al.* (2010) and Geenty *et al.* (2014). The design and management of the MLA RF is briefly reported by Hebart and Brien (2018).

Table 1. Summary statistics for lamb survival and indicator traits including the number of additional records included (+N) since Brien *et al.* (2010), the number of sires, dams and number of contemporary groups (N_CG). Trait units are presented within parentheses

Trait	N	+N	Mean	Median	s.d.	Sires.	Dams	N_CG.
LS0 (0/1)	25,638	-	0.93	1	0.26	739	11,567	41
LS3 (0/1)	26,313	11,121	0.88	1	0.32	739	11,653	41
LS7 (0/1)	26,677	-	0.86	1	0.34	739	11,685	41
LSW (0/1)	26,709	11,517	0.85	1	0.35	739	11,694	41
BWT (kg)	27,240	12,352	4.8	4.7	1.1	739	11,721	41
ML (cm)	10,213	-243	10.0	10	1.2	269	5,101	18
CRL (cm)	19,442	8,991	45.8	46	5.0	402	7,994	32
THO (cm)	10,210	-244	39.0	39	3.9	269	5,102	18
RT (°C)	18,695	8,881	39.1	39.2	0.9	438	7,927	33
OBV (1-5)	22,997	9,472	2.3	2	1.1	715	9,777	40
BCS (1-7)	23,251	8,699	2.5	2	1.0	478	9,116	39
MBS (1-5)	13,114	827	2.5	2	1.1	281	5,987	23
LE (1-5)	17,468	7,442	1.1	1	0.5	538	7,272	38
BLT (sec)	16,382	7,719	8.5	3	16.6	402	7,394	32
STD (sec)	12,550	3,752	16.1	6	26.2	289	6,091	24
CONT (sec)	12,253	3,682	29.0	17	33.0	289	5,938	24
UDD (sec)	3,613	-109	70.4	60	46.4	204	2,439	16
FOLL (sec)	6,645	-199	57.2	44	41.5	205	3,754	16

Statistical analysis. Variance components were estimated for all traits using the restricted maximum likelihood (REML) method within an animal model in ASREML-R 4.2 (Butler *et al.* 2023). A deep pedigree was utilised; 171,363 individuals comprising 18,326 and 66,212 unique sires and dams respectively. All traits were analysed on the observed scale. All models included fixed effects of contemporary group defined by cohort (year by site; 41 levels), sex (male or female), sire breed type (3 levels), dam breed type (4 levels), age of dam (1-8 years), type of birth (single, twin or multiple) and the interaction between cohort and age (days) as a co-variate. Data was filtered to retain complete records across all fixed effects. Birth time (0, 1 or 2 indicating the trait was observed at birth, within 4hrs of birth or >4hrs of birth respectively) was fitted as an additional covariate for RT, OBV, BCS, MBS, LE and timed behaviour traits. Direct additive, maternal genetic, maternal permanent environmental, litter and genetic group variances were estimated. Litter was defined as the combination of year of birth x dam id. Genetic group variance was fitted via a Q matrix constructed for 147 genetic groups (Swan *et al.* 2014).

RESULTS AND DISCUSSION

Mean lamb survival was highest at day 0 (93%, Table 1) and decreased with age to 85% by weaning. The biggest decrease in mean lamb survival was between 0 and 3 days (5% decrease) indicating that most lamb losses occur up to 3 days of age, with marginal decreases in survival thereafter till weaning.

Table 2. Estimates of additive genetic (V_a), maternal genetic (V_m), maternal permanent environmental (V_{mpe}), litter (V_c), genetic group (V_{gg}) and phenotypic (V_p) variance components as well as direct (h^2) and maternal (m^2) heritability estimates and common litter effects(c^2)

Trait**	V_a	V_m	V_{mpe}	V_c	V_{gg}	V_e	V_p	h^2	m^2	c^2
LS0	0.001	0.001	0.001	0.015	0.002	0.042	0.062	0.01	0.01	0.25
LS3	0.001	0.002	0.004	0.017	0.001	0.068	0.094	0.01	0.02	0.19
LS7	0.001	0.003	0.004	0.016	0.002	0.079	0.105	0.01	0.03	0.16
LSW	0.001	0.003	0.005	0.016	0.002	0.087	0.112	0.01	0.02	0.14
BWT	0.100	0.138	0.025	0.092	0.147	0.322	0.825	0.12	0.17	0.11
ML	0.269	0.004	0.000*	0.381	0.000*	0.289	0.942	0.29	0.00	0.40
CRL	3.639	1.367	0.000*	4.296	1.627	6.935	17.864	0.20	0.08	0.24
THO	1.206	0.905	0.131	1.955	0.730	5.359	10.285	0.12	0.09	0.19
RT	0.042	0.000*	0.017	0.343	0.006	0.799	1.208	0.03	0.00	0.28
OBV	0.082	0.000*	0.011	0.634	0.022	0.272	1.021	0.08	0.00	0.62
BCS	0.354	0.025	0.000*	0.271	0.040	0.209	0.898	0.39	0.03	0.30
MBS	0.044	0.217	0.000*	0.770	0.256	0.020	1.307	0.03	0.17	0.59
LE	0.006	0.002	0.000*	0.110	0.004	0.063	0.185	0.03	0.01	0.59
BLT	5.277	4.706	0.206	103.38	7.616	158.96	280.15	0.02	0.02	0.37
STD	17.810	9.044	5.201	271.01	6.534	353.74	663.34	0.03	0.01	0.41
CONT	20.126	201.44	14.203	544.47	188.30	226.29	1194.8	0.02	0.17	0.46
UDD	102.58	0.000*	168.35	732.63	0.000*	935.56	1939.1	0.05	0.00	0.38
FOLL	49.956	27.074	89.238	1228.8	34.761	197.67	1627.5	0.03	0.02	0.76

*indicates variance components that hit the boundary due to negligible variation. **Trait units can be found in Table 1. ***Standard errors range 0.01-0.05 for h^2, m^2 and c^2 estimates.

Direct (V_a) and maternal genetic (V_m), maternal permanent environmental (V_{mpe}), litter (V_c) and genetic groups (V_{gg}) variances were able to be estimated separately for each trait. Direct heritability (h^2) estimates were low for lamb survival traits (0.01; Table 2) in line with reviewed estimates by Safari *et al.* (2005b). Indicator traits also generally had low h^2, except for BWT (0.12), ML (0.29), CRL (0.20), THO (0.12) and BCS (0.39). Estimates of h^2 were all equal to or smaller than those reported in Brien *et al.* (2010), except for BCS, though total phenotypic variances (V_p) reported were very similar (e.g. for Ls3 and LsW). Given the greater number of variance components estimated herein compared to Brien *et al.* 2010 (especially the separation of maternal components as well as V_c), it is reasonable to assume that V_a may have been over-estimated in the prior study. Estimates of maternal heritabilities (m^2) were low and equal to or greater than h^2 for lamb survival traits (Table 2), similar to estimates by Safari *et al.* (2005b), though not all of the papers reporting m^2 analysed in this review disentangled maternal genetic and permanent environmental variances, including Brien *et al.* 2010. Estimates of m^2 were highest for BWT (0.17), MBS (0.17) and CONT (0.17). Although small, V_{mpe} estimates accounted for 2.3%, 3.8%, 3.9% and 4.1% of V_p for LS0, LS3, LS7 and LSW, respectively. Estimates of V_{gg} were generally small (accounting for up to 9% of V_p the variance) except for BWT, MBS and CONT (accounting for 17.8%, 19.6% and 15.8%, respectively).

Common litter effects (c^2) accounted for the largest amount of variation across each trait. For lamb survival traits, c^2 appeared to decrease with age with 0.25 reported for LS0 compared to 0.14 for LSW (Table 2), consistent with Everett-Hincks *et al.* (2014) who reported c^2 values of 0.13 and

0.09 for LS3 and LSW respectively. Structural traits (ML, CRL and THO) reported c^2 of 0.40, 0.24 and 0.19 respectively while scored traits (VIG, MBS, LE) averaged c^2 of approximately 0.60. Timed traits generally returned c^2 values of approximately 0.40 except for FOLL which had the highest c^2 of the dataset at 0.76. These c^2 values remained equally high or higher for all traits when BWT or BWT quadratic (for lamb survival) co-variates, fit within type of birth, were added to the model (data not shown) indicating that the larger values observed are not capturing an effect of BWT.

CONCLUSION

Estimates of direct genetic and maternal heritabilities were updated for 18 traits including lamb survival and related characters. Direct and maternal heritabilities for lamb survival traits are confirmed as low when derived from a linear mixed model analysis. Common litter effects accounted for the greatest amount of variation across all traits (11% to up to 76%). Several indicator traits have heritability estimates well above those for lamb survival traits and may have potential value for use in selection programs to increase genetic gain, depending on the magnitude of their genetic correlation with lamb survival (yet to be reported from data used in this study) and their ease of recording.

ACKNOWLEDGEMENTS

We are grateful to Daniel Brown of AGBU for assistance with providing data from two sources, firstly the Information Nucleus program of the Cooperative Research Centre for Sheep Industry Innovation and secondly, the MLA Resource Flock. We acknowledge the contributions of the many research and technical staff from research agencies involved and support provided by Australian sheep breeders. Finally, this study is part of the MDC project 'Fit to Lamb' P.PSH.1399, led by Murdoch University, administered by Meat & Livestock Australia, with funding provided by the Australian government, Murdoch University, the University of Adelaide, Australian Wool Innovation Ltd, Animal Health Australia and CSIRO.

REFERENCES

Brien F.D., Hebart M.L., Smith D.H., Hocking Edwards J.E., Greeff J.C., Hart K.W., Refshauge G., Bird-Gardiner T.L., Gaunt G., Behrendt R., Robertson M.W., Hinch G.N., Geenty K.G. and van der Werf J.H.J. (2010) *Anim. Prod. Sci.* **50**: 1017.

Butler D.G., Cullis B.R., Gilmour A.R., Gogel, B.G. and Thompson R. (2023) 'ASReml- R Reference Manual Version 4.2.' VSN International Ltd., UK.

Cloete S.W.P., Misztal I. and Olivier J.J. (2009) *J. Anim. Sci.* **87**: 2196.

Dwyer C.M., Conington J., Corbiere F., Holmoy I.H., Muri K., Nowak R., Rooke J., Vipond J. and Gautier J.M. (2016) *Animal* **10**: 449.

Everett-Hincks J.M. and Cullen N.G. (2009) *J. Anim. Sci.* **87**: 2753.

Everett-Hincks J.M., Lopez-Villalobos N., Blair H.T. and Stafford K.J. (2005) *Livest. Prod. Sci.* **93**: 51.

Everett-Hincks J.M., Mathias-Davis H.C., Greer G.J., Auvray B.A. and Dodds K.G. (2014) *J. Anim. Sci.* **92**: 2885.

Hebart M.L. and Brien F.D. (2018) *Proc. World Cong. Genet. Appl. Livest. Prod.* **11**: 703.

Hinch G.N. and Brien F.D. (2014) *Anim. Prod. Sci.* **54**: 656.

Safari E., Atkins K.D., Fogarty N.M. and Gilmour A.R. (2005a) *Proc. Assoc. Advmt. Anim. Breed. Genet.* **16**: 28.

Safari E., Fogarty N.M. and Gilmour A.R. (2005b) *Livest. Prod. Sci.* **92**: 271.

SIL (2024) Genetic Trends. Retrieved on 7 February 2024, from https://www.sil.co.nz/files/1702244282_NZGE122023_InfoGTGraphs_DP.pdf

Swan A.A., Brown D.J. and van der Werf J.H.J. (2016) *Anim. Prod. Sci.* **56**: 87.

Proc. Assoc. Advmt. Anim. Breed. Genet. 26: 264-267

AN EX-POST COST-BENEFIT ANALYSIS OF THE OVITA LAMB SURVIVAL PROJECT

K.M. McRae[1], J.M. Everett-Hincks[1], S.-A. Newman[1], W.E. Bain[1], G. Greer[1], H. Baird[1], H. Manley[1], G. Shackell[1], S.M. Clarke[1], K.G. Dodds[1], N. Amyes[2], N. Jopson[3], M. Lee[4], P.L. Johnson[1], S.J. Rowe[1] and J.C. McEwan[1]

[1] AgResearch Ltd., Invermay Agricultural Centre, Mosgiel 9092, New Zealand
[2] AgResearch Ltd., Ruakura Agricultural Centre, Hamilton 3214, New Zealand
[3] AbacusBio Ltd., Dunedin 9016, New Zealand
[4] B+LNZ Genetics, Dunedin 9016, New Zealand

SUMMARY

Improving lamb survival has both economic and animal welfare benefits for sheep farmers. However, genetic improvement is often considered of little practical benefit because of the low trait heritability and recording limitations, especially under extensive management. The Ovita Lamb Survival research project improved industry recording and genetic evaluation of lamb survival from 2003. To quantify research impacts, an ex-post cost-benefit analysis was undertaken using genetic trends of direct and maternal lamb survival breeding values from 1995 until 2023. In 2023, the average direct breeding value for lamb survival in dual-purpose ram breeding flocks was 1.98% higher than in 1995, and the maternal breeding value was 1.10% higher than in 1995. Conservative gene flow calculations indicate that genetic improvements are likely responsible for an additional 337,000 commercial lambs weaned in 2023, worth $54 million per year to the New Zealand economy. Accounting for inflation, genetic lags and the timing of the research investment (2003-2012), and a discount rate of 8%, the net present value was estimated at $164 million. These results provide strong evidence of the returns from the Ovita investment.

INTRODUCTION

The number of lambs born is a critical component of productivity and profitability on New Zealand sheep farms. Despite advances in management practices, lamb survival remains a significant issue for farmers (Greer *et al.* 2015). Improving lamb survival not only impacts economic returns but also addresses animal welfare concerns.

Lamb survival is a complex trait, influenced by direct and maternal genetic and environmental effects. Therefore, genetic improvement is often considered of little practical benefit because of the low trait heritability and recording limitations, especially under extensive management (Morris *et al.* 2000). Ovita was a joint research venture between AgResearch and Beef + Lamb New Zealand (2003-2013). One workstream commencing in 2003 was to examine ways to improve lamb survival through the identification of genetic markers. This project resulted in the collection of 103,000 detailed lamb survival and mortality records from 38 New Zealand performance-recorded flocks (Kerslake *et al.* 2005; Everett-Hincks and Duncan 2008; Everett-Hincks *et al.* 2014) and resulted in improvements in the recording and genetic evaluation for this trait, progressively implemented from 2012. Along with this research, there were several technological developments, including pregnancy scanning, foetal aging and full and partial DNA parentage. Genomic selection was implemented in 2017, and single-step genomic evaluation in 2019, all increasing the accuracy of the survival breeding values.

When lamb survival and mortality records were used to estimate genetic parameters, the total (direct + maternal) heritability estimates were low (0.007±0.002 to 0.058±0.003) for survival traits and moderate (0.38±0.01) for lamb birth weight (Everett-Hincks *et al.* 2014). For the traits studied, the heritability attributed to maternal effects (expressed when the ewe lamb produces progeny) accounted for at least half to two-thirds of the total genetic variation, indicating that genetic progress could be made by selecting ewe lambs based on their estimated breeding values for lamb survival. Additionally, it was established that accurate recording of survival to weaning was more important than component lamb mortality traits in an effective breeding program. As a result, a revised lamb survival trait was developed and implemented into Sheep Improvement Limited (SIL), the national performance recording and genetic evaluation system (Vanderick *et al.* 2015). Lamb survival (SUR) is currently included in both maternal (NZMW) and terminal (NZTW) indexes, with survival maternal (SURM) additionally included in NZMW (Santos *et al.* 2015).

MATERIALS AND METHODS

Genetic trends in direct (SURgBV) and maternal (SURMgBV) lamb survival from 1995 until 2023 were obtained from nProve (formerly SIL) for NZ dual-purpose maternal flocks (NZGE v6; GE analysis 42120). Numbers of breeding ewes, lambs tailed (docked) from ewes, and all processing and export data were obtained from the Beef + Lamb New Zealand Economic Service.

A simplified gene flow model was used, assuming that rams were first used as rising two-year-olds and kept for three years, the realised value of SURgBV in a commercial flock from the ram team was the average breeding value (BV) of stud rams born 3-5 years prior, halved (as rams only account for half of the genetic improvement in progeny). This then flows onto daughters, with the realised value of SURgBV in commercial ewes, the average BV of the sire team used 5-7 years prior, halved, assuming a replacement rate of approximately 33% (Farrell 2020). The direct BV of the commercial flock in any given year was calculated as the sum of the ram and ewe BVs. The assumption is that the majority of recognised profit from lambs born is in the subsequent calendar year. The model for SURMgBV in commercial ewes follows the same pattern, with the realised value being half the average BV of stud rams born 5-7 years prior. The total increase in lamb survival from BVs was calculated by summing the direct BV and the maternal BV within a given year.

To calculate the extra number of lambs weaned, the following equation was used:

$$Extra\ lambs\ weaned = ewes\ mated\ \times \frac{lambs\ born}{ewes\ mated} \times commercial\ BV$$

Lambs born to ewes mated were calculated using industry data (B+LNZ Economic Service). Estimates of average survival vary in the literature; however, in this model, an average survival of 0.88 from birth to weaning was used (Everett-Hincks *et al.* 2014) to back-calculate the number of lambs born. Survival from tailing to weaning was assumed to be 0.98 (Byrne *et al.* 2012), and industry tailing numbers were adjusted accordingly.

The number of carcasses processed per year was estimated from total lamb meat expressed on a carcass weight equivalent, divided by the average export lamb weight. The value of lamb meat and co-products (including skins) per carcass was calculated by dividing the Free on Board (FOB) total value by the estimated number of carcasses (B+LNZ Economic Service) and adjusted for inflation (Reserve Bank of New Zealand Inflation Calculator). This was used to calculate the value of the additional lambs weaned per year, in which the lambs were processed using 2023 as the base year.

Ovita investment was obtained from internal AgResearch records and adjusted for inflation as above ($4.2 million in 2023 dollars). The net present value (NPV, using a discount rate of 8%) was estimated for the period 2003-2023 (the period from when Ovita investment started).

RESULTS AND DISCUSSION

In the 2023 birth year, the average SURgBV in Dual-purpose ram breeding flocks was 1.98% higher than in 1995, and SURMgBV was 1.10% higher than in 1995 (Figure 1). Values did not change markedly between 1995 and 2003, when the Ovita program began. In Terminal flocks, where there is no selection for SURMgBV, the average value was 0.06% compared to 1995. The average

SURgBV for terminal flocks, which is included in the terminal worth selection index, has increased by 0.95%.

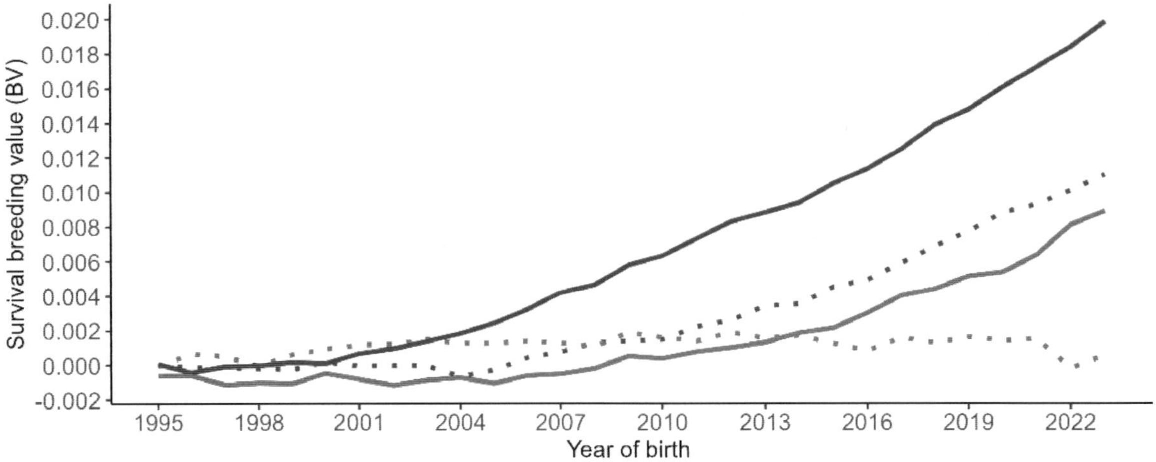

Figure 1. Genetic trends in SURgBV (solid line) and SURMgBV (dashed line) in B+LNZ Genetics recorded dual-purpose (blue) and terminal (red) flocks by year of birth

In 2023, New Zealand grazed 14.8 million ewes, with 127% lambs tailed to ewes mated. A conservative gene flow model suggested that an extra 337,000 lambs were weaned in 2023, because of genetic improvements in lamb survival since 1995 (Figure 2). Including the value of lamb meat and co-products (including skins), this is worth $53 million per year to the New Zealand economy. Accounting for genetic lags, changes in NZ ewe numbers, and the timing of the research investment, the Net Present Value of the investment is $164 million using a discount rate of 8%.

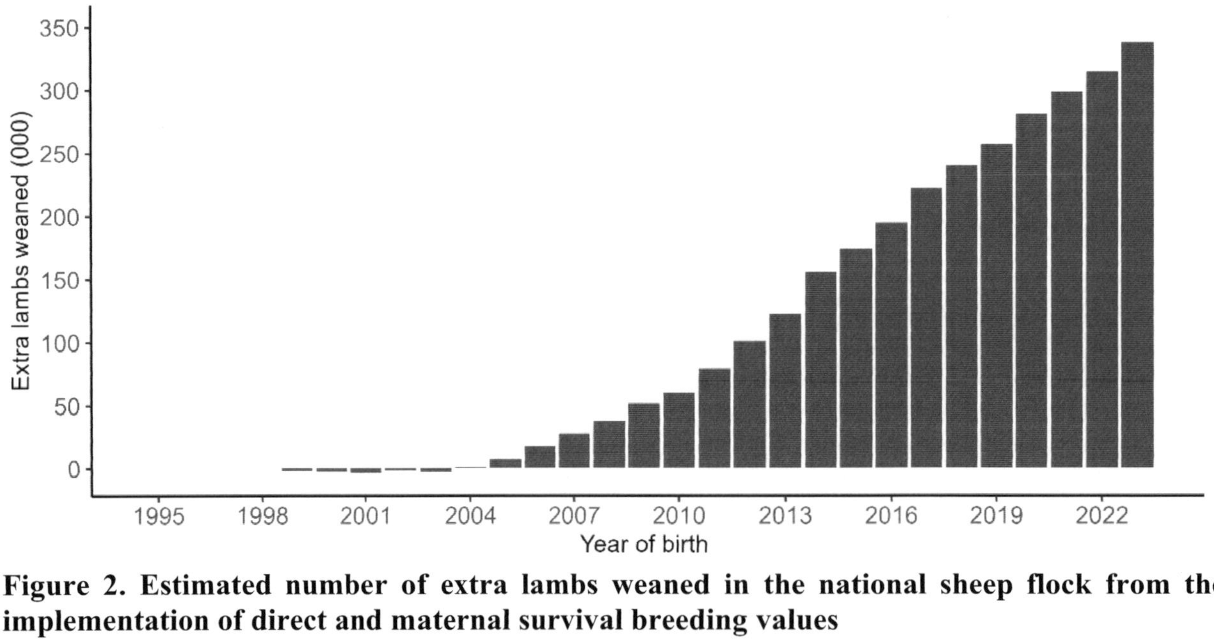

Figure 2. Estimated number of extra lambs weaned in the national sheep flock from the implementation of direct and maternal survival breeding values

These results have several observations and caveats. The first is that the results presented here do not include the difference in average direct and maternal survival BVs of all stud ram lambs born, and those of the rams sold to commercial farmers. That data is not available and will have varied over time as recording, selection indices, and breeding values improved. However, this benefit is likely to be substantial. On the converse, B+LNZ recorded flocks probably only account for some 90% of rams sold in New Zealand. A second interesting observation is that when expressed on a per carcass basis, the inflation-adjusted value of lamb meat and co-products (including skins) has increased on average by $1.50/year over the 1995-2004 period. This reflects heavier carcass weights, increased carcass processing within New Zealand and a shift to exporting chilled rather than frozen meat. This partially compensates for the lower numbers of ewes wintered.

Another caveat is that when comparing genetic trends with industry production data, such as lambs tailed to ewes mated, is that these data are not adjusted for changes in the number of animals in the various land classes over time. This is required to more accurately calculate the proportion of increased farm productivity that is attributable to genetics. Secondly, accurate pregnancy scanning data is not recorded, even though this is collected by many commercial farmers, meaning that industry figures on lamb survival are not available.

Ascribing changes in genetic gain over time to specific changes in the genetic evaluation is also difficult. Direct survival BLUP breeding values were first available in 2001, and maternal breeding values in 2005, with genomic breeding values in the 2017-2019 period. However, their inclusion, parameters used, and index weighting have varied over time. Perhaps more importantly, the survival trait was poorly recorded until 2012, when existing data were subjected to a forensic examination and strict data filters were implemented. This was also accompanied by increasing use of DNA parentage and pregnancy scanning data. Genomic selection is expected to best assist in sex-limited traits, expressed late in life, and notably, maternal survival breeding values are now increasing more rapidly.

In undertaking this evaluation, several future opportunities were identified to improve lamb survival more rapidly; perhaps the most important is that whole ram breeding flock genotyping, coupled with accurate recording of the trait, will greatly increase the rate of genetic progress in maternal lamb survival. This is shown in increases in individual pedigree-recorded flocks, which have been two to four-fold greater than the industry average.

CONCLUSION

In conclusion, this study demonstrates that an inflation-adjusted $4.2M investment in a low heritability trait has yielded substantial returns driven by improved trait recording and increased utilisation of genomics. Further, it highlights the potential for even greater benefits and supports expanding this approach to a wider variety of traits.

REFERENCES

Byrne T.J., Ludemann C.I., Amer P.R. and Young M.J. (2012) *Livest. Sci.* **144:** *20.*

Farrell L.J., Kenyon P.R., Tozer P.R., Ramilan T. and Cranston L.M. (2020) *Agric. Syst.* **184:** 102888.

Greer A.W., Corner-Thomas R.A., Logan C.M., Kenyon P.R., Morris S.T., Ridler A.L., Hickson R.E. and Blair H.T. (2015) *New Zeal. J. Agric. Res.* **58:** 359.

Morris C.A., Hickey S.M. and Clarke J.N. (2000) *New Zeal. J. Agric. Res.* **43:** 515.

Kerslake J.I., Everett-Hincks J.M. and Campbell A.W. (2005) *Proc. New Zeal. Soc. Anim. Prod.* **65:** 13.

Everett-Hincks J.M. and Duncan S.J. (2008) *Open Vet. Sci. J.* **2**: 55.

Everett-Hincks J.M., Mathias-Davis H.C., Greer G.J., Auvray B.A. and Dodds K.G. (2014) *J. Anim. Sci.* **92:** 2885.

Santos B.F.S., McHugh N., Byrne T.J., Berry D.P. and Amer P.R. (2015) *J. Anim. Breed. Genet.* **132:** 144.

Vanderick S., Auvray B., Newman S.-A., Dodds K.G., Gengler N. and Everett-Hincks J.M. (2015) *J. Anim. Sci.* **93:** 3765.

Proc. Assoc. Advmt. Anim. Breed. Genet. 26: 268-271

GENETIC PARAMETERS OF GESTATION LENGTH WITH EARLY IN LIFE GROWTH TRAITS IN AN AUSTRALIAN WAGYU POPULATION

M. Facy[1,2], A. Lloyd[1,3] and J. Grose[1]

[1] 3D Genetics, Pukawidgi, Bukkulla, NSW, 2360 Australia
[2] Davies Livestock Research Centre, University of Adelaide, Roseworthy, SA, 5371 Australia
[3] Evolved Projects Pty Ltd, Geelong, VIC, 3220 Australia

SUMMARY

Wagyu selection has typically ignored the trait of gestation length with selection focus primarily on carcase quality. As a result, the gestation length in Wagyu is one of the longest of the beef cattle breeds averaging 287.5 days. The objective of this study was to estimate the heritability of gestation length and its correlation with weight traits in an Australian Fullblood Japanese Black (Wagyu) population. Gestation length was found to have a heritability of 0.37, with only weak correlations with early in life weight traits. The maternal heritability of gestation length of 0.04 was much lower than the direct heritability reflecting that this is a trait of the calf. The genetic correlation between the maternal effects was considerably higher than that between additive and maternal effects (0.28 with birth weight and 0.81 with yearling weight). Due to the moderate heritability of gestation length and in general favourable correlations with weight traits, it would be possible to select simultaneously on both gestation length and early in life growth traits.

INTRODUCTION

Wagyu or Japanese Black cattle are well known for their high marbling characteristics, with a large selection emphasis on this trait in Australia. However, due to selection being heavily focused on marbling, there may have been detrimental effects on other traits. One of these traits is gestation length, with Wagyu having one of the longest gestation lengths among beef cattle breeds of 287.5 days (Ibi *et al.* 2008). These longer gestations result in detrimental effects on both the calf and the cow due to increased rates of dystocia and a shorter time for the cow to recover after calving prior to rebreeding, which can affect pregnancy rates (Stachowicz *et al.* 2023). The objective of this study was to estimate genetic parameters for gestation length and investigate the relationship between gestation length and early in life growth liveweight traits (birth weight, weaning weight and yearling weight).

MATERIALS AND METHODS

The phenotypic and genotypic data were provided by 3D Genetics, a Fullblood Wagyu Bull breeding program located in Northern New South Wales running ~2000 breeders. The breeding program started in 1997 and has been growing every year since. The phenotypic data include gestation length (GL), birth weight (BW), weaning weight (WW) and yearling weight (YW). A description and summary of traits are included (Table 1).

Genotyping was conducted over 20 years, initially with microsatellites and then over the last decade utilising SNP genotypes; as a result, multiple SNP chip arrays have been utilised. These arrays include Illumina 777k, Illumina GGPLD V3 30K, Illumina GGPLD V4 30K, Illumina ICB 50K, Illumina GGP 100K and Weatherby's Scientific Versa50K. There were 10,830 SNPs common across all chips with 8,675 animals, which were imputed up to GGPLD30K (21,791 SNPs) using Minimac4 (Howie *et al.* 2012). Genotype data was filtered to exclude duplicate animals and SNPs with less than 1% minor allele frequency. The heterozygosity fraction (Het) was calculated as the proportion of imputed heterozygote SNPs for each individual animal.

Table 1. Description and summary of traits included in the analysis

Trait	Acronym	Description	Mean	Standard deviation	No. of records	No. of Unique Dams
Gestation length	GL	Time interval between conception and parturition for Artificial Insemination	286.9	5.59	2,434	1,438
Birth weight	BW	Weight within 24 hours of birth	28.6	4.45	4,957	2,006
Weaning weight	WW	Weight at age 121-300 days	151.4	39.75	6,026	2,169
Yearling weight	YW	Weight at age 301-500 days	307.6	69.55	4,306	1,663

Statistical Analysis. Genetic parameters were estimated through a series of univariate and bivariate models using ASREML-R (Butler *et al.* 2023). The genomic relationship matrix was calculated using VanRanden's first method (VanRaden 2008) which included 8,675 animals in the GRM for analysis.

Dam joining number, dam class, sex, contemporary group and heterozygosity fraction (Het) were included as fixed effects when appropriate for the trait (Table 2). Dam class was split into heifers for 2-year-old cattle; mature for 3 to 10-year-old cattle and old for cattle older than 10 years old. Birth group was defined as birth year and birth windows which was split into early, mid and late birthing windows (A, B and C). Birth group was used to account for age differences within each birth year. The number of contemporary groups ranged from 32 for birth weight to 187 for yearling weight (Table 2). The univariate and bivariate models had the additive and maternal effects fitted to calculate the heritability and maternal effect for all traits except for YW. When appropriate, genetic correlations were also estimated between the maternal components of the traits.

Table 2. Fixed effects, contemporary-group definition for each trait included in the analysis

Trait	Fixed effects	Contemporary group	Number of contemporary groups
GL	Sex, Dam joining number, Het	Birth year, Gestation length management group	55
BW	Sex, Dam class, Het	Birth group, birth weight management group	32
WW	Sex, Dam class, Het	Birth group, birth and weaning weight management group	86
YW	Dam class, Het	Birth group, sex, weaning and yearling weight management group	187

RESULTS AND DISCUSSION

Gestation length (GL) was moderately heritable (0.37) in this population (Table 3). This heritability estimate was similar to other published estimates, which ranged from 0.27 to 0.53 (Baco *et al.* 1998; Ibi *et al.* 2008; Oyama *et al.* 2004), with Oyama *et al.* (2011) estimating a weighted mean of 0.35. The maternal heritability was low (0.04) with no permanent maternal environmental variance (Table 3). The low maternal heritability was consistent with previous studies (0.04 to 0.14; Ibi *et al.* 2008; Oyama 2011), with Oyama *et al.* (2011) estimating a weighted mean of 0.09. The lower maternal heritability would indicate that the dam has little influence over GL but rather is directly impacted by the additive genetic effects of the calf.

The heritability of the liveweight traits ranged from 0.21 to 0.46, with all having low standard errors (Table 3). WW had the lowest heritability (0.21) but the highest maternal heritability (0.24). The various liveweight trait heritability estimates agreed with published estimates (Ibi *et al.* 2008; Lopez *et al.* 2020; Mehrban *et al.* 2021; Oyama 2011).

Maternal effects were evident for BW and WW with maternal heritabilities of 0.08 and 0.24, respectively (Table 3). WW had the largest maternal heritability of all the traits; in breeding programs that WW maternal effect is referred to as MILK. This would indicate that the ability of the mother has a large effect on WW and should be selected for in a breeding program. It is well recognised within the commercial Wagyu industry that Wagyu dams have poor milking ability which can have a detrimental effect on calf survival (Australian Wagyu Association 2025). YW did not have any maternal genetic variance in this population, so it was removed from the model.

Table 3. Additive, maternal and residual estimates alongside heritability and direct maternal heritability with standard errors in brackets

Trait	Additive variance	Maternal genetic variance	Residual variance	Heritability	Maternal heritability
GL	9.96	0.90	15.86	0.37 (0.03)	0.04 (0.02)
BW	7.50	1.31	7.42	0.46 (0.02)	0.08 (0.01)
WW	154.63	170.99	395.59	0.21 (0.02)	0.24 (0.02)
YW	352.14	-	776.27	0.31 (0.02)	-

The genetic correlation between GL and BW was 0.25 (Table 4), which is a weak, positive correlation. The phenotypic correlation was also weak and positive correlation (0.25, Table 4). The estimates herein are lower than those estimated by Lopez *et al.* (2020) in Hanwoo cattle of 0.51 for genetic correlation, however; the phenotypic correlation is similar (0.21). These correlations between GL and BW would indicate, a small increase in BW would occur as GL increases. BW is treated as an optimal trait in breeding programs, as such that BW is controlled to reduce calving issues; however, not lowering BW that it starts to effect calf survivability. This is evident in the Australian Wagyu Association Breeder Feeder Index placing a non-linear economic value on BW (Australian Wagyu Association 2025). Therefore, the genetic correlation between GL and BW could be favourable or unfavourable depending on the population mean. If the population mean is small for BW, the correlation would be favourable; however, if the population mean is large for BW, the correlation would be unfavourable. Even though the correlation is weak, the impacts GL could have on BW need to be considered in selection.

The genetic correlation between BW maternal and GL was uncorrelated at 0.02 however with a large standard error (Table 4); meaning these two traits have no impact on each other. Despite this low correlation, there was a high positive genetic correlation between BW maternal and GL maternal of 0.81 indicating strong association between the two traits. Despite the correlation being high between those two traits the maternal heritabilities were low (0.04 for GL and 0.08 for BW). As the heritability is low for the maternal component of BW and GL this has little impact in a breeding program however should be still considered in selection decisions.

The genetic correlation between GL and both WW (-0.09) and YW (-0.09) were both weak and negative. (Table 4). These estimates are similar to Lopez *et al.* (2020) of -0.09 for both genetic and phenotypic correlation. These correlations between WW and YW with GL were favourable but weak. This weak association between the traits means GL could be selected independently of weight traits in a breeding program.

Table 4. Genetic and phenotypic correlations for weight traits with gestation length for both additive and maternal effects.

Trait	Genetic correlation Additive	Genetic correlation Maternal	Phenotypic correlation
BW Additive	0.25 (0.06)	0.26 (0.19)	0.25 (0.05)
BW Maternal	0.02 (0.12)	0.81 (0.20)	
WW Additive	-0.09 (0.08)	-0.25 (0.07)	-0.11 (0.04)
WW Maternal	0.12 (0.21)	0.28 (0.21)	
YW Additive	-0.09 (0.07)	0.12 (0.21)	-0.15 (0.03)

The genetic correlation between WW maternal (often referred to as MILK) and GL was 0.12 for additive and 0.28 maternal (Table 4). These maternal genetic correlations are weak but positive, meaning a small increase in GL would occur as WW maternal increases. The Wagyu breed is well known for displaying poor milking ability and longer gestation length. These weak, positive maternal genetic correlations should be considered in a breeding program as they are unfavourable and selection on one trait could cause them to move in the wrong direction. WW maternal and GL are both components in a Wagyu breeding program that breeders are concerned about and even though the correlation is weak still need to be considered in breeding programs.

CONCLUSION

Gestation length can increase birthing difficulty due to increased birth weight and decrease pregnancy rates. The direct heritability and maternal heritability of gestation length (GL) was estimated to be 0.37 and 0.04 respectively. These heritability estimates demonstrate it is possible to make genetic gain by selecting for reduced gestation length. Further, the genetic and phenotypic correlations were in general weak with early in life weight traits which indicates it is possible to select simultaneously for higher growth and shorter gestation length.

REFERENCES

Australian Wagyu Association (2025) 'Breeder-Feeder Index: New AWA Selection Index', https://www.wagyu.org.au/breeder-feeder-index.

Baco S., Harada H. and Fukuhara R. (1998) *Anim. Sci. Tech.* **69**: 7.

Butler D.G., Cullis B.R., Gilmour A.R., Gogel B.G. and Thompson R. (2023) 'ASReml-R Reference Manual Version 4.2'. VSN International Ltd., Hemel Hempstead, HP2 4TP, UK.

Howie B., Fuchsberger C., Stephens M., Marchini J. and Abecasis G.R. (2012) *Nat. Genet.* **44**: 955.

Ibi T., Kahi A.K. and Hirooka H. (2008) *J. Anim. Sci.* **79**: 297.

Lopez B.I., Santiago K.G., Seo K., Jeong T., Park J.-E., Chai H.-H., Park W. and Lim, D. (2020) *Animals* **10**: 1083.

Mehrban H., Naserkheil M., Lee D.H. and Ibáñez-Escriche N. (2021) *Anim. Biosci.* **34**: 824.

Oyama K. (2011) *J. Anim. Sci.* **82**: 367.

Oyama K., Katsuta T., Anada K. and Mukai F. (2004) *Anim. Sci.* **78**: 195.

Stachowicz K., Ooi E. C. and Amer, P. (2023) Interbull Bulletin **59**: 171.

VanRaden P.M. (2008) *J. Dairy Sci.* **91**: 4414.

Proc. Assoc. Advmt. Anim. Breed. Genet. 26: 272-275

ACROSS-BREED ANALYSIS OF MATERNAL TRAITS IN NEW ZEALAND BEEF CATTLE

J. Kang[1, 2], D.J. Garrick[3], M.A. Lee[1], F. Weik[1] and J.A. Archer[1]

[1] Beef + Lamb New Zealand Genetics, Dunedin, Otago, 9016 New Zealand
[2] Charles Perkins Centre, The University of Sydney, Camperdown, NSW, 2006 Australia
[3] The Helical Company Limited, Rotorua, 3076 New Zealand

SUMMARY

This study assessed breed differences in the maternal traits body condition score (BCS), mature cow weight (MCWT) and hip height (HIP) using data from the Beef + Lamb New Zealand Genetics Beef Progeny Test. Age, breed effects and heterosis were estimated to inform the development of an across-breed genetic evaluation system. The results showed that breed differences exist, while positive and significant heterosis effects suggested that crossbred animals may exhibit improved maternal performance. Age effects indicated a steady increase in HIP and MCWT with age, until reaching a plateau at around 7-8 years. BCS followed a non-linear trajectory, peaking at two years before decrease upon first calving then gradually increasing with age until a decline after seven years.

INTRODUCTION

Improving maternal performance is key to enhancing the profitability of cow-calf operations in many commercial beef herds. Three maternal traits, BCS, MCWT and HIP play pivotal roles in production efficiency. MCWT, reflecting the combined weight of muscle, fat, bone, and internal organs, is an important determinant of maintenance feed requirements. While heavier cows tend to be associated with increased weaning weights of calves, they also incur higher maintenance costs due to greater energy demands. HIP, closely linked to frame size, provides insight into growth potential and mature weight, contributing to structural efficiency, and in combination with weight can assist in differentiating between cows which are heavier. This is mainly due to a larger frame-size (generally viewed as undesirable) from cows which are heavier due to increased muscle and fat with moderate frame size. BCS, as an independent measure of energy reserves and fat cover, is phenotypically important for reproductive success. It is especially important during periods of feed shortage when cows must rely on stored body reserves, and cows which naturally maintain higher condition score are generally viewed as being easier (and/or lower cost) to manage.

Exploiting genetic variation within and across breeds for maternal traits requires knowledge of age variation, breed differences, and heterosis effects. The focus in this study was BCS, MCWT and HIP on animals representing Angus and Hereford cattle breeds. By understanding both direct and maternal breed effects on these traits, this work will help to establish options to better utilise genetic resources available, including crossbreeding, in order to enhance the productivity and efficiency of the New Zealand beef industry.

MATERIALS AND METHODS

Data A total of 9,066 animals from the Beef + Lamb New Zealand Genetics beef progeny test were analysed, mainly comprising purebred Angus (5,989) and Hereford (1,515). Crossbreeding has taken place by mating sires from Angus, Hereford, and other breeds (including Stabilizer, Simmental, Shorthorn, Charolais, Limousin, Gelbvieh, and South Devon) with Angus and Hereford dams (Table 1). Since 2021, diallelic crosses between Angus and Hereford have been introduced specifically to study the effects of crossbreeding.

Table 1. Overview of the data structure for cows in the beef progeny test, with numbers representing the count of female animals that have maternal traits (MCWT, BCS, and HIP) recorded

Timeline				**2015 - 2020**				**2021 - 2024**		**Total**
Herd		A	B	C	D	E	F		G	
Dam breed		Angus	Angus	Angus	Hereford	Angus	Angus	Hereford	Angus	
	Angus	2,018	1,471	1,162	71	430	303	36	605	6,096
	Hereford	42	122	-	1,143	-	46	372	-	1,725
	Stabilizer	-	104			-				104
	Simmental	178	137	-	7			-		322
Sire breed	Shorthorn	2	1			-				3
	Charolais	82	5	-	4			-		91
	Limousin	6	6			-				12
	Gelbvieh	2	3			-				5
	South Devon	5				-				5
Total		2,335	1,849	1,162	1,225	430	349	408	605	8,363

Cows aged two-year-old and older were measured for MCWT, BCS and HIP (Table 2). BCS were visually assessed by either a trained scorer or farmer (trained and regularly calibrated to expert scorers) on a scale of one to ten (Beef + Lamb New Zealand 2017). Repeated measurements on MCWT and BCS were taken at three timepoints within each year: prior to mating, at weaning and prior to calving.

Table 2. Summary of phenotypic records on maternal traits

Trait	**BPT cows with known breed**				**Crossbred BPT cows**			
	Number of records	Number of cows	Range	Mean (SD)	Number of records	Number of cows	Range	Mean (SD)
BCS (Score)	62,538	7,774	3 - 10	6.9 (0.93)	3,004	504	4 - 9	7.06 (0.80)
HIP (cm)	7,253	4,636	115 - 158	129 (4.30)	650	480	118 – 140	128.4 (4.07)
MCWT (kg)	66,434	8,314	264 - 820	525.8(81.06)	4,216	858	297 – 776	491.5 (71.42)

Analyses Univariate animal models were fitted to estimate both breed effects and genetic parameters:

$$y = \mathbf{X}b + \mathbf{Z}a + \mathbf{P}c + e \qquad (1)$$

where y is an $n \times 1$ column vector of phenotypes, where n represents the number of phenotypes; X is an incidence matrix for fixed effects (b), Z is an $n \times q$ incidence matrix for both additive genetic effects (a) of q animals represented in the pedigree, P is an $n \times k$ incidence matrix for permanent environment effects (c) of k dams with one or more maternal performance measures; and e is the column vector of random residual effects. Fixed effects were the age of animals (in years), contemporary groups (i.e., recording herd x date x management group), breed composition covariates (i.e., Angus, Hereford and Other) of animals and their dams (i.e., Dam - Angus, Dam - Hereford and Dam – Other for any unspecified breed composition), and the effects of heterosis (0 for purebred animals and 1 for first crosses).

Basic assumptions of mixed model equations were applied here, such as $E[y] = Xb$ and $var\begin{pmatrix} a \\ c \\ e \end{pmatrix} = \begin{pmatrix} A\sigma_a^2 & 0 & 0 \\ 0 & I\sigma_c^2 & 0 \\ 0 & 0 & I\sigma_e^2 \end{pmatrix}$, where I is the identity matrix of size k for permanent environmental effects and size n for residual effects and the pedigree relationship matrix (A) was constructed using pedigree information spanning three generations on the sire side and five generations on the dam side. Variance component estimation was achieved using ASREML 4.2 (Gilmour 2019).

RESULTS AND DISCUSSION

The lowest heritability (0.26) and repeatability (0.39) among the traits evaluated was for BCS. In comparison, MCWT showed a moderate heritability of 0.50 and a repeatability of 0.78. HIP was the most heritable trait, with an estimated heritability of 0.64 and a repeatability of 0.75. These estimates (Table 3) are consistent with the findings reported by Weik *et al.* (2021) using a subset of this dataset which did not include more recently collected data including records from the diallelic cross.

Table 3. Variance component estimates (σ_a^2= additive genetic variance, σ_{pe}^2= permanent environmental variance, σ_e^2= residual variance), heritabilities (h^2± SE) and repeatabilities (t ± SE) from univariate animal models

Trait	σ_a^2	σ_{pe}^2	σ_e^2	h^2	t
BCS	0.13	0.07	0.29	0.26 (0.02)	0.39 (0.01)
HIP	8.12	1.31	3.20	0.65 (0.03)	0.75 (0.01)
MCWT	1168.54	513.66	509.24	0.53 (0.02)	0.77 (0.01)

The breed estimates suggest that, on average, Angus animals had lower MCWT, BCS, and HIP compared to the other breeds (Table 4). However, no significant differences were observed between Angus and Hereford breed effects for these traits, likely due to the fact that only herd F with 757 animals had both dam breeds in the same contemporary groups. Among maternal breed effects, Hereford dams contributed to higher MCWT (+7.66 kg), whereas other breeds had a negative effect (-5.83 kg) compared to Angus, while maternal breed differences in BCS and HIP were negligible. Overall, breed differences in maternal traits were small relative to the mean values (a 0.5 cm difference in HIP between Angus and Hereford, with a mean of 128.9 cm). Notably, heterosis effects were larger (0.75 for HIP), accounting for approximately 26% of the genetic standard deviation, highlighting the potential benefits of crossbreeding for improving maternal traits.

Table 4. Estimated direct and maternal breed (Angus as reference) effects and heterosis effect (standard error) for maternal traits

Trait	Direct breed effects		Maternal breed effects		Heterosis
	Hereford	Other	Hereford	Other	
BCS	0.08 (0.04)	0.12 (0.04)	0.05 (0.03)	0.02 (0.02)	0.04 (0.02)
HIP	0.50 (0.50)	1.18 (0.35)	0.02 (0.36)	-0.72 (0.23)	0.75 (0.17)
MCWT	-1.50 (2.85)	6.11 (2.61)	7.66 (2.61)	-5.83 (1.77)	7.50 (1.69)

Regarding age effects, HIP was found to increase steadily until approximately seven years of age (Table 5). Likewise, MCWT continued to increase until cows reached nine years of age. In contrast, BCS followed a non-linear trajectory: cows at two years of age were well-conditioned but experienced a rapid decline in BCS upon first calving. Although BCS gradually improved with advancing age, it began to decline again after seven years.

Table 5. Estimated age effects (standard error) for maternal traits

Trait	2	3	4	5	6	7	8	9	10+
BCS	0.61 (0.03)	-0.10 (0.02)	-0.11 (0.01)	0 (0)	0.11 (0.01)	0.16 (0.01)	0.15 (0.01)	0.07 (0.02)	-0.05 (0.02)
HIP	0 (0)	2.22 (0.22)	4.33 (0.30)	4.85 (0.31)	4.95 (0.31)	5.21 (0.31)	5.24 (0.32)	5.33 (0.34)	5.14 (0.35)
MCWT	-109.3 (1.31)	-64.1 (0.90)	-30.28 (0.48)	0 (0)	22.17 (0.45)	34.38 (0.60)	40.74 (0.77)	41.89 (0.94)	41.92 (1.17)

CONCLUSION

This study provides information to aid in the development of an across-breed genetic evaluation system relevant to the NZ beef-cattle population. The estimated breed effects and heterosis highlights the importance of incorporating breed differences into a multi-breed genetic evaluation framework. Furthermore, the positive heterosis estimates emphasise the potential advantages of crossbreeding. Future work should focus on expanding the dataset of crossbred animals to enable more accurate estimation of these effects.

ACKOWLEDGEMENT

This project was funded through the "Informing New Zealand Beef" programme, supported by Beef + Lamb New Zealand and the Ministry for Primary Industries. We also acknowledge the farmers participated in the Beef + Lamb New Zealand Genetics Beef Progeny Test for contributing data to this research.

REFERENCES

Beef + Lamb New Zealand (2017) Beef cow body condition scoring.

Gilmour A.R. (2019) *J. Anim. Breed. Genet.* **136**: 262.

Weik F., Hickson R. E., Morris S. T, *et al.* (2021) *Animals* **11**: 2509.

MULTI-TRAIT SELECTION WITH DESIRED GAINS AND NON-LINEAR INDEXES

J.H.J. van der Werf

School of Environmental & Rural Science, University of New England, Armidale, NSW, 2351 Australia

SUMMARY

A linear selection index for multiple traits is not satisfactory in many practical cases, such as when selecting for traits that have an optimum phenotypic value with a nonlinear profit function. This paper will discuss various ways to select for a breeding objective that includes a trait with an optimum phenotype. The study shows that, contrary to general belief, a linear updated index does not always achieve a desirable result due to correlated responses in multiple trait selection. A simple derivation of weights for a restricted index is proposed to deal with the problem.

INTRODUCTION

Multiple trait selection based on a linear index can be problematic when there is uncertainty about economic weights or when some traits have economically optimal values for their phenotype. With optimal traits there may be a tendency to use a non-linear index to avoid selecting animals that deviate much from the optimal value. How to select for traits with a non-linear profit function has been addressed many times, and it is commonly accepted that a linear index is still a satisfactory approach, provided that index weights are regularly updated depending on how far the current population mean is away from the trait optimum (Goddard 1983; Dekkers *et al.* 1995; Martin-Collado *et al.* 2016). Martin-Collado *et al.* (2016) considered a case study where they compared linear indexes with non-linear indexes and linear updated indexes. They also looked at a linear index where there is no longer a weight for the phenotypic values above the optimum, which they referred to as Non-Linear-Flat (NFL Index). Valuing phenotypic values of individual selection candidates differently is a departure from the common view on index selection, where the selection weights depend on the population average rather than individual phenotypes. However, with optimum traits it is tempting to use a selection method where individuals that deviate from the optimum are penalized. Real world breeding objectives involve multiple traits and whatever weights are used for traits with an optimal value will also affect responses in other traits that may be linked to profit linearly. Equally, the correlation with other objective traits and their index weights will be important to ensure that an optimum trait is not pulled away from its optimum. This could be the case if the optimum trait has a strong correlation with other objective traits and has relatively less value. To ensure that trait responses are optimal, a restricted index, which applies a linear weight, could be used when the optimum trait mean is near its optimum. The aim of this paper is to explore trait responses for various selection strategies in a two-trait breeding objective when one trait has an economic optimum. Four selection strategies are compared, being the linear index, the linear updated index, the penalized phenotype index and the restricted linear updated index.

MATERIALS AND METHODS

A simulation study was used for a population of 2,000 progeny born in each generation. Results were averaged over 200 replicates. Males (20%) and females (80%) were selected based on an index using individual phenotypes for two traits. Two traits were considered; one with a linear profit function and a trait with a nonlinear profit function. Phenotypic standard deviations (PSD) are 1, hence trait responses are expressed in units of PSD. We assumed equal heritabilities of 0.25 and genetic and residual correlations were either 0 or 0.5 and 0.3, respectively. Random normal deviates were used in R to simulate breeding values and phenotypes for the individuals in each of 30 generations. We did not account for relationships between individuals, inbreeding or the Bulmer effect. The profit function was equal to $v_1.x_1+v_2.x_2 + c.x_2^2$, where x_i is a phenotypic value for trait i. Following Martin-Collado *et al.* (2016), we derived that $c = -v_2/(2.x_{2Opt})$, were x_{2Opt} is the optimum phenotypic value for trait 2 relative to the value in generation 0 ($x_{2(0)}$) and v_2 is the derivative of the profit equation for trait 2 at $x_{2(0)}$. We assume equal economic values of 1 for v_1, v_2 and the value for x_{2Opt} was also 1. These values could be varied to explore the relative importance of each trait and the distance of x_2 from its optimum. We used four selection strategies.

1. The linear index was derived from selection index theory, with the index weights derived as $b = \mathbf{P}^{-1}\mathbf{G}v$, where **P** is the 2x2 variance-covariance matrix among an individual's phenotypes and **G** = is the covariance matrix between phenotypes and breeding values. The vector v contains economic values derived in generation 0 as $v = [v_1, v_2]$. Individuals are selected based on b'x, where x is a vector of phenotypes $[x_1, x_2]$.
2. The linear updated index was updated every generation using the economic values as derivatives of the profit function, i.e. $v = [v_1, v_2 + 2c\bar{x}_2]$, where $\bar{x}_2$ is the population mean of x_2 at the current generation.
3. The penalized phenotype index was constructed the same way as the linear updated index, but with x_2 replaced by x_{2d}, the deviation of x_2 from the optimum value: $x_{2d} = -abs(x_{2Opt} - x_2)$.
4. The restricted linear updated index follows the updated index but once at the optimum uses a restricted index to keep the response for trait 2 equal to zero. At that point, the economic value for trait 2 in the updated index would be zero, which is the slope of the profit function at the optimum value. However, the restricted index weight for trait 2 requires a non-zero values to counterbalance a correlated response. The restricted index weight is derived below.

Desired gain index. A response to selection can be predicted for each of the breeding objective traits as a vector $d_g = \mathbf{V}v/\sqrt{(v'\mathbf{V}v)}$, where **V** is the variance-covariance matrix of the estimated breeding values (EBVs) for the objective traits for an individual, and v is vector with economic values of the objective traits. The matrix **V** can be derived from selection index theory as $\mathbf{G}.\mathbf{P}^{-1}\mathbf{G}$ and v'**V**v is the variance of the index, which can also be written as b'**P**b. However, it is important to note that the selection response only depends on the covariance matrix among EBVs (**V**) and the weights used for these EBVs in the index (v). Now let d_{desg} be a vector of desired responses. Then the vector with weights to achieve these responses would be derived as $v_{desg} = \mathbf{V}^{-1}d_{desg}.\sqrt{(v'_{desg}.\mathbf{V}.v_{desg})}$. The last term is a scalar (variance of the index) and can be ignored as it is only important to know relative weights. In our case, with two traits, the response for the second trait needs to be zero, i.e. $d_{desg} = [z, 0]$. Therefore, taking the ijth elements of the inverse of **V** (v^{ij}) gives desired gain weights $v_{desg} =[v^{11}.z, v^{22}.z]$, or relative weights are $[1, v^{21}/v^{11}]$.

RESULTS

The response to selection based on the four index strategies is shown in Fig.1 for the case where there are no genetic correlations between the traits. The response for the linear index is linear for both traits. When the optimum trait is not correlated with the other traits, the updated index strategy is an adequate method to keep the phenotypic mean of the non-linear trait at its optimal value. The response of the restricted linear updated index is identical to the response of the linear updated index. When the response for the non-linear trait is slowing down near the optimum value, the response for the other objective trait can increase and shows a higher response compared to the linear index. The penalised phenotype strategy also converges toward the optimum phenotype value but at a slower rate than the updated indexes.

Proc. Assoc. Advmt. Anim. Breed. Genet: 26: 276-279

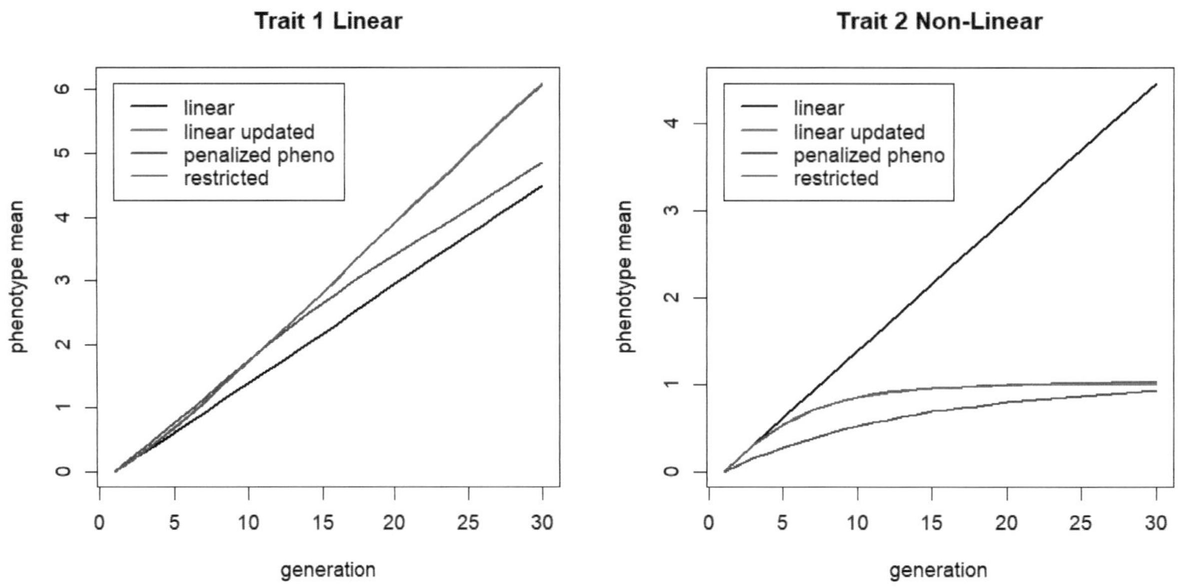

Figure 1. Response to selection for two traits that are related to profit in a linear and non-linear way. The non-linear trait has an optimal value at 1 phenotypic standard deviation above the current mean. The traits have equal economic value in generation 0. Responses are compared for four different index selection strategies (see main text for detail) for the case where both genetic and environmental correlations are 0

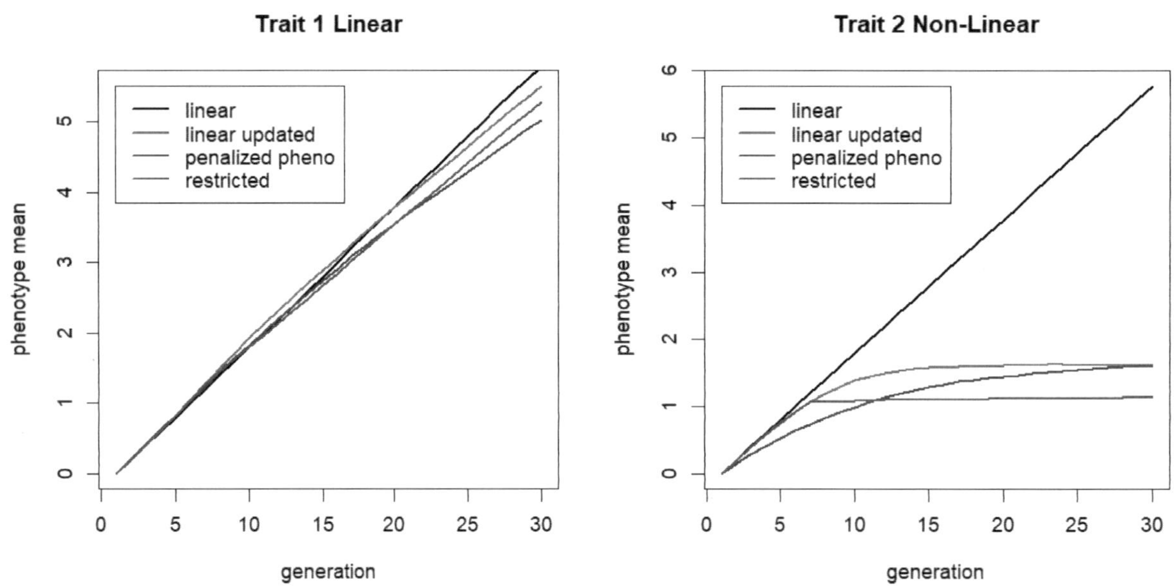

Figure 2. Response to selection for two correlated traits related to profit in a linear and a non-linear way. The non-linear trait has an optimal value at 1 phenotypic standard deviation above the current mean. The traits have equal economic value in generation 0. Responses are compared for four different index selection strategies (see main text for detail) for the case where genetic and environmental correlations are 0.5 and 0.3, respectively

When the two traits are correlated the updated index method does not keep the nonlinear trait at its optimum value while with the restricted index approach the trait 2 response does converge to its optimum (Figure 2). However, the overall profit of these two indices is very similar as for the updated index the cost of overshoot in trait 2 is compensated by more genetic gain for trait 1.The penalized phenotype index also converges to the same value as the linear updated index strategy, but at a slower rate. A stronger genetic correlation increases the value at which the updated linear index approach converges to. Similarly, if the economic value of trait 1 is relatively higher, then the convergence value for the optimum trait is also higher, i.e. further above the optimum value.

DISCUSSION

Several papers have suggested that the updated linear index is an adequate method to select for optimum traits. This is true for single trait selection or if the optimum trait is not correlated with the other objective traits. However, in multiple trait selection, updated economic values do not always guarantee traits to converge to an optimum value. A restricted index was used to keep the nonlinear traits at its optimal value, and restricted index weights were derived from the variance covariance matrix of EBVs for the objective traits. Although this was illustrated in a two-trait example, the method can easily be extended to more traits (e.g. see Van der Werf and Clark 2022).

Martin-Collado *et al.* (2016) illustrated updated linear indexes for a sheep scenario and found the method to be suitable. However, in their case the genetic correlation was close to zero and the non-linear trait had relatively a higher economic value. In most cases, the other objective traits combined will have more value and if these traits combined have a significant correlation with the nonlinear trait, we cannot expect the updated linear index to converge to the optimum for traits with an optimal value. The higher phenotypic value that the response converges to depends on the correlation and the economic value of the other traits, i.e. it is a function of the pull from the residual objective traits. The derivative of the profit function for the nonlinear trait at its optimum is zero, but in most cases this economic weight is not sufficient to counterbalance a correlated response. The restricted index used values that were designed for that purpose, i.e. they are effectively the implicit economic values used in desired gains indexes. Constrained selection indexes for optimal traits were also proposed by Goddard (1983) and Dekkers *et al* (1995) but less straightforward compared to the method proposed in this paper.

The penalized phenotype index method showed a slower response than the linear index method but ultimately converges to the same phenotypic mean. The time to convergence could be increased could be increased by penalising the deviations more strongly (i.e. $P_{2d} = -abs(P_{2Opt} - P_2).\lambda$, where a higher λ-value (>1) is a penalty). However, this is at the expense of the response for trait 1. The penalized phenotype selection method is overall less effective than a linear index, and a mate allocation strategy might be a more efficient way to avoid extreme phenotypes.

CONCLUSION

In multiple trait selection a linear updated selection index method is not adequate for traits with an optimum value as it does not sufficiently control the correlated response with other objective traits. We derived restricted index weights that ensure phenotypic optimal values in multiple trait selection.

REFERENCES

Dekkers J.C.M., Birke P.V. and Gibson J.P. (1995) *Anim. Sci.* **61**:165.

Goddard M.E. (1983) *Theor. Appl.Genet.* **64**: 339.

Martin-Collado D., Byrne T.J., Visser B. and Amer P.R. (2016) *J. Anim. Breed. Genet.* **133**: 476.

van der Werf J.H.J. and Clark S.A. (2022) *Proc.World Congr.Genet.Applied to Livest.Prod.* **12**: 1975.

Proc. Assoc. Advmt. Anim. Breed. Genet: 26: 280-283

DIGITAL TWIN: MODELLING BREEDING PROGRAM APPROACHES IN AUSTRALIAN WAGYU

J. Copley[1], B. Hayes[2], G. Moser[1], B. Engle[3], K. Villiers[2] and M. Kelly[1]

[1] Australian Agricultural Company Ltd, Newstead, QLD, 4006, Australia
[2] Centre for Animal Science, Queensland Alliance for Agriculture and Food Innovation, St Lucia, QLD, 4072 Australia
[3] USDA, ARS, U.S. Meat Animal Research Center, Clay Center, NE, 68933 USA

SUMMARY

We developed a 'digital twin' of AACo's Wagyu seedstock breeding program utilising the genomicSimulation software. We aimed to find the optimal selection pressure and reproductive technology to rapidly meet the objective of a completely polled herd while optimising genetic gain for key traits and managing inbreeding. Our results demonstrated that a polled Wagyu herd could be achieved within 8 years using a combination of invitro fertilisation (IVF) and natural service (NS) in a breeding program which applied 100% weighting to polled. However, we demonstrated that applying this level of selection pressure came at a significant cost to genetic progress for other traits. A 60% weighting to polled selection was able to achieve a polled herd within 10 years whilst meeting the other objectives.

INTRODUCTION

Breeding programs in beef cattle are challenging to evaluate in a timely manner. Whilst genomic technologies allow breeding decisions to be made earlier-in-life and has accelerated the rate of genetic gain in AACo's herd, substantial risk is still attached to breeding programs. Therefore, we sought to create a 'digital twin' of AACo's breeding program using the genomicSimulation software (Villiers *et al.* 2022). This software allowed for proposed breeding programs utilising different: scenarios, selection pressures, technologies and target traits to be fully tested before implementation. The digital twin is a detailed representation of AACo's Wagyu breeding program where the aim is to develop a polled pure-bred Wagyu herd. Selection for polled was balanced with selection for a Wagyu production index (WAGIN). Inbreeding, measured as Runs of Homozygosity (ROH), was also evaluated. The 'twin' aimed to find the optimal breeding program to achieve a completely polled herd in the shortest time using a combination of natural service (NS), invitro fertilisation (IVF) and artificial insemination (AI).

MATERIALS AND METHODS

Genomic resources. The founding population for the digital twin was a group of 2,500 genotyped animals, each with a Bovine TropBeef (50K) SNP array imputed to 770K. SNP effects for the modelled production traits were taken directly from AACo's internal prediction pipeline. These SNP effects are updated as required, an assumption that was also carried into the digital twin. The WAGIN production index is comprised of several production traits and is specifically designed for selection of Wagyu sires. The poll allele within the Wagyu population was modelled as a single additional SNP at the map location of the known Celtic mutation; chromosome 1: 2429319 bp (Randhawa *et al.* 2020). The effect of the poll allele was set to 1 and the effect of the horn allele to 0.

Computational resources. The digital twin simulations utilised the genomicSimulation package in R (version 0.2.4.003) (Villiers *et al.* 2022). GenomicSimulation was designed for the simulated stochastic crossing and selection of real genotypes, which combined with known SNP effects allowed for the calculation of genomic breeding values (GBVs) of individual animals. Simulation was conducted in a unix environment using the resources of the University of Queensland's HPC.

Breeding program assumptions. The digital twin was used to model three different breeding programs: NS, AI and IVF, and investigated the outcomes of placing different selection emphasis on polled vs WAGIN. NS assumed that all heifers and mature cows were mated using natural service with a pregnancy rate of 85%. Bulls were joined at 3% of the female herd, with offspring per bull being capped at 20% of all calves. Females were replaced at 25% per annum, replacement heifers being selected on estimated genetic merit. Bull selection was conducted in two steps, firstly 36% were randomly culled, remaining candidates were then selected on genetic merit to meet a 33% annual replacement rate. Virgin bulls were mated to heifers only, after which they were placed within a common pool of mature bulls. The AI scenario assumed that all heifers were AI'd as yearlings, with 50% of AI resulting in pregnancy, the remainder of the available females were mated using natural service. Two different IVF scenarios were modelled, IVF5 assuming 5 embryos per donor and IVF10 assuming 10. Both selected 350 mature cows as recipients for embryo's donated by either the 65 (IVF5) or 35 (IVF10) heifers of highest genetic merit, all remained females were natural mated. The selection of sires, donors and replacement heifers was subject to different levels of selection intensity for polled (polled weight): 0, 0.4, 0.6 and 1. Polled weight was balanced against production traits, expressed using WAGIN. Selection candidates were ranked by combining rescaled GBVs for polled and WAGIN.

Inbreeding. Inbreeding was defined using runs of homozygosity (ROH). ROHs are long continuous segments in the genome which, due to their length, are assumed to come from a common ancestor (Meyermans *et al.* 2020). ROHs were detected using the sliding window approach in PLINK 1.9 (Chang *et al.* 2015). Both the GBVs and ROHs values presented were the mean value of 25 replicates per generation, necessary due to individual matings in natural service scenarios not being predetermined.

RESULTS AND DISCUSSION

A range of different 'poll weights' were tested within the digital twin, ranging from 0, indicating no selection pressure for polled, to 1, indicating 100% selection pressure on polled, ignoring production traits. We have chosen to present only a subset of these poll weights in the analysis. The Polled Frequency presented in Figure 1. are the combined sum of PP and HP animals divided by number of animals times 2. All breeding programs, even NS were able to achieve a 100% Polled Frequency when applying 100% weighting to Polled. However, our results clearly demonstrated that applying only a 60% weighting to polled would achieve a similar result. This is due to the low number of polled animals available in the base population. Within this herd, there is simply no point in applying selection pressure beyond 60% for polled. Unsurprisingly, programs using intensive breeding tools such as AI and IVF were able to achieve a polled herd in less time. The results of the Polled Frequency analysis carry implications for the WAGIN gain results in Figure 2. As expected, applying 100% selection intensity for Polled meant that genetic gain for WAGIN was zero for 10 years, until a completely polled herd was achieved. Conversely, genetic gain for WAGIN under a 60% selection intensity for polled was still reasonable under all breed programs.

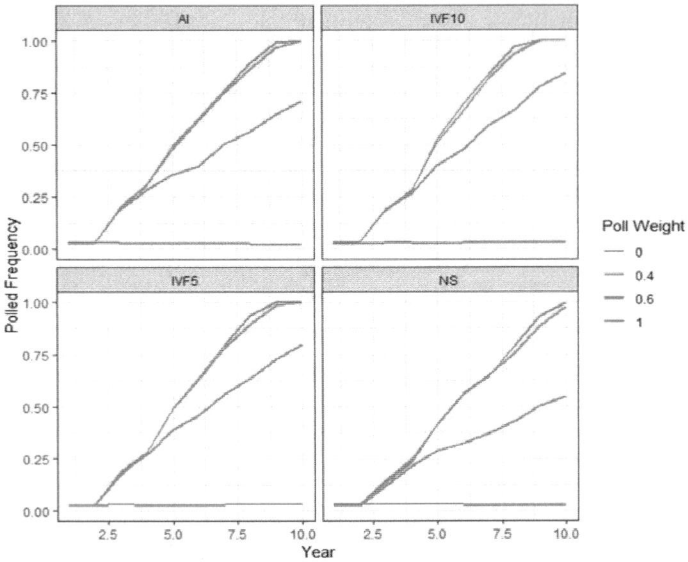

Figure 1. Genetic gain for Polled Frequency

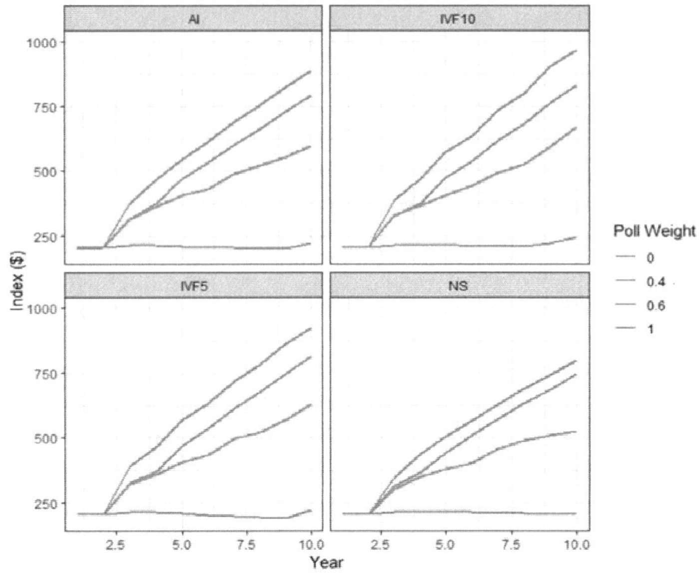

Figure 2. Genetic gain for WAGIN

IVF and AI assisted breeding were, as expected, able to provide superior rates of genetic gain. The opportunity cost of the forgone genetic gain for WAGIN is difficult to determine. It must account for the value of the genetic gain for production traits in real terms and balance this against the value of increased polled frequency. Such analysis is beyond the scope of this paper but must assign a monetary value to polled through factors including decreased mortality from dehorning, market access from improved animal welfare.

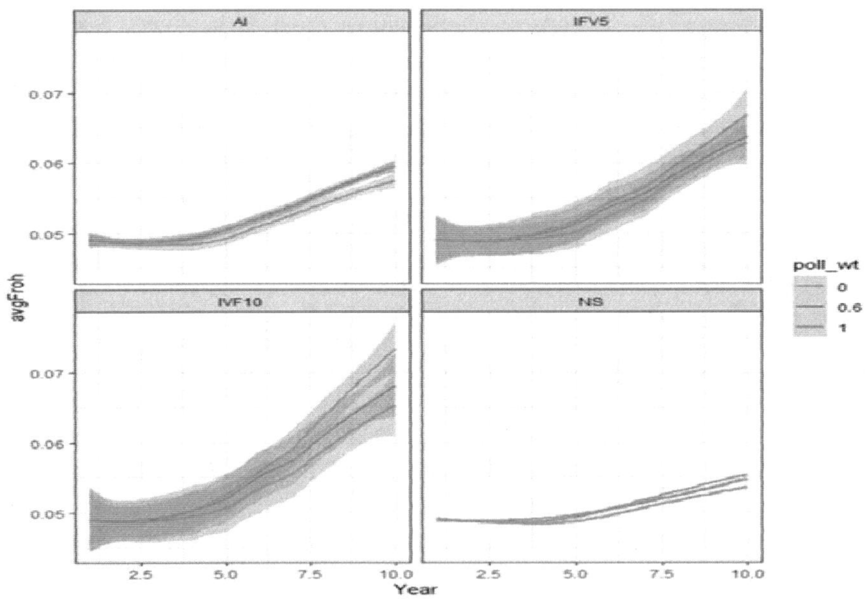

Figure 3. Inbreeding rate for different breeding program scenarios

Another important factor when considering the cost benefit analysis of the respective breeding programs is the projected level of inbreeding. Our previous results clearly demonstrated the superiority of IVF and AI programs in achieving the breed objectives, but this is accompanied by a higher level of inbreeding (Figure 3). Compared to the NS breeding program where ~5.5% of the genome was found within ROHs, the IVF10 program resulted in ~6.5-7.4% within ROHs (Figure 3). Whilst inbreeding depression could not be modelled in this scenario, it has been observed in dairy cattle previously (Pryce *et al.* 2014) that a 1% increase in inbreeding was associated with 1-1.5% reduced milk, fat and protein yields as well as declining fertility. Increasing selection weight on polled did increase inbreeding, however the type of breeding program was found to be more impactful. For example, the inbreeding in the NS program at polled weight = 1 was lower than the AI program at polled weight = 0. This is a direct function of the small effective population size of polled animals available for selection as well the limited number of animals inducted as AI sires and IVF donors. The use of technologies like AI and IVF, which accelerate the rate of genetic gain for target traits, are not without drawbacks.

CONCLUSION

The results of the simulations indicated that AACo achieving a polled Wagyu herd is possible within a 10-year time frame. However, selection for polled must be finely balanced against selection for WAGIN. We concluded that polled weight in the breeding program should not exceed 0.6, doing so does not increase the proportion of polled animals in the herd and would only serve to compromise genetic gain for WAGIN. AI and IVF breeding programs performed better than NS in achieving the breeding objectives sooner but did result in higher rates of inbreeding. Maintaining genetic diversity is especially important for Wagyu breeding programs, thus upcoming iterations of the digital twin analyses will incorporate optimal contribution selection and be able to balance multiple traits with rates of inbreeding.

ACKNOWLEDGEMENTS

Meat and Livestock Australia contributed financial support for this work under the Animal Wellbeing Extension and Adoption Partnership, P.PSH.1371. The USDA is an equal opportunity provider and employer.

REFERENCES

Chang C.C., Chow. C.C., Tellier L.C., Vattikuti S., Purcell S.M. and Lee J.J. (2015) *Gigascience.* **4**: 8.

Meyermans R., Gorssen W., Buys N. and Janssens. S. (2020) *Genomics* **21**: 94.

Pryce J.E., Haile-Mariam M., Goddard M.E. and Hayes B.J., (2014) *Genet. Sel. Evol.* **46**: 71.

Randhawa G.S., Soltysiak M.P.M., El Roz H., de Souza C.P.E., Hill K.A. and Kari L. (2020) *PloS One* **15**: e0232391.

Villiers K., Dinglasan E., Hayes B.J. and Voss-Fels K.P. (2022) *G3 Genes Genom. Genet.* **12**: jkac216.

Proc. Assoc. Advmt. Anim. Breed. Genet. 26: 284-286

ROUTINE DIAGNOSIS OF GENETIC EVALUATIONS AND AUTO STANDARDISATION TO NON-LINEAR COST FUNCTION

X. Zhang, F. Hely, K. Stachowicz and P. Amer

AbacusBio, Level 5, 333 Princes Street Dunedin 9016, New Zealand

SUMMARY

Routine genetic or genomic evaluations are often conducted multiple times a year. When there is a new distribution of estimated breeding values, non-linear cost functions that have been developed will often need to be adjusted or recalculated. The purpose of this paper is to present routine methods to 1) diagnose and display whether the genetic evaluations have deviated from the past course, and 2) standardise the genetic evaluations of a new population so that the previously developed non-linear cost function can be applied to the standardised evaluations. Automation of this routine could expedite the decision on whether the new data still suits the already-developed non-linear cost function and avoid potential errors created by manually recalibrating the non-linear cost functions incorrectly.

INTRODUCTION

The index value, or cost function, of a measured trait with linear economic weights is simply the product of its economic value in trait units and its estimated breeding value (EBV; Falconer and Mackay 1996; Hazel 1943). For traits with non-linear economic weights, the shape of the non-linear cost function (NLCF) and its position relative to the range of estimated breeding values adds a complexity into the index formulation (Quinton *et al* 2019). For example, a cost, C(\$/trait unit) for a trait, could be connected to EBV via a conditional NLCF as below

$$C = \begin{cases} \beta_0 + \beta_1(g - T), & g \le T \\ \beta_0 + \beta_1(g - T) + \beta_2(g - T)^{\gamma}, & g > T \end{cases} \qquad (1)$$

where β_0, β_1 and β_2 are intercept, linear and nonlinear (curve) coefficients, respectively, g_i is the EBV, T is the threshold in g scale where the economic value changes from linear to non-linear, and γ is the exponent. Values of C, g and T are known. The NLCF is only nonlinear when $g > T$. β_0, β_1, β_2 and γ determine the shape of the NLCF. β_0 and β_1 are easy to solve by taking two pairs of g and C observations into the linear part, i.e. $g \le T$. For β_2 and γ, given two sets of observations of g and C, we can now form two equations

$$C_1 = \beta_0 + \beta_1(g_1 - T) + \beta_2(g_1 - T)^{\gamma}$$

and

$$C_2 = \beta_0 + \beta_1(g_2 - T) + \beta_2(g_2 - T)^{\gamma}$$

Since the value of β_0 merely shifts the curve up or down but does not change the shape, we can simplify the above by assuming the cost is 0 at T, then $\beta_0 = 0$, and the above becomes

$$C_1 - \beta_1(g_1 - T) = \beta_2(g_1 - T)^{\gamma}$$

and

$$C_2 - \beta_1(g_2 - T) = \beta_2(g_2 - T)^{\gamma}$$

We can solve for γ by taking the ratio of the above two equations

$$\frac{C_1 - \beta_1(g_1 - T)}{C_2 - \beta_1(g_2 - T)} = \left(\frac{g_1 - T}{g_2 - T}\right)^{\gamma}$$

Then taking the log of both sides and rearranging we have

$$\gamma = \frac{log\big(C_1 - \beta_1(g_1 - T)\big) - log\big(C_1 - \beta_1(g_2 - T)\big)}{log(g_1 - T) - log(g_2 - T)}$$

Then we can solve $\beta_2$using either of the initial equations, e.g.

$$\beta_2 = \frac{C_1 - \beta_1(g_1 - T)}{(g_1 - T)^\gamma}$$

When the distribution of estimated breeding values changes due to a new evaluation method, a new genetic base, pre-selection, or the trait is evaluated in a different environment for a new cohort of candidates, the original coefficients including T_i may not hold anymore. Thus, the NLCF needs to be adjusted or recalculated (Burndon 1990; Quinton *et al.* 2019). This is required to ensure that the index still penalises or incentivises the right individuals in the right degree based on where they are on the estimated breeding value scale.

A set of economic weights is usually calculated to last for multiple years unless there is a major change in the genetic/genomic evaluation results described above, when an update of the index may be triggered (Cole and VanRaden 2018). However, there is often a lack of communication between genetic evaluation providers and those developing or maintaining the index regarding whether and how the evaluation results have changed between runs, sometimes due to data security or IP protection. It would be useful to make a diagnosis tool as part of the evaluation pipeline, allowing index experts to detect major changes in the evaluation so that they can decide the next appropriate step, e.g. to change the NLCF, standardise the EBV, or to understand more about the evaluation results before making a decision. The purpose of this paper is to present a routine method to diagnose changes in EBV and to standardise these when the latest EBV doesn't deviate from the past distribution.

ROUTINE DESCRIPTION

The routine will consist of two parts. The diagnostics part and the auto-standardisation part.

Diagnosis. The first diagnostic is to verify whether the evaluation distribution changed. The EBV of every trait in each evaluation run will go through the routine for a series of descriptive analyses including summary statistics and tests for a normal distribution. The results will be recorded in a report, including the number of animals, mean, median, SD, min, max, outlier values, histograms, results from Shapiro-Wilk or Kolmogorov-Smirnov test of normality, and ranking of animals.

The second diagnostic is to check if there is a genetic base shift. When we have accumulated multiple evaluation results, then for a set of reference animals that have EBV across all evaluations, we can run correlation, pairwise t-test (when EBV are normally distributed), pairwise F-test and ANOVA, or Kruskal-Wallis H test (when using genotypes) to check if the latest EBV have deviated from the original EBV for which the NLCF was created. If genetic progress in the EBV is expected, then we can run a linear regression on the EBV means across evaluations, leaving out the latest mean, then compare the observed latest mean with the predicted mean, to see if the observation is within the expected range, say, within 2 SD of the predicted mean. The formula to calculate the standard deviation of a predicted value, $\hat{y}_i$, as a linear regression is:

$$SD(\hat{y}_i) = s\sqrt{1 + \frac{1}{n} + \frac{(x_i - \bar{x})^2}{\sum_{i=1}^{n}(x_i - \bar{x})^2}}$$

where $s = \sqrt{\frac{\sum_{i=1}^{n}(y_i - \hat{y}_i)^2}{n-2}}$ is the standard error of $\hat{y}$, n is the number of observations, x_i is the *i*-th predictor, and $\bar{x}$ is the mean of all predictors. Then we repeat the analysis for EBV for every trait. Lastly, we can check the rank correlation of animals overlapped across evaluations, especially base or proven animals.

The third diagnostic is to check if the EBV threshold, T, has changed. The relationship between the non-linear cost changing point and the phenotype threshold usually stays stable, whereas the corresponding T to the phenotype threshold can change across evaluations. When the phenotype threshold is available to both the initial and the latest datasets, we can run a linear regression between phenotype and EBV of the initial data and apply the same coefficients to the latest data to estimate T in the latest dataset. If phenotypes of at least one dataset are not available, we can still run a linear regression between the EBV values of the 2 sets of data to estimate new T from this regression. In addition, we can check the proportion of individuals with EBV beyond the threshold where the cost is non-linear. The threshold will be recorded from the latest NLCF, along with other parameters.

If there has been a distribution change, genetic base shift or a change of threshold is detected, the routine programme will show a warning and flag the out-of-range EBV, so that index experts can decide if they want to communicate with evaluation experts about the cause and impact of the changes, and if they need to re-design the NLCF. These diagnostics can run every time a new set of EBV is available, and the results will be recorded for the comparison.

Auto standardisation. When an index review is needed, commonly every 3-5 years, the historical diagnosis results can be used to help with decision making. If the vector of the latest set of EBV, **g**, has passed the aforementioned diagnostics, e.g. their distribution has not changed significantly, or the evaluation expert has confirmed they are homogenous, then we can simply standardise them to the initial scale, $\mathbf{g}_0$, for which the NLCF was created.

$$\mathbf{g}_{new} = \frac{(\mathbf{g} - \mu_{\mathbf{g}}) \cdot \sigma_{\mathbf{g}_0}}{\sigma_{\mathbf{g}}} + \mu_{\mathbf{g}_0} \tag{2}$$

where $\mathbf{g}_{new}$ is the vector of standardised latest EBV; $\mu_{\mathbf{g}}$ and $\sigma_{\mathbf{g}}$ are the mean and standard deviation of the latest EBV, and $\mu_{\mathbf{g}_0}$ and $\sigma_{\mathbf{g}_0}$ are the mean and standard deviation of EBV at the initial evaluation where the NLCF was made. The standardisation formula parameters will be recorded for future reference.

Auto fitting of NLCF. Finally, after diagnosis and auto standardisation of EBV, we fit $\mathbf{g}_{new}$ into equation (1) to obtain the corresponding new cost for **g** and record the new NLCF.

CONCLUSION

We have described a routine diagnostics pipeline to first validate new trait EBV against historical trait EBV, then use an automation pipeline to perform a series of calculations, ultimately generating new non-linear cost values. By streamlining and simplifying the selection index review process, this pipeline provides timely warnings to index experts when data deviates from expectations, aiding them in making informed decisions about necessary adjustments to the NLCF.

REFERENCES

Burndon R.D. (1990) *Theor. Appl. Genet.* **79**: 65.

Cole J.B. and VanRaden P.M. (2018) *J. Dairy. Sci.* **101**: 3686.

Falconer D.S. and Mackay T.F.C. (1996) 'Introduction to Quantitative Genetics' 4th Edition Longman Group Ltd.

Hazel L.N. (1943) *Genetics* **28**: 476.

Quinton C.D., Amer P.R., Byrne T.J., Archer J.A., Santos B. and Hely F. (2019) *Proc. Assoc. Advmt. Anim. Breed. Genet.* **23**: 47.

Proc. Assoc. Advmt. Anim. Breed. Genet. 26: 287-290

MICRON PREMIUMS AND THEIR ROLE IN MERINO BREEDING PROGRAMS

S.F. Hatcher[1,2], A.E. Casey[3], S.F. Walkom[4], D.J. Brown[4] and A.A. Swan[4]

[1] Makin Outcomes Pty Ltd, PO Box 8358, Orange East, NSW, 2800 Australia
[2] Charles Sturt University, Locked Bag 588, Wagga Wagga, NSW, 2678 Australia
[3] Advanced Breeding Services, 4 Fraser Place, Orange, NSW, 2800 Australia
[4] Animal Genetics Breeding Unit*, University of New England, Armidale, NSW, 2350 Australia

SUMMARY

Micron premiums (MP) are an appropriate means of incorporating drivers of the value of fine wool into Merino breeding programs. Changes in the relative value of wool to meat since 2003 resulted in significant changes in Merino breeding objectives and the demographics of the Australian sheep flock. Over the same period, the average MP for fine wool (< 19 microns) have been greater than 10% regardless of the variation in the Eastern Market Indicator and an increased proportion of fine wool in the Australian clip. However, due to high heritability of fibre diameter (FD) MERINOSELECT indexes are adjusted to reflect a reduced MP to produce more balanced indexes. While individual Merino breeders can increase the emphasis they place on FD when making ram selection and purchase decisions, MPs are not routinely reported making it problematic for Merino breeders to balance the value of focussing on genetic gain in FD and the suitability of MERINOSELECT indexes in meeting their flock's breeding objective.

INTRODUCTION

Micron premiums (MP), defined as the percentage increase in price per kg for a reduction of one micron, are a measure value of changing fibre diameter (FD) relative to fleece weight (FW) (e.g. the 19 MP is the difference between the price of 19 µm and 20 µm wool, divided by the price of 19 µm wool, expressed as a percentage). Expressing MP as a percentage provides a constant relative value over time, irrespective of the absolute price. MPs provide Merino breeders with the means to link the market value for wool with the breeding objective and selection emphasis used in their breeding programs. Due to the economically unfavourable genetic correlation (0.3) between FD and FW (Safari *et al.* 2005), Merino breeding objectives using MPs were based on a desired gains approach and generally ranged between 3 to 20% with simultaneous improvement in FD and FW possible within this range (Swan *et al.* 2007). Producers wishing to place emphasis on increasing FW while maintaining FD would use a 3% MP, those wishing to reduce FD and maintain FW would use a 15% MP. Producers placing an equal emphasis on FW and FD would use a MP midway of that range. MP-based breeding objectives were routinely used in within- and across-flock selection for Merinos up to the early 2000s. However significant changes have occurred in the landscape for Merino wool production enterprises driven by variation in the relative value of meat to wool with consequent changes in the demographic structure of the flock. As a result, the contribution of FD to Merino breeding objectives began to decline (Swan *et al.* 2017). This paper explores the value of MPs over the past 20 years together with changes in the fibre diameter composition of the Australian wool clip to ascertain the importance of MPs in 'modern' breeding objectives for Merino sheep.

THE CHANGING LANDSCAPE OF MERINO PRODUCTION

Changing demographic of the Australian sheep flock. In 2003 income from the sale of wool

* A joint venture of NSW Department of Primary Industries and Regional Development and the University of New England

comprised 66.0% of the total receipts for specialist Australian sheep producers, however by 2022 this had reduced to just 38.0%, with income from sheep sales comprising 61.9% of total receipts (ABARES 2024). Higher meat prices have led to wethers being sold as younger animals into the lamb or store sheep market rather than retained for wool production. In 2003, 43.7% of total Australian sheep numbers were ewes (ABS 2004) by 2023 this had increased to 58.6% (MLA 2023). In addition, large numbers of Merino ewes are now mated to rams from meat breeds to generate progeny with high value lamb carcases (Swan *et al.* 2007).

Evolution of Merino breeding objectives. Significant changes have occurred in Australian Merino breeding enterprises over the past 20 years. Up to the early 2000s, FD was the dominant trait in the Merino Production Plus (MPP) index, however this began to change from 2005 when the influence of yearling bodyweight (YWT) began to increase (Swan *et al.* 2017). From 2010, the emphasis on clean FW and YWT increased substantially with FD making a small negative contribution, due to a positive trend in the trait (Swan *et al.* 2017). The relative economic contribution of FD to the current wool focussed MERINOSELECT indexes is 21 and 17% for the Fine Wool (75:25 wool to meat income ratio) and Wool Production (71:29) indexes respectively (MLA 2024). There has also been a recent industry directive for all MERINOSELECT indexes to target animal welfare, in particular reducing breech wrinkle (MLA 2024), which has unfavourable genetic correlations with FD (Hatcher and Preston 2018).

MICRON PREMIUMS SINCE 2003

MPs for fine (17, 18 & 19 micron) and medium (20, 21 and 22 micron) Merino fleece wool were calculated from the daily micron price guides (MPGs, average of the Northern and Southern selling centres) reported by the Australian Wool Exchange (AWEX) in their National Market reports from the 1 July 2003 to the 26 June 2024 (Table 1) representing the past 20 wool selling seasons. The data was also spilt into the past 10 and past 5 selling seasons. Over the past 20 wool selling seasons the 17MP, 18MP and 19MP averaged 11.7, 11.1 and 10.4% respectively. The average of the medium wool MPs (20MP, 21MP and 22MP) were about a third to a quarter the size of the fine wool MPs.

Table 1. Average and range of MP for fine and medium Merino fleece wool for the past 20 (from 2003/04), 10 (from 2013/14) and 5 (from 2019/20) wool selling seasons to 2023/24

MP	Past 20 seasons *2003/24 to 2023/24*			Past 10 seasons *2013/14 to 2023/24*			Past 5 seasons *2019/20 to 2023/24*		
	Ave	Min	Max	Ave	Min	Max	Ave	Min	Max
17	11.7	0.5	73.4	9.8	0.5	28.9	15.5	0.7	28.9
18	11.1	-5.9	31.6	10.5	1.6	28.8	14.9	2.2	28.8
19	10.4	-0.7	37.1	8.0	-0.4	26.4	11.1	-0.4	26.4
20	3.9	-1.0	12.7	3.0	-0.9	10.4	4.1	-0.2	10.4
21	2.8	-7.6	10.8	2.0	-7.6	10.8	2.4	-2.0	10.8
22	2.7	-3.3	13.8	1.7	-3.3	10.3	3.9	1.4	9.2

MPs were reasonably consistent despite significant variation in the overall market level of Australian wool (Eastern Market Indicator, EMI) sold at auction (Figure 1) and increases in the weight of fine (17.5 and 18.5 μm) wool tested (a proxy for wool production) since 2003/04 (Figure 2). This highlights the consistent high value placed upon fine wool by the global wool trade, regardless of the overall level of the Australian market and volume of fine wool.

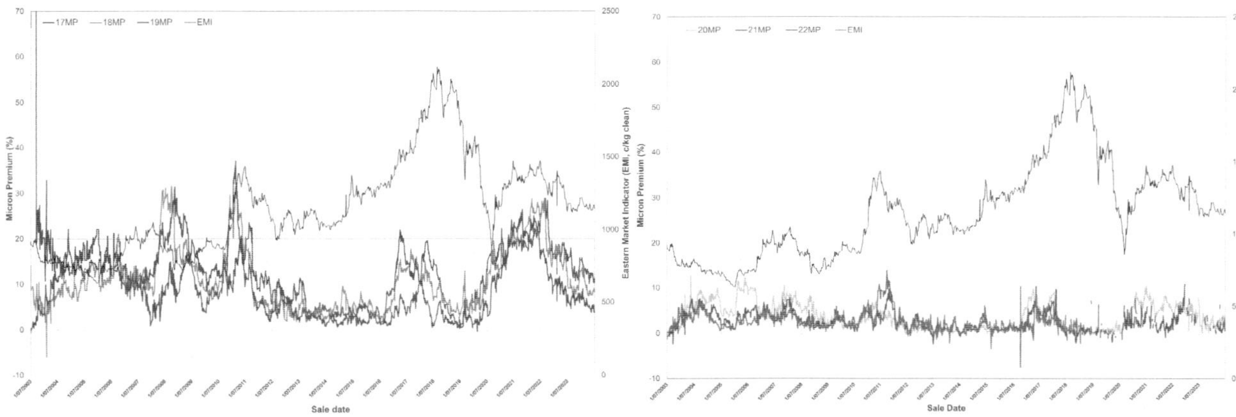

Figure 1. Variation in a) fine and b) medium wool MPs over the past 20 wool selling seasons compared with the Eastern Market Indicator (EMI, c/kg clean) wool.

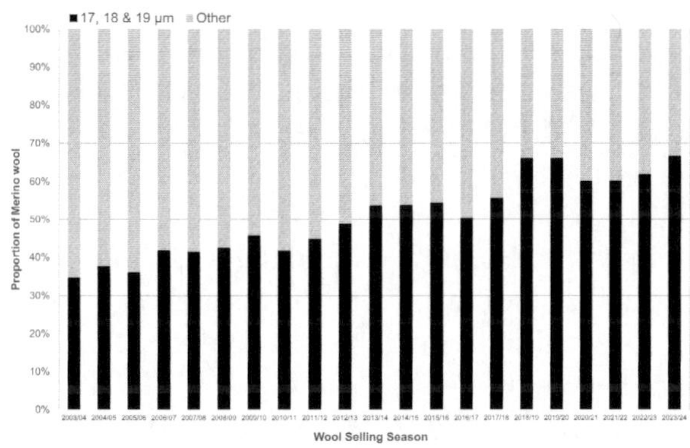

Figure 2. Variation in the weight of fine (≤ 19.5 µm) wool relative to broader Merino wool tested (AWTA Key Test Data) over the past 20 wool selling seasons (Source: AWTA Analytics. Accessed 29 September 2024).

HOW ARE MPS ACCOMODATED IN MERINOSELECT INDEXES?

The standard MERINOSELECT indexes were designed for four general production systems that cover the broad range of breeding directions used in the Merino industry (MLA 2024). These range from those strongly focussed on reducing fibre diameter (Fine Wool) to those with a strong focus on increasing wool production (Wool Production) along with those that balance sheepmeat and wool production (Sustainable Merino) and those with a strong focus on sheepmeat production (Merino Lamb) (MLA 2024). Given the current high MPs for fine wool, incorporating the current FD price relationships into the suite of MERINOSELECT industry indexes would place a higher relative emphasis on reducing FD. As FD is a highly heritable trait, the net effect would approach single trait selection for FD regardless of the emphasis placed on other traits in any of the MERINOSELECT indexes. This would limit genetic gain in other traits that make an economic contribution to each index. As genetic trends across the MERINOSELECT database indicate most Merino breeders have historically placed a high emphasis on increasing FW while maintaining FD (Swan *et al.* 2017), the MERINOSELECT indexes are adjusted to express a reduced MP as with historic indexes, 50% for Fine Wool and 40% for Wool Production (Figure 3) compared to the prices paid for FD.

CONCLUSION

The development of industry selection indexes is extremely challenging in the current market environment with emphasis on an increased range of wool, meat and reproduction traits plus the desire to include welfare traits. While the latter often have little to no market value, they are included

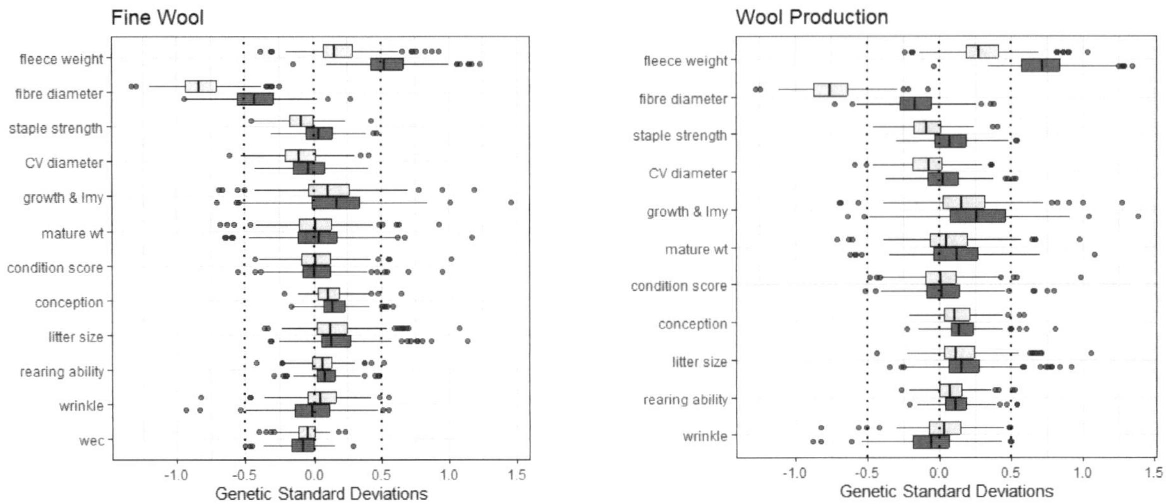

Figure 3. Range in relative selection advantages (superiority of top 10% to average) realised from selection on the current index (purple) or when full MP realised within the same index (yellow) for the Fine Wool and Wool Production indexes across MERINOSELECT member flocks in their 2023 drop progeny.

as indicators in the Sheep Sustainability Framework and the Australian Wool Sustainability Scheme. The genetic relationships, often unfavourable, between traits adds to the challenge. Variation in flock structure, current production level of individual flocks and genetic relationships between traits within a flock means the relative economic value of traits in MERINOSELECT indexes is a sound starting point but not a perfect fit for every Merino enterprise. As the emphasis on FD in MERINOSELECT indexes does not reflect the high MPs for fine wool, individual Merino breeders require easy access to MPs and the relative FD emphasis in MERINOSELECT indexes to make informed decisions on the index they choose to use on genetic gain in FD and the need to apply independent emphasis on traits to make progress towards their flock's breeding objective. Simple response predictions tools that allow the impact of changing trait weightings are required to make it straightforward to assess the impact of changing trait weightings, in particular MPs.

REFERENCES

Australian Bureau of Agricultural and Resource Economics and Sciences (2024) Farm Data Portal. https://www.agriculture.gov.au/abares/data/farm-data-portal Accessed 24-Sep-2024.

Australian Bureau of Statistics (2004) Agricultural Commodities, Australia 2002/03 Çatalogue no. 7121. Australian Bureau of Statistics, Canberra, ACT Australia.

Hatcher S. and Preston J.W.V (2018) *Small Ruminant Res.* **164**: 48.

Meat & Livestock Australia (2023) Industry projections 2023. Australian sheep – July update. Meat & Livestock Australia, North Sydney, Australia.

Meat & Livestock Australia (2024) MERINOSELECT indexes – A ram breeder's guide. Meat & Livestock Australia, North Sydney, Australia

Safari E., Fogarty N.M. and Gilmour A.R. (2005) *Livest. Prod. Sci.* **92**: 271.

Swan A.A., Banks R.G., Brown D.J. and Chandler H.R. (2017) *Proc. Assoc. Advmt Anim. Breed. Genet.* **22**: 365.

Swan A.A., van der Werf J.H.J. and Atkins K.D. (2007) *Proc. Assoc. Advmt. Anim.Breed. Genet.* **17**: 483.

Proc. Assoc. Advmt. Anim. Breed. Genet: 26: 291-294

COST-BENEFIT OF GENOTYPING COMMERCIAL MERINO SHEEP

G. Rose[1], S. Martin[2], D.J. Brown[1], L.P. Kahn[3] and J.H.J. van der Werf[3]

[1] Animal Genetics Breeding Unit*, University of New England, Armidale, NSW, 2351 Australia
[2] SheepMetrix, Young, NSW 2594 Australia
[3] School of Environmental & Rural Science, University of New England, Armidale, NSW 2351 Australia

SUMMARY

Genomic predictions of genetic merit are commonly used in the seedstock industry but application to commercial livestock systems is lagging in comparison. Therefore, this paper aims to show the cost-benefit of using genomic information in culling and selection decisions of commercial Merino sheep. The benefit of such information depends on the achieved selection superiority and the number of discounted expressions of that superiority in the future. We compared a moderately accurate genomic test with average accuracy of 57% across all traits with a highly accurate test with average accuracy of 80% across all traits. The break-even cost of genotyping males in a multiplier flock for use as sires in a commercial flock was AU$258 (moderate accuracy) and AU$377(high accuracy) and for ewes in the multiplier flock it was AU$107 (moderate accuracy) and AU$156 (higher accuracy). The break-even cost of selecting replacement ewes in a commercial flock is between AU$14 and AU$21. These results indicate that there is opportunity for cost-effective genotyping in commercial Merino flocks, especially where home-bred rams are selected from a multiplier flock.

INTRODUCTION

Genomic testing has rapidly been adopted in the seedstock industry, with over 0.5 million animals genotyped in the Sheep Genetics database (MLA 2024). Genomic prediction is becoming more accurate as more animals with measured phenotypes also have genotypes. While most of the genomic testing completed is currently in seedstock breeder flocks, there is potential to extract value from genomic testing in commercial flocks. Genomic predictions have already been used to predict genetic merit of animals at a flock level, such as the current Flock Profile prediction (Swan *et al.* 2018) for commercial Merino flocks. Commercial producers use a Flock Profile to help benchmark where their flock sits within industry and use the results to choose better genetic merit rams and track genetic progress. Individual prediction of the genetic merit of commercial sheep without recorded phenotypic information could also be useful to rank animals within flock, e.g. for decisions around culling or mating of ewes for flock replacement, or for selecting ewes or rams in multiplier flocks to improve the genetic merit of sires used in commercial flocks. Genotyping of commercial cattle is already utilised in the beef and dairy industries, but the value proportion has not been explored in sheep. Therefore, the aim of this paper was to develop a cost-benefit analysis of genomic prediction in commercial Merino's to underpin potential adoption of the technology in the sheep industry.

MATERIALS AND METHODS

The value of using genomic prediction in decision making was modelled for a commercial Merino sheep flock. This modelling estimated the cost-benefit and break-even costs of genotyping to get a genomic breeding value (GBV). Commercial GBV products are not yet available in sheep,

---

*A joint venture of NSW Department of Primary Industries and Regional Development and the University of New England

so we compared tests with assumptions about accuracies to show the value of potential future products. We used selection index theory to predict selection differentials of selected ewes or rams using the Sustainable Merino (SM) index (MLA 2024) as an objective. Selection accuracy was compared for three different information sources:

1. A postweaning phenotype for weight, fibre diameter and clean fleece weight;
2. A moderately accurate genomic test, similar to the current expected accuracy (equivalent to having five progeny measured, average accuracy across 20 traits = 57%); and
3. A more accurate genomic test, as to be expected in the next few years (equivalent to 20 progeny measured, average accuracy across all traits = 80%)

Other traits included weight (4 traits), fibre diameter (3) and clean fleece weight (3) at different age stages, staple strength (3), condition score, faecal egg count, number lambs weaned, lean meat yield, intramuscular fat, dag and breech wrinkle scores. We predicted future expression of selection superiority of ewes or rams in their descendants, or of their own phenotype using a deterministic geneflow method. The benefit of selective culling depended on:

1. The predicted difference, how much better are the animals that are kept?
2. The number of future expressions of that superiority, which in turn depends on:
    a. The number of future phenotypic expressions of the animals itself
    b. The number of expressions in future offspring and their descendants
    c. When these expressions are realised, as future benefits should be discounted

Future expression of the selection superiority of selected parents was modelled with the gene flow method (Hill 1974). Three scenarios of selection were considered: 1) selection of replacement ewes in a commercial flock, 2) selection of males born in a multiplier flock, to be used as sires in a commercial flock, and 3) selection of ewes within the multiplier flock.

Selection index theory was used to predict selection superiorities, from the regression of the breeding objective on the information that is used to predict it. These predictions were calculated using MERINOSELECT genetic parameters for 20 traits (not all trait results presented) and economic values (pers com Sheep Genetics).

Table 1 shows the proportion of selection superiority in 1-year old offspring of commercial rams and ewes resulting from selection of: 1) commercial flock replacement ewes, 2) multiplier males to be used in the commercial tier, and 3) multiplier replacement dams. Considering a time horizon of 20 years, the cumulative discounted expression (CDE) is the sum of these expressions after discounting them at 5% based on year of expression. When calculated over a 20-year time horizon and per breeding ewe in the flock the CDE for these three scenarios is 0.676, 0.721 and 0.446, respectively.

Table 1- Proportion of selection superiority in 1-year old offspring of commercial ewes resulting from selection of 1) commercial replacement ewes, 2) multiplier males to be used in the commercial tier, and 3) multiplier replacement dams

Selection strategy	**Year after selection**									
	1	2	3	4	5	6	7	8	9	10
Replacem. Ewes Comm.	0.00	0.11	0.10	0.10	0.10	0.10	0.10	0.04	0.05	0.04
Multiplier Males	0.00	0.18	0.17	0.17	0.04	0.05	0.05	0.05	0.05	0.04
Multiplier Dams	0.00	0.00	0.00	0.02	0.04	0.05	0.05	0.06	0.06	0.06

Value of selecting better replacement females in a commercial flock. We assumed a flock of 1,000, breeding ewes with a weaning rate of 1.2 lambs per breeding ewe, an annual attrition rate of

10%, and a culling age of 7 years for the ewes. This resulted in a replacement rate of 23.7% of hogget ewes entering the breeding flock. Hence 237 ewes were selected at hogget age from 432 candidates (selection rate of 54%), if only 80% of 540 hoggets was genotyped or measured for phenotype.

Economic values accounted for multiple expressions and discounts for traits expressed later in older animals. Therefore, we calculated the value of selection based on the genetic value of ewes expressed at 2 years old. The value of selecting better replacement ewes is then equal to i.(S_p *CDE_p + $S_g CDE_g$).EconVal per replacement ewe, where i = 0.72 is the selection intensity, S_p and S_g are vectors with the phenotypic and genetic selection differential for each trait, respectively, CDE_p and CDE_g are the phenotypic and genetic cumulative discounted future expressions of traits in the entire commercial flock (0.81 × 1,000 and 0.676 × 1,000, respectively), and EconVal is a vector with the economic value for each trait.

Value of selecting better multiplier rams and dams. We assumed a commercial flock of 8,000 ewes with rams sourced from a multiplier flock of 400 breeding females. Rams for the commercial flocks were kept for 3 matings. Annually, 59 rams were selected from 173 candidates (34%) and 95 ewes from 173 candidates (55%) that were measured or genotyped in the multiplier flock. The CDE in the commercial flock of 8,000 ewes is calculated as 0.721 × 8,000 = 5771. Selecting replacement ewes for the multiplier flock results in CDE in the commercial flock of 8,000 ewes of 0.446 × 8,000 = 3,568. The cost benefit analysis estimated the break-even cost of testing. The break-even cost was the profit per selected animal multiplied by the number of selected animals and divided by the number of animals that were DNA-tested.

RESULTS AND DISCUSSION

Table 2 shows the benefit of selection, expressed in dollar value per selected ewe. The example in Table 2 had 213 selected ewes expressing the benefit while there were 432 ewes genotyped. The results show that the break-even cost for obtaining information is modest, ranging between AU$7 and AU$22, which is lower than the current genotyping cost. It should be stated that the difference between phenotypic selection and genomic selection is the largest selection differential and benefit that can be obtained with genomic selection. There is also a shift towards more benefit for hard to measure traits, such as reproduction and other traits related to sustainability. For example, with selection on phenotype, 65% of the benefit is due to improvement of wool traits and 35% is from reproduction traits indirectly through selection on live weight, whereas with accurate genomic selection these relative contributions are 27% and 73%, respectively.

Table 2. Benefit per selected replacement ewe of three selection strategies for replacement females in a commercial flock, and the break-even cost of collecting the information used for selection, using 20 traits

Selection criteria	Benefit per selected ewe	Break-even cost
Selection on phenotype	AU$14.19	AU$7.01
Selection on genotype (~57% accurate)	AU$ 28.66	AU$14.16
Selection on genotype (~80%accurate)	AU$43.69	AU$21.58

The break-even cost is quite insensitive to several assumed parameters, in particular weaning rate and replacement rate. A weaning rate of 1.5 would give a break-even cost of genotyping of AU$14.10, rather than AU$14.16 as in Table 2 (weaning rate 1.2) as the extra benefits per ewe are offset by more genotyping per selection candidate, i.e. there are more selection candidates, and this outweighs the greater selection intensity. The proportion of selection candidates that is genotyped may vary, with a lower proportion genotyped giving lower costs but also lower selection intensity.

Genotyping a larger proportion is more beneficial with similar break even cost for any genotyped proportion of ewes >75%. Table 2 shows clearly that the accuracy of the genomic test has a large effect on the break-even cost of genotyping.

Table 3 shows the break-even cost for selecting males or replacement females in the multiplier flock, respectively. The break-even cost is a lot higher for the multiplier because the much larger number of animals that benefit compared to the number of animals genotyped. Furthermore, the selection intensity in the rams selected in the multiplier flock for use in the commercial flock is higher, which also leads to a different optimal proportion of males genotyped. This optimal proportion of multiplier males genotyped was around 50-60%, with a break even cost of AU$270 whereas the break-even cost with 34% genotyped was AU$257. Selecting of males in the multiplier has a higher break-even cost compared to selecting females because fewer males were selected from potential candidates (34% in males compared to 55% in females).

Table 3. Benefit of three selection strategies for selecting rams and ewes from the multiplier and the break-even cost of collecting the information used for selection

Selection criteria	Selection in rams	Selection in ewes
Selection on phenotype	AU$ 141.46	AU$ 58.78
Selection on genotype (~57% accurate)	AU$ 257.73	AU$ 107.09
Selection on genotype (~80% accurate)	AU$ 376.61	AU$ 156.49

CONCLUSIONS

These results indicate that there is opportunity for cost-effective genotyping in commercial Merino flocks. This opportunity seems to be attractive at this stage to commercial producers that use a multiplier flock to breed rams for their commercial flocks based on the current cost of genotyping. These benefits do not include other potential advantages of having genotyped animals, for example, being able to inform potential buyers of commercial ewes the genetic potential of animals. Despite this, this research shows the potential benefits that commercial genotyping products have for the Merino sheep industry in Australia.

ACKNOWLEDGMENTS

Funding for this work was from MLA-MDC project P.PSH.1483 and through funding from the Australian Government's Future Drought Fund.

REFERENCES

Hill W.G. (1974) *Anim. Sci.* **18**: 117.

Meat and Livestock Australia (2024) MERINOSELECT indexes A ram breeder's guide. https://www.sheepgenetics.org.au/globalassets/sheep-genetics/getting-started/sheep-genetics---asbvs-and-indexes/merinoselect-indexes-breeder-2024.pdf

Swan A.A., Gurman P.M., Boerner V., Brown D.J., Clarke S., Gore K., Granleese T. and van der Werf J.H.J. (2018) *Proc. World Cong. Genet. Appl. Livest. Prod.* **11**: 346.

Proc. Assoc. Advmt. Anim. Breed. Genet: 26: 295-298

A COMPARISON OF PROGENY OF SELECTED AND UNSELECTED SIRES WITHIN THE SETTING OF A COMMUNITY-BASED BREEDING PROGRAMME CATERING TO WOMEN SMALLHOLDER GOAT REARERS IN BIHAR INDIA

S. Abeykoon[1], P. Amer[2], P. Wilcox[1] and C. Nimbkar[3]

[1] Department of Mathematics and Statistics, University of Otago, North Dunedin, Dunedin 9016 New Zealand
[2] AbacusBio Limited, Central Dunedin, Dunedin, 9016 New Zealand
[3] Nimbkar Agricultural Research Institute, Phaltan, Maharashtra, 415523 India

SUMMARY

Phenotypic goat data is generated as part of a community-based breeding programme (CBBP) to select breeding sires at the age of 100 days according to a score derived from combining multiple selection criteria within a multifaceted women's empowerment project. This analysis primarily aimed to utilise this phenotypic data to establish the effect of selecting sires within the CBBP on average daily gain (ADG) up to 100 days. A secondary aim was to establish the effect birth year of kids fathered by unselected sires had on ADG thereby removing the influence of the CBBP. Analysis of the data using multiple linear regression models revealed an overall statistically significant result that progeny of selected sires had a superior mean ADG 13.6% ($P < 0.05$) higher than the mean ADG of kids of unselected sires. This trend did not translate unanimously to village specific results. Analyses of birth year effects of kids of unselected sires on ADG showed an incremental underlying trend in growth rate increasing by 2.4g/day ($P < 0.05$) each year. Though some of the village specific results are inconclusive, overall results support the wide popularity of using a CBBP as one component in a women's empowerment programme to provide economic benefits through goat rearing despite limitations in an analysis utilising a dataset designed for sire selection.

INTRODUCTION

Goat rearing provides support to approximately 70% of landless, marginal, and small farm households in rural areas of the Bihar state of India (Schurink *et al.* 2022). Goats, being smaller and easier to tend to than other livestock, are usually reared by women (Barooah *et al.* 2020) who live with their families in small villages of several hundred people interspersed among fields of row and tree crops. They commonly keep 1–2 breeding dams with their associated litters as a relatively liquid supplementary source of income (Kumar *et al.* 2018).

Traditionally, most sires were unselected roaming bucks that had the potential to breed with most does in a village over an extended period (Schurink *et al.* 2022) causing a considerable risk of some sires mating with their own progeny. Resultant inbreeding in ruminants can lead to stunted growth, reduced fertility, and higher mortality rates (Leroy 2014). To ameliorate this situation and to increase goat productivity through genetic improvement, the Aga Khan Foundation (AKF) introduced a community-based goat breeding programme (CBBP) as part of a women's empowerment programme called Project Mesha which also included other initiatives such as training women in animal husbandry to provide preventive healthcare for goats and providing entrepreneurial opportunities through tending for breeding bucks (Nimbkar *et al.* 2021).

Primarily, this study aims to establish if there is an effect of selecting sires based on a score derived from combining multiple selection criteria on the average daily gain (ADG) in grams per day of progeny in a CBBP through the analysis of a dataset designed for selecting sires. Secondarily, it aims to use the same dataset to uncover the underlying trend of a kid's birth year on its ADG without the influence of the CBBP.

MATERIALS AND METHODS

Programme management. Project Mesha is a multifaceted women's empowerment program which targets goat keeper women who are members of the self-help groups within the Bihar Rural Livelihood Mission (JEEViKA) (Schurink *et al.* 2022). Key components of Project Mesha include: pashu sakhis who are women trained in animal husbandry and deliver fee-based animal preventive healthcare such as vaccinations, deworming, castration, breeding expertise, and encouragement of performance recording to assist genetic evaluation (Schurink *et al.* 2022); creating entrepreneurial opportunities for women via becoming breeding buck keepers; and the development and day-to-day management of a CBBP (Nimbkar *et al.* 2021).

The CBBP was initiated in 2018, 2 years after the commencement of Project Mesha. Individual animal-level data from these villages, including pedigree, phenotypic and dam health data, were recorded and stored using the Dtreo flexible cloud database system (https://dtreo.io/). Kids were selected as potential breeding sires based on having a desirable combined score derived from the following selection criteria: fast-growth in the first 100 days; being a twin; and having dams with a lifetime record of good health and liked by their owners (a subjective score provided by the owner at doe evaluation). Primary selection was made by the Mesha field team using data in the form of Dtreo generated reports of selection indexes for male kids and observation for absence of phenotypic defects (Savoia *et al.* 2023). If a kid was selected, its owner's consent was obtained to keep it uncastrated until the point of secondary selection occurring at the age of 6 to 8 months (Nimbkar *et al.* 2021). Project Mesha purchases selected sires at a price premium over the market value for meat and transports them to villages at least 20 km away.

Statistical analysis. Kids included in the analysis: i) had a minimum of three weight records; ii) had a known sex either 'male' or 'female'; and iii) were not identified as outliers with consecutive weights resulting in ADGs 3 standard deviations or greater deviating than the average. This totalled 2,002 kids including 960 from selected sires.

Four linear regression models were used for the analysis of sire identity effect. These models were fitted to data from all 2,002 kids.

In Models 1–4 (Table 1) Y refers to the ADG estimates (g/day) obtained by linear regression; β_0 is the intercept term; K_h is the sire identity effect (h = 1 and 2) (1, unselected; 2, selected); V_i is the village effect (i = 1–18); S_j is the sex effect (j = 1 and 2) (1, Female; 2, Male); L_k is the litter size (k = 1–4); W_l is the season of birth effect (l = 1 and 2) (1, Dry (November to May), 2, Rainy (June to October)); and P_m is the dam parity effect (m = 1–3) (1, First Parity (1), 2, Second Parity (2), 3, Third Parity or above ($\geq$ 3)). Lastly, e is a term representing the residual error. Additionally, Model 3 incorporates a new main effect: an unordered nominal calendar year of birth effect Q_n (n = 1–6) (1, Year 2018, 2, Year 2019, 3, Year 2020, 4, Year 2021, 5, Year 2022, 6, Year 2023) and Model 4 incorporates a continuous calendar year of birth effect Q.

Table 1. Models to analyse sire identity effect on ADG (g/day)

Model number	Model
1	$Y_{hijklm} = \beta_0 + K_h + V_i + S_j + L_k + W_l + P_m + e_{hijklm}$
2	$Y_{hijklm} = \beta_0 + (K_h \times V_i) + S_j + L_k + W_l + P_m + e_{hijklm}$
3	$Y_{hijklmn} = \beta_0 + (K_h \times V_i) + S_j + L_k + W_l + P_m + Q_n + e_{hijklmn}$
4	$Y_{hijklm} = \beta_0 + (K_h \times V_i) + S_j + L_k + W_l + P_m + Q + e_{hijklm}$

A further 2 linear regression models, Models 5 and 6 (Table 2), were designed to analyse the year effect on ADG of progeny fathered by unselected sires (1,042 kids). Variables were defined as in Models 1–4.

Table 2. Models to analyse birth year effect on ADG (g/day)

Model number	Model
5	$Y_{ijklmn} = \beta_0 + V_i + S_j + L_k + W_l + P_m + Q_n + e_{ijklmn}$
6	$Y_{ijklm} = \beta_0 + V_i + S_j + L_k + W_l + P_m + Q + e_{ijklm}$

RESULTS AND DISCUSSION

Primary objective – sire identity effect. Model 1 results analysing the difference in means show kids generated by selected sires grew 13.6% faster ($P < 0.05$) compared to kids of unselected sires. Model 2 village-specific sire identity effects translated into statistically significant ($P < 0.05$) improvements in mean ADG of progeny of selected sires for villages 3, 5, 8, 9, 11, and 13.

When categorical or continuous year effects were included (Models 3 and 4 respectively), both models resulted in the same statistically significant villages for sire identity effect. Villages 3, 5, and 8 results showed increased selected sire effects ($P < 0.05$). Villages 1, 2, and 10 results showed decreased selected sire effects ($P < 0.05$). Villages 9 and 11 which lost statistical significance from Model 2 ($P > 0.05$) had progressively increasing proportions of selected sires throughout analysed years potentially leading to unstable data structures which could promote multicollinearity.

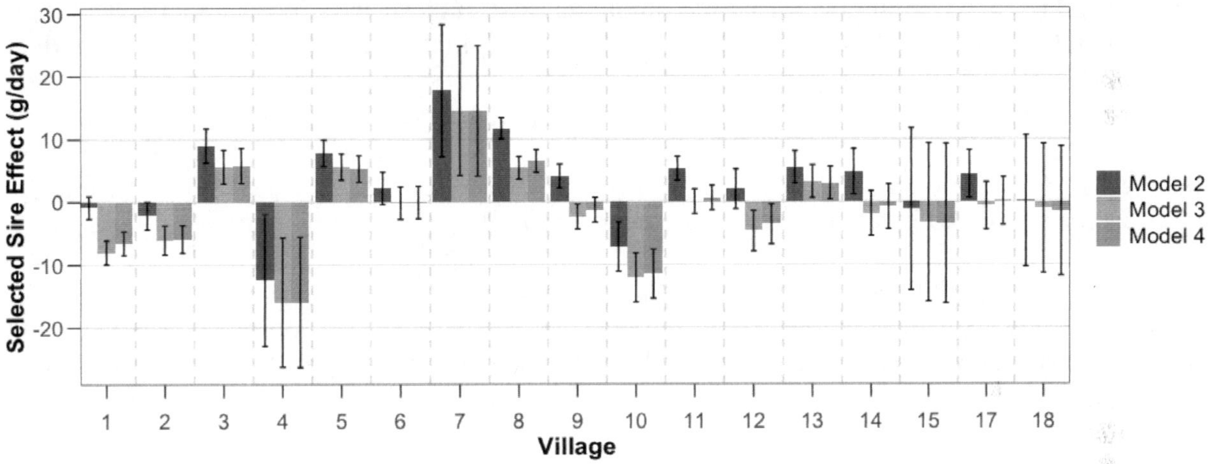

Figure 1. Mean selected sire effect on ADG (g/day) ± SE of recorded kids in Villages 1–18 in accordance with Models 2, 3, and 4

Secondary objective – birth year effect. Birth year effect on ADG was analysed using the kids of unselected sires only. Model 5 with the categorical year effect rendered the birth year effect on ADG insignificant ($P > 0.05$) for all years analysed, 2018 until 2023, when compared to the mean ADG of the reference year, 2018. Results from Model 6 designed to illustrate a continuous year effect on ADG indicated a statistically significant ($P < 0.05$) linearly increasing ADG with a gradient of 2.4g/day for each additional year within the study (standard error = 0.6 g/day).

Variability in ADG among progeny in this study was expected, as ADG was not the sole criterion for selecting young bucks as sires. In addition to growth potential, sires were chosen from healthy mothers deemed desirable by their owners and twinning was selected for (Nimbkar *et al.* 2021). This simultaneous incorporation of multiple traits within a breeding programme will mean that there are differences in rates of genetic gain for individual traits.

This analysis faced certain data structure limitations, i.e. using a dataset designed to select sires for a case control study and having gaps in progeny trait records. Future analyses could improve robustness of model outcomes by acknowledging confounding factors but that would not be practical when considering the primary purpose of this data is to select sires.

Despite these limitations in analysis, overall results from Model 1 are complementary to the substantial practical success the project has encountered. Currently, 150 villages have a selected breeding buck in residence, generating business income for women goat keepers and pashu sakhis. Some women, now recognised as 'buck entrepreneurs,' have expanded their operations to manage multiple breeding bucks. By sourcing sires from distant villages, the project effectively mitigates inbreeding, with each buck servicing approximately 20 does per month. Assured availability of high-quality breeding bucks has encouraged more women to engage in goat rearing, stimulating rural economic activity and enhancing women's financial independence which were the main considerations in this women's empowerment project.

CONCLUSION

Analysis of a dataset created to select sires for the CBBP component of a programme designed to create economic opportunities for women goat keepers indicated an overall significant increase in ADG for kids born to selected sires. This result didn't translate unanimously to village specific results. Analysis of only kids of unselected sires, and thus removing the influence of the CBBP, showed kids born in later years had statistically significant improved growth rates when treating the birth year effect as a continuous variable implying that factors like project management or environmental factors could also have contributed to improving progeny growth rates. Individual birth year effects on ADG fluctuated meaning more data is necessary to consider extrapolating this incremental trend. This analysis faced data structure limitations and considered only one of multiple breeding objectives of the CBBP. Other benefits of the project included economic opportunities and avoidance of goat inbreeding due to the CBBP leading to healthier progeny thereby improving the lives of women involved in this women's empowerment initiative.

ACKNOWLEDGEMENTS

Project Mesha and all goat recording activities as part of the Community-Based Breeding Programme were coordinated by the Aga Khan Foundation with funding from the Gates Foundation.

REFERENCES

Barooah B., Sarkar R., Dubey P. and Bagai A. (2020) 'Project Mesha formative evaluation report'. https://www.3ieimpact.org/sites/default/files/2020-05/Project-Mesha-Formative-Research-Report.pdf Accessed on 31 Mar. 2025.

Kumar D., Kumar R. and Sharda K. (2018) *J. Pharmacogn. Phytochem.* **7**: 2329.

Leroy G., (2014) *Anim. Genet.* **45**: 618.

Nimbkar C., Amer P., Schurink A., Sawhney T. and Azam R. (2021) *Proc. Assoc. Advmt. Anim. Breed. Genet.* **24**: 272.

Savoia S., Santos B., Liebergreen S., Ooi E.C., Kumar S., Teviotdale M., Amer P. and Byrne T. (2023) *Int. Comm. Anim. Rec. Tech. Ser.* **27:** 219.

Schurink A., Kumar S., Chakraborty R., Sawhney T., Nimbkar C. and Amer P. (2022) *Proc. World Cong. Genet. Appl. Livest. Prod.* **12**: 1897.

Proc. Assoc. Advmt. Anim. Breed. Genet. 26: 299-302

OPTIMAL BREEDING PROGRAMS FOR LAO NATIVE GOATS IN SMALLHOLDER SYSTEMS

S.V. Le[1], S. de las Heras-Saldana[2], P. Alexandri[2] and J.H.J. van der Werf[1]

[1] School of Rural & Environmental Science, University of New England, Armidale, NSW, 2351 Australia
[2] Aninal Genetics Breeding Unit*, University of New England, Armidale, NSW, 2351 Australia

SUMMARY

This study used stochastic simulation to evaluate breeding strategies aimed at improving body weight of Lao native goats in smallholder systems while maintaining genetic diversity. Simulation reflected both traditional breeding practices and alternative strategies, including random mating, selection on phenotype, and rotational mating scenarios at herd and village level. After 20 mating cycles, traditional practices (random mating at herd level) resulted in limited body weight improvement and a high inbreeding level (0.216). Selection and rotational mating showed improvement in body weight while also affecting inbreeding level. The rotation strategy achieved the highest body weight gain (1.2 phenotypic standard deviation, PSD, units) and a relatively low inbreeding level (0.071). The genetic gain and inbreeding levels were 0.899 PSD and 0.077, respectively, for village-selection and 0.69 PSD and 0.24, respectively, for herd-selection. Overall, rotational mating significantly increased the body weight while maintaining low inbreeding, making it a promising breeding strategy for Lao farmers. For successful implementation, this approach will require pedigree and animal performance recording and active farmer participation.

INTRODUCTION

Genetic improvement of indigenous livestock provides an important strategy to improve the living standard of farmers in developing countries. Lao native goats play an important role in the livelihoods of smallholder farmers, contributing up to 46% of total household income (Walkden-Brown *et al.* 2024). Like other indigenous breeds, Lao native goats have small body sizes and a slow growth rate but they are adapted to local environments and management practices. These goats are primarily raised in smallholder systems with herd sizes ranging from 1 to 37 goats (Walkden-Brown *et al.* 2024). Based on spatial data, Le (2024) observed that goats in Laos graze on communal lands or are kept on private property, with males used either from their herd or exchanged with other herds. Some gene flow was detected between herds both within and across villages (Le *et al.* 2023; Le *et al.* 2024). Currently, no formal breeding program has been introduced or implemented for Lao native goats in smallholder systems and farmers lack systematic evaluation of productivity (Gray *et al.* 2019). Several studies have explored different breeding strategies for smallholder systems, the cooperative village or village-based breeding programs (CBBPs) closely align with the observed management practices in Laos (Gizaw *et al.* 2014; Le 2024). To assess the potential of CBBPs in this context, we simulated various breeding scenarios in which sires and dams were selected either randomly or based on an observed phenotype for body weight. These sires were used either within their herds or exchanged with other herds within and between villages. When sires are selected from the same herd or different herds, females could be mated with siblings or relatives, which would increase the inbreeding rate. The objective was to identify the most effective breeding programs for improving body weight while maintaining genetic diversity in smallholder systems, providing recommendations to Lao farmers on sustainable breeding practices.

---

* A joint venture of NSW Department of Primary Industries and Regional Development and the University of New England

MATERIALS AND METHODS

Breeding objective. Lao native goats have a small body size and are bred for meat production. Kids are light at birth weighing between 1.1 to 1.5kg and reach 9.7kg at weaning at around 3 months of age. Traders, abattoirs and restaurant owners offer higher prices to Lao farmers when their goats reach 20kg weight at around a year old (Walkden-Brown *et al.* 2024). As a result, Lao farmers are increasingly motivated to improve the productivity of their goats by enhancing their body weight to meet the growing demand from traders exporting to Vietnam (Walkden-Brown *et al.* 2024).

Breeding scenarios. Different breeding scenarios were simulated in R using a custom script to reflect both traditional breeding practices and alternative strategies at herd and village levels. The base population for each herd contained 22 unrelated animals (20 females and 2 males) and 20 females were mated to one of two males in the two mating rounds thereafter, with an average of two progeny per female per mating, ranging from 1 to 3 offspring. Each village had three herds. The simulation ran for 20 mating cycles, assuming a 50:50 sex ratio among offspring. The total number of animals simulated for one breeding program at the herd and village level was therefore 822 and 2,466, respectively. The herd of the offspring was assigned based on the herd of their dams. Starting from the third mating cycle, the mating strategy varied depending on the specific breeding scenario. Each breeding scenario was replicated 1,000 times to track patterns of genetic gain and inbreeding level.

Heritability (h^2) for body weight was set at 0.2, with offspring survival from birth to mating age at 0.8 and dam survival after each mating round at 0.9. Phenotypes were sampled as the sum of the breeding value and a random residual effect, with variance equal to h^2 and ($1-h^2$), respectively. The phenotypic variance was assumed to be 1 and $\sigma_a^2 = h^2$. Breeding values in the base population were sampled randomly for $N(0, \sigma_a^2)$, whereas breeding values in subsequent generations were simulated as a parental average breeding value with an additional Mendelian sampling term. The variance due to Mendelian sampling decreased with increased inbreeding (Van der Werf and De Boer 1990).

$$\mathrm{var}(\Phi) = \frac{1}{2}[1 - \frac{1}{2}(F_s + F_d)]\sigma_a^2$$

where $\mathrm{var}(\Phi)$ is a variance of the trait; F_s and F_d are inbreeding coefficients for sire and dam, respectively; σ_a^2 is the additive genetic variance before selection.

From the third mating cycle onward, dams and sires were selected either randomly or based on their phenotype, depending on the breeding scenario. For each herd, two male offspring from the (i - 2) mating cycle were selected, while 20 dams were selected from the dams of the (i – 1) mating cycle and the female offspring from the (i – 2 mating cycle) (Figure 1). Mating occurred either randomly or in a rotational system, depending on the scenario.

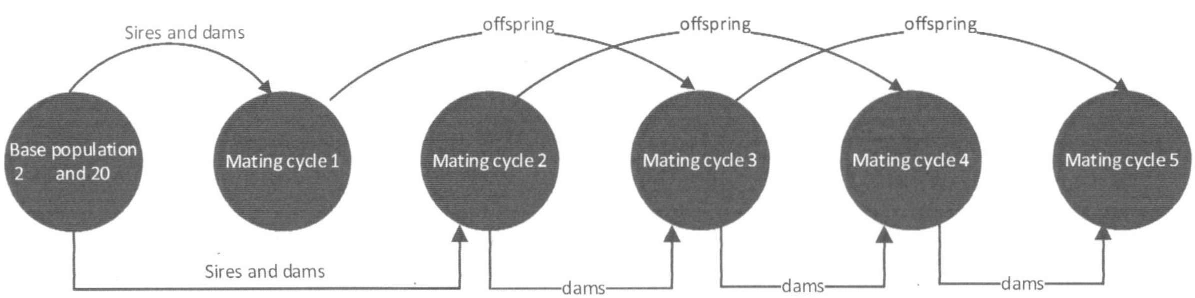

Figure 1. The diagram of the first five mating cycles for herd mating breeding scenarios

Definition of the breeding scenarios. *Herd-random.* Sires and dams are randomly selected in a herd and they are mated randomly.

Herd-selection. Sires and dams are selected based on their phenotype for each herd. The two heaviest sires and 20 heaviest dams are mated randomly.

Village-random. This was a simulation for three herds within a village, and sires and dams were randomly selected. From the third mating cycle, two sires and 20 dams were selected randomly from each herd and these six sires were randomly allocated to these 60 females.

Village-selection. This was similar to village-random, but sires and dams are selected based on their phenotype.

Village-rotation. This was similar to village-selection. Instead of random mating, a mating rotation was applied, where selected males from herd 1, herd 2, and herd 3 were mated with selected females in herd 2, herd 3, and herd 1, respectively.

Data analysis. In each breeding scenario, the body weight improvement was calculated as the average body weight across all the animals born from each mating cycle and then averaged across 1,000 replicates. Similarly, the inbreeding coefficient was averaged over all progeny born in each mating cycle, and then averaged across 1,000 replicates.

RESULTS AND DISCUSSION

The changes in body weight and inbreeding level of simulated Lao native goats over 20 mating cycles at herd and village levels are displayed in Figures 2A and 2B, respectively. In terms of breeding strategies, the rotation approach led to the greatest improvement, followed by selection, while random mating showed little effect on body weight. The maximum bodyweight improvements achieved after 20 mating cycles were for; village-rotation 1.2 PSD units, village-selection 0.9 PSD units, herd-selection 0.69 PSD units, herd-random 0.26 PSD units, and for village-random 0.108 PSD units.

Although the inbreeding levels increased in all breeding scenarios, the inbreeding levels for both breeding scenarios at herd level were higher than those at village level. The rate of inbreeding at the 20th mating cycle for herd-selection and herd-random scenarios were 0.24 and 0.216, respectively, while the inbreeding levels at village levels were 0.077, 0.071, and 0.029 for selection, rotation, and random mating strategies, respectively.

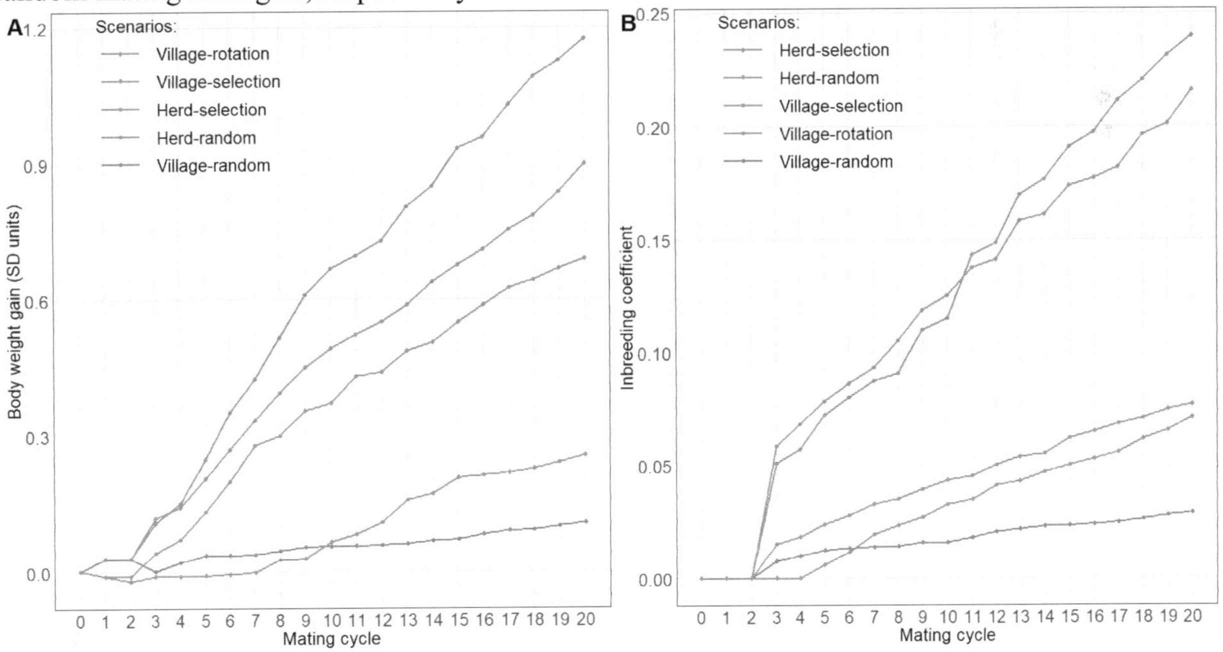

Figure 1. Average body weight gain (A) and inbreeding level (B) across mating cycles under all breeding scenarios

The rotational program at the village level appears to be a promismng strategy to improve the bodyweight of Lao native goats while maintaining a low level of inbreeding. For practical implementation, rotational breeding within a village or between villages could form a foundation for genetic improvement.

The findings in this study were in alignment with previous studies. For example in Ethiopia, Gizaw *et al.* (2014) reported significant genetic improvement in Menz sheep in village breeding programs, with birth weight and weights at three and six months increasing by 0.49 to 2.46 kg after three generations. Similarly in Sudan, Omer *et al.* (2022) found the highest genetic gain in milk yield performance of Butana cattle under the rotational use of bulls between herds compared to the use of herd bulls.

Lao farmers mainly rely on local grass and fodder to feed their goats (Gray *et al.* 2019). However, feed availability is characterised by seasonal shortages, low nutritional content and limited energy and protein (Xayalath *et al.* 2021). Supplementation is rarely practised in smallholder systems (Walkden-Brown *et al.* 2024). Therefore, improving the genetic potential for weight gain in Lao native goats must be accompanied by forage production initiatives and supplementary feeding practices. Additionally, since Lao goats typically graze on communal lands for 6 to 8 hours daily and bucks are generally uncastrated (Walkden-Brown *et al.* 2024), the implementation of a rotational breeding strategy requires specific management adjustments. Breeding bucks should be kept separate, and non-breeding bucks castrated to prevent undesired mating.

Future studies could explore more advanced genetic evaluation tools involving a larger number of herds per village, such as Best Linear Unbiased Prediction (BLUP), and genomic BLUP, to estimate the breeding values of economically important traits for Lao goats. Genetic improvement can also be achieved through within-breed selection or crossbreeding with exotic breeds. Boer goats or Bach-Thao goats are available in Laos; however, there is limited documentation on their performance and development in the region. Further studies should evaluate the effectiveness of crossbreeding strategies to improve the productivity of Lao native goats.

CONCLUSION

Lao native goats have small body sizes, and traditional breeding practices in Laos showed minimal body weight improvement over time while increasing the inbreeding rate. A rotational mating strategy, an alternative breeding program that involves selecting bucks based on phenotype ranking and exchanging them between herds, could improve the body weight of Lao goats while maintaining genetic diversity.

ACKNOWLEDGEMENTS

The first author thanks Australian Commonwealth Government for funding through the Research Training Program scholarship at the University of New England, Armidale, NSW, Australia.

REFERENCES

Gizaw S., Getachew T., Goshme S., *et al.* (2014) *Animal* **8**: 1249.
Gray G.D., Walkden-Brown S.W., *et al.* (2019) *ACIAR LPS/2016/027 Final Report.*
Le S. (2024) PhD. Thesis, University of New England.
Le S., de las Heras-Saldana S., *et al.* (2023) *Proc. Assoc. Advmt. Anim. Breed. Genet.* **25**: 250.
Le S., de Las Heras-Saldana S., Alexandri P., *et al.* (2024) *J. Anim. Breed. Genet.* **141**: 531.
Omer E.A., Hinrichs D., Addo S. and Roessler R. (2022) *J. Dairy Sci.* **105**: 5261.
van der Werf J.H.J. and De Boer I. (1990) *J. Dairy Sci.* **68**: 3124.
Walkden-Brown S., Colvin A., Millar J., *et al.* (2024) *ACIAR LS17/034 Final Report.*
Xayalath S., Mujitaba M.A., Ortega A.D.S.V. and Rátky J. (2021) *Acta Agr Debr* **55**: 227.

Proc. Assoc. Advmt. Anim. Breed. Genet: 26: 303-306

IDENTIFICATION OF RELATIONSHIPS AND PEDIGREE SIMPLIFICATION USING GRAPH THEORY

M.H. Ferdosi and D. Johnston

Animal Genetics Breeding Unit*, University of New England Armidale, NSW, 2351 Australia

SUMMARY

Reliable pedigree and genomic data are essential for single-step genomic best linear unbiased prediction. Discrepancies between pedigree and genomic relationships must be identified and resolved before analysis. While parent-progeny relationships are typically verified using genomic data, other relationships, such as those between progeny and grandparents or half-sibs, are often unchecked. The numerator relationship matrix and genomic relationship matrix further obscure the precise nature of individuals' relationships. For example, half-sibs and grandparent-progeny pairs appear identical in the numerator relationship matrix. This article identifies these relationships among individuals in the numerator relationship matrix that are inconsistent with the genomic relationship matrix, applying graph theory to identify the precise type of their relationships and simplify pedigrees to pinpoint and correct errors. The results demonstrate that graph theory algorithms can accurately identify pedigree relationship types and extract any individual's ancestors, descendants, or other relationships from the pedigree.

INTRODUCTION

The single-step genomic best linear unbiased prediction (ssGBLUP) integrates genomic and pedigree information to estimate breeding values accurately (Legarra *et al.* 2014). A critical component of quality control in this process is ensuring alignment between the genomic relationship matrix (GRM) and numerator relationship matrix (NRM). Pedigree errors are common in industry data, and some can be identified and corrected through DNA parentage verification and discovery. However, certain relationships, such as progeny-grandparent or cousin relationships, are largely unchecked (Connors *et al.* 2017). Incomplete pedigrees often lead to GRM values exceeding NRM values. However, when NRM values are significantly higher than GRM values (by more than 0.25, with GRM values near zero), pedigree errors are likely the cause. The scatter plot of NRM versus GRM values highlights these over-specified relationships. However, correcting these issues using additional information, such as herd and multi-sire grouped mating, requires identifying the precise relationship type. For instance, NRM values between individuals and their grandparents are identical to those among half-sibs, i.e., 0.25. The graph theory in mathematics considers structures called graphs, which consist of vertices (nodes) connected by edges (links) to analyse relationships, connectivity, and networks. Graph theory is used in this paper to resolve pedigree inconsistencies and precisely identify the nature of relationships in the pedigree that are not supported by relationships defined by the NRM and GRM. In addition, applying this theory simplifies the pedigree by extracting the ancestors or descents of individuals with issues to facilitate locating and fixing the pedigree errors.

---

* A joint venture of NSW Department of Primary Industries and Regional Development and the University of New England

MATERIALS AND METHODS

Test pedigree. A small pedigree with 15 individuals was created to test the graph theory approach. The R package visPedigree (Luan 2018) was used to plot this pedigree (Figure 1).

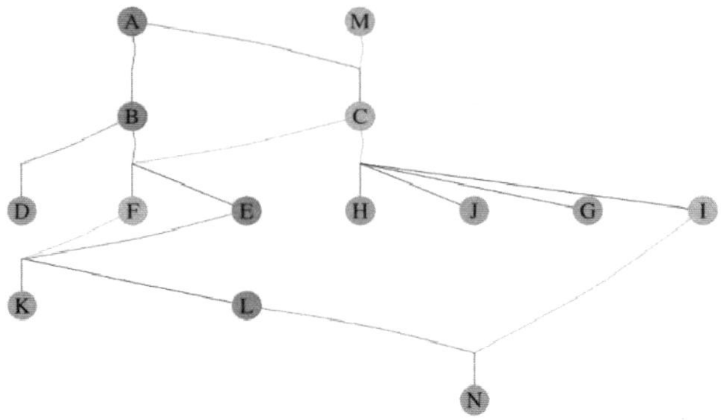

Figure 1. A simple pedigree used to check the graph theory algorithm for the identification of various relationships. Dark blue, golden yellow and olive-green shows, sire, dam and individuals with unknown sex, respectively

Simulated genotype and pedigree. This study utilised QMSim v2 (Sargolzaei and Schenkel 2009) and its first example to generate pedigrees and genotypes, producing a population spanning 10 generations. Genotypes from generations 8 and 10 were used. In contrast, those from generation 9 were omitted to demonstrate how inconsistencies between NRM and GRM can reveal pedigree errors, particularly where parent-progeny relationships cannot be verified genomically. In each generation, 20 males were randomly mated with 400 females, with each female producing two progenies. The genome consisted of 30 chromosomes, each spanning 100 cM and containing 333 markers. The sires of two individuals in generation 8 were altered to simulate pedigree errors.

Genomic and pedigree relationship matrices. The GRM was built using VanRaden first method (VanRaden 2008), and the NRM was built using the pedigree package (Coster 2022) in R for animals with genotypes.

Identification of the relationships among individuals. The package igraph in R (Csardi and Nepusz 2006), was used for graph theory analysis. In this study, we use graph theory to identify Parent-Offspring (PO), Full-sibs (FS), Half-sibs (HS), Progeny-Grandparent (PG), and Cousins (CO) relationships and their extended relationships. The pedigree data typically has three columns for individuals, sires, and dams. The first step in using graph theory to analyse pedigrees was to convert the pedigree into a direct graph, i.e., each parent-offspring relationship was represented as a directed graph where the 'edge' indicates a flow from one node to the other. The directed graph identified groups of full-sibs, half-sibs, grandparents in common, cousins, and more. The closely related individuals can be identified with the neighbourhood function, and since the pedigree is converted to a directed pedigree, the neighbour incoming function can identify the parents, and the outgoing function can identify the progenies of specific individuals. The intersect function was used to find common parents, which can be used to distinguish full sibs from half-sibs. These procedures can be extended with recursive functions to identify the great-grandparents, other ancestral relationships, and even further ones. So, the pedigree of an individual and its ancestral individuals can easily be extracted and visualised.

RESULTS AND DISCUSSION

The test pedigree was used to assess the methodology. The relationships for some individuals are shown in Table 1. In this table, although A—D and H—G relationships were 0.25 and cannot be differentiated in NRM, the precise type of their relationships has been identified with graph theory, i.e., A—D and H—G were Progeny-Grandparent and Half-sibs, respectively.

Table 1. Description of the relationships in the test pedigree (Figure 1) based on the numerator relationship matrix (NRM)

ID1	ID2	NRM Value	Relationship Type
A	B	0.5	Parent-Offspring
A	D	0.25	Progeny-Grandparent
H	G	0.25	Half-sibs
B	J	0.125	Uncle/Aunt
G	D	0.0625	Cousins
E	F	0.625	Full sibs
A	N	0.375	Progeny-GreatGrandParent

Figure 2 displays all individuals whose NRM values exceeded their GRM values by more than 0.25, highlighting potential pedigree issues. As indicated by the blue and olive colours, most of these individuals were half-sibs or cousins. By applying graph theory to extract ancestral or descendant pedigrees, individuals with incorrect pedigrees can be identified for further investigation. The individual with the most frequent inconsistency between NRM and GRM was the '6109' (the replaced sire with '6104'). Additionally, individuals '8361' and '8362' exhibited the highest differences between NRM and GRM values. Figure 3 presents a simplified pedigree where grandsire' 6104' was replaced with '6109', revealing that '8361' and '6109' stood out. This approach, when combined with additional information such as flock profiles or multiple-sire mating pedigrees, enhances the efficiency and accuracy of pedigree correction. While other methods, such as recursive algorithms, can also identify these relationships, graph theory algorithms are already well-developed and optimised, making them highly feasible for implementation in scripting languages like R (R Core Team 2021).

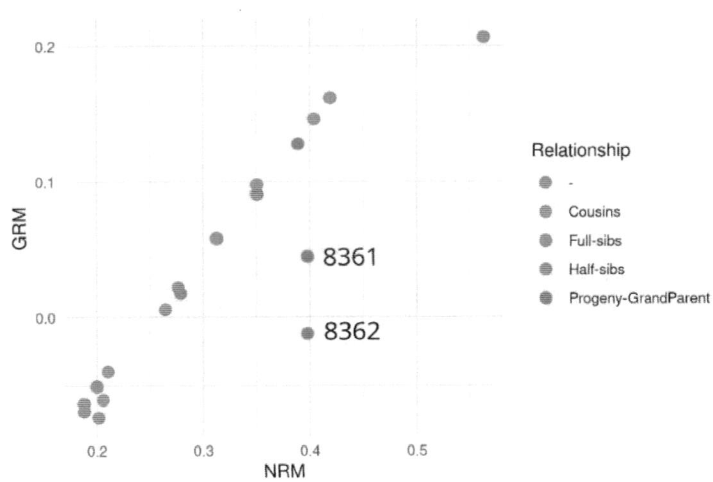

Figure 2. Scatter plot with relationships from the GRM and NRM of simulated data where the pedigree relationships were higher than genomic relationships. The dash symbol represents undetermined relationships. Individuals 8361 and 8362 shows deviation in NRM in Comparison to the GRM

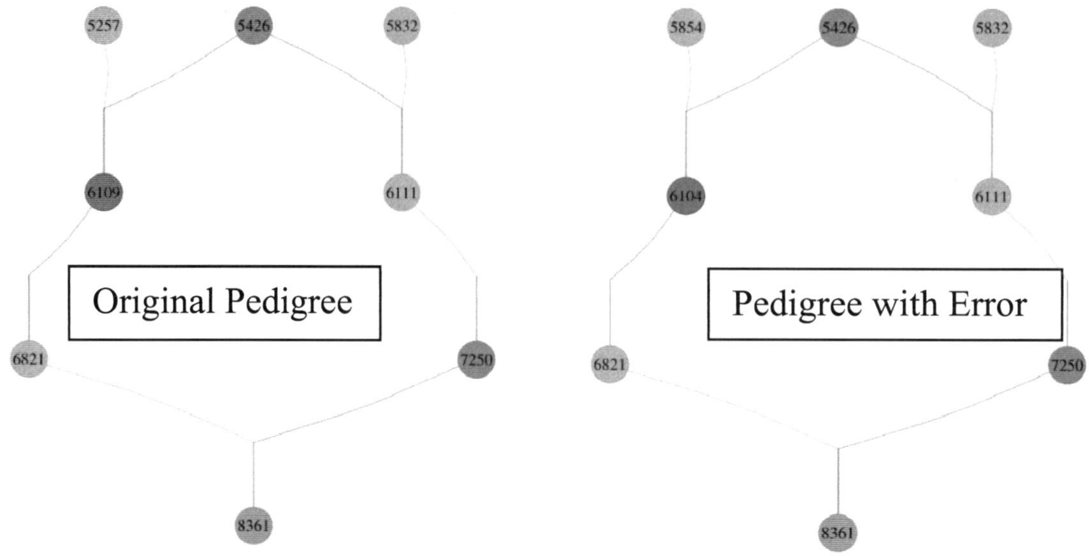

Figure 3 The pedigree shows grandsire '6104' incorrectly assigned to its progeny. Dark blue, golden yellow and olive-green shows, sire, dam and individuals with unknown sex in the pedigree

CONCLUSION

In this study, we applied graph theory to analyse pedigree relationships, specifically identifying cases where the NRM values exceeded the GRM values. Using the same framework, we demonstrated how to simplify the pedigree by isolating a sire's progeny with incorrect pedigree assignment. This approach enables the rapid identification and resolution of pedigree discrepancies using additional data from breed societies and breeders. Correcting pedigree errors and improving the alignment between NRM and GRM values increases the reliability of genomic prediction accuracy. Moreover, the availability of advanced graph theory algorithms and libraries in popular scripting languages further streamlines pedigree analysis, making it more accessible and efficient.

ACKNOWLEDGEMENTS

This study was supported by Meat and Livestock Australia project L.GEN.2204.

REFERENCES

Connors N., Cook J., Girard C., Tier B., Gore K., Johnston D. and Ferdosi M. (2017) *Proc Assoc Advmt Anim Breed Genet.* **22**: 317.
Coster A. (2022) 'pedigree' (*The Comprehensive R Archive Network*).
Csardi G. and Nepusz T. (2006) *InterJournal* **1695**: 1.
Legarra A., Christensen O.F., Aguilar I. and Misztal I. (2014) *Livest Sci* **166**: 54.
Luan S. (2018) 'visPedigree: A package for tidying and drawing animal pedigree'.
R Core Team (2024) 'A Language and Environment for Statistical Computing.' (*R Foundation for Statistical Computing*: Vienna, Austria).
Sargolzaei M. and Schenkel F.S. (2009) *Bioinformatics* **25**: 680.
VanRaden P.M. (2008) *J. Dairy Sci.* **91**: 4414.

Proc. Assoc. Advmt. Anim. Breed. Genet. 26: 307-310

ASSESSING THE GENETIC DIVERSITY OF AUSTRALIAN GOATS

P. Alexandri[1], T. Atkinson[2], D.J. Brown[1], K. Gore[1], K. Smith[3], S.F. Walkom[1] and T. Granleese[3]

[1] Animal Genetics Breeding Unit*, University of New England, Armidale, NSW, 2351 Australia
[2] NSW Department of Primary Industries and Regional Development, Dubbo, NSW, 2830 Australia
[3] Western Local Land Services, Dubbo, NSW, 2830 Australia
[4] NSW Department of Primary Industries and Regional Development, Grafton, NSW, 2460 Australia

SUMMARY

Assessing genetic diversity and population structure of Australian goats is the first step towards establishing a genomic reference population to enable a multibreed genetic evaluation. This study explored the genetic structure of Australian goat breeds, focusing mostly on Rangeland, Boer, and Kalahari Red goats. Results revealed clear genetic distinctions between Boer, Rangeland, dairy, and fibre breeds. Rangeland goats exhibited multiple ancestries but formed distinct groups, suggesting they are a separate breed. Australian Boer goats clustered with global Boer populations, while Australian Kalahari Red goats were genetically similar to Boers, indicating little differentiation. The results highlight the genetic makeup of Australia's diverse goat population and lay the groundwork to facilitate multibreed genetic evaluations, ultimately improving breeding strategies and accelerating genetic progress across Australian goat breeds.

INTRODUCTION

Currently little is known about the genetic structure of Australian goat breeds. Goats were first brought to Australia by European settlers with the first goats arriving onboard the first fleet (January 1788). Since then, multiple introductions have been recorded. Goats adapted to the arid conditions of the Australian environment and expanded through most of the continent without human management, resulting in a population of feral goats of unknown genetic composition, commonly referred to as Rangeland goats (Kijas *et al.* 2013). The Australian Rangeland goat population plays a crucial role in supporting the country's goat meat industry, an export sector valued at $235 million annually, but the majority of goat meat is sourced from harvesting operations that collect goats in a semi-feral state. Within the Measured Goats project (Granleese *et al.* 2023), Rangeland goats along with Boer and Kalahari breeds will be used to establish a genomic reference population to assess the genetic structure and diversity of Australian meat goat breeds and estimate heterosis effects to be used for updating the KIDPLAN genetic evaluation. Therefore, this study aimed to deliver an initial characterisation of the genetic structure of Australian goat breeds, focusing on understanding their population structure, and between breed relationships.

MATERIALS AND METHODS

The Measured Goats project (Granleese *et al.* 2023) provided genotype data from 2,588 goats selected within the project and genotyped with a 70K goat specific panel (https://neogenaustralasia.com.au/ggp-goat-70k/). Genotyped animals of known breed came predominantly from Boer and Rangeland breeds but individuals from Kalahari Red, Australian

---

* A joint venture of NSW Department of Primary Industries and Regional Development and the University of New England

Contender and Savanna breeds were also included. To better represent the genetic diversity of the Australian goat population, previously published genotype data from Kijas *et al.* (2013) were also included. This dataset comprises 183 genotypes from Australian Boer, Angora, Cashmere, and Rangeland goats. Additionally, the Goat ADAPTmap Project (Stella *et al.* 2018) provided genotypes from a broad range of goat breeds worldwide, enabling the study of goat diversity on a global scale. The original ADAPTmap dataset includes 4,653 animals from 169 populations across 35 countries, spanning Europe, West Asia, North America, South America, and Oceania (Colli *et al.* 2018). Initial analysis showed only distant relationships between a big number of international goat breeds and the Australian goat population. Therefore, for the purpose of this study, only genotypes from breeds present in Australia were considered.

To merge genotypes, SNP profiles from each of the above data sets were mapped to the latest goat genome assembly (ARS1.2, Brickhart *et al.* 2017). This step was crucial, given some of the data came from the first version of the Illumina goat 50K SNP chip (Tosser-Klopp *et al.* 2014).

Table 1. Number of unrelated individual goat genotypes (N) available after quality control per breed and country or continent. Study indicates previously published genotypes or data generated within this study

Breed	N	Country / Continent	Source of genotype data
Anglo Nubian	4	Australia	Measured Goats Project
Australian Brown	2	Australia	
Boer	14	Australia	
British Alpine	5	Australia	
Contender	13	Australia	
Kalahari Red	4	Australia	
Pygmy	2	Australia	
Rangeland	334	Australia	
Savanna Goats	4	Australia	
Boer	2	USA	
Alpine	87	Europe	Colli *et al.* (2018)
Boer	5	Australia	
Boer	84	Worldwide	
Cashmere	7	Australia	
Nubian	41	Worldwide	
Rangeland	33	Australia	
Saanen	81	Worldwide	
Toggenburg	13	Africa	
Angora	3	Australia	Kijas *et al.* (2013)
Boer	5	Australia	
Cashmere	3	Australia	
Cashmere	2	New Zealand	
Rangeland	6	Australia	
Total	754		

Genotype quality control was applied to exclude SNPs on sex chromosomes and unmapped locations, SNPs with minor allele frequency (MAF) lower than 0.05 and a genotyping rate lower than 90%. Individuals with more than 10% missing SNPs were also excluded.

To avoid bias in the analysis due to closely related animals, individual genotypes were selected within breed to maximise genetic distance. A genomic relationship matrix using all available genotypes was constructed based on the approach described by Yang *et al.* (2011) and only individuals with an average genomic relationship below 0.2 were retained for the analysis.

The genetic distance between individuals of different breeds was estimated using the VCF2Dis software (https://github.com/BGI-shenzhen/VCF2Dis). The results were used to construct a Neighbor-Joining (NJ) phylogenetic tree with FastMe2.0 (Lefort *et al.* 2015), which was visualised using the iTOL program (Letunic and Bork 2024).

RESULTS AND DISCUSSION

The final genotypes dataset after quality control and genomic relationship filtering included 754 unrelated individuals from 16 breeds (Table1) genotyped in 27,876 SNPs. Number of animals, breed, country and data origin are detailed in Table 1.

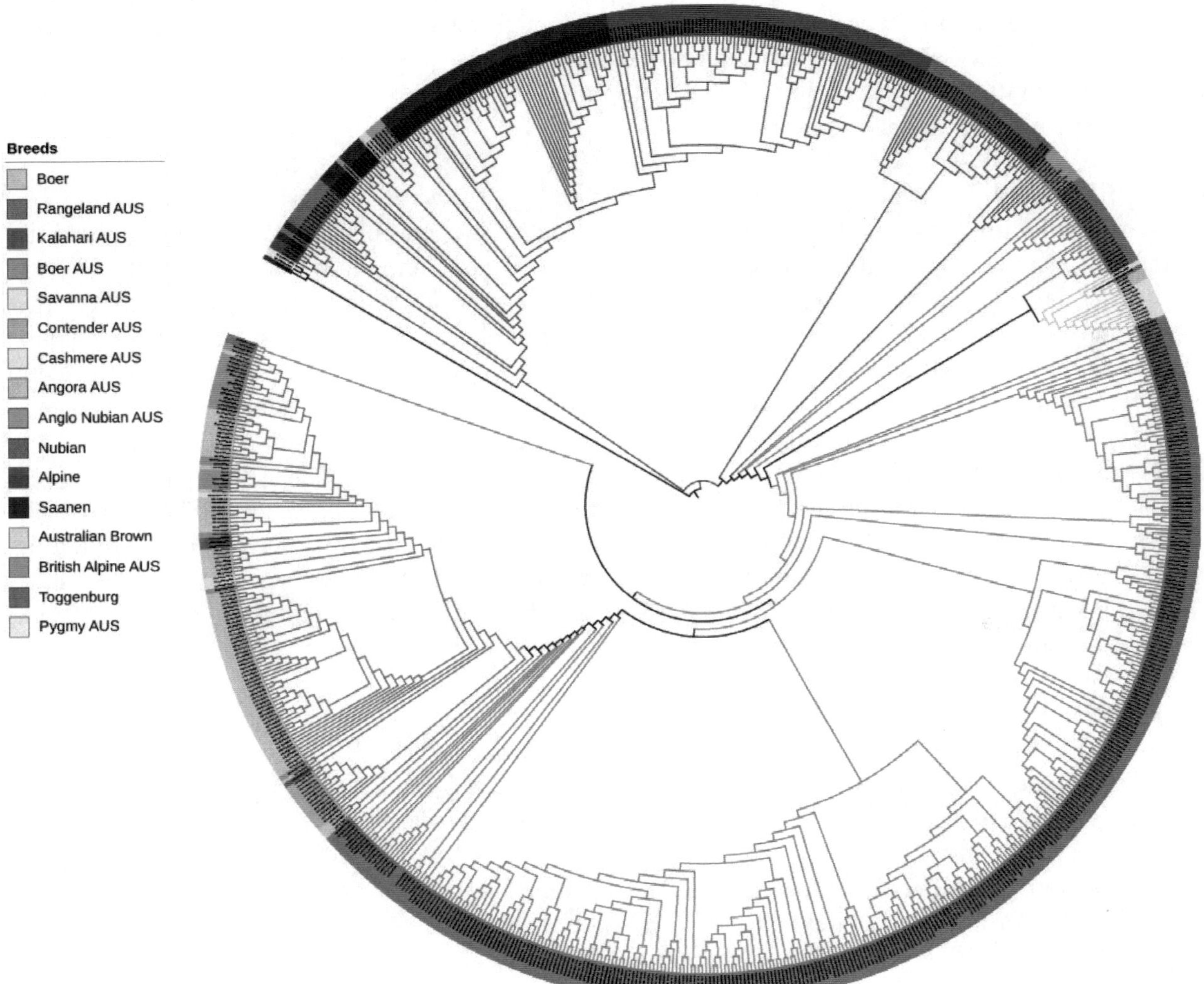

Figure 1. Phylogenetic relationships between breeds presented on Table 1 using 27,876 SNPs

Phylogenetic analysis showed the relationship between each breed as well as between individuals within each breed. The 754 individuals were grouped into 17 branches representing different breeds and within breed populations (Figure 1). The main branches included Boer, Rangeland, Australian fibre breeds (Angora and Cashmere) and Australian and worldwide dairy breeds indicating a clear genetic division existing between Boer, fibre and dairy breeds as well as some Rangeland populations. A closer look within the Rangeland individuals reveals several well defined groups. These groups indicate the existence of a separate Rangeland breed but also the diverse origin of the Rangeland goat population as several individuals appear to cluster closer to Boer, fibre or dairy groups respectively. Australian Boer goats belong to the well defined Boer group which includes worldwide Boer population but also populations with Boer ancestry such as Australian Contender and Australian Savanna goats. The Kalahari Red individuals clustered within the Boer branch but not in separate and defined groups indicating little to no differentiation between Australian Boer and Kalahari populations. More analysis is under way to describe the genetic structure of the Australian goat population better using admixture approaches. This analysis will also produce allele frequencies to define the breed composition for genotyped individuals of unknown breed.

Currently, only a small number of goats are performance recorded annually (predominantly Boer, Granleese *et al.* 2023), with KIDPLAN breeding values calculated by Sheep Genetics (Meat & Livestock Australia). Expanding the scope of genetic recording to include a broader range of goats and more detailed performance data is critical to advancing the Australian goat industry. Estimating the genetic diversity and breed structure of Australian goats is the first step towards a genomic reference population that will facilitate the development of a multibreed genetic evaluation, fostering genetic improvement across various breeds.

CONCLUSION

First results towards establishing an Australian goat reference population have highlighted the genetic differentiation existing between Australian Boer, Rangeland, dairy and fibre goats. Despite their potential multiple ancestries, Rangeland goats appear to be a separate breed clearly distinct from other populations. Australian based Kalahari Red goats are genetically similar to Boer goats, indicating they can be treated as the same breed. Ongoing studies aim to refine genetic structure and breed composition, aiming to enhance genetic evaluations and breeding strategies.

ACKNOWLEDGEMENTS

The Measured Goats in the Rangelands project is an initiative co-funded by NSW Department of Primary Industries and Meat and Livestock Australia Donor Company. The authors give thanks to the efforts of the Condobolin Agricultural Research and Advisory Station manager and their staff, the skilled team of technicians and scientists. The authors also thank Dr James Kijas for supplying a subset of the Australian goat genotypes used in the analysis

REFERENCES

Brickhart D.M., Rosen B.D, Koren S. *et al.* (2017) *Nat. Genet.* **49**: 643.
Colli L., Milanesi M., Talenti A., *et al.* (2018) *Genet. Sel. Evol.* **50**: 58.
Granleese T., Mortimer S.I., Atkinson T., *et al.* (2023) *Proc. Assoc. Advmt. Anim. Breed. Genet.* **25**: 230.
Kijas J.W., Ortiz J.S., McCulloch R., *et al.* (2013) *Anim. Genet.* **44**: 325.
Lefort V., Desper R. and Gascuel O. (2015) *MBE* **32**: 2798.
Letunic I. and Bork P. (2024) *Nucleic Acids Res.* **52**: W78.
Stella A., Nicolazzi E.L., Van Tassell, C.P., *et al.* (2018) *Genet. Sel. Evol.* **50**: 61.
Tosser-Klopp G., Bardou P., Bouchez O., *et al.* (2014). *PLOS ONE* **9**: e862.
Yang J., Lee S.H., Goddard M.E. and Visscher P.M. (2011) *Am. J. Hum. Genet.* **88**: 76.

Proc. Assoc. Advmt. Anim. Breed. Genet. 26: 311-314

BREED COMPARISONS IN COMBINED LAMBPLAN: GENETIC GROUP AND LINKAGE ASSESSMENT

A.J. McMillan, P. Alexandri, P.M. Gurman, D.J. Brown, K.L. Bunter, K.P. Gore, S. de las Heras-Saldana, L. Li, S.F. Walkom and A.A. Swan

Animal Genetics Breeding Unit*, University of New England, Armidale, NSW, 2351 Australia

SUMMARY

In response to changing industry selection practices and a desire for a simpler genetics landscape, Sheep Genetics is developing a Combined LAMBPLAN evaluation that merges the Maternal and Terminal sire analyses. A key challenge, especially when combining multi-breeds for genetic evaluation, is ensuring fair comparisons across diverse genetic backgrounds. To address this, genetic groups have been redefined and assessed for prediction accuracy, and across-flock linkage has been evaluated to determine the degree of connectedness and robustness of prediction across the combined population. Results show that the new genetic groups effectively capture population structure and that strong linkage supports robust across-flock comparisons. These findings provide key evidence for implementing the Combined LAMBPLAN evaluation to improve the accuracy and relevance of Australian Sheep Breeding Values (ASBVs).

INTRODUCTION

Genetic evaluations play a critical role in improving the productivity and efficiency of livestock industries. In Australia, Sheep Genetics (SG) has traditionally conducted separate evaluations for Maternal and Terminal sheep breeds (Brown *et al.* 2007). However, as breeding objectives shift, and producers adopt flexible breeding strategies and regularly incorporate genetics from diverse sources they have a need to assess animals across these two analyses. A growing number of animals from the separate analyses are common between the two evaluations. An increase in the number of composite animals, and the use of outside breed sires reflect a reduced reliance on strict breed classifications and an emphasis on performance-based selection (McMillan *et al.* 2023).

To better capture genetic relationships across these populations and improve evaluation accuracy, a combined analysis is being developed, integrating the LAMBPLAN Maternal and Terminal analyses into a Combined LAMBPLAN genetic evaluation (Walkom *et al.* 2025). A key challenge in this transition is ensuring that animals from a diversity of breeds and environments are fairly treated.

In both current LAMBPLAN analyses, genetic groups account for breed differences and genetic trends in the base populations (Brown *et al.* 2007). Genetic group formation is based on breed and time periods. The current LAMBPLAN analysis utilises a multi-trait estimate of sire linkage across flocks (Brown *et al.* 2007). This linkage value measures the flocks' connectedness and gives breeders the confidence to compare animals across flocks and breeds. This linkage is created by using sires across multiple flocks and time periods. In addition, the Information Nucleus Flock (van der Werf *et al.* 2010), other resource flocks, and breeder groups provide strong comparisons through direct head-to-head comparisons.

This study aims to (1) calculate genetic group accuracies in the new combined analysis and (2) evaluate across flock linkage in the new Combined LAMBPLAN genetic evaluation. By quantifying these factors, we provide insights into the effectiveness of the combined evaluation to deliver accurate Australian Sheep Breeding Values (ASBVs) across a broader population.

---

* A joint venture of NSW Department of Primary Industries and Regional Development and the University of New England

MATERIALS AND METHODS

The data for these analyses was taken from a combined dataset of the Maternal and Terminal analyses built by Sheep Genetics in November 2024, as per Walkom *et al.* (2025). The analyses will focus on two well-recorded trait groups, weight (WT) and ultrasound measured eye muscle depth (EMD), at three age stages: weaning, post-weaning and yearling (wwt, pwt, ywt, wemd, pemd, yemd) and the less recorded carcase trait of abattoir measured intramuscular fat (imf)

Genetic Group Definition. With the initiative to combine the analyses, particularly regarding data reduction described by Walkom *et al.* (2025), a new set of 92 genetic groups were defined to reflect this new dataset, using the same classifiers: breed and time periods. Efforts were made to collapse groups where appropriate to ensure computational efficiency this reduced the number of genetic groups from 111 and 94 respectively in the current Maternal and Terminal analysis.

Genetic Group Accuracy Estimation. To evaluate the accuracy of genetic group estimates, a reduced version of the full mixed model was considered, incorporating only contemporary groups and genetic groups. The model is expressed $as\ y = X\mathrm{B} + ZQg + \cdots + e$ where: y is the vector of observations, X is the design matrix for fixed effects (B), Z is the incidence matrix linking records to genetic group effects, Q is the incidence matrix relating individuals to genetic groups, g is the vector of genetic group effects, and e is the residual error term. To obtain the prediction error variance (PEV) matrix for genetic groups, the equations corresponding to contemporary groups were absorbed into the genetic groups' equations, resulting in the following expression:

$$PEV = \left[Q'Z'ZQ + I\frac{\sigma_e^2}{\sigma_{gg}^2} - Q'Z'X(X'X)^{-1}XZQ\right]^{-1} \sigma_e^2$$

where σ_e^2 is the residual variance and σ_{gg}^2 represents the genetic group variance. The accuracy of genetic group estimates was computed using the diagonal elements of the PEV matrix, with the accuracy for each genetic group calculated as: $\sqrt{}(1 - \frac{PEV_{ii}}{\sigma_{gg}^2})$, where PEV_{ii} represents the prediction error variance of the genetic group.

Linkage Analyses. The routine linkage analyses were conducted on the new combined data following procedures found in Huisman *et al.* (2005). This is a sire-based analysis conducted at a flock level for the phenotypic data recorded from 2019- 2024. The linkage for the weight and eye muscle depth traits were assessed within the trait but across stages, with linkage for intramuscular assessed as a single trait.

RESULTS AND DISCUSSION

Genetic Group Accuracy Estimation. Figure 1 presents the genetic group accuracy for the newly defined groups, demonstrating the effectiveness of estimating genetic group effects within the combined LAMBPLAN dataset where adequate data is available. The weight traits show very high accuracy, benefiting from extensive records across most genetic groups. Only 5% of the weight genetic groups have an accuracy of prediction under 85%. In contrast, for eye muscle depth traits approximately 40% of genetic groups have an accuracy of prediction under 85 %. This is due to generally fewer performance records for eye muscle depth across stages, resulting in a poorer ability to predict the genetic groups. However, post-weaning muscle depth, which is very well recorded, follows a similar pattern to that observed for weight traits. Intramuscular fat, the trait with the least amount of recording, has 48% of the genetic groups above 85% prediction accuracy. The major breeds tend to have more recording and, thus higher accuracy. The minor breeds tend to have less data, and their genetic group solutions with low prediction accuracy will be regressed to the mean.

These findings confirm that the new genetic group definitions provide robust estimates particularly for the breeds with more representation in the analyses. The high accuracy for key traits and breeds underscores the effectiveness of the updated framework in supporting reliable genetic evaluations across the diverse combined LAMBPLAN dataset.

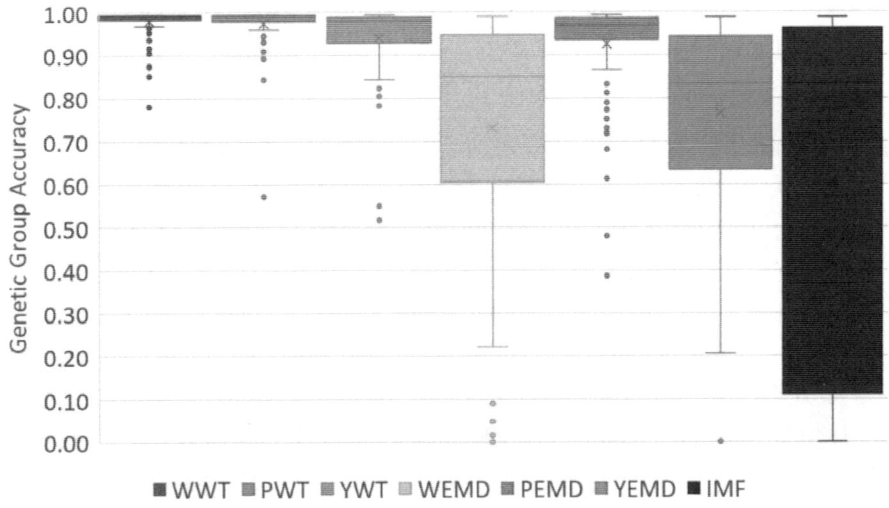

Figure 1. Accuracy of genetic group effects for a range of traits in the Combined LAMBPLAN dataset

Linkage Analyses. Table 1 summarises the linkage analyses for the WT, EMD, and IMF trait groups for the Combined LAMBPLAN dataset.

Table 1. Summary of linkage for three trait groups in the Combined LAMBPLAN dataset

	WT	EMD	IMF
Total flocks	758	708	36
Average number of linked flocks (NOLF)	19.4%	19.6%	33.5%
Average linkage to linked flocks	69%	68%	56%
Linked flocks	728	690	35

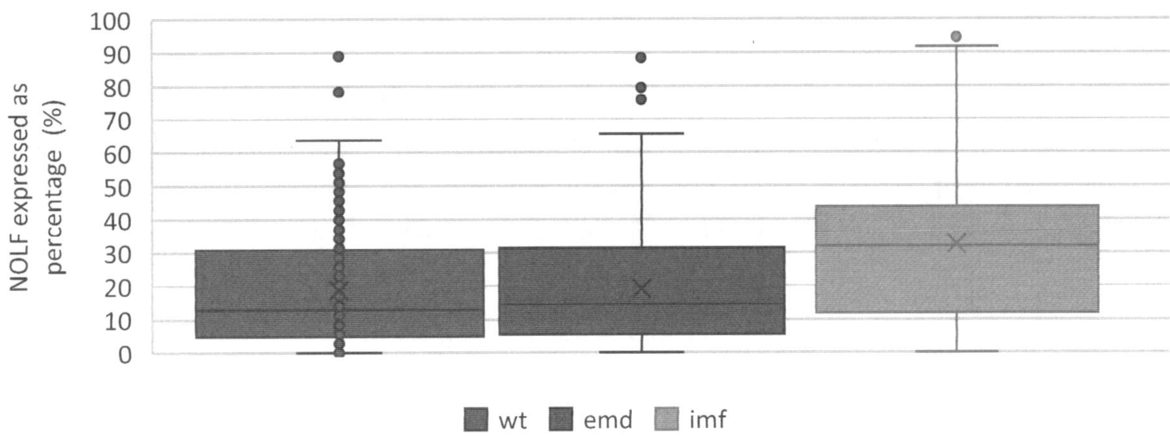

Figure 2. Distribution of individual flocks' number of linked flocks (NOLF) in combined LAMBPLAN dataset

Figure 2 presents the distribution of number of linked flocks (NOLF) for individuals flocks. The data reveal a broad range, indicating that most flocks maintain connections with a substantial portion of the population. Flocks, on average, are linked to approximately 20% of the total flocks for the weight and eye muscle depth groups and 30% for those flocks recording IMF, which are fewer. The Information Nucleus Flock and MLA Resource Flocks are key by linking over 80% of the flocks. This is a positive outcome regarding the utility of the combined analysis and value of the structured resource flocks.

The heatmaps of flock linkage not shown here reveal that although historical patterns reflecting the distinction between maternal and terminal animals are traceable, the resource flocks intersect these divisions. This connectivity results in high linkage for the resource flocks, with proportions ranging from 78% to 88% for weight, 75% to 88% for eye muscle depth, and 83% to 94% for intramuscular fat. Similar patterns exist for organised breeder groups, which have high linkage within these groups but also links to the wider industry.

The presence of these highly connected flocks facilitates strong genetic links between what might otherwise be more isolated groups. This suggests that while some historical divisions exist, a clear genetic divide between maternal and terminal animals is not obvious, reinforcing the validity of comparisons across these groups. Overall, results confirm the connectedness within the population, supporting the integrity of genetic evaluations and ensuring broad connectivity among flocks.

While this paper focussed on weight and carcase traits. Further work is being conducted for the other trait groups in particular reproduction which is essential for breeders with a maternally focused breeding objective.

CONCLUSION

This study demonstrated the ability of the Combined LAMBPLAN analysis to accurately estimate genetic group effects from a diverse breed background and the ability to compare ASBVs between the vast majority of flocks. Given these results, the Combined LAMBPLAN analysis will have the ability to provide well-estimated ASBVs that are comparable across previous analysis divides, providing breeders with the ability to accurately select animals from a broader gene pool.

ACKNOWLEDGEMENTS

This research was funded by Meat and Livestock Australia project L.GEN.2204. The authors gratefully acknowledge the teams behind the Sheep CRC for Sheep Industry Innovation Information Nucleus and MLA Resource flock as well as individual industry flock owners involved in flock management and data collection.

REFERENCES

Brown D.J., Huisman A.E., Swan A.A., Graser H.U., Woolaston R.R., Ball A.J., Atkins K.D. and Banks R.G. (2007) *Proc. Assoc. Advmt. Anim. Breed. Genet.* **17**:187.

Huisman A.E., Tier B. and Brown D.J. (2005) *Proc. Assoc. Advmt. Anim. Breed. Genet.* **16**: 250.

McMillan A.J., Walkom S.F. and Brown, D.J. (2023) *Proc. Assoc. Advmt. Anim. Breed. Genet.* **25**: 31.

van der Werf J.H.J., Kinghorn B.P. and Banks R.G. (2010) *Anim. Prod. Sci.* **50**: 998.

Walkom, S.F, Alexandri, P, Brown, D.J., Bunter, K.L., Connors, N.K, de las Heras-Saldana, S., Gore, K. P., Gurman, P.M., Li, L., McMillan, A.J., Miller, S.P., Bradley, P.E. and Swan, A.A. (2025) *These proceedings*

Proc. Assoc. Advmt. Anim. Breed. Genet. 26: 315-318

USING GENOTYPING BY SEQUENCING TO MONITOR THE GENETIC DIVERSITY OF AUSTRALIAN HONEY BEES

P. Alexandri[1], S.P. Miller[1], E.A. Frost[2], N.C. Chapman[2], S.F. Walkom[1], R.G. Banks[1] and K.L. Bunter[1]

[1] Animal Genetics Breeding Unit*, University of New England, Armidale, NSW, 2351 Australia
[2] NSW Department of Primary Industries and Regional Development, Paterson, NSW, 2421 Australia

SUMMARY

European honey bees in Australia are largely genetically isolated due to strict border controls. They contribute significantly to the economy through honey production and pollination, yet national efforts for their genetic improvement have historically been limited. This study estimated the genetic diversity of Australian bees using low-pass genome sequencing on 711 samples from 26 locations, provided by breeders participating in the national PlanBee project. Most genotypes were obtained from pooled drone samples. Results showed average observed heterozygosity (Ho: 0.20) was lower than the expected one. Bias tended towards low Ho, with some pooled drone samples conversely exhibiting unexpectedly high values. These high values were potentially attributed to multiple queen ancestries of the drone pool. Low FST values (0-0.07) between sampling locations indicated minimal population structure, likely due to gene flow through the exchange of genetic material within the PlanBee project. Findings suggest a need for broader sampling and better documentation of hive history and queen lineage.

INTRODUCTION

European honey bees were initially introduced to Australia from Western Europe with later introductions from Eastern Europe and the Mediterranean. These introductions resulted in an admixed population (Chapman *et al.* 2016), which has subsequently been isolated due to geographical isolation, strict border control protocols, and limited importation of bee genetic material. European honey bees contribute to the Australian economy both directly through honey production and indirectly by providing pollination services for various crops (Chapman *et al.* 2022). Despite their importance, limited attempts have been made to genetically improve bee populations. Furthermore, previous studies of Australian bee diversity used a limited number of areas and markers (i.e. 95 single nucleotide polymorphisms -SNPs, Chapman *et al.* 2016; Chapman *et al.* 2019). Accurately benchmarking and monitoring the genetic diversity of Australian bees is important because bees are kept in highly variable environments influencing their productivity and viability. The recent arrival of the Varroa mite (*Varroa destructor*) in Australia poses an additional challenge as loss of genetic diversity may limit the potential to select for varroa mite resistance (Büchler *et al.* 2010). This work aimed to assess genetic diversity in the Australian bee population using low-pass genome sequencing data, with a focus on identifying population structure and gene flow between different bee-keeping locations.

MATERIALS AND METHODS

Queen bee tissues and pooled drone larvae and pupae tissues were collected from 26 different locations or apiaries across Australia. The number of drones for pooled samples ranged from 5 to

* A joint venture of NSW Department of Primary Industries and Regional Development and the University of New England

10. A total of 1235 samples were submitted to Gencove (https://gencove.com/) for DNA extraction and low-pass sequencing at two different depths (4x and 10x) using a commercially developed method. DNA was successfully extracted and sequence data were obtained for 717 samples.

Table 1. Number of drones (N_D), queens (N_Q) and total number of samples (N) from each location retained after quality control

State	Locations	N_D	N_Q	N
WA	1-10	41		41
NSW	11-16	211	152	374
SA	17-19	18		18
QLD	20	30		30
TAS	21-24	219		219
VIC	25-26	40		40
Total	26	559	152	711

During the initial analysis of sequencing data, paired-end reads were aligned to the *Apis mellifera* genome Amel_HAv3.1 (GCF_003254395.2, Wallberg *et al.* 2019) using Sentieon Driver's BWA-MEM implementation (Freed *et al. unpublished data*). At this stage, 6 samples were excluded because of low (<60%) alignment. Haplotype calling was performed using Sentieon's `Haplotyper` with genome variant call format (gVCF, with information for both variant and non-variant positions) output per sample. Joint variant calling for processed haplotypes was performed using Sentieon's `GVCFtyper` with a maximum number of 12 alternate alleles reported. Resulting variant call format (vcf) files containing 2,827,009 variable sites were combined and further quality control was applied to remove variants with a missing rate of more than 10%, Minor Allele Frequency (MAF) lower than 5% as well as variants deviating from Hardy – Weinberg Equilibrium ($P < 0.0005$). After quality control 711 (Table 1) samples genotyped in 48,874 SNPs could be used for further analysis. To estimate levels of genetic diversity in the sampled locations, heterozygosity was estimated with PLINK 2.0 (Chang *et al.* 2015). The existence of more than one distinct population was assessed using pairwise genetic differentiation (F_{ST}) using the method implemented in Arlequin 3.5.2 (Excoffier *et al.* 2010).

RESULTS AND DISCUSSION

Observed queen heterozygosity values for each location were estimated either directly from the queen's genotype or indirectly using pooled drone genotypes (Figure 1). The average expected heterozygosity in the data was 0.23. Across all locations, observed heterozygosity was 0.20 and ranged from 0.04 to 0.50 (Figure 1). For the queen genotypes only, the average observed heterozygosity was 0.21 and it ranged between 0.15 and 0.47. For pooled drone genotypes, the highest 10% of heterozygosity values were observed in samples with more than 6 individuals, indicating a higher number of heterozygous SNPs in relatively larger pooled samples of drones. Including more individuals in the pooled sample may give a better representation of the queen's genome and provide the opportunity to capture more of the queen's alleles. However, higher heterozygosity in pooled drone genotypes can also be a result of the drones descending from more than one queen. Petersen *et al.* (2017) reported drones of different ancestry were present in pooled samples that came from hives where the queens had recently been replaced, highlighting the importance of recording the hive history and the queen's age or time in that hive before sampling.

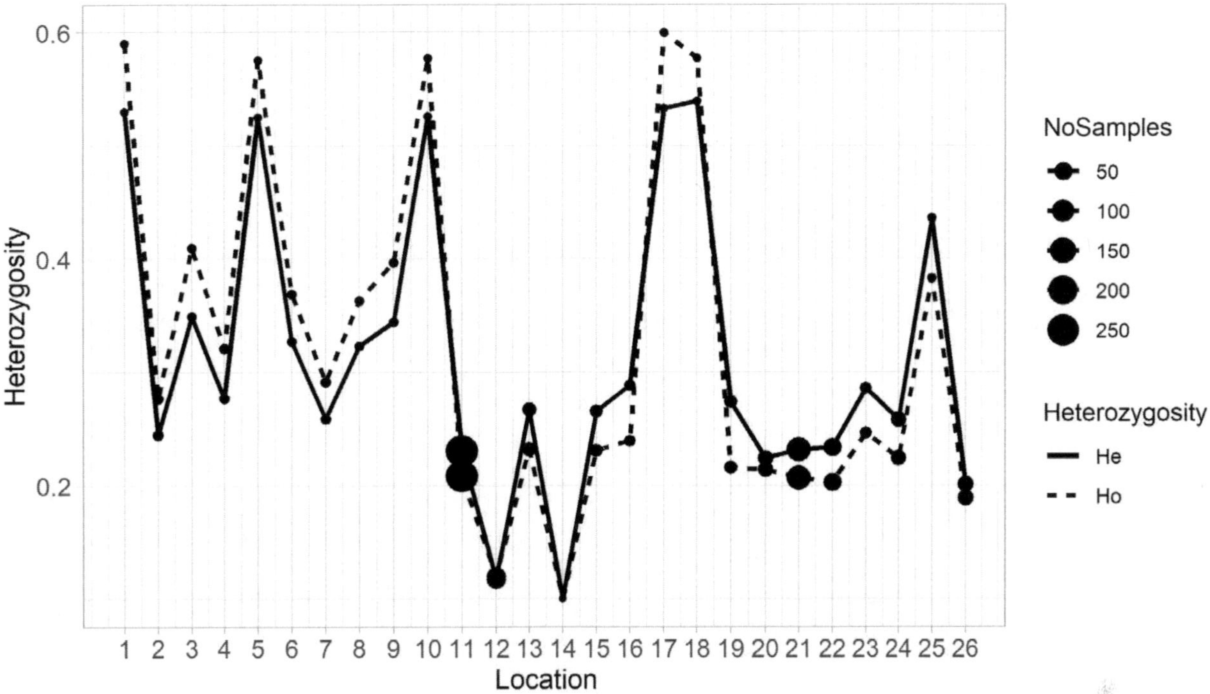

Figure 1. Observed (Ho) and expected (He) heterozygosity values for each location. Dot sizes indicate the number of samples from each location. Location codes correspond to Table 1

Previous studies have demonstrated the hybrid origin of Australian bees (Chapman *et al.* 2016) and the contribution of management and admixture to increased diversity in bee populations has been well described (Harpur *et al.* 2012). However, most samples in this study had lower heterozygosity than expected, potentially due to lower levels of genetic diversity within the sampled populations. A similar study in the New Zealand bee population revealed that although the overall levels of diversity were high enough to ensure long-term viability, within regions and companies the diversity was significantly reduced (Petersen *et al.* 2021). Moreover, low heterozygosity for Australian bees as a result of the sampling process cannot be excluded and more diverse sampling may be needed. Pairwise F_{ST} values estimated for all locations ranged from 0 to 0.07 (Figure 2), indicating no distinct subpopulations. Western Australian bee samples were expected to be more differentiated based on previous results using microsatellites (N. Chapman, unpublished data), reflecting a ban prohibiting importations to WA from other states. Genomic data used in this study captured greater variation on the genome and therefore would be more efficient at detecting population differentiation than using a small number of microsatellite markers. Further, several potential sub-populations (e.g. states) were represented by few samples, and some breeders were known to exchange genetic material between locations. Therefore, the low differentiation between Western Australia and the other sites likely reflects the sampling process and the existence of gene flow through queen transfer from Western Australia to Eastern states during the PlanBee project.

CONCLUSION

The study found that the observed heterozygosity (0.20) across sampling locations was slightly lower than expected (0.23). Low FST values (0–0.07) between location indicate minimal population differentiation and suggest a more homogeneous population due to gene flow from queen movement. The lower-than-expected heterozygosity may reflect high homozygosity but also limited sample diversity, highlighting the need for broader sampling. Sufficient drones per pool are required for better representation of the queen's genome, but better documentation of hive history and queen lineage is also needed to avoid bias from drones originating from multiple queens.

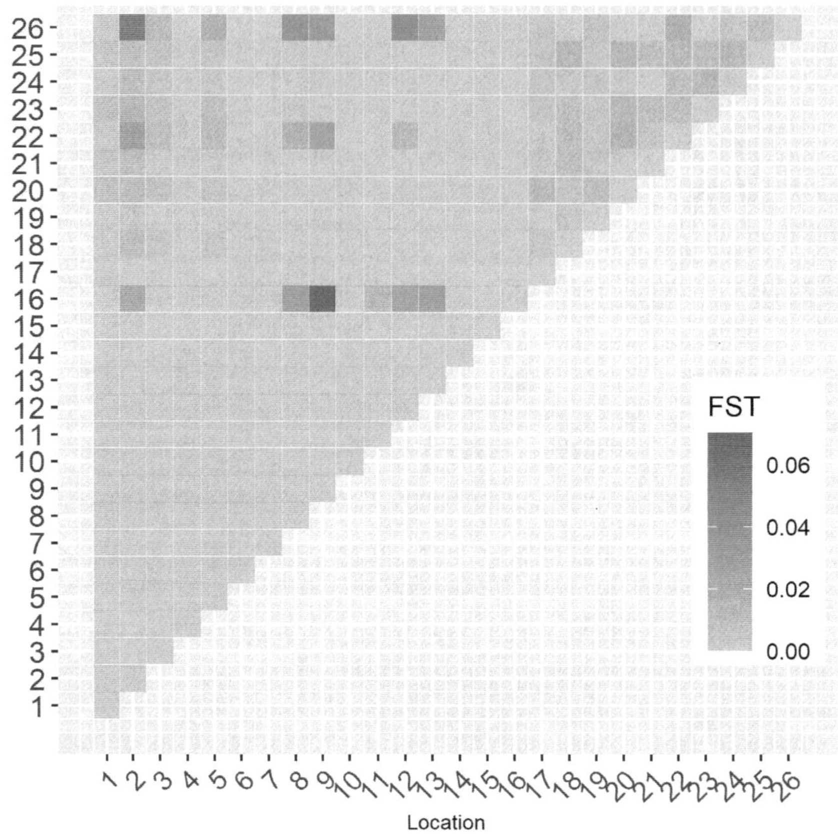

Figure 2. Pairwise genetic differences (FST) between locations estimated for all sequenced samples. Location codes correspond to Table 1

ACKNOWLEDGEMENTS

PlanBee project was supported by AgriFutures Australia through funding from the Australian Government Department of Agriculture, Water and the Environment as part of its Rural R&D for Profit program, participating research institutions and industry.

REFERENCES

Büchler R., Berg S. and Le Conte Y. (2010) *Apidologie* **41**: 393.

Chapman N.C., Harpur B.A., Lim J., *et al.* (2016) *Apidologie* **47**: 26.

Chapman N.C., Sheng S., Lim J., *et al.* (2019) *Apidologie* **50**: 28.

Chapman N.C., Frost E.A. and Banks R.G. (2022) *Proc. World Cong. Genet. Appl. Livest. Prod.* **12**: 2512.

Chang C., Chow C.C., Tellier L.C.A.M., *et al.* (2015) *GigaScience*, **4**: s13742–015–0047–8.

Excoffier L. and Lischer H.E.L. (2010) *Mol. Ecol. Resour.* **10**: 564.

Harpur B.A., Minaei S., Kent C.F. and Zayed A. (2012) *Mol. Ecol.* **21**: 4414.

Petersen G.E.L., Fennessy P.F., Van Stijn T.C.,*et al.* (2017). *Proc. Assoc. Advmt. Anim. Breed. Genet.* **22**: 217.

Petersen G.E.L., Fennessy P.F. and Dearden P.K. (2021) *Anim. Prod. Sci.* **61**: 1856.

Wallberg A., Bunikis I., Vinnere Petterson O., *et al.* (2019) *BMC Genomics* **20**: 275.

Proc. Assoc. Advmt. Anim. Breed. Genet. 26: 319-322

BREED-OF-ORIGIN OF ALLELES AND GENOMIC BREED COMPOSITION ESTIMATION IN NEW ZEALAND CROSSBRED DAIRY CATTLE

Y. Wang, A.M. Winkelman and B.L. Harris

Research and Development, Livestock Improvement Corporation, Hamilton 3240, New Zealand

SUMMARY

An animal's genomic breed composition (GBC) is the partition of its genome based on the genetic inheritance from its ancestors or ancestral breeds. Understanding the genomic breed composition for crossbreed animals is valuable for estimating heterosis, tracing breeding history, assessing genetic contributions from foundational breeds, and guiding selection decisions in crossbreeding programs. In this study, genomic breed composition and breed of allele origin (BoA) of New Zealand crossbred dairy cattle were estimated using the R package *optiSel*. Genomic information was available for 145,022 Holstein-Friesian Jersey crossbred animals to determine the breed composition. On average, 95.6% of the animals' alleles could be assigned to a breed in the reference. A correlation of 0.805, 0.896, and 0.889 was achieved between pedigree and genomic-based breed composition of Friesian, Holstein, and Jersey in the crossbred animals, respectively. Although pedigree information was not utilised in the calculation process, a clear pattern of haplotype inheritance can be observed from the breed of allele origin information.

INTRODUCTION

Estimating crossbred animals' breed composition offers numerous benefits, such as predicting heterosis, understanding target breeds' population structure and breeding history, and guiding management decisions in future crossbreeding programs. Traditionally, pedigree information has been used to determine breed composition, though the reliability of these estimates can be affected by missing, inaccurate, or incomplete records in the pedigree. Moreover, pedigree-based approaches rely solely on an animal's ancestry and cannot account for genomic features like crossovers or meiotic assortment. Consequently, the information derived from pedigree data represents an expected value and does not capture deviations from this expectation.

Vandenplas *et al.* (2016) developed a method that relies solely on genomic data to trace haplotypes in crossbred animals and assign each allele within the haplotype to its breed of origin. With the breed-specific origin of alleles, the genomic breed composition of each individual can be calculated simply by dividing the number of alleles from each breed by the total number of alleles. Similarly, Eiríksson *et al.* (2021) developed the AllOr method based on comparing haplotypes in overlapping windows. However, both methods required the purebred populations in the reference to have a high level of genetic purity and minimal to no introgression from other breeds, and the targeted crossbred animal to have a simple crossbred history, which does not apply to the New Zealand dairy population since crossbred sires were widely used since 2006 (Harris 2022). This study aimed to assign the breed origin of alleles of crossbred animals and evaluate their breed composition using genomic information using R package *optiSel* (Wellmann 2019).

MATERIALS AND METHODS

Animals and genotypes. The purebred animals in the reference population consist of 312 Friesian (FRI), 4110 Holstein (HOL), and 39,012 Jersey (JER) animals. Animals classified as purebred Jersey have a pedigree-based breed composition of Jersey above 87.5%. Animals classified as purebred Friesian or Holstein have a pedigree-based breed composition of Holstein or Friesian above 75%. The target population of this study consists of 145,022 Holstein-Friesian Jersey crossbred animals. They are categorised into three groups based on the pedigree quality indicated by animals' official indicator type: High (79,689 animals) – both parents are DNA parentage tested; Medium (57,019 animals) – mating indicates that either parent is possible; Low (8314) - no mating date, mating dates indicate more than one possible parent, negative parentage test, or any other reasons that may lead to uncertain parent. Among the whole population, 55% of the population has high pedigree quality, 39.3% of the population has medium pedigree quality, and 5.7% of the population has low pedigree quality. The predictions were estimated based on Pearson's correlation between the breed composition estimated based on pedigree and genomic information, and the correlation was estimated within each group accordingly.

The genotype data used in this study was part of the data used for routine national animal evaluation. Animals from both reference and target populations were genotyped in panels with different densities and imputed to a filtered Illumina50k level. Each animal's phase information was generated from the phasing or imputation process using Beagle 5.3 (B. L. Browning, 2018; 2021).

Assigning alleles to breed of origin. Alleles were assigned to their breed of origin in the target population by comparing haplotypes from purebred animals with those from crossbred animals. Frequencies were calculated for each allele within a defined haplotype segment for each breed group in the reference population. The breed origin of each allele was identified by selecting the breed with the highest allele frequency from the previous step. This method ensured that only segments with matching markers in both haplotypes were counted, thus accurately estimating segment frequency at each marker. Details of the mathematical notation can be found in Wang *et al.* (2017). The analysis was performed chromosome by chromosome, and the overall genomic breed composition was subsequently calculated by summing the alleles from each reference breed and dividing by the total number of alleles.

RESULTS AND DISCUSSION

Genomic breed composition estimation. The average assignment rate of alleles was 95.6% across the target population. The rate of assignment positively correlated with pedigree quality: for animals with high pedigree quality, the average assignment rate was 96.1%; for those with medium pedigree quality, it decreased to 95.3%, and for animals with the lowest pedigree quality, it dropped to 92.6%. Similarly, the correlation between pedigree-based and genomic-based breed composition improved with higher pedigree quality. For example, the correlation between Friesian pedigree-based breed composition (pFRI) and genomic-based breed composition (gFRI) was highest at 0.863 for animals with high pedigree quality, decreased to 0.822 for animals with medium pedigree quality, and dropped further to 0.389 for those with the lowest pedigree quality. The Friesian breed correlation values were lower than those for Holstein and Jersey because the purebred Friesian reference population was smaller than the reference populations for these two breeds.

For the Holstein breed proportion, based on pedigree (pHOL) and genomic (gHOL) data, the correlation was 0.932 for animals with high pedigree quality, 0.911 for those with medium quality, and 0.447 for animals with the lowest pedigree quality. Results for the Jersey breed were similar, with the correlation between Jersey pedigree (pJER) and genomic (gJER) measures at 0.939 for animals with high pedigree quality, 0.909 for medium, and 0.406 for low pedigree quality. Outliers were often detected among animals with a pJER value of $1/2^n$, which is illustrated in Figure 1. This may be due to the incorrect assignment of purebred Jersey ancestors (parent, grandparent, or great-grandparent) for these animals.

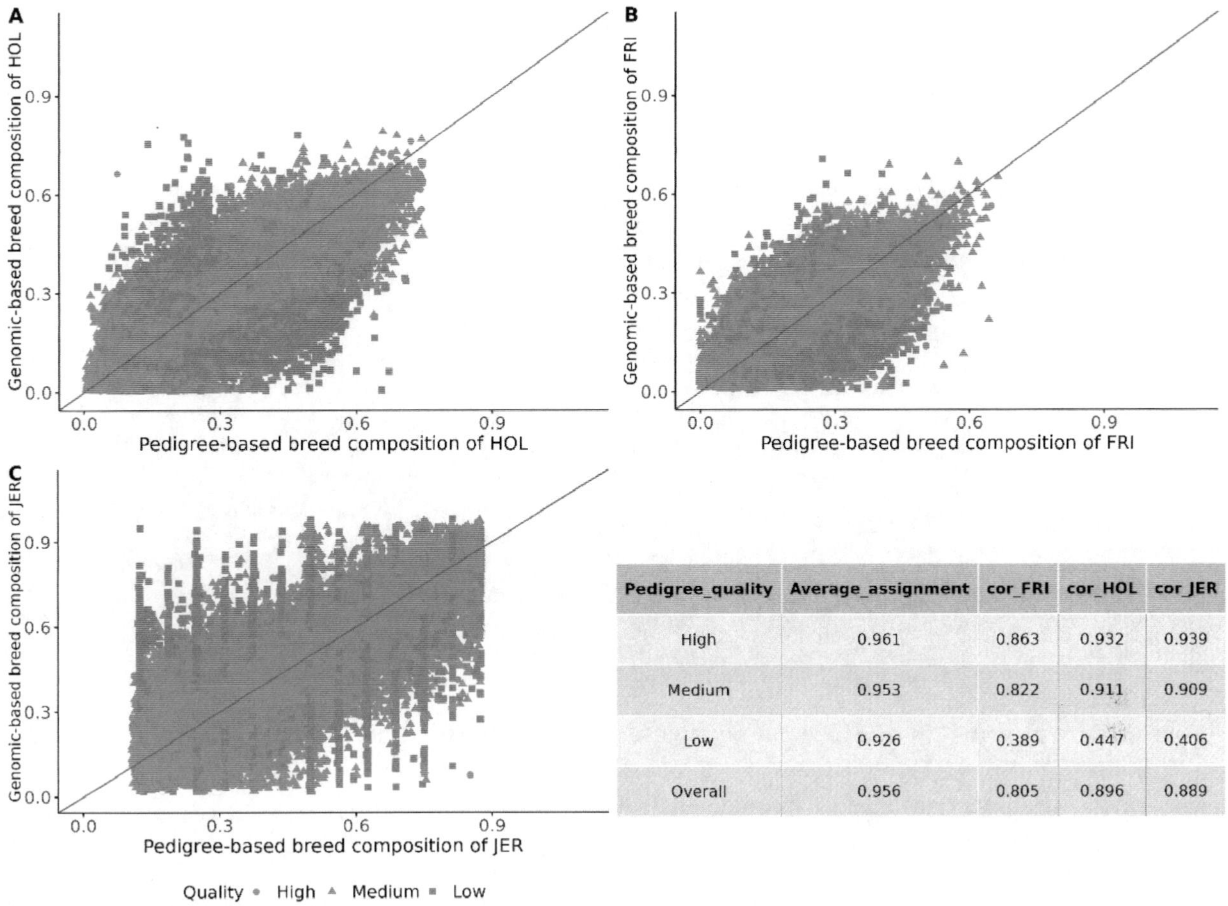

Pedigree_quality	Average_assignment	cor_FRI	cor_HOL	cor_JER
High	0.961	0.863	0.932	0.939
Medium	0.953	0.822	0.911	0.909
Low	0.926	0.389	0.447	0.406
Overall	0.956	0.805	0.896	0.889

Figure 1. The relationship between pedigree and genomic-based breed composition of Holstein (A), Friesian (B), and Jersey (C) in the target population

Haplotype inheritance in an example trio family. In the target population, one chromosome of each pair was predominantly assigned to the paternal haplotype, while the other chromosome was assigned to the maternal haplotype. Figure 2 illustrates this with a representative example of haplotypes of chromosome 25 from a trio family, including crossbred parents and ten full-sib offspring. Although pedigree information was not incorporated in the *optiSel* package for assigning the breed origin of alleles, the haplotype inheritance pattern is still clearly observable. Some offspring inherited a haplotype directly from one parent without crossover, such as offspring 2, whose haplotype 2 was directly inherited from the sire's haplotype 2. Crossovers are also evident; for instance, offspring 5's haplotype 2 combines segments from dam haplotype 2 in the first half and dam haplotype 1 in the latter half. Additionally, some alleles remain unassigned, as seen in the last segment of offspring 2's haplotype 2. This could be due to low genetic diversity among breeds or insufficient marker density at the chromosome's end to accurately determine alleles' breed origin.

In theory, full-sib progeny within the same family should have identical breed compositions based on pedigree information. In this example trio family, the sire has a pFRI of 0.258, and the dam has a pFRI of 0.148. Therefore, all full-sib progeny have an expected pFRI of 0.203, calculated as the average of the sire's and dam's values. Similarly, for the Holstein and Jersey proportions, each full-sib progeny has a pHOL of 0.333, the average of the sire's pHOL (0.598) and the dam's pHOL (0.068), and a pJER of 0.46, the average of the sire's pJER (0.136) and the dam's pJER (0.784). However, based on genomic information, the GBC of the progeny showed variation. The gFRI ranged from 0.146 to 0.202, with an average of 0.172. For the gHOL, values ranged from 0.301 to

0.361, averaging 0.328. The gJER ranged from 0.457 to 0.5, with an average of 0.477. The values showed deviation with pedigree based breed compositions, resulting from crossing-overs and chromosomal assortments taking place during meiosis.

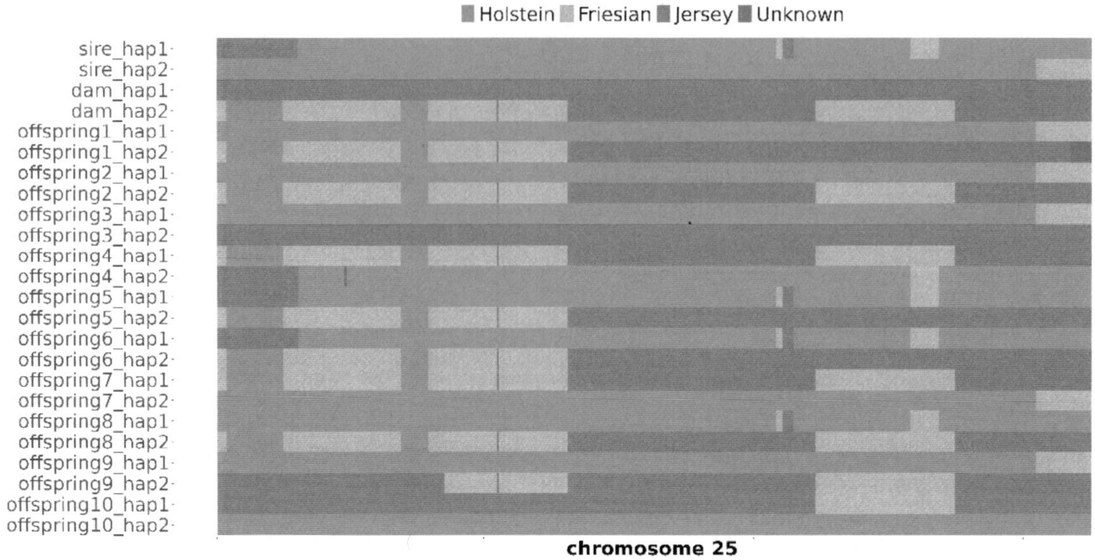

Figure 2. Breed-of-origin of alleles in a representative example trio family with sire, dam, and ten full-sib offspring. Each three-breed cross individual is represented in two rows, one representing the paternal and one representing the maternal haplotype

CONCLUSION

The breed-of-origin of alleles in Holstein-Friesian-Jersey crossbred dairy cattle was determined empirically using R package *optiSel*, achieving a 95.6% assignment rate without requiring pedigree information. High correlations were achieved between pedigree and genomic-based breed composition, especially for animals with medium to high pedigree quality. Based on the breed of origin of alleles, a clear haplotype inheritance pattern can be observed even the pedigree information was not utilized. Understanding the breed-of-origin of alleles enables future genomic prediction models to incorporate breed-specific effects of SNP alleles, leading to a potential increase in prediction accuracy.

ACKNOWLEDGEMENTS

The NZ Ministry of Primary Industries financially supported this study through the SFF Futures Program: Resilient Dairy–Innovative Breeding for a Sustainable Dairy Future (grant number PGP06-17006).

REFERENCES

Harris B.L. (2022) *JDS communications* **3**: 152.

Eiríksson J.H., Karaman E., Su G. and Christensen O.F., (2021) *Genet. Sel. Evol.* **53**: 84.

LIC and DairyNZ. (2022) New Zealand Dairy Statistics 2022-23, Hamilton, New Zealand.

Vandenplas J., Calus M.P., Sevillano C.A., Windig J.J. and Bastiaansen J.W. (2016) *Genet. Sel. Evol.* **48**: 61.

Wang Y., Segelke D., Emmerling R., Bennewitz J. and Wellmann R., (2017). *G3-Genes Genom. Genet.* **7**: 4009.

Wellmann R. (2019) *BMC Bioinformatics* **20**: 1.

ASCERTAINING THE BREED COMPOSITION OF AN ADMIXED SHEEP POPULATION USING GENOMIC INFORMATION: A SIMULATION STUDY

S. Sahoo[1], M.H. Ferdosi[1], J.H.J. van der Werf[2] and S. de las Heras-Saldana[1]

[1] Animal Genetics Breeding Unit,* University of New England, Armidale, NSW 2351 Australia
[2] School of Environmental & Rural Science, University of New England, Armidale, NSW 2351 Australia

SUMMARY

The aim of this study was to evaluate the reliability of estimating the breed composition of an admixed population using simulated data. Comprehending breed composition will help to discern the animal's ancestry and to define breed groups in genetic evaluation. The simulation was performed in two steps: initially, a historical population was simulated and then a recent population comprising two subpopulations was simulated. Haplotypes of these subpopulations were coded differently and selected randomly and independently for 5 generations, followed by a cross between them in generation 6, with further inter-crossing for the last 4 generations. True breed composition (TBC) was calculated from coded haplotypes for the crossbred population and later compared with results obtained from Admixture and BreedComp algorithms. BreedComp provided more accurate breed composition estimates than Admixture in this study. These findings suggest that the approach can be extrapolated to real data to assign animals into genetic groups to improve the prediction accuracy of breeding values.

INTRODUCTION

Animal breeders strive to select the best animals to genetically improve breeding populations (Niehoff *et al.* 2024). Due to selection and admixture, many populations are heterogeneous, which needs to be accounted for in genetic evaluations. It is thereby important to identify genetic groups to account for the genetic variation in admixed populations. Genetic groups or unknown parent groups (UPGs) can be assigned to animals of unknown parentage to account for differences in genetic merit (Westell *et al.* 1988). Different methods have been implemented to define genetic groups, based on birth year, region or flocks of origin, or generation from pedigree records. However, this approach is problematic when the pedigree is incomplete (Masuda *et al.* 2022). With the advent of genomic data, a potentially powerful alternative method for breed of origin is available (Eiriksson *et al.* 2022), but little is known about the accuracy of using genomic data to define genetic groups for admixed populations.

Since genomic data became available, modified versions of best linear unbiased prediction (BLUP) such as genomic BLUP (GBLUP) and single-step GBLUP have been used to improve the estimation of breeding value (Legarra *et al.* 2009). Previous studies have reported that the complexity of the relationship matrix increased with the integration of pedigree and genomic data. Due to different base populations being used for the numerator relationship matrix (**A**) and genomic relationship matrix (**G**), incompatibility between pedigree and genomic data produced biased estimates of breeding value (Masuda *et al.* 2022). A generalised UPG, also known as the metafounders (MF) model, was developed to ensure the compatibility between pedigree and genomic relationships by adjusting **A** to match **G,** balancing and accounting for relationships between and within base populations of **A** and **G** (Legarra *et al.* 2015; Himmelbauer *et al.* 2024).

---

* A joint venture of NSW Department of Primary Industries and Regional Development and the University of New England

However, incomplete pedigree in real data makes it difficult to properly define MF, as an accurate definition of relationships is needed. In this context, accurate estimation of breed composition based on genotype data could enhance the definition of genetic groups in admixed populations and improve compatibility with pedigree data.

Gurman *et al.* (2017) compared the estimates of clustering approaches such as Admixture (Alexander *et al.* 2009) and BreedComp (Boerner and Wittenburg 2018) with the breed composition estimated using pedigree data of Australian sheep breeds in which BreedComp provided slightly better estimates than Admixture. Similarly, findings by Boerner and Wittenburg (2018) highlighted that BreedComp performed better in crossbred animals artificially generated from real genotyped data of 11 different cattle breeds. However, pedigree-based breed assignment can be inaccurate due to missing pedigree records, affecting its comparison with genomic-based breed assignment methods.

Simulation studies can be used to determine the accuracy of these methods to estimate breed composition using genomic data, as tracing the alleles passed to subsequent generations using haplotype information can aid in determining the accurate breed contribution from either parent. Although simulation studies have been used to determine breed composition with clustering algorithms (Himmelbauer *et al.* 2024), the accuracy of the clustering methods using genomic data to evaluate its efficiency is largely unexplored.

Hence, this study used simulated data to assess the reliability of estimating breed composition based on genomic information to infer the parental contribution to each animal in an admixed population.

MATERIALS AND METHODS

Simulation overview. A historical sheep population was simulated using QMSim (Sargolzaei and Schenkel 2009). The genome consisted of 26 chromosomes with 49,400 SNPs, a mutation rate of 2.5×10^{-5}, an effective population size (Ne) of 100, and a crossover interference of 25cM. A total of 2000 animals were simulated descending from 20 sires and 1000 dams.

The historical population was divided into populations A and B using custom code in R. To track back the ancestral populations, alleles in haplotypes were recoded as -1 and 1 for population A, and -2 and 2 for population B, respectively. All 1,000 dams in both populations were selected along with 54 males in population A and 26 males in population B to create a difference in effective population size between these two populations. A recombination function was added to the population using the hsphase R package (Ferdosi *et al.* 2014). Mating within populations A and B was conducted randomly and independently for 5 generations.

In generation 6, 26 sires from population B were mated with 1000 dams from population A. The resulting cross population was used to simulate four more generations of inter-crosses, using 54 sires and 1,000 dams from generations 7 to 10.

From animals in generation 5 to 10 (n = 1000 per generation), the haplotypes were recoded (0, 1) format and converted to genotypes to estimate the breed composition using supervised Admixture (K = 2) (Alexander *et al.* 2009) and BreedComp (Boerner 2017) analysis. The true breed composition (TBC) was measured based on counts of alleles at each locus coded according to the breed of origin. The accuracy of the estimates was calculated as the difference between TBC with estimates obtained from Admixture and BreedComp for the crossbred generations 6 to 10, and these differences were plotted using Python v3.11 (Python Software Foundation 2023). Population parameters, such as average inbreeding coefficient and effective population size for crossbred population, were calculated using *--het* in Plink (Purcell *et al.* 2007) and SNeP software (Barbato *et al.* 2015), respectively.

RESULTS AND DISCUSSION

Effective population size and inbreeding coefficients. Inbreeding coefficients (F) and effective population size (Ne) were estimated for generations 5 to 10 (Table 1). Ne for population A at generation 5 (264) is higher than that of population B (167) because of fewer sires being used in population B, which also explains the higher F-value in population B. In generation 6, there was an increase in Ne as a result of the cross between populations A and B, which were mated randomly for 5 generations. As F-value was calculated using observed and expected homozygotes, a negative estimate of F explained the increase in the level of heterozygosity (Purcell *et al.* 2007) as observed in generation 6. From generation 7 to 10, Ne declined in the crossbred population as it followed inter se mating within the population leading to an increase in inbreeding coefficients.

Table 1. Estimation of inbreeding coefficient (F) and effective population size (Ne) from generation 5 to 10

Generation	5 (PopA)	5 (PopB)	6	7	8	9	10
F	0.0084	0.0234	-0.024	0.0007	0.0015	0.0085	0.0116
Ne	264	167	338	329	310	308	292

Breed composition estimates. Density plot depicts the distribution for the difference between TBC and estimates from Admixture and BreedComp (Figure 1). BreedComp exhibited a lower sigma value ($\sigma = 0.08$) than Admixture ($\sigma = 0.13$) revealing that the difference between TBC and estimates of BreedComp were closer to zero as compared to Admixture results, indicating slightly more accurate estimates. Similar findings were reported by (Gurman *et al.* 2017; Boerner and Wittenburg 2018) for Australian sheep and cattle breeds respectively, where BreedComp was able to perform better than Admixture.

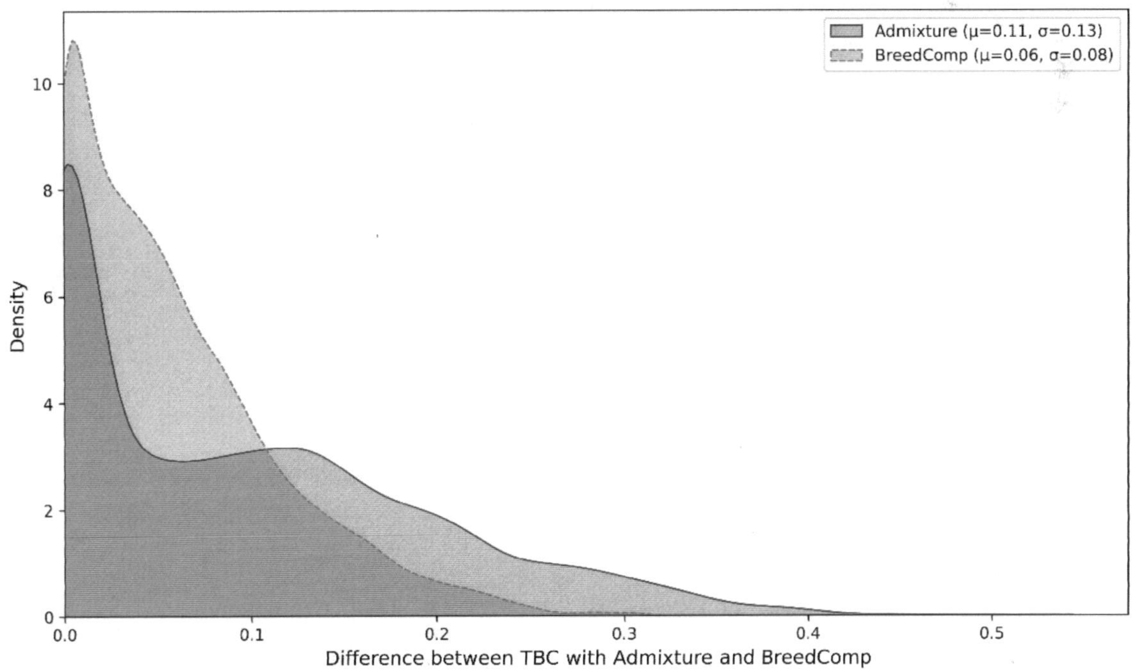

Figure 1. Density Plot illustrating the difference of TBC with estimates of Admixture and BreedComp for crossbred population

As Ne estimates were calculated on the basis of linkage disequilibrium (LD) (Barbato *et al.* 2015), a decrease in Ne estimates from generation 7 to 10 (Table 1) reflects a high level of LD in the crossbred population. Neither Admixture nor BreedComp algorithms account for LD. However, BreedComp relaxes marker independence by accounting for linear dependencies between markers, possibly explaining why BreedComp could provide better estimates of breed composition compared to Admixture results (Boerner and Wittenburg 2018).

The BreedComp algorithm can determine breed composition, enabling proper grouping of animals based on genomic data. Reliable breed composition estimates will help in clustering similar animals into MFs, which can be achieved by defining them based on genotype data. This approach helps overcome the limitation of missing and inaccurate pedigree records. Further research will be needed to investigate applying this method to estimate the breed composition of real animals based on genomic data to assign animals properly into MFs to test if this improves the accuracy of the prediction of breeding value which will enhance the overall productivity of the farm.

CONCLUSIONS

This study provided insights into the use of Admixture and BreedComp algorithms to estimate reliability for evaluating breed composition in admixed sheep populations using genomic information. BreedComp provided more accurate estimates of breed composition than Admixture in this study. The application of algorithms like BreedComp will help in precise assignment of animals to define metafounders based on genomic data, thereby improving the accuracy of breeding value in genetic evaluation.

ACKNOWLEDGMENTS

Shweta Sahoo acknowledges the financial support provided by the UNE-IPRA scholarship.

REFERENCES

Alexander D.H., Novembre J. and Lange K. (2009) *Genome Res* **19**: 1655.

Barbato M., Orozco-terWengel P., Tapio M. and Bruford M.W. (2015) *Front. Genet.* **6**: 109.

Boerner V. (2017) *Proc. Assoc. Advmt. Anim. Breed. Genet.* **22**: 97.

Boerner V. and Wittenburg D. (2018) *Front. Genet.* **9**: 185.

Eiriksson J.H., Byskov K., Su G., Thomasen J.R. and Christensen O.F. (2022) *J. Dairy Sci* **105**: 5178.

Ferdosi M.H., Kinghorn B.P., van der Werf J.H.J., Lee S.H. and Gondro C. (2014) *BMC Bioinformatics* **15**: 172.

Gurman P.M., Swan A.A. and Boerner V. (2017) *Proc. Assoc. Advmt. Anim. Breed. Genet.* **22**: 341.

Himmelbauer J., Schwarzenbacher H., Fuerst C. and Fuerst-Waltl B. (2024) *J. Dairy Sci.* **107**: 8170.

Legarra A., Aguilar I. and Misztal I. (2009) *J. Dairy Sci.* **92**: 4656.

Legarra A., Christensen O.F., Vitezica Z.G., Aguilar I. and Misztal I. (2015) *Genetics* **200**: 455.

Masuda Y., VanRaden P.M., Tsuruta S., Lourenco D.A.L. and Misztal I. (2022) *J. Dairy Sci.* **105**: 923.

Niehoff T.A.M., Ten Napel J., Bijma P., Pook T., Wientjes Y.C.J., Hegedus B. and Calus M.P.L. (2024) *Genet. Sel. Evol.* **56**: 41.

Purcell S., Neale B., Todd-Brown K., Thomas L., Ferreira M.A., Bender D., Maller J., Sklar P., de Bakker P.I., Daly M.J. and Sham P.C. (2007) *Am. J. Hum. Genet.* **81**: 559.

Python Software Foundation. 2023. Python Language Reference, version 3.11. *Python Software Foundation*, Wilmington, DE. Available at https://www.python.org

Sargolzaei M. and Schenkel F.S. (2009) *Bioinformatics* **25**: 680.

Westell R.A., Quaas R.L. and Van Vleck L.D. (1988) *J. Dairy Sci.* **71**: 1310.

Proc. Assoc. Advmt. Anim. Breed. Genet. 26: 327-330

INBREEDING MEASURES FOR AUSTRALIAN WAGYU - METHODS AND IMPLICATIONS

D.P. Garrick[1,2,3], D.J. Garrick[1,2] and C.F. Teseling[4]

[1] The Helical Company Limited, Rotorua, New Zealand
[2] Theta Solutions LLC, Lacey WA, USA
[3] A.L. Rae Centre for Genetics and Breeding, School of Agriculture and Environment, Massey University, New Zealand
[4] Australian Wagyu Association, Armidale, NSW, 2350 Australia

SUMMARY

This study analysed inbreeding patterns in the Australian Wagyu population using three complementary approaches: traditional pedigree-based coefficients, and genomic approaches based on runs of homozygosity (ROH), or runs of heterozygosity (ROHet). Data included a pedigree of >677,000 animals including 330,000 genotyped on the GGP100K platform. Results revealed that genomic inbreeding estimated from (ROH) consistently tracked about 10% higher than pedigree-based inbreeding, perhaps reflecting ancient inbreeding not captured in pedigree analysis. A steady increase in both metrics was observed from 2000-2024. These findings provide a foundation for implementing genomic-based mate selection strategies to optimise both genetic gain and maintenance of diversity in Wagyu breeding programs.

INTRODUCTION

The genetic architecture of modern cattle reflects their unique domestication history, with global populations descending from just two or three domestication events approximately 10,000 years ago (Pitt *et al.* 2018). These ancient population bottlenecks established a foundational level of inbreeding inherent in all contemporary cattle breeds. Against this historical backdrop, the intensification of modern breeding programs has heightened the importance of managing genetic diversity within specific populations, preventing inbreeding depression, and limiting the emergence of deleterious recessive alleles that could impact animal fitness, health, or welfare.

The Australian Wagyu population represents a compelling case study in genetic diversity management, having originated from a limited number of foundation sires exported from Japan in the 1990s. This relatively recent bottleneck, combined with the population's continued closure to outside genetics, creates unique challenges for maintaining genetic diversity while pursuing intensive selection for desired traits. Understanding the patterns and extent of inbreeding in this population is crucial for its long-term viability and continued genetic improvement.

Traditional approaches to monitoring inbreeding have relied on pedigree analysis, which provides expected relationships between animals based on their common ancestors. However, pedigree-based inbreeding coefficients are limited to the extent of pedigree recording that defines the known founders assumed to be unrelated and non-inbred, and cannot capture the actual segments of DNA inherited from common ancestors, nor can they identify specific genomic regions under selection pressure. The advent of high-density genomic data has enabled more precise characterisation of realised inbreeding through multiple complementary approaches.

This study characterises genetic diversity within the Australian Wagyu population using three distinct but complementary approaches: traditional pedigree-based inbreeding coefficients, runs of homozygosity (ROH), and runs of heterozygosity (ROHet). By comparing these metrics, we aim to provide a comprehensive assessment of inbreeding patterns and their implications for breeding program management. This research extends beyond previous work investigating X chromosome heterozygosity (Garrick *et al.* 2023) in the Australian Wagyu population.

MATERIALS AND METHODS

The data used in this study was a 2024 extract from the comprehensive Australian Wagyu Association (AWA) Helical database. The extract contains >591,500 SNP genotype samples originating from 951 different kinds of panels as defined by the genotype chip, version, manifest and service provider. The corresponding recorded pedigree of >677,000 animals included registered Japanese Black, Akaushi, Composite, and commercial animals.

Genotype samples were discarded from analyses if they had a raw call rate lower than 95% or if < 10% or > 90% of their SNPs were homozygous. SNPs were excluded if the minor allele frequency (MAF) was less than 15%. This higher than usual threshold was selected after exploratory analyses and serves to mitigate the effects of the smaller Japanese Black effective population size (Zhang and Banks 2021) and the founder bottleneck which exists outside of Japan. For this analysis, only animals genotyped on GGP100K chips were considered, amounting to 329,695 individuals with 64,180 informative SNPs after MAF filtering. A more strictly defined subset of that population was created by limiting animals to those with a recorded date of birth that were AWA registered Japanese Black (not Japanese Red, Akaushi, composite, or unregistered commercial animals), and that consisted of 94,279 individuals.

Traditional pedigree-based inbreeding coefficients were computed from the complete pedigree using an algorithm suitable for large populations (Meuwissen and Luo 1992).

Runs of homozygosity (ROH) were computed by assessing runs of homozygous alleles (SNP is AA or BB) with the following criteria:

1. Max gap between SNPs was 1,000,000 base pairs (1Mb)
2. Minimum ROH length was 500,000 base pairs (500kb)
3. Minimum number of SNPs in a run was 30 SNPs
4. Minimum SNP density was one SNP per 125,000 base pairs
5. 1 missing SNP was allowed within a run
6. 1 heterozygous SNP was allowed within a run

An ROH coefficient representing genomic inbreeding was computed as the length of the genome in ROH / 2.9 Gb which is the assumed length of the bovine genome. This inbreeding metric provides more detailed information than pedigree-based methods, which cannot account for ancient inbreeding events or missing pedigree information.

Runs of heterozygosity (ROHet) were computed by assessing runs of heterozygous alleles (SNP is AB) with the following criteria:

1. Minimum length of 250kb (250,000 base pairs)
2. Containing at least 15 SNPs
3. Maximum gap allowed between SNPs was 1Mb (1,000,000 base pairs)
4. 3 homozygous SNPs (AA or BB) were allowed within a run
5. 2 missing SNPs were allowed within a run

An ROHet coefficient was computed as the length of the genome in ROHet / 2.9 Gb. ROHet represent regions where an animal shows sustained heterozygosity which may in some cases be an artifact of the SNP chip design but in general should be indicative of genetic diversity or historical admixture in the context of analysing registered Japanese Black Wagyu individuals.

The Helical software system (Garrick *et al.* 2025) implements the analytical methods described above and was used to extract the data from the AWA Helical database, perform necessary quality controls, and to undertake the analyses.

RESULTS AND DISCUSSION

Founder animals in the pedigree were assumed non-inbred and unrelated so therefore comprise the base for pedigree inbreeding. Pedigree-based inbreeding coefficients averaged across each birth year are steadily increasing within the Japanese Black population. Observing Figure 1, genomic inbreeding via ROH follows a nearly identical trend to pedigree-based inbreeding on the same individuals but with the mean inbreeding at each time point being roughly 10% higher than the pedigree inbreeding. This mean difference represents the ancient inbreeding existing in those founder pedigree animals which can be captured in descendants via genomic analysis. The variability from 2000-2006 may be attributed to the limited number of animals born in this period who had been genotyped on the GGP100K chip.

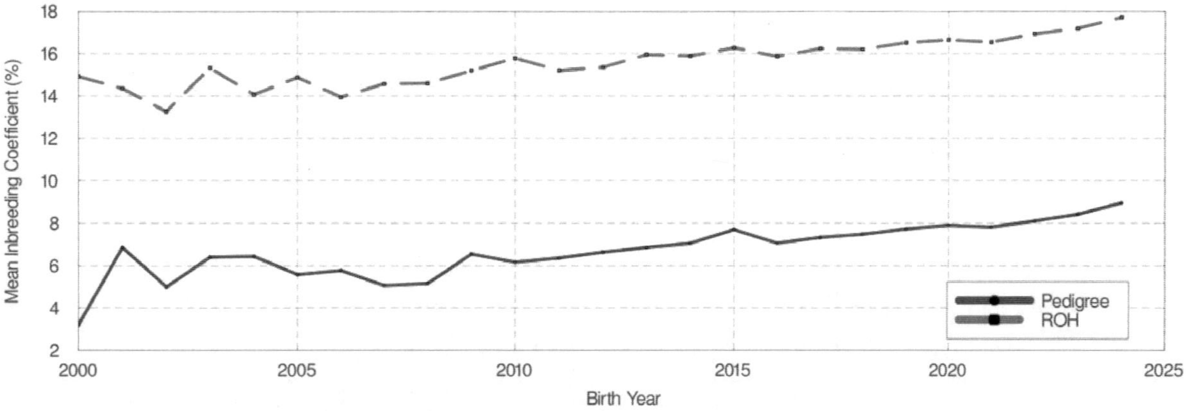

Figure 1. Mean inbreeding % by birth year for AWA registered Japanese Black Wagyu using a pedigree-based method (left axis) and genomic inbreeding (right axis)

Figure 2 shows that while there is a strong relationship between genomic and pedigree inbreeding, there also exists a wide distribution of genomic inbreeding (ROH) estimates for animals with the same pedigree inbreeding estimate. This distribution presents opportunities to utilise precision breeding to significantly improve the management of inbreeding over the long term, by assessing the specific genomic regions of homozygosity between potential mates.

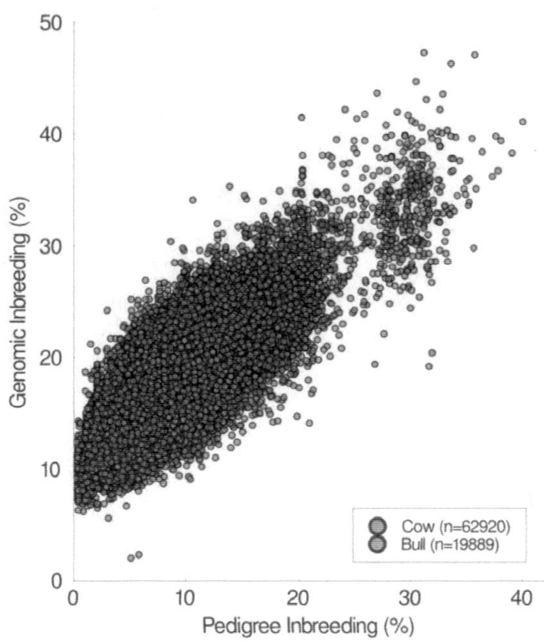

Figure 2. Comparison of genomic inbreeding % via ROH vs. pedigree inbreeding % for AWA 2020-2024 born genotyped registered Japanese Black Wagyu

Figure 3 visualises the distribution of ROHet versus ROH for the same set of animals. A negative correlation between ROH and ROHet is observed indicating that for increasing levels of ROH, the proportionate runs of heterozygosity decreases, however ROHet alone is not a good indicator for genomic inbreeding. For a given level of ROH, there exists a distribution of animals with widely varying ROHet. For example, animals with 20% ROH have between 5-9% of ROHet. These runs of heterozygosity present opportunities for mate selection which further account not just for animals with differing regions of ROH, but also to select mates with complementary regions of ROHet within the genome.

The increased uptake and interest in mate-allocation software reflects these opportunities in practice. Extensions of such tools to incorporate individual ROH and/or ROHet profiles could potentially target matings that maximise beneficial homozygosity in trait-specific regions while minimising overall ROH across the genome.

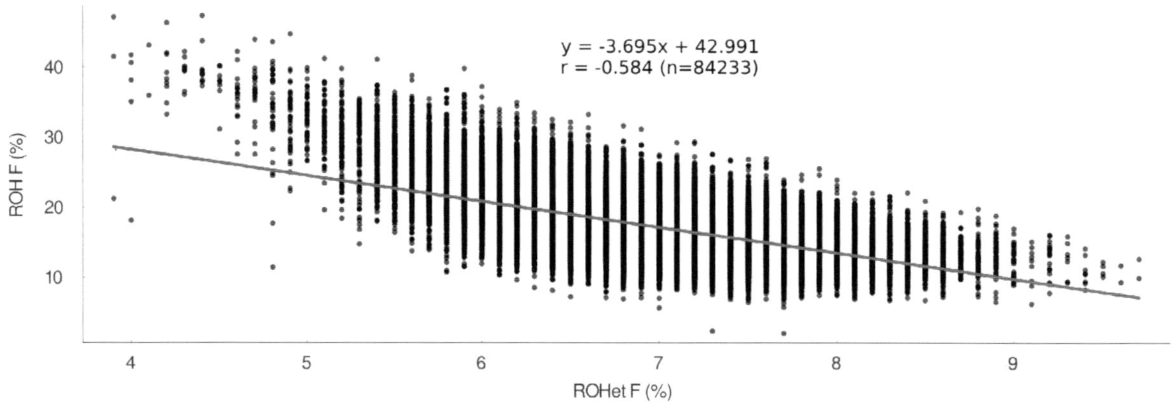

Figure 3. ROH versus ROHet for AWA 2020-2024 born genotyped registered Japanese Black Wagyu

CONCLUSION

The Japanese Black population registered with the Australian Wagyu Association shows increasing levels of inbreeding by birth year. Trends in genomic inbreeding estimates using an ROH analysis align well with trends in pedigree-based estimates but present new opportunities for mating decisions which optimise inbreeding and genetic gain within specific regions of the genome. While rising inbreeding warrants monitoring, the presence of heterozygous genome segments provides opportunities for maintaining genetic diversity. Future breeding decisions should focus on balancing genetic gain with maintenance of diversity, utilising specific ROH and ROHet segment information to guide mate selection with precision breeding. Implementation of these genomic tools in breeding programs could help optimise long-term genetic improvement while managing inbreeding accumulation.

REFERENCES

Garrick D.P., Garrick D.J. and Howse J.E. (2025) *Proc. Assoc. Advmt. Anim. Breed. Genet.* **26**: *These proceedings.*

Garrick D.P., Handcock, R.C. and Teseling, C. (2023) *Proc. Assoc. Advmt. Anim. Breed. Genet.* **25**: 226.

Meuwissen T. and Luo Z. (1992) *Genet. Sel. Evol.* **24**: 305.

Pitt D., Sevane N., Nicolazzi E.L., MacHugh D.E., Park S.D.E., Colli L., Martinez R., Bruford M.W. and Orozco-terWengel P. (2018) *Evol. Appl.* **12**: 123

Zhang, Y.D. and Banks, R. (2021) *Proc. Assoc. Advmt. Anim. Breed. Genet.* **24**: 451.

ABSENCE OF C>T POLYMORPHISM IN THE INTRON 1 (5′-UTR) REGION OF THE *MTHFR* GENE IN INDIAN BARBARI GOATS

A. Kumar, S.P. Singh, D. Sharma, G. K. Gupta, Pooja, A. Chauhan, V. Aditya and B. Kumari

Department of Animal Genetics and Breeding, College of Veterinary Science & Animal Husbandry, UP Pandit Deen Dayal Upadhyaya, Pashu Chikitsa Vigyan Vishwavidyalaya Evam Go-Anusandhan Sansthan (DUVASU), Mathura-281001 (India)

SUMMARY

The methylenetetrahydrofolate reductase (*MTHFR*) gene plays a crucial role in one-carbon metabolism and has been investigated for its association with production traits in livestock. This study aimed to examine a previously reported polymorphism in the *MTHFR* gene (rs641815079) in 272 adult female Barbari goats maintained at the University Goat Farm, Department of Physiology, DUVASU, Mathura, using PCR-RFLP with the *BsaH*I restriction enzyme.

The results revealed a monomorphic banding pattern, where all individuals exhibited the CC genotype, confirming the absence of C>T polymorphism at this site. Sanger sequencing further validated the presence of the allele, leading to the formation of the *BsaH*I restriction site. Due to the monomorphic nature of the population, no association study with milk production traits was conducted. These findings suggest that this region of the *MTHFR* gene does not contribute to genetic variation in Barbari goats, emphasising the need for alternative genetic markers for selection programs.

INTRODUCTION

Genetic selection plays a significant role in improving milk yield, composition, and metabolic efficiency in dairy animals. The *MTHFR* gene, a key regulator of one-carbon metabolism and DNA methylation, has been linked to various productive traits in cattle and goats. Several studies have demonstrated that polymorphisms within this gene can impact milk yield, protein percentage, and metabolic efficiency in ruminants. Previous research in Xinong Saanen and Guanzhong dairy goats (An *et al.* 2015; Hou *et al.* 2015) identified three genotypes (CC, CT, and TT) at the g.1951C>T locus (rs641815079) of the *MTHFR* gene. The rs641815079 SNP, located at NC_030823.1:g.40362342C>T, has been previously associated with milk composition traits in other dairy goat breeds (An *et al.* 2015). However, there is a lack of studies investigating *MTHFR* polymorphisms in Indian goat breeds, particularly Barbari goats, which are known for their high adaptability and moderate milk production. The objectives of this study were to genotype the *MTHFR* rs641815079 variant using PCR-RFLP in Barbari goats and determine allele and genotype frequencies in the studied population.

MATERIALS AND METHODS

Animal selection and sample collection. A total of 272 adult female Barbari goats from the University Goat Farm, DUVASU, Mathura were included in the study. Blood samples (5 mL) were collected via jugular venipuncture using EDTA-coated vacutainers and stored at −20°C for further analysis. Ethical approval for this study was obtained from the Institutional Animal Ethics Committee (IAEC), DUVASU, Mathura, under the Committee for Control and Supervision of Experiments on Animals (CCSEA), Government of India (Approval No. V-11011(13)/6/2022-CPCSEA-DADF, dated 09 June 2022).

DNA extraction and PCR amplification. Genomic DNA was extracted using the phenol-chloroform method and assessed for quality using 1% agarose gel electrophoresis and NanoDrop spectrophotometry.

The targeted SNP, rs641815079, is a known variant in goats and corresponds to the T→C substitution at genomic coordinate NC_030823.1:g.40362342C>T based on the ARS1.2 goat reference genome. The ARS1.2 reference genome lists the C allele at this position. A 546 bp fragment within non-coding region of the *MTHFR* gene was amplified using PCR as per the PCR condition mentioned in Table 1, with the following previously reported primer set (An *et al.* 2015)

F 5'-CCGAACATCTGTTGACCTC-3'

R 5'-AGGAAGAAGGCTGGTGAG-3'

PCR amplification was performed in a 25 µl reaction volume containing 12.5 µl of 2X PCR Master Mix (EmeraldAMP GT PCR master mix, Takara Bio, RR310A), 0.5 µl each of 10 pmole/µl forward and reverse primers, 1.0 µl of genomic DNA (~100 ng), and 10.5 µl of nuclease-free water.

Table 1. PCR conditions for the amplification of the *MTHFR* gene

Steps		1	2			3	4
		Initial denaturation	35 Cycles of			Final extension	Storage
MTHFR	Temp.	94°C	94°C	58°C	72°C	72°C	12°C
	Time	5 min	30 sec	30 sec	30 sec	8 min	∞

Amplified fragments were visualized using 1.0% agarose gel electrophoresis, and the results were documented using a gel documentation system (Figure 1).

***MTHFR/BsaH*I PCR-RFLP Assay.** The PCR products were digested using the *BsaH*I restriction enzyme following standard protocols. The digested fragments were separated by 2.0% agarose gel electrophoresis, stained with ethidium bromide, and visualized under a UV transilluminator.

Sanger sequencing. To confirm the presence of the restriction site for the *BsaH*I enzyme and to validate the monomorphic nature of the rs641815079 locus, PCR products from three randomly selected Barbari goats were purified and subjected to Sanger sequencing. The obtained sequences were aligned using standard bioinformatics tools to detect the presence of the T→C substitution, which forms the recognition site for *BsaH*I.

RESULTS AND DISCUSSION

***MTHFR/BsaH*I PCR-RFLP assay.** The *BsaH*I/PCR-RFLP assay revealed only one banding pattern (CC genotype, 426 bp & 120 bp) (Figure 2). CT and TT genotypes were absent, confirming the monomorphic nature of the Barbari goat population for this variant in the studied population. The study found that all screened animals lacked variation for the *MTHFR* rs641815079 variant, with all individuals exhibiting the CC genotype. Consequently, no association study with milk production traits was conducted.

***MTHFR/BsaH*I DNA sequencing.** Revealing that all sequenced were homozygous for the allele of the C allele of the rs641815079 variant. Sanger sequencing confirmed the PCR-RFLP results (Figure 3). The C allele, which is also present in in the ARS1.2 reference genome, the recognition sequence for *BsaH*I, explaining the uniform digestion pattern in all individuals. Due to the monomorphic nature of the population, no association study with milk production and quality traits were conducted.

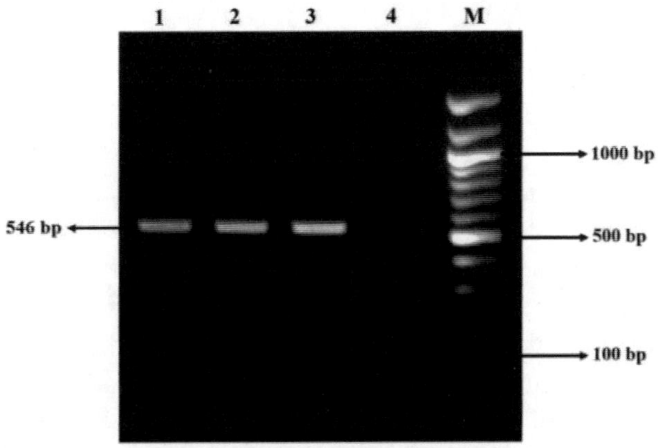

Figure 1. Agarose (1.0%) gel electrophoresis showing a 546 bp PCR products in all lanes (1-4), M= Marker (100bp ladder)

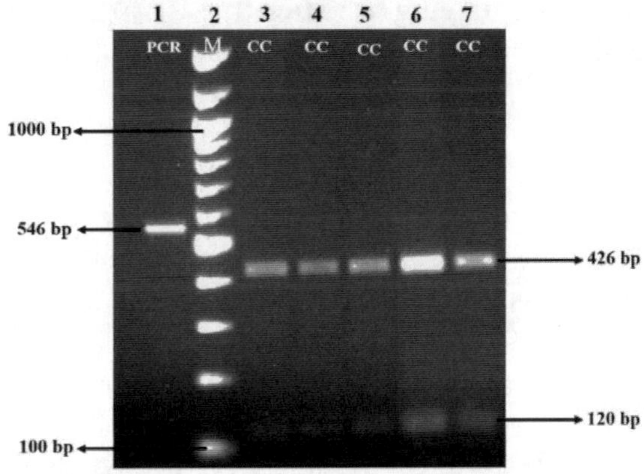

Figure 2. *MTHFR/BsaH*I PCR-RFLP assay showing genotype pattern in 2.0% agarose gel; Lane 1: Undigested PCR product, 2: Marker (100 bp ladder), 3-7: CC genotype (426 & 120 bp)

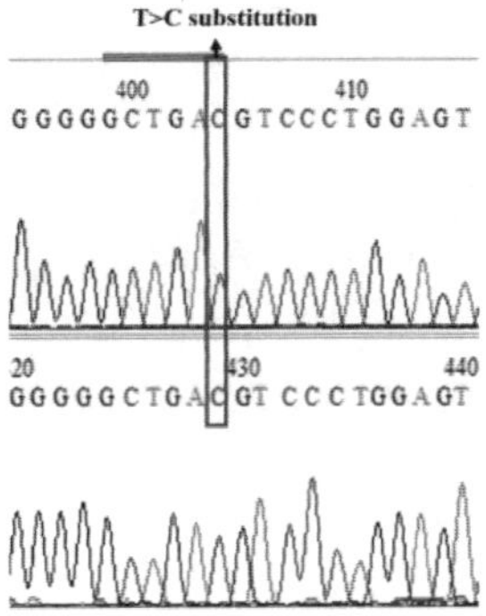

Figure 3. DNA sequencing showing the rs641815079 C allele highlighted by the purple box in the GCTGAC recognition site of *BsaH*I enzyme

Comparison with previous studies. An *et al.* (2015, 2016) and Hou *et al.* (2015) found several polymorphic variants in Xinong Saanen and Guanzhong dairy goats and reported significant associations with milk protein content for some of these variants, including the rs641815079 variant.

The frequency of the C allele for the rs641815079 variant was 0.72, whereas in our study, it was 1.00, indicating a fixed allele in Barbari goats.

Implications of monomorphism. The absence of observed genetic variation for the rs641815079 variant at the *MTHFR* locus in this study may be attributed to the relatively small population size, limiting the detection of rare alleles. To validate these findings, a larger and more diverse sample of Barbari goats should be screened. Although monomorphism was observed at the investigated locus, the original study plan included assessing the association of this polymorphism with various milk production and quality traits. These included daily milk yield, fat percentage, solids-not-fat (SNF), protein content, and lactose percentage. If polymorphism had been detected, statistical analyses would have been conducted to evaluate its association with these economically important traits in Barbari goats. It is also important to note that the rs641815079 variant is likely a marker and not necessarily a causal variant. In other goat breeds, such as Xinong Saanen and Guanzhong, other polymorphisms within the *MTHFR* gene (e.g., *c.*2494G>A, *g.2244A>G*) have shown significant associations with milk protein traits (An *et al.* 2015; *Hou et al.* 2015; An *et al.* 2016). Future studies in Barbari goats should consider exploring these additional variants or identifying novel SNPs through targeted resequencing. A genome-wide association study (GWAS) could also be more effective in identifying whether the *MTHFR* region or other genomic loci are linked to milk production and composition traits in this breed.

CONCLUSION

This study investigated the rs641815079 variant in the non-coding region of the *MTHFR* gene in Barbari goats using PCR-RFLP with the *BsaH*I restriction enzyme. The amplified region (546 bp) was monomorphic, with all individuals exhibiting the CC genotype. Sanger sequencing confirmed the results of the PCR-RFLP assay. Due to monomorphism, no association study with milk production traits was performed.

ACKNOWLEDGEMENTS

The authors express gratitude to the Honourable Vice chancellor, DUVASU, Mathura, for providing research facilities. Special thanks to the officer, In charge and staff at the University Goat Farm for assistance in sample collection and data recording.

REFERENCES

An X., Song Y., Hou J., Han P., Peng J., Zhang L., Wang J. and Cao B. (2015) *Small Rumin. Res.* **130**: 76.

An X., Song Y., Hou J., Wang S., Gao K. and Cao B. (2016) *Anim. Genet.* **47**: 499.

Hou J., Song Y., Gao T., Yue L. and Cao B. (2015) *PLoS One*. **10**: e0133015.

Proc. Assoc. Advmt. Anim. Breed. Genet. 26: 335-338

PRELIMINARY GENETIC CORRELATIONS OF MERINO MEAT COLOUR TRAITS WITH LEAN MEAT YIELD AND EATING QUALITY

S.I. Mortimer[1], S.M. Fowler[2], B.W.B. Holman[3], T.I.R.C. Alvarenga[1], D.L. Hopkins[2, 4], K.L. Egerton-Warburton[5], J.L. Smith[6], B.C. Hine[6] and A.A. Swan[7]

[1] NSW Department of Primary Industries and Regional Development, Armidale Livestock Industries Centre, Armidale, NSW, 2351 Australia
[2] NSW Department of Primary Industries and Regional Development, Centre for Sheep and Red Meat Development, Cowra, NSW, 2794 Australia
[3] NSW Department of Primary Industries and Regional Development, Wagga Wagga Agricultural Institute, Wagga Wagga, NSW, 2650 Australia
[4]Canberra, ACT, 2903 Australia
[5] NSW Department of Primary Industries and Regional Development, Orange Agricultural Institute, Orange, NSW, 2800 Australia
[6] CSIRO, Agriculture and Food, F.D. McMaster Laboratory, Armidale, NSW, 2350 Australia
[7] Animal Genetics and Breeding Unit*, University of New England, Armidale, NSW, 2351 Australia

SUMMARY

Genetic correlations were estimated among retail colour stability and fresh meat colour traits assessed on Merino lamb loins, as well as genetic correlations of the colour traits with lean meat yield and objective eating quality traits of intramuscular fat and shear force. The meat colour traits were confirmed to be heritable. Improving intramuscular fat or shear force is expected to lead to less redness and faster rate of browning of the meat during retail display. Selection for lean meat yield is expected to improve slightly retail colour stability.

INTRODUCTION

Of the MERINOSELECT standard indexes, the Merino Lamb index has been developed to improve profitability of a self-replacing production system where income is mostly derived from lamb production and some from adult wool production (MLA 2024). In addition to improving fleece weight, reproduction and growth, the index aims to achieve genetic gains in lean meat yield and eating quality in line with their economic values. Lean meat yield combined with growth and together with eating quality traits contribute approximately 20% of the index's economic response.

While there is no information on the cost of meat discolouration during retail display for Australian lamb meat, its colour influences decisions made by consumers and retailers. Consumers use meat colour to assess the quality and freshness of meat cuts at the point of purchase. For retailers, reduced display life of meat products from browning over time lowers their value due to discounting and downgrading of products. Impacts on retail colour stability of lamb cuts following selection for lean meat yield or eating quality are unknown, unlike the situation for fresh meat colour traits where favourable genetic relationships have been reported (Mortimer *et al.* 2017b, 2018). Using data from carcases of the wether progeny of Australian Wool Innovation's Merino Lifetime Productivity (MLP) project, this study reports genetic correlation estimates among retail colour stability and fresh meat colour traits assessed on loins and of the meat colour traits with lean meat yield and objectively measured eating quality traits of intramuscular fat and shear force.

---

* A joint venture of NSW Department of Primary Industries and Regional Development and the University of New England

MATERIALS AND METHODS

The traits for this study were recorded on carcases produced from 2017- and 2018-born wethers at the Macquarie and New England sites of the MLP project (Ramsay *et al.* 2019), with pre-slaughter procedures described by Mortimer *et al.* (2021). After 24 h of chilling post-slaughter, the fresh meat colour traits of relative redness (*a**, CFA), yellowness (*b**, CFB) and lightness (*L**, CFL) were measured on the cut surface of the left *longissimus lumborum* (LL, loin) at the 12th rib of each carcase as described by Newell *et al.* (2020). A sub-sample from the LL of each carcase was collected and aged for 5 days to measure retail colour stability traits (Holman *et al.* 2021). Colourimetric measurements of *a** (RCA), *b** (RCB) and *L** (RCL) were recorded initially and at 1-day intervals over 3 days, with values after 2 days of simulated retail display defined as the trait of interest. The oxymyoglobin/metmyoglobin ratio (RCR), a measure of discolouration of meat, was estimated by dividing the captured light reflectance at a wavelength of 630 nm by that at a wavelength of 580 nm. From other sub-samples of the LL of each carcase, the objective eating quality traits of intramuscular fat (IMF, %) and shear force (SF5, N) were measured, using methods as described by Holman *et al.* (2019) and Fowler *et al.* (2014), respectively. Lean meat yield (LMY, %) was predicted as first described by Gardner *et al.* (2010), but with a variant developed for the Merino breed (GE Gardner, personal communication). Summary statistics for each trait are shown in Table 1.

Statistical analysis. Variance and covariance components for each trait were estimated using ASReml (Gilmour *et al.* 2015). Phenotypic variances and heritabilities were estimated from pedigree only univariate analyses fitting an animal model. Fixed effects included in the model were birth-rearing type, age of dam, age at measurement, Merino wool type (ultra/superfine, fine/fine-medium, medium/strong) and contemporary group (as defined in the MERINOSELECT data base, modified for slaughter group). For the objective eating quality traits, hot carcase weight was included as a covariate. A direct genetic effect of animal was fitted as the random effect. A maternal permanent environmental effect and a sire by site interaction were also included as random effects but were found to be not significant for all traits. Genetic and phenotypic correlations among the traits were estimated from bivariate analyses.

Table 1. Summary statistics for meat colour, objective eating quality and lean meat yield traits and estimates of phenotypic variance (σ^2_P) and direct heritability (h^2)

Trait	Records	Mean	SD	Range	σ^2_P (SE)	h^2 (SE)
CFA	1446	20.5	1.4	14.2-25.2	1.89 (0.08)	0.23 (0.08)
CFB	1446	3.2	1.1	-1.0-7.1	0.78 (0.04)	0.52 (0.10)
CFL	1446	35.2	2.2	28.9-44.1	3.69 (0.17)	0.45 (0.10)
RCA	1437	16.5	1.8	10.2-22.7	2.98 (0.12)	0.09 (0.05)
RCB	1437	15.9	1.6	8.9-22.1	2.31 (0.10)	0.11 (0.06)
RCL	1437	39.2	2.9	30.6-48.8	5.56 (0.28)	0.59 (0.12)
RCR	1437	3.3	0.6	1.9-5.6	0.26 (0.01)	0.11 (0.06)
IMF	1437	4.28	1.33	0.81-10.8	1.34 (0.07)	0.87 (0.10)
SF5	1438	25.4	5.6	13.0-58.4	27.28 (1.33)	0.52 (0.11)
LMY	683	57.6	2.6	46.8-64.4	4.60 (0.34)	0.47 (0.17)

RESULTS AND DISCUSSION

Except for RCL, heritability estimates for the retail colour stability traits recorded after 2 days of simulated retail display were less than 0.15 (Table 1) and were similar to estimates reported by Mortimer *et al.* (2017b). Heritabilities were higher (estimates all greater than 0.20) for the fresh meat colour traits than published estimates for Merino loins (Greeff *et al.* 2008; Mortimer *et al.* 2017b). With all estimates greater than 0.40, heritability estimates for LMY, IMF and SF5 were much higher (0.29±0.11, 0.58±0.11 and 0.10±0.09, respectively) than estimates presented by Mortimer *et al.*

(2017a, b). Compared with the earlier reports, our estimates of additive genetic variances for LMY and the fresh meat colour and objective eating quality traits were much higher.

Genetic correlations between retail colour stability traits and LMY tended to be low and favourable though not significantly different from zero (Table 2). Estimates of LMY with RCR (0.10±0.39), RCA (0.18±0.41) and RCL (0.28±0.24) suggest that selection for LMY would not increase the rate of discolouration of Merino loins during retail display or unfavourably lower meat redness and lightness. Genetic correlations of the retail colour stability traits with IMF and SF5 were generally moderate to high, with the sign of estimates being consistent with the strong negative genetic correlation between IMF and SF5 (-0.79±0.08, Table 2). Selection yielding favourable increases in IMF or decreases in SF5 is expected to lead to greater meat discolouration (-0.71±0.18 and 0.66±0.20, respectively) and less red meat (-0.49±0.21 and 0.53±0.24, respectively), although lighter meat colour (0.69±0.08 and -0.51±0.14, respectively). These estimates were consistent with those derived from a larger multi-breed population (Mortimer *et al.* 2014).

Genetic correlations of LMY with fresh meat colour traits were low and not significantly different from zero. Previously, stronger estimates also with large standard errors had been reported, which suggested selection to increase LMY would reduce meat redness and increase meat lightness (Mortimer *et al.* 2018). The genetic correlations of the fresh meat colour traits with IMF and SF5 were generally consistent with estimates reported by Mortimer *et al.* (2017b), with favourable responses expected to occur following selection to improve the objective eating quality traits.

Table 2. Genetic (below diagonal) and phenotypic (above diagonal) correlations among meat colour, objective eating quality and lean meat yield traits

	CFA	CFB	CFL	RCA	RCB	RCL	RCR	IMF	SF5	LMY
CFA		0.55	-0.44	0.19	-0.02	-0.10	0.13	0.08	-0.16	0.06
CFB	0.57 (0.14)		0.43	0.13	0.14	0.33	0.04	0.41	-0.30	-0.05
CFL	-0.03 (0.22)	-0.06 (0.24)		-0.05	0.17	0.50	-0.09	0.35	-0.19	-0.05
RCA	0.36 (0.30)	-0.19 (0.25)	-0.47 (0.24)		0.67	-0.21	0.95	-0.17	0.00	0.17
RCB	0.12 (0.31)	0.58 (0.23)	0.68 (0.22)	0.13 (0.38)		n.e.[1]	0.60	0.03	-0.05	0.16
RCL	0.00 (0.21)	0.70 (0.11)	0.92 (0.05)	-0.17 (0.25)	n.e.		-0.29	0.34	-0.17	0.04
RCR	0.18 (0.30)	-0.43 (0.21)	-0.61 (0.20)	0.96 (0.03)	0.02 (0.37)	-0.34 (0.22)		-0.23	0.03	0.15
IMF	0.24 (0.17)	0.73 (0.08)	0.74 (0.09)	-0.49 (0.21)	0.28 (0.23)	0.69 (0.08)	-0.71 (0.18)		-0.38	-0.17
SF5	-0.32 (0.19)	-0.69 (0.10)	-0.69 (0.12)	0.53 (0.24)	-0.30 (0.26)	-0.51 (0.14)	0.66 (0.20)	-0.79 (0.08)		0.08
LMY	0.16 (0.30)	0.07 (0.25)	-0.03 (0.26)	0.18 (0.41)	-0.16 (0.39)	0.28 (0.24)	0.10 (0.39)	-0.12 (0.20)	-0.20 (0.25)	

[1] n.e., not estimable

These results are preliminary, as the analyses were based on a small Merino data set and did not fit a genetic group effect. With many genetic parameter estimates associated with relatively large standard errors, more accurate estimates are required for all traits to extend understanding of the implications of selection using the MERINOSELECT Merino Lamb index for meat colour traits. While the genetic correlation estimates among each of the retail colour traits, the fresh colour traits and among LMY, IMF and SF5 were in reasonable agreement with literature estimates, some estimates differed markedly. For example, the estimates of LMY with IMF (-0.12±0.20) and SF5 (-0.20±0.25) were weaker and in the case of SF5 the estimate was negative rather than positive.

CONCLUSION

The heritability estimates confirm that genetic variation in retail display and fresh meat colour traits exists and can be exploited by Merino breeding programs. Where selection programs focus on improving intramuscular fat or shear force, less redness and faster rate of browning of the meat during retail display is expected to occur. Selection for lean meat yield is expected to improve slightly retail colour stability. If analyses of more data confirm these relationships, responses in these traits from selection on the Merino Lamb index should be predicted and monitored.

ACKNOWLEDGEMENTS

We acknowledge the co-investment by Meat & Livestock Australia in funding the conduct of this study (MLA P.PSH.1908 and L.EQT.1908). Australian Wool Innovation Limited is the owner of pedigree and fixed effect data from the MLP Project used by this study. Staff are thanked at: NSW DPIRD Cowra for sample collection and measurement; NSW DPIRD, Trangie, and CSIRO, Armidale, for sheep management; and Fletcher International Exports for their support.

REFERENCES

Fowler S.M., Schmidt H., van de Ven R., Wynn P. and Hopkins D.L. (2014) *Meat Sci.* **97**: 597.

Gardner G.E., Williams A., Siddell J., Ball A.J., Mortimer S., Jacob R.H., Pearce K.L., Hocking Edwards J.E., Rowe J.B. and Pethick D.W. (2010) *Anim. Prod. Sci.* **50**: 1098.

Gilmour A.R., Gogel B.J, Cullis B.R. and Thompson R. (2015) ASReml User Guide Release 4.1. VSN International Ltd, Hemel Hempsted, UK.

Greeff J.C., Safari E., Fogarty N.M., Hopkins D.L., Brien F.D., Atkins K.D., Mortimer S.I. and van der Werf J.H.J. (2008) *J. Anim. Breed. Genet.* **125**: 205.

Holman B.W.B., Bailes K.L., Meyer R.G. and Hopkins D.L. (2019) *J. Food Sci. Technol.* **56**: 3957.

Holman B.W.B., Diffey S.M., Logan B.G., Mortimer S.I. and Hopkins D.L. (2021) *Food Anal. Methods* **14**: 697.

MLA (2024) https://www.sheepgenetics.org.au/globalassets/sheep-genetics/getting-started/sheep-genetics---asbvs-and-indexes/merinoselect-indexes-breeder-2024.pdf accessed 30 January 2025.

Mortimer S.I., Smith J.L., Hine, B.C., Fowler, S.M., Holman, B.W.B., Hopkins D.L., Egerton-Warburton K.L. and Swan A.A. (2021) *Proc. Assoc. Advmt. Anim. Breed. Genet.* **24**: 255.

Mortimer S.I., Fogarty N.M., van der Werf J.H.J., Brown D.J., Swan A.A., Jacob R.H., Geesink G.H., Hopkins D.L., Hocking Edwards J.E., Ponnampalam E.N., Warner R.D., Pearce K.L. and Pethick D.W. (2018) *J. Anim. Sci.* **96**: 3582.

Mortimer S.I., Hatcher S., Fogarty N.M., van der Werf J.H.J., Brown D.J., Swan A.A., Jacob R.H., Geesink G.H., Hopkins D.L., Hocking Edwards J.E., Ponnampalam E.N., Pearce K.L. and Pethick D.W. (2017a) *J. Anim. Sci.* **95**: 2385.

Mortimer S.I., Hatcher S., Fogarty N.M., van der Werf J.H.J., Brown D.J., Swan A.A., Jacob R.H., Geesink G.H., Hopkins D.L., Hocking Edwards J.E., Ponnampalam E.N., Warner R.D., Pearce K.L. and Pethick D.W. (2017b) *J. Anim. Sci.* **95**: 4260.

Mortimer S.I., van der Werf J.H.J., Jacob R.H., Hopkins D.L., Pannier L., Pearce K.L., Gardner G.E., Warner R.D., Geesink G.H., Hocking Edwards J.E., Ponnampalam E.N., Ball A.J., Gilmour A.R. and Pethick D.W. (2014) *Meat Sci.* **96**: 1016.

Newell M.T., Holman B.W.B., Refshauge G., Shanley A.R., Hopkins D.L. and Hayes R.C. (2020) *Small Rumin. Res.* 1**92**: 106235.

Ramsay A.M.M., Swan A.A. and Swain B.C. (2019) *Proc. Assoc. Advmt. Anim. Breed. Genet.* **23**: 512.

GENETIC PARAMETER ESTIMATION OF CARCASE QUALITY PHENOTYPES MEASURED WITH CAMERAS IN AUSTRALIAN WAGYU

E. de Bruyn, K.L. Moore and S.P. Miller

Animal Genetics Breeding Unit*, University of New England, Armidale, NSW, 2351 Australia

SUMMARY

Computer image analysis is increasingly important in carcase evaluations to determine pricing in beef markets. Images captured by cameras can accurately predict carcase traits such as marble score and eye muscle area. As technology develops, older camera models are being replaced as newer versions are developed. This study estimated genetic parameters for carcase traits using the MIJ-30 camera and compared it to estimates based on the older Mirror-type camera. Traits investigated were hot standard carcase weight, AUS-MEAT marble score, HK-333 Japanese mirror-type camera fineness index and MIJ-30 fineness index. Univariate animal models were used to estimate variance components and heritabilities. Bivariate analysis was done to make pairwise comparisons between traits and estimate correlations. Traits were moderately heritable (0.33 – 0.58). The genetic correlation between the fineness index of the two cameras was 0.55 ± 0.19, indicating that measures from each camera should be treated as separate traits for genetic evaluation. This work supports the need to continuously validate newer carcase evaluation technologies to better understand evolving traits and how to incorporate them into genetic evaluations.

INTRODUCTION

In Australia, the value of a carcase is predominantly based on its weight. However, carcase components are often marketed based on their predicted eating quality (Pethick *et al.* 2021). After harvest, AUS-MEAT chiller assessments are carried out to grade carcasses and evaluate expected eating quality. In Wagyu carcases, much emphasis is placed on marbling as an indicator of quality. A shift in assessment from evaluating marbling as a standalone measure, towards incorporating other quality measures to assess carcase pricing, such as eye muscle shape and marbling fineness, has been slowly adopted. Consideration for these factors is given with downstream markets and use in mind. This was investigated in Japanese Black and Crossbred steers in Japan (Kuchida *et al.* 2003; Takeo *et al.* 2016). A significant price difference was shown between high and low marbling fineness carcases. Especially in lower marble score (Japanese body marble scores 4 – 2) carcases. The greatest price difference was observed for carcases graded as Japanese body marble score 2, with a ¥140 per kg difference (roughly about 1.73 AUD per kg) between low and high marbling fineness. Allowing factors other than weight and visual grading to be considered for carcase pricing is expected to become a routine practice as retailers and restaurants attempt to provide consistent and high-quality eating experiences to their customers. Therefore, it is important to estimate their genetic relationship with already established systems such as AUS-MEAT assessed traits.

The Australian Wagyu Association (AWA) uses carcase cameras to routinely collect carcase data, including marble score, marble fineness, and eye muscle area, which is genetically evaluated by BREEDPLAN (AWA, www.wagyu.org.au). The HK-333 Japanese mirror-type camera was initially used by AWA to record carcase traits, starting in 2008. Traits recorded include eye muscle area, marbling percentage, marbling fineness, fatness, marbling particle proportionality, and muscle and fat colour. Genetic parameters for the mirror-type camera carcase traits were estimated and found to be moderate to highly heritable, along with high genetic correlations with corresponding

* A joint venture of NSW Department of Primary Industries and Regional Development and the University of New England

AUS-MEAT carcase measurements (Zhang *et al.* 2015; Zhang and Banks 2019). Since 2018, AWA has used the compressed lightweight MIJ-30 (Meat Image Japan) camera to replace the mirror-type camera. The camera uses a cloud-based system to predict carcase grading (MIJ, www.mij-labo.co.jp).

This study aimed to estimate the genetic parameters of hot standard carcase weight, AUS-MEAT marble score, the MIJ-30 fineness index and the HK-333 Japanese mirror-type camera fineness index and their genetic relationship. These genetic parameters will be used to assess how the MIJ-30 fineness index trait can best be included in genetic evaluations so that fineness measurements from both technologies can contribute to genetic evaluations.

MATERIALS AND METHODS

Carcase traits. Carcase traits included were hot standard carcase weight (kg), AUS-MEAT marble score (0-10), Mirror-type camera fineness index (count/cm^2) and MIJ-30 camera fineness index. The Mirror-type camera analysed images captured using the HK-333 imaging device and BeefAnalyserII software developed in Japan. The HK-333 fineness index counts the number of intramuscular fat streaks per cm^2 in the ribeye area, limited to particle sizes ranging from 0.01 to 0.5 cm^2 (Zhang *et al.* 2015; Lee *et al.* 2018). The MIJ-30 camera uses BeefAnalyserII software and measures the fineness index by dividing the sum of the circumference of the marbling flecks by the square root of the eye muscle area with a greater result indicating finer marbling [*Translated from Japanese to English using Google Translate*] (Shimabukuro *et al.* 2022).

Data access and editing. Pedigree and carcase data were extracted from the AWA BREEDPLAN database for 32,605 animals with recorded carcase traits from 1,661 unique sires. The data was manipulated using R software version 3.6.0 (R Core Team 2023) using packages readr and dplyr. Outliers were removed and defined as observations with more than three standard deviations from the mean. The site where the eye muscle area was measured varied between contemporary groups. Carcase camera measurements were taken at the 5th/6th rib, 6th/7th rib, 10th/11th rib and 12th/13th rib sites, which was accommodated in the analyses through the contemporary group (CG) effect in the model. Other factors accounted for in CG include pre-harvest factors (such as herd, year, sex and management group), harvest date and harvest group. After data editing, 2,104 animals with the mirror-type camera fineness index remained, with 325 unique sires represented. Of these, one sire had more than 100 progeny, 62 sires had less than 100 but more than 10 progeny, 162 sires had less than 10 progeny, and 100 sires had single progeny. For the MIJ-30, 6,384 animals with fineness index remained after data editing, and 689 unique sires were represented. Of these, three sires had more than 100 progeny, 188 had less than 100 but more than 10 progeny, 367 had less than 10 progeny, and 131 had single progeny.

Statistical models. The animal model was applied to estimate genetic parameters. Variance components and heritabilities were estimated by REML using ASREML software (Butler *et al.* 2023). CG was fitted as a fixed class effect. Hot standard carcase weight was adjusted for harvest age, by fitting harvest age as a linear and quadratic covariate. All other carcase traits were adjusted for carcase weight, by fitting carcase weight as a linear and quadratic covariate. A three-generation pedigree was used for the analyses. Bivariate analysis was performed in pairs to estimate genetic correlations between the reported traits.

RESULTS AND DISCUSSION

Table 1 displays descriptive statistics for the investigated carcase traits. Table 2 shows the genetic parameters for the estimated carcase traits. All the traits were moderately heritable, estimates were 0.43, 0.58, 0.33 and 0.49 for hot standard carcase weight, AUS-MEAT marble score, mirror-type camera fineness index and MIJ-30 fineness index, respectively. Despite differences in dataset size these heritabilities are similar to estimates obtained using a subset of this data in 2022, no estimates for the MIJ-30 fineness index were reported (G. M. Jeyaruban, personal communication, 28 January 2025). A larger number of records were available for the current study, with 53.02% and 53.73% more records for hot standard carcase weight and AUS-MEAT marble score, respectively. The heritability for the MIJ-30 fineness index was estimated by Tamagawa *et al.* (2024) for different muscles which ranged from 0.49 to 0.69. The heritability estimate for MIJ-30 fineness index falls within the lower end of the range of heritabilities provided from the reported study.

Table 1. Descriptive statistics for carcase traits from the Australian Wagyu Association

	No. Sires	N	Mean	sd	Min	Max
Hot standard carcase weight (kg)	1543	25,154	424.40	49.05	266.80	580.0
AUS-MEAT marble score (0-10)	1485	23,665	7.89	1.69	3.0	12.0
Mirror-type camera fineness index	325	2,104	2.59	0.60	0.94	4.65
MIJ-30 fineness index	689	6,384	5.59	1.05	1.53	9.86

Low genetic and phenotypic correlations were estimated between hot standard carcase weight and the other reported carcase traits. Genetic and phenotypic correlations between hot standard carcase weight with the mirror-type camera fineness index and the MIJ-30 fineness index, respectively, were estimated to be not significantly different from zero. This is not unexpected as the carcase quality traits were adjusted for carcase weight. However, the models could not converge fully, which could be because the carcase quality traits were adjusted for carcase weight and thus have genetic correlations with carcase weight that were close to zero. When adjusting all the traits for harvest age, non-zero correlations (ranging from -0.13 to 0.07) between carcase weight and the other traits were obtained, and the models completely converged (data not shown). AUS-MEAT marble score was positively correlated (phenotypic and genetic) with both fineness indices. The AUS-MEAT marble score was highly correlated genetically with the MIJ-30 fineness index (0.83 ± 0.03). Based on the definition provided for the MIJ-30 fineness index, it was expected that a greater marble score would coincide with a higher fineness index. That is because as more fat streaks are fitted into the same area of the eye muscle in increasing marble score, the total fat streaks' circumference also increases, which increases the fineness index. Also, with the same number of marbling streaks but with more area covered in marbling per streak, the marbling score will increase, but so will the circumference of the streaks. Hence, there is a direct connection between the marbling score and the MIJ-30 camera fineness index.

Table 2. Phenotypic variances (V_p), genetic correlations (standard error) above the diagonal, phenotypic correlations (standard error) below the diagonal and heritabilities (standard error) on the diagonal for carcase traits from the Australian Wagyu Association

	V_p	Hot standard carcase weight	AUS-MEAT marble score	Mirror-type camera fineness index	MIJ-30 fineness index
Hot standard carcase weight (kg)	1150.50	0.43 (0.03)	0.003 (0.05)	0.02 (0.15)*	-0.05 (0.08)*
AUS-MEAT marble score (0-10)	2.17	-0.001 (0.01)	0.58 (0.03)	0.33 (0.13)	0.83 (0.03)
Mirror-type camera fineness index	0.23	-0.01 (0.04)*	0.33 (0.03)	0.33 (0.09)	0.55 (0.19)
MIJ-30 fineness index	0.62	-0.02 (0.02)*	0.63 (0.01)	0.25 (0.06)	0.49 (0.05)

* The model did not fully converge.

The mirror-type camera fineness index positively correlates with the MIJ-30 camera fineness index phenotypically (0.25 ± 0.06) and genetically (0.55 ± 0.19). Kuchida *et al.* (2018) reported a high phenotypic correlation (0.954) between the two cameras. However, that study is not comparable, as Kuchida *et al.* (2018) used the same index trait definition for both cameras, while this current study defines each camera's fineness index differently. No comparable literature on the genetic correlation between the two index traits could be found. The moderate genetic correlation indicates that these traits are different and should be treated as such in genetic evaluations. Reporting for the carcase marbling fineness index EBV from the evaluation should be on the MIJ-30 scale as this is the current technology for measurement, and the mirror-type camera traits will decline in significance as this technology has been superseded and no longer used for measurement in Australian Wagyu. These two traits share a similar trait name, i.e. fineness index, which can give the impression that they are similar traits. However, comparison of the definitions clearly indicate that the way the two cameras measure the fineness index trait is different. The two fineness index measures are expected to be related as 'number of fat streaks' is a contributing metric to the mirror-type camera fineness index, and the sum of the circumference of all fat streaks is a metric in the MIJ-30 camera fineness index.

CONCLUSIONS

The high genetic correlation between the AUS-MEAT marble score and the MIJ-30 fineness index aligns with the MIJ-30 trait definition, as more fat streaks correspond to a higher marble score. The fineness index of the mirror-type camera and the MIJ-30 camera are heritable. The two traits are moderately correlated, indicating that they should be treated as different traits in a genetic evaluation. As newer technology becomes available, future research should continue to validate the recorded traits and how to accurately incorporate them into genetic evaluations.

ACKNOWLEDGEMENTS

The authors acknowledge receiving the Dr Michael J Bradfield Memorial Scholarship, co-funded by the University of New England and the Agricultural Business Research Institute. Furthermore, the Australian Wagyu Association and the Agricultural Business Research Institute are acknowledged for providing access to the BREEDPLAN data.

REFERENCES

AWA. www.wagyu.org.au. Wagyu Breedplan. Australian Wagyu Association.

Butler D.G., Cullis B.R., Gilmour A.R., Gogel B.J. and Thompson, R. (2023) 'ASReml-R Reference Manual Version 4.2'. SN International Ltd, Hemel Hempstead, UK.

Kuchida K., Kikuchi A., Kato K., Hidaka S., Suzuki M. and Miyoshi S. (2003) *J. Jpn. Soc. Anim. Sci.* **74**: 23.

Kuchida K., Sakaguchi Y., Kano A., Goto Y. and Komine H. (2018) *Proc. Int. Cong. Meat. Sci. Tech.* **63**: 13.

Lee B., Yoon S., Lee Y., Oh E., Yun Y.K., Kim B.D., Kuchida K., Oh H.K., Choe J. and Choi Y.M. (2018) *Korean. J. Food. Sci. An.* **38**: 606.

MIJ. www.mij-labo.co.jp. Products | MIJ Labo.

Pethick D.W., Hocquette J.-F., Scollan N.D. and Dunshea F.R. (2021). *Animal* **15**: 100356.

Shimabukuro M., Kano A., Komine H. and Kuchida K. (2022) *J. Jpn. Soc. Anim. Sci.* **93**: 125.

Takeo A., Asa R., Hagiya K. and Kuchida K. (2016) *J. Jpn. Soc. Anim. Sci.* **87**: 253.

Tamagawa Y., Takahashi M., Hagiya K. and Kuchida K. (2024) *Anim. Biosci.* **37**: 1526.

Zhang Y. and Banks R.G. (2019) *Proc. Assoc. Advmt. Anim. Breed. Genet.* **23**: 83.

Zhang Y., Tier B. and Banks R. (2015) *Proc. Assoc. Advmt. Anim. Breed. Genet.* **21**: 409.

Proc. Assoc. Advmt. Anim. Breed. Genet: 26: 343-346

MULTIBREED GENOMIC SELECTION FOR SENSORY EATING QUALITY OF LAMB USING CONSUMER ASSESSMENTS

D.J. Brown[1], S.Z.Y. Guy[1], S.I. Mortimer[2], L. Pannier[3], P. McGilchrist[4], D. Pethick[3] and A.A. Swan[1]

[1] Animal Genetics Breeding Unit*, University of New England, Armidale, NSW, 2350 Australia
[2] NSW Department of Primary Industries and Regional Development, Armidale, NSW, 2351 Australia
[3] School of Agricultural Science, Murdoch University, Perth, WA, 6150 Australia
[4] School of Environmental and Rural Science, University of New England, Armidale, NSW, 2351 Australia

SUMMARY

This paper reports updated genetic parameter estimates for correlations between consumer eating quality and carcase traits for Australian sheep. Seven consumer sensory-assessment traits (tenderness, flavour, juiciness, smell, overall liking, star rating and MQ4 index) were collected on loin and topside cuts ($n \sim 6{,}500$). Heritability estimates ($\hat{h}^2$) were low to moderate for these traits ($0.09 < \hat{h}^2 < 0.26$), with the topside cut having slightly more genetic variation than the loin. Genetic correlations ($\hat{r}_g$) between sensory traits were high within cuts ($\hat{r}_g > 0.96$) and across cuts ($\hat{r}_g > 0.73$). There were moderate to high genetic correlations between the consumer sensory and the objective eating quality traits (intramuscular fat and shear force), and low to weakly negative correlations between the sensory traits and carcase lean meat yield. These estimates will be used to revise selection indexes and develop breeding values for consumer eating quality.

INTRODUCTION

Breeding programs can balance the antagonistic relationship between yield and eating quality using modified selection indexes. These indexes are based on breeding objectives where an economic value is defined for consumer-assessed eating quality (Swan *et al.* 2015). Balanced selection for carcase yield and eating quality was further enhanced from 2017 underpinned by substantial investment in genomic reference populations. Genomically enhanced breeding values are now available for carcase yield traits and objectively measured eating quality traits.

Genetic correlations between consumer eating quality and selection traits (carcase yield and objective measures of eating quality) are required to develop selection indexes. Current eating quality indexes are based on correlations estimated from consumer eating quality trials on approximately 1,500 animals from multiple breeds. The continued collection of this information has resulted in almost four-fold the amount of both consumer and carcase data available across all sheep breeds. This paper reports updated genetic parameter estimates between consumer eating quality and carcase traits for Australian sheep breeds. These estimates will be used to further revise selection indexes, and potentially develop new multi-trait breeding values for consumer eating quality.

MATERIALS AND METHODS

Data were available for crossbred lambs recorded in Australian sheep reference populations across multiple years and data sources; mainly from the Sheep CRC Information Nucleus Flock (lambs born in 2009 and 2010) (van der Werf *et al.* 2010), and the MLA Resource Flock (lambs

* A joint venture of NSW Department of Primary Industries and Regional Development and the University of New England

born across 2017-2021). Lambs were processed at an average age of 266 ± 92 days (mean ± SD), with an average hot carcase weight of 23.7 ± 4.1 kg and carcase subcutaneous fat of 4.4 ± 2.4 mm. Carcase composition and objective eating quality traits examined were hot carcase weight (kg), lean meat yield (%) assessed from computed tomography scanning and dual energy x-ray, loin intramuscular fat (%) measured by chemical analysis and loin shear force at 5 days of aging (N).

Consumer eating quality traits were collected from a representative subset of animals from each project. There were 979 sires represented by an average of 6.9 progeny per sire. Samples were collected, prepared, cooked and tasted by consumers according to the protocol described by Pannier *et al.* (2025). Briefly, five slices from the topside (*Musculus semimembranosus*) and loin (*M. longissimus lumborum*) muscles from each carcase were prepared and grilled under standardised conditions. Samples were allocated to untrained consumers according to a 6x6 Latin square design, who scored each sample for tenderness, juiciness, flavour, smell and overall liking between zero and 100 (100 being most preferred). Consumers also assessed each sample according to quality grades (star rating) of 'unsatisfactory' (2 star), 'good every day' (3 star), 'better than every day' (4 star) or 'premium' (5 star) (Pannier *et al.* 2025). Each loin and topside were eaten 10 times, with the average across the 10 consumer responses utilised for each muscle. An MQ4 value was also calculated according to Pannier *et al.* (2025) by averaging the consumer scores across traits within cuts (MQ4) and across all traits for both cuts (Overall MQ4).

Table 1. Summary statistics for eating quality (consumer-assessed and objectively measured) and carcase composition traits

Trait		*n*	Mean	Min	Max	CV (%)
Loin (0 to 100)	Tenderness	6,429	70.4	27.6	94.2	14.2%
	Flavour	6,429	68.5	39.2	93.0	12.1%
	Juiciness	6,429	64.9	28.1	94.8	14.9%
	Overall liking	6,429	69.3	37.8	95.0	12.6%
	Star rating[A]	6,103	3.7	2.5	5.0	9.9%
	MQ4	6,429	68.3	37.3	91.8	12.6%
Topside (0 to 100)	Tenderness	6,486	47.0	5.0	93.3	26.1%
	Flavour	6,486	54.7	5.0	89.8	16.4%
	Juiciness	6,486	51.0	6.0	87.3	21.0%
	Overall liking	6,486	52.0	6.0	89.6	19.4%
	Star rating[A]	6,161	3.0	2.0	4.4	12.2%
	MQ4	6,486	51.2	6.0	90.0	19.0%
Overall MQ4		6,749	59.7	30.8	90.0	13.2%
Hot carcase weight (kg)		45,356	23.7	9.2	47.0	17.2%
Lean meat yield (%)		11,184	56.3	38.8	70.7	8.1%
Intramuscular fat (%)		40,921	4.4	1.1	14.4	27.5%
Shear force (N)		37,116	32.3	10.8	107.0	33.7%

[A] Star rating is scored on a 4-point scale from 2 to 5.

The above traits were analysed in a series of bivariate models ASReml (Gilmour *et al.* 2021) using a mixed model, fitting a fixed effect of contemporary group defined as flock, management group, year of measurement, sex, and kill group sub-classes (as well as laboratory for shear force and intramuscular fat) and random effects of animal (additive genetic effect based on the pedigree) and genetic groups defined by breed to account for different base animal breeds. Additional fixed effects of birth type, rearing type, age of dam and age at measurement were also fitted. Intramuscular fat and shear force were also corrected for hot carcase weight. Heterosis was not fitted in the model.

RESULTS AND DISCUSSION

On average, the loin samples scored higher for eating quality compared to the topside samples, though there was generally more variation in the topside cuts (Table 1). Smell was not heritable for both cuts and thus no results were shown. Heritability estimates were moderate for all the other consumer eating quality traits (0.09 to 0.20 for the loin, 0.12 to 0.26 for the topside), with the topside generally having higher estimates (Table 2). Heritability estimates were high for carcase weight, intramuscular fat and lean meat yield, and moderate for shear force. The variance attributed to breed was 0.75 to 2.3 times that of the within-breed variance for loin consumer scores, carcase weight and lean meat yield, while for the other traits, it tended to be small relative to the genetic variance.

Consumer-assessed eating quality exhibited genetic variation, with the highest heritability estimated for topside tenderness which agrees with multi-breed estimates reported by Mortimer *et al.* (2015).

Table 2. Additive genetic ($\hat{\sigma}_a^2$), phenotypic ($\hat{\sigma}_p^2$), genetic group ($\hat{\sigma}_{GG}^2$) and heritability estimates ($\hat{h}^2$)(± standard error), averaged across bivariate analyses of eating quality and carcase traits

Trait		$\hat{\sigma}_a^2$	$\hat{\sigma}_p^2$	$\hat{\sigma}_{GG}^2$	$\hat{h}^2$
Loin (0 to 100)	Tenderness	16.8 ± 2.77	85.3 ± 1.56	11.1 ± 4.69	0.20 ± 0.03
	Flavour	5.2 ± 1.59	60.7 ± 1.09	5.8 ± 2.60	0.09 ± 0.03
	Juiciness	10.1 ± 2.31	81.0 ± 1.46	6.5 ± 3.10	0.12 ± 0.03
	Overall liking	8.6 ± 1.90	66.5 ± 1.20	7.1 ± 3.13	0.13 ± 0.03
	Star rating	0.02 ± 0.00	0.12 ± 0.00	0.01 ± 0.01	0.16 ± 0.03
	MQ4	9.6 ± 1.89	63.4 ± 1.15	7.4 ± 3.20	0.15 ± 0.03
Topside (0 to 100)	Tenderness	31.4 ± 4.33	122.9 ± 2.25	2.9 ± 2.48	0.26 ± 0.03
	Flavour	9.0 ± 2.06	74.2 ± 1.33	0.00 ± 0.00	0.12 ± 0.03
	Juiciness	10.9 ± 2.71	93.8 ± 1.68	1.4 ± 1.41	0.12 ± 0.03
	Overall liking	16.2 ± 2.89	90.9 ± 1.64	0.6 ± 1.04	0.18 ± 0.03
	Star rating	0.03 ± 0.00	0.12 ± 0.00	0.00 ± 0.00	0.23 ± 0.03
	MQ4	16.2 ± 2.68	81.8 ± 1.48	0.8 ± 1.10	0.20 ± 0.03
Overall MQ4		11.9 ± 1.78	53.0 ± 0.95	3.5 ± 1.89	0.22 ± 0.03
Hot carcase weight (kg)		3.3 ± 0.12	6.7 ± 0.05	7.5 ± 1.60	0.49 ± 0.02
Lean meat yield (%)		2.4 ± 0.22	6.0 ± 0.09	3.0 ± 0.90	0.47 ± 0.03
Intramuscular fat (%)		0.5 ± 0.02	0.9 ± 0.01	0.1 ± 0.04	0.62 ± 0.02
Shear force (N)		21.7 ± 1.27	82.3 ± 0.68	3.2 ± 1.32	0.27 ± 0.01

Figure 1 presents estimates of genetic correlations ($\hat{r}_g$). The consumer eating quality traits were highly correlated with each other within cuts ($\hat{r}_g > 0.96$; SEs ~ 0.03), as well as across cuts ($0.73 < \hat{r}_g < 0.97$; SEs ~ 0.10). Consumer eating quality traits of the different muscles were genetically similar, and thus, selection for better eating quality for one muscle will still result in an improvement in eating quality in other muscles. Moderate to strong genetic correlations were observed between consumer eating quality traits and objective eating quality traits. Shear force had strong negative (favourable) correlations with loin and topside traits ($-0.88 < \hat{r}_g < -0.64$). Meanwhile, intramuscular fat had a slightly stronger positive relationship with the eating quality traits from the loin ($0.44 < \hat{r}_g < 0.63$) compared to the topside ($0.28 < \hat{r}_g < 0.43$). The relationships between consumer eating quality traits and lean meat yield were weak to negligible in both the loin and topside ($-0.21 < \hat{r}_g < -0.01$). The correlation between intramuscular fat and shear force was moderate at -0.47 ± 0.03.

Consumer eating quality traits were generally more highly correlated with shear force than intramuscular fat, particularly in the loin. However, both traits are very useful selection criteria in breeding programs. With technological advances in objective carcase measurements, price signals are now emerging for intramuscular fat measured at processing. Therefore, future indexes for all breeds should directly include eating quality in the breeding objective as an economic trait.

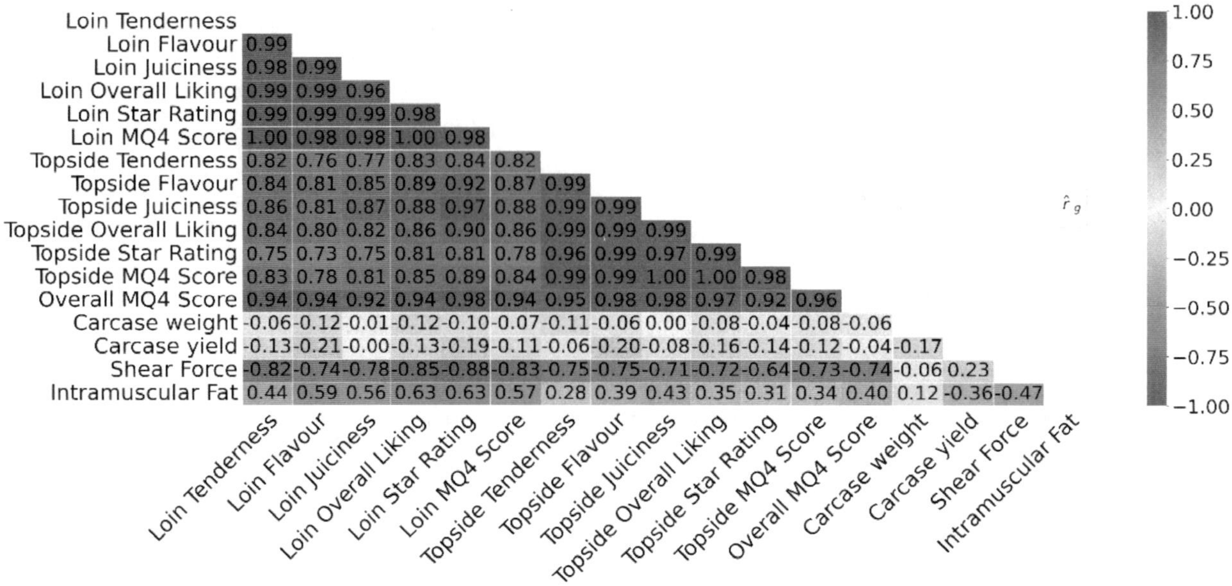

Figure 1. Genetic correlation estimates between eating quality and carcase composition traits (all SEs ~0.05)

The re-estimated genetic parameters between topside overall liking and lean meat yield align with previous estimates reported by Swan *et al.* (2015) and Guy *et al.* (2022). However, notably strong genetic correlations were evident between loin eating quality and shear force in the current study, which were similar to those reported by Guy *et al.* (2022) in Terminal sire breeds. Both these were significantly higher than those reported by Mortimer *et al.* (2015). Since this study used more data, there is greater confidence in the precision of estimates of genetic correlation between these traits. Due to the changes in correlations with the updated data, it will be necessary to review indexes that include eating quality, along with predictions of response to selection. Future work will explore the development of breeding values for consumer eating quality in multi-trait models.

ACKNOWLEDGEMENTS

This work was funded by MLA Project L.GEN.1704 and L.GEN.2204. The authors are grateful for the staff involved in collecting this data, funded through MLA L.GEN.1814 and B.SGN.0144.

REFERENCES

Gilmour A.R., Gogel B.J., Cullis B.R., Welham S.J. and Thompson R. (2021) 'ASReml SA User Guide Release 4.2' VSN International Ltd, Hemel Hempstead, HP1 1ES, UK.

Guy S.Z.Y., Mortimer S.I., Pannier L., McGilchrist P., Brown D.J., Pethick D. and Swan, A.A. (2022) *Proc. World Cong. Genet. Appl. Livest Prod.* **12**: 3000.

Pannier L., Tarr G., Pleasants T., Ball A., McGilchrist P., Gardner G.E. and Pethick D.W. (2025) *Meat Sci. In Press*

Mortimer S.I., Swan A.A, Pannier L., Ball A.J., Jacob R.H. *et al.* (2015) *Proc. Assoc. Advmt. Anim. Breed. Genet* **21**: 209.

Swan A.A., Pleasants T. and Pethick D. (2015) *Proc. Assoc. Advmt. Anim. Breed. Genet* **21**: 29.

van der Werf J.H.J, Kinghorn B.P. and Banks R.G. (2010) *Anim. Prod. Sci.* **50**: 998.

Proc. Assoc. Advmt. Anim. Breed. Genet: 26: 347-350

ESTIMATED FREQUENCIES OF MYOSTATIN MUTATIONS (DOUBLE MUSCLING) IN AUSTRALIAN TROPICALLY ADAPTED BEEF BREEDS

K.L. Moore[1], D.J. Johnston[1] and T. Grant[2]

[1] Animal Genetics Breeding Unit*, Armidale, NSW, 2351 Australia
[2] Queensland Department of Primary Industries, Toowoomba, QLD, 4350 Australia

SUMMARY

Myostatin mutations have been identified in many beef breeds and have been associated with increased carcase yields, reduced fat, increased calving difficulty, delayed puberty and reduced female fertility. This study aimed to estimate the allele frequencies of myostatin mutations in Australian tropically adapted beef breeds. Nine myostatin variants (viz. C313Y, D182N, E226X, E291X, F94L, NT419, NT821, Q204X and S105C) were assessed in Brahman (n=2,554), Droughtmaster (n=2,188), and Santa Gertrudis (n=904) cattle that were part of the Repronomics research project. Results showed the myostatin variants NT821 and F94L were segregating in the Droughtmaster and Santa Gertrudis breeds but not in the Brahman. In total, 17.4% and 26% of Droughtmaster and Santa Gertrudis animals carried at least one copy of a myostatin allele for either variant. The frequency of the NT821 allele was 0.08 and 0.12, and the F94L allele was 0.01 and 0.02, respectively, for Droughtmaster and Santa Gertrudis animals. Hardy-Weinberg equilibrium was observed for both NT821 and F94L in both breeds. However, there were fewer than expected two copy NT821 animals for both breeds. With no variants detected in Brahman, the myostatin variants likely originated from the *Bos Taurus* influence of the Droughtmaster and Santa Gertrudis breeds. The allele frequencies of NT821 in the Droughtmaster and Santa Gertrudis breeds, along with the literature reports of increased calving difficulties associated with NT821, indicate that breeders should monitor and manage the presence of the NT821 myostatin variant. Further research should be undertaken to quantify the size of the effect that having one and two copies of the mutation has across all economically important traits.

INTRODUCTION

The myostatin gene influences muscle production and is responsible for 'double muscling' in beef cattle (Fiems 2012; McPherron and Lee 1997). Research has identified many variants of the gene including C313Y, D182N, E226X, E291X, F94L, NT419, NT821, Q204X and S105C. The mutation variants are either deletions or substitutions, but all result in making the myostatin gene inactive (Dunner *et al.* 2003). In an extensive review, Fiems (2012) reported double-muscled animals had increased carcase yields with reduced fat but have also been linked to increased calving difficulty and reduced female fertility. However, limited studies have considered myostatin mutations in tropically adapted beef breeds. Vankan *et al.* (2010) reported that the F94L mutation was present in Australian Droughtmaster animals but not Brahman, and the NT821 mutation has been identified in Australian Santa Gertrudis animals (O'Rourke *et al.* 2012). Neither study estimated the breed allele frequencies of these myostatin variants. This study aimed to estimate the allele frequencies of myostatin mutations in Australian Brahman, Droughtmaster and Santa Gertrudis breeds.

MATERIALS AND METHODS

Animal Data. Myostatin genotypes were obtained for Brahman (n=2,554), Droughtmaster

---

* A joint venture of NSW Department of Primary Industries and Regional Development and the University of New England

(n=2,188), and Santa Gertrudis (n=904) animals. Animals were genotyped on Neogen's GGP_TropBeef chip which included nine myostatin variants: C313Y, D182N, E226X, E291X, F94L, NT419, NT821, Q204X and S105C. Animals were born between 2015 and 2023 and were part of the Repronomics project, an extensive reference population targeting female reproduction traits and collecting economically important phenotypes from birth to slaughter (Johnston *et al.* 2017). DNA samples were collected on all project animals born and the dataset included calves that died at or shortly after birth. Brahman, Droughtmaster and Santa Gertrudis sires were selected to represent the wider breed, especially if the sire was influential and had limited female reproduction information available. There were 104, 83 and 41 sires represented in the dataset, respectively, for Brahman, Droughtmaster and Santa Gertrudis. Sires were mated to cows of the same breed, except that at the start of the Repronomics project some Droughtmaster and Santa Gertrudis sires at the Brian Pastures herd were mated to base Tropical Composite cows from the previous Beef CRC genetics project (Barwick *et al.* 2009). Myostatin genotypes were unavailable for the base cows at the start of the project, but approximately 50% of the sires were genotyped for myostatin. Female calves produced during the project were retained in the cow herd and were the dams of future generations. Cows were only culled if they failed to wean a calf.

Statistical analyses. Allele frequencies were calculated based on the number of animals with 0, 1 or 2 copies of the myostatin alleles. A chi-square test was used to test for Hardy-Weinberg equilibrium based on the allele frequencies calculated from the current study and tested against a critical value of 3.84, based on 1 degree of freedom and $P<0.05$.

RESULTS AND DISCUSSION

No myostatin variants were shown to be segregating in the Brahman, but both the NT821 and F94L myostatin variants were present in the Droughtmaster and Santa Gertrudis (Table 1). Of the Droughtmaster and Santa Gertrudis animals, 17.4% and 26% carried at least one myostatin allele from either variant (Table 2) with similar allele frequencies across males and females (results not shown). Myostatin variants C313Y, D182N, E226X, E291X, NT419 and S105C were shown not to be segregating in any breed, with all genotyped animals having no copies of these myostatin variants. However, one Droughtmaster sire was a heterozygote for Q204X, with 40 out of 66 genotyped progeny also being heterozygous for Q204X. No other animals in the dataset were found to have a copy of the Q204X mutation. Ryan *et al.* (2023) identified that the Q204X mutation was present in 27% of Irish Charolais and 8% of Limousins.

Table 1. Frequency of myostatin alleles for the NT821 and F94L myostatin variant in Brahman, Droughtmaster and Santa Gertrudis animals

	NT821					F94L				
Breed	**0**	**1**	**2**	**N animals**	**Freq(p)**	**0**	**1**	**2**	**N animals**	**Freq(p)**
Brahman	2,554	0	0	2,554	0.000	2,554	0	0	2,554	0.000
Droughtmaster	1,852	331	5	2,188	0.078	2,138	48	0	2,186	0.011
Santa Gertrudis	697	199	8	904	0.119	872	32	0	904	0.018

The estimated gene frequency of the NT821 mutation was 0.08 for Droughtmaster and 0.12 for Santa Gertrudis animals (Table 1). Animals with at least one NT821 allele were sired by 50 and 32 different sires, respectively, for Droughtmaster and Santa Gertrudis. Hardy-Weinberg equilibrium was tested, and for Droughtmaster (X^2=5.91) and Santa Gertrudis (X^2=2.23), the NT821 myostatin variant was not in Hardy-Weinberg equilibrium (critical value=3.84, $P<0.05$, df=1) for Droughtmaster, while the Santa Gertrudis subset did not depart significantly from Hardy-Weinberg equilibrium. For Droughtmaster, 13 homozygous NT821 animals were expected, and just five homozygotes were observed. A similar pattern was observed for Santa Gertrudis, with 13 homozygous NT821 animals expected and eight homozygotes observed. It is not yet understood why fewer homozygotes were observed; further research is required, but it may be due to embryo loss, myostatin resulting in a higher number of calf deaths due to dystocia (and calves were not found to obtain a DNA sample) or myostatin being linked to poor reproductive success (Repronomics cows that fail to wean a calf are culled).

Table 2. Number and (frequency) of myostatin genotypes for combinations of NT821 and F94L myostatin variants in Droughtmaster and Santa Gertrudis animals

		Droughtmaster					Santa Gertrudis		
		F94L					F94L		
		0	1	2			0	1	2
NT821	0	1,805 (0.826)	46 (0.021)	0 (0.000)	NT821	0	669 (0.740)	28 (0.031)	0 (0.000)
	1	328 (0.150)	2 (0.001)	0 (0.000)		1	195 (0.216)	4 (0.004)	0 (0.000)
	2	5 (0.002)	0 (0.000)	0 (0.000)		2	8 (0.009)	0 (0.000)	0 (0.000)

The frequency of the F94L allele was 0.01 for Droughtmaster and 0.02 for Santa Gertrudis animals (Table 1). Animals with at least one F94L allele were sired by 21 and 13 different sires, respectively, for Droughtmaster and Santa Gertrudis. Hardy-Weinberg equilibrium was tested, and for both Droughtmaster (X^2=0.26) and Santa Gertrudis (X^2=0.29), the F94L myostatin variant was in Hardy-Weinberg equilibrium (critical value=3.84, P<0.05, df=1).

The NT821 and F94L variants have been identified in many beef breeds (Dunner *et al.* 2003: Ryan *et al.* 2023), but in particular, the NT821 variant has been associated with Belgian Blue and F94L with Limousin breeds. A sizeable Irish study showed that 99% of purebred Limousins had at least one copy of F94L, and 99% of purebred Belgian Blue had at least one copy of NT821 (Ryan *et al.* 2023). The Irish study considered 12 *Bos Taurus* breeds and reported that although both variants were associated with calving difficulty, the impact was smaller for F94L and greatest for NT821. The effect of one and two copies of the NT821 mutation was an increase in calving difficulty scores of 0.13 and 0.37 for calves with myostatin and an increase of 0.05 and 1.30 for cows with myostatin, respectively. The same study also reported that F94L and NT821 resulted in heavier, more muscled and leaner carcases. Calves with one and two copies of the F94L mutation yielded 7.6kg and 20.6kg heavier carcases than zero copy animals, respectively, and 0.5 and 1.6 scores higher for conformation, respectively. One copy of F94L resulted in carcases 0.26 scores leaner than zero copy animals. However, two copies of F94L animals were shown to have fat scores 1.21 higher than zero copy animals, which is against the reported trend of leaner carcases. Calves with one and two copies of the NT821 mutation yielded 16.6kg and 23.3kg heavier carcases than zero copy animals, respectively, and 1.3 and 2.6 scores higher for conformation, respectively. Calves with one and two copies of the NT821 mutation were leaner than zero-copy animals by 0.98 and 3.85 fat scores. Although an association with reduced female reproduction has been reported (Fiems 2012), no studies have reported the impact of one or two copies of the mutation. These effects reported in the literature highlight the need for further research to fully understand the impact of these myostatin variants in Australian tropical beef breeds.

Droughtmaster and Santa Gertrudis were developed as composites with Brahman and *Bos Taurus* origins. In developing both composites, the predominant *Bos Taurus* breed was Shorthorn. Other breeds may also contribute to the genetic makeup of Droughtmaster and Santa Gertrudis from the grading-up process used in some herds to build numbers. Given that no myostatin variants were detected in the Brahman population of this study, it is probable that the NT821 and F94L variants detected originated from the *Bos Taurus* content of the composites. Both variants were detected in the Irish purebred Shorthorn populations, with 1% and 7% of Irish Shorthorns having one or more copies of the F94L and NT821 variants, respectively (Ryan et al., 2023). In the present study, Droughtmaster and Santa Gertrudis had 2% and 3% of animals with one or more copies of the F94L variant, respectively. The number of Droughtmaster and Santa Gertrudis animals with one or more copies of NT821 was much higher than that observed in Irish Shorthorns (and the other Irish breeds considered, apart from Belgian Blue), with 15% and 23% of animals having one or more copies of NT821, respectively. Selection for more muscled animals is likely to lead to increases in the frequency of NT821 over time, in the absence of considering the myostatin status of animals.

CONCLUSIONS

This study showed that none of the myostatin variants considered in this study were segregating in the Brahman population, but NT821 and F94L were present in the Droughtmaster and Santa Gertrudis populations. F94L was present at lower frequencies, and literature reports that it increases muscling but has minimal effect on calving difficulty. NT821 was segregating at higher frequencies and has been reported to increase muscling and calving difficulties. The increase in meat yield may be advantageous but needs to be balanced with the risk of calving difficulties and reduced female reproduction observed in other studies. As most research has been undertaken on *Bos Taurus* breeds, further research is required to investigate the impact of these myostatin mutations on the tropically adapted beef breeds in Australia. Droughtmaster and Santa Gertrudis breeders should monitor and manage the presence of the NT821 myostatin variant. To inform the potential impact on future breeding programs and genetic evaluation, further research will be undertaken to quantify the size of the effect that having one and two copies of the mutation has across all economically important traits.

ACKNOWLEDGEMENTS

Thank you to Neogen for providing myostatin genotypes for the project animals. The authors acknowledge funding from MLA (projects B.NBP.0759, P.PSH.1221, and L.GEN.2007) and the contributions of the project collaborator institutions: AGBU, Queensland Department of Primary Industries, Northern Territory Department of Primarily Industries and Fisheries, University of Queensland and QAAFI. Thank you to the collaborating organisations and the large team collecting DNA samples.

REFERENCES

Barwick S.A., Johnston D.J., Burrow H.M., Holroyd R.G., Fordyce G., Wolcott M.L., Sim W.D. and Sullivan M.T. (2009) *Anim. Prod. Sci.* **49:** 367.

Dunner S., Miranda M.E., Amigues Y., Cañón J., Georges M., Hanset R., Williams J. and Ménissier F. (2003) *Genet. Sel. Evol.* **35**:103.

Fiems L.O. (2012) *Animals* **2**: 472.

Johnston D.J., Grant T.P., Schatz T.J., Burns B.M., Fordyce G. and Lyons R.E. (2017) *Proc. Assoc. Advmt. Anim. Breed. Genet.* **22:** 385.

McPherron A.C. and Lee S.J. (1997) *Proc. Natl. Acad. Sci. U.S.A.* **94**:12457.

O'Rourke B.A., Greenwood P.L., Arthur P.F. and Goddard M.E. (2012) *Anim. Gen.* **44**: 86.

Ryan C.A., Purfield D.C., Naderi S. and Berry D.P. (2023) *J. Anim. Sci.* **101:** 1.

Vankan D.M., Waine D.R. and Fortes M.R.S. (2010) *Animal* **4.4**: 530.

Proc. Assoc. Advmt. Anim. Breed. Genet: 26: 351-354

IDENTIFICATION OF GENES HABOURING MUTATIONS AFFECTING EATING QUALITY TRAITS IN BEEF CATTLE USING BAYESIAN GENOMIC PREDICTION METHODS

M. Forutan[1], H. Aliloo[2], S. Clark[2], P. McGilchrist[2] and B. Hayes[1]

[1] Queensland Alliance for Agriculture and Food Innovation, The University of Queensland, Brisbane, QLD, 4072 Australia
[2] School of Environmental and Rural Science, University of New England, Armidale, NSW, 2350 Australia

SUMMARY

Methods such as Genome-Wide Association Studies (GWAS) and Bayesian genomic prediction (BayesR) are commonly employed to enhance understanding of the genetic architecture of complex traits, identify genetic variants associated with these traits, and assist in breeding decisions and market allocation. This study aims to uncover genes harbouring variants influencing eating quality traits, including tenderness, juiciness, flavour, overall liking, and meat quality score (MQ4) in a large and diverse population of *Bos taurus indicus* cattle from Australia, the USA, and Ireland. The analysis involved 7,380 young males and females with phenotypic data and genotypes imputed up to 709,768 SNP (Illumina HD array). The BayesR approach was applied with a chain length of 40,000 iterations and a burn-in of 5,000 iterations. Notably, the findings highlight 324 SNPs with exceptionally high posterior inclusion probabilities (PIP > 0.9999 quantile for each trait), linked to 100 candidate genes. Among these, shared genetic signals across most of the traits within or close to genes such as *CAPN1, CAST, bta-mir-2407,* and *CCDC171* underscore their pivotal roles in meat quality across diverse populations. These insights contribute significantly to the global effort to enhance meat quality through genomics-driven cattle breeding programs.

INTRODUCTION

With global meat consumption projected to increase by 2030, especially in developing countries, the beef industry faces both a significant opportunity and a challenge to meet growing consumer demands. The Organisation for Economic Co-operation and Development (OECD) forecasts that this rise in demand will drive a need for more tailored beef products, opening avenues for beef consumer segmentation. Different criteria and/or perceptions are used to define meat quality (including sensory, nutritional, technological or hygienic quality (Geletu *et al.* 2021) and sustainability (Font-I-Furnols 2023)). Meat tenderness is considered the main sensory trait and is heavily influenced by consumer preferences and market demand (Warner *et al.* 2022).

Understanding the genetic basis of beef quality traits is essential for improving the efficiency and precision of cattle breeding programs. These traits are polygenic, shaped by many genes with small effects, as well as environmental factors (Arikawa *et al.* 2024), making them challenging to accurately identify and predict. The majority of studies have focused on using GWAS models to identify genomic regions associated with meat quality related traits like tenderness, important for beef quality in cattle populations (Medeiros de Oliveira Silva *et al.* 2017; Arikawa *et al.* 2024) but limited studies have utilized Bayesian methods for association analyses. Classical models, like GWAS, fit each SNP separately which can limit precision and statistical power due to multiple testing. On the other hand, Bayesian methods implicitly account for population structure and the multiple-testing problem inherent in classical single-marker GWAS (Wolc and Dekkers 2022).

This study leverages a genetically diverse dataset of *Bos taurus indicus* cattle from Australia (AUS), the USA, and Ireland (IR) to identify potential causal variants underlying eating quality traits. Results enhance understanding of genetic variation influencing beef cattle eating quality.

MATERIALS AND METHODS

Data collection. A total of 7,384 animals (3197, 1297 and 2900 from the USA, IR and AUS respectively) had 5 consumer meat eating quality traits; tenderness (TENDER), flavour (FLAVOUR), juiciness (JUICY), overall liking (OVERALL) and meat quality score (MQ4), where MQ4 is formed by weighting the four sensory scores (Watson *et al.* 2008). These records belong to 156 groups and grill cooking method. Each group had at least 5 animal records. In total across three countries there were 4552 males, 1625 steers, and 1188 females. Samples aged from 3 to 52 days with average (±SD) of 13.4 (3.1). A total of 5,532 animals were hormone growth promoter (HGP) free. Animals with carcass weights outside 2 times the interquartile range for each country were removed. Relaxed thresholds based on the distribution of 99.9% of the records were used to remove the possible outliers for 5 eating quality traits.

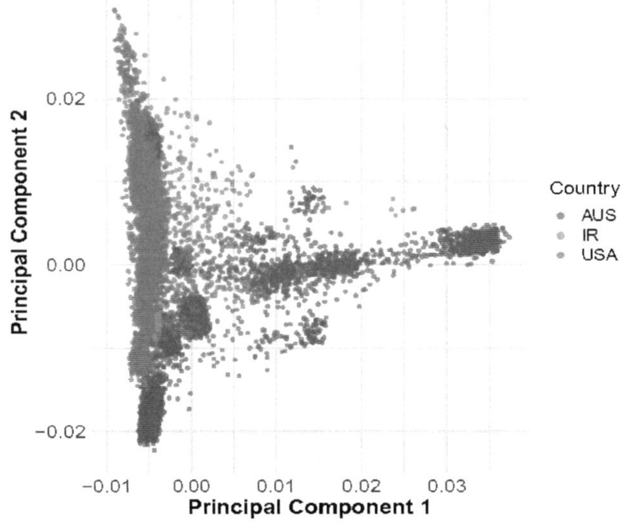

Figure 1. Principal component analysis (PCA) results for genotyped animals (Legends indicate the contributing countries, reflecting the genetic diversity of the dataset.)

Genotyping. Low-density genotypes from AUS (50k and 100k SNP chips) and imputed high-density genotypes from the USA and IR were available. Genotypes were imputed up to 709,768 SNPs (bovine high-density (HD) array) using findhap4 software (VanRaden *et al.* 2013). Imputation was performed separately for each country, utilizing a reference set of 4506 cattle from relevant breeds genotyped with the Bovine HD array (Hayes *et al.* 2023).

Bayesian genomic prediction. In the BayesR approach (Erbe *et al.* 2012), the following prediction equation was used to estimate the SNP effects for each of the five meat quality traits.

$$\boldsymbol{y} = \boldsymbol{Xb} + \boldsymbol{Zg} + \boldsymbol{e}$$

where $\boldsymbol{y}$ is a vector of phenotypes, $\boldsymbol{b}$ includes the estimates of fixed class effects of contemporary group, SEX and HGP treatments, and covariates of days aged after slaughter, carcass weight, first 4 principal components (PCAs) from genotypes, heterosis, and country, $\boldsymbol{g}$ is a vector of $\boldsymbol{m}$ SNP effects with distribution $\boldsymbol{g} \sim \boldsymbol{N}(\boldsymbol{0}, \boldsymbol{\sigma_i^2})$. The heterosis was defined as the regression of the trait on proportion of heterozygote loci across all loci for each animal. The genetic variance of the trait $(\boldsymbol{\sigma_g^2})$ was assumed $\boldsymbol{\sigma_i^2} = \{\boldsymbol{0},\ \boldsymbol{10^{-4}\sigma_g^2},\ \boldsymbol{10^{-3}\sigma_g^2},\ \boldsymbol{10^{-2}\sigma_g^2}\}$. This setup allows the BayesR model to have a more flexible SNP effect distribution which is a mixture of four possible normal distributions, all with a mean of 0 but with different variances. $\boldsymbol{Z}$ is a matrix $\boldsymbol{n\ x\ m}$ of standardized genotypes and $\boldsymbol{e}$ is a vector of random residuals with $\boldsymbol{e} \sim \boldsymbol{N}(\boldsymbol{0}, \boldsymbol{\sigma_e^2})$. The posterior distributions were sampled using Marcov Chain Monte Carlo (MCMC) with Gibbs Sampling in GCTB (Zeng *et al.* 2018) with 40,000 iterations of which the first 5,000 are discarded as burn-in.

Functional analysis. We conducted a follow-up study identifying genes impacting eating quality in cattle. Top SNPs with Posterior Inclusion Probability (PIP) values above the 0.9999 quantile were used, focusing on strong associations. Genes within 10 kb of selected SNPs were analysed using *Bos taurus* ARS-UCD1.2 assembly as a reference for gene ontology enrichment analysis. Functional enrichment analysis was performed using DAVID 6.8 (Huang *et al.* 2009), with *Bos taurus* as background.

RESULTS AND DISCUSSION

Population overview. Figure 1 visualizes breed structure using PCA of imputed genotypes in GCTA (Yang *et al.* 2011b). In this figure, PC1 separates *Bos taurus* from *Bos indicus* animals, with Australian animals distributed along this axis. PC2 distinguishes different *Bos taurus* breeds, with Irish, some USA, and Australian animals lying along this axis. Animals used in this study were commercial animals and did not have an accurate breed assigned, so the labels in the figure refer to the contributing country rather than breed identities. (Figure 1).

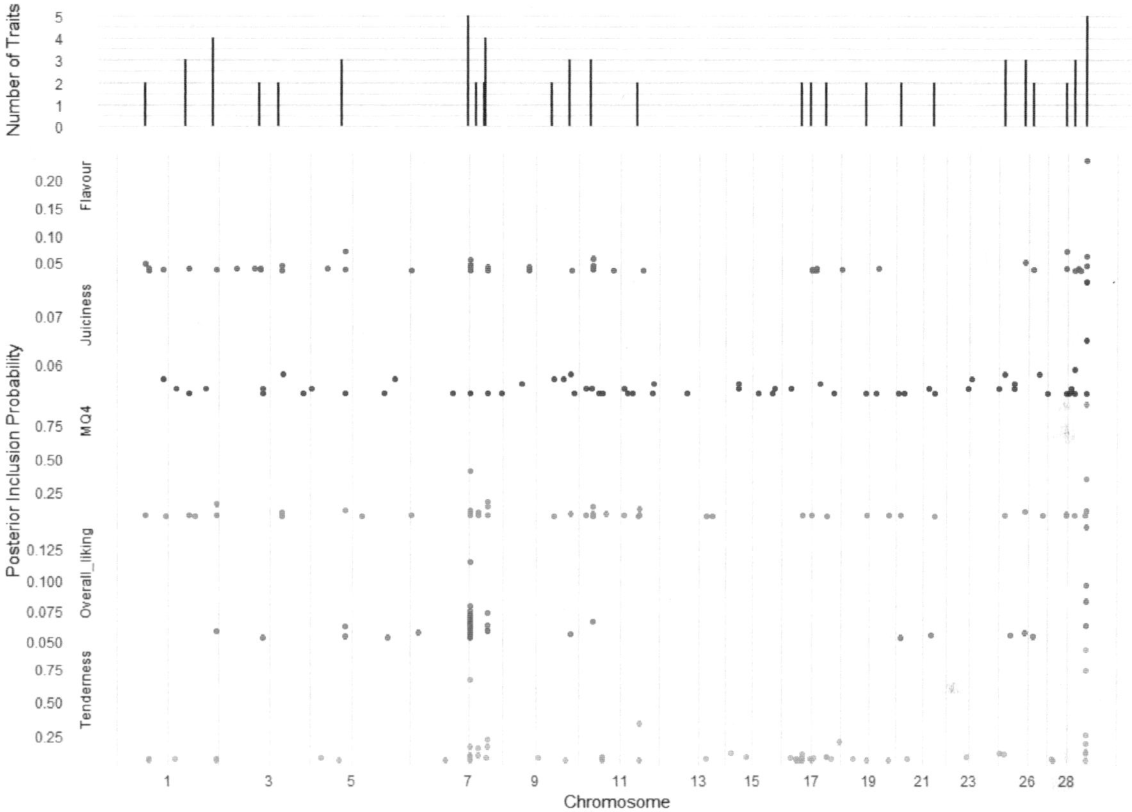

Figure 2. A bird's-eye view of the Manhattan plots showing only the top SNP (PIP>0.9999) for each of the meat quality traits. The lollipop heights in the top track are proportional to the number of traits each SNP is associated with.

Identification of potential candidate genes. Potential candidate genes were identified from top SNPs with PIP values above the 0.9999 quantile, yielding 324 SNPs across five traits. These SNPs clustered on several chromosomes, with notable associations to CAST and CAPN1 genes on chromosomes 7 and 29. Some SNPs were linked to multiple traits (Figure 2).

A total of 100 candidate genes located within a window of 10 kb upstream or downstream of the TOP-SNPs, including both novel genes and those previously reported in the literature. Among the novel candidates, *ANXA5* (Wang *et al.* 2024) and *BNC2* (Tan *et al. Unpublished data*) were highlighted for their potential roles in lipid metabolism and energy balance, contributing to marbling and tenderness. Additionally, *bta-mir-2407* was identified as a potential lipid related differentially expressed miRNAs (Li *et al.* 2022), further influencing meat quality. Several previously known genes were also confirmed, such as *CAPN1* and *CAST*, which are well-established for their involvement in post-mortem proteolysis and tenderness (Arikawa *et al.* 2024), as well as *TTLL5*, and *DOCK2,* which are recognized for their role in omega-3 and omega-6 fatty acids profile in muscle, and actin cytoskeleton remodelling, respectively (Lemos *et al.* 2016). The most strongly enriched term identified by Gene Ontology (GO) enrichment analysis of genes associated with eating quality was “GTPase activator activity” (*p-value* $< 8.0 \times 10^{-5}$).

CONCLUSION

This study provides a comprehensive overview of the genetic basis for meat eating quality traits in beef cattle by using relatively large and diverse population of *Bos taurus indicus*. The identification of novel genes, alongside the validation of established markers like *CAPN1* and *CAST*, highlights both new opportunities and robust evidence for improving eating quality through genetic selection.

REFERENCES

Arikawa L.M., Mota L.F.M., Schmidt P.I., Frezarim G.B., Fonseca L.F.S., Magalhães A.F.B., Silva D.A., Carvalheiro R., Chardulo L.A.L. and de Albuquerque L.G. (2024) *Meat Sci.* **209**: 109402.

Erbe M., Hayes B., Matukumalli L., Goswami S., Bowman P., Reich C., Mason B. and Goddard M. (2012) *J. Dairy Sci*. **95**: 4114.

Font-I-Furnols M. (2023) *Foods.* **12**: 2144

Forutan M., Lynn A., Aliloo H., Clark S.A., McGilchrist P., Polkinghorne R. and Hayes B.J. (2023) *Front. Genet.* **14**: 1089490.

Geletu U.S., Usmael M.A., Mummed Y.Y. and Ibrahim A.M. (2021) *Vet. Med. Int.* **2021**: 7340495.

Hayes B.J., Copley J., Dodd E., Ross E.M., Speight S. and Fordyce G. (2023) *Genet. Sel. Evol*. **55**: 71.

Huang D.W., Sherman B.T. and Lempicki R.A. (2009) *Nat. Protoc.* **4**: 44.

Lemos M.V., Chiaia H.L.J., Berton M.P., Feitosa F.L., Aboujaoud C., Camargo G.M., Pereira A.S., Albuquerque L.G., Ferrinho A.M. and Mueller L.F. (2016) *BMC Genomics* **17**: 1.

Li Y., Yang M., Lou A., Yun J., Ren C., Li X., Xia G., Nam K., Yoon D. and Jin, H. (2022) *Sci. Rep.* **12**: 5926.

Medeiros de Oliveira Silva R., Bonvino Stafuzza N., de Oliveira Fragomeni B., Miguel Ferreira de Camargo G., Matos Ceacero T., Noely dos Santos Gonçalves Cyrillo J., Baldi F., Augusti Boligon A., Zerlotti Mercadante M.E. and Lino Lourenco D. (2017) *PloS One* **12**: e0169860.

VanRaden P., Null D., Sargolzaei M., Wiggans G., Tooker M., Cole J., Sonstegard T., Connor E., Winters M. and Van Kaam J. (2013) *J. Dairy Sci*. **96**: 668.

Wang W., Ma C., Zhang Q. and Jiang Y. (2024) *J. Proteom.* **298**: 105129.

Warner R.D., Wheeler T.L., Ha M., Li X., Bekhit A.E.-D., Morton J., Vaskoska R., Dunshea F.R., Liu R.and Purslow P. (2022) *Meat Sci.* **185**: 108657.

Watson R., Polkinghorne R. and Thompson J.M. (2008) *Aust. J. Exp. Agric.* **48**: 1368.

Wolc A. and Dekkers J.C. (2022) *Genet. Sel. Evol*. **54**: 31.

Yang J., Lee S.H., Goddard M.E. and Visscher P.M. (2011a) *Am. J. Hum. Genet.* **88**: 76.

Yang J., Manolio T.A., Pasquale L.R., Boerwinkle E., Caporaso N., Cunningham J.M., De Andrade M., Feenstra B., Feingold E. and Hayes M.G. (2011b) *Nat. Genet.* **43**: 519.

Zeng J., De Vlaming R., Wu Y., Robinson M.R., Lloyd-Jones L.R., Yengo L., Yap C.X., Xue A., Sidorenko J. and McRae A.F. (2018) *Nat. Genet.* **50**: 746.

Proc. Assoc. Advmt. Anim. Breed. Genet. 26: 355-358

GENETIC CORRELATIONS BETWEEN MATURE BODY CONDITION, WEIGHT, AND HIP HEIGHT AND EARLY FAT DEPTH TRAITS IN AUSTRALIAN ANGUS CATTLE

M. Samaraweera[1], H. Aliloo[2], C.J. Duff[1] and S.A. Clark[2]

[1] Angus Australia, 86 Glen Innes Road, Armidale, NSW, 2350 Australia
[2] School of Environmental and Rural Science, University of New England, Armidale, NSW, 2350 Australia

SUMMARY

The objective of this study was to estimate genetic correlations between mature cow traits: body condition (MBC), weight (MCW), hip height (MCH), and live animal ultrasound scan and carcass fat depth traits: P8 fat (P8) and rib fat (RIB) in Australian Angus cattle. Genetic parameters were estimated using univariate animal models, and up to four records per cow were used in the analyses for MBC, MCW, and MCH. The heritability estimates for MBC, MCW, and MCH were 0.21 ± 0.01, 0.43 ± 0.01, and 0.55 ± 0.03, respectively. The strongest genetic correlation was observed between the MBC and MCW (0.87 ± 0.01), followed by MCW and MCH (0.83 ± 0.01). Moderate genetic correlations were observed between MBC and ultrasound scan P8 fat (0.35 ± 0.03) and rib fat (0.31 ± 0.03) traits measured on heifers and steers. Genetic correlations between MBC and carcass P8 fat (0.19 ± 0.08) and rib fat (0.30 ± 0.15) were low to moderate. Progress from selection for MBC based on live animal ultrasound scan and carcass fat depth traits measured on young cattle would be slow due to low genetic correlations. Since MBC is adequately heritable, this trait could be used as a direct measure to improve cow condition through selection.

INTRODUCTION

Mature cow body condition and size (height and weight) are important traits influencing the productivity and reproductive efficiency of mature cows. However, mature cow traits are generally recorded less frequently than early life traits due to practical challenges. Therefore, selection decisions for mature cow body composition traits, such as body condition, are often based on similar traits, like fat depth measurements taken at early ages.

Angus Australia has been publishing an estimated breeding value (EBV) for cow weight, body condition and height through the TransTasman Angus Cattle Evaluation (TACE). Understanding the genetic correlations between traits measured in young animals, such as those obtained via ultrasound scans and direct carcass measurements, and mature cow body condition and mature cow size is important for genetic improvement in cow herd productivity. The objective of this study was to estimate the genetic correlations between mature cow traits, body condition, hip height, and body weight, and live animal ultrasound scan and carcass fat depth traits and carcass weight using data from Australian Angus.

MATERIALS AND METHODS

Data. All data including phenotypes and pedigrees used for this study were extracted from the Angus Australia database. Mature cow body condition (MBC, recorded as a score), mature cow weight (MCW, measured in kg), and mature cow hip height (MCH, measured in cm) were measurements taken at the time of weaning of its calf. Mature body condition scores were collected using a zero to five scoring guide where score 0 represents emaciated body condition, and score 5 represents extremely fat body condition (Angus Australia 2024). For analysis, body condition scores were renumbered from 1 to 6. Data cleaning criteria for MCW, as outlined by Graser *et al.* (2005), were applied to define MBC and MCH. Data cleaning for MBC, MCH, and MCW resulted in 37,329, 13,640, and 177,931, records, respectively.

The ultrasound scan traits used for this study were P8 fat (P8, measured in mm) and rib fat (RIB, measured in mm). All ultrasound scan traits were separately analysed as bull (B) or heifer and steer (H) traits. Records of carcass fat depth at p8 rump site (CP8, measured in mm), carcass rib fat (CRIB, measured in mm), and carcass weight (CWT, measured in kg) were used in the analyses.

Statistical analyses. A repeatability animal model including an animal permanent environmental effect was used for all three mature cow traits. The model was also fitted with quadratic effects of the animal age (for MBC, MCH, and MCW) and quadratic effects of calf age (for MBC) as covariates. Contemporary group and other fixed effects fitted in the univariate animal models for scan and carcass traits were as described by Graser *et al.* (2005). Sire by herd was fitted as an additional random effect for scan traits. The genetic correlations between traits were estimated by bivariate animal models. The variance components were estimated using pedigree-based models implemented in airemlf90 (Misztal *et al.* 2018).

RESULTS

Most of the animals had a favourable body condition score, with an average score of 3.3 ± 0.77 (Table 1). On average, mature cows were 134 cm in height and weighed 572 kg (Table 1). All mature cow traits were measured on average at an age of 1,422 days with an average calf age at weaning of 195 days. There was a substantially higher number of records available for scan traits compared to carcass traits (Table 2).

Table 1. Descriptive statistics of mature body condition (MBC), mature cow height (MCH), and mature cow weight (MCW)

Trait	Unit	No. of records	No. of animals with records	Mean	SD	Min	Max
MBC	Score	37,329	23,047	3.3	0.77	1	6
MCH	cm	13,640	9,871	134	5	117	155
MCW	kg	177,931	97,797	572	97	176	1004

Table 2. Summary statistics of scan and carcass traits used in the study

Trait[1]	Unit	No. of records	Mean	SD	Min	Max
P8_B	mm	194,104	5.0	2.5	1	15
P8_H	mm	184,428	7.5	3.5	1	21
RIB_B	mm	194,571	4.0	1.9	1	11
RIB_H	mm	184,249	5.8	2.6	1	16
CP8	mm	11,862	21.1	6.6	3	46
CRIB	mm	4,279	16.4	6.2	3	41
CWT	kg	12,086	443	64	214	611

[1]Scan traits: B and H refers to bull and heifer scanned traits, P8: live-scan P8 fat, RIB: live-scan rib fat; Carcass traits: CP8: carcass fat depth at p8 rump site, CRIB: carcass rib fat, CWT: dressed carcass weight.

Mature cow traits have moderate to high heritability, with the highest heritability observed for MCH (0.55 ± 0.03), followed by MCW (0.43 ± 0.01), and MBC (0.21 ± 0.01) (Table 3). Both MCH (0.73) and MCW (0.74) were more repeatable than MBC (0.37). The heritability estimates of heifer scan traits were higher than those for bull scan trats (Table 3). The heritability estimates for scan fat depth traits ranged from 0.23 ± 0.01 for RIB_B to 0.40 ± 0.01 for P8_H. The heritability for CWT (0.49 ± 0.03) was greater than those for carcass fat depth traits, with the lowest for CRIB (0.35 ± 0.05).

Mature cow height is strongly associated with MCW ($r_G = 0.83 \pm 0.01$). The genetic correlation between MBC and MCW in this study was also strongly positive (0.87 ± 0.01). The genetic correlation between MBC and MCH was low (0.26 ± 0.03).

Table 3. Heritability estimates with standard deviations ($h^2 \pm SD$), additive genetic (σ_a^2), animal permanent environmental (σ_{pe}^2), sire by herd (σ_{sxh}^2), phenotypic (σ_p^2) variances for mature cow and scan traits based on univariate analyses

Trait[1]	σ_a^2	σ_{pe}^2	σ_{sxh}^2	σ_p^2	$h^2 \pm SD$
MBC	0.057	0.044		0.272	0.21 ± 0.01
MCH	7.360	2.517		13.450	0.55 ± 0.03
MCW	1067	755		2469	0.43 ± 0.01
P8_B	0.484		0.048	1.778	0.27 ± 0.01
P8_H	1.530		0.117	3.840	0.40 ± 0.01
RIB_B	0.199		0.021	0.852	0.23 ± 0.01
RIB_H	0.733		0.072	1.987	0.37± 0.01
CP8	12.98			26.94	0.48 ± 0.03
CRIB	6.82			19.48	0.35 ± 0.05
CWT	491			1001	0.49 ± 0.03

[1]Traits and units are as given in Table 1 and Table 2.

The genetic correlations between mature cow traits and scan and carcass fat depth traits and CWT are given in Table 4. Accordingly, MBC and heifer and steers scan (P8_H, RIB_H) or the carcass fat depth traits (CP8 and CRIB) were lowly to moderately genetically correlated. The genetic correlation between MBC and RIB_B (0.11 ± 0.05) is lower than that between MBC and RIB_H (0.31 ± 0.03). As expected, CWT and MCW (0.39 ± 0.05) and MCH (0.21 ± 0.09) were positively genetically correlated.

Table 4. Genetic correlations between mature cow traits and carcass and scan traits from bivariate analyses

Traits[1]	MBC	MCH	MCW
P8_B	0.08 ± 0.04	-0.19 ± 0.05	-0.22 ± 0.02
P8_H	0.35 ± 0.03	-0.28 ± 0.03	-0.11 ± 0.01
RIB_B	0.11 ± 0.05	-0.20 ± 0.05	-0.23 ± 0.02
RIB_H	0.31 ± 0.03	-0.23 ± 0.03	-0.10 ± 0.01
CP8	0.19 ± 0.08	-0.16 ± 0.09	-0.30 ± 0.05
CRIB	0.30 ± 0.15	0.01 ± 0.15	-0.13 ± 0.10
CWT	0.05 ± 0.09	0.21 ± 0.09	0.39 ± 0.05

[1]Traits and units are as given in Table 1 and Table 2.

DISCUSSION

There is sufficient additive genetic variation for effective selection of MBC, MCH, and MCW. Comparable heritability estimates have been reported in the literature. For example, heritability estimates from 0.16 ± 0.03 (Granleese and Clark 2019) to 0.21 (Johnston *et al.* 1996) for MBC, from 0.44 ± 0.05 (Granleese and Clark 2019) to 0.58 ± 0.06 (Donoghue *et al.* 2018) for MCH, and 0.43 (Granleese and Clark 2019; Johnston *et al.* 1996) for MCW in Australian Angus.

The high genetic correlations between MCW and MCH observed in our study (0.83 ± 0.01) indicates that selection for higher finishing weights and the associated correlated response for MCW could lead to an increase in MCH. A high genetic correlation between MBC and MCW in this study (0.87 ± 0.01) is expected given that MBC was not accounted for in the models for MCW. The genetic correlation between MBC and MCW in this study is greater than the previous studies by Granleese and Clark (2019) (0.62 ± 0.07), Donoghue *et al.* (2018) (0.74 ± 0.11 at weaning at parity two), and Johnston *et al.* (1996) (0.49). Placing selection emphasis on increasing CWT could lead to higher MBC due to positive genetic correlations between CWT and MCW (0.39 ± 0.05) as well as between MCW and MBC (0.87 ± 0.01).

Body condition, P8, and RIB measures reflect the fat cover thickness of an animal. However, low to moderate genetic correlations between MBC and heifer or steer scan (P8_H, RIB_H) or the carcass (CP8 and CRIB) fat depth traits measured on young animals in this study indicate that they are not genetically the same trait in young and mature animals. This also consistent with the biological differences in subcutaneous fatty tissue development in young and old animals (Robelin 1981). Low to moderate genetic correlations between MBC and scan fat depth similar to our study have been reported in the literature (Weik *et al.* 2022; Pitchford 2023). Results of our study suggest that optimising MBC based on scan and carcass fat depth traits measured on young cattle is possible, but a large change in the genetic merit of fat depth traits is required to achieve a small change in the genetic merit of MBC. Therefore, measuring MBC as a direct measure is easier and more cost effective than measuring scan fat depth in mature cows.

Targeting a higher rib fat or P8 fat at young and finishing cattle would decrease MCH and MCW as indicated by the negative genetic correlations obtained in this study. Similar negative genetic correlations between MCW and scan P8 (-0.28 ± 0.05), RIB (-0.32 ± 0.05), and CRIB (-0.40 ± 0.10) were reported by Weik *et al.* (2022). Similarly, the genetic correlations between MCH and scan P8 (-0.31 ± 0.05), and RIB (-0.32 ± 0.05), CRIB (-0.28 ± 0.10) were also negative (Weik *et al.* 2022). These negative genetic correlations are advantageous if the goal is to select for smaller mature cow size, thereby reducing feed requirement for maintenance and potentially allowing a higher stocking density on pasture. However, smaller cow weights lead to reduced returns for cull cows. Therefore, selection targets for MCW as well as MBC, and MCH should be aligned with the breeding objective.

CONCLUSION

There is sufficient genetic variation to directly select for mature cow traits. Fat depth traits measured in young animals have low genetic correlations with MBC, resulting in slow genetic progress for MBC if selection decisions for MBC are based on live animal ultrasound scan and carcass fat depth traits. Therefore, collecting MBC as a direct measure would be easy and cost effective rather than measuring scan fat depth in mature cows.

ACKNOWLEDGEMENTS

The authors thank Angus Australia members for collecting and submitting the data.

REFERENCES

Angus Australia (2024) https://www.angusaustralia.com.au/education/breeding-and-genetics/collection-guidelines-for-angus-research/). (Accessed November 2024).

Donoghue K.A., Lee S.J., Parnell P.F. and Pitchford W.S. (2018) *Anim. Prod. Sci.* **58**: 145.

Granleese T. and Clark S.A. (2019) *Proc. Assoc. Advmt. Anim. Breed. Genet.* **23**: 95.

Graser H.U., Tier B., Johnston D.J. and Barwick S.A. (2005) *Aust. J. Exp. Agric.* **45**: 913.

Johnston D.J., Chandler H. and Graser H.U. (1996) *Aust. J. Agric. Res.* **47**: 1251.

Misztal I., Tsuruta S., Lourenco D.A.L., Masuda Y., Aguilar I., Legarra A. and Vitezica Z. (2018) 'Manual for BLUPF90 family of programs', University of Georgia, Athens, USA.

Pitchford W.S. (2023) *Anim. Prod. Sci.* **63**: 957.

Robelin J. (1981) *J. Lipid Res.* **22**: 452.

Weik F., Hickson R.E., Morris S.T., Garrick D.J. and Archer J.A. (2021) *Animal* **12**: 25.

Proc. Assoc. Advmt. Anim. Breed. Genet. 26: 359-362

JOINT MODELING OF MEAN AND DISPERSION INCREASES THE ACCURACY OF RANDOM REGRESSION ON SIMULATED GROWTH CURVES

C.L. Nel[1,2] and A.R. Gilmour[3]

[1] Directorate: Animal Sciences: Western Cape Department of Agriculture, Elsenburg, 7607, South Africa
[2]Department of Animal Sciences, Stellenbosch University, Matieland, 7602, South Africa
[3]11 Holman Way, Orange, NSW, 2800 Australia

SUMMARY

We considered the use of hierarchical generalized linear models (HGLM) to model a smooth trend for residual variance in a random regression analysis. The traditional RR model uses a step function to model the change in residual variance over time. We compared the approaches on their ability to recover input parameters from two simulated growth datasets with varying inputs of dispersion parameters. Solutions from the RR analysis using the step function were susceptible to changes in dispersion, while results from the HGLM analysis were stable and more accurate. The application to real growth datasets with more complicated features should be considered in the future.

INTRODUCTION

In random regression (RR) analysis on a longitudinal axis, say time, just as genetic variance changes smoothly along the axis, the residual variance/dispersion (σ_e^2) also changes smoothly with time. The latter is often approximated with using a step-function that divides the axis (e.g. age, days in milk) into intervals and estimating σ_e^2 within each interval. This easy solution does not model the dispersion smoothly, and the number of intervals, as well as the boundaries, is arbitrary.

In growth data of some species, the effect of scale with increasing age can lead to a sharp increase in dispersion during early stages of the s-shaped growth curve. Following RR results, it has been postulated that severe scale effects can dominate the analysis despite using a step function (Apiolaza *et al.* 2000; Nel *et al.* 2025). The issue is notable, since heterogeneity of residuals linked to factors in the mixed model can affect other solutions such as animal genetic effects (Hill 1984).

Joint modelling of the mean and dispersion was first proposed in the context of 'normal regression' (Aitkin 1987). In this setting, the computing procedure iterates between two generalized linear models (GLM): a model for the mean ($E(y)$) and a model for dispersion ($E(e_i^2) = \phi_i$), where ϕ_i is analysed as a gamma variable with a log-link function. More recently, this framework was extended to include random variables in the mean model as a hierarchical GLM (HGLM; Lee and Nelder 1996), and further extensions (e.g. double HGLM) have become of considerable interest in animal breeding (Rönnegård *et al.* 2010).

In this paper we evaluate the functionality of HGLM to model both dispersion and genetic variance smoothly in the context of RR along the longitudinal axis. It was proposed that the HGLM machinery could be beneficial compared to traditional RR models in enabling (1) a smooth trend for dispersion of residuals along the longitudinal trajectory, and (2) allowing the solutions to the mean model to be appropriately weighted according to the GLM estimates derived from the dispersion model. We compared the HGLM process to a traditional RR model based on the ability to recover parameters from a simulated growth dataset.

MATERIALS AND METHODS

Simulated data. All simulations were done using custom scripts in R. The design of the dataset roughly reflected that seen for an ostrich growth dataset (Nel *et al.* 2025) but augmented for simplicity. The simulated data consisted of a population recorded monthly for live weight (LW) as a longitudinal trait during the first nine months of age (t).

The population consisted of five discrete generations of 1000 progeny each. In each generation, 200 males and 200 females were randomly selected and assigned into pairwise breeding groups, each pair producing five full-sib progenies. The first generation was considered the base population with no records for LW. The final phenotype dataset thus consisted of 4000 individuals recorded for LW nine times. The pedigree contained 5000 identities, the progeny of 800 sires and dams.

For animal i recorded at age $j = 1, \dots, 9$, the phenotype for LW was simulated as: $y_{ikj} = \mu_j + cg_k + pe_i + a_{ij} + e_{ij}$ where μ_j is the mean for age j following a third-degree polynomial curve (Figure 1) and cg_k is a contemporary group effects of $k = 1, \dots, 80$ levels (detail not shown). The animal permanent environmental (PE) variance (σ^2_{pe}) was constant across ages : $pe_i \sim N(0, \sigma^2_{pe})$ but both the additive genetic variance $a_{ij} \sim N(0, \sigma^2_{aj})$ and residual variance $e_{ij} \sim N(0, \sigma^2_{ej})$ varied with age. The structure of this heterogeneity was introduced as follows:

Additive genetic effects were defined for each individual as linear (first order) reaction norms (RN) formulated to include both scale and rank type changes (Falconer 1990) across the age trajectory. This was achieved indirectly by first sampling breeding values at the start ($j = 1$; $\sigma^2_{a1} = 5$) and at the end ($j = 9$; $\sigma^2_{a9} = 15$) with correlation of 0.5 between breeding values (a_1,a_9) and calculating RN components intercept (a_0) and slope (a_x) after standardising j to values between -1 and 1.

The residual variances were defined two ways. In Sim_I, the dispersion of residuals was as defined $\phi_j = \exp(-0.52 + 1.73j - 0.13j^2)$, a second-degree polynomial trend that first showed a sharp increase in residual variance, but declined as the mean growth rate plateaus. In Sim_II, the curve $\phi_j = \exp(1.21 - 0.21j + 0.07j^2)$ was designed to reflect a sharp increase in dispersion of variance with age (Figure 1).

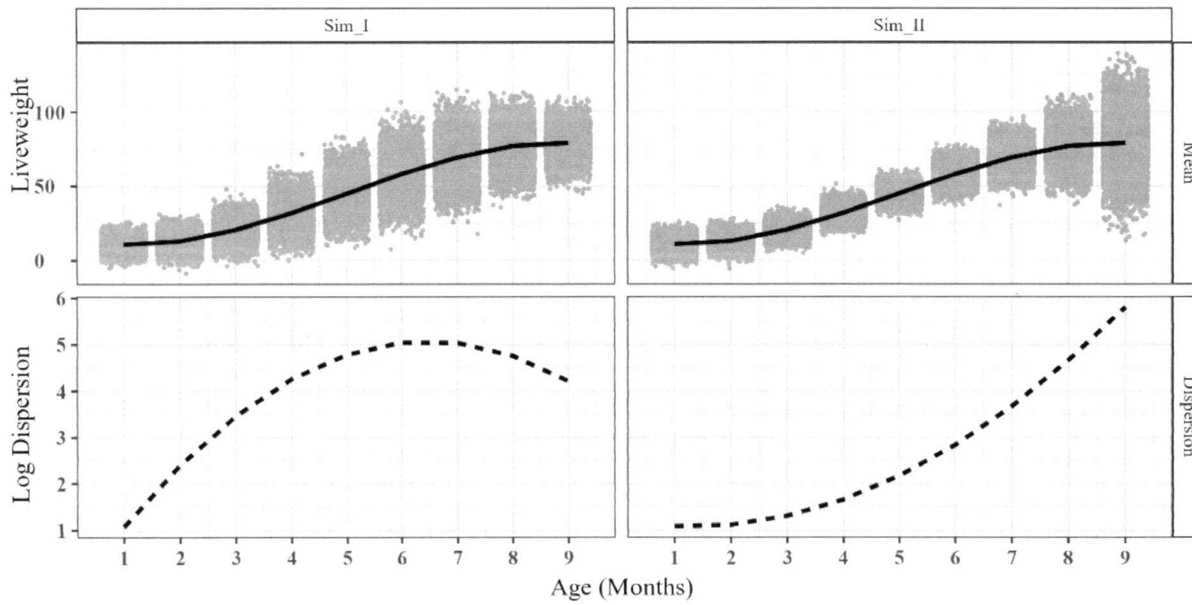

Figure 1. The distribution of LW phenotypes (gray) surrounding the mean growth trend (solid, black) as determined by the respective dispersion inputs (dashed, black) to simulations I and II

Statistical analysis. The data was analysed using the ASREML V4.2 software (Gilmour *et al.* 2021). For the traditional RR analysis, the residual was modelled as a step function in three intervals

(1-3, 4-6, 7-9 months). This design deliberately allowed an extent of change in $\phi_i$within intervals, a notable artefact of the data in Nel *et al.* (2025), but most likely true for the majority of previous RR analyses on growth data using a step function. The HGLM procedure uses the same mean model but weighted by inverse variances predicted by a dispersion model fitted simultaneously. In both cases, the linear model terms for the mean model were specified according to the simulation model design: $y = Xb + Za_0 + Za_x + Zpe + e$ assuming a normal distribution with an identity link but with heterogenous time dependent residual variance. Model testing was not a deliberate part of the study. The dispersion model in the HGLM analysis analysed the squared residuals from the mean model ($E(e_i^2) = \phi_i$) assuming a gamma distribution with a log link: $\log(\phi) = Xb$ where this *Xb* models a curvilinear (quadratic polynomial) trend. The appropriate covariance functions were used to find solutions for genetic variance (σ^2_{aj}) and accuracy of prediction ($cor(\widehat{ebv}, ebv)$) at selected age points, *j*. For each scenario, the simulation and analysis were repeated 20 times, and a mean of these results was used as the final estimate of either RR or HGLM.

RESULTS AND DISCUSSION

Mean and dispersion modelling. Both analyses showed reasonable convergence behaviour for both datasets (Sim_I and II), although RR usually converged in fewer iterations compared to HGLM. For HGLM, the dispersion model predicted both simulation $\log(\phi)$ trends with a very high accuracy with errors too small for it to be visually discernible from the true curves shown in Figure 1. The only example we found of HGLM applied to predict a smooth residual in RR was a dispersion model for lactation based on days in milk (Jaffrezic *et al.* 2000). To our knowledge, it has not been applied to growth curves, where the proportional change in the dispersion parameter ϕ is most likely of much greater magnitude than that expected for lactation data.

Variance parameters. It was apparent that RR analysis was susceptible to changes in dispersion, which varied depending on $\log(\phi)$ (Table 1). For Sim_I, with high dispersion at intermediate ages, the RR analysis was prone to overestimating genetic variance at the RN intercept component ($\sigma^2_{a_0}$) and underestimate variance of the RN slope ($\sigma^2_{a_x}$) – the latter by a considerable margin. This would explain the overestimation of the genetic correlation between breeding values at the ends of the trajectory ($p(a_1, a_9) = 0.85 > 0.5$; Table 1), which shows a partial failure in capturing reranking effects (Falconer 1990). In Sim_II, RR overestimated $\sigma^2_{a_x}$, likely due to high dispersion at the final age points. In both scenarios, the HLGM process was less affected by the curves of $\log(\phi)$, and solved the mean model with estimates very close to the true values – but perhaps also slightly underestimating the animal PE effect (Table 1).

Table 1. True values and solutions for variance parameters following RR and HGLM analysis. *The estimated values are the mean following 20 simulations

Parameters	True Values	Sim_I*		Sim_II*	
		RR	HGLM	RR	HGLM
σ^2_{pe}	10	9.02	9.73	9.63	9.64
σ^2_{a0}	7.15	8.64	7.28	8.00	7.31
σ^2_{ax}	1.89	0.88	1.75	2.59	1.88
$p(a_0, a_x)$	0.56	0.80	0.55	0.60	0.54
$p(a_1, a_9)$	0.50	0.83	0.54	0.43	0.51

Genetic variance at different ages and accuracy of prediction. The RR analysis estimated $\sigma^2_{a_j}$ of Sim_I close to the real values for first ($j = 1$) and last ($j = 9$) age points, but slightly higher values at intermediate ages (Table 2). For Sim_II, it overestimated $\sigma^2_{a_j}$ for 9 months of age by a notable margin, an indication of overestimating scale effects. HGLM results better aligned with the known input values which also resulted in a higher accuracy of prediction, most particularly for Sim_I (Table 2). Generally observing lower accuracy values at the ends of the trajectory was expected, since errors in estimating RN slope components are accentuated as distance to the intercept increases.

The computation in HGLM includes iterative updates of w_i, the ith diagonal element of the weight matrix $W$of the mixed model equations, by the inverse of solutions ϕ_i from the dispersion model. In the current design, a benefit of this smooth adjustment from the HGLM process was clear, and, as argued by Rönnegård *et al.* (2010), the possibly important aspects of leverage should not be overlooked in animal breeding analysis with high heterogeneity. It should be noted, however, that the highly accurate modelling of $\log(\phi)$ would have been very important to the correct adjustment of W, and the process of model testing was ignored in this study.

Also, in real datasets, $\log(\phi)$ trends are unlikely to be well specified by inflexible polynomials. In a separate analysis, dispersion models making use of cubic smoothing splines also converged with accurate solutions. This is expected to be the most powerful application of HGLM to real data where both the magnitude and the change in $\log(\phi)$ are unknown.

Table 2. True values and solutions for genetic parameters and prediction accuracy following RR and HGLM analysis. *The estimated values are the mean following 20 simulations

Age	True Values	Genetic Parameters				Accuracy of EBVs			
(j)		Sim_I*		Sim_II*		Sim_I*		Sim_II*	
	$\sigma^2_{a_j}$	RR	HGLM	RR	HGLM	RR	HGLM	RR	HGLM
1	4.98	4.58	5.12	5.23	5.20	0.63	0.69	0.70	0.70
3	5.36	6.28	5.55	5.65	5.55	0.70	0.72	0.74	0.74
5	7.16	8.64	7.28	8.00	7.31	0.71	0.72	0.75	0.76
7	10.38	11.67	10.32	12.30	10.48	0.68	0.71	0.75	0.77
9	15.03	15.36	14.67	18.54	15.07	0.66	0.69	0.74	0.76

CONCLUSION

In the presence of sharp changes in dispersion, the genetic solutions from RR analysis can be affected by the choice of the step function. Analysis within the HGLM framework, in turn, can have more accurate genetic solutions as found here. However, the design of the simulated dataset was highly simplistic, and the process needs to be tested on real data.

REFERENCES

Aitkin M. (1987) *J. R. Stat. Soc. C-App.* **36**: 332.

Apiolaza L.A., Gilmour A.R. and Garrick D.J. (2000) *Can. J. Forest. Res.* **30**: 645.

Falconer D.S. (1990) *Gen. Res.* **56**: 57.

Gilmour A.R., Gogel B.J., Cullis B.R., Welham S.J. and Thompson A.N. (2021) ASReml User Guide Release 4.2 Functional Specification. www.vsni.co.uk.

Hill W.G. (1984) *Anim. Sci.* **39**: 473.

Jaffrezic F., White I.M.S., Thompson R. and Hill W.G. (2000) *J. Dairy Sci.* **83**: 1089.

Lee Y. and Nelder J.A. (1996) *J. R. Stat. Soc. B.* **58**: 619.

Nel C.L., Gilmour A.R., Muvhali P.T., Cloete S.W.P., Kekana M. and Engelbrecht A.E. (2025) *Brit. Poultry. Sci.* (Online Early).

Rönnegård L., Felleki M., Fikse F., Mulder H.A. and Strandberg E. (2010) *Genet. Sel. Evol.* **42**: 8.

Proc. Assoc. Advmt. Anim. Breed. Genet. 26: 363-366

ACCURACY OF GENOMIC PREDICTIONS FOR BEEF EATING QUALITY TRAITS

H. Aliloo[1], M. Forutan[2], P. McGilchrist[1], B. Hayes[2] and S. Clark[1]

[1] School of Environmental and Rural Science, University of New England, Armidale, NSW, 2351 Australia

[2]Queensland Alliance for Agriculture and Food Innovation, University of Queensland, Brisbane, QLD, 4072 Australia

SUMMARY

Eating quality is the primary factor influencing consumer purchasing decisions for beef products. Consumer-derived sensory eating quality traits capture the actual eating experience and are directly aligned with consumer expectations. Accurate genomic prediction of these traits can enable selective breeding, allowing for continuous genetic improvement and consistent consumer satisfaction. In this study we used an international dataset from Australia, the USA and Ireland to estimate genetic parameters for five meat eating quality traits and to evaluate the accuracy of genomic estimated breeding values under different cross-validation scenarios. Heritability estimates ranged from 0.19 to 0.31 for Australia, 0.07 to 0.20 for the USA, 0.09 to 0.17 for Ireland, and 0.14 to 0.22 for the combined dataset. Prediction accuracies for all traits were moderate when using the Australia-only reference population but increased to high accuracies when international data were included. This highlights the value of incorporating international data into the reference population, creating a larger and more diverse dataset, and ultimately improving prediction accuracies.

INTRODUCTION

Eating quality is a key driver of beef purchasing decisions, as it directly influences consumer satisfaction and perceived value for money. Factors such as tenderness, flavour, juiciness, and overall liking play a crucial role in shaping consumer preferences and purchasing behaviours. Meat Standards Australia (MSA) has developed an index to grade beef carcasses based on their expected eating quality, using chemical and physical characteristics of the meat (Watson *et al.* 2008). The MSA index incorporates carcass traits such as marbling, ossification and hang method to predict the expected satisfaction grade of individual beef cuts. These factors, along with the proposed cooking method, are considered to influence the eating quality of the meat. Consumer-derived sensory eating quality traits, on the other hand, reflect the actual eating experience and align more closely with consumer expectations. These traits are typically assessed through structured sensory evaluation protocols, often involving trained consumer panels or large-scale untrained consumer testing, where key attributes of consumer satisfaction are quantified using standardized scoring systems.

Despite their importance, sensory eating quality traits are costly and difficult to measure, and direct assessments on selection candidates are not feasible as they require animal slaughter. Consequently, genetic improvement for these traits in cattle breeding programs is hindered by the scarcity of direct phenotypic data. Genomic selection offers an opportunity to provide accurate estimates of genetic merit for eating quality traits, enabling more efficient genetic improvement. Genomic selection incorporates the genomic relationship between animals in a reference set, comprising both genotyped and phenotyped individuals, and those in a target set with only genotypes to predict the genetic merit of the latter (Meuwissen *et al.* 2001). However, the accuracy of genomic estimated breeding values remains a key factor in determining the overall effectiveness of breeding strategies.

This study aimed to estimate genetic parameters and evaluate the accuracy of genomic prediction for consumer-derived eating quality traits using an international dataset from Australia, the USA, and Ireland. By comparing different reference sets, we assessed the potential benefits of multi-country data for enhancing the reliability of genomic evaluations in the Australian population.

MATERIALS AND METHODS

Data. Striploin muscle samples were grilled and evaluated by 10 consumers following the MSA consumer testing protocols (Watson *et al.* 2008). Each consumer assessed the samples for tenderness (TENDER), juiciness (JUICY), flavour (FLAVOR) and overall liking (OVERALL), scoring them on a scale from 0 to 100. A meat quality score (MQ4) was then calculated for each sample based on the optimum linear discriminating function developed by Watson *et al.* (2008). To minimise the impact of extreme values, the highest and lowest two scores for each sample were excluded, and the remaining six were averaged. After quality control, the final dataset comprised 3,526 records from Australia (AUS), 3,197 from the USA, and 1,297 from Ireland (IRE) for analysis. The genotypes were obtained using a variety of different medium density panels and were then imputed up to 700K separately for each country using appropriate reference sets. In total, there were 4,928, 3,230 and 1,312 genotyped animals 4072 available from AUS, USA and IRE, respectively.

Statistical model. A mixed linear univariate animal model was fitted separately for each of the 5 meat eating quality traits from each country using BLUPF90+ (Misztal *et al.* 2014):

$$\mathbf{y} = \mathbf{Xb} + \mathbf{Zu} + \mathbf{e}$$

where **y** is the vector of phenotypes; **b** includes the estimates of fixed cross-classified contemporary group, sex and hormonal growth promotant effects, and covariates of days aged after slaughter, carcass weight, first 4 principal components and heterosis; **u** is the vector of genomic estimated breeding value (GEBV) of animals; and **e** is the residual term. **X** and **Z** are incident matrices relating observations to fixed and random effects, respectively. It was assumed that $v(u) = G\sigma_a^2$ where G was the genomic relationship matrix based on VanRaden (2008) and σ_a^2 is the additive genetic variance. The full dataset was used to estimate variance components and heritabilities for each trait within each country. Additionally, to estimate variance components and heritabilities across all three countries, all records for each eating quality trait were combined into a single dataset. A fixed cross-classified country effect was added to account for differences between countries.

Cross-validation. To assess the genomic evaluation accuracies for AUS animals, three 5-fold cross validation strategies were implemented. In the first strategy (RANKED1), AUS animals were sorted based on the first principal component (PC1) and assigned into one of five groups of approximately equal size to maintain a balanced genetic representation in each group. In the second strategy (RANKED2), AUS animals were sorted based on PC1 and divided into five groups of approximately equal size. The top 20% of animals were placed in the first group, the next 20% in the second group, and so on. This approach created more distinct genetic clusters within the validation sets. In the third strategy (RANDOM), animals were randomly assigned into five groups of approximately equal size, without considering any genetic or temporal structure. In each iteration, one group's phenotypes were masked while the remaining four were used to estimate GEBVs. Accuracy was evaluated using LR method (Legarra and Reverter 2018) by comparing GEBVs from whole-data models with those from partial-data cross-validations. Accuracies were averaged across the five validation groups, and the standard error was calculated as STD/√5. The predictive ability of the model for AUS animals was assessed using two reference sets, one with only AUS animal and another that included animals from USA and IRE in addition to AUS cohort (AUS_USA_IRE).

RESULTS AND DISCUSSION

Variance components and heritabilities. Table 1 presents the estimates of variance components and heritabilities (± standard errors) for the five meat quality traits, obtained from a single-trait GBLUP model. Among all countries, TENDER exhibited the highest additive genetic variance, resulting in the highest heritability estimates across the five traits. Heritabilities were generally highest in AUS compared to USA and IRE, with the largest difference observed for TENDER. The lowest heritabilities were found for JUICY in AUS and IRE, and for FLAVOR in the USA. When combining data from all three countries (AUS_USA_IRE), the heritability estimates were intermediate, falling between the individual within-country estimates. Forutan *et al.* (2025) applied the BayesR approach using the same dataset and obtained similar estimates for variance components and heritabilities. Our results indicate that the additive genetic component makes a significant contribution to the variability of meat eating quality traits, suggesting that genetic improvement through selective breeding is feasible.

Table 1. Variance components and heritabilities (± SE) for five meat quality traits estimated for each country separately (AUS, USA, IRE) and for the combined dataset (AUS_USA_IRE)

Trait	Variances	AUS	USA	IRE	AUS_USA_IRE
	σ_a^2	56.47 ± 8.91	28.18 ± 5.99	21.24 ± 9.53	34.62 ± 3.97
TENDER	σ_e^2	124.48 ± 7.15	111.74 ± 5.36	109.69 ± 8.92	121.78 ± 3.50
	h^2	0.31 ± 0.05	0.20 ± 0.04	0.16 ± 0.07	0.22 ± 0.02
	σ_a^2	26.92 ± 6.49	22.47 ± 5.93	10.53 ± 7.64	20.33 ± 3.15
JUICY	σ_e^2	117.57 ± 5.72	124.32 ± 5.60	101.66 ± 7.58	120.38 ± 3.10
	h^2	0.19 ± 0.04	0.15 ± 0.04	0.09 ± 0.07	0.14 ± 0.02
	σ_a^2	22.63 ± 4.95	7.65 ± 3.50	16.00 ± 7.22	13.99 ± 2.27
FLAVOR	σ_e^2	84.59 ± 4.27	95.21 ± 3.75	85.03 ± 6.80	90.94 ± 2.28
	h^2	0.21 ± 0.04	0.07 ± 0.03	0.16 ± 0.07	0.13 ± 0.02
	σ_a^2	34.74 ± 6.32	13.67 ± 4.25	17.59 ± 7.95	21.06 ± 2.82
OVERALL	σ_e^2	97.15 ± 5.24	101.86 ± 4.26	92.82 ± 7.47	101.15 ± 2.68
	h^2	0.26 ± 0.04	0.12 ± 0.04	0.16 ± 0.07	0.17 ± 0.02
	σ_a^2	34.99 ± 6.07	15.16 ± 3.85	16.75 ± 7.26	34.62 ± 3.97
MQ4	σ_e^2	89.89 ± 4.97	82.10 ± 3.65	83.85 ± 6.79	121.78 ± 3.50
	h^2	0.28 ± 0.05	0.16 ± 0.04	0.17 ± 0.07	0.22 ± 0.02

Accuracy of GEBVs. Figure 1 illustrates accuracy, slope and bias obtained from the 5-fold cross-validations for the five meat quality traits, using AUS-only or AUS_USA_IRE reference populations. For the AUS-only reference, JUICY had the highest prediction accuracy (0.69) among all traits and across all validation strategies, while the other four traits had similar accuracies, ranging from 0.62 to 0.65. Among the different validation strategies, RANKED2 consistently produced the lowest prediction accuracies for all traits (0.44 – 0.51), whereas the other strategies yielded similar results, with accuracies ranging from 0.62 to 0.69. Adding international data from USA and IRE to the reference population improved the prediction accuracies for all traits and across all validation strategies, highlighting the value of a larger and more diverse reference population. The prediction accuracies followed a similar trend to the AUS-only reference, with RANKED2 producing the lowest accuracies, ranging between 0.59 and 0.66, and the other two strategies achieving similar accuracies, ranging between 0.83 and 0.89. Considering the RANDOM validation strategy, the accuracies from the AUS_USA_IRE reference population showed notable improvements compared to the AUS-only reference, increasing by 0.21 for TENDER (from 0.62 to 0.83), by 0.20 for JUICY (from 0.69 to 0.89), by 0.25 for FLAVOR (from 0.64 to 0.89), by 0.23 for OVERALL (from 0.64 to 0.87), and by 0.23 for MQ4 (from 0.63 to 0.86). These findings underscore the benefits of incorporating international data to enhance prediction accuracies for beef eating quality traits in Australia.

The lower accuracy in RANKED2 is expected, as animals from one breed were used to predict the genetic merit of a different breed, representing the lower bound of prediction accuracy. In contrast, RANKED1, which split animals from different breeds between reference and validation sets, improved accuracy by better capturing within-breed variation. The poor performance of RANKED2 highlights that due to significant breed diversity in commercial populations, genetic merit predictions from one breed background cannot accurately translate to another.

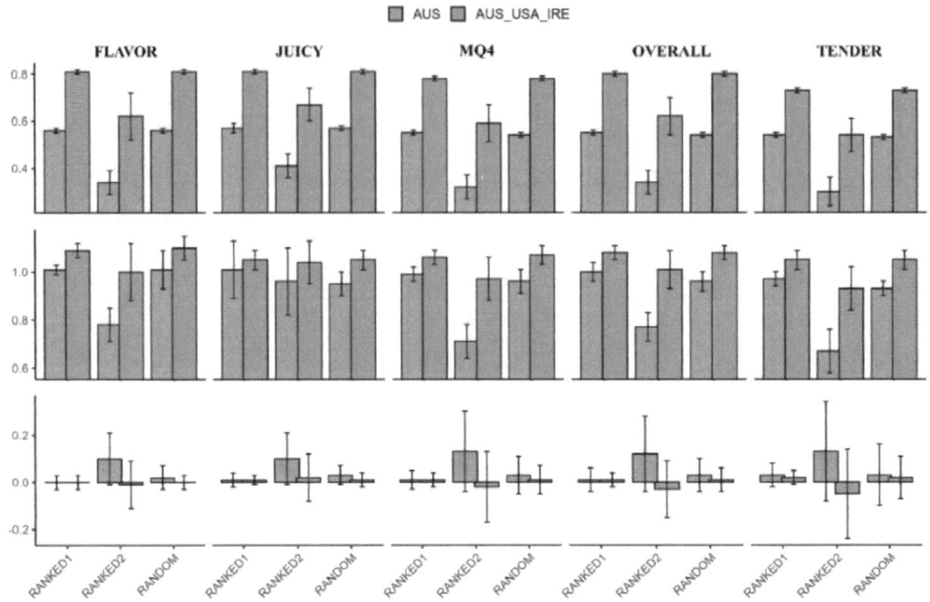

Figure 1. Accuracy (top), slope (middle) and bias (bottom) for five eating quality traits across different cross-validation scenarios, using AUS-only or AUS_USA_IRE reference sets

CONCLUSION

This study highlights the importance of a diverse reference population for genomic prediction of eating quality traits in commercial beef cattle. Incorporating international data improves the accuracy of genetic merit predictions by creating a larger reference with greater breed diversity.

REFERENCES

Forutan M., Aliloo H., Clark S.A., McGilchrist P. and Hayes B.J. (2025) *Proc. Assoc. Advmt. Anim. Breed. Genet.* **26:** *These proceedings.*

Legarra A. and Reverter A. (2018) *Gent. Sel. Evol.* **50**: 1.

Meuwissen T.H., Hayes B.J., and Goddard M.E. (2001) *Genetics* **157**: 1819.

Misztal I. Tsuruta S., Lourenço D., Aguilar I., Legara A. and Vitezica Z. (2014) Manual for BLUPF90 family of programs. [accessed February 28, 2022]. https://nce.ads.uga.edu/wiki/lib/exe/fetch.php?media=blupf90_all.pdf.

VanRaden P.M. (2008) *J. Dairy Sci.* **91**: 4414.

Watson R., Polkinghorne R. and Thompson J.M. (2008) *Aust. J. Exp. Ag.* **48**: 1368.

Proc. Assoc. Advmt. Anim. Breed. Genet. 26: 367-370

YEARLING WEIGHT AND HORN TRAITS IN SOUTH AFRICAN MERINO LINES DIVERGENTLY SELECTED FOR AND AGAINST NUMBER OF LAMBS WEANED

S.W.P. Cloete[1], C.L. Nel[1,2] and J.J.E. Cloete[3]

[1] Department of Animal Sciences, Stellenbosch University, Private Bag X1, Matieland, 7602, South Africa

[2] Directorate: Animal Sciences, Department of Agriculture, Western Cape Government, Private Bag X1, Elsenburg, 7607, South Africa

[3] Elsenburg Agricultural Training Institute, Private Bag X1, Elsenburg 7607, South Africa

SUMMARY

Ovine horn phenotypes have been studied extensively but are still not fully understood. This study used data from a divergent selection experiment for (H-Line) and against (L-Line) number of lambs weaned per ewe mated (NLW) in two South African Merino lines to study the effect of selection line on yearling weight (YW) and horn traits. Mixed model analyses allowed the estimation of fixed effects and genetic parameters for YW, horn length (HL) and horn base circumference (HC) in ram yearlings. H-Line ewes were more likely to be polled (0.651 vs. 0.353; Chi^2=46.8) with a lower frequency of horns than L-Line ewes (0.032 vs. 0.323; Chi^2=128.4). HC in both sexes and HL in ewes were independent of selection line with YW as a covariate. Single- and three-trait heritability estimates were respectively 0.54 and 0.58 for YW, 0.42 and 0.45 for HL, and 0.44 and 0.53 for HC. Genetic correlations among traits ranged from 0.66 between YW and HL to 0.77 between HL and HC. Studies on horn traits in South African Merinos could provide motivation for a greater focus on the elimination of horns, as in other major sheep-producing countries.

INTRODUCTION

In wild ungulates, horns are important for defence against predators and for determining an individual's social status in a group (Simon *et al.* 2022). The utility of horns in domestic sheep is less clear as it is often associated with health and management issues, as well as injuries to other sheep and human handlers (Hatcher *et al.* 2019; Simon *et al.* 2022). Horned Soay sheep had greater reproductive success, which was traded off against a shorter lifespan (Johnston *et al.* 2013). However, evidence linking horns to reproductive success is scarce (Simon *et al.* 2022).

Against this background, we studied horn traits in a South African Merino resource flock divergently selected for and against reproduction with successful results (Cloete *et al.* 2004). Given that horns are often dimorphic between sexes in sheep (Simon *et al.* 2022), rams and ewes were considered separately.

MATERIALS AND METHODS

Experimental site and animals. The well-recorded Elsenburg Merino flock (Cloete *et al.* 2004; Nel *et al.* 2024) provided material for this study. The flock was established in 1986 and used a common base population to establish a divergent selection experiment involving two lines selected for (High or H-Line) or against (Low or L-Line) number of lambs weaned per ewe mated. The two lines were maintained in a single flock, except when being mated in single-sire groups during summer (January-February) to lamb in winter (June-July) annually. Details regarding the composition, selection regimes and management of the flock are provided in Cloete *et al.* (2004) and Nel *et al.* (2024) and will not be elaborated here.

Recordings. This study hinged on detailed horn phenotypes recorded from 2016 to 2023 on yearling ewe (N=791) and ram (N=843) progeny that were available for selection as replacements. In addition, yearling weights (YW) at an average (SD) age of 334 (29) days were also recorded.

COVID-19 restrictions prohibited the recording horns of the 2019 drop in May 2020, but YW was available (N=1847). Horn phenotypes were scored as polled, knobs, scurs and horns in both sexes (Pickering *et al.* 2009; Duijvesteijn *et al.* 2018). Horn length (HL) was measured from the base to the tip of the horn in rams and all ewes with horns or measurable scurs (Figure 1) using a standard tape measure. Horn base circumference (HC) was also measured in those individuals where it was feasible. Assuming symmetric horns, the record was consistently obtained from the right horn/scur by the same person. Progeny (N=60) and grand progeny (N=74) of a polled industry ram that was used in 2021 and 2022 to link the flock to industry resources (Nel *et al.* 2024) were excluded from all analyses. Using 2300 Merino progeny in the Sheep CRC information nucleus flock, Hatcher *et al.* (2019) showed that a polled ram mated to horned ewes will have 49% polled ram and 82% polled ewe progeny. The corresponding figures for a horned ram were respectively 6 and 18%, demonstrating the profound impact such a ram could have.

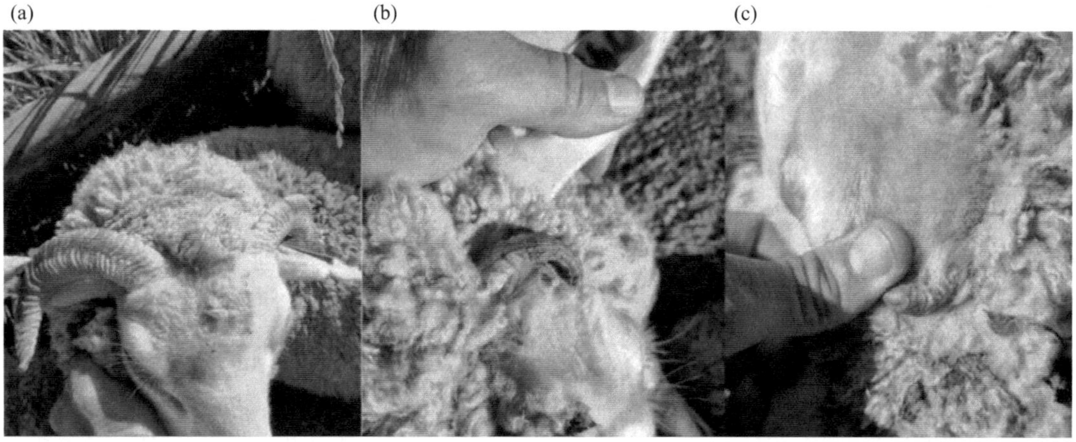

Figure 1. Yearling ewes with horns (a), measurable scurs (b) and scurs not measured (c)

Statistical analysis. Frequencies derived from the presence/absence of horn phenotypes of ewe yearlings were compared between lines using the online Chi²-calculator of Preacher (2001). YW, HL and HC of rams were analysed using a linear mixed model analyses in ASReml V4.2 (Gilmour *et al.* 2021). The analysis included the fixed effects of selection line (H- or L-Line), dam age (2-6+ years), birth type (single or multiple) and birth year as defined. Yearling age and YW (only for horn traits) were included as linear covariates. The additive random effect of animal was added to the model to enable the estimation of genetic and phenotypic (co)variance components and ratios. Birth year was excluded from the parallel analysis on HL and HD of 74 horned ewe yearlings as the horn phenotype was absent in the H-Line (N=20) in two out of seven years.

RESULTS AND DISCUSSION

Horn phenotypes within ewe selection lines. H-Line ewe yearlings were almost twice as likely to be polled than their L-Line contemporaries (P<0.01; Table 1). The presence of knobs and scurs was independent of the selection line (P>0.20) and added up to around 32% of the phenotypes when combined. L-Line yearlings were about 10 times more likely to be horned than their H-line contemporaries. Polled ewes contributed 0.459 to the 2448 Merinos sampled by Duijvesteijn *et al.* (2018), a proportion that was intermediate compared to the Elsenburg selection lines. At around 50% of the ewes sampled, knobs and scurs combined were more common than in Table 1. At 0.036, the proportion of horned ewes was in the same range as in the H-Line, but much lower than in the L-Line. The relative proportions of horn phenotypes in 35 Icelandic ewes were 0.743 polled, 0.171 scurs, and 0.086 horns (Simon *et al.* 2024). Although these figures deviated quite considerably from those in Table 1, it would not be too deviant from H-Line proportions if ewes with knobs were regarded as polled. Simon *et al.* (2024) did not detail how knobs were dealt with. It is not yet known if the results in Table 1 are related to selection taking place in the resource flock, as Simon *et al.* (2019) did not find an association of horns with reproduction in livestock species they reviewed. It is unclear if the observed line effects are conclusively the result of the divergent selection practiced in the flock since 1986, or an effect of random genetic drift. Out of 72 ewes with scurs in the H-Line, 17 were measurable. All 13 scurs in L-Line were unmeasurable, not allowing comparisons in this line. A two-sample t-test assuming unequal variances was used to test the hypothesis that horn traits in the H-Line were not different (P>0.05). The alternative hypothesis that horns had larger dimensions than scurs was accepted (HL: t =5.15; HC: t =5.70; P<0.01; Table 1).

Table 1. Horn length (HL) and horn base circumference (HC) of measurable horns and scurs (only in the H-Line) and the relative proportions of horn phenotypes (record numbers in brackets), within the divergently selected H- and L-Lines in yearling Elsenburg Merino ewes

Horn/Scur size, ewe numbers and horn phenotype	HL (cm)	**H-Line** HC (cm)		**L-Line**
Total number of ewes			624	167
Polled	NA	NA	0.651 (406)	0.353 (59) **
Knobs	NA	NA	0.202 (126)	0.246 (41) NS
Scurs	7.5 (17)	6.3	0.115 (72)	0.078 (13) NS
Horns	20.1 (20)	11.3	0.023 (20)	0.323 (54) **

** Significant (P<0.01); NS Not significant (P>0.05); NA Not applicable

Environmental and genetic effects on YW and horn traits in rams. The low incidence of horns in general and H-Line ewes in particular compromised genetic analyses on horn traits across sexes. Apart from being absent in two years, the numbers of H-Line ewe yearlings with horns ranged from 1 to 7 in the other years, while the corresponding range in the L-Line was from 6 to 10. Both H-Line ewes and rams were heavier than their L-Line contemporaries (Table 1). When analysed with YW as a covariate, HD in both sexes as well as HL in ewes were independent of selection line (P>0.05; Table 2). HL was still slightly higher in H-Line rams after the correction (P<0.05).

Table 2. Least-squares means (±SE) depicting the effect of selection line (H-Line vs L-Line) on yearling weight, horn length and horn circumference of Merino ewe and ram yearlings

Sex and level	**Yearling weight (kg)**	**Horn length (cm)**	**Horn circumference (cm)**
Ewe yearlings	**	NS	NS
H-Line	32.5±0.2	17.6±1.6	10.7±0.5
L-Line	28.1±0.3	20.0±0.9	10.1±0.3
Ram yearlings	**	*	NS
H-Line	35.4±0.4	35.6±0.2	16.5±0.1
L-Line	31.4±0.5	34.7±0.3	16.5±0.1

* Significant (P<0.05); ** Significant (P<0.01); NS Not significant (P>0.05)

Single-trait heritability estimates were 0.54±0.09 for YW, 0.42±0.09 for HL and 0.44±0.10 for HD, the latter traits analysed with YW as a covariate. The three-trait analysis yielded slightly higher heritability estimates ranging from 0.45 for HL to 0.58 for YW (Table 3). Adjusting for size could have reduced between family variation in HL and HC in single-trait analyses, although significance could not be demonstrated. The estimated heritability for YW is higher than a previous figure of 0.38 on the same flock (Cloete *et al.* 2005). However, the previous study also reported a maternal genetic variance ratio of 0.09. Genetic correlations ranged from 0.67 between YW and HL to 0.77

between HL and HC. Phenotypic correlations were also positive. It was not unexpected that YW, HL and HD was positively correlated at both the genetic and phenotypic levels, as horn traits could also be size-related. However, the direct impact of genetic variation in horn traits of domestic sheep is not apparent, as the overall objective is to breed towards polled sheep, for reasons outlined by Hatcher *et al.* (2019). In contrast, horns have an intrinsic economic value as part of the trophy-hunting industry in wild ungulates. Sim and Coltman (2019) reported somewhat lower heritability estimates than those in the present study of 0.33 for HL and 0.36 for HD in Dall's sheep (*Ovis dalli dalli*) that were harvested routinely. Heritable variation in ovine horn traits thus seems available.

Table 3. Phenotypic variances, as well as heritability (bold on the diagonal), genetic correlations (below diagonal) and phenotypic correlations (above diagonal) for yearling weight, horn length and horn base circumference of yearling Merino rams in the H- and L-Lines from a three-trait genetic analysis

Effect and level	Phenotypic variance	Yearling weight	Horn length	Horn circumference
Yearling weight (kg)	31.54	**0.58±0.10**	0.64±0.03	0.62±0.03
Horn length (cm)	26.81	0.67±0.09	**0.45±0.10**	0.64±0.03
Horn circumference (cm)	2.850	0.70±0.09	0.77±0.08	**0.53±0.10**

CONCLUSIONS

The incidences of polledness and horns differed markedly between ewes from the H- and L-Lines. After adjusting for the bigger skeletal size of H-Line ewes and rams, HD in both sexes and HL in ewes were independent of selection line. Apart from the efforts of some individual breeders, there is no concerted effort to breed for polled phenotypes in South African Merinos. It would be of value to demonstrate that selection for reproduction could aid in achieving this goal. Further studies on the inheritance of horns in South African sheep are thus warranted.

REFERENCES

Cloete S.W.P., Gilmour A.R., Olivier J.J. and Van Wyk J.B. (2004) *Anim. Prod. Sci.* **44**: 745.

Cloete S.W.P., Olivier J.J., Van Wyk J.B., Schoeman S.J. and Erasmus G.J. (2005) *Proc. Assoc. Advmt. Anim. Breed. Genet.* **16**: 24.

Duijvesteijn N., van der Werf J.H.J. and Kinghorn B.P. (2024) *Genet. Sel. Evol.* **56**: 61.

Gilmour A.R., Gogel B.J., Cullis B.R., Welham S.J. and Thompson A.N. (2021) ASReml User Guide Release 4.2 Functional Specification. VSN International Ltd, Hemel Hempsted, UK.

Hatcher S., Gardner G.E., Gill S., Lee S., Swan A.A. and van der Werf J.H.J. (2019) Sheep CRC Report 13_30, prepared as part of the Sheep CRC Program 2007-2014.

Johnston S.E., Gratten J., Berenos C., Pilkington J.G., Clutton-Brock T.H., Pemberton J.M. and Slate J. (2013) *Nature* **502**: 93.

Nel C.L., Gurman P., Swan A.A., van der Werf J., Snyman M., Dzama K., Olivier W., Scholtz A. and Cloete S. (2024) *J. Anim. Breed. Genet.* **141**: 65.

Pickering N.K., Johnson P.L., Auvray B., Dodds K.G. and McEwan J.C. (2009) *Proc. Assoc. Advmt. Anim. Breed. Genet.* **18**: 88.

Preacher K.J. (2001) https://quantpsy.org/chisq/chisq.htm.

Sim Z. and Coltman D.W. (2019) *Front. Genet.* **10**: 959.

Simon R., Drögemüller C. and Lühken G. (2022) *Genes* **13**: 832.

Simon R., Elísabetardóttir K. and Lühken G. (2024) *Arch. Anim. Breed.* **67**: 237.

Proc. Assoc. Advmt. Anim. Breed. Genet. 26: 371-374

MICROBIABILITY AND HERITABILITY OF DEER SEASONALITY TRAITS

J.L. Clarke[1,2], T.P. Bilton[1], H. Henry[1], K.M. McRae[1], F. Booker[1], B. Veenvliet[1], W. Bain[1], J.A. Ward[1], B. Thompson[1], D. Stevens[1], J.C. McEwan[1], B.J. Perry[1], M.A. Black[2] and S.J. Rowe[1]

[1] AgResearch Invermay Agricultural Centre, Private Bag 50034, Mosgiel, 9053, New Zealand
[2] Department of Biochemistry, University of Otago, Dunedin, 9016, New Zealand.

SUMMARY

Deer are seasonal animals that experience changes in growth and other traits throughout the year. Rumen microbial communities (RMC) are affected by diet and host genetics and have been shown to influence phenotypic traits. The aim of this project was to determine the variance explained by the rumen microbiome (microbiability), and the animals' genetics (heritability) on seasonal growth rates and immune responses. Rumen contents from 630 Red Deer and Wapiti cross deer were collected, processed and sequenced to generate a RMC profile of the rumen microbiome. From the RMC, a microbial similarity matrix between individuals was constructed and used a as random effect matrix in a linear mixed model to estimate trait variation attributed to the RMC. This was compared to models using only genomics, and a joint model using both host genomics and the RMC. Estimates of microbiability and heritability for these traits were similar across models, where winter growth rate and log of carbohydrate larval antigen (LnCARLA) displayed negligible microbiability. This indicates that while the RMC explains a significant amount of variation in growth, it does not rank animals at times when feed intake and growth is low and has no association with disease resistance. Therefore, caution should be taken when using RMC for trait predictions in seasonal animals.

INTRODUCTION

Molecular phenotypes are becoming more prevalent as predictors in breeding schemes. One such predictor in ruminants is rumen microbial communities (RMC). Understanding the behaviour of these communities in individuals is important in assessing the utility of RMC as a tool for prediction. Deer are unlike other agricultural ruminant species as they are very seasonal animals, with many traits, such as weight, growth, antler growth and reproduction, fluctuating annually (Berlioz *et al.* 2017). For example, growth traits in winter often fluctuate for deer where feed is limited, and often deer stop growing or lose weight over winter (Thompson *et al.* 2023). This has been observed in farmed deer even with diet supplementation. Growth of deer is important for agricultural practices to meet venison demand in spring where increased growth rates are usually associated with increased carcass mass. The RMC in both deer and other ruminants are controlled by a host of factors such as sex, breed, season and host genome, and have been associated with physiological mechanisms, most importantly such as immune response, disease resistance and feed efficiency (Hess *et al.* 2022). Due to associations to diet and feed efficiency, it is likely that the deer RMC is associated with other growth and physiological traits. The effect of the RMC on important traits is largely undocumented in deer and may lead to increased understanding of deer and aid in selection processes. The amount of variation of important traits attributed to the RMC is vital to understanding the role microbes play in predicting key phenotypes. This study aims to analyse the amount of variation for key growth traits attributed to the deer host genome and the deer RMC through different methods.

MATERIALS AND METHODS

Experimental animals and protocols applied in this study were approved by the AgResearch Ruakura (Hamilton, NZ) Animal Ethics committee (approvals 14906, 14907 and 14981).

Animals and phenotypes. This study consisted of 630 Red Deer (*Cervus elaphus*) and Wapiti cross deer (*Cervus canadensis*) from four different cohorts (i.e. birth years) across New Zealand (Table 1). SNP data was generated for individuals using genotyping by sequencing (GBS) and sex, breed and year of birth (YOB) were also recorded. Traits of interest were growth rate (GR), including autumn GR, winter GR and spring GR, hot carcass weight (HCW), preslaughter weight (PRSLW) and log of carbohydrate larval antigen (Ln CARLA). Growth rates were computed as the difference in weight (kilograms) over a specified period. GR was taken from January (post weaning weight at approximate 2 months of age) to October (preslaughter at approximately 11 months of age). Autumn GR (February to May) measured growth from approximately 3 to 6 months of age, Winter GR (May to August) from 6 to 9 months of age, and Spring GR (August to October) from 9 to 11 months of age. Live weights were recorded at approximately the same time each year (within a few days).

Table 1. Distribution of individuals across breed, sex and collection year cohort

Breed	Sex	Collection Year (Birth Year)				Total
		2019 (2018)	2021 (2019)	2022 (2021)	2023 (2022)	
Red Deer	male	113	10	144	101	368
	female	-	9	1	-	10
Wapiti cross	male	-	-	72	60	132
	female	-	-	60	60	120
Total		113	19	277	221	630

Rumen microbial counts and profiles. Rumen samples were collected from the animals at the time of slaughter and were preserved using the TNx2 solution (Budel *et al.* 2022). Microbial DNA was sequenced using restriction enzyme reduced representation sequencing (Hess *et al.* 2020). Sequences were processed through the reference-based pipeline described in Hess *et al.* (2020) using the genome taxonomy database (GTDB) (Parks *et al.* 2021) to classify microbial sequences at the genera level. The microbial count matrix was created using relative abundance of genera per sequence and filtered so that only genera that had a relative abundance greater than 0.01% within at least one cohort were retained. The microbial counts were converted to a relative abundance matrix (i.e., counts were converted into proportions per sample), log10 transformed and scaled per genera within each cohort and breed, then converted into a microbial relationship matrix (MRM) as described in Hess *et al.* (2020).

Animal genotyping. To investigate heritability of growth and disease resistance traits a genomic relationship matrix (GRM) was created. This GRM was computed in KGD (Dodds *et al.* 2015) using VanRaden method 1 with non-missing SNPs for each matrix entry and assuming missing data is at random. SNPs with a call rate of 70% were retained.

Statistical models. The following three models were fitted to each of the traits:

1) $y_{ijkl} \sim YOB_j + Sex_k + Breed_l + m_i + e_{ijkl}$
2) $y_{ijkl} \sim YOB_j + Sex_k + Breed_l + g_i + e_{ijkl}$
3) $y_{ijkl} \sim YOB_j + Sex_k + Breed_l + m_i + g_i + e_{ijkl}$

where y_{ijkl} is the trait, i indexes the individual, YOB, Sex, Breed are fixed effects, $m_i \sim N(0, \sigma_m^2\mathbf{M})$ where **M** is the MRM, $g_i \sim N(0, \sigma_g^2\mathbf{G})$ where **G** is the GRM, e_{ijkl} is the residual error term such that $e_{ijkl} \sim N(0, \sigma_e^2\mathbf{I})$and **I** is the identity matrix. The microbial, genetic and residual variances are σ_m^2, σ_g^2, and σ_e^2 respectively, and the microbiability and heritability are $\sigma_m^2/\ \sigma_t^2$ and $\sigma_g^2/\ \sigma_t^2$ where σ_t^2

is the total variance. All models were fitted in ASReml v4.2 (Gilmour *et al.* 2015). Microbiability is the amount of variation of the trait explained by the RMC.

RESULTS AND DISSCUSSION

There were very minor differences between the heritability and microbiability estimates in the joint model (model 3) compared to the single models (models 1 and 2) (Table 2). The RMC did not explain any of the variation across Winter GR as the microbiability estimate was negligible across models 1 and 3. This indicates that over winter when deer don't grow or consume as much food the RMC shuts down and therefore cannot be used to indicate growth. Genomic estimation of growth can still occur over this period, given the high heritability estimates across all models.

Interestingly, microbiability of Autumn GR was much higher than that of Winter GR. As the RMC is heavily influenced by diet (Rowe *et al.* 2019), one would expect the growth rates closest to the time of RMC collection would have a higher microbiability. Alternatively, the microbiability of Autumn GR may be higher than winter GR estimates due too little to no feed availability in winter compared to other seasons. Therefore, Winter GR has a lower microbiability estimate when compared to other growth measures as the RMC are not acting at all in this time, therefore this period is dissimilar from all other metrics.

The microbiability estimates for the seasonal and overall growth rates are largely centred around 0.10, where estimates ranged from 0.07 ± 0.041 to 0.14 ± 0.046 in model 1 and 0.09 ± 0.041 to 0.13 ± 0.044 in model 3 across the traits GR, Spring GR, Autumn GR and PRSLW. This indicates that although many of these time points are far removed from the sample date, the RMC still has an effect or association with growth in New Zealand deer.

Table 2. Microbiability and heritability estimates of growth and seasonality traits across the three models

Trait	MRM (model 1)		GRM (model 2)		MRM + GRM (model 3)			
	m^2	SE	h^2	SE	m^2	SE	h^2	SE
GR	0.10*	0.039	0.68*	0.079	0.11*	0.037	0.65*	0.078
AutumnGR	0.07*	0.041	0.38*	0.089	0.09*	0.041	0.39*	0.087
WinterGR	0.00	0.000	0.66*	0.081	0.04	0.036	0.66*	0.081
SpringGR	0.14*	0.046	0.42*	0.085	0.13*	0.044	0.39*	0.082
HCW	0.02	0.029	0.59*	0.089	0.05	0.031	0.73*	0.078
PRSLW	0.13*	0.043	0.65*	0.083	0.12*	0.038	0.60*	0.083
LnCARLA	0.00	0.026	0.52*	0.083	0.00	0.000	0.52*	0.083

*Estimates that are significant to 5%
m^2 = microbiability, h^2 = heritability, SE = standard error

LnCARLA and HCW also had a very small and not significant microbiabilities but high heritability estimates for the single and joint models. This is interesting as the traits with the lowest microbiability estimates (Winter growth rate and LnCARLA) had very high heritability estimates, whereas traits with lower heritability estimates (Spring GR and Autumn GR) had higher microbiability estimates. This indicates that microbes and the host genome may be explaining different aspects of the variation across the traits. The heritability estimates were significantly higher than the microbiability estimates, demonstrating that while the RMC explains a significant amount of variation across most traits, it does not compare to using genomic data.

LnCALRA had a very low microbiability estimate in both models (0.00 ± 0.026 and 0.00 ± 0.00 for models 1 and 3 respectively), indicating that the deer RMC does not provide information on parasite resistance as measured here. This is not consistent with associations between the RMC and chronic wasting disease investigated in literature (Minich *et al.* 2021). Thus, implying that the RMC may protect individuals from disease through other mechanisms than LnCARLA.

In rumen related traits such as feed intake in sheep, it has been shown that adding RMC information improves genomic prediction (Rowe *et al.* 2019), however this study has seen that in deer, the RMC adds very little explained variation compared to genomic models for growth and disease resistance traits. Therefore, the RMC may not improve prediction as much as expected. Discretion should be used when using RMC in growing animals that express seasonal growth such as deer.

CONCLUSION

For growth traits the host genome is much more predictive than the RMC. The RMC explains a significant amount of the variation in the growth traits, but the RMC is impacted by seasonal growth and caution should be used when selecting a periodic phenotype. Furthermore, RMC samples may yield more accurate representation of growth traits when taken in tandem of growth measurements, rather than at slaughter.

ACKNOWLEDGEMENTS

Financial support was provided via the "Mapping the New Zealand Ruminotype landscape" and AgResearch's Strategic Science Investment Funding (SSIF) from the Ministry of Business, Innovation and Employment (MBIE). We thank Deer Industry New Zealand for support and additional funding of the overall PhD project.

REFERENCES

Berlioz E., Azorit C., Blondel C., Sierra Tellado Ruiz M. and Merceron G. (2017) *Hyst. Ita. J. Mamm.* **28**: 222.

Budel J.C.C, Hess M.K., Bilton T.P., Henry H., Dodds K.G., Janssen P.H., McEwan J.C. and Rowe S.J. (2022) *Anim. Microbiome* **4**: 39.

Dodds K.G., McEwan J.C., Brauning R., Anderson R.M., van Stijn T.C., Kristjánsson T. and Clarke, S.M. (2015) *BMC Genomics* **16**: 1047.

Gilmour A.R., Gogel B.J., Cullis B.R. and Thompson R. (2015) ASReml User Guide Release 4.2 structural specification. VSN International Ltd, Hemel Hempsted, UK.

Hess M., Zetouni L., Hess A., Budel J., Dodds K.G., Henry H., Brauning R., McCulloch A., Hickey S.M., Johnson P., Elmes S., Wing J., Bryson B., Knowler K., Hyndman D., Baird H., McRae K., Jonker A., Janssen P., . . . Rowe S.J. (2022) *Genet. Sel. Evol* **55**: 53.

Hess M.K., Rowe S.J., Van Stijn T.C., Henry H.M., Hickey S.M., Brauning R., McCulloch A.F., Hess A.S., Kirk M.R., Kumar S., Pinares-Patiño C., Kittelmann S., Wood G.R., Janssen P.H. and McEwan, J.C. (2020). *PLoS One* **15**: e0219882.

Minich D., Madden C., Evans M.V., Ballash G.A., Barr D.J., Poulsen K.P., Dennis P.M. and Hale V.L. (2021) *Sci. Rep.* **11**: 13218.

Parks, D. H., Chuvochina, M., Rinke, C., Mussig, A. J., Chaumeil, P.-A. and Hugenholtz, P. (2021) *Nucl Aci Res*, **50**: D785.

Rowe S.J., Hess M., Zetouni L., Hickey S., Brauning R., Henry H., Flay H., Budel J., Bryson B., Janssen P., Jonker A. and McEwan J. (2019) *Proc. Trop. Agri. Con.* **36**: 177.

Thompson B., Stevens D., Ward J., Dodds K. and Asher G. (2023) *Proc. N.Z. Soc. Anim. Prod.* **82**: 50.

CHARACTERISATION OF SHORT TANDEM REPEATS (STRs) IN CATTLE USING LONG READ SEQUENCING

Q.H. Tran[1,2], I.M. MacLeod[1,2], T.V. Nguyen[1], J. Wang[1]and A.J. Chamberlain[1,2]

[1] Agriculture Victoria, Centre for AgriBioscience, Bundoora, VIC, 3083 Australia
[2] School of Applied Systems Biology, La Trobe University, Bundoora, VIC, 3083 Australia

SUMMARY

Short tandem repeats (STRs), also known as microsatellites, are highly variable DNA mutations that were widely used as genetic markers in cattle populations. However, previous studies focused only a small set of known STRs, and our understanding of STRs at the whole-genome level and their potential functional impact remains limited. This study is the first to examine STRs across the entire genome in Holstein and Jersey breeds using long-read sequencing. We investigated the number of alleles in multiallelic STRs, their allele lengths, their annotation, and the diversity of STRs between populations. Our study characterised over 352,000 STRs and provided evidence suggesting that there may be strong negative selection pressure acting on STRs that cause frameshift mutations in protein coding regions. This study provides a database of STRs at the whole-genome level that serves as a valuable resource for both functional studies and population structure analysis in cattle.

INTRODUCTION

Short tandem repeats (STRs) also known as microsatellites, are DNA regions characterized by the repetition of 1 to 6 base pair (bp) motifs. STRs are highly variable with multiple allele lengths (i.e. repeat motifs). In humans, STRs have been identified as causal variants for both complex and monogenic traits (Tanudisastro *et al.* 2024). However, unlike SNPs, which have been well characterized at sequence level across many populations, there are limited studies on STRs in cattle at the whole-genome scale. This is likely due to the lack of standardised bioinformatic tools for genotyping STRs, as each tool has its own strengths and limitations. Additionally, available short-read sequence data can be unreliable for detecting STRs longer than the read lengths, because sequence reads cannot cover the entire repeat region (Tanudisastro *et al.* 2024). This study is the first to explore whole-genome STRs in a cattle population using long-read sequencing data. LongTR, a recently developed tool, was chosen for STR genotyping because it demonstrated high performance among STR genotyping tools for long-read data (Ziaei Jam *et al.* 2024). Polymorphic STRs across the entire genome were identified in 108 Holstein and Jersey cattle. This work provides an overall insight of STR diversity and their potential functional impact on the cattle genome.

MATERIALS AND METHODS

Fifty Holstein and 58 Jersey cattle were sequenced using an Oxford Nanopore Technology (ONT) PromethION sequencer with R9.4.1 or R10.4.1 flow cells, achieving an average read coverage of 25X. The FAST5 files were basecalled using either Guppy v6.1.7 or Dorado v0.7.0 in super high accuracy mode. The output sequence reads were filtered using FiltLong (https://github.com/rrwick/Filtlong) and then aligned to the ARS-UCD 2.0 reference genome (Rosen *et al.* 2020) using Minimap2 (Li 2018). Prior to STR genotyping, a reference list of STRs was created from the reference genome using Tandem Repeat Finder (Benson 1999). Only autosomal STR loci with repeat motifs ranging from 2 to 6 bp were selected for the predefined list. This list, along with the 108 aligned sequence files, were input to LongTR with the parameter settings --min-reads 4 (minimum total reads required to genotype a locus) and --max-tr-len 10000 (maximum STR length of the reference allele). The VCF output from LongTR was then further filtered, removing STR loci with a total number of informative reads (DP in INFO fields) that were too low (< 1500) or too high

(> 3000), where the thresholds were chosen based on the observed genome-wide distributions. Additionally, individual genotypes with a quality score (Q) < 0.9, number of valid reads per animal (DP in Format fields) < 8, or an alternate pattern marked as <DEL> (deletion) were set to missing. Finally, only polymorphic loci with a missing genotype rate of less than 20% were retained. To assess the genotyping accuracy, Mendelian consistency was calculated for 2 parent–offspring trios in our population. The filtered STRs were annotated using Variant Effect Predictor (VEP) (McLaren *et al.* 2016) version 113 with the parameters --symbol and --biotype. To evaluate the population structure based on genome-wide STRs, the average pairwise genetic differentiation (weighted Fst) between the two breeds, Holsteins and Jerseys, was calculated for all STR within a window size of 2,000,000 bp and a sliding window of 10,000 bp using VCFtools (Danecek *et al.* 2011).

RESULTS AND DISCUSSION

STR genotyping. A reference list of 836,981 autosomal STR loci from the reference genome was created. After genotyping these STR in 108 animals using LongTR, 352,554 polymorphic STRs were retained after quality filtering. The average Mendelian consistency of the STR genotypes in two parent-offspring trios, was 74% for the raw unfiltered STR and increased to 80% after filtering. The number of STRs per chromosome generally corresponded with the chromosome length, except for chromosomes 19 and 25, which had fewer than expected STRs (Figure 1a). Di- and pentanucleotide repeats were the most common STRs, representing 39.2% and 21.4%, respectively (Figure 1b). Most STRs (84.4%) had between 5 to 15 alleles, with dinucleotide repeats showing the highest variability (Figure 1c). This observation is consistent with other studies in humans and cattle using short-read data and different bioinformatic tools, which may suggest that dinucleotide repeats have a higher mutation rate (Bhati *et al.* 2023; Shi *et al.* 2023).

We calculated the length of the major (i.e. most frequent) allele at each STR locus. The length of the major allele typically ranged from 20 to 40 bp, while dinucleotide STRs showed more variability in major allele length, which aligned with the higher number of alleles (Figure 1d). Overall, only 0.5% of major alleles were longer than 150 bp (the common length of reads in short-read sequencing), while the maximum lengths of major STRs for di-, tri-, tetra-, penta-, and hexanucleotide repeats were 2,175, 3,227, 1,994, 2,682, and 1,937 bp, respectively. Furthermore, 19,115 STR loci had at least one allele longer than 150 bp, accounting for 5.4% of all STR loci and these are likely to be better characterised using long read versus short read sequence. Dinucleotide STRs had the highest proportion of loci with at least one allele longer than 150 bp (10% of total dinucleotide STR loci). For other repeat motifs, the proportion ranged from 2.1% to 3.2%. The maximum allele lengths recorded for di-, tri-, tetra-, penta-, and hexanucleotide repeats were 16,753, 7,569, 8,788, 11,311, and 6,733 bp, respectively. The long repeat expansions with rare frequency are of interest, because they could be potential pathogenic variants, and they are difficult to detect in studies using short-read data.

STR annotation. The VEP annotation found that most STRs were in intergenic (48.9%) and intronic regions (42.8%). Only 0.3% of STRs (1,147 loci) were in coding regions. For non-coding regions, the proportion of STRs by repeat motif corresponded to the overall distribution of STRs in the genome (Figure 1e). However, in coding regions, trinucleotide STRs dominated, comprising 66% of all coding-region STRs, followed by hexanucleotide STRs at 17% (Figure 1e). Conversely, there were a considerably lower proportions of di-, tetra-, and pentanucleotide STRs in coding regions, suggesting that these may be under strong negative selection because they cause frameshifts in the DNA codons with a much higher risk of detrimental impact on protein function compared to tri- or hexanucleotide STRs.

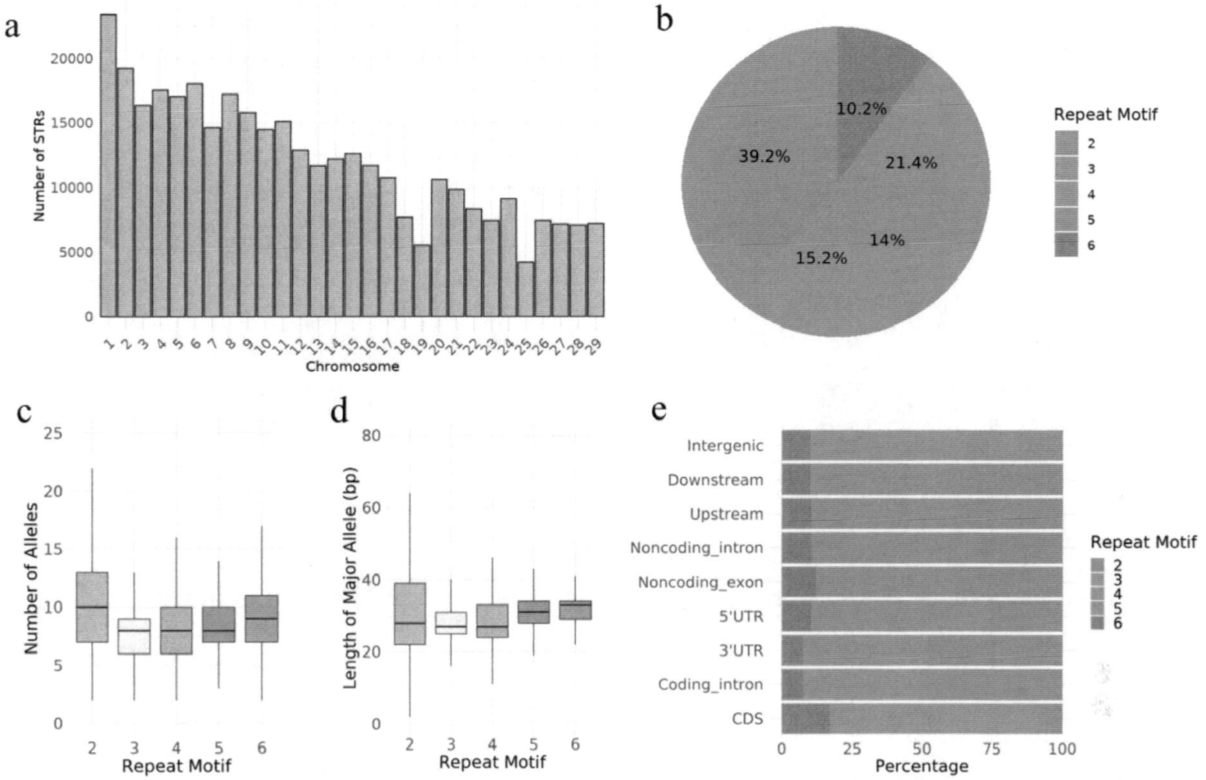

Figure 1. Polymorphic STR characteristics. a. Number of STR by chromosomes. b. Proportion of STR by repeat motif size. c Distribution of number of alleles per STR by repeat motif size. d Distribution of length of major allele by repeat motif size. e Percentage of STRs of repeat motif size within regions of different function

Population structure. In the past STRs have traditionally been used to investigate population differences in cattle. However, previous studies were limited to a small set of known STRs. In this study, we examined the use of genome-wide STRs to assess population structure. We calculated pairwise estimates of genomic differentiation (Fst) between the Holstein and Jersey populations using STRs across chromosome regions. We identified two regions with the most highly differentiated STRs (average Fst > 0.15) (Figure 2). The first region was from 40 to 43 Mb on chromosome 7, and the second was from 26 to 28.7 Mb on chromosome 19. For example, in Jerseys, the major allele (82% of all alleles) for the STR at position 41,775,200 on chromosome 7 had 7 repeats of the AGC motif, while the major allele (68% of alleles) for Holsteins at this STR had 4 repeats of AGC. Interestingly, the region from 40 to 43 Mb on chromosome 7 was reported as part of a run of homozygosity (ROH) island across United States, New Zealand and Australian Jersey populations based on SNP data indicating potential selection pressure across this region (Howard *et al.* 2015). While regions with high Fst do not necessarily implicate STR as causal variants, it is of interest to further investigate STR functional impact and associations with key dairy traits.

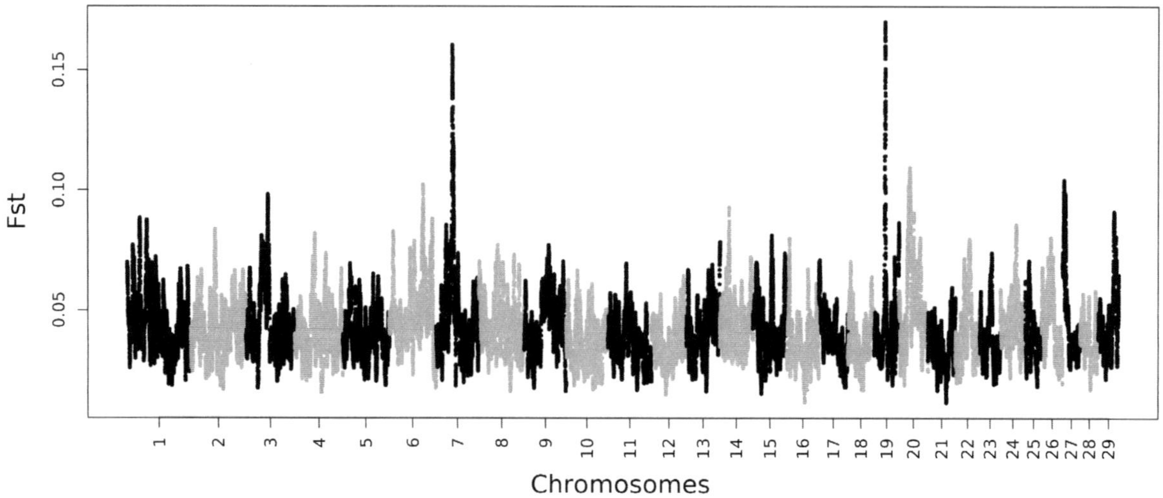

Figure 2. Genome-wide Manhattan plot of average Fst between Holstein and Jersey populations

CONCLUSIONS

This study is the first to examine the diversity of STRs in a cattle population at the whole-genome scale using long-read sequence data. While the study found that challenges still exist in accurately genotyping STRs from whole-genome sequence data, we demonstrated that large numbers of STRs were accurately genotyped and many were highly polymorphic. The considerably higher number of tri- and hexanucleotide STRs and lower number of di-, tetra-, and pentanucleotide STRs in coding regions reflected the potential functional impact of STRs and suggests selection pressure against frameshift causing mutations. The high variability of STRs also made them a powerful approach for identifying genomic regions that may be under different selective forces in the two different breeds.

ACKNOWLEDGEMENTS

The authors acknowledge financial support from the DairyBio program, which is jointly funded by Agriculture Victoria, Dairy Australia and The Gardiner Foundation.

REFERENCES

Benson G. (1999) *Nucleic Acids Res.* **27**: 573.

Bhati M., Mapel X.M., Lloret-Villas A. and Pausch H. (2023) *Genetics* **225**: iyad161.

Danecek P., Auton A., Abecasis G., *et al.* (2011) *Bioinformatics* **27**: 2156.

Howard J.T., Maltecca C., Haile-Mariam M., Hayes B.J. and Pryce J.E. (2015) *BMC Genomics* **16**: 187.

Li H. (2018) *Bioinformatics* **34**: 3094.

McLaren W., Gil L., Hunt S.E., Riat H.S., Ritchie G.R.S., Thormann A., Flicek P. and Cunningham F. (2016) *Genome Biol.* **17**: 122.

Rosen, B.D., Bickhart, D.M., Schnabel, R.D., *et al.* (2020) *GigaScience* **9**: 1.

Shi Y., Niu Y., Zhang P., *et al.* (2023) *Nat. Commun.* **14**: 2092.

Tanudisastro H.A., Deveson I.W., Dashnow H. and MacArthur D.G. (2024) *Nat. Rev. Genet.* **25**: 460.

CONSIDERATIONS IN IMPUTATION OF SKIM SEQUENCED DATA

M.S. Tahir[1], J. Wang[1], A.J. Chamberlain[1,2], C.M. Reich[1], B.A. Mason[1] and I.M. MacLeod[1,2]

[1] Agriculture Victoria, Centre for AgriBioscience, Bundoora, VIC, 3083 Australia
[2] School of Applied Systems Biology, La Trobe University, Bundoora, VIC, 3083 Australia

SUMMARY

Skim sequencing genotypes have become a cost-effective alternative to standard SNP array genotypes for genomic studies. However, compared to deep sequencing, skim sequencing suffers from randomly missing or inaccurate genotypes. The true genotypes can often be recovered with imputation using high-quality sequenced reference population genotypes. We skim-sequenced 293 cattle at 1x depth at an estimated cost of 36 AUD per sample. We aligned the skim sequence data of each animal to the bovine reference genome and imputed it to the whole genome sequence using GLIPMPSE2. Two imputed genotypes metrics (dosage: probability of called genotype and best-guess: 0, 1, or 2), called by GLIMPSE2, were compared to the real genotypes of the same animals with custom XT-50K SNP bead-chip to calculate per-SNP and per-animal imputation accuracy. The per-SNP and per-animal imputation accuracy with dosage genotypes were found to be 0.96 and 0.97, respectively. Sequencing depth of whole genome, imputation chunks, and 5kb windows around variants was determined for all animals and compared to the per-animal imputation accuracy. The correlation between the mean genome sequencing depth and imputation accuracy (per-animal) across all animals was weak (0.04). However, the difference in the mean sequencing depths of the whole genome, imputation chunks, and 5kb windows around variants between the animals with highest and lowest imputation accuracy was significant. The findings of this study show that skim sequence genotypes imputed with deeper sequenced reference population were highly accurate and cost-effective. Optimising the depth of skim sequencing can further improve the imputation accuracy.

INTRODUCTION

Skim sequencing, also known as low-coverage (< 3X) whole genome sequencing, has emerged as a cost-competitive alternative to standard SNP array genotyping (Li *et al.* 2021). It can capture the existing array-based SNPs as well as new variants in individual animals. A disadvantage of skim sequencing is missing information for some random genomic regions. This results in either sporadic missing genotypes or inaccurate genotypes. Missing information in skim sequence can be recovered with imputation using a high-quality deep-sequenced reference population (Pasaniuc *et al.* 2012).

Imputation methodology and landscape of sequencing depth can impact the final accuracy of the skim-sequence imputed genotypes (Al Bkhetan *et al.* 2019). For example, Beagle (Browning and Browning 2007) is highly accurate for imputation of variant genotypes to whole-genome sequence but showed relatively low imputation accuracy when used with skim-sequenced samples (Daetwyler *et al.* 2021) since it was not developed for such data. New imputation tools like GLIMPSE2 (Rubinacci *et al.* 2023) and QUILT (Davies *et al.* 2021) have been developed specifically to impute skim-sequenced data. Both tools are reported to show similar imputation accuracy, but GLIMPSE2 is computationally more efficient compared to QUILT (Rubinacci *et al.* 2023).

In livestock, it is important to maintain continuity between new and old genotyping platforms so that genetic evaluations relying on a particular variant panel are not negatively affected. In this study, GLIMPSE2 was used to impute skim-sequenced data of 293 cattle to the whole genome sequence (WGS) level. The empirical accuracy of imputation was tested for the variants that overlapped a custom XT-50K variant panel used for genetic evaluations in the Australian dairy industry (Xiang *et al.* 2021). We also tested the impact of skim-sequencing depth of target samples on imputation accuracy.

MATERIALS AND METHODS

A total of 293 (142 Holstein, 148 Jersey, and 3 crossbred) animals were skim-sequenced to an average 1x coverage costing 36 AUD per sample. Sequences were aligned to the bovine reference genome using BWA (Li *et al.* 2009). The aligned bam files were used as input for the imputation pipeline. Run8 of the 1,000 bull genomes project was used as the imputation reference with 4,118 animals of various *Bos. taurus* breeds. GLIMPSE2, with default parameters, was used to impute genotypes for 39,457,248 variants. It splits the reference genotypes into imputation chunks, phases the target samples using pre-phased information from the split chunks of the reference panel, imputes the missing genotypes, and then ligates the imputed chunks of the target samples. We also tested non-default values for parameters like imputation chunk size 18 Mb (default=4 Mb), and effective population sizes 1,000 and 10,000 (default=100,000).

GLIMPSE2 provides two metrics of imputed genotypes: 1) dosage, the estimated alternative allele dosage (a continuous metric of probability of genotype called), and 2) best-guess, the genotype called (0, 1, or 2) based on estimated allele dosage. The per variant and per animal empirical accuracy of imputation was calculated as a correlation between imputed genotypes (dosage and best-guess) and quality-filtered real genotypes of 32,614 variants (32,045 SNPs and 569 indels) genotyped for the same 293 animals using a custom XT-50K bead-chip. The 17 animals with the highest ($\geq$ 0.99) and 17 animals with the lowest (< 0.75) imputation accuracy (per animal) were selected as two groups to check the impact of sequencing depth on imputation accuracy.

GLIMPSE2 provided the average sequence depth of all imputed chromosome chunks for each animal. In addition, BedTools (Quinlan *et al.* 2010) coverage function was used to generate the variant window sequence depth (average sequence depth of 5kb window around each variant compared). These parameters were compared for animals with high and low imputation accuracy.

RESULTS AND DISCUSSION

The mean variant and animal-based imputation accuracies are summarized in Table 1. The per variant mean imputation accuracy of skim-sequenced data was similar to previously reported imputation accuracies from standard genotyping arrays (Nguyen *et al.* 2021; Nguyen *et al.* 2024). Imputation accuracy by GLIMPSE2 in this study was better when compared to the findings of Lamb *et al*, 2023, who also used GLIMPSE but on low-coverage Oxford Nanopore sequencing data.

Table 1. Mean empirical accuracy of imputation per variant and per animal for dosage or best guess genotypes, using different chunks and effective population sizes

	Mean Accuracy							
	Chunk: 4Mb Ne:1000		Chunk: 4Mb Ne: 10,000		Chunk: 4Mb Ne: 100,000		Chunk: 18 Mb Ne=1000	
Genotypes	Variant	Animal	Variant	Animal	Variant	Animal	Variant	Animal
Dosage	0.962	0.976	0.957	0.974	0.952	0.971	0.960	0.972
Best-Guess	0.955	0.974	0.953	0.971	0.950	0.970	0.957	0.970

Consistent per variant and per animal imputation accuracies were observed when using different imputation chunk sizes (4 Mb and 18 Mb), however, imputation with 4 Mb chunk size was computationally efficient. A slight but consistent increase in imputation accuracy was observed for decreasing effective population sizes (*Ne*) which is consistent with the *Ne* being much lower in cattle. Per chromosome imputation accuracy varied a little. Variants with low minor allele frequency

(MAF) had slightly higher mean imputation accuracy than those with higher MAF (Figure 1a and b). Mean imputation accuracy for Holstein and Jersey was 0.99 and 0.95, respectively. Imputation accuracy of 22 animals (Jersey) was less than 0.9. Lower accuracy of Jersey animals may be due to the smaller number of Jersey animals in the reference population compared to the Holstein. Out of 32,614, there were 154 variants with imputation accuracy less than 0.8 (randomly spread genome-wide).

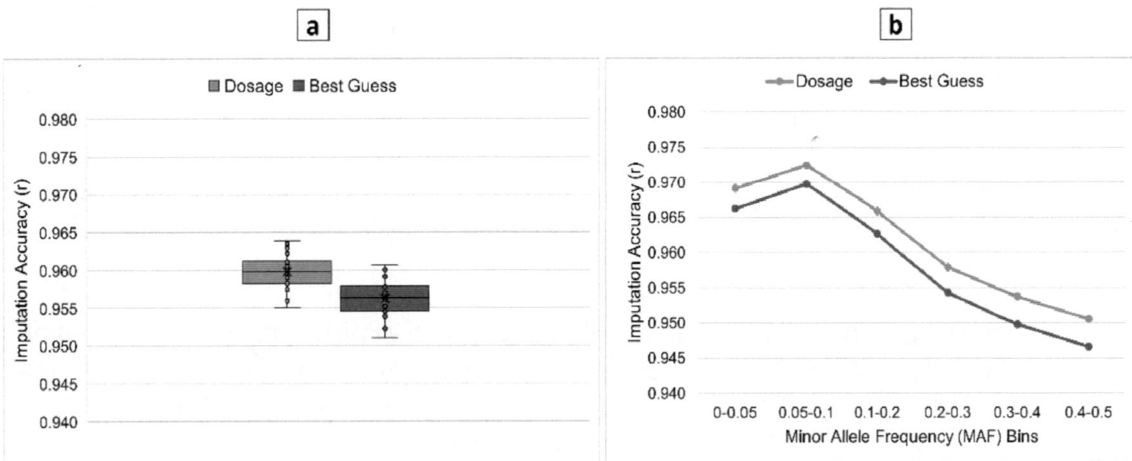

Figure 1. Mean accuracy of imputation across chromosomes and MAF bins. a: mean imputation accuracy of variants on each chromosome (best guess and dosage genotypes). **b:** mean imputation accuracy of variants grouped by MAF (best guess and dosage genotypes).

A weak correlation of 0.04 was found between the average skim sequence depth of animals and their imputation accuracy. The correlation between average sequence depths of imputation chunks (4 Mb) and imputation accuracies of animals was 0.1. The correlation between the average variant window (5 kb) sequence depth and imputation accuracy of animals was only 0.08. However, a distinct difference (p-value < 0.001) was observed in the average sequence depths of animals with the highest and lowest imputation accuracies (Figure 2a). Similarly, significant differences were observed when comparing the average imputation chunk sequencing depth (p-value $< 2.2\text{x}10^{-16}$) and the variant window sequence depth (p-value $< 2.2\text{x}10^{-16}$) between high and low imputation accuracy animals (Figure 2b and c).

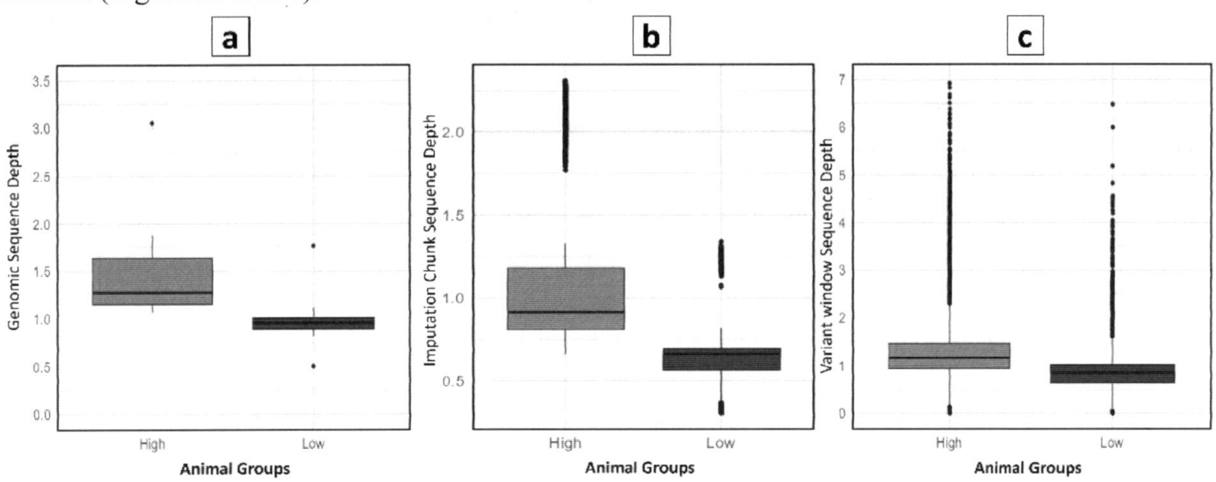

Figure 2. Comparison showing different measures of average sequence depth in animals with high or low imputation accuracy. a: genome sequence depth. **b:** sequence depth of imputation chunks. **c:** variant window sequence depth.

These differences between high and low imputation accuracy animals suggest that sequencing depth is associated with the accuracy of imputation as expected. However, this association could not be established at the population level (293 animals). It might be because of the interplay between the stochastic nature of the software algorithm and the landscape of the sequence information around each variant site within animals that impacts correct haplotyping and therefore imputation (Al Bkhetan *et al.* 2019 and Wragg *et al.* 2024). The higher average sequencing depth of high imputation accuracy animals (~1.5x) suggests that skim sequencing the samples with at least 1.5x depth can improve the overall imputation accuracy. The cost of skim sequencing assay with 1.5x depth is estimated to still be less than the cost of genotyping with a standard 50K SNP chip. Another alternative can be combining the skim sequence genotypes with "target-capture", a targeted sequencing of specific variants for higher coverage genotypes. This is of value for important SNPs such as known deleterious mutations that may be missed or have low accuracy in skim sequence genotypes.

CONCLUSION

The imputation accuracy of the skim-sequenced dataset was similar to that obtained by imputing standard SNP genotypes to WGS level. Our results suggest optimizing sequence depth versus costs. Combining skim sequencing with target capture may further improve the imputation accuracy.

REFERENCES

Al Bkhetan Z., Zobel J., Kowalczyk A., Verspoor K. and Goudey B., (2019) *BMC Bioinformatics* **20**: 1.

Browning S.R. and Browning B.L. (2007) *Am. J. Hum. Genet.* **81**: 1084.

Daetwyler H.D., Li J., Vander Jagt C.J., MacLeod I.M., Pickrell J., Vasquez M., Hoff J. and Chamberlain A.J. (2021) *Proc. Assoc. Advmt. Anim. Breed. Genet.* **24**: 78.

Davies R.W., Kucka M., Su D., Shi S., Flanagan M., Cunniff C.M., Chan Y.F. and Myers S. (2021) *Nat Genet.***53**: 1104.

Lamb H.J., Nguyen L.T., Copley J.P., Engle B.N., Hayes B.J. and Ross E.M. (2023) *BMC Biol.* **21**: 286.

Li H. and Durbin R. (2009) *Bioinformatics*. **25**: 1754.

Li J.H., Mazur C.A., Berisa T. and Pickrell J.K. (2021) *Genome Res.* **31**: 529.

Nguyen T.V., Bolormaa S., Reich C.M., Chamberlain A.J., Medley A., Schrooten C., Daetwyler H.D. and MacLeod I.M. (2021) *Proc. Assoc. Advmt. Anim. Breed. Genet.* **24**: 267.

Nguyen T.V., Bolormaa S., Reich C.M., Chamberlain A.J., Vander Jagt C.J., Daetwyler H.D. and MacLeod I.M. (2024) *Genet. Sel. Evol.* **56**: 72.

Pasaniuc B., Rohland N., McLaren P.J., Garimella K., Zaitlen N., Li H., Gupta N., Neale B.M., Daly M.J., Sklar P. and Sullivan P.F. (2012) *Nat. Genet.* **44**: 631.

Quinlan A.R. and Hall I.M. (2010) *Bioinformatics* **26**: 841.

Rubinacci S., Hofmeister R.J., Sousa da Mota B. and Delaneau O. (2023) *Na.t Genet.* **55**; 1088.

Wragg D., Zhang W., Peterson S., Yerramilli M., Mellanby R., Schoenebeck J.J. and Clements D.N. (2024) *Genet. Sel. Evol.* **56**: 6.

Xiang R., MacLeod I.M., Daetwyler H.D., de Jong G., O'Connor E., Schrooten C., Chamberlain A.J. and Goddard M.E. (2021) *Nat. Commun.* **12**: 860.

THE FEASIBILITY OF AN IMPUTATION REFERENCE POPULATION FOR STRUCTURAL VARIATION IN CATTLE

A.J. Chamberlain[1,2], T.V. Nguyen[1], J. Wang[1], X. Wang[3], C.J. Vander Jagt[1] and I.M. MacLeod[1,2]

[1] Agriculture Victoria, Centre for AgriBioscience, Bundoora, VIC, 3083 Australia
[2] School of Applied Systems Biology, La Trobe University, Bundoora, VIC, 3083 Australia
[3] Northwest A&F University, Yangling, China

SUMMARY

To understand the full impact of structural variation on traits important to cattle industries it would be desirable to generate a reference population for the imputation of large numbers of structural variants (SV) into existing populations with detailed phenotypic records for the traits of interest. This pilot study investigates two characteristics of SV that could impact their imputation; the precision of SV calling and their linkage disequilibrium (LD) with known single nucleotide polymorphisms (SNP) in the genome. Results indicate there are a number of SV that are called with low to zero standard deviation in length and starting position. Also, the pattern of LD between SV and SNP was similar to that between SNP and other SNP. Combined these results indicate that it is feasible to impute many SV into large existing populations with SNP genotypes which will enable exploration of their impact on traits important to cattle industries.

INTRODUCTION

Structural variants (SV) can be large insertions or deletions (INDEL, >50 base pairs), inversions, translocations, copy number variations or segmental duplications. Human studies estimate that SV together occupy a proportion of the genome that is equal to or greater than that of single nucleotide polymorphisms (SNP) and small INDEL (Feuk *et al.* 2006; Ho *et al.* 2020) and cause greater diversity at the nucleotide level than any other form of genetic variation (Chaisson *et al.* 2019). Multiple studies in cattle have demonstrated that SV impact classic mendelian traits, quantitative traits and gene expression (Kadri *et al.* 2014; Rothammer *et al.* 2014; Lee *et al.* 2021).

The full impact of SV on traits important to cattle industries has not yet been explored. It is therefore desirable to generate a large reference population that would enable imputation of SV into large populations with phenotypic records for the traits of interest, as has been done extensively with SNP and small INDEL from the 1000 Bull Genomes project (Hayes and Daetwyler 2019). To this end, the Bovine Long Read Consortium (BovLRC, Nguyen *et al.* 2023a) was formed to characterise structural variation at population scale as well as generate an imputation reference panel for SV. For this we developed a process (Figure 1) that was computationally efficient and scalable. It requires partners to pre-process sequences to smaller summary files using a custom built nextflow pipeline (Nguyen *et al.* 2023b), which standardises data processing and avoids sharing of large sequence or alignment files. Following joint calling of SV as well as SNP and then imputation of sporadic missing genotypes the reference database would be distributed to all partners for imputation into their own populations.

We previously assessed the mendelian consistency of SV called in cattle with the same pipeline used in this study (Nguyen *et al.* 2023b). The accuracy of SV imputation would depend on the accuracy of SV characterisation and genotype calling in the reference population. Genotype calling accuracy of SV is influenced by multiple factors; the aligner, genotyping method, depth of coverage, sequence data type, SV size and precision of breakpoints among others (Duan *et al.* 2022). Determining SV breakpoints is particularly challenging, and SV detection methods often generate SVs with imprecise breakpoints (Kosugi *et al.* 2019). Duan *et al.* (2022) showed that modifying SV breakpoints resulted in decreased F1 scores for five SV detection software, where F1 is the harmonic mean of precision and recall. They also found F1 scores varied with SV length. The degree of linkage disequilibrium (LD) between the SV and SNP in the reference population also impacts the ability to impute SV. Therefore, this pilot study investigates 1) the precision of SV breakpoints and 2) LD between SV and known SNP in two breeds to determine the feasibility of an imputation reference population for SV.

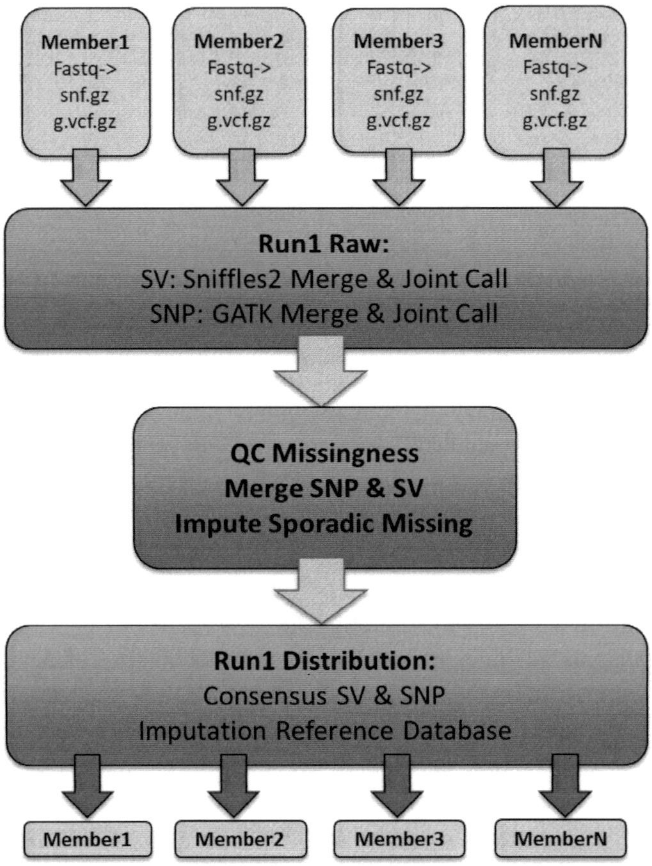

Figure 1. A process for the characterisation of structural variation in the bovine genome at population scale and the subsequent generation of an imputation reference panel for SV. Note fastq -> snf/g.vcf.gz is undertaken with nextflow pipeline nf-EXPLOR which uses Sniffles2 to call SV and Clare3 to call SNP within individuals. Green objects represent steps undertaken by participating organisations and blue those undertaken by Agriculture Victoria.

MATERIALS AND METHODS

DNA sequencing, read processing and genotype calling. High molecular weight DNA was extracted from semen, liver tissue or whole blood using Gentra Puregene kit (Qiagen). Sequencing libraries were prepared using ligation sequencing kit v9 or v10 (Oxford Nanopore Technology, ONT) according to manufacturer's instructions and sequenced on R9.4.1 or R10.4.1 flowcells on a PromethION (ONT). Super high accuracy base calling was undertaken with Guppy v6.1.7 or Dorado v0.7.0. and reads with q-score greater than 7 retained for analysis. All samples were processed as outlined in Figure 1. Briefly nf-EXPLOR (https://github.com/tuannguyen8390/nf-EXPLOR) was used to trim poor quality bases from reads using FiltLong (https://github.com/rrwick/Filtlong), map filtered reads to the bovine genome ARS-UCD2.0 (Rosen *et al.* 2020) using Minimap2 (Li 2018) and detect SV with Sniffles2 v2.6.1 (Sedlazeck *et al.* 2018) for each individual. Sniffles2 was used to subsequently merge SV from all animals and re-genotype. SV larger than 3Mb were removed. Genotypes with quality score less than 5 were set to missing, then SV with greater than 20% missing genotypes were removed.

SV breakpoint analysis. Long read sequences from 108 animals, 50 Holstein and 58 Jersey, were generated and processed as outlined above. From the variant call format (VCF) file the standard deviation (SD) of the SV length and start position was extracted for 30,668 deletions and 37,486 insertions, which were then plotted in frequency histograms with bins of 20bp.

Linkage disequilibrium analysis. Long read sequences for 41 animals, 19 Holstein and 22 Jersey, were used to discover SV (Chamberlain *et al.* 2023) as outlined above. The insertions and deletions were then called in 93 Holstein and 105 Jersey with short read sequences, that had greater than 10x coverage, using Paragraph v2.0 (Chen *et al.* 2019). Short read sequences were publicly available and taken from Run 8 of the 1000 Bull Genomes Project. SV genotypes were then merged with SNP genotypes called from the same short read sequences. LD was calculated as r^2 between all pairs of SV (2,513) and SNP (972,276) on chromosome 1 with minor allele frequency > 0.01 within windows of 500kb using PLINK v1.9 (Chang *et al.* 2015). Similarly, LD was calculated between pairs of SNP however, only a random 10% of SNP were used for computational efficiency. LD decay was then plotted as the mean r^2 within bins of 1000bp.

RESULTS AND DISCUSSION

The accuracy of SV imputation would depend on the accuracy of SV characterisation and genotype calling in the reference population. Key characteristics of SV are their length and position in the genome, both of which are challenging to estimate because of the need to call SV within individuals and then merge and genotype across the population. For this reason, many software, including Sniffles2, estimate the standard deviation of length and position in merged datasets, where low standard deviation reflects consistent calling across the population. Figure 2 A-D show frequency histograms of the SD of length and starting position of 30,668 deletions and 37,486 insertions discovered in 108 animals Though there were SV with large SD the majority had very low to zero SD for both length and position. Based on the results of Duan *et al.* (2022) such SV are likely to be genotyped with high precision and recall, something further studies will confirm. This indicates that there are a significant number of SV that Sniffles v2.6.1, the variant detection software used, was able to call consistently, and likely accurately, within this population. It is these SV that will be good candidates to impute into larger populations.

LD between the SV and SNP in the reference population is required for an accurate imputation. It is common practice for large target populations, with detailed phenotypic records for the traits of interest, to be genotyped with low density (~50K) SNP panels which are subsequently imputed to full sequence using reference populations like the 1000 bull genomes dataset. Therefore, we assessed the LD between SV and SNP genotypes called from whole genome short read sequence data from Run 8 of the 1000 bull genomes project. Figure 2E shows the decay in LD with distance between SV and SNP was similar to that between SNP with SNP, indicating that SV could be imputed with similar accuracy to SNP, provided that genotyping of SV is highly accurate.

CONCLUSION

Results from this pilot study in two breeds indicate that a significant proportion of SV could be imputed into large target populations with SNP genotypes and phenotypic records for traits of interest. Further work is required to test the accuracy of imputation. However, imputation of SV will enable a more detailed understanding of the impact of SV on the cattle genome and phenotypes of interest to cattle industries.

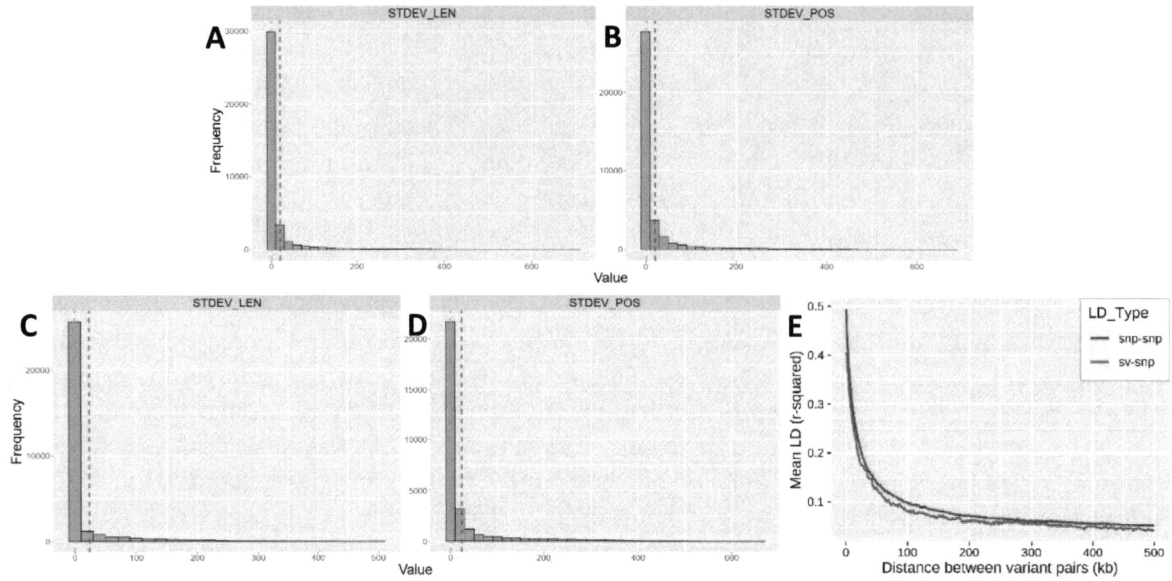

Figure 2. Frequency histograms of the standard deviation of length and position of insertions (A and B respectively) and deletions (C and D respectively) detected from long read sequence of 108 animals from 2 breeds. Red and green lines indicate the mean and median respectively. E) Mean linkage disequilibrium (LD as r^2) between variant pairs, either SV (insertions and deletions only) with SNP (blue) or SNP with other SNP (red) within 500kb on chromosome 1.

ACKNOWLEDGEMENTS

The authors thank DairyBio, a joint venture project between Agriculture Victoria, Dairy Australia and The Gardiner Foundation, for funding.

REFERENCES

Chaisson M.J.P., Sanders A.D., Zhao X., *et al.* (2019) *Nat. Commun.* **10**: 1784.

Chamberlain A.J., Nguyen T., Wang J., *et al.* (2023) *Proc. Assoc. Advmt. Anim. Breed. Genet.* **25**: 274.

Chen S., Krusche P., Dolzhenko E., *et al.* (2019) *Genome Biol.* **20**: 291.

Chang C.C., Chow C.C., Tellier L.C.A.M., *et al.* (2015) *GigaScience* **4**: s13742–015–0047–8.

Duan X., Pan M. and Fan S. (2022) *BMC Genomics* **23**: 324.

Feuk L., Carson A.R. and Scherer S.W. (2006) *Nat. Rev. Genet.* **7**: 85.

Hayes B.J. and Daetwyler H.D. (2019) *Annu. Rev. Anim. Biosci.* **7**: 89.

Ho S.S., Urban A.E. and Mills, R.E. (2020) *Nat. Rev. Genet.* **21**:171.

Kadri N.K., Sahana G., Charlier C., *et al.* (2014) *PLoS Genet.* **10**: e1004049.

Kosugi S., Momozawa Y., Liu X., *et al.* (2019) *Genome Biol.* **20**: 117

Lee Y.-L., Takeda H., Costa Monteiro Moreira G., *et al.* (2021) *PLoS Genet.* **17**: e1009331.

Li H. (2018) *Bioinformatics* **34**: 3094.

Nguyen T.V., Vander Jagt C.J., Wang J., *et al.* (2023a) *Genet. Sel. Evol.* **55**: 9.

Nguyen T.V., Wang J., Chamberlain A.J., *et al.* (2023b) *Proc. Assoc. Advmt. Anim. Breed. Genet.* **25**: 202.

Rosen B.D., Bickhart D.M., Schnabel R.D., *et al.* (2020) *GigaScience* **9**: 1.

Rothammer S., Capitan A., Mullaart E., *et al.* (2014) *Genet. Sel. Evol.* **46**: 44.

Sedlazeck F.J., Rescheneder P., Smolka M., *et al.* (2018) *Nat Methods* **15**: 461.

Proc. Assoc. Advmt. Anim. Breed. Genet. 26: 387-390

INSIGHTS FROM POPULATION-SCALE LONG-READ SEQUENCING: STRUCTURAL VARIANT CHARACTERISATION AND ANNOTATION

T. Nguyen[1], J. Wang[1], A. Chamberlain[1,2] and I. Macleod[1,2]

[1] Agriculture Victoria, Centre for AgriBioscience, Bundoora, VIC, 3083 Australia
[2] School of Applied Systems Biology, La Trobe University, Bundoora, VIC, 3083 Australia

SUMMARY

This study presents an analysis of structural variants (SVs) in 108 dairy cattle (50 Holstein and 58 Jersey) using high quality long-read Oxford Nanopore Technologies sequence data. Analysis of SV size distribution across allele frequencies revealed distinct evolutionary patterns, with larger SVs predominantly occurring at lower frequencies. A larger proportion of the longer insertions show low allele frequency compared to deletions, suggesting possible differential selective pressures and potentially because long insertions are harder to identify than deletions. The study validated large breed-specific known SV including HH0, HH5, and POLL, demonstrating the reliability of our approach. In investigating the relationship between the severity of SV impact, SV allele frequency and length, we found that the larger the variant the more likely the mutation (alternate allele) had a predicted high impact. Furthermore, among deletions, the longer and less common variants showed a larger proportion of predicted high impact effects compared to more common SV. This population-scale long-read reference dataset provides valuable new insights into the genomic architecture of dairy cattle and establishes a foundation for understanding the functional impact of SVs in cattle breeding.

INTRODUCTION

Although there have been limited studies of structural variants (SVs) in livestock, it is likely that they play a crucial role in the genetic diversity and adaptability of livestock (Nguyen *et al.* 2023). SV are defined as variants longer than 50 bp of DNA sequence: this includes deletions (DEL), insertions (INS), duplications (DUP), inversions (INV), and translocations. Several recent cattle studies have identified SV that cause embryonic lethality (Schütz *et al.* 2016), influence physical traits (Rothammer *et al.* 2014), and disease resistance (Lee *et al.* 2021). Therefore, gaining a better understanding of SVs at population scale is essential for breeding programs aimed at improving livestock because it is likely that on average SV will have a higher impact on gene function and regulation compared to single nucleotide polymorphisms (SNPs) (Lee *et al.* 2023). Long-read sequencing technologies have emerged as a powerful tool for population-scale studies of SVs compared to traditional short-read sequencing that struggles to accurately resolve complex genomic regions due to the limited read length. Long-read sequencing provides longer contiguous sequences of DNA, which enhances the ability to identify and characterize SVs with greater precision (Amarasinghe *et al.* 2020). A long-read reference population is crucial to study the SV diversity within and between cattle breeds, and to enable research to identify SVs that may impact economically important traits.

MATERIALS AND METHODS

DNA sequencing. 50 Holstein and 58 Jersey animals were selected (total = 108), avoiding full and half sib relationships to maximise diversity. High molecular weight DNA was extracted from semen, liver tissue or whole blood using Gentra Puregene DNA extraction kit (Qiagen). Sequencing libraries were prepared using ligation sequencing kit V9 or V10 or 14 (Oxford Nanopore Technologies) according to manufacturer's instructions and sequenced on R9.4.1 or R10.4.1

flowcells on PromethION P24 (Oxford Nanopore Technologies). Super high accuracy basecalling was performed with either Guppy v6.1.7 or Dorado v0.7.0.

Data analysis. Reads were quality trimmed using FiltLong (https://github.com/rrwick/Filtlong accessed December 2022) with default settings. Filtered reads were then mapped to ARS-UCD2.0 reference genome (Rosen *et al.* 2020) using Minimap2 (Li 2018). For discovery and genotyping of SV, Sniffles2 v.2.5.3 (Sedlazeck *et al.* 2018) was used to detect SVs for each sample and subsequently to merge and re-genotype SVs from all individuals. We only considered SVs on chromosome 1-29 of less than 3 Mb length, and a minor allele count > 0. Genotypes with quality score < 5 were set to missing and variants with > 20% missing genotypes removed. Variant Effect Predictor (VEP) software (McLaren *et al,* 2016) was deployed to annotate the SVs using Ensembl cattle annotation release 112 (exported from ARS-UCD1.3), having identical autosomal coordinates to ARS-UCD2.0).

RESULTS AND DISCUSSION

Across all Holstein animals, the mean and median raw sequence read coverage was 25.2X and 24.9X, respectively, with read length N50 values of 26,184bp (mean) and 23,666bp (median). The Jersey samples yielded comparable sequencing metrics, with mean and median coverage of 22.7X and 24.6X, respectively, and read N50 values of 22,366bp (mean) and 20,830bp (median). These sequencing metrics indicate high-quality data suitable for downstream genomic analyses.

Four distinct types of SVs were identified across the joint genotyping of all 108 sequences in Figure 1A. INS were the most abundant of the total SVs (55.5%), followed closely by DEL (45.3%). INV were much less frequent while DUP were the least common. This relative abundance of SV types is in keeping with population scale studies in humans (Collins *et al.* 2020). Here we focus mainly on results for INS and DEL. The length distribution of INS and DEL variants (Figure 1B) revealed that both were most common in the shorter length ranges (<10kb), while SV > 20kb were much less common. Additionally, a peak for both DEL and INS was observed between 100 – 200 bp, and a strong INS peak between 7,000 to 9,000 bp. These peaks likely represent different types of transposable elements and have been reported in cattle, human and mouse studies (Collins *et al.* 2020; Ferraj *et al.* 2023).

Using this dataset, we confirmed the presence of several known SVs, including two Holstein specific embryonic lethal DEL, Brachyspina/HH0 (Charlier *et al.* 2012) and HH5 (Schütz *et al.* 2016), as well as a known DUP which causes the so called "Friesian" polled (POLL) phenotype (Rothammer *et al.* 2014). The aligned sequence reads across the HH0 deletion region are visualized in Figure 1C, for one animal genotyped as 0/1 (carrier) and one free of the deletion (0/0). No other Holstein or Jersey samples were found to carry the deletion. In addition to HH0, two Holstein carriers of the 138 kb HH5 DEL were identified. As seen in Figure 1D, the HH5 DEL is more difficult to directly visualise because it is much longer than most of the sequenced reads. Finally, in Figure 1E, we confirmed a known POLL carrier (Jersey bull).

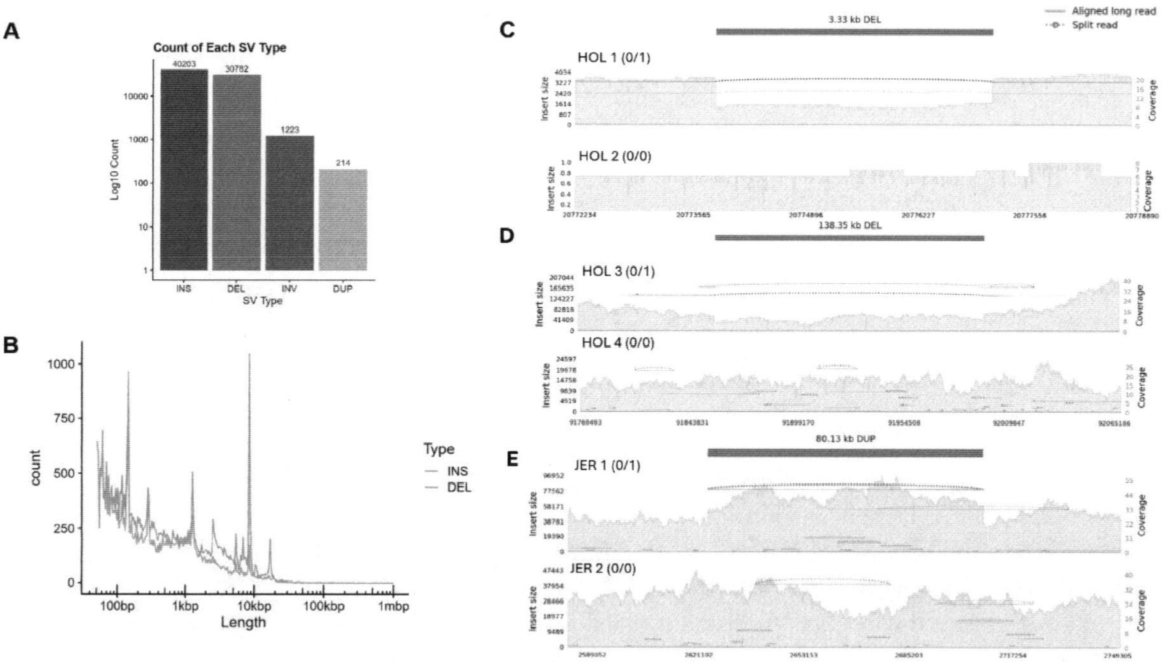

Figure 1. Structural variant discovery in 108 cattle: (A) distribution of types (INS: insertions; DEL: deletions; INV: inversions; DUP: duplications) and (B) length distribution of INS and DEL. Samplot coverage profiles visualising aligned reads for individual animals across the breakpoints for known recessive lethal DEL HH0 (C), HH5 (D) and POLL DUP (E) (carrier genotypes labelled 0/1 and homozygous reference as 0/0). The Y-axes show estimated length of split reads "insert size" (left) and the read coverage (right). The x axis is the position along the chromosome.

Using all discovered DEL and INS, the relationship between SV size, alternate allele frequency (AF) distribution, and VEP predicted "HIGH" functional impact of SV on genes was investigated to gain insights into possible selection pressures (Figure 2). Within each length category in Fig 2, the SV with low AF (<0.15) were more common than higher allele frequencies (AF>0.15) and overall ~40% of DEL and INS fell in the <1 kb with AF < 0.15 category. Additionally, as SV length increased, the proportion of low frequency variants of each length category increased e.g. 57% of all INS < 1kb length were low AF, while 97% of all INS > 10kb length were low AF, and similarly for DEL the respective proportions were 52% and 75%. This potentially indicates that there has been negative selection pressure operating on longer SV because they are more likely to be deleterious while the smaller variants may be better tolerated and less likely to negatively affect fitness of the animal. Among DEL, there was further support for this hypothesis because we observed strong enrichment for "HIGH" impact variants in the longer and low AF categories, particularly for SV > 10kb with AF < 0.15 (Fig 2: red colouring). Although this pattern was not observed for INS, it is important to note that several technical and biological factors could explain this: (i) VEP software may have inherent bias in predicting functional impacts of different SV types (e.g. predicting the result of a gene deletion is easier than predicting the impact of inserting a new piece of DNA into a gene), (ii) technical challenges in precise alignment of insertion events due to low coverage and/or bias of the reference genome, and (iii) the sequence content of insertions may require more complex interpretation frameworks for specific types of insertion, such as retrotransposons).

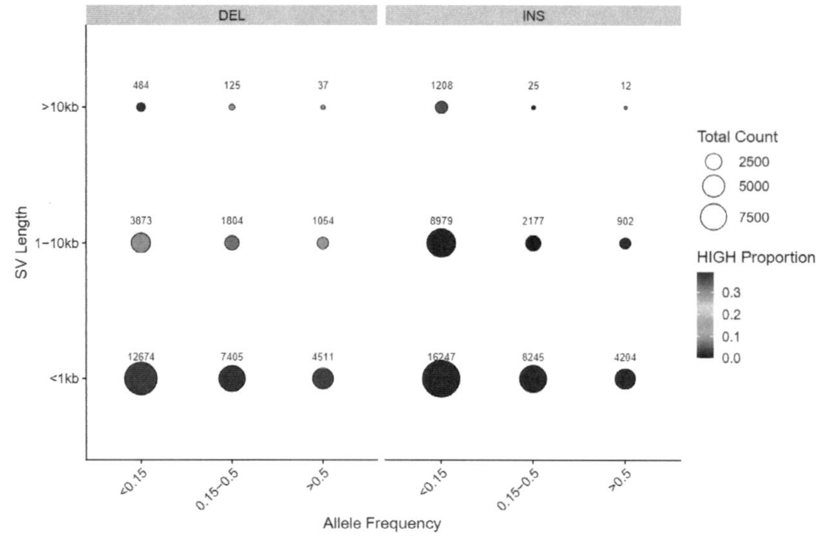

Figure 2. Proportion of Variant Effect Predictor HIGH impact deletions (DEL) and insertions (INS) by variant length and alternate allele frequency categories. Circle size indicates total variant count in each category and the number above each circle is the exact total. Colour gradient indicates the proportion of variants (0.0 to 0.4) in each category that were predicted HIGH functional impact. The Y-axis shows SV length in three bins and the X-axis displays three allele frequency bins.

CONCLUSION

This population-scale analysis of dairy cattle SVs using long-read sequencing revealed several key findings: 1) SV length distributions differed between INS and DEL , 2) a high proportion of long DEL predicted to have a HIGH functional impact were at low AF , 3) evolutionary constraints may be operating on larger SVs as evidenced by their low allele frequencies, and 4) known breed-specific variants were successfully detected. This reference dataset establishes a foundation for investigating the role of structural variation in livestock genetics and breeding applications.

ACKNOWLEDGEMENTS

The authors thank DairyBio, a joint venture project between Agriculture Victoria, Dairy Australia and The Gardiner Foundation, for funding.

REFERENCES

Amarasinghe S.L., Su S., Dong X., *et al.* (2020) *Genome Biol.* **21**: 30.
Charlier C., Agerholm J.S., Coppieters W., *et al.* (2012) *PLoS One* **7**: e4308.
Collins R.L., Brand H., Karczewski K.J., *et al.* (2020) *Nature* **581**: 444.
Ferraj A., Audano P.A., Balachandran P., *et al.* (2023) *Cell Genomics* **3**: 100291.
Lee Y.L., Bosse M., Takeda H., *et al.* (2023) *BMC Genomics* **24**: 225.
Lee Y.L., Takeda H., Costa Monteiro Moreira G., *et al.* (2021) *PLoS Genet.* **17**: e1009331.
Li H. (2018) *Bioinformatics* **34**: 3094.
McLaren W., Gil L., Hunt S.E., *et al.* (2016) *Genome Biol.* **17**: 122.
Nguyen T.V., Vander Jagt C.J., Wang J., *et al.* (2023) *Genet. Sel. Evol.* **55**: 9.
Rosen B.D., Bickhart D.M., Schnabel R.D., *et al.* (2020) *GigaScience* **9**: 1.
Rothammer S., Capitan A., Mullaart E., *et al.* (2014) *Genet. Sel. Evol.* **46**: 44.
Schütz E., Wehrhahn C., Wanjek M., *et al.* (2016) *PLoS ONE* **11**: e0154602.
Sedlazeck F.J., Rescheneder P., Smolka M., *et al.* (2018) *Nat. Methods* **15**: 461.

Proc. Assoc. Advmt. Anim. Breed. Genet. 26: 391-394

EXPLORING BLOOD BIOMARKERS AS POTENTIAL INTERMEDIATE PHENOTYPES FOR TRANSITION COW DISEASES: A FOUNDATION FOR FUTURE MULTI-OMICS INTEGRATION

A. Kudriashova[1,2], A.J. Chamberlain[1,3], J.A.A. McArt[4], J.E. Hemsworth[1], C. M. Reich[1] and M.E. Goddard[1,2]

[1] Agriculture Victoria, AgriBio, Centre for AgriBioscience, Bundoora, VIC, 3083 Australia
[2] School of Agriculture, Food and Ecosystem Sciences, University of Melbourne, VIC, 3647 Australia
[3] School of Applied Systems Biology, La Trobe University, Bundoora, VIC, 3083 Australia
[4] Department of Population Medicine and Diagnostic Sciences, College of Veterinary Medicine, Cornell University, Ithaca, NY

SUMMARY

Transition cow health is essential for dairy cows' productivity and longevity. Current breeding methods inadequately address transition period challenges. Blood biomarkers like β-hydroxybutyrate (BHB), non-esterified fatty acids (NEFA), glucose, and urea assess cows' metabolic status and provide insights into their health. This study examined the association between common blood biomarkers and transition diseases in dairy cows, identifying them as potential intermediate phenotypes. Blood samples from 200 clinically ill transition cows were compared with healthy controls. Serum biomarkers were quantified, and statistical analysis identified statistically significant differences between diseased and control groups. Linear models assessed the impact of diagnosis and other factors on biomarker concentrations. Results indicated that clinically ill cows exhibited patterns of systemic inflammation, with elevated globulins, reduced albumin, and altered lipid and liver markers. Energy metabolism biomarkers (BHB, NEFA, glucose) were affected by metabolic disorders, while inflammatory biomarkers (albumin, albumin-globulin ratio, globulin) were associated with infection-related diseases. These findings suggest that serum biomarkers related to energy metabolism and inflammation are valuable for predicting disease in transition dairy cows.

INTRODUCTION

Given the elevated occurrence of illnesses during the postpartum period, the health of transition cows is crucial for dairy cows' productivity, lifespan, and survival. Current breeding approaches do not adequately address the physiological challenges of the transition period, such as energy deficit, immune suppression, and calcium homeostasis. Genetic improvement of transition cow health could rely on records of cow disease. However, metabolic disorders generally have low heritability and are hard to collect on enough cows to generate accurate EBVs (Pryce *et al.* 2016). Intermediate phenotypes, also known as endophenotypes, might be more heritable and easier to collect than records of disease and better describe the underlying biology than the clinical symptoms of specific disorder or complex phenotypes such as overall health or productivity. Endophenotypes provide a functional link between genetic information and complex traits, potentially enhancing the accuracy of selection in breeding programs. Blood biomarkers such as BHB, NEFA, glucose, and urea are widely used to assess cows' metabolic status and provide accurate measurable insights into metabolic health. Their higher heritability compared to health data accessible from farms and veterinary records makes blood biomarkers more desirable phenotypes (Luke *et al.* 2019). Further, these biomarkers can be used to identify gene variants associated with disease susceptibility and resilience.

The present study was performed to explore associations of common blood biomarkers of health in postpartum dairy cows and to identify these biomarkers as potential intermediate phenotypes of transition cow diseases.

METHODOLOGY

Blood samples were collected from 200 clinically ill dairy cows within 70 days after calving and matched with one or two healthy control animals from the same herd with similar calving dates. The study included 15 commercial dairy farms in Gippsland, Australia, all using grazing-based feeding systems supplemented with forage and concentrates during milking. Blood samples were collected from a coccygeal vessel, incubated at ~20°C for 1.5 hours, and centrifuged at 1,500 x g for 15 minutes; serum was stored at -80°C. Serum biomarkers listed in Table 1 were quantified using a Chemwell 2910 auto-analyser (Awareness Technology, USA) at AgriBio, Melbourne, Australia. Transition cow diseases included ketosis (n =13), milk fever (n = 50), mastitis (n = 51), metritis (n = 18), endometritis (n = 38), retained foetal membrane (n = 34), subclinical hypocalcemia (n = 178), and subclinical hyperketonemia (n = 74). Several animals had multiple diseases. A total of 331 samples were included in the control group.

Statistical analyses were conducted using R version 4.4.2 (2024-10-31). Descriptive statistics (mean, standard deviation, standard error, minimum, and maximum) were calculated for each biomarker in the "Control" and "Diseased" groups (Table 1). Welch's t-test determined statistically significant differences in biomarker levels between the groups. Density plots were generated to assess normal distribution, group overlap, and outliers for each biomarker (Figure 1). BHB distribution was normalised using log-transformation.

Outliers were identified in the Diseased group for urea, magnesium, calcium, and glucose. Since these extreme values reflected true physiological states associated with disease, they were retained in the analysis but underwent proportional winsorization to reduce their influence without compromising data integrity and eliminating their relative order. For each biomarker, values exceeding a predefined upper threshold (upper normal range) were proportionally scaled down. The winsorized value was computed as:

$$X_{win} = T_{upper} + \left(X_{orig} - T_{upper}\right) \times scalefactor$$

where X_{orig} - original biomarker value, T_{upper} upper threshold, scale factor - proportion by which extreme values above the threshold are adjusted, and X_{win}new winsorized biomarker value. A scale factor of 0.3 reduced extreme values to 30% of their deviation beyond the upper threshold.

A fixed effect model was fitted to assess the effect of disease diagnosis, farm identity, breed, parity, weeks after calving, sample date, haemolysis, and treatment.$\text{Biomarker level} = \beta_0 + \beta_1(Control) + \beta_2(Diagnosis) + \cdots + \beta_k(Covariates) + \varepsilon$, where β_I represents the estimated effect size of each predictor and ε represents the residual error.

RESULTS AND DISCUSSION

Figure 1 and Table 1 summarise differences between the "Control" and "Diseased" groups. Mean magnesium, NEFA, total protein, and triglycerides concentrations showed no statistically significant differences. Higher biomarker concentrations in comparison to "Control" were identified for BHB, bilirubin, globulin, urea and glucose, and lower range for albumin-globulin ratio and albumin. Calcium, while showing lower concentrations for sick animals, also had outliers in the upper range.

Overall statistical and visual analyses indicated that clinically ill cows' metabolic profiles follow patterns of systemic inflammation, including elevated globulins, reduced albumin, altered lipid metabolism (cholesterol and NEFA), and liver markers like bilirubin. These patterns suggest that the "Diseased" group exhibits a systemic inflammatory response rather than localised stress. Globulins increase during inflammation as they contain antibodies and inflammatory proteins. Reduced albumin and albumin-globulin (A/G) ratio occur as the liver prioritises Acute Phase Protein synthesis over albumin. Changes in lipid metabolism during systemic inflammation are characterised by reduced cholesterol concentrations. Elevated bilirubin indicates liver stress or red blood cell breakdown during inflammation. Higher BHB levels suggest an energy imbalance, commonly seen in inflammatory conditions with negative energy balance.

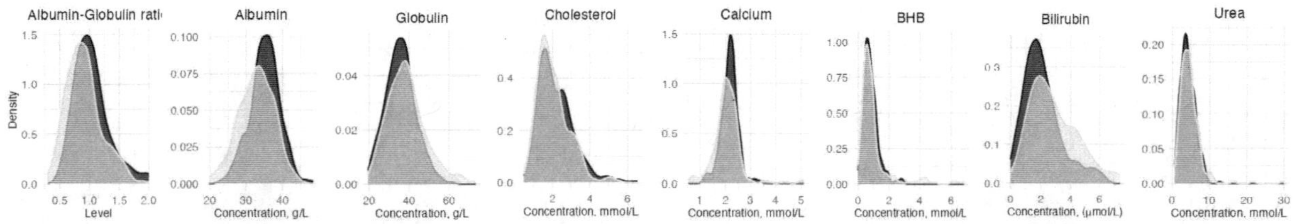

Figure 1. Density distribution histogram of diseased (yellow) and control (blue) groups
Albumin-globulin ratio, albumin, globulin, cholesterol, calcium, BHB, bilirubin, urea (from left to right).

Table 1. Statistical analysis of biomarker differences between control and diseased groups

Biomarker	Diagnosis	Mean	SD	Min	Max	SE	p-value	Normal range
A/G	Control	1.04	0.30	0.50	2.00	0.02	<0.001	0.70 - 1.1
	Sick	0.91	0.29	0.30	1.90	0.02		
Albumin	Control	35.16	4.12	23.00	48.50	0.23	<0.001	25.0 - 38.0 (g/L)
	Sick	32.82	4.77	19.80	43.90	0.33		
BHB	Control	0.82	0.48	0.00	2.88	0.03	<0.05	< 1.2 (mmol/L)
	Sick	1.03	1.14	0.00	6.85	0.08		
Bilirubin	Control	2.14	1.31	0.00	6.33	0.07	<0.001	0.00 - 24.0 (µmol/L)
	Sick	2.88	1.58	0.00	7.35	0.11		
Urea	Control	4.22	1.89	0.70	12.70	0.10	<0.05	2.1 - 10.7 (mmol/L)
	Sick	4.71	3.04	1.10	29.9	0.21		
Calcium	Control	2.16	0.33	1.14	3.37	0.02	<0.05	2.00 - 2.75 (mmol/L)
	Sick	2.07	0.55	0.55	5.10	0.04		
Cholesterol	Control	2.20	0.93	0.48	6.51	0.05	<0.05	1.73 - 7.73 (mmol/L)
	Sick	2.04	0.84	0.65	5.34	0.06		
Globulin	Control	35.80	7.46	19.50	61.50	0.41	<0.001	30.0 - 45.0 (g/L)
	Sick	38.80	9.71	20.80	73.60	0.68		
Glucose	Control	3.25	0.49	1.30	4.50	0.03	<0.001	2.5 - 4.2 (mmol/L)
	Sick	3.72	1.29	1.70	9.90	0.09		
Magnesium	Control	0.99	0.16	0.40	1.62	0.01	0.87	0.74 - 1.44 (mmol/L)
	Sick	0.99	0.29	0.29	2.89	0.02		
NEFA	Control	0.74	0.29	0.16	1.72	0.02	0.46	0.2 - 0.8 (mmol/L)
	Sick	0.76	0.32	0.12	1.59	0.02		
Total protein	Control	70.96	7.23	45.40	93.30	0.40	0.36	60.0 - 85.0 (g/L)
	Sick	71.62	9.07	51.80	102.80	0.64		
Triglycerides	Control	0.20	0.09	0.02	0.68	0.01	0.26	0.00 - 0.5 (mmol/L)
	Sick	0.21	0.08	0.08	0.52	0.01		

Linear models were fitted to assess the effect of disease diagnosis on serum biomarkers, suggesting that these biomarkers serve as indicators of distinct physiological states. Statistically significant associations ($P < 0.05$) are shown in a Table 3. Magnesium was strongly associated with RFM, subclinical hypocalcaemia with low cholesterol concentrations, and hyperketonemia with increased urea concentrations. Milk fever demonstrated a positive association with glucose in untreated animals while clinical signs persisted. Both subclinical conditions were associated with positive NEFA and bilirubin levels. Overall, the linear model results highlighted that energy metabolism biomarkers (BHB, NEFA, glucose) were strongly influenced by metabolic disorders, while inflammatory biomarkers albumin, albumin-globulin ratio, and globulin, as well as NEFA, were associated with infection-related diseases.

Table 2. Linear model p-values for biomarker-disease associations

Biomarker	Adj r^2	Metritis	Endometritis	Mastitis	Milk fever	Ketosis	RFM	HK	HCa
Albumin	0.39	0.004		0.030					
Urea	0.34							0.012	
Magnesium	0.20	0.022					0.007		
Cholesterol	0.52								0.012
BHB	0.57					<0.001		<0.001	
Globulin	0.40			0.009					
Bilirubin	0.31					0.028		0.001	0.001
NEFA	0.34		0.048					0.012	0.009
Glucose	0.38				<0.001	<0.001		0.020	0.036
Calcium	0.64				<0.001				<0.001
AG	0.42	0.043		0.027					

Cell colours represent the effect's direction: red for positive (increased biomarker concentrations), blue for negative (decreased concentrations), and white for non-statistically-significant effects, with statistically significant effect p-values displayed within each cell. HK denotes subclinical hyperketonemia, HCa subclinical hypocalcaemia, and RFM retained foetal membrane.

CONCLUSION

These results highlight that certain serum biomarkers have the potential to be used as intermediate phenotypes, linking the biological changes in metabolic and inflammatory pathways to clinical disease (ketosis, RFM, and milk fever). Elevated NEFA and BHB, and decreased glucose concentrations, suggest a systemic energy deficit, reinforcing their potential as predictive markers. Decreased A/G ratio and increased globulin concentrations indicate a systemic inflammatory response, suggesting that protein markers can serve as early indicators of infection. Our findings demonstrate that serum biomarkers related to energy metabolism and inflammation can serve as valuable intermediate phenotypes for disease prediction in transition cows

ACKNOWLEDGEMENTS

This study was undertaken as part of the DairyBio program, which is jointly funded by Dairy Australia, the Gardiner Foundation, and Agriculture Victoria and is a PhD research project at the University of Melbourne.

REFERENCES

Luke T.D.W., Nguyen T.T.T., Rochfort S., Wales W.J., Richardson C.M., Abdelsayed M. and Pryce J.E. (2019) *J. Dairy Sci.* **102**: 11142.

Pryce J.E., Parker Gaddis K.L., Koeck A., Bastin C., Abdelsayed M., Gengler N., Miglior F., Heringstad B., Egger-Danner C., Stock K.F., Bradley A.J. and Cole J.B. (2016) *J. Dairy Sci.* **99**: 6855.

Proc. Assoc. Advmt. Anim. Breed. Genet. 26: 395-398

USING MULTI-OMICS DATA TO IDENTIFY CANDIDATE GENES FOR MILK LACTOSE PERCENTAGE IN DAIRY CATTLE

M. Ghoreishifar[1,2] R. Xiang[1,3], T. Lopdell[4], M. Littlejohn[4], A. Chamberlain[1,2], J. Pryce[1,2] and M. Goddard[1,3]

[1] Agriculture Victoria Research, AgriBio Centre for AgriBioscience, Bundoora, VIC, 3083 Australia
[2] School of Applied Systems Biology, La Trobe University, Bundoora, VIC, 3083 Australia
[3] Faculty of Science, University of Melbourne, VIC, 3010 Australia
[4] Livestock Improvement Corporation, Private Bag 3016, Hamilton, 3240 New Zealand

SUMMARY,

Genome-wide association studies (GWAS) have successfully identified many quantitative trait loci (QTL) associated with complex traits in humans and livestock. However, pinpointing the causal variants and target genes through which these QTL influence phenotypes remains challenging. Transcriptome data from mammary tissue were analysed using genetic score omics regression (GSOR). This method correlates observed gene expression to genetically predicted phenotypes and is used to find associations between gene expression and genomic breeding values (GEBVs). GSOR identified 706 genes whose expression in mammary glands is significantly associated with GEBVs estimated for milk lactose percentage (FDR $\leq$ 10%). We observed a significant co-occurrence between the identified genes and GWAS signals for lactose percentage reported in an independent study (P=2.5E-10; odds-ratio=3.8). The co-occurring genes with GWAS signals enriched for ion transport Gene Ontology (GO) term included *ATP6V0A2*, *DNAH10*, *LRRC8B*, *LRRC8C*, *KCNJ2*, *P2RX4*, and *SLC34A2*. These findings introduce known and novel candidate genes related to the regulation of milk lactose, showing their involvement in the ion transport mechanism and supporting the importance of the osmoregulatory function of lactose in milk.

INTRODUCTION

Genome-wide association studies (GWAS) have successfully identified many quantitative trait loci (QTL) associated with complex traits in humans and livestock. However, pinpointing the causal variants and target genes through which these QTL influence phenotypes remains challenging. This difficulty arises due to the small effect sizes of most variants, linkage disequilibrium (LD) among nearby variants, and the fact that most QTLs are located in non-coding regions of the genome. QTL in non-coding regions are believed to affect a phenotype by regulating the expression of their target genes; in such cases, they are referred to as expression QTL (eQTL).

Three types of evidence are used to identify the genes through which eQTL operates: (1) genes located near the most significant GWAS variant, (2) genes whose expression is correlated with the trait based on the GSOR method (unpublished data), and (3) genes whose physiological roles are related to the trait. Although none of these types of evidence is definitive on its own, their convergence on the same genes provides convincing evidence supporting the correct identification of genes (*unpublished data*).

The current study has three primary objectives: (1) to identify genes whose observed expression levels in the mammary gland of NZ dairy cows are significantly associated with their estimated genomic breeding values (GEBVs) for milk lactose percentage, (2) to investigate the extent of co-occurrence between the associated genes and signals obtained from an independent GWAS on lactose percentage, and (3) to determine whether the co-occurring genes (with GWAS signals) enriched in a specific Gene Ontology (GO) term.

MATERIALS AND METHODS

Phenotypic data. Phenotypic data for test-day lactose percentage was provided by DataGene Pty Ltd (Melbourne, Australia). Outliers deviating ±3 SD of the mean phenotypic value were excluded. Test-day records were included if the cow's age at calving was between 18-25 months and days in milk (DIM) between 5-315 days. The final data contained ~4.99 million test day records for ~ 477 K cows. Using ASReml (Gilmour *et al.* 2022), adjusted phenotypes were estimated and averaged for each cow (i.e. effect of Cow) using the model proposed by Khansefid *et al.* (2023):

$$y_{ijklm} = \mu + H_iTD_j + M_k + pol(DIM, 8) + pol(Age, 2) + Cow_l + e_{ijklm}$$

where, y_{ijklm} is the test-day record for lactose percentage, μ is the effect of overall mean, H_iTD_j is the effect of the ith herd and jth test date; M_k is the effect of the kth calving month; $pol(DIM, 8)$ and $pol(Age, 2)$ are the regression coefficients of Legendre polynomials of order 1–8 for DIM and of order 1–2 for age at calving in months; Cow_l and e_{ijklm} are the random effect of the lth cow and the random residual term, respectively.

Genotype data. Genotype data were available for 81,658 Australian cows, of which 79% were Holstein, 16% Jersey and 5% Aussie Red. Genotypes were imputed to whole genome sequences using Run9 of the 1000 Bull Genomes project (Daetwyler *et al.* 2014). Variants with minor allele frequency (MAF) < 0.01 and genotype frequencies departing from Hardy-Weinberg equilibrium (*P* < 1E-8) were excluded. LD pruning was performed to exclude variants that are in high LD (r^2 > 0.95). These procedures retained 1,181,628 variants for subsequent analyses. Using genotypes and phenotypes of 81,658 cows described above, we trained a BayesR3 (Breen *et al.* 2022) model to estimate prediction equations (SNP effects) for lactose percentage. The model was as follows: $\boldsymbol{y} = \boldsymbol{Xu} + \boldsymbol{Vg} + \boldsymbol{e}$, where $\boldsymbol{y}$ is vector of adjusted phenotypes; $\boldsymbol{X}$ is an incidence matrix, **u** is a vector of fixed effect including breed (three levels); **V** is the coded genotype; **g** is a vector of SNP effects; and **e** is the residual term. BayesR3 model was run using the default parameters. The estimated SNP effects were used to calculate $\widehat{GEBV}_{cis}$ (in the gene expression data) using nearby variants (1 Mb) to a specific gene.

Gene expression data and GSOR. Expression measures of ~12.3 K genes in mammary tissue from ~350 New Zealand (NZ) cows (including Holstein Friesian, Jersey and their crosses) were used. Imputed whole genome sequence variants were available for these cows. The processing of samples, RNA extraction, library preparation, and RNA sequencing were described in detail in Littlejohn *et al.* (2016). The same set of sequence variants as described for the Australian data were kept for the NZ animals. To perform GSOR, we used a method called genetic score omics regression, abbreviated to GSOR (unpublished data). This method estimates the significance of the correlation between the expression level of a gene and the GEBVs derived from nearby variants of that gene ($\widehat{GEBV}_{cis}$). The following model was used:

$$\widehat{GEBV}_{cis} = b\Omega + g + e$$

where, $\widehat{GEBV}_{cis}$ is a vector calculated using the effect estimated for variants located in the 1 Mb region surrounding the gene, **b** is the regression coefficient of the $\widehat{\mathbf{GEBV}}_{cis}$ on Ω, which is a vector representing gene expression measures; and **g** is a vector of random polygenic effects with $\mathbf{g} \sim \mathrm{N}(0, \mathbf{G}\sigma_g^2)$, where **G** is the genomic relationship matrix estimated from genome-wide SNPs, and σ_g^2 is the additive genetic variance explained by genome-wide SNPs; $\boldsymbol{e}$ is the vector of residuals with $\boldsymbol{e} \sim N(0, \boldsymbol{I}\sigma_e^2)$, where **I** is an identity matrix and σ_e^2 is residual variance. To account for multiple testing, we applied Benjamini-Hochberg correction and genes with $FDR \leq 0.10$ were deemed significant.

Do GSOR significant genes overlap with GWAS signals? To test this hypothesis, we investigated the overlap (co-occurrence) between the GSOR significant genes, whose expression levels are correlated with $\widehat{\mathbf{GEBV}}_{cis}$, and a GWAS on lactose concentration previously reported in the study by (Lopdell *et al.* 2017). GWAS results were downloaded and lifted over to the new reference genome (ARS-UCD 1.2), and SNPs with $P \leq 1E-8$ was regarded as significant (i.e., GWAS signals).

To find the degree of overlap between the GSOR significant genes and GWAS signals, we partitioned the genome into non-overlapping windows of 100 Kb. We counted the number of windows containing at least one GWAS signal, and a total number of GSOR significant genes located within these windows. These quantities were compared against a total number of windows and total number of GSOR significant genes. We used a Fisher Exact test to investigate significance of our findings at alpha ≤ 0.05.

Functional annotation analyses. We conducted functional annotation analyses for the GSOR significant genes that co-occur with GWAS signals, using the total number of GSOR significant genes as the background. We used DAVID bioinformatic tool (Sherman *et al.* 2022) for this analysis with UniProt Keyword annotation, and terms with FDR ≤ 0.05 were considered significant.

RESULTS AND DISCUSSION

The correlations between the expression of 12,237 mammary genes across cows and $\widehat{\mathbf{GEBV}}_{cis}$ for lactose percentage were tested using GSOR. Of the tested genes, 704 significant associations were found at FDR ≤ 10%. However, similar to GWAS, GSOR is also susceptible to false positives (unpublished data). Therefore, we assessed the overlap between GSOR significant genes and GWAS signals by evaluating the extent of co-occurrence within non-overlapping windows. The results are presented in Table 1 and suggest there is a meaningful biological agreement between the GWAS findings and the GSOR results. For example, using 100 Kb windows, there were 24,861 non-overlapping windows, and 706 GSOR significant genes. We found 321 windows containing GWAS signals, and 35 GSOR significant genes located within these windows (co-occurred with GWAS signals), resulting in a *p*-value of 2.5E-10 (odds ratio=3.8).

Table 1. Fisher Exact Test of co-occurrence between GSOR significant genes and GWAS signals using non-overlapping windows of different sizes

Window size (Kb)	N of Windows	N of GSOR significant genes	N of windows containing GWAS signal(s)	N of GSOR genes in windows containing GWAS signal(s)	*P*-value (odds-ratio)
100	24,861	706	321	35	2.5E-10 (3.8)
500	4,984	706	186	57	3.1E-6 (2.16)

GSOR can be confounded by LD where the correlation between gene expression and trait can be caused by SNP marker(s) instead of a causal variant (*unpublished data*). Such a correlation does not indicate a causal link between the expression of the gene and the trait. Therefore, additional sources of evidence could help prioritize the causal gene-trait associations at GWAS loci.

If GSOR significant genes co-occurring with GWAS signals include causal genes, they are expected to show enrichment for biologically relevant GO terms. We tested this hypothesis by investigating the enrichment of GO terms for the genes that show a GSOR correlation and are near GWAS signals. Our results showed the biological terms Transport, and its child term Ion transport are significantly enriched (Table 2) with 11 and seven genes involved in these GO terms, respectively.

Our findings are consistent with a previous study on milk lactose traits (Lopdell *et al.* 2017), that highlighted the role of membrane transport genes, including *LRRC8C*, *P2RX4*, KCNJ2 and *ANKH* as key modulators of milk lactose content. In addition to the first three genes, our study also identified novel genes including *ATP6V0A2*, *DNAH10*, *LRRC8B* and *SLC34A2* which are involved in Ion transport biological process. These genes are expected to influence osmotic balance through modulation of ion concentrations in milk.

Table 2. Enrichment of GSOR significant genes co-occurred with GWAS signals compared to background genes (i.e., the total GSOR significant genes)

Category	Term	Genes	*P* value	FDR
Biological process	Transport	ATP6V0A2, RAB5C, DNAH10, LRRC8B, LRRC8C, KCNJ15, KCNJ2, P2RX4, P2RX6, SLC34A2, SLC50A1	2.64E-4	5.5E-3
Biological process	Ion transport	ATP6V0A2, DNAH10, LRRC8B, LRRC8C, KCNJ2, P2RX4, SLC34A2	7.4E-4	8.2E-3
Molecular function	Ion channel	LRRC8B, P2RX4, P2RX6, LRRC8C, KCNJ15, KCNJ2	1.5E-4	2.5E-3

CONCLUSION

Our study identified 704 genes from mammary gland, whose expression levels were associated with milk lactose percentage. Of these genes, a significant proportion showed co-occurrence with GWAS signals reported in an independent study. Enrichment analysis of these co-occurring genes highlighted transport and ion transport processes, implicating both known and novel membrane transport genes in regulating osmolality and influencing milk lactose concentration. These findings offer valuable insights into the genetic basis and biological regulation of lactose percentage.

ACKNOWLEDGEMENTS

The authors acknowledge DataGene for providing the Lactose data, as well as Dr Tuan Nguyen and Dr Iona MacLeod for imputation of sequence genotypes. We also appreciate the financial support from DairyBio, a joint venture of Dairy Australia, The Gardiner Foundation, and Agriculture Victoria (Melbourne, Australia).

REFERENCES

Breen E.J., MacLeod I.M., Ho P.N., Haile-Mariam M., Pryce J.E., Thomas C.D., Daetwyler H.D. and Goddard M.E. (2022) *Commun. Biol.* **5**: 661.

Daetwyler H.D., Capitan A., Pausch H., Stothard P., Van Binsbergen R., Brøndum R.F., Liao X., Djari A., Rodriguez S.C., Grohs C. and Esquerré D. (2014) *Nat. Genet.* **46**: 858.

Gilmour A.R., Gogel B.J., Cullis B.R., Welham S.J., Thompson R., Butler D., Cherry M., Collins D., Dutkowski G. and Harding, S. (2014) ASReml User Guide Release 4.1. VSN International Ltd, Hemel Hempsted, UK.

Khansefid M., Pryce J.E., Shahinfar S., Axford M., Goddard M.E. and Haile-Mariam M. (2023) *Anim. Prod. Sci.* **63**: 1031.

Littlejohn M.D., Tiplady K., Fink T.A., Lehnert K., Lopdell T., Johnson T., Couldrey C., Keehan M., Sherlock R.G., Harland C. and Scott A. (2016) *Sci. Rep-UK* **6**: 25376.

Lopdell T.J., Tiplady K., Struchalin M., Johnson T.J., Keehan M., Sherlock R., Couldrey C., Davis S.R., Snell R.G., Spelman R.J. and Littlejohn M.D. (2017) *BMC Genom.* **18**: 1.

Sherman B.T., Hao M., Qiu J., Jiao X., Baseler M.W., Lane H.C., Imamichi T. and Chang W. (2022) *Nucleic Acids Res.* **50**: W216.

Proc. Assoc. Advmt. Anim. Breed. Genet. 26: 399-402

EVIDENCE OF CHROMOSOME X ASSOCIATION WITH KNOBBED ACROSOMES

B.S. Firewski[1], W.L.A. Tan[1], T. Melo[2], L.R. Porto-Neto[3], A. Reverter[3], M. McGowan[4] and M.R.S. Fortes[1]

[1] The University of Queensland, School of Chemistry and Molecular Bioscience, St Lucia, QLD, 4072 Australia
[2] Universidade Estadual de Sao Paulo (UNESP), Campus Jaboticabal, Brazil
[3] CSIRO Agriculture & Food, St Lucia, QLD, 4067 Australia
[4] The University of Queensland, School of Veterinary Sciences, Gatton, QLD, 4343 Australia

SUMMARY

Sperm morphological defects are an important component trait of bull fertility. This paper reports a genome-wide association study targeting Knobbed Acrosomes (KA), a relatively common sperm morphological defect in tropical bulls. Phenotypes of bulls in a multi-breed population were available for analysis, and 25% had the KA defect in 1% or more of its sperm cells. Of genotyped bulls, 2,183 were classified as cases (KA ≥ 1%) and 3,657 were controls (no KA defects observed). Associated SNPs were aggregated to define QTL boundaries. Genes mapped to QTL were subject to functional annotation and enrichment analysis, aiming at identifying plausible candidate genes for KA. The estimated heritability for KA presence was 0.16. Significant associations defined a QTL on the X chromosome, containing eight candidate genes: *AMOT*, *ATRX*, *CYSLTR1*, *GPR174*, *HTR2C*, *LDHA*, *LPAR4*, and *P2RY10*. Selection against KA sperm defect may be possible and could improve fertilisation rates in tropical herds.

INTRODUCTION

The standardised Bull Breeding Soundness Evaluation (BBSE) measures traits relevant to male fertility (Entwistle *et al.* 2003). Sperm morphological defects are evaluated as part of BBSE. The proportion of sperm with morphological defects identified in a sample is an important criterion in BBSE. Bulls may be deemed unfit for mating if high percentages of sperm morphological defects are identified. One such defect is knobbed acrosomes (KA): the acrosome of the sperm is thickened or ridged, thus affecting the sperm's ability to bind to the zona pelucida, which is detrimental to fertilisation. Furthermore, in bulls that present with KA sperm, non-KA sperm's ability to form zygotes and bind to the zona pelucida is also affected (Thundathil *et. al.* 2000). Therefore, it is possible that the impact of KA presence is currently underestimated. In this study, we aimed to investigate the genetics underpinning the presence of KA defects in tropical bulls.

MATERIALS AND METHODS

Animals and phenotypes. BBSE records from 6,063 bulls comprising six different breeds were included in this study. Two breeds were research herds from the Cooperative Research Centre for Beef Genetic Technologies (Beef CRC) consisting of Brahman and Tropical Composite bulls. The remaining four breeds obtained from industry herds were Santa Gertrudis, Droughtmaster, Ultra black, and Belmont Red. This population of bulls and their BBSE phenotypes were first described in a previous paper (Porto Neto *et al.* 2023). Sperm morphology was observed before as a combined trait (i.e. percentage of normal sperm), but the occurrence of KA was not investigated previously.

The distribution of the percentage of KA-affected sperm (%KA) within this data was heavily skewed (i.e., not normally distributed). Therefore, we explored a case versus control GWAS design, by classifying bulls as affected (case) or unaffected (control). Cases were bulls with 1% or higher presence of sperm with KA defects. Controls were bulls with 0% KA (no KA sperm found on BBSE evaluation).

Genotypes and GWAS. Most animals were genotyped at ~ 50K. A reference panel that utilised Beef CRC and industry animals genotyped at higher density (~700K) were used to impute animals to higher density. The animals used in the reference population were representative of the bulls used in this study, as reported before (Porto-Neto *et al.* 2023). The reference population was phased using Eagle 2 (v2.4.1) and then used to impute the 50K genotypes using Minimac3 for autosomes and Minimac4 for Chromosome X. All SNP with imputation $r^2 > 0.8$, a call rate > 0.85 and a minor allele frequency > 0.05 were retained for further analysis. This resulted in a total of 661,037 SNPs.

The association between SNP and trait was calculated using EMMAX methods (Kang 2010). This method uses the following model iteratively for each SNP *j*:

$$y = X\beta + z_j g_j + Za + e$$

where y is a vector describing phenotypes for all individuals, X is the matrix for fixed effects, including continuous covariates (body weight and scrotal circumference) and categorical effects (contemporary groups defined by year and breed), β is the vector of coefficients for the fixed effects, z_j is the vector of genotypes SNP *j*, g_j is the fixed additive effect of SNP *j*, Z is the matrix of genotypes for all SNP, a is the random additive breeding value of each animal and assumed to follow a normal distribution with zero mean and variance $G\sigma_a^2$ where G is the genomic relationship matrix (GRM) computed using Method 1 of VanRaden (2008), and e represents the random residual effect assumed to follow a normal distribution with zero mean and variance $I\sigma_e^2$. The threshold of significance for SNP association with the presence of KA defect was set to $P \leq 10^{-8}$. We also considered SNPs with $P \leq 10^{-6}$ to define QTL boundaries. Methods for aggregating SNPs and defining QTL boundaries were described previously (Van den Berg *et al.* 2016; Fortes *et al.* 2020).

RESULTS AND DISCUSSION

Of all the bulls with BBSE records, 25% had KA ≥ 1%. However, less than 1% of bulls would have failed the BBSE test due to KA defects. In the industry, when KA is higher than 30%, the bull would be considered not fit for mating and would have failed the BBSE test according to clinical veterinarian practices. Thus, less than 1% of bulls in this multi-breed population are considered to have a clinical fertility problem due to the presence of KA alone. However, the implications of KA occurrence in Australian beef bulls will be considered more relevant or less severe, depending on whether this sperm defect is regarded as compensable. If KA is a compensable sperm defect, as suggested by industry standards (Fordyce *et al.* 2006), the <1% of the population with KA occurrence may seem of little consequence. However, sperm with KA also affect the fertilisation capacity of non-KA sperm in the same animal (Barth 1986; Thundathil *et al.* 2000). Thus, the consequences of KA occurrence are currently unknown. In the present dataset, 25% of bulls had KA occurrence in their ejaculate sperm. In other words, one-quarter of studied bulls may have lower fertility due to KA occurrence.

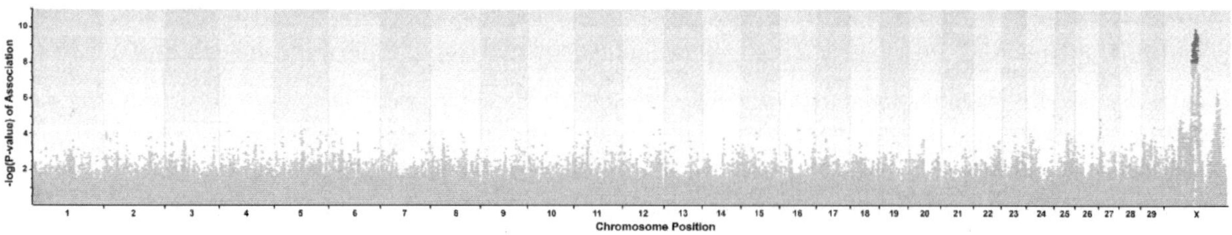

Figure 1. Manhattan plot showcasing genome-wide association results for the sperm morphological defect known as Knobbed Acrosomes. Note the significant associations identified in chromosome X. SNP marked in red are considered significantly associated (-log(*P*-value) > 8).

The heritability estimate for KA presence as a binary trait was 0.16. Among genotyped bulls, 2,183 were classified as cases (KA ≥ 1%) and 3,657 were controls. The association study points to two QTL regions in the X chromosome (Figure 1). Previous studies identified sex-linked recessive inheritance patterns for KA occurrence in Friesian bulls (Donald and Hancock 1953). Charolais bulls are disproportionally affected by KA presence and severely affected bulls have lower fertility (Barth 1986). Barth (1996) also observed pedigree evidence for a recessive inheritance of KA defects in Charolais bulls. Our results corroborate the view that KA inheritance might be sex-linked and point more specifically to QTL regions on the X chromosome.

Table 1. Quantitative trait loci boundaries and details

SIG	START (bp)	END (bp)	SIZE (bp)	PEAK POS	N SNP
10^{-8}	X:63.859.185	X:74.316.807	10.457.622	69.611.250	989
10^{-6}	X:61.130.332	X:77.980.200	16.849.868	69.611.250	1252

SIG = significance in *P*-value, bp = base pairs, POS= position, N SNP= number of SNPs.

The results point to large QTL regions on chromosome X (Table 1). Inside the defined QTL boundaries, we discovered eight candidate genes: *AMOT*, *ATRX*, *CYSLTR1*, *GPR174*, *HTR2C*, *LDHA*, *LPAR4*, and *P2RY10*. Of these genes, *ATRX* is a promising functional candidate. *ATRX* has been linked to Sertoli cell proliferation, and Sertoli-mediated spermatogenesis (Bagheri-Fam *et al.* 2011). Previous GWAS review and prioritisation analyses have linked *ATRX* with bull fertility traits (Fonseca *et al.* 2018). Knockout mice demonstrate that spermatogenesis defects may result from the impaired interaction between the faulty ATRX protein and the androgen receptor (AR), supporting *ATRX* as a key candidate gene (Bagheri-Fam *et al.* 2022).

The other candidate genes mentioned above cannot be excluded from future work. *CYSLTR1* expression can aid in cases of testicular inflammation (Awad *et al.* 2023). *GPR174* codes for a molecule involved with spermatogenesis due to its role in testosterone signalling (Tsutsui *et al.* 2011). A recent study linked *LPAR4* expression to testicular function and immunity (Dai *et al.* 2024). The conditional deletion *LDHA* in Sertoli cells disrupts spermiogenesis in mice by impairing lactate production, highlighting a mechanism critical for sperm development (Zhang *et al.* 2022). In short, there is literature evidence to support a role in spermatogenesis for 5 of the positional candidates found in this study: *ATRX, CYSLTR1, GPR174, LPAR4,* and *LDHA*. They might act independently or in a coordinated manner to affect spermatogenesis. The peak SNP is an intergenic variant (rs110162429) mapped to a large copy number variant, and its closest annotated feature is an enhancer. Their hypothetical impact on KA occurrence needs further investigation.

When the acrosome of the sperm is thickened or ridged, the sperm's ability to bind to the zona pelucida of the oocyte is compromised (Thundathil *et al.* 2000). Not binding to the zona pelucida impairs fertilisation. Thus, this research contributes to understanding acrosome-mediated infertility, a problem reported across mammalian species, including cattle and humans (Thundathil *et al.* 2000; Aitken *et al.* 1990; Moretti *et al.* 2005). The candidate genes suggested herein need further research to validate their association and elucidate their roles in acrosome-mediated infertility.

CONCLUSION

The occurrence of KA sperm is a heritable and relatively common sperm defect in tropical bulls, with significant associations mapped to the X chromosome. Selective breeding against KA occurrence could enhance fertility rates, especially considering that 25% of bulls in this multi-breed population had KA-affected sperm. The notion that KA defects might be inherited in a recessive

Mendelian fashion further contributes to hopes of identifying risk (or causative) alleles that could aid in discarding bulls with the deleterious genotype.

ACKNOWLEDGMENTS

This work had support from CSIRO, the University of Queensland, and Meat & Livestock Australia. The authors thank the beef producers who contributed to this project (L.GEN.1818).

REFERENCES

Aitken R.K.L., Bolton V. and Hargreave T. (1990) *Fertil. Steril.* **54**: 701.

Awad A.M., Elshaer S. L., Gangaraju R., Abdelaziz R.R. and Nader M.A. (2023) *Int. Immunopharmacol.* **125**: 111127.

Bagheri-Fam S.A.A., Svingen T., Combes A.N., Sinclair A.H., Koopman P. and Harley V. (2011) *Hum. Mol. Gen.* **20**: 2213.

Bagheri-Fam S., Alankarage D., Frost E.R. and Harley V.R. (2022) *Data Brief* **42**: 108230.

Barth A. (1986) *The Can. Vet. J.* **20**: 379.

Dai P., Ding M., Yu J., Gao Y., Wang M., Ling J., Dong S., Zhang X., Zeng X. and Sun X. (2024) *Toxics* **12**: 342.

Donald H.P. and Hancock J.L. (1953) *J. Agric. Sci.* **43**: 178.

Entwistle K. and Fordyce G. (2003) Evaluating and reporting bull fertility. Australian Assocaition of Cattle Veterinarians. Indooroopilly, Qld.

Fonseca P.A.D., dos Santos F.C., Lam S., Suárez-Vega A., Miglior F., Schenkel F.S., Diniz L.D.F., Id-Lahoucine S., Carvalho M.R.S. and Cánovas A. (2018) *J. Anim. Sci.* **96**: 4978.

Fordyce G.E.K., Norman S., Perry V., Gardiner B. and Fordyce P. (2006) *Theriogeniol.* **66**: 1140.

Fortes M.R.S., Porto-Neto L.R., Nguyen L.T., Freitas A.C., Melo T.P., Scalez D.C.B.S., Hayes B., Raidan F.S.S., Reverter A. and Boe-Hansen G.B. (2020) *Gen. Sel. Evol.* **52**: 46.

Hancock J.L. (1949) *Vet. Rec.* **61**: 308.

Kang H.M., Sul J.H., Service S.K., Zaitlen N.A., Kong S., Freimer N.B., Sabatti C. and Eskin E. (2010) *Nature Gen.* **4**: 348.

Moretti E.C.G., Scapigliatti G., Cosci I., Sartini B. and Baccetti B. (2005) *J. Submicro. Cyt. Pathol.* **37**: 297.

Porto-Neto L.R., Alexandre P.A., Hudson N.J., Bertram J., McWilliam S.M., Tan A.W.L., Fortes M.R.S., McGowan M.R., Hayes B.J. and Reverter A. (2023) *PLoS One.* **18**: e0279398.

Thundathil J.M.R., Palasz A., Barth A.D. and Mapletoft R. (2000) *Theriogeniol.* **54**: 921.

Tsutsui K., Bentley G.E., Bedecarrats G., Osugi T., Ubuka T. and Kriegsfield L.J. (2011) *Front. Neuroendocrinol.* **31**: 284.

Van den Berg I.B.D. and Lund M. S. (2016) *J. Dairy. Sci.* **99**: 8932.

VanRaden P.M. (2008) *J. Dairy Sci.* **91**: 4414.

Proc. Assoc. Advmt. Anim. Breed. Genet. 26: 403-406

PRELIMINARY INVESTIGATION OF THE GENETICS OF SHEDDING IN MATERNAL SHEEP

S.F. Walkom[1], J.R. Keiller[2], M.P. Kelly[3] and D.J. Brown[1]

[1] Animal Genetics Breeding Unit* University of New England, Armidale, NSW, 2351, Australia
[2] Cashmore Nudies, Cashmore, VIC, 3305, Australia
[3] Low Footprint Lamb, Croxton East, VIC, 3301, Australia

SUMMARY

There has been increasing interest from commercial sheep producers to shift to shedding breeds such as the Dorper and Wiltshire Horn, which has triggered the development of shedding composite lines. This study analysed the genetic expression of spring shedding in adult ewes using a single-step animal model, from which potential marker effects could be identified. Early investigations into the genetic architecture of shedding support previous literature which suggests that the trait is highly heritable and controlled by a dominant gene effect. However, further research of major SNP effects within this population and the multi-breed population within the Sheep Genetics LAMBPLAN is required.

INTRODUCTION

The viability of sheep production systems can be impacted by the varying prices of commodities, such as labour and feed, and proceeds from wool and meat (Rose *et al.* 2014). This is evident in self-replacing meat focussed flocks where the value of the high micron wool is outweighed by the increasing cost to remove the wool, shearing and crutching, meaning wool is considered a net loss to the production system. This has resulted in commercial sheep producers shifting to shedding breeds and triggering the development of shedding composite lines throughout the Australian sheep meat industry. In these cases, the breeders are looking to take advantage of desirable 'Terminal' and 'Maternal' genetic characteristics of the major non-shedding breeds/composites while also introducing shedding attributes. Consequently, both the commercial and seedstock sectors are interested in breeding from animals with a known genetic capacity to shed their fleece. Shedding animals are considered easy-care animals with not only the demand for shearing greatly reduced but also a reduction in the need for preventative treatment practices.

Assessed through visual shedding scores, wool-shedding ability exhibited moderate to strong genetic variation in UK, and American flocks (Pollot 2011; Matika *et al.* 2013; Vargas Juardo *et al.* 2019) with heritability estimates ranging from 0.26 to 0.54. Preliminary analysis of an earlier subset of the shedding data utilised within this study, reported similar heritability estimates and showed strong genetic correlations between shedding scored at different ages of recording (Guy *et al.* 2021). This aligned with the findings of Vargas Juardo *et al.* (2019) and Vargas Juardo *et al.* (2016) who showed strong correlations between lamb and adult shedding assessments, and that shedding scores were highly repeatable across shedding events in adult sheep, respectively. This is not surprising with Pollot (2011) reporting that the mode of inheritance in shedding characteristics matches that observed of a dominant major gene effect. Thus, Matika *et al.* (2013) proposed that seasonal fleece shedding should be regarded as the ancestral trait where the dominant nature of the target locus suggests that the derived trait, that is, fleece retention, is caused by the recessive mutation.

---

* A joint venture of NSW Department of Primary Industries and Regional Development and the University of New England

This study continues the analysis of shedding in two composite shedding flocks, previously reported by Guy *et al.* (2021), by analysing spring shedding expression in adult ewes using a single-step animal model, from which potential marker effects were identified.

MATERIALS AND METHODS

Shedding Data. Shedding was scored annually in late spring, between October and December, providing repeat records across 10 years within the two genetically linked composite shedding flocks in western Victoria. The flocks were developed from a composite maternal flock of primarily Coopworth origin, with the shedding characteristics infused via Dorper, Wiltipoll and Wiltshire Horn sires.

As per Guy *et al.* (2021), the wool-shedding score was based on the Sheep Genetics (2019) 1-5 scale. The shedding score refers to the animal's skin area covered in wool at the time of assessment. An animal with a score of 1 does not currently have wool longer than 1cm on any part of the body, it may have grown wool and completely shed, or it may be a hair sheep. An animal with a score of 5 is completely covered in wool (excluding the head and lower limbs). The three middle scores refer to the proportion of the body covered in wool and are independent of the pattern in which shedding is expressed. A score of 2 should reflect ~1-32% wool coverage, score 3 ~33-66% and score 4 ~67-99% wool coverage.

Statistical analysis. The analysed shedding traits were based on the expression of shedding of their first adult fleece (grown over winter) in a spring lambing flock. Three shedding traits were analysed within the study, 1) SS, shedding score (1 to 5 scale); 2) CS, fleece is Completely Shed (score 1 vs. scores 2-5); and 3) FF, No shedding evident (Full Fleece, score 5 vs. scores 1-4). The trait based on the shedding score was also analysed with a repeatability model, where repeat records (Table 1) were available across adult shedding periods (repeat across years).

Variance components and breeding values were estimated from a single-step genomic BLUP model incorporating genomic and pedigree information with a lambda of 0.75 (Legarra *et al.* 2014), using the WOMBAT linear model program (Meyer 2007). Shedding records were available for 6,226 animals, from two flocks, with an associated pedigree of 36,595 animals, with genotypes available for 11,435 animals. Animal information, pedigree and genotypes were sourced from the Sheep Genetics LAMBPLAN database (Brown *et al.* 2007). After the removal of SNPs based on QC checks, a combined imputed SNP genotype for 60,867 markers was constructed for all animals. In this preliminary study, breed effects were not accounted for in the genomic relationship matrix.

The fixed effects model included birth and rearing type (11, 21, 22, 31, 32, 33), sex (M, F), the number of days since the winter solstice (days, indication of days into the shedding period) as a linear covariate, and contemporary group. The contemporary group was defined as a concatenated term that included property, year of birth, year of shedding, and period of scoring (21-day window of recording, Table 1).

Marker effects. Wombat (Meyer 2007) was used to estimate marker effects for the shedding traits. Predicted marker effects were obtained as 'back solved solutions', through a linear transformation of the predicted breeding values 'single step' genomic BLUP, taking advantage of relationships between individuals and information from un-genotyped animals as per Aguilar *et al.* (2019). The derived probabilities were plotted as a Manhattan plot against the genome position of each SNP an arbitrary p-value of equal or less than 1.0×10^{-5} (blue line in figure 1) was considered as a threshold to identify regions of potential significance.

RESULTS AND DISCUSSION

The heritability of shedding score (1-5) within this population was moderate to high at 0.58 ± 0.03 (Table 1). This aligns with the previous literature estimates from pedigree-based animal models for populations of shedding composite sheep. Estimates ranged from 0.80 in UK Easycare sheep

(Matika *et al.* 2013), 0.47 to 0.59 in a US Romanov, White Dorper and Katahdin composites in the US (Vargas Jurado *et al.* 2020) and 0.54 in a NZ Wiltshire and Wiltshire-composite flock (Costilla *et al.* 2024). In support of the previous findings by Vargas Juardo *et al.* (2016), the shedding score was moderate to highly repeatable across adult shedding events, 0.54 ± 0.02 (Table 1). The heritability of the binary CS trait was lower at 0.17 (Table 1), which suggests that there may still be genetic variation in shedding beyond just that associated with a single dominant gene effect. The FF trait was explored to test the Matika *et al.* (2013) proposal that the fleece retention is caused by a recessive mutation and thus the threshold between scores 4 and 5 would be triggered by the dominant gene. However, whilst convergence was achieved the linear multifactorial model has limitations for this trait, where the low proportion of expression (4%, Table 1) exacerbates the binary nature, producing a heritability of 0.13 ± 0.02.

Table 1. Data summary and parameter estimates from the univariate analysis of spring shedding traits in composite shedding sheep

	First winter shedding			Repeat adult shedding
Trait	Full Fleece (1/0)	Completely Shed (1/0)	Shedding Score (1-5)	Shedding Score (1-5)
Animals	4,345	4,345	4,345	6,226
Records	4,345	4,345	4,345	10,629
CGs	31	31	31	147
Trait mean	0.04	0.26	2.46	2.55
Phenotypic Variance	0.03 ± 0.00	0.17 ± 0.00	1.12 ± 0.03	1.10 ± 0.02
Heritability	0.13 ± 0.02	0.46 ± 0.03	0.58 ± 0.03	0.45 ± 0.02
Repeatability				0.54 ± 0.02

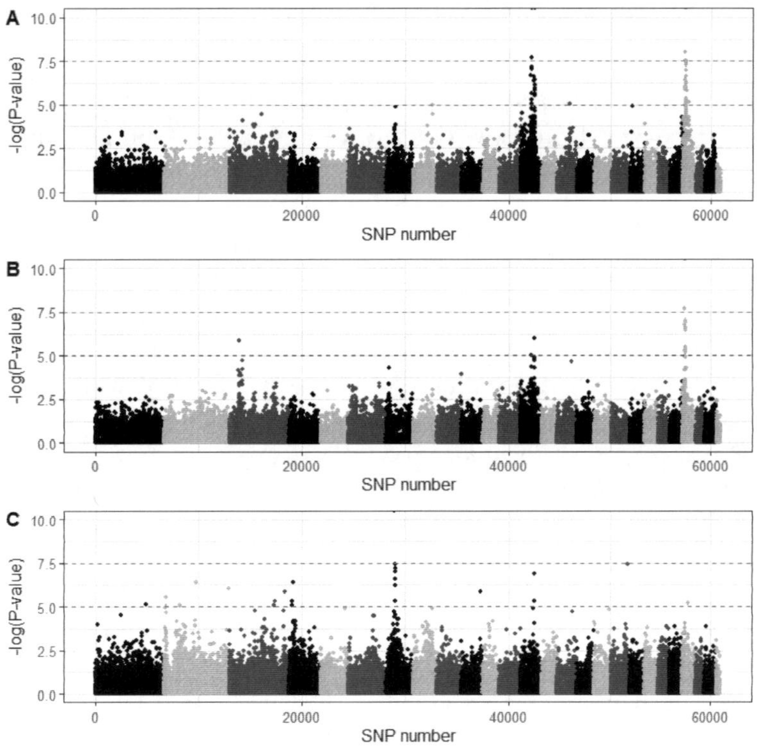

Figure 1. Significance of marker effects from back solved solutions from single-step univariate analysis for Shedding traits recorded on the animal's first adult fleece (A; Shedding score (SS, 1-5), B; Completely Shed (CS, 1/0), C; Full Fleece (FF, 1/0)

Estimates of the significance of marker effects back-solved from the single-step G-BLUP are presented in Figure 1. When shedding was analysed using the SS trait, regions of interest were identified on chromosomes 7, 8, 13, 15, 19 and 23. Marker effects from the CS trait reduced the significance of regions on chromosomes 13 and 23. A peak on chromosome 3 was also observed for the CS trait. A GWAS study of 200 Easycare-shedding sheep (Matika *et al.* 2013) noted a significant SNP on chromosome 3 but discounted it due to the low allele frequency. The small proportion of fully fleeced animals in this study limited the ability to describe the genetics of shed vs. fleeced animals. However, a noticeable peak was observed on chromosome 7 for the FF trait.

Early investigations into the genetic architecture of shedding in these composite shedding flocks align with the previous literature proposals for a dominant major gene effect, with several potential chromosomal regions identified. However, as with the study by Matika *et al.* (2013), the shedding gene in this population has been infused via a limited number of sire lines, thus noticeable regions of significance may reflect breed effects not necessarily the causative shedding gene. Additional analysis regarding the location of significant SNPs along the genome within this population and the multi-breed population within the Sheep Genetics LAMBPLAN are required, but also whether different chromosomal regions are associated with shedding in the wool breeds like the Wiltipol compared to the more hair breeds such as the Dorper.

CONCLUSION

Early investigations into the genetic architecture of shedding in an Australian composite shedding population support previous literature that suggests that the trait is highly heritable and controlled by a dominant gene effect. Consequently, genetic selection can proceed effectively even though marker effects are yet to be identified. Further research on the impact of significant SNPs for shedding within the Australian sheep population is required.

ACKNOWLEDGEMENTS

The research within this project (P.PSH.1421) was funded via the Meat and Livestock Australia Donor Company and contributions from Low Footprint Lamb and Cashmore Nudies.

REFERENCES

Aguilar I., Legarra A., Cardoso F., Masuda Y., Lourenco D. and Misztal I. (2019) *Genet. Sel. Evol.* **51**: 28.

Brown D.J., Huisman A.E., Swan A.A., Graser H.U., Woolaston R.R., Ball A.J., Atkins K.D. and Banks R.G. (2007) *Proc. Assoc. Advmt. Anim. Breed. Genet.* **17**: 187.

Costilla R., McIntyre S., Newman S.A., Stanley A., Campbell D., Lee M.A., Archer J.A., Dodds K.G. and Johnson P.L. (2024) *NZ J. of Agric. Res.* **67**: 402.

Guy S., Kellier J. and Brown D.J. (2021) *Anmi. Prod. Sci.* **61**: 82.

Legarra, A., Christensen O. F., Aguilar I. and I. Misztal (2014) *Livest. Sci.* **166**: 54.

Matika O., Bishop S.C., Pong-Wong R., Riggio V. and Headon D.J. (2013) *Anim. Gene.* **44**: 742.

Meyer, K. (2007) *J. Zhejiang Uni. Sci. B.* **8**: 815.

Pollott G. E. (2011) *J. Anim. Sci.* **89**: 2316.

Rose G., Mulder H.A., Thompson A.N., van der Werf J.H.J. and van Arendonk J.A. (2014) *Ag. Sys.* **131**: 94.

Sheep Genetics (2019) https://www.sheepgenetics.org.au/updates/news/visual-shedding-score-guidelines/

Vargas Jurado N., Leymaster K.A., Kuehn L.A. and Lewis R.M. (2016) *J. Anim. Breed. Gene.* **133**: 396.

Vargas Jurado N., Kuehn L.A. and Lewis R.M. (2020) *J. Anim. Breed. Gene.* **137**: 365.

Proc. Assoc. Advmt. Anim. Breed. Genet. 26: 407-410

GENETIC EVALUATION OF TAIL LENGTH IN AUSTRALIAN SHEEP

M.J. Hodge[1], D.J. Brown[2] and S.F. Walkom[2]

[1] Sheep Genetics, Meat & Livestock Australia, Armidale, NSW, 2350 Australia
[2] Animal Genetics Breeding Unit*, University of New England, Armidale, NSW, 2351 Australia

SUMMARY

Sheep tail docking is common practice to reduce the incidence of flystrike and stained fleece, yet it is associated with poor animal welfare. This study aims to explore genetic variation in tail length, which may offer the potential to reduce sheep tail length via genetic selection. Tail length is moderate to highly heritable (0.35±0.01-0.67±0.03) with low (0.14±0.01) to moderately high (0.37±0.01) genetic correlations with body weight. These results indicate tail length may be reduced via genetic selection, offering a welfare-friendly alternative to tail docking.

INTRODUCTION

Flystrike cost the Australian sheep industry in excess of $300 million in 2022 (Shephard *et al.* 2022). Tail length is a key indicator of breech flystrike, as long tails are associated with faecal- and urine-stained fleece which attract the sheep blowfly (James 2006). Currently, sheep producers dock tails at marking using either a hot knife or rubber ring (Grant 2004). However, both methods are associated with poor animal welfare and can negatively impact lamb growth rates. Whist retaining an appropriate tail length (approx. third or fourth tail joint) prevents rectal prolapse and cancer (Woodruff *et al.* 2023), public expectations of sheep welfare may result in tail docking being banned. Therefore, exploring welfare-friendly, cheaper alternatives like genetic selection to reduce sheep tail length, with minimal health and production impacts is important.

Past research found tail length to be moderate to highly heritable (0.38 – 0.68; Greeff *et al.* 2015; Johnson *et al.* 2023; Oberpenning *et al.* 2023; Teubes *et al.* 2023). Strong positive genetic correlations between tail length and body length, tail length and body weight at marking (Greeff *et al.* 2015; Teubes *et al.* 2023). However, these studies were focussed on single flocks of Merino sheep. Past studies also used varied in their method of phenotyping, hence, it is also unclear whether a subjective tail length score (1-5) is adequate, or whether an objective measurement (cm) may be more appropriate to define the tail length phenotype. This study investigated tail length across multiple flocks and breeds to quantify genetic variation in tail length in mixed breeds of sheep and phenotyping protocols, and its correlation with body weight.

MATERIALS AND METHODS

Phenotypes. Tail length and body weight records on 57,760 lambs from the Information Nucleus and MLA Resource Flock (van der Werf *et al.* 2010) and industry flocks were sourced from the Sheep Genetics database. Breeds represented in industry flocks included commercial shedders and a maternal composite from a Coopworth base. The dataset included phenotypes measured two ways – 1) assessed as a 1 to 5 score of tail length relative to the hock, or 2) measured in cm from the tip of the tail to the base (Greeff *et al.* 2015). Data editing removed sites with low numbers of animals phenotyped (n<500), animals with non-genetic influences on tail length performance (i.e., tail bitten off), where the average age of the contemporary group was <14 days and >85 days (to align with commercial sheep marking periods). Further data was removed including tail length records taken

* A joint venture of NSW Department of Primary Industries and Regional Development and the University of New England

at birth due to low record counts and when measured tail length records were outside of 4 standard deviations (SD) from the mean. The final dataset included animals from 6 flocks, measured across 12 years comprising 35,383 records for assessed tail length score and 8,964 records for measured tail length (Table 1). Where an animal had multiple body weights taken, the weight measured closest to the date of tail measurement (n=44,347) was used for the bivariate analysis.

Table 1. Summary statistics of tail length phenotypes used in the analysis

Trait	Assessed (score 1-5)	Measured (cm)	Scaled trait (1-5)
Flocks	6	2	6
Animal count	35,383	8,964	44,347
Sire breeds	12	25	27
CGs	344	74	416
Mean age at tail (SD)	29.17 (17.58)	30.76 (8.51)	45.45 (17.77)
Mean tail length (SD	2.82 (0.69)	24.77 (0.411)	2.94 (0.71)
Mean age at weight (SD)	51.04 (19.77)	31.39 (9.06)	47.32 (19.79)

Statistical analysis. To evaluate whether tail length can be adequately phenotyped using a subjective assessment (score 1-5) only (i.e., no measured (cm) trait), a third tail length trait was created (n=44,347) based on all records with both an assessed score and measured tail length. To create the scaled trait, the distribution of measured and assessed data was first evaluated, then a normalisation formula was applied to transform measured tail data to a scaled score (1-5) value that followed a normal distribution. A univariate animal model using a pedigree of 100,648 animals over 18 and 22 generations on the dam and sire side, respectively, was performed using ASReml (version 4.2) (Gilmour *et al.* 2021) for the three tail length traits. Fixed effects included lamb contemporary group (defined as site of measurement, sex, year of birth, and date of measurement), birth type (n=3), age in days at tail measurement (n=123), and age of the dam was fit as a categorical variable (n=12). A second univariate analysis was performed with all prior noted effects, as well as body weight as a covariate. The full model fit for each tail trait was

$$y = Xb + Za + e$$

where y, b, a, and e are the vectors of observed traits of animals, fixed effects, direct additive genetic effects, and residual effects, respectively, and X and Z are the incidence matrixes for the fixed and direct additive genetic effects, respectively. Bivariate analysis was performed between the tail traits and body weight. The fixed effects previously defined were fitted with additional fixed effects for body weight (age at body weight measurement (n=138), rear type (n=3), and a random maternal genetic effect). Maternal permanent environment and sire breed effects were found to be low and not significantly different from 0, respectively, (results not reported) and excluded from the model.

RESULTS AND DISCUSSION

Heritability. Variance components for the tail length traits with and without body weight fitted as a co-variate are shown in Table 2. Heritability estimates ranged from 0.34 to 0.67 which are similar to those reported in the literature (Greeff *et al.* 2015; Johnson *et al.* 2023; Oberpenning *et al.* 2023). The heritability of measured tail length was significantly higher than both the 1-5 score estimates. This is likely due to different progeny- and sire-breeds and phenotyping protocol (Table 1), as well as any reduced accuracy resulting subjective assessment compared with objective measurement. As more sire breeds were represented in the progeny with measured tail length, there was greater variation in measured tail length which may explain its higher heritability. Both the 1-5 assessed and scaled traits were associated with a lower heritability compared to the measured trait (Table 2). The scaled trait also had a high number of sire breeds represented therefore, it may be possible that the lower associated heritability for both assessed and scaled (1-5) tail length is a result of the reduced variation from a subjective assessed score compared to an objective measured trait. Greeff *et al.* (2015) reported similar heritability estimates for both tail length and reported no significant influence of maternal permanent environmental effects. Further, when body weight at lamb marking was fit as a covariate the tail length heritability was reduced.

Table 2. Genetic parameter estimates and standard errors (± SE) for tail length traits

Tail measurement	Weight fit as covariate	Phenotypic variation ± SE	Genetic variation ± SE	Heritability ± SE
Assessed (1-5 score)		0.44±0.00	0.18±0.01	0.41±0.01
Measured (cm)	No	9.59±0.16	6.46±0.36	0.67±0.03
Scaled trait (1-5)		0.37±0.00	0.13±0.01	0.35±0.01
Assessed (1-5 score)		0.43±0.00	0.18±0.01	0.41±0.01
Measured (cm)	yes	8.56±0.15	5.56±0.32	0.65±0.03
Scaled trait (1-5)		0.37±0.00	0.13±0.01	0.34±0.01

Genetic correlations. Phenotypic correlations ranged from 0.24 to 0.51, and genetic correlations ranged from 0.14 to 0.37, depending on the method of tail length phenotyping (Table 3). Correlations reported in the present study align with Oberpenning *et al.* (2023), however are lower than those reported by Greeff *et al.* (2015). This may be due to Greeff *et al.* (2015) measuring body weight and tail length measurement on the same day compared to the present study where this was not necessarily the case. Age at phenotyping (both tail length and body weight) was fit in the model to try to account for this. While the present study did not have enough records to evaluate the relationship between tail length and body length, past research by Greeff *et al.* (2015) noted body weight and body length adjustments at lamb marking removed any genetic correlation between tail and body length. Hümmelchen *et al.* (2025) reported small differences in body weight between short and long tailed groups. Genetic correlations between assessed and measured tail length measurement (cm) were able to be estimated in the present study due to singularities resulting from insufficient data. This warrants further investigation.

Table 3. Genetic and phenotypic correlations (± SE) for tail length and body weight

Body weight	Assessed (1-5 score)	Measured (cm)	Scaled trait (1-5)
Genetic correlation ±SE	0.14±0.01	0.37±0.01	0.16±0.01
Phenotypic correlation ±SE	0.24±0.03	0.51±0.04	0.31±0.03

Recommendations. Reducing tail length via genetic selection may be a viable, welfare friendly alternative to tail docking as it is moderate to highly heritable. However, the heritability of assessed tail length was approximately 20% lower than measured tail length. Given there was no significant difference the heritability estimates of assessed tail length and the scaled trait, it may be a viable option to develop breeding values based on assessed tail length in industry flocks (enabling greater levels of trait recording due to simplicity) and the scaled traits based on measured tail length in resource flocks. Despite the unfavourable genetic correlation between tail length and body weight, sheep producers may make gains in both traits by identifying and selecting animals which perform favourably for both traits or using a selection index appropriately weighting both traits. The relationship between tail length, body length, and carcase structure requires further investigation, as past studies have reported carcase abnormalities in shorter tailed sheep (James 2006). Assessment of tail length as a 1-5 score, with tail length relative to the hock (Table 4) as recommended by Greeff *et al.* (2015) could be a viable option for industry sheep flocks to genetically reduce tail length.

Further, it would be worthwhile assessing and measuring tail length in MLA research flocks to allow for future investigation into the relationship between both traits. While the dataset used in this study is larger than past studies, it is still relatively small, and it would therefore be worthwhile re-estimating genetic parameters following the collection of more tail length measurements according to the protocol suggested. It would therefore be worthwhile recommending measuring body weight at the same time as phenotyping for tail length for any future re-estimates of genetic parameters.

Table 4. Recommended scoring guide for measuring tail length

Score	Tail length
1	Tail is half way from the base of the tail to the hock
2	Tail is three quarters of the way from the base of the tail to the hock
3	Tail is at the hock
4	Tail is longer than the hock
5	Tail is towards the pasterns/very long

CONCLUSION

The moderate to high heritability of assessed and measured tail length indicates that genetic selection offers a welfare friendly alternative to tail docking. However, the unfavourable genetic correlation with liveweight needs to be accounted for in breeding programs. Phenotyping tail length using an assessment (1-5 score) provides a simple, quick measure of tail length for sheep breeders interested in genetically reducing tail length. Further work is required to estimate genetic correlations between tail length and additional important to sheep production traits like carcase composition and quality, wool quality and production, and reproduction.

ACKNOWLEDGEMENTS

The authors gratefully acknowledge the Sheep CRC for Sheep Industry Innovation Information Nucleus and MLA Resource flock, and individual industry flocks involved in and data collection.

REFERENCES

Gilmour A.R., Gogel B.J., Cullis B.R., Welham S.J. and Thompson R. (2021) ASReml User Guide Release 4.2, V.I. Ltd, Editor. 2021, Hemel Hempstead: UK. HP2 4TP.

Grant C. (2004) *Appl. Anim. Behav. Sci.* **87***:*255.

Greeff J.C., Karlsson L.J.E. and Schlink A.C. (2015) *Proc. Assoc. Advmt. Breed. Genet.* **21**: 237.

Hümmelchen H., Wagner H., Brügemann K,. König, S. and Wehrend A. (2025) *Vet. Med. Sci.* **11**: 70138.

James P.J. (2006) *Aust. J. Exp. Agric.* **46***:* 1.

Johnson P.L., Scobie D.R., Dodds K.G., Powdrell S.J.H., Rowe S.J. and McRae, K.M. (2023) *J. Anim. Sci.* **101**: 161.

Oberpenning J., Bohlouli M., Engel P., Hümmelchen H., Wagner H., Wehrend A. and König S. (2023) *J. Anim. Breed. Genet.* **140**: 132.

Shephard R., Ware J.W., Blomfield B. and Neithe G. (2022) Final Report Project B.AHE.327. Meat & Livestock Australia, North Sydney.

Teubes M., Cloete S.W.P. Dzama, K. and Scholtz A.J. (2023) *Anim. Prod. Sci.* **63**: 1043.

Woodruff M., Munoz C., Coleman G., Doyle R. and Barber S. (2023) *Animals* **13**: 963.

van der Werf J.H.J., Kinghorn B.P. and Banks R.G. (2010) *Anim. Prod. Sci.* **50**: 998.

ASSESSING GENETIC VARIABILITY IN FLYSTRIKE RESISTANCE ACROSS EXPRESSION LEVELS IN AUSTRALIAN MERINO SHEEP

E. Dehnavi[1], A.A. Swan[1], B.C. Swain[2], J.C. Greeff[3], J.L. Smith[4], G. Burbidge[5], A.M.M. Ramsay[6], T.L. Bird-Gardiner[7] and D.J. Brown[1]

[1] Animal Genetics Breeding Unit*, University of New England, Armidale, NSW, 2351 Australia
[2] BCS Agribusiness, Gartmore, Gunnedah, NSW, 2380 Australia
[3] Department of Primary Industries and Regional Development, Perth, WA, 6000 Australia
[4] CSIRO Agriculture and Food, Armidale, NSW, 2350 Australia
[5] Burbidge Farms, Tarcutta, NSW, 2652 Australia
[6] Stenhouse Consulting, Brunswick East, VIC, 3057 Australia
[7] Department of Primary Industries and Regional Development, Agricultural Research Centre, Trangie, NSW, 2823 Australia

SUMMARY

This work aimed to provide guidance to breeders on the level of flystrike expression required to achieve effective selection and to develop genomic reference data for future prediction. Genetic parameters and breeding values were compared across different expression thresholds (1% to 10%) using data sourced from 21 research and industry breeder flocks. Among the thresholds examined (1% to 10%), a 2.5% threshold was adequate for excluding data with no genetic variation for breech strike while retaining sufficient data for analysis. Previously, a 5% threshold was considered effective for capturing data with adequate variation, primarily from research flocks with minimal flystrike prevention practices. Lowering this threshold to 2.5% enables greater breeder participation by making it more achievable to obtain information for selection purposes. As flystrike resistance has proven to be heritable, breeding for genetic resistance offers a sustainable, long-term solution, helping to reduce flystrike. Additionally, this change supports a broader breeding approach to reduce flystrike, potentially minimising reliance on chemical treatments and management practices.

INTRODUCTION

The Australian sheep industry faces significant financial and animal welfare challenges due to flystrike in sheep, a common and debilitating infection (Shephard *et al.* 2022). Research has demonstrated flystrike resistance is heritable, making genetic resistance a promising and sustainable solution for long-term control through breeding programs (Smith *et al.* 2009; Dehnavi *et al.* 2023). Establishing a well-designed sheep reference population is essential for developing genomically enhanced Australian Sheep Breeding Values (ASBVs; Brown *et al.* 2010), particularly given the varying incidence rates of flystrike across different environments (Bird-Gardiner *et al.* 2013; Greeff *et al.* 2014; Smith *et al.* 2009). This study aims to estimate genetic parameters using data collected from a range of Australian Merino sheep populations, including research and seedstock flocks, with documented cases of flystrike across varying expression levels. The findings aim to guide breeders on the adequate level of flystrike expression required to facilitate the effective selection and development of a reference population. The Flystrike Genomics Reference Flock project (funded by Australian Wool Innovation) is a three-year project aimed at increasing the phenotypic and genomic information needed to establish a flystrike ASBV.

---

* A joint venture of NSW Department of Primary Industries and Regional Development and the University of New England

MATERIALS AND METHODS

Data. The dataset comprised 39,998 breech strike (BRS) records (1/0) from young sheep (up to yearling age), including 20,806 ewes and 19,192 wethers and rams, alongside 26,136 genotypes with 61,998 markers on these animals. This dataset was collected from 21 research and industry seedstock flocks. The pedigree consisted of 79,484 animals from 6,805 sires and 36,597 dams. Data filtering was applied across a range of expression thresholds ranging from 1% to 10% per flock, year of birth and sex (Table 1). A minimum of 20 animals per contemporary group (defined as flock, year of birth and sex) were included. At each expression threshold, the data was partitioned into two subsets: The *Excluded* subset, consisting of records below the threshold level due to low incidence rates within each of the flock-year of birth-sex groups, and the *Included* subset, containing records where the incidence rate was equal to or greater than the threshold levels.

Statistical Analyses. Analyses were conducted using single-step REML in BLUPF90 software (Misztal *et al.* 2014). Initial bivariate analyses of subsets (*Excluded* and *Included*) revealed non-significant genetic correlations, particularly at lower thresholds (data not shown). Consequently, separate univariate analyses were conducted for each of the *Excluded* and *Included* subsets at each threshold level of expression. BRS records (1/0) were standardised (mean=0, SD=1) for strike incidence rate within the flock, year of birth, and sex to account for heterogeneous variance (Dehnavi *et al.* 2024b). Linear mixed models included contemporary groups (flock, year of birth, and sex) and the birth-rear type interaction as fixed effects, with the animal's direct genetic effect fitted as a random effect. Variance components estimated from single-step REML analysis were used to compute breeding values.

Genotype data were incorporated using the H matrix in single-step genomic best linear unbiased prediction (Misztal *et al.* 2014). Heritability estimates were obtained for each subset across the different threshold levels. Because of the non-significant genetic correlation between *Excluded* and *Included* subsets (data not shown), Pearson correlations were calculated to compare research breeding values (RBVs) of common sires and their progeny between the two subsets, contrasting RBVs derived from the *Excluded* data with those obtained from the *Included* subset (Figure 1).

Table 1. Summary of the number of sires (Nsire) and their progeny count (Nprog) categorised by expression threshold (Exc: *Excluded*, Inc: *Included*). The proportion of sires in common across subsets and their respective progeny counts (Link%) are provided in parentheses

Count\ Threshold	1%	2%	2.5%	3%	4%	5%	8%	10%
NsireExc	348	523	597	653	718	756	825	880
(Link%)	(32%)	(43%)	(46%)	(43%)	(34%)	(28%)	(27%)	(27%)
NsireInc	760	698	676	625	520	454	392	351
(Link%)	(15%)	(32%)	(41%)	(45%)	(46%)	(47%)	(56%)	(67%)
NprogExc	4,358	4,898	6,064	4,256	4,615	4,333	5,009	5,084
(Link%)	(39%)	(35%)	(39%)	(23%)	(21%)	(18%)	(18%)	(16%)
NprogInc	4,093	6,925	7,949	6,342	6,345	5,270	5,503	4,725
(Link%)	(15%)	(28%)	(34%)	(31%)	(38%)	(37%)	(53%)	(60%)

RESULTS AND DISCUSSION

As the threshold increased from 1% to 10%, a significant reduction in the number of records available for the *Included* subset was observed. The number of sires in the *Excluded* subset (NsireExc) rises steadily, peaking at 880 sires at the 10% threshold. Conversely, the number of sires in the *Included* subset (NsireInc) decreases with increasing threshold levels, while their linkage percentage exhibits an inverse pattern, rising from 15% linkage at the 1% threshold to 67% linkage at the 10% threshold. These patterns highlight the trade-off between data quantity and the genetic linkage captured between the two subsets (Table 1).

Figure 1 illustrates the number of records in the *Excluded* (Exc) and *Included* (Inc) subsets across various expression thresholds, along with relevant heritability estimates (h^2). The heritability for the entire unfiltered dataset was low at 0.05 ± 0.01. However, as the thresholds increased, heritability estimates for the high-incidence subset (*Included*; Inc) improved significantly, reaching a maximum of 0.17 ± 0.02 at the 10% threshold. This suggests that applying higher expression thresholds enhances heritability but reduces the number of animals in the analysis.

These heritability estimates are lower than those reported in previous studies, such as 0.32 ± 0.11 (Smith *et al.* 2009) and 0.57 ± 0.13 (Greeff *et al.* 2014) for breech strike using different models and data transformation. However, Greeff *et al.* (2021) later reported a heritability of 0.07 ± 0.01 using a linear model on logarithm-transformed breech strike counts. These differences may stem from variations in trait definition, modelling approaches, scaling and environmental and management factors influencing trait expression. Notably, none of the earlier studies applied thresholds on expression levels.

Higher incidence rates are often associated with increased heritability on the liability scale when running a threshold model (Dahlqwist *et al.* 2019). This aligns with the observed trend in this study, where applying higher expression thresholds led to greater heritability estimates. This occurs because a greater proportion of individuals exceeding the expression threshold leads to higher observable genetic variance. However, this relationship is complex, as heritability estimates are influenced by both the proportion of individuals exceeding the threshold and the underlying genetic variance, which may not increase linearly with incidence (Dahlqwist *et al.* 2019).

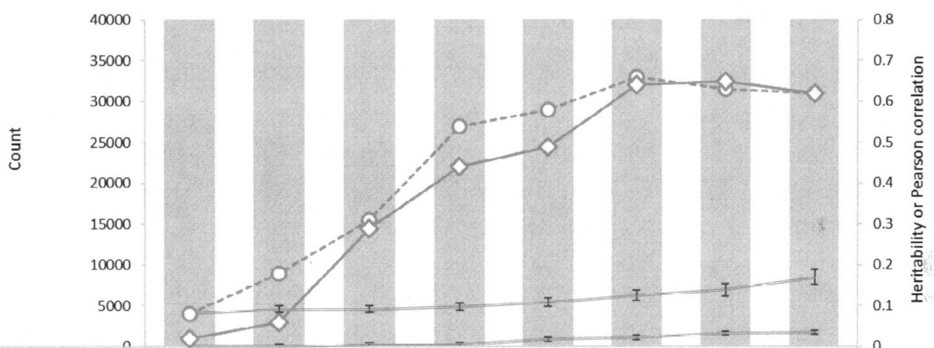

Threshold (%)	1	2	2.5	3	4	5	8	10
N (Exc)	11,403	14,184	15,681	18,837	22,350	25,519	29,263	31,845
N (Inc)	28,595	25,814	24,317	21,161	17,648	14,479	10,735	8,153
h2 (Exc)	0.00	0.00	0.00	0.01	0.02	0.02	0.03	0.03
h2 (Inc)	0.08	0.09	0.09	0.10	0.11	0.13	0.14	0.17
RBVs' cor common prog	0.08	0.18	0.31	0.54	0.58	0.66	0.63	0.62
RBVs' cor common sire	0.02	0.06	0.29	0.44	0.49	0.64	0.65	0.62

Figure 1. Number of records (orange and blue bars) and heritability (double orange and blue lines) in *Excluded* (Exc) and *Included* (Inc) subsets. The subsets represent different levels of expression thresholds (1%, 2%, 2.5%, 3%, 4%, 5%, 8%, and 10%). The Pearson correlation was assessed between the RBVs of common sires and their progeny (RBVs' corr common prog or sire) derived from *Excluded* data with those derived from *Included* data (solid and dotted green lines)

Figure 1 also highlights the comparison of breeding values of common sires and their progeny between the two subsets (solid and dotted green lines). A comparison was conducted between the RBVs of common sires and their respective progeny across the two (*Excluded* and *Included*) subsets. At lower thresholds, the weaker pedigree linkage between the subsets, and the low heritability resulted in more diverse RBV predictions, as indicated by the lower Pearson correlation (solid and dotted green lines in Figure 1). Increasing the threshold levels strengthened the pedigree linkage, improving the Pearson correlation of RBVs for common individuals between the two subsets.

Although heritability estimates remain adequate within the *Included* subset (Inc; 0.08 to 0.09), the low linkage and reduced correlation between RBVs of common sires at these thresholds (1% to 2%) highlight the importance of incorporating high-incidence records. While a threshold of 5% appears reasonable (Dehnavi *et al.* 2024a), lowering the threshold to 2.5% is a viable compromise to enable a great level of participation. Below this level, heritability and genetic variance are minimal (with phenotypic variance approaching 1), indicating that excluding the low-incidence subset (*Excluded*) would preserve genetic information and allow for accurate selection decisions based on genetic merit. This adjustment balances the number of excluded records, prediction accuracy, and genetic variation while encouraging breeders to actively use breeding values for selection for flystrike resistance.

CONCLUSION

Identifying the most appropriate threshold for flystrike expression remains an important consideration to ensure the most accurate breeding values are provided to breeders and enable future genomic selection. The selected threshold impacts both heritability estimates and the correlations of predicted breeding values significantly. Higher thresholds improve both heritability estimates and the accuracy of breeding value predictions in sheep selection programs. However, to encourage breeders to participate effectively in direct selection for flystrike, a threshold of 2.5% appears optimal, as it excludes records with minimal genetic variation while maintaining selection efficiency for more breeders. To ensure broader applicability, this approach will be evaluated further for body strike in young sheep and records of both breech and body flystrike across later ages.

ACKNOWLEDGEMENTS

This analysis was funded by Australian Wool Innovation Ltd (Project ON-00775). It used data provided by the Sheep CRC Information Nucleus Flocks (INF), Merino Lifetime Productivity Project (MLP), Commonwealth Scientific and Industrial Research Organisation (CSIRO), the Department of Primary Industries and Regional Development Western Australia (DPRID), the Australian Merino Sire Evaluation Association (AMSEA) and breeder flocks including Burbidge farms. Acknowledgement is also made to the Australian Government, who support the research, development, and marketing of Australian wool.

REFERENCES

Bird-Gardiner T.L., Brown D.J., Smith J. and Mortimer S.I. (2013) *Proc. Assoc. Advmnt. Anim. Breed. Genet.* **20**: 183.

Brown D.J., Swan A.A. and Gill J.S. (2010) *Anim. Prod. Sci.* **50**: 1060.

Dahlqwist E., Magnusson, P.K.E., Pawitan Y. and Sjölander, A (2019) *Hum. Genet.* **138**: 425.

Dehnavi E., Swan A.A. and Brown D.J. (2024a) 'Milestone report' AWI ON-00775.

Dehnavi E., Swan A.A., Burbidge G. and Brown D.J. (2024b) *Proc. Aus. Assoc. Anim. Sci.* **35:** 70.

Dehnavi E., Swan A.A., Ramsay A.M.M., Bird-Gardiner T.L. and Brown D.J. (2023) *Proc. Assoc. Advmnt. Anim. Breed. Genet.* **25**: 354.

Greeff J.C., Karlsson L.J.E. and Schlink A.C. (2014) *Anim. Prod. Sci.* **54**: 125.

Greeff J.C., Schlink A.C. and Karlsson L.J.E. (2021) *Anim. Prod. Sci.* **61**:1893.

Misztal I., Tsuruta S., Lourenco D.A.L., Aguilar I., Legarra A. and Vitezica Z. (2014) 'Manual for BLUPF90 family of programs'.

Shephard, R., Webb Ware, J., Bloomfield B. and Niethe, G. (2022) Final report of project B.AHE.0327, Meat and Livestock Australia Ltd.

Smith J.L., Brewer H.G. and Dyall T. (2009) *Proc. Assoc. Advmnt. Anim. Breed. Genet.* **18**: 334.

Proc. Assoc. Advmt. Anim. Breed. Genet. 26: 415-418

THE ROLE OF CONFORMATION TRAITS IN EVALUATING HEAT TOLERANCE IN DAIRY CATTLE

L.M. Jensen[1,2], M. Haile-Mariam[1], S. Bolormaa[1] and J.E. Pryce[1,2]

[1] Agriculture Victoria Research, AgriBio, Bundoora, VIC, 3083 Australia
[2] School of Applied Systems Biology, La Trobe University, Bundoora, VIC, 3083Australia

SUMMARY

Dairy cows are highly susceptible to heat stress, and this has a detrimental effect on milk production, reproductive performance, and animal welfare. These adverse effects will grow as dairy producing regions continue to be impacted by climate change. To address this, Australia implemented a genetic evaluation for heat tolerance in 2017. The objective of this study was to assess the relationships between the heat tolerance breeding values and some conformation traits in Australian Holstein and Jersey cows. The heat tolerance slope, which is the decline in milk, fat, and protein yield as the temperature-humidity index (THI) increases was calculated using a random regression model for heat tolerance, was correlated with seven conformation traits. Angularity had the strongest genetic and phenotypic correlation with heat tolerance slope traits. However, this could be an artifact of angularity's relationship to mean milk production, as heat tolerance phenotypes and milk production traits are strongly negatively correlated. Conformation traits are related to heat tolerance proportionally to how they are related to milk yield and therefore are not good candidates proxy traits for more complex heat tolerance traits.

INTRODUCTION

Acute and chronic exposure to high ambient temperatures causes reduced milk production, reduced fertility, and increased morbidity (Hansen 2009; Wheelock *et al.* 2010; Ouellet *et al.* 2019). A recent study has shown the world has likely surpassed the 2°C threshold of the 2015 Paris agreement (Esper *et al.* 2024); the focus to help cattle adapt to warmer climate conditions is now more important than ever. In pasture-based systems, like Australia, implementing permanent, long-term mitigation strategies such as genetic selection for heat tolerance is important to ensuring continued milk production as temperatures continue to increase. Heat tolerance is defined as the cow's ability to maintain production during heat stress. Various phenotypes related to heat tolerance have been shown to be heritable such as respiratory rate, rectal temperature (Dikmen *et al.* 2012) and milk yield (Ravagnolo *et al.* 2000; Nguyen *et al.* 2016). In fact, the rate of decline in milk yield under heat stress has been used to calculate breeding values in several countries with Australia being the first to implement these in the national genetic evaluation (Nguyen *et al.* 2017). While these methods allow for genetic selection of heat tolerance, the phenotype is inextricably entwined with milk production which has a negative genetic correlation with milk yield.

Linear type classification (conformation) data provides a large repository data set within the Australian dairy cattle population. Many registered cows are classified at least once in their lifetime, providing data on the cows' physical characteristics by highly trained group of evaluators. Currently, twenty-four linear traits are measured and recorded in both the Holstein and in the Jersey classification systems.

Previous studies have reported negative genetic correlation between body depth and many health and fertility traits, and it may be inferred that body depth may have an antagonistic relationship with heat tolerance given the positive genetic correlations between heat tolerance and health traits (Haile-Mariam *et al.* 2004). In part this may be due to the increased surface area to mass ratio (SA:M) found in cattle with smaller body depth. Sensible heat loss, or the physical transfer of heat without a phase change, is thought to be dependent on an animal's surface area per unit of body weight (Hansen

2004). Brahman, *Bos indicus*, have a significantly longer epidermis than Angus, *Bos taurus*, and this may play a role in the improved heat tolerance of Brahman cattle (Mateescu *et al.* 2023). Conformation traits could be indicative of heat tolerance as they pertain to more difficult to measure phenotypes such as SA:M. This study investigated the usefulness of conformation traits to evaluate heat tolerance in Holstein and Jersey cattle.

MATERIALS AND METHODS

Conformation data. Data on 115,490 Australian Holsteins and 41,889 Australian Jerseys cows born between 1999 and 2021 were used in this study. Classification records for primiparous cows were retrieved from the DataGene database. Conformation traits, Table 1, thought to be related to heat tolerance were selected with pin set used as a comparison, given the hypothesis that it lacked relationships to any perceived biological process involved in heat tolerance. Conformation traits are scored on a linear scale of 1-9, except pin set which is scored according to biological extremes with 5 serving as the mid-point and 1 and 9 the extremes in opposite directions.

Table 1. Conformation trait record numbers, mean, and standard deviation for Holstein and Jersey cattle

Trait	Holstein			Jersey		
	No. Records	Mean	SD	No. Records	Mean	SD
Body Depth	115260	6.13	1.25	41878	6.20	0.85
Body Length	*n.a.*	*n.a.*	*n.a.*	40756	6.08	1.16
Stature	115458	6.51	1.40	41892	5.83	1.03
Chest Width	115490	5.64	1.17	41894	5.55	0.91
Pin Width	115490	6.30	1.27	41879	5.50	0.94
Angularity	115490	5.89	1.15	41843	6.58	0.85
Pin Set	115490	4.09	1.30	41889	5.34	0.83

Production Data. For each cow, the slope (HTS) and intercept (HTI) of the random regression model for milk, fat, and protein yield as THI increases were calculated as per the current Australian breeding value for heat tolerance described in Nguyen et al. (2016). The model includes the second-order orthogonal polynomial THI interacting with a fixed regression coefficient of milk, fat, or protein yield on THI and the random regression coefficient on heat load in addition to several fixed effects to account for environmental variability. Temperature-humidity index (THI) was calculated using the equation: $THI = T_{db} + 0.36T_{dp} + 41.2$, where T_{db} is dry-bulb temperature and T_{dp} is dew-point temperature (Yousef 1985).

Statistical analysis. A tri-variate pedigree-based REML analysis was used to calculate the genetic and phenotypic correlations between each of the conformation traits and heat tolerance slope (HTS) and heat tolerance intercept (HTI) traits. The variance components were estimated using ASReml (v. 4.2.1) fitting an animal model. HTS and HTI for milk, fat, and protein yields were both adjusted for year-season of calving. Confirmation traits were adjusted for age at classification, days in milk at classification, and the classifier-round by herd effect by including them in an animal model analysis.

RESULTS AND DISCSSION

The genetic correlations between HTS milk and the conformation traits ranged from -0.01 (chest width) to -0.26 (angularity) in Holsteins (Table 2) and 0.01 (chest width) and -0.38 (angularity) in Jerseys (Table 3). The genetic correlations between conformation traits and HTS fat or HTS protein were similar to HTS milk in Holsteins. However, in Jersey, the genetic correlations between conformation traits and HTS fat or HTS protein were greater than HTS milk. Body depth had a genetic correlation of -0.15 with HTS milk and -0.25 with HTS fat (Table 3). Perhaps this is due to Jerseys producing more fat and protein in the milk; therefore, reduced energy consumption from heat stress effects milk composition more than volume for Jerseys.

The genetic correlations between HTI milk and the conformation traits ranged from 0.01 (chest width) and 0.23 (angularity) for Holstein and from 0.02 (chest width) and 0.37 (angularity) for Jerseys (not presented). These results align with the negative genetic correlations between HTI and HTS of -0.96, --.97, and -0.97 for Holsteins and of -0.97, -0.99, -0.98 for Jerseys for milk, fat, and protein respectively.

Table 2. Conformation trait heritability and genetic correlation with Heat Tolerance Slope for Holstein cows

Trait	h^2	Milk r_g*	Fat r_g	Protein r_g
Body Depth	0.27	-0.13 ± 0.03	-0.17 ± 0.03	-0.11 ± 0.03
Stature	0.34	-0.16 ± 0.03	-0.11 ± 0.03	-0.20 ± 0.03
Chest Width	0.17	-0.01 ± 0.04	-0.05 ± 0.04	-0.04 ± 0.04
Pin Width	0.29	-0.17 ± 0.03	-0.15 ± 0.03	-0.18 ± 0.03
Angularity	0.17	-0.26 ± 0.03	-0.27 ± 0.04	-0.24 ± 0.04
Pin Set	0.33	-0.05 ± 0.03	0.01 ± 0.03	-0.03 ± 0.03

* r_g = genetic correlation

Table 3. Conformation trait heritability and genetic correlation with Heat Tolerance Slope for Jersey cows

Trait	h^2	Milk r_g*	Fat r_g	Protein r_g
Body Depth	0.23	-0.15 ± 0.03	-0.25 ± 0.03	-0.18 ± 0.03
Body Length	0.26	-0.19 ± 0.03	-0.25 ± 0.03	-0.25 ± 0.03
Stature	0.37	-0.18 ± 0.03	-0.28 ± 0.03	-0.25 ± 0.03
Chest Width	0.14	0.02 ± 0.04	-0.18 ± 0.04	-0.07 ± 0.04
Pin Width	0.19	-0.13 ± 0.04	-0.24 ± 0.04	-0.20 ± 0.04
Angularity	0.20	-0.37 ± 0.03	-0.33 ± 0.4	-0.38 ± 0.04
Pin Set	0.27	-0.06 ± 0.03	-0.06 ± 0.03	-0.06 ± 0.03

* r_g = genetic correlation

The relationship between HTS and conformation traits in this study indicates that conformation traits had weak to moderate relationships with heat tolerance traits in both Holstein and Jersey cows, and the strongest correlations were obtained with angularity (range: -0.24 to -0.38) (Table 2 and 3). Angularity describes the angle, openness, and spring of rib that cattle possess with the ideal rib having adequate spacing between the ribs, angled back from the chest towards the flank, and springing out from the vertebrae when viewed from behind. The negative correlation with HTS indicates that more angular cows are more susceptible to the effects of heat stress. Given the positive genetic correlation of 0.48 between angularity and milk yield (Berry *et al.* 2004), the relationship between HTS and angularity could simply be a by-product of the relationship with milk production. This could be tested using the HTS adjusted for HTI (heat tolerance independent from milk production). Despite evidence of body size differences impacting heat tolerance in beef cattle, this may largely be due to differences between *Bos indicus* and *Bos taurus* animals or different breeds (Mateescu *et al.* 2023; Madhusoodan *et al.* 2019). Given that dairy cattle selection mainly happens within breed, conformation trait differences may not have enough genetic variation to be strong indicators of HTS. Moreover, milk production and the associated metabolic heat production are strongly correlated with many conformation traits, and that relationship may be stronger than that between conformation traits and heat tolerance mechanisms.

CONCLUSION

Angularity had the greatest genetic correlation with heat tolerance slope, -0.24 and -0.38 with in Holstein and Jersey respectively. Considering there was only low to medium correlation between heat tolerance and conformation traits and that this is possible an indirect influence of milk production, the conformation traits are not deemed good candidates to use as a proxy trait for heat tolerance. These results also show that the selection for conformation traits or HTS does have any intended negative effect.

ACKNOWLEDGEMENTS

This work would not have been possible without the support and data sharing from Holstein Australia and DataGene. The DairyBio research program and the Gardiner Foundation contributed financial support for this work.

REFERENCES

Berry D. P., Buckley F., Dillon P., Evans R.D. and Veerkamp R.F. (2004) *Irish J. Agr. Food Res.* **43**: 161.
Dikmen S., Cole J.B., Null D.J. and Hansen P.J. (2012) *J. Dairy Sci.* **95**: 3401.
Esper J., Torbenson M. and Büntgen U. (2024) *Nature* **631**: 94.
Haile-Mariam M., Bowman P.J. and Goddard M.E. (2004) *Aust. J. Agr. Res.* **55**: 77.
Hansen P.J. (2009) *Philos. T. Roy. Soc. B* **364**: 3341.
Hansen P.J. (2004) *Anim. Reprod. Sci.* **82**: 349.
Madhusoodan A.P., Sejian V., Rashamol V.P., Savitha S.T., Bagath M., Krishnan G. and Bhatta R. (2019) *J. Anim. Behav. Biometeorol.* **7**: 104.
Mateescu R.G., Sarlo Davila K.M., Hernandez A.S., Nunez Andrade A., Zayas G.A., Rodriguez E.E., Dikmen S. and Oltenacu P.A. (2023) *Front. Genet.* **14**: 1107468.
Nguyen T.T.T., Bowman P.J., Haile-Mariam M., Nieuwhof G.J., Hayes B.J. and Pryce J.E. (2017) *J. Dairy Sci.* **100**: 7362.
Nguyen T.T.T., Bowman P.J., Haile-Mariam M., Pryce J.E. and Hayes B.J. (2016) *J. Dairy Sci.* **99**: 2849.
Ouellet V., Cabrera V.E., Fadul-Pacheco L. and Charbonneau É. (2019) *J. Dairy Sci.* **102**: 8537.
Ravagnolo O., Misztal I. and Hoogenboom G. (2000) *J. Dairy Sci.* **83**: 2120.
Wheelock J.B., Rhoads R.P., VanBaale M.J., Sanders S.R. and Baumgard L.H. (2010) *J. Dairy Sci.* **93**: 644.
Yousef M.K. (1985) 'Stress Physiology in Livestock' CRC Press, Boca Raton.

Proc. Assoc. Advmt. Anim. Breed. Genet. 26: 419-422

LINEAR ANALYSIS OF RUMEN THERMOLOGGER DATA TO COMPARE TOLERANCE TO HEAT CHALLENGE IN EIGHT SHEEP BREEDS

C.L. Nel[1,2], K. Mafunga[2], S.K. Maloney[3], T.S. Brand[1,2], K. Dzama[2], P.G. Theron[2] and S.W.P. Cloete[2]

[1] Directorate: Animal Sciences, Western Cape Department of Agriculture, Elsenburg, 7607 South Africa

[2] Department of Animal Sciences, Stellenbosch University, Matieland, 7602 South Africa

[3] School of Human Sciences, The University of Western Australia, Crawley, WA, 6009 Australia

SUMMARY

The aim of this study was to analyse rumen temperature (RT) in eight sheep breeds recorded for a four-month period during a hot and arid summer in Mediterranean South Africa. Challenges in linear mixed model analysis of the automatically recorded rumen thermologger data included the effect of drinking, a high level of autocorrelation between frequent observations, and a visible lag period between the hourly recorded temperature humidity index (THI) and RT. Marked differences in RT existed between breeds at high THI, with lowest to highest RT predicted for the Namakwa Afrikaner, Dorper, Meatmaster, Dormer, Ile de France, Merino, South African Meat Merino, and Dohne Merino. The low ranking of the composite and purely indigenous breeds supported the hypothesis that these breeds have a higher capacity to tolerate heat stress.

INTRODUCTION

In South Africa (SA), a large proportion of sheep producers favour exotic Merino-type breeds for intensive and extensive meat and wool production. These breeds produce large carcasses and, at varying levels, an additional income through wool. However, they are expected to be less heat-tolerant, based on respiration rate (Cloete *et al.* 2021). The indigenous sheep breeds of South Africa are known for their distinctly short, hairy coats, slender frames, and fat tails – attributes often assumed related to adaption to high heat load (Molotsi *et al.* 2019). However, standardised testing that includes indigenous South African breeds is rare, and when it has been done, was conducted over short exposure times. Moreover, the review by McManus *et al.* (2020) concluded that variance in heat tolerance is not well explained by the hair-type vs. wool-type criteria. Considerable variation could also exist within these groups, depending on the history of each breed.

A comparison among breeds requires a measure of animal strain in response to different levels of heat stress. Recently, the use of automated rumen thermologgers was proven to provide a reliable proxy for core temperature in sheep (Vesterdorf *et al.* 2022) after already having shown potential for use in cattle (Liang *et al.* 2013). This method delivers automated repeated measurements at a very high density (~ every 10 mins) over extended periods (~ months or years). However, the high-density form of the data differs from that usually evaluated by linear mixed models by, most notably, not qualifying as independent observations.

Here we report on a preliminary analysis and results after using this method to compare eight well-known sheep breeds over a hot and dry summer season in Mediterranean SA. Although other approaches to interpret these data exist (Vesterdorf *et al.* 2022), the use of the linear mixed model framework was a deliberate decision and some of the challenges encountered in this process are discussed from the point of view that similar data could in future substantiate a genetic analysis. From this data, across breed prediction curves including exotic, composite, and indigenous breeds are presented to contribute to the current gap of similar datasets.

MATERIALS AND METHODS

Study site and period. The experiment was conducted at the Langgewens research farm near Moorreesburg in the Western Cape Province of SA. The farm is situated in the winter-grain cropping-pasture area, which is known for hot, dry summers with maximum temperatures commonly exceeding 30°C. Data recording commenced in early summer for roughly four months from 21 November 2023 to 4 March 2024.

Genetic resources. The eight sheep breeds included the Merino, Dohne Merino, South Africa Meat Merino (SAMM), Ile de France (IDF), Dormer, Dorper, Meatmaster, and the Namakwa Afrikaner (NA). For each breed, eight entire ram lambs were randomly screened from within the Langgewens multibreed sheep flock (Cloete *et al.* 2021). This flock is managed as a single group throughout the year, thus ensuring a standardised background for individuals from different breeds. The lambs were born during April – July of 2023, and the average (± SD) age of rams at the start of the experiment was 168 ± 18 days.

Data. Climate data recorded at a weather station installed close to the paddocks were used to derive an hourly measure of heat stress using the temperature humidity index (THI) as defined by LPHSI (1990). Once each month, the live weight (LW; Kg), fat depth (FATD; in mm), and coat depth (COATD; in mm) were measured on each animal.

Rumen temperature (RT) was recorded using custom made thermologgers (Bryn O Morgan Industries, Perth), which were built into rumen capsules. The loggers were programmed to record RT every 5 minutes and calibrated against a certified platinum sensor thermometer (CENTER 376) before and after deployment in the animals. Each logger recorded 34,327 measurements over the experimental period. Following Vesterdorf *et al.* (2022), any measurement that showed an absolute change > 0.40°C from the previous measurement was flagged along with the next 11 measurements (in total, one hour) as being associated with drinking. In contrast to Vesterdorf *et al.* (2022), these values were set to missing instead of imputing an average. The remaining datapoints were used to calculate an average RT value for every hour, to reduce the level of pseudo-replication and because climatic data were only available at hourly intervals. Lastly, the total dataset spanning across four months was subset to four periods within each month spanning seven days chosen manually to: (1) include both low and high average THI values; (2) be separated by at least 20 days from another replicate period; and (3) not including the dates of animal handling. Finally, the data consisted of 64 animals recorded for four replicates (Month) each with 168 (7 days x 24) hourly RT measurements.

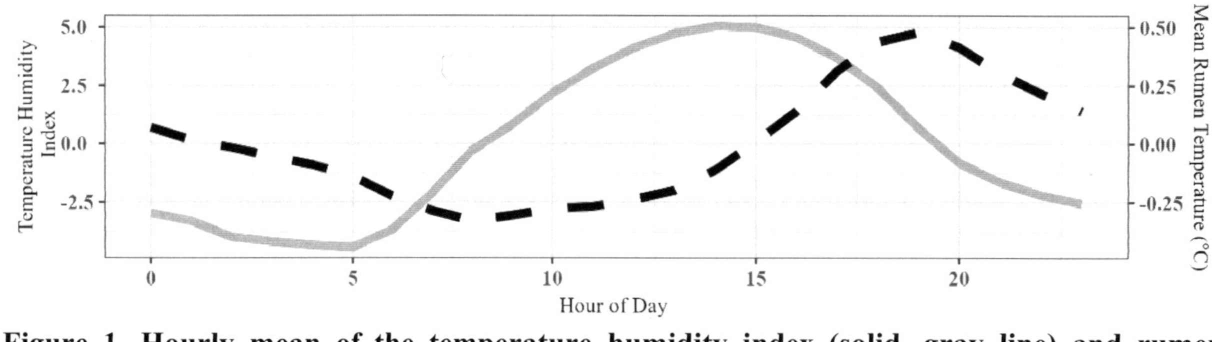

Figure 1. Hourly mean of the temperature humidity index (solid, gray line) and rumen temperature (dashed, black line) as recorded across the four months of the experiment. Values were scaled to zero by the overall mean(s)

The average raw means for a typical daily 24-hour cycle are shown in Figure 1. Several issues arise when analysing these data. Firstly, the co-dependence on a daily cycle causes an issue when both THI and an hour-of-day term are included in the model. Ignoring this term might not be ideal, since there could be factors other than THI, such as rumination (Liang *et al.* 2013), that can affect the diurnal rhythm of RT. Secondly, a marked lag effect is present on almost all days with THI reaching its peak at approximately 2pm, when the average RT is still increasing to peak later at around 5pm. This complicates any regression of $RT_i \sim THI_i$ where i is hour of the day. To overcome this issue, the row-axis linking records was redefined to allow for different starting points for each daily 24-point cycle. The nominated starting point was chosen based on the respective lowest values according to the raw means in Figure 1, being 5am and 8am, respectively. Thus, a regression $RT_i \sim THI_i$ was defined as, for example, the effect of THI at 2pm on the RT at 5pm.

Statistical analysis. Followed by building a linear mixed model using ASREML V4.2 software (Gilmour *et al.* 2021). Fixed effects were tested using conditional Wald F-statistics and included Breed (8 levels), the environmental variable THI (15:32) and the animal measurements LW (29:69 kg), FATD (0.09:0.53 mm) and COATD (8.3:55.6 mm) as linear covariates. The addition of random terms followed, using changes in Log-Likelihood values compared to the nested model. Random effects included animal ID (64 levels) as well as animal within month compounded (ID.Month: 248 levels) as identity terms. The model design also included testing of a first order autoregressive structure (AR1) accounting for the temporal autocorrelation between samples t+1 and t where t = 1 to t =168 within a Month. Finally, prediction of RT according to variation in THI was allowed to follow a curvilinear trend modelled by a cubic smoothing spline with eight knot points nested within Breed.

RESULTS AND DISCUSSION

Model definition. The fixed effects in the model included a significant interaction between THI x Breed ($P < 0.01$), while the covariates LW, FATD, and COATD did not affect RT ($P > 0.05$). Random terms that were retained included ID ($P < 0.01$) and ID.Month ($P < 0.05$). As expected, measurements taken at adjacent timepoints were strongly correlated ($P < 0.01$) with the AR1 autocorrelation parameter p = 0.91. However, it is not yet certain whether this is the best possible approach to incorporate high density data into a linear mixed model framework. The study of Liang *et al.* (2013) used daily means of THI and RT, where some information was lost but with reduced dependence between observations. A study of genetic analysis of comparable datasets (vaginal thermometers in sows) using a repeatability model exists (Freitas *et al.* 2023), but it is not clear how dependence between adjacent measurements was accounted for.

From preliminary analyses, it is notable that using splines to model a daily RT cycle (i.e. a prediction of the trend in Figure 1) also produced informative results as long as THI was not modelled. A viable alternative might be to categorise days according to mean temperature (cool, mild, hot, etc.) and to obtain these spline solutions within each category for each breed. Alternatively, excluding THI and estimating AR1 for each breed could be a good indicator of resilience to repeated changes in heat load (Ghaderi Zefreh *et al.* 2024). Using this approach, groups with a greater fluctuation from the general mean of RT will have higher measures of temporal autocorrelation over the observed period.

Breed response curves to THI. The predicted change in RT for each of the eight sheep breeds is shown in Figure 2. A slow initial increase in RT with increased heat load accords with the trends shown by Liang *et al.* (2013). At THI between 25 – 27, there was a curvilinear increase in RT in all breeds, but high variability for predicted RT when THI exceeded 30. At the highest THI, the RT (in °C) for Dohne Merino (40.5 ± 0.1), SAMM (40.4 ± 0.1), and Merino (40.4 ± 0.1) were the highest, while the IDF (40.3 ± 0.1), Dormer (40.2 ± 0.1) and Meatmaster (40.2 ± 0.1) trended intermediately, while the Dorper (40.0 ± 0.1) and Namakwa Afrikaner (39.9 ± 0.1) had the lowest RT values. These differences could be judged almost negligible at the absolute level, but it is important to consider the extremely narrow distribution of the RT data (CV = 1.3% in the pruned dataset). Based on eye temperature and respiration rate, Cloete *et al.* (2021) similarly showed that the Namakwa Afrikaner (indigenous) and composites with indigenous parent breeds (Meatmaster and Dorper) were better able to manage heat stress than were purely exotic Merino type breeds. It is of interesting to observe a similar result based on RT data with the contention that this capacity for heat tolerance is not explained by any of the measurable differences in LW, FATD, or COATD. However, some response curves, such as that of the IDF, is difficult to explain, and more information is needed to comprehend re-ranking of breeds across different THI levels.

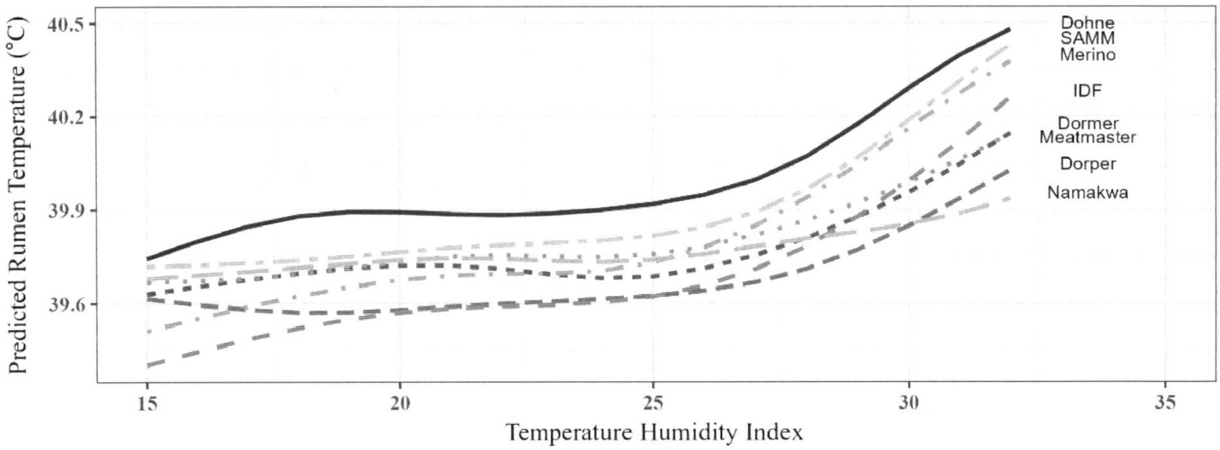

Figure 2. Predicted rumen temperature for eight sheep breeds across the observed temperature humidity index

CONCLUSION

Incorporating high density thermologger data into the standard linear mixed model framework creates challenges, but a large-scale genetic analysis remains feasible in future. On the breed level, indigenous types appear to have a quantifiable advantage under high heat load that is not explained by weight or coat depth. This points to an adaptive developmental history of breed genetics, but a larger dataset would be helpful in confirming or refuting these results.

ACKNOWLEDGEMENTS

The authors would like to acknowledge the important contributions to planning and field work of Ms Resia Swart and the working team of the Langgewens research farm. Also, Dr Luoyang Ding for assistance with logger preparation and calibration. The Western Cape Agricultural Research Trust provided financial support to the work.

REFERENCES

Cloete S.W.P, Van Der Merwe D.A., Steyn S., Muller A., Melgar J., Cornwallis C.K. and Brand T.S. (2021) *Proc. Assoc. Advmt. Anim. Breed. Genet.* **24**: 62.

Freitas P.H.F., Johnson J.S., Wen H., Maskal J.M., Tiezzi F., Maltecca C., Huang Y., DeDecker A.E., Schinckel A.P. and Brito L.F. (2023) *Genet. Sel. Evol.* **55**: 65.

Ghaderi Zefreh M., Pong-Wong R. and Doeschl-Wilson A. (2024) *Animal* **18**: 101248.

Gilmour A.R., Gogel B.J., Cullis B.R., Welham S.J. and Thompson A.N. (2021) ASReml User Guide Release 4.2 Functional Specification. www.vsni.co.uk.

Liang D., Wood C.L., McQuerry K.J., Ray D.L., Clark J.D. and Bewley J.M. (2013) *J. Dairy Sci.* **96**: 5072.

LPHSI (1990) Agricultural Engineering Technology Guide, Clemson University, SC, USA.

Molotsi A.H., Dube B. and Cloete S.W.P. (2019) *Diversity* **12**: 14.

Vesterdorf K., Beatty D.T., Barnes A. and Maloney S.K. (2022) *Anim. Prod. Sci.* **62**: 1671.

THE EFFECT OF TREE SHADE AND BREED ON HEAT STRESS INDICATORS IN DORMER AND SOUTH AFRICAN MEAT MERINO EWES AND LAMBS

K. Mafunga[1], C.L. Nel[1,2], K. Dzama[1] and S.W.P. Cloete[1]

[1] Department of Animal Sciences, Stellenbosch University, Private Bag X1, Matieland, 7602, South Africa

[2] Directorate: Animal Sciences, Department of Agriculture, Western Cape Government, Private Bag X1, Elsenburg, 7607, South Africa

SUMMARY

The provision of natural tree shade on the respiration rate (RR) and rectal temperature (RT) of ewes and lambs was studied during the perinatal period. Pregnant Dormer and South African Meat Merino (SAMM) ewes were randomly allocated to 18 paddocks, 9 with access to tree shade (SH) and 9 without access to shade (NS) prior to lambing in March-April 2023. The RR and RT were recorded in the afternoon from pregnant ewes randomly selected from the SH and NS paddocks on three occasions with an expected high ambient temperature. Newborn lambs were recorded for RR and RT at noon within 24 hours of birth. Micro-climate data were obtained from loggers in the paddocks. Minimum temperature for SH paddocks was higher than in the NS paddocks, the opposite being true for maxima. Ewes and lambs in SH paddocks had a lower RR and RT compared to the animals in NS paddocks, the impact accentuated at a higher THI in lambs. Dormer ewes breathed faster than their SAMM contemporaries, but this breed effect was reversed in lambs (both $P < 0.05$). Ewe traits were repeatable, with values of 0.32 for RT and 0.16 for RR. Albeit in a small study, with heat stress not observed on some days, it seems that tree shade may improve animal welfare by reducing detrimental impacts of high THI on hot days during autumn lambing. Selective breeding for heat tolerance may also be contemplated in the current flock, if supported by further research.

INTRODUCTION

There is accordance among scientists that the global climate will become more unstable in future (Blignaut *et al.* 2009), including more frequent and severe droughts and an increase in global temperatures. Heat stress can impact on ovine production and reproduction (Van Wettere *et al.* 2021) particularly compromising more vulnerable groups, such as neonatal lambs (Cloete *et al.* 2021).

Against this background we studied the provision of tree shade to two typical South African (SA) sheep breeds, the SA Meat Merino (SAMM) and the Dormer during an autumn lambing season. We hypothesised that access to tree shade would mitigate the impact of heat as suggested by the literature (De *et al.* 2020; Pent *et al.* 2021). Since both breeds originate from ovine germplasm from temperate regions, we did not expect marked differences between breeds for their reaction to heat. We could also derive repeatability estimates for heat stress indicator traits in ewes.

MATERIALS AND METHODS

Experimental site and animals. The study took place under Mediterranean conditions on the Elsenburg Research Farm near Stellenbosch, in the Western Cape of SA, and used the foundation flocks of both breeds (Dormer: Muller *et al.* 2020; SAMM: Muller *et al.* 2023). The Dormer is the primary terminal sire breed in SA and the SAMM is the main dual-purpose breed (meat and wool). The breeds were maintained in the same flock, except when mated in single-sire groups during spring 2022 to lamb in autumn 2023. The site receives 625 mm of rain p.a., 78% expected from April to September, and is characterised by cool and wet winters (minima around 5°C) and hot, dry summers (maxima around 30°C). Grazing consisted of irrigated kikuyu, irrigated and dryland lucerne, and dryland oat fodder crops.

Recordings. Microclimate data for relative humidity and ambient temperature were obtained from Tinytag® data loggers placed in the paddocks, while a nearby official weather station supplied macroclimatic data. Weather station data were used to calculate the daily temperature-humidity index (THI) throughout the lambing period. Recently shorn pregnant ewes of both breeds (166 in total) were randomly allocated within breed to 9 irrigated kikuyu lambing paddocks with access to tree shade (SH) or 9 paddocks without shade (NS), 14 days before lambing in March. Five or 6 Dormer and 3 or 4 SAMM ewes were allocated to each paddock. For a subsample of ewes, the respiration rate (RR, by counting side movements for 30s) and rectal temperature (RT, using a clinical thermometer) were recorded at noon 3 times on days when the maximum temperature was expected to be >30 ˚C. Upon lambing, birth weights (BW) of 216 lambs were recorded within 24 h of birth (Muller *et al.* 2020; 2023), and RR and RT were recorded simultaneously during routine lambing rounds from around 12:00. At tail-docking, on an average (SD) age of 11.4 (2.6) days, a sample of neck wool was clipped from individual ewes and analysed for cortisol concentration (CC), as an indicator of heat stress (Nejad *et al.* 2014).

Statistical analysis. Ewe RR and RT were subjected to linear mixed model analyses involving the fixed effects of breed (Dormer vs. SAMM), shade treatment (SH vs. NS) and ewe age (2 to 6 years). There was no attempt to model ambient climate, as recording was reserved for hot days. The random effect of animal was added to the model to account for the repeated sampling of ewes. A fixed model involving the same fixed effects was employed for CC. Likewise, a fixed, linear model was used to analyse lamb data, fitting the effects of breed, shade treatment, sex (ram or ewe), age of dam (2 to 6 years) and birth type (single or multiple). Lamb data were recorded under a variety of climate conditions, allowing fixed linear and random spline effects of the weather station THI to be fitted and interacted with shade treatment. All analyses proceeded in ASReml V4.2 (Gilmour *et al.* 2021). The random between-ewe variance was used to estimate repeatability of RT and RR in ewes.

RESULTS AND DISCUSSION

Micro-climate. The maximum temperature and the THI at the maximum temperature were greater in NS paddocks than in SH paddocks (P<0.01; Table 1). This trend was reversed for minimum temperatures (P<0.01). Previous studies also reported that trees moderated the micro-climate of the surrounding area, being cooler during the day and warmer at night (Cloete *et al.* 2021).

Table 1. Two-sample t-test results for micro-climate data obtained from Tinytag® data loggers in shade (SH) and no-shade (NS) paddocks for the period around lambing

Climate variable	SH	NS	T-value
Maximum ambient temperature (˚C)	21.9	23.6	-13.4*
THI at Maximum ambient temperature	20.7	21.9	-13.7*
Minimum ambient temperature (˚C)	13.3	12.9	3.3*
THI at Minimum ambient temperature	13.3	12.9	3.7*

* Significant (P<0.01); THI – Temperature-humidity index

Birth weight and heat stress indicators in neonatal lambs. Dormer lambs were 11% heavier at birth than their SAMM contemporaries (P<0.01), but BW was unaffected by shade treatment (P>0.05; Table 1). Cloete (1993) similarly reported a heavier BW for SAMM lambs than Dormers, although the same result was not observed in a more recent study (Cloete *et al.* 2021). Cloete (1993) related the smaller size of Dormer lambs to its role as a terminal sire breed on smaller Merino ewes. Sex, birth type and dam age effects on BW were as reported Muller *et al.* (2020; 2023). Lambs in NS paddocks had a higher RT (P<0.05) and a 19.1% higher RR (P<0.01) than those in SH paddocks. These results are consistent with literature pertaining to tree shade (Cloete *et al.* 2021; Pent *et al.*

2021; Dos Santos *et al.* 2021). The RT of Dormer lambs were slightly higher than those of SAMM lambs (P<0.05), but the RR of SAMM lambs was 11.9% higher than that of Dormer lambs. Birth type was the only other fixed effect to affect an indicator trait, with RR being 9.9% faster in singles than in multiples (respective geometric means 107.2 vs. 97.5 breaths/min).

Table 2. Least-squares means (±SE) depicting the effects of shade treatment (shade vs no-shade) and breed (Dormer vs SA Meat Merino) on birth weight and the heat stress indicators (rectal temperature and log-transformed respiration rate (geometric mean)) of lambs

Effect and level	Birth weight (kg)	Rectal temperature (˚C)	Respiration rate/min
Shade treatment	0.08	*	**
Shade	5.10±0.08	39.3±0.1	4.54±0.04 (93.7)
No-shade	5.25±0.10	39.5±0.1	4.72±0.05 (111.6)
Breed	**	*	*
Dormer	4.90±0.09	39.5±0.1	4.57±0.05 (96.6)
SA Meat Merino	5.45±0.09	39.3±0.1	4.68±0.05 (108.1)

* Significant (P<0.05); ** Significant (P<0.01)

Both RT and RR were affected by the interaction of shade treatment with linear and spline components of the THI (P<0.01; Figure 1). Both traits were independent of the shade treatment at THI <21. Treatment differences were accentuated as THI increased, to the extent that the RR of NS lambs exceeded that of SH lambs by a factor of 1.8 (respective geometric means of 238.2 vs. 132.9; P<0.01). These results conform to previous studies, indicating that the benefit of tree shade becomes appreciably smaller when heat stress is low, such as on cooler days (Cloete *et al.* 2021), during cooler seasons (Pent *et al.* 2021) or cooler times of the day (Dos Santos *et al.* 2021).

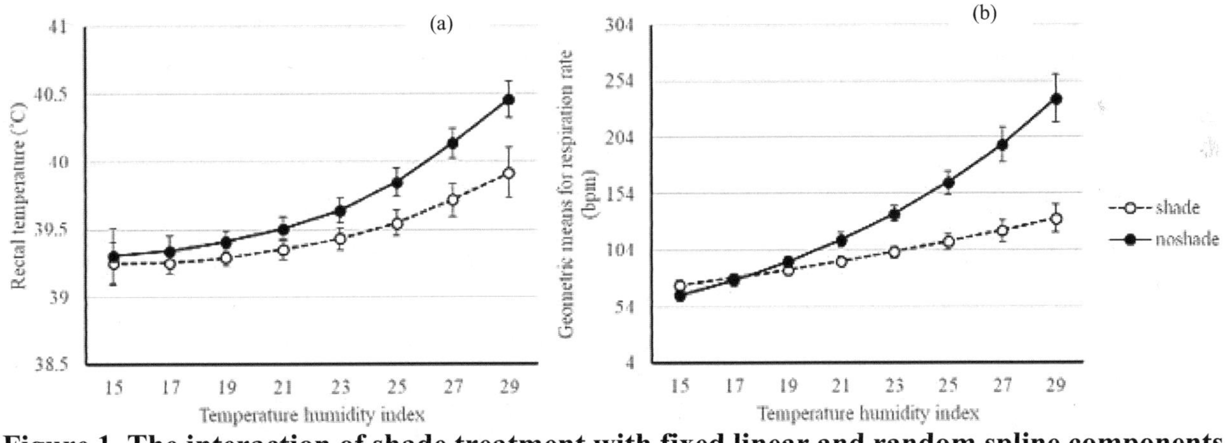

Figure 1. The interaction of shade treatment with fixed linear and random spline components of the temperature humidity index for rectal temperature (a) and respiration rate (b) of newborn lambs. Each mean is accompanied by the appropriate standard error

Heat stress indicators and wool cortisol concentrations in ewes. The RT of ewes in NS paddocks was 1.3% higher (P<0.01) than in SH paddocks (Table 3). Likewise, the RR of ewes in NS paddocks exceeded that for SH paddocks by 72.8% (P<0.01). These results are consistent with those of other studies reporting the impact of tree shade in mature sheep under hot conditions (De *et al.* 2020; Pent *et al.* 2021; Dos Santos *et al.* 2021). The RT was independent of breed (P>0.10), but Dormer ewes had a 12.8% higher RR (P < 0.05). It is possible that Dormer ewes relied more on RR to maintain core temperature than their SAMM contemporaries. The backfat thickness of Dormers exceeds the mean for SAMMs by 35% at hogget slaughter (Cloete *et al.* 2004), which could help to partly explain the breed differences. Wool CC was independent of breed but trended higher in ewes grazing NS paddocks (P=0.06). Both RT and RR were repeatable in ewes, at respectively 0.32±0.08 and 0.16±0.08. Previous short-term studies also suggested that RT was repeatable, but that RR was also affected by reranking of animals between cool and hot conditions (Steyn 2022).

Table 3. Least-squares means (±SE) depicting the effects of shade treatment (shade vs no-shade) and breed (Dormer vs SA Meat Merino) on birth weight and the heat stress indicators (rectal temperature and log-transformed respiration rate (geometric mean)) in ewes

Effect and level	Rectal temperature (°C)	Respiration rate/min	Cortisol (pg/mg)
Shade treatment	**	**	0.06
Shade	39.1±0.1	4.54±0.04 (118.3)	2.80±0.02
No-shade	39.6±0.1	4.72±0.05 (204.2)	3.23±0.02
Breed	0.19	*	0.13
Dormer	39.4±0.1	4.57±0.05 (167.6)	2.85±0.01
SA Meat Merino	39.3±0.1	4.68±0.05 (149.3)	3.18±0.02

* Significant (P<0.05); ** Significant (P<0.01)

CONCLUSION

Tree shade ameliorated the effects of heat stress in both pregnant ewes and neonatal lambs by markedly reducing RR on hot days. Access to tree shade during autumn lambing is expected to contribute to sheep welfare under the Mediterranean conditions. Many sheep farming areas in SA have hotter summers than the Western Cape, warranting further studies in more challenging environments. RR varied between breeds, possibly in association with the smaller size of Dormer lambs and an increased fatness of Dormer ewes. Derived within-season repeatability estimates should be followed up with longer-term across-season studies, while data should be accumulated to allow preliminary heritability estimates for heat stress indicator traits.

REFERENCES

Blignaut J., Ueckermann L. and Aronson, J. (2009) *S. Afr. J. Sci.* **105**: 61.

Cloete J.J.E., Hoffman L.C., Cloete S.W.P. and Fourie, J.E. (2004) *S. Afr. J. Anim. Sci.* **34**: 44.

Cloete S.W.P. (1993) *S. Afr. J. Anim. Sci.* **23**: 38.

Cloete S.W.P., Muller A., Steyn S., Van der Merwe D.A., Nel C.L., Cloete S., Kruger A.C.M. and Brand, T.S. (2021) *J. Thermal Biol.* **99**: 103024.

De K., Sharma S., Kumawat P.K., Kumar D., Mohapatra A. and Sahoo A. (2020) *J. Vet. Behav.* **40**: 103.

Gilmour A.R., Gogel B.J., Cullis B.R., Welham S.J. and Thompson A.N. (2021) ASReml User Guide Release 4.2 Functional Specification. VSN International Ltd, Hemel Hempsted, UK.

Muller A., Brand T.S., Scholtz A.J., Kruger A.C.M. and Cloete, S.W.P. (2020) *Small Rumin. Res.* **191**: 106181.

Muller A., Brand T.S., Scholtz A.J. and Cloete, S.W.P. (2023) *Small Rumin. Res.* **228**: 107091.

Nejad J.G., Lohakare J.D., Son J.K., Kwon E.G., West J.W. and Sung K.I. (2014) *Animal* **8**: 1.

Pent G.J., Fike J.H. and Kim I. (2021) *Agrofor. Syst.* **95**: 21.

Santos M.L.P.D., Dada J.M.V., Muniz P.C., Nunes-Zotti M.L.A., Barros F.R.A.D. and Vieira F.M.C. (2021) *Small Rumin. Res.* **228**: 107091.

Steyn S. (2022) MSc (Agric) Thesis, University of Stellenbosch.

Van Wettere W.H.E.J., Kind K.L., Gatford K.L., Swinbourne A.M., Leu S.T., Hayman P.T., Kelly J.M., Weaver A.C., Kleemann D.O. and Walker S.K. (2021) *J. Anim. Sci. Biotechnol.* **12**: 1.

Proc. Assoc. Advmt. Anim. Breed. Genet: 26: 427-430

PRELIMINARY ANALYSIS OF IMMUNE COMPETENCE TRAITS IN NORTHERN AUSTRALIAN TROPICALLY ADAPTED BEEF BREEDS

K.L. Moore[1], D.J. Johnston[1], T. Grant[2] and J. Croaker[3]

[1] Animal Genetics Breeding Unit*, Armidale, NSW, 2351 Australia
[2] Queensland Department of Primary Industries, Toowoomba, QLD, 4350 Australia
[3] Australian Brahman Breeders' Association, Rockhampton, QLD, 4700 Australia

SUMMARY

Immune competence traits (cell-mediated and antibody-mediated immune responsiveness) were evaluated in northern Australian tropically-adapted beef breeds. A total of 784 Brahman, Droughtmaster, and Santa Gertrudis steers were recorded at weaning for the two traits. The development of a statistical model showed year of birth and breed were significant ($P<0.05$) for both immune competence traits. For the cell-mediated immune response, the starting (day 0) live weight and the baseline skin thickness ratio was also significant. Heritability estimates were low to moderate for both traits, with 0.18 and 0.22 for cell-mediated and antibody-mediated immune responses, respectively. These preliminary results indicate that genetic variation exists for these immune competence traits in tropically adapted beef breeds. Further research is required to determine if selecting for immune competence in tropical breeds also improves profit by improving production, health (i.e. reduced reliance on antibiotics to prevent or treat disease) and welfare traits.

INTRODUCTION

Disease has welfare, production and cost implications for beef enterprises. Immune competence traits have been developed for the genetic evaluation of general disease resistance in the Angus Australia national genetic evaluation (Hine *et al.* 2019). Animals with higher immune responses are expected to better resist disease challenges, resulting in fewer losses and improved performance. In Angus, the traits have been estimated to be moderately heritable, with estimates ranging between 0.25 and 0.42 (Hine *et al.* 2019; Reverter *et al.* 2021a; Reverter *et al.* 2021b). Reverter *et al.* (2021a) found immune competence traits generally had small to moderate genetic correlations with production traits. The strongest genetic correlation indicated that animals with low antibody-mediated immune responses had higher weaning (r_g=-0.35) and yearling weights (r_g=-0.39). Hine *et al.* (2021, 2022) compared animals with low and high immune responses and, although not significant due to the study's size, reported animals with low immune responses had higher mortality and disease rates in the feedlot. Immune competence has not been evaluated in northern Australian tropically adapted beef breeds. This paper investigates if the immune competence traits are heritable in Australian tropically adapted beef breeds.

MATERIALS AND METHODS

Animal Data. Immune competence traits were recorded on 784 northern Beef Information Nucleus (BIN) steers from three tropically adapted beef breeds (Brahman n=342, Droughtmaster n=307, and Santa Gertrudis n=135). The northern BIN steers were the males born in the Repronomics project, an extensive reference population targeting female reproduction traits and collecting many phenotypes from birth to slaughter (Johnston *et al.* 2017). Animals were born across two years at two Queensland Department of Primary Industries research herds: Spyglass Beef Research Facility (110 km north of Charters Towers, QLD) and Brian Pastures Research Facility

* A joint venture of NSW Department of Primary Industries and Regional Development and the University of New England

(18 km east south-east of Gayndah, QLD).

Four Repronomics cohorts had immune competence traits measured at weaning over 14 days, following the protocols described by Hine *et al.* (2019). In brief, two traits were measured: cell-mediated immune response and antibody-mediated immune response. On day 0 of the test, animals were removed from their dam, weighed and vaccinated with the Zoetis bovine Ultravac 7in1 vaccine. The magnitude of delayed-type hypersensitivity responses in the skin (caudal fold) under the tail was used to assess cell-mediated immune response. On day 14, a baseline skin thickness measurement was taken with callipers on the tail's caudal fold on both sides. Steers were injected with a small amount of saline on one side (control site) and 7in1 vaccine on the other (test site). Two days later, the skin thickness was measured at the control and test sites, and the cell-mediated immune response was calculated as; cell-mediated immune response = 100 X log(skin thickness at the test site / skin thickness at the control site). The antibody-mediated immune response to a component of the Ultravac 7in1 vaccine was tested via a blood sample taken on day 14. The production of anti-tetanus toxoid serum IgG1 was used to assess the antibody-mediated immune response, and it was reported as 100 X the optical density units (corrected for plate-to-plate variation) measured in the serum. Live weights were recorded on days 0, 14 and 16 of the immune competence test. Minimal data edits were undertaken with just one animal removed because the measured cell-mediated immune response was an extreme outlier (more than two standard deviations higher than the next highest record).

Statistical analyses. Statistical models were developed with a pooled breed dataset, using PROC MIXED in SAS (SAS Institute, 2007); sire was fitted as random, and fixed effects were tested with step-wise elimination until only significant ($P<0.05$) terms remained. The initial model considered as class effects the birth cohort (a concatenation of herd and year of birth) and sire breed, and as linear and quadratic covariates, the day 0 live weight for both traits and the baseline skin thickness ratio for the cell-mediated immune response trait. In addition, all first-order interactions were tested for significance. In Repronomics, cows were naturally mated for their first two parities, and for subsequent parities, were AI mated. In the current dataset, there was confounding between sire and mating type, and thus also dam age. Therefore, dam age was not considered in statistical models. The least squares means were calculated for significant fixed effects for each trait.

Genetic parameter and heritability estimates were obtained for each trait from a univariate mixed linear animal model using ASReml (Gilmour *et al.* 2021). The significant fixed effects identified above were fitted for all traits except day 0 live weight. For day 0 live weight, the model fitted cohort and sire breed as class effects, animal age as a covariate, and animal and the dam permanent environment as random effects. Three generations of pedigree were used, and 80 sires were represented with an average half-sib family size of 9.7, ranging from 1 to 42.

RESULTS AND DISCUSSION

On day 0 of the immune competence test, steers were, on average, 186 days old and 212.5kg. Cell-mediated and antibody-mediated immune responses were normally distributed, with averages of 28.93 100 X log(mm) and 90.61 100 X optical density units, respectively (Table 1). Variation was recorded in cell-mediated and antibody-mediated immune responses, with coefficients of variation of 29.8% and 18.8%, respectively.

Cohort (i.e. herd year) and sire breed were significant for both traits, with day 0 live weight and baseline skin thickness also significant for the cell-mediated immune response. Cohort least squares means indicated no difference in the immune responses at the two herd locations within the same year, but that birth cohorts born in 2023 had lower immune responses compared with the 2022 birth cohorts. Sire breed least squares means are presented in Table 2 and indicate that Santa Gertrudis animals had significantly higher cell-mediated immune response than Droughtmaster and Brahman, which were not significantly different from each other. For antibody-mediated immune response,

Santa Gertrudis and Brahman animals were not significantly different from each other but had significantly higher antibody-mediated immune responses than Droughtmaster. Linear covariate solutions showed that animals with increased cell-mediated immune response had decreased day 0 live weight and increased baseline skin thickness. The negative relationship between cell-mediated immune response and day 0 live weight aligns with the genetic correlations reported by Reverter *et al.* (2021a). Sire solutions (results not shown) showed variation in immune responses between sires. There was as much variation in sire solutions within a breed as there was across breeds, with all breeds having sires with higher and lower solutions for immune competence traits.

Table 1. Summary statistics, phenotypic (V_p) variance and heritability estimates for immune competence traits recorded on tropically adapted beef breeds

Trait	N	Mean	Std	Min	Max	V_p	h^2
Age at test day 0 (days)	783	186.0	21.9	118	228		
Weight at test day 0 (kg)	782	212.5	33.1	118	316	453.11	0.58 (0.12)
Skin thickness at test day (log(mm))	783	-0.003	0.03	-0.11	0.09		
Cell-mediated immune response (100 X log(mm))	783	28.93	8.61	5.65	55.22	63.03	0.18 (0.09)
Antibody-mediated immune response (100 X optical density units)	783	90.61	17.02	22.77	127.67	276.91	0.22 (0.09)

Table 2. Breed least squares means (standard error) for cell-mediated and antibody-mediated immune response for tropically adapted beef breeds

Trait	Brahman	Droughtmaster	Santa Gertrudis
Cell-mediated immune response (100*log(mm))	28.37 (0.59)[a]	26.95 (0.62)[a]	35.41 (0.94)[b]
Antibody-mediated immune response (100*optical density units)	94.07 (1.23)[b]	86.41 (1.30)[a]	92.10 (1.96)[b]

Subscripts within trait represent least squares means significantly different based on 95% confidence intervals.

Preliminary variance component estimates (Table 1) showed that immune competence traits had low to moderate heritability estimates of 0.18 (0.09) and 0.22 (0.09) for cell-mediated and antibody-mediated immune responses, respectively. Day 0 live weight was estimated to have a strong heritability of 0.58 (0.12). Maternal effects could not be fitted and tested for significance for all traits due to the current data structure. Therefore, the day 0 live weight heritability may be inflated as it will likely include maternal effects. Heritability estimates for cell- and antibody-mediated immune response from the current study were lower than those reported for Angus. Cell-mediated immune response heritability in Angus was estimated to range between 0.25 (Reverter *et al.* 2021b) and 0.31 (Reverter *et al.* 2021a), with a phenotypic variance of 63.64 (Reverter *et al.* 2021a), which was very similar to phenotypic variance estimates in the tropically adapted beef breeds in the current study. Antibody-mediated immune response heritability in Angus was estimated to range between 0.32 (Hine *et al.* 2019) and 0.42 (Reverter *et al.* 2021a), with a phenotypic variance of 541.45 (Reverter *et al.* 2021a), which was much higher than the phenotype variance estimated in the current study. The studies in Angus cattle adjusted the immune competence traits by age, whereas in this study, the traits were weight-adjusted. However, variance components when the traits were age-adjusted (results not shown) were similar to those from weight-adjusted traits. The antibody-mediated immune response coefficient of variation in the study by Reverter *et al.* (2021a) was much higher (32.0%) than the current study (18.8%), corresponding to the higher phenotypic variance estimated.

A further cohort of animals will be recorded for immune competence, and carcase data will be recorded for all animals. This expanded dataset will then be used to increase the accuracy of the heritability estimates and estimate the genetic relationships with carcase traits of economic importance. Survival and health records, where available, will also be considered to understand the relationship between immune competence traits and health and welfare traits. These preliminary estimates indicate that the traits will respond to genetic selection and, if shown to be genetically related to traits of economic importance or improved health and welfare outcomes, may be useful for the genetic improvement of tropically adapted beef breeds.

CONCLUSIONS

This preliminary study showed that the immune competence traits were heritable in tropical beef breeds and would respond to genetic selection. However, further research with a larger dataset is required to increase the accuracy of the heritability estimates and to determine if genetic selection for immune competence will lead to improved animal production, health and welfare. An additional year of immune competence data is planned, and steers will be backgrounded, finished, and carcase data collected. With the larger dataset, the availability of carcase records and female reproduction records from sisters, the genetic correlations between immune competence and carcase and female reproduction traits will be estimated. Survival and health records, where available, will also be considered.

ACKNOWLEDGEMENTS

The authors acknowledge funding from MLA (projects B.NBP.0759, P.PSH.1221, P.PSH.0743, P.PSH.0774 and L.GEN.2201) and the contributions of Australian Brahman Breeders' Association Limited, Droughtmaster Australia and a consortium of Santa Gertrudis breeders. Thank you to the collaborating organisations and the large team involved in designing, running, and collecting data. Thank you to Brad Hine and Aaron Ingham for their work in developing the protocols for recording immune competence and to CSIRO for collecting the immune competence phenotypes.

REFERENCES

Gilmour A.R., Gogel B.J., Cullis B.R., Welham S.J. and Thompson R. (2021) 'ASReml SA User Guide Release 4.2' VSN International Ltd, Hemel Hempstead, HP1 1ES, UK.

Hine B.C., Bell A.M., Niemeyer D.O.M., Duff C.J., Butcher N.M., Dominik S., Ingham A.B. and Colditz I.G. (2019) *J. Anim. Sci.* **97**: 4053.

Hine B.C., Bell A.M., Niemeyer D.O.M., Duff C.J., Butcher N.M., Dominik S., Porto-Neto L., Li Y., Reverter A., Ingham A.B. and Colditz I.G. (2021) *J. Anim. Sci.* **99**: 1.

Hine B.C., Niemeyer D.O.M., Bell A.M., Reverter A. and Ingham A.B. (2021) *MLA final report L.GEN.1817.*

Johnston D.J., Grant T.P., Schatz T.J., Burns B.M., Fordyce G. and Lyons R.E. (2017) *Proc. Assoc. Advmt. Anim. Breed. Genet.* **22**: 385.

Reverter A., Hine B.C., Porto-Neto L., Li Y., Duff C.J., Dominik S. and Ingham A.B. (2021a) *J. Anim. Sci.* **99**: 1.

Reverter A., Hine B.C., Porto-Neto L., Alexandre P.A., Li Y., Duff C.J., Dominik S. and Ingham A.B. (2021b) *Anim. Prod. Sci.* **61**: 1919.

SAS Institute (2007) SAS user's guide in statistics, 9th edition. SAS Institute Inc., Cary, NC, USA.

ASSOCIATIONS BETWEEN IMMUNE COMPETENCE BREEDING VALUES AND HEALTH TRAITS IN AUSTRALIAN ANGUS CATTLE

L.A Mowbray[1], M. Samaraweera[1], C. Duff[1], B.C. Hine[2], A. Reverter[3], P.A. Alexandre[3], L.R. Porto-Neto[3] and A.B. Ingham[3]

[1] Angus Australia, 86 Glen Innes Road, Armidale, NSW, 2350 Australia
[2] CSIRO, Agriculture and Food, F.D. McMaster Laboratory, Armidale, NSW, 2350, Australia
[3] CSIRO, Agriculture and Food, Queensland Bioscience Precinct, St Lucia, QLD, 4067Australia

SUMMARY

Health and resilience traits provide breeders with an opportunity to target genetic selection for improved disease resistance and optimal welfare outcomes. The objective of this study was to explore the relationship between three immune competence Genomic Estimated Breeding Values (GEBV) for Angus cattle and novel Pinkeye and Worm Egg Count (WEC) phenotypes. Animals categorised with a low Cell-Mediated Immune Response GEBV had more cases of corneal ulceration during Pinkeye outbreaks ($P = 0.015$). A trend for more corneal ulceration was observed in animals categorised with a low ImmuneDEX GEBV, though this was not significant. A trend for a higher WEC was observed in animals categorised as low ImmuneDEX but was not statistically significant. Angus Australia and CSIRO will continue to collect data to investigate these associations further.

INTRODUCTION

Immune competence breeding values provide a genetic prediction of an animal's ability to mount an adaptive immune response. Selection based on immune competence traits could benefit the beef industry by reducing the impact of disease, improving animal welfare, enhancing productivity and reducing reliance on antimicrobials and anthelmintics to treat disease amid growing resistance issues. Three immune competence GEBVs: Cell-Mediated Immune Response (Cell-IR), Antibody-Mediated Immune Response (Ab-IR), and ImmuneDEX (IDEX) have been produced by Angus Australia, in collaboration with CSIRO. At present, only the IDEX trait, a combined proxy trait for Cell-IR and Ab-IR (Reverter *et al.* 2021), is published as part of the fortnightly Trans-Tasman Angus Cattle Evaluation and was developed to enable selection of animals with a balanced ability to mount both antibody and cell-mediated responses.

Pinkeye (Infectious Bovine Keratoconjunctivitis) and intestinal roundworms (Nematodes) are two common diseases that significantly impact cattle productivity and profitability in Australia (Shephard *et al.* 2022).

A pinkeye scoring guideline for the collection of pinkeye phenotypes was recently developed by CSIRO and Angus Australia due to its welfare and economic impact on beef cattle (Angus Australia 2024). The objective of this study was to explore the relationship between genetic predictions for IDEX and the two component traits (Cell-IR, Ab-IR) and novel Pinkeye and Worm Egg Count (WEC) phenotypes in Angus cattle.

MATERIALS AND METHODS

Data. Pinkeye scores were collected from 17 cohorts of Angus weaners during, or shortly after, a Pinkeye outbreak in 2024 using Angus Australia's 0 to 4 score Pinkeye grading system. A total of 1,996 animals were scored for pinkeye in both eyes. Of these, 1,757 had genotypes. Due to the low frequency of scores 1 to 4 (Table 1), animals were assigned to categories of 'Corneal Ulceration' (scores 0-1) or 'No Corneal Ulceration' (scores 2-4).

Individual faecal samples were collected per rectum from 19 cohorts of Angus Sire Benchmarking Program steers and heifers at weaning (14 cohorts) and at feedlot induction (5 cohorts). Animals sampled at feedlot induction had received only one lifetime anthelmintic drench more than 6 months prior to sampling. Individual WECs were conducted on each animal (n=2,072), along with pooled larval cultures for each cohort. Of the total number of animals sampled 1,841 had genotypes.

GEBVs were extracted for each immune trait from the Angus Australia database. Phenotype measuring protocols for the immune competence traits are outlined in Hine *et al.* (2019), while Reverter *et al.* (2021) reports the most recent heritability estimates as 0.31±0.06 (Cell-IR), 0.42±0.06 (Ab-IR) and 0.37±0.06 (IDEX). The animals were categorised into Low (L), Middle (M), or High (H) groups for each immune competence GEBV based on their rank in the sample population.

Table 1. Description, frequency and categorisation of Pinkeye Scores of the sample population.

Score	Description	Left Eye	Right Eye	Category
0	No evidence of disease	1605	1624	No Ulceration
1	Tearing	14	13	No Ulceration
2	Mild to moderate corneal ulceration	18	5	Ulceration
3	Severe corneal ulceration	9	12	Ulceration
4	Corneal scarring	111	103	Ulceration
	Total	1757	1757	

Analyses. For analysis of pinkeye incidence data, contingency tables were created to compare the prevalence of corneal ulceration observed in the H, M, and L groups for each immune competence GEBV. Pearson's chi-squared test was applied to each table using Microsoft Excel to check for significant differences in the distribution of corneal ulceration among the groups.

The WEC data were log-transformed using the formula log(WEC +100), and normality was assessed based on a normal probability test. Residual WEC (rWEC) were then calculated using models that accounted for contemporary group and sex. Contemporary groups were formed by concatenating herd, cohort, and observed year, resulting in 19 groups with average of 97 animals each. Residual WEC and significance of differences in WEC among L, M, and H groups were assessed. Group means and differences were evaluated using p-values adjusted with a Bonferroni correction using the R software package Agricolae (de Mendiburu 2023).

Logistic regression was used to predict the probability of worm infestation based on each immune competence GEBV. For this analysis, WEC data were converted to a binary trait where WEC >0 (eggs present) expressed as 1 and WEC =0 (eggs absent) as 0.

RESULTS AND DISCUSSION

Pinkeye. The prevalence of Pinkeye cases across the 17 cohorts ranged from 2.4% to 42.6% and was 12.6% overall. This is consistent with recent estimations of Pinkeye incidence across Australian beef herds (Kneipp 2020). However, the low incidence of disease made inference from the Pinkeye scoring system challenging, leading to the conversion of pinkeye scores into a binary system for analysis.

Table 2 shows that the L group for Cell-IR GEBVs had a higher proportion of corneal ulceration (16%) compared to the M and H (11% each) Cell-IR GEBV groups (P = 0.015). The IDEX GEBV also demonstrated a trend for more cases of corneal ulceration in the L group than the M and H groups, but this was not statistically significant (P = 0.23). The H group of Ab-IR GEBV had the highest proportion of corneal ulceration cases when compared to M and L Ab-IR GEBV groups. The trends in the Cell-IR and IDEX traits align with previous observational studies suggesting a reduced ability to mount an immune response is associated with increased disease susceptibility. For instance, Hine *et al.* (2021) found that low immune competence Angus steers had a higher proportion of health-related mortalities at the feedlot.

Results from the current study suggest that the ability of an animal to mount a cell-mediated immune response may play a role in pinkeye incidence and, more specifically, severity. Cell-mediated immune responses such as conjunctival macrophages and cytokine production from corneal epithelial cells may play a role at the ocular surface (Knop and Knop 2007). Further studies will be required to confirm if such factors influence pinkeye incidence and severity in L Cell-IR GEBV animals by delaying the immune response at the ocular surface.

Kneipp *et al.* (2020), estimated that Pinkeye treatment costs the Australian cattle industry $AU 9.67million annually. Therefore, if genetic selection can help reduce the incidence and severity of pinkeye, significant economic benefits could be expected for industry.

Table 2. Descriptive statistics for High (H), Middle (M) and Low (L) immune response Genomic Breeding values (GEBV) and outcomes on corneal ulceration from Pinkeye

GEBV	Group	n	Mean GEBV ± SD	Min	Max	Corneal Ulceration	Proportion
	H	586	-0.06 ± 0.12	-0.21	0.44	70	12%
IDEX	M	585	-0.30 ± 0.05	-0.39	-0.21	72	12%
	L	586	-0.53 ± 0.11	-0.94	-0.39	80	14%
	H	586	0.37 ± 0.78	-0.59	3.69	65	11%
Cell-IR[1]	M	585	-1.31 ± 0.42	-2.02	-0.59	64	11%
	L	586	-3.08 ± 0.84	-6.37	-2.02	93	16%
	H	586	-2.14 ± 2.60	-5.35	10.87	81	14%
Ab-IR	M	585	-7.27 ± 1.16	-9.30	-5.35	66	11%
	L	586	-12.29 ± 2.33	-20.72	-9.31	75	13%

[1] *Values for corneal ulceration differ significantly (P = 0.015)*

Roundworms. The overall worm burden in this dataset was low with a mean WEC of 152 (range: 0 - 3,600) and the frequency distribution right-skewed. WEC data was log transformed to improve the normality of data before analysis. Though not statistically significant, the means of rWEC were higher in low immune response GEBV groups as compared to high immune response GEBV groups (Figure 1). The trends are similar to findings from Hine *et al.* (2022), which reported immune competence traits were favourably genetically correlated with WEC (-0.19 ± 0.23) in Merino sheep.

Differences of having a worm burden or not between animals categorized as L and H for IDEX, Cell-IR, and Ab-IR were not significant, with odds ratios of 0.86 (95% CI = 0.67-1.12), 0.85 (95% CI = 0.66-1.09), and 0.90 (95% CI = 0.69-1.16), respectively. A limitation of this study was that 30% of the sample population had zero WEC. Considering the sporadic nature of egg shedding in cattle (Parkinson *et al.* 2019), this could also mean many false negatives exist in the dataset, increasing the likelihood of a type II error (incorrectly accepting the null hypothesis). Therefore, a larger sample size will be needed to improve the robustness of analysis and accurately estimate associations between immune competence and WEC in beef cattle.

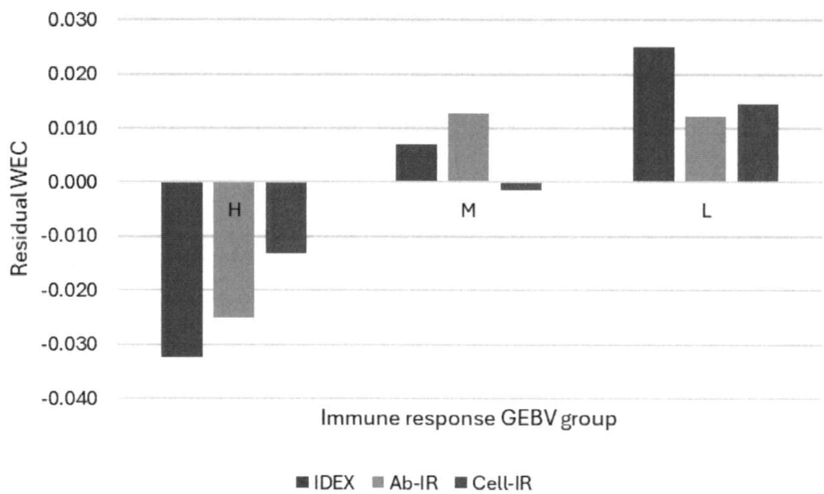

Figure 1. Means of residual worm egg counts (rWEC) in Low (L), Middle (M) and High (H) groups for ImmuneDEX (IDEX), antibody-mediated immune response (Ab-IR), cell-mediated immune response (Cell-IR) GEBV groups

CONCLUSION

The trends observed in this study support the hypothesis that animals with lower immune competence are associated with increased corneal ulceration from Pinkeye and higher WEC. These results serve as a foundation for continued recording of disease phenotypes in order to more accurately characterise the associations between immune competence GEBVs and disease incidence in Angus cattle.

REFERENCES

Angus Australia (2024) https://www.angusaustralia.com.au/news/collecting-pinkeye-scores-2024. (Accessed December 2024).

de Mendiburu F (2023). agricolae: R package version 1.3-7, https://CRAN.R-project.org/package=agricolae (Accessed December 2024).

Hine B.C., Acton G.A., Elks D.J., Niemeyer D.D.O., Bell A.M., Colditz I.G., Ingham A.B. and Smith J.L. (2022) *Animal* **16**: 100544.

Hine B.C., Bell A.M., Niemeyer D.D., Duff C.J., Butcher N.M., Dominik S., Ingham A.B. and Colditz I.G. (2019) *J. Anim. Sci.* **97**: 4053.

Hine B.C., Bell A.M., Niemeyer D.D., Duff C.J., Butcher N.M., Dominik S., Porto-Neto L.R., Li Y., Reverter, A., Ingham A.B. and Colditz I.G. (2021) *J. Anim Sci.* **99**: 1.

Kneipp M., Govendir, M., Laurence M. and Dhand N.K. (2021) *Prev. Vet. Med.* **187**: 105232.

Knop E. and Knop N. (2007) *Chem. Imm. Al.* **92**: 36.

Parkinson, T.J., Vermunt, J.J., Malmo J. and Laven R. (2019) 'Diseases of cattle in Australia' pp 221-242, 2nd ed. Massey University Press, Auckland.

Reverter, A., Hine B.C., Porto-Neto L., Li Y., Duff C.J., Dominik S. and Ingham A.B. (2021) *J. Anim. Sci.* **99**: 1.

Shephard R., Webb Ware J., Blomfield B. and Niethe G. (2022) B.AHE.0327: Priority list of endemic diseases for the red meat industry - final report, Meat & Livestock Australia, North Sydney.

Proc. Assoc. Advmt. Anim. Breed. Genet. 26: 435-438

INCLUDING RESILIENCE IN THE BREEDING OBJECTIVE OF SUSTAINABLE MERINO SHEEP

E.G. Smith[1], S.F. Walkom[2], A.A Swan[2] and S.A. Clark[1]

[1] School of Environmental and Rural Science, University of New England, Armidale, NSW, 2350 Australia
[2] Animal Genetics Breeding Unit*, University of New England, Armidale, NSW, 2351 Australia

SUMMARY

This study examined the effect of including an indicator trait of resilience (log variance of fibre diameter variation (FD_Lnvar)) into a current industry selection index designed to produce sustainable Merino sheep in Australia. A desired gains index was used to estimate the relative contribution of resilience to the index when either maintaining or improving (decreasing) FD_Lnvar by a quarter or half genetic standard deviation over ten years, compared to the standard index. The relative contribution of resilience needed, to achieve these goals, was shown to be 3.5, 12.4 and 25.3% respectively. Adjusting the index to maintain resilience had little impact on production traits, but prioritising higher FD_Lnvar responses reduced gains in key production traits including clean fleece weight and growth.

INTRODUCTION

General resilience describes an animal's ability to respond and recover from unspecified environmental disturbances (Colditz and Hine 2016). In sheep, general resilience can be assessed using the variability in fibre diameter measured along the wool staple, using a trait definition such as the log variance of deviations in fibre diameter (FD_Lnvar). Animals with larger trait estimates for FD_Lnvar are potentially more susceptible to environmental stresses and may have poorer health outcomes (Smith *et al.* submitted). The current MERINOSELECT indexes, released in 2024, include several traits related to susceptibility to disease and resilience including wrinkle, dag, worm egg count and condition score (Sheep Genetics 2024). However, as much as 80% of the genetic variation in FD_Lnvar is not currently explained by the traits included in this evaluation (Smith *et al.* submitted). There is interest in including resilience traits like FD_Lnvar in breeding objectives, but its economic value is hard to quantify due to the difficulty in quantifying its economic impact. A desired gains index is often useful for exploring the implications of including new traits in selection indexes where the economic benefits of improving a trait are unknown. This paper assesses the impact of adding FD_Lnvar to the Sustainable Merino (SM) index for Australian Merino sheep. A desired gains approach estimated the relative index contribution needed to maintain or improve resilience by 0.25 or 0.50 genetic SD over ten years. The effect of including resilience on genetic gain in other traits was also examined.

MATERIALS AND METHODS

Breeding objective. The SM index is based on a self-replacing ewe flock, producing 17-19μm wool and selling lambs post-weaning, off-shears. In this system, income from the production of wool and meat are approximately equal (46:54). The index prioritises genetic improvement of clean fleece weight, growth, lean meat yield (LMY) and reproduction (conception, litter size and rearing ability) while maintaining fibre diameter and staple strength. There is also a focus on reducing worm egg

* A joint venture of NSW Department of Primary Industries and Regional Development and the University of New England

count, dag, wrinkle and mature weight (Sheep Genetics 2024). Together, the SM index includes 14 traits (Table 1), with fleece traits assessed at post-weaning, hogget and adult stages. To assess the impact of including resilience, three SM index variations were compared to the base index: (1) SM+R (zero change in FD_Lnvar), (2) SM+R0.25 (improving FD_Lnvar by a quarter) and (3) SM+R0.50 (improving FD_Lnvar by half a genetic SD). A desired gains approach was used to calculate the relative contribution of resilience to the index needed to guide selection towards improving FD_Lnvar to align with these three objectives.

Desired gains index. A desired gains index calculates trait values indirectly using index-weighting factors based on a predetermined desired or restricted amount of genetic gain for one or more traits. Here the method uses the equation of Pešek and Baker (1969): $\boldsymbol{b} = \boldsymbol{P}^{-1}\boldsymbol{G}(\boldsymbol{G}'\boldsymbol{P}^{-1}\boldsymbol{G})^{-1}\,\boldsymbol{d}$, where $\boldsymbol{b}$ is a vector of weighting factors for the selection index, $\boldsymbol{P}$ is the phenotypic variance-covariance matrix, $\boldsymbol{G}$ is the covariance matrix of the phenotypic and genotypic values and $\boldsymbol{d}$ is the vector of relative desired gains used to realise the breeding goals. The genetic parameters (phenotypic and genetic correlation and heritabilities) were sourced from Sheep Genetics (2024). The heritability of FD_Lnvar was assumed to be 0.11 and correlations between FD_Lnvar and traits in SM are shown in Table 1. The weighting factors for each trait in the index were transformed into standard trait values (STV) by multiplying the trait weighting factor by its genetic standard deviation. The relative contribution (%) of each trait to the index was calculated as: $\frac{|STV_j|}{\sum_{j=1}^{24}|STV_j|} \times 100$, where $|STV_j|$ was the absolute STV of trait j. The sign of the relative contribution was adjusted to indicate the direction of trait improvement.

Table 1. Genetic and phenotypic correlations between FD_Lnvar and traits included in Sustainable Merino (Smith *et al.* submitted). Correlations are averaged across age stages

Trait	Genetic correlation	Phenotypic correlation
Clean fleece weight	-0.10	-0.01
Fibre diameter	-0.05	-0.13
Fibre diameter CV	-0.07	0.08
Staple strength	-0.01	-0.05
Growth	-0.11	-0.07
Lean meat yield	0.08	-0.01
Mature weight	-0.05	-0.07
Condition score	-0.23	-0.02
Conception[1]	*-0.05*	*0.01*
Litter size[1]	*-0.05*	*0.01*
Rearing ability[1]	*-0.05*	*0.01*
Wrinkle	0.07	-0.01
Worm egg count	0.01	-0.05
Dag	0.08	0.01

[1]Assumed based on an informed estimate.

The genetic gain per generation from a single round of selection using a selection intensity of one was calculated as, $\boldsymbol{g} = \boldsymbol{b}'\boldsymbol{G}/\boldsymbol{\sigma_I}$, where $\boldsymbol{g}$ is a vector of genetic gain in each trait and $\boldsymbol{\sigma_I}$ is the standard deviation of the index. Annual genetic gain was derived by multiplying $\boldsymbol{g}$ by the ratio of selection intensity $\boldsymbol{i}$ and generation interval **L.** A generation interval of 3 years for males and 4 years for females was used, with selection intensities of 2.06 for males (5% selected) and 0.80 for females (50% selected). Selection was 65% index-based and 35% visual. A 30% reduction in response was assumed over 10 years due to a decrease in genetic variance. The number of half-sibling records available for each trait was: 30 for post-weaning fleece traits, weight, worm egg count, resilience, early post-weaning wrinkle and dag; 30 for hogget fleece and weight traits and lean meat yield; 7 for adult fleece traits and mature weight; and 5 for reproduction traits and condition score.

RESULTS AND DISCUSSION

Figure 1 shows the relative contribution of each trait to the index as a proportion of the total index value. Achieving zero change in FD_Lnvar required a relative contribution of -3.5%, similar to worm egg count or fibre diameter within the SM+R index. Achieving greater genetic gain in FD_Lnvar, equivalent to a quarter (SM+R0.25) or half (SM+R0.50) of a genetic SD required larger relative trait contributions of -12.4 and -25.3%, respectively. In the latter index, the increased emphasis on FD_Lnvar resulted in significant reductions in selection pressure on other traits including clean fleece weight, fibre diameter, conception and rearing ability compared to SM. For SM+R0.50 the relative trait values were around halved for all traits in the index compared to the SM. These higher contributions were necessary to achieve the specific reduction in FD_Lnvar due to the low heritability of FD_Lnvar and weak genetic correlations between FD_Lnvar and other traits in the breeding objective. This study relied on genetic parameters for FD_Lnvar estimated from a limited dataset (Smith *et al.* submitted), therefore, there is some uncertainty regarding the direction and strength of the genetic correlations. Future work could explore the sensitivity of the relative contribution to changes in the genetic correlations similar to Safari *et al.* (2006).

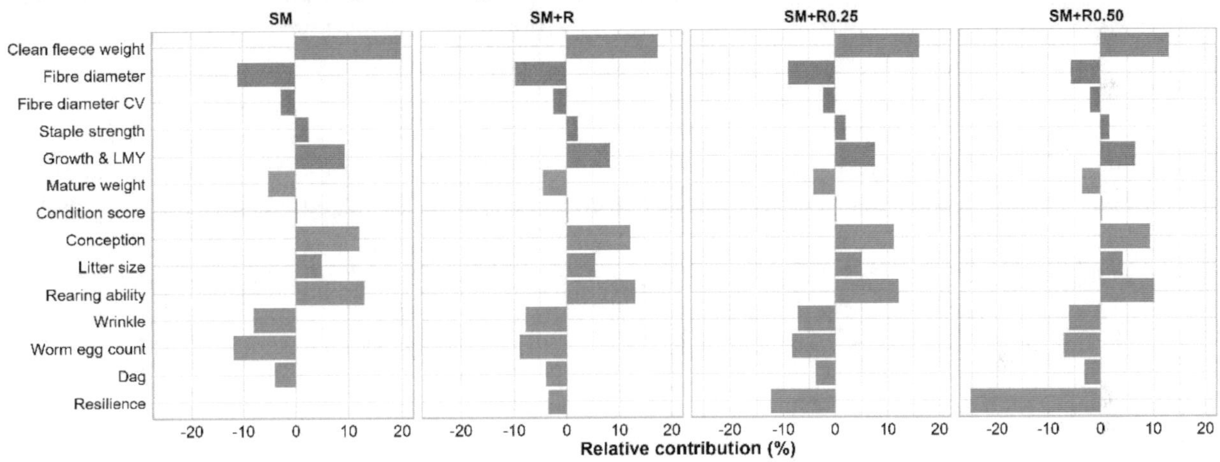

Figure 1. Relative contribution (%) of each trait to the index of Sustainable Merino (SM), including resilience (FD_Lnvar) maintained (SM+R), improving by 0.25 (SM+R0.25) or 0.5 (SM+R0.50) of a genetic SD. Traits are summed across age stages

The impacts of including selection pressure on FD_Lnvar on genetic gain in other production and health traits relative to the SM index are shown in Figure 2. The SM+R index results in an average decline in the response of 12% compared to SM. Genetic gain in wrinkle was, however, improved beyond SM in all indexes that included resilience, due to favourable correlated responses with FD_Lnvar. Despite this improvement, both SM+R0.25 and SM+R0.50 resulted in proportionally higher losses in average response of 16% and 28%, respectively, compared to SM. The response in condition score was not different between the SM and SM+R0.25 and slightly higher between SM and SM+R0.50 index, owing to its low relative contribution to the index and moderate favourable correlations with FD_Lnvar. The total index value decreased by 13, 17 and 25% in the SM+R, SM+0.25 and SM+R0.50 indexes, respectively. The decline in response in the former two indexes is possibly acceptable, although attempting to improve FD_Lnvar by half a genetic SD is unlikely to be economically viable within this breeding objective. Nonetheless, higher levels of resilience (FD_Lnvar) may be warranted in specific contexts, such as in reaction to more extreme production environments or where policy changes concerning animal welfare pose a threat to market access. Further, the current selection on the SM index results in a slight favourable decrease in FD_Lnvar, which suggests that the current SM index is unlikely to be detrimental to resilience.

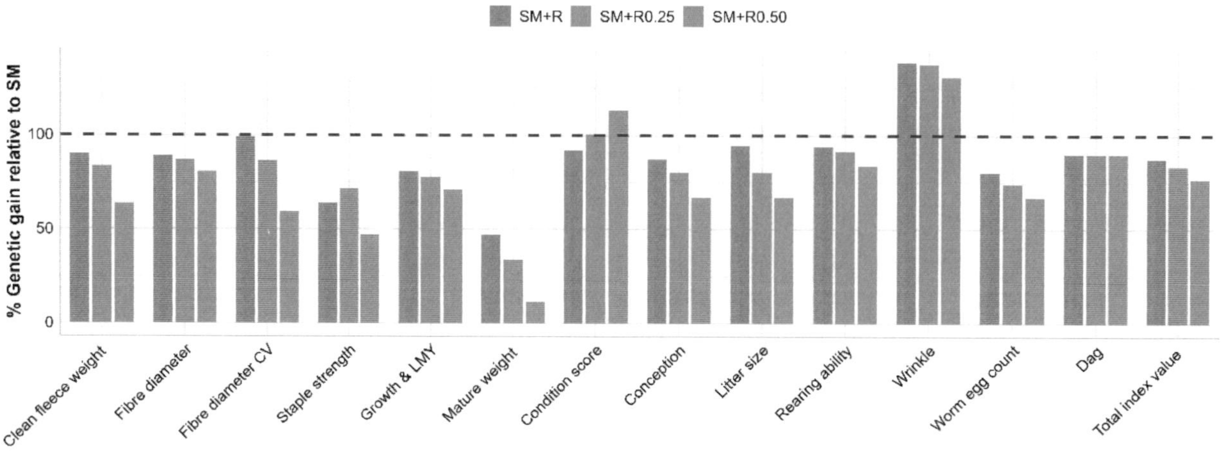

Figure 2. The proportion of genetic gain over 10 years in each trait relative to the Sustainable Merino (SM). Scenarios include maintaining resilience (SM+R) and improving by 0.25 (SM+R0.25) or 0.5 (SM+R0.50) of a genetic SD. Responses are averaged across age stages

In this study, desired gains for FD_Lnvar were chosen subjectively to illustrate the impact of selecting for resilience within the existing SM breeding objective. Future work could focus on optimising these desired gains to balance improvements in resilience with improvements in other traits. Alternatively, other methods for valuing resilience could be explored, such as a bioeconomic model. Bioeconomic models account for the complex interactions between genetics, physiology, management practices and economic parameters, and therefore may provide a more realistic assessment of how resilience traits influence profitability. These models are also better suited to evaluate uncertainty, which is important particularly for traits like resilience which affect a system's ability to adapt to disturbances.

CONCLUSION

This study demonstrated that maintaining or increasing resilience by 0.25 or 0.50 of a genetic SD in the SM index required relative contributions of -3.5, -12.4 and -25.3% respectively. Greater emphasis on resilience within these indexes led to a decline in genetic response for other important production traits, highlighting the trade-offs associated with including FD_Lnvar in the index. Further research is needed to refine the value of resilience before considering integrating FD_Lnvar into industry breeding programs.

ACKNOWLEDGEMENTS

The authors thank the Meat and Livestock Australia funded project L.GEN.2204 for the development of the Sheep Genetics Index and background IP used in this study.

REFERENCES

Colditz I.G. and Hine B.C. (2016) *Anim. Prod. Sci.* **56**: 1961.
Pešek J. and Baker R. (1969) *Can. J. Plant. Sci.* **49**: 803.
Safari E., Fogarty N. and Gilmour A. (2006) *Aust. J. Exp. Agric.* **46**: 283.
Sheep Genetics (2024) Sheep Genetics.
Smith E.G., Walkom S.F., Waters D.L. and Clark S.A. (submitted) *Gen. Sel. Evol.*

Proc. Assoc. Advmt. Anim. Breed. Genet. 26: 439-442

LATEST DEVELOPMENTS IN GENOMIC TOOLS FOR ANGUS AUSTRALIA: SteerSELECT, HeiferSELECT and BreedCHECK

A. Reverter[1], C. Duff[2], M. Samaraweera[2], P.A. Alexandre[1], B.C. Hine[3], A.B. Ingham[1], L.A Mowbray[2] and L.R. Porto-Neto[1]

[1] CSIRO, Agriculture and Food, QBP, St Lucia, QLD, 4067 Australia
[2] Angus Australia, 86 Glen Innes Road, Armidale, NSW, 2350 Australia
[3] CSIRO Agriculture and Food, F.D. McMaster Laboratory, Armidale, NSW, 2350 Australia

SUMMARY

Angus SteerSELECT, HeiferSELECT, and BreedCHECK are genomic tools developed through a partnership between Angus Australia and the CSIRO. These tools are designed to deliver genomic estimated breeding values (GEBV) that enhance decision-making regarding the optimal feeding strategies for feedlot cattle (Angus SteerSELECT), the selection of replacement heifers (HeiferSELECT), and the assessment of breed composition (Angus BreedCHECK) in commercial Angus beef breeding enterprises. Central to these tools is a newly established SNP map comprising 61,105 autosomal markers. The implications of these SNPs on various growth, health, and fertility traits will be examined, along with the allele frequencies observed in a multi-breed population.

INTRODUCTION

A reference population of genotyped and phenotyped animals allows for the reliable estimation of SNP effects that can be used to compute indirect predictions of genomic estimated breeding values (GEBV) for commercial animals that have neither progeny nor phenotypes. These indirect predictions have been widely tested in livestock populations (Hidalgo *et al.* 2023).

Commercially available, Angus SteerSELECT, HeiferSELECT, and BreedCHECK are customised genomic tools designed to provide commercial Angus producers with tools to improve the genetic merit and management of animals in their herds. Specifically, Angus HeiferSELECT enables predictions of the breeding potential of cattle that have not had a calf, and SteerSELECT predicts the productive performance of cattle that have not had any phenotypes recorded. Both tests can be performed at any age allowing management decisions to be made early based on the potential of each individual animal.

With regular monitoring to ensure their relevance, this study aims to showcase the technical advancements implemented in Angus SteerSELECT, HeiferSELECT and BreedCHECK, including larger reference populations and denser SNP marker map.

MATERIALS AND METHODS

Angus SteerSELECT. Angus SteerSELECT is a tool to identify the genetic merit of Angus steers (87.5% Angus content or greater; see Angus BreedCHECK later in this section), providing data to inform management decisions aimed at sorting steers into the most appropriate finishing path and end market. Results of the initial validation were reported by Hine *et al.* (2021), with cross-validation estimates of bias, dispersion and accuracy of genomic predictions presented in Alexandre *et al.* (2021), and the potential value of incorporating commercial data into the reference population examined in Reverter *et al.* (2023).

Early implementations of Angus SteerSELECT relied on a reference population of 3,993 steers with feedlot measures including average daily gain (ADG) and daily feed intake (DFI), and 5,890 steers with carcase measures including carcase weight (CWT), eye muscle area (EMA), MSA marbling score (MMS), ossification (OSS), and rib fat depth (RIB). In comparison, the latest development uses a reference population of 5,267 (or 32% increase) and 6,997 (or 19% increase) steers with feedlot and carcase measurements, respectively.

Angus HeiferSELECT. Coupled with bull selection, the targeted selection of replacement females is crucial for maximising genetic gain in commercial beef enterprises. The decision of which heifers to retain for breeding, and which to cull, potentially affects the profitability of the herd for years to come (Wathes *et al.* 2014). Angus HeiferSELECT is an advanced genomic tool developed to inform the selection of replacement heifers in commercial Angus breeding operations. Alexandre *et al.* (2022) validated Angus HeiferSELECT genomic predictions for birth weight (BW), weaning weight (WW), yearling weight (YW) and mature cow weight (MCW) based on the animals' self-performance as well as the average performance of their progeny. More recently, Alexandre *et al.* (2023) demonstrated the opportunity to include fertility indicator traits, including age at first calving (AFC), days to calving (D2C), and pregnancy test outcome (PREG) in the Angus HeiferSELECT trait repertoire as correlated traits to calving ease (CE), the trait reported in Angus HeiferSELECT.

Combined, Angus SteerSELECT and HeiferSELECT provide information for 14 traits across five categories: (1) Body weights: BW, WW, YW, and MCW; (2) Feedlot performance: ADG and DFI; (3) Carcase: CWT, EMA, MMS, OSS, and RIB, (4) Immune competence: ImmuneDEX (IDEX); and (5) Fertility: Milk (as maternal component of WW) and CE (presented in birth weight units of kgs). For the immune competence phenotypes, measurement protocols can be found in Hine *et al.* (2019), and the most recent genetic parameter estimates in Reverter *et al.* (2021).

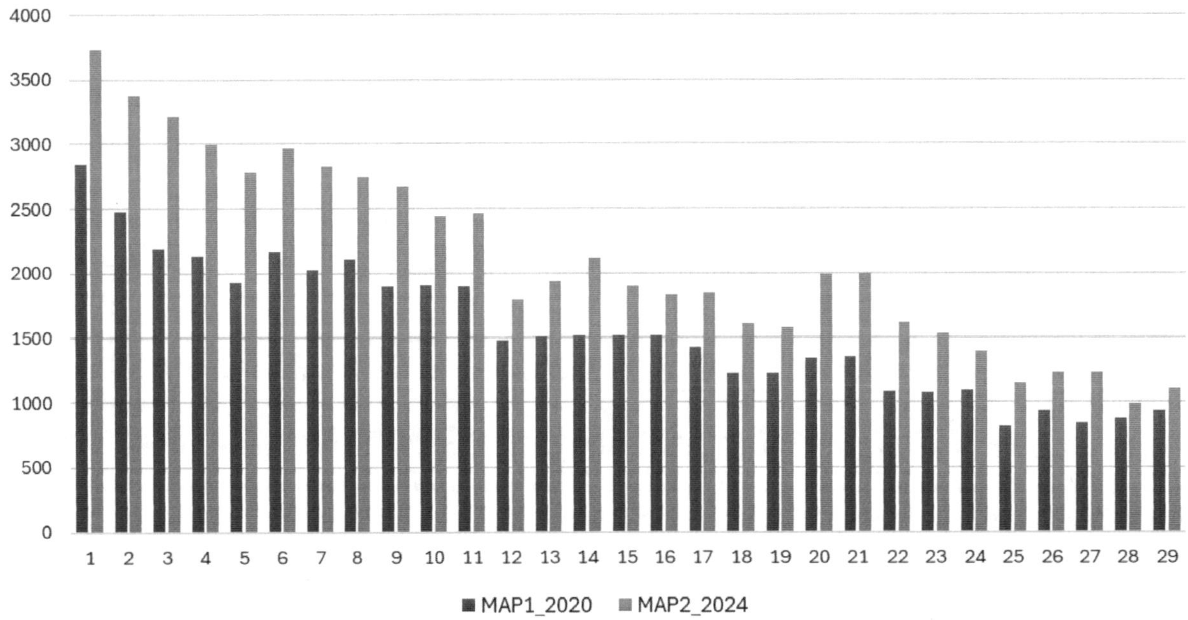

Figure 1. Number of SNP across the 29 bovine autosomal chromosomes used in the original (MAP1_2020) and the most recent (MAP2_2024) developments of Angus SteerSELECT, HeiferSELECT and BreedCHECK genomic tools

Angus BreedCHECK. Based on numerical approaches outlined in Reverter *et al.* (2020), Angus BreedCHECK is a genomic tool that predicts breed composition, reporting on relative contribution of 11 breeds with a focus on Angus content. Following its initial implementation (V1), based on a reference population of 1,398 cattle across 11 breeds, Duff *et al.* (2023) compared the breed composition values from Angus BreedCHECK on a large industry dataset of genotyped animals, to their known breed background. The latest development (V2) of Angus BreedCHECK is based on a reference population of 2,743 cattle, i.e. almost doubling population size used in V1, across the same

11 breeds. Comparisons between V1 and V2 were based on B-allele frequencies and percent absolute error (%Error) in the Angus content for cattle in five registries including the Angus Herd Book Register (HBR), Angus Performance Register (APR), Angus Commercial Register (ACR), Angus HeiferSELECT (AHS) and Multi Breed Register (MBR).

Anticipating higher genomic prediction accuracies and central to the advancements implemented in the three genomic tools (Angus SteerSELECT, HeiferSELECT, and BreedCHECK), is the use of a larger SNP map that grew by 35% from 45,364 (namely, MAP1_2020) to 61,105 autosomal SNPs (MAP2_2024).

RESULTS AND DISCUSSION

Figure 1 shows the number of SNP per chromosome (BTA) for the previous (MAP1_2020) and most recent (MAP2_2024) implementation of Angus SteerSELECT, HeiferSELECT, and BreedCHECK genomic tools. All chromosomes experienced an increase in the number of SNP ranging from 12% in BTA 28 (from 875 to 984) to 49% in BTA 22 (from 1,088 to 1,622).

Table 1. Number of records (N), mean and standard deviation (SD) and estimates of genetic variance (Vg) and heritability (h^2) for the 14 traits across the 5 categories considered in the development of Angus SteerSELECT and HeiferSELECT genomic tools

Trait, Unit	N	Mean	SD	Vg	h^2
Category 1 – Body Weights					
BW, kg	100,375	36.03	5.05	6.33	0.412
WW, kg	94,440	245.52	52.55	97.55	0.203
YW, kg	70,203	402.43	73.93	270.27	0.299
MCW, kg	13,236	544.34	81.87	847.60	0.374
Category 2 – Feedlot Performance					
ADG, kg/day	5,267	1.619	0.352	0.015	0.208
DFI, kg/day	5,267	14.502	2.265	0.812	0.316
Category 3 – Carcase Characteristics					
CWT, kg	6,979	431.34	66.17	425.22	0.404
EMA, cm^2	4,742	90.06	10.71	31.32	0.443
MMS, score	6,942	502.36	127.92	3,637.68	0.347
OSS, score	4,787	147.31	18.30	58.39	0.315
RIB, mm	4,388	17.02	6.06	6.97	0.313
Category 4 – Immune Competence					
IDEX, score	5,319	-0.003	1.139	0.37	0.303
Category 5 – Fertility					
Milk, kg	94,440	245.52	52.55	65.36	0.116
PREG, wk	6,774	12.81	7.34	9.15	0.267

Listed in Table 1 are the summary statistics and estimates of genetic variance and heritability for the 14 traits included in Angus SteerSELECT and HeiferSELECT genomic tools. Despite the larger sample size, figures in Table 1 are very close to those reported for previous implementations (Hine *et a*l. 2021; Alexandre *et al.* 2021; Reverter *et al.* 2023). Heritability estimates were moderate and ranged from 0.116 ± 0.013 for Milk (i.e. WWT maternal) to 0.443 ± 0.023 for EMA.

To aid producers make optimal multi-trait selection decisions, following derivation by Quinton *et al.* (2021), traits in the Angus SteerSELECT and HeiferSELECT tools are modelled into economic selection index being cow-calf value, feedlot and carcase value and total breeding value.

Table 2 shows the %Error of Angus BreedCHECK in the genomic assignment to Angus between old and new versions. Reflecting very consistent results between the two versions, 96.2% of cattle in the largest HBR registry had a %Error < 1%. The biggest discrepancy was for the multi-breed registry (MBR) known to contain many Brangus cattle and for which the %Error was 2.8% on average, with 76.6% of animals having a %Error < 5%.

Table 2. Absolute percent error (%Error) in the genomic assignment to Angus between old and new versions of Angus BreedCHECK: number of records (N), mean, standard deviation (SD), minimum, and percent of animals with %Error < 1% and < 5%

Registry	N	Mean	SD	Min.	Max.	<1%	<5%
HBR	210,114	0.2	1.1	0	52	96.2	96.7
APR	112,646	0.3	1.3	0	61	94.0	95.5
ACR	7,100	0.6	2.8	0	61	90.2	93.2
AHS	312	0.6	1.9	0	16	89.1	92.0
MBR	4,193	2.8	3.4	0	56	42.2	76.6

CONCLUSION

Angus SteerSELECT, HeiferSELECT, and BreedCHECK are genomic tools designed to assist decision-making regarding the optimal management and feeding strategies for steers, the selection of replacement heifers, and the assessment of genomic breed composition in commercial Angus beef breeding enterprises.

ACKNOWLEDGEMENTS

The assistance of Gavin Purtell from CSIRO and Christopher de Crespigny from Angus Australia is gratefully acknowledged. CSIRO acknowledges the Traditional Owners of the lands that we live and work on across Australia and pays its respects to Elders past and present.

REFERENCES

Alexandre P.A., Li Y., Hine B.C., Duff C.J., Ingham A.B., Porto-Neto L.R. and Reverter A. (2021) *Genet. Sel. Evol.* **53**: 77.

Alexandre P., Porto-Neto L., Hine B., Ingham A., Samaraweera M., Duff C. and Reverter A. (2022) *Proc. World Cong. Genet. Appl. Livest. Prod.* **12**: 1930.

Alexandre P., Porto-Neto L.R., Hine B.C., Samaraweera M., Byrne A.I., Ingham A., Duff C. and Reverter A. (2023) *Proc. Assoc. Advmt. Anim. Breed. Genet.* **25**: 67.

Duff C.J., Samaraweera A.M., Byrne A.I., Ingham A.B., Alexandre P.A., Porto-Neto L.R. and Reverter A. (2023) *Proc. Assoc. Advmt. Anim. Breed. Genet.* **25**: 326.

Hidalgo J., Lourenco D., Tsuruta S., Bermann M., Breen V. and Misztal I. (2023) *J. Anim. Sci.* **101**: skad355.

Hine B.C., Bell A.M., Niemeyer D.O.D., Duff C.J., Butcher N.M., Dominik S., Ingham A.B. and Colditz I. (2021) *J. Anim. Sci.* **97**: 4053.

Hine B.C., Duff C.J., Byrne A., Parnell P., Porto-Neto L., Li Y., Ingham A.B. and Reverter A. (2021) *Anim. Prod. Sci.* **61**: 1884.

Quinton C.D., Archer J.A., Amer P.R., Harburg S, Petersen G., Byrne A., Duff C. and Parnell P. (2021) *Proc. Assoc. Advmt. Anim. Breed. Genet.* **24**: 308.

Reverter A, Hudson N.J, McWilliam S, Alexandre P.A, Li Y, Barlow R, Welti N, Daetwyler H, Porto-Neto L.R and Dominik S. (2020). *J. Anim. Sci.* **98**: skaa337.

Reverter A., Hine B.C., Porto-Neto L., Alexandre P.A., Li Y., Duff C.J., Dominik S. and Ingham A.B. (2021) *Anim. Prod. Sci.* **61**: 1919.

Reverter A., Porto-Neto L., Hine B.C., Alexandre P.A., Samaraweera M., Byrne A.I., Ingham A.B. and Duff C.J. (2023) *Anim. Prod. Sci.* **63**: 947.

Wathes D.C., Pollott G.E., Johnson K.F., Richardson H. and Cooke J.S. (2014) *Animal* **8**: 91.

Proc. Assoc. Advmt. Anim. Breed. Genet. 26: 443-446

2025 NATIONAL BREEDING OBJECTIVE REVIEW FOR AUSTRALIAN DAIRY CATTLE: INSIGHTS FROM STAKEHOLDER CONSULTATION

T.T.T. Nguyen[1], M. Axford[1], P. Thurn[1], L. Monks[1], G. Nieuwhof[1], P. Williams[1], D. Watson[2], H. McLaren[3] and M. Shaffer[1]

[1] DataGene Ltd., AgriBio, 5 Ring Road, Bundoora, VIC, 3083 Australia
[2] Down To Earth Research, Scarborough, WA, 6019 Australia
[3] Dairy Australia, HWT Tower, Southbank, VIC, 3006 Australia

SUMMARY

DataGene conducts a comprehensive review of the National Breeding Objective (NBO) every five years to ensure breeding cows meet the evolving needs of Australian dairy farmers. A key component of an NBO review is to explore perceptions, attitudes, and usage of Australian breeding values and indices among farmers and herd improvement (HI) industry personnel. This review employed a mixed-method approach, including a quantitative survey of 217 dairy farmers and 38 HI personnel, alongside 20 qualitative in-depth interviews with dairy farmers. The survey specifically targeted dairy farmers using artificial insemination (AI). The results were further discussed at 28 stakeholder events involving 138 participants, providing additional perspectives to inform breeding strategies. The Balanced Performance Index (BPI) remains highly influential across regions, breeds, and production systems. While there is clear demand for a pasture-based index, opinions on a Total Mixed Ration (TMR) index remain polarised. Farmers in hot and humid climates raised interest in a region's specific index. Jersey breeders demonstrated strong reliance on existing Australian indices, though a breed-specific index was deemed non-essential. These findings, together with economic and genetic reviews, will guide the refinement of breeding tools to better align with industry needs and enhance farmer engagement.

INTRODUCTION

The Australian Dairy National Breeding Objective (NBO) sets breeding priorities to improve net farm profit and ensure Australian dairy herds meet the future needs of farmers and the industry. To remain effective, the NBO must adapt to changing dairy business requirements, advancements in breeding technologies, new knowledge and consumers' preferences. DataGene reviews the NBO and its associated indices every five years, with additional reviews conducted as necessary. The current NBO focuses on increasing farm profitability while driving sustainability outcomes. This is delivered through three key breeding indices: Balanced Performance Index (BPI), Health Weighted Index (HWI) (Axford *et al.* 2021) and Sustainability Index (SI) (Nguyen *et al.* 2023). The BPI is an economic index that reflects most farmer preferences. It drives net profit through a balance of functionality, type and yield. The HWI puts extra emphasis on health and fertility. The SI fast-tracks the reduction in greenhouse gas emission intensity. These indices are critical tools for ranking bulls, cows and herds, enabling farmers to select superior genetics that deliver long-term productivity, profitability, and sustainability.

A key component of an NBO review is understanding perceptions, attitudes usage of genetics, breeding indices and traits among dairy farmers and herd improvement (HI) industry personnel. To gather the necessary insights, DataGene commissioned Down To Earth Research (DTER) to implement a mixed-method approach. A nationwide online survey explored: i) Current use of Australian indices, ii) The influence of Australian indices on breeding decisions, iii) Attitudes towards genetics and breeding, and iv) Trait preferences and future genetic priorities.

To delve deeper into emerging themes, qualitative in-depth interviews were conducted with dairy farmers. These discussions focused on perceived need for indices tailored to specific production systems, regions, breeds, and future genetics priorities and requirements. The findings were discussed at a series of stakeholder meetings, providing additional perspectives into breeding strategies. This paper presents the findings of the online survey, in-depth interviews and stakeholder engagement meetings. These findings will inform the refinement of breeding tools to adapt to industry changing needs and enhance farmer engagement.

MATERIALS AND METHODS

Online survey. An online survey was designed using a structured questionnaire and was 'live' between late June and early July 2024. Key aspects of the methodology include:

- A survey notification email was sent to Dairy Australia levy payers (with an email address available) and DataGene's Genemail members informing them of the survey, benefits of participation and a unique link to complete the online survey.
- Survey promotion was undertaken by Dairy Australia and DataGene which included a generic link to enable completion of the survey.
- In total 296 dairy farmers and industry personnel participated in the online study. Once partial and poor-quality surveys were removed, 255 high quality surveys remained, 217 completed by dairy farmers and 38 by HI industry personnel.
- Online survey data was extracted and analysed by senior DTER consultants using IBM Statistical Package for Social Sciences (SPSS) and Q research software.

In-depth interviews. Initially, a series of 10 in-depth interviews were conducted with dairy farmers who completed an online survey and agreed to provide additional insights. Two interviews were conducted with dairy farmers representing each of the following segments: TMR, Jersey as main breed in the herd, pasture-based farms, Holstein as main breed in the herd, and sub-tropical farms. Additionally, to capture feedback from a different segment of farmers than DataGene may typically engage with, interviews targeted respondents with commercial dairy farms rather than those with registered cattle. To gain further understanding from the TMR farmers, where the survey participation was low, an additional 10 in-depth interviews were carried out. Key aspects of the in-depth interview process included:

- Interviews were completed during August 2024 and averaged approximately 20 minutes in duration.
- Respondents were typically passionate about herd genetics, represented a variety of locations, production and feeding systems and did not have registered cattle in their herd.
- To ensure open, honest discussion and feedback was provided to DTER, respondents were guaranteed full confidentiality, and a discussion guide was used rather than a structured questionnaire.

Stakeholder meetings. To gain additional perspectives to inform breeding strategies, the results of the online survey and the in-depth interviews were presented and discussed at 28 events, engaging 138 participants including bull companies, resellers, breed organisations, farmers, industry, and research.

RESULTS AND DISCUSSION

While the results reflect insights from a broad cross-section of the industry, the survey specifically targeted dairy farmers using artificial insemination (AI). This approach may introduce some bias, as participants with strong opinions – whether positive or negative – toward herd genetics may be more likely to respond. Consequently, the data represents the perspectives of respondents and may not fully reflect the views of the entire dairy farming population.

Breeding profile of milking herds. Nationally, 54% of dairy farmer respondents have at least some registered cattle and 46% are solely commercial dairy farms. Holstein are the most common breed of cattle on respondent farms (61%). This is reflected across all regions, except among DairyTas respondents, where crossbred cows dominate. In total, Jerseys are the main breed of cows on 18% of respondent farms, crossbreds (17%) and red breeds (13%). The results are consistent with Dairy Australia survey data. Note that while the main breed of cow was requested, respondents were able to select more than one main breed.

Use of indices and breeding values. The majority of dairy farmers (88%) and HI industry personnel (92%) use at least one Australian indices or breeding value when selecting AI sires. The BPI is the most commonly used, followed by the HWI and SI. Index usage varies by breed and herd size, with 100% of Jersey breeders using indices compared to 72% of other breeds, and higher usage among larger herds. Feed systems influence preferences, with HWI favoured by farmers feeding less than 1.5 tonnes of grain. Additionally, 76% of farmers and 84% of HI personnel use Australian Breeding Values (ABVs), while 71% of both groups rely on the Australian production index (ASI). These findings highlight the widespread influence of Australian indices, particularly the BPI, across diverse regions, breeds, and farming systems.

Degree of influence of indices. When selecting AI sires, 43% of dairy farmer respondents say the BPI has a large influence on their decision making with a further 36% indicating it has little influence. Among HI industry respondents, 61% say the BPI has a large influence and 29% a little influence when choosing semen. Notably, HWI influences AI selection for a significantly greater proportion of respondents with commercial dairy farms than those with registered cattle (68% and 53% respectively).

Attitudes towards breeding indices for specific systems. Online survey and in-depth interview findings provide some evidence that there is a demand for a seasonal pasture-based index among those implementing a pasture-based feed system. Both the small number of farmers implementing a TMR and HI personnel support the need for a TMR index. Some in-depth interview respondents see the benefit of splitting housed versus pasture-based systems, while others believe cows will perform regardless, provided they are fed sufficiently. Unsurprisingly, while there is little demand among southern and western farmer respondents for a northern Australian index, there is demand in the sub-tropical region.

Perceived need for a Jersey specific index. Survey and in-depth interview results suggest that the current Australian indices are having substantial impact on AI sire selection among respondents whose main breed of cattle is Jersey and while there may be some interest in developing a Jersey specific index However, may be a ‘nice to have’ rather than ‘need to have’.

Trait preferences. When selecting AI sires, mammary system, daughter fertility and protein production (kg) are rated most important traits by dairy farmers and HI personnel. Dairy farmer respondents with registered cattle place a significantly higher emphasis on overall type than those without registered cattle. This is reversed for calving ease and feed saved. Farmers in seasonal calving/pasture-based systems rank fertility as a higher priority than other groups. Subtropical farmers place a higher emphasis on heat tolerance compared to the overall respondent average.

Future needs. While seasonal conditions, financial conditions and other factors may change perceptions in future, survey results suggest animal health, calf vitality, lameness and beef-on-dairy are key areas that genetics may assist dairy farmers. Robotic milking and beef-on-dairy appear to be less of a concern. However, these may represent opportunities for genetic improvement and are likely to be more highly rated among those with robotic dairies and selling dairy beef.

Between August and October 2024, DataGene conducted meetings with stakeholders from bull companies, resellers, breed organisations, farmers, industry, and research. These sessions aimed to present findings from the survey and phone interviews, gather feedback on key themes, and gain new insights. Feedback largely aligned with previous results, highlighting four main themes:

- Production remains key for farm income, explaining the unpopularity of negative milk L bulls ABV. Exacerbating this, the negative weighting applied to milk litres in the BPI formula (reflecting feed cost) is poorly understood.
- Simplicity is preferred, as farmers favour a manageable number of traits applicable across feeding systems.
- Feed Saved ABV is not well understood, lacks credibility due to low reliability, and requires extension support for wider adoption. However, feed efficiency was viewed to be important, further highlighting the confusion around this trait.
- Support for the base change was noted.

CONCLUSION

In conclusion, the stakeholder consultation for the 2025 NBO review, conducted by DataGene through an online survey, in-depth interviews, and stakeholder meetings, revealed the following key insights. Stakeholders expressed strong support for the Balanced Performance Index (BPI). They also showed clear support for a seasonal calving/pasture-based index. There was high support for a specific index tailored for hotter regions. Views on the need for a specific index for TMR herds were polarised. Support for a Jersey-specific index remained limited. These findings, together with economic and genetic reviews, will help guide the refinement of breeding tools to adapt to industry needs.

ACKNOWLEDGEMENTS

The authors wish to acknowledge the valuable contributions of the survey respondents, interviewees, and participants of the stakeholder meetings. Special thanks are extended to DataGene's Genetic Evaluation Standing Committee, Jo Newton (DairyBio), and the Dairy Australia and DataGene teams for their promotional support. We also gratefully recognise the technical support provided by John Droppert (Dairy Australia), the Dairy Farm Monitor Project team, and Natalie Nelson (Agriculture Victoria).

REFERENCES

Axford M., Santos B., Stachowicz K., Quinton C., Pryce J.E. and Amer P. (2021) *Anim. Prod Sci.* **61**: 1940.

Nguyen T.T.T., Richardson C.M., Post M., Amer P., Nieuwhof G.J., Thurn P. and Shaffer M. (2023) *Anim. Prod Sci.* **63**: 1126.

EXPLORING ADOPTION FACTORS, AND PREFERENCES TO THE USE OF BEEF GENETIC TOOLS IN NORTHERN AUSTRALIA.

P. Menchon[1], J.K. Manning[1], D.L. Swain[1,2] and A. Cosby[1]

[1] Institute for Future Farming Systems, School of Health, Medical and Applied Sciences, CQUniversity Australia, Rockhampton, QLD, 4701 Australia
[2] TerraCipher, 337 Laurel Bank Rd., Alton Downs, QLD, 4702 Australia

SUMMARY

Genetic improvement of beef cattle has emerged as a fundamental priority in Northern Australia. This study aims to identify the factors that influence decision-making regarding the use of genetic tools by Northern Australian beef farmers. The genetic tools considered are BREEDPLAN estimated breeding values (EBVs), TACE, IGS, Leachman, Ingenity and genomic profiles, as well as indices that integrate these genetically assessed traits, such as $Indexes. Australia beef farmers. Surveys were collected from beef farmers across Northern Australia and logistic regression analyses and Chi-square tests were performed. Age was the only demographic factor that had a significant effect on the adoption of genetic tools for seedstock and commercial farmers. For seedstock farmers, each additional year of age increased the odds of adopting genetic tools by 6.4%. however, for commercial farmers, each additional year decreases the odds by 5.2%. For commercial farmers who collect phenotypic data, the likelihood of their adoption of genetic tools increases. Understanding genetic tools and the impact on their business is a key motivation for adoption.

INTRODUCTION

Beef farmers in Northern Australia have historically been slower to adopt genetic tools than their southern counterparts. This region faces several challenges such as extreme climate, poor market access and exposure to ticks that influence production, and economic viability of livestock enterprises (Bell and Sangster 2023). The size of farms in remote areas alongside low connectivity also challenges technology adoption. Genetic improvement offers a solution to some of these production challenges. Despite recent RD&E efforts promoting their use among northern Australian farmers (Bell and Sangster 2023; Greenwood *et al.* 2018; MLA 2022; NABRC 2022), adoption of genetic tools and technology remains a priority. To address this, the diffusion of existing technologies through participatory extension programs with farmers (Nettle *et al.* 2022; van der Werf 2023) is an opportunity to increase awareness and adoption of genetic improvement in commercial herds. This research aims to enhance the understanding of the factors that motivate northern Australian beef farmers to use genetic improvement tools in their herds and to identify their data collection preferences.

MATERIALS AND METHODS

This research was conducted during 2024 through online surveys in Qualtrics with seedstock and commercial farmers from Northern Australia. This research was approved by the CQUniversity Australia Human Research Ethics Committee (approval number 24882). The survey comprised 32 questions and responses were anonymous. The survey targeted Northern Australian beef farmers segmenting commercial and seedstock producers. To determine the sample size that achieved representativeness across Northern Australia, the formula proposed by Daniel and Cross (2018) was used $n=[(z^2.p.q)/d^2]$, where n= sample size, z = desired confidence level, p = proportion of a characteristic of the population to be sampled, the number of beef farmers related to the total number of farms in northern Australia, q= (1-p), and d= desired precision. Farmer proportions were extracted from the Australian Bureau of Agricultural and Resource Economics and Sciences (2023) and

Government of Western Australia (2021). This corresponds to p values of 0.675 and q of 0.325. Considering a 90% desired precision (d=0.1), the sample size to be collected should be 84 across Northern Australia. Seedstock and commercial beef farmers were classified according to the use of genetic tools in their herds into two sub-categories, Adopters (A) and Non-Adopters (NA).

Statistical analysis. Firstly, binary logistic regressions were used to identify the variables that influence the adoption of genetic tools by beef farmers. The dependent variables are qualitative and categorised (Adopter; Non-Adopter). Commercial farmers were classified as "Adopters" if, within the past five years, they engaged in breeding programs, purchased heifers with genetic evaluations, or acquired bulls or used AI from genetically evaluated bulls, provided they also met the first two criteria. Seedstock farmers were deemed "Adopters" if they participated in breeding programs (e.g., BREEDPLAN) and sold bulls or heifers with EBV data. It is proposed to follow the model used by Quddus (2022) $Y=Logit(p)=Ln(p/(1-p))=\alpha+Z'\delta$., where Y takes the value of one if the beef cattle farmer uses genetic tools in their herd (Adopters); the probability that the farmer adopts any of the genetic tools being p, (1- p) is the probability that the farmer does not use genetic tools (Non-Adopters), Z is the vector of the independent variables, δ is the coefficient to be estimated and α is a constant. The independent variables used in the model were demographic and productive factors (e.g. age, gender, education level, role on-farm, number of employees, herd size, breeding systems, breeding methods, tropical/temperta breed and data collection). The odds ratio *(Exp(B))* quantifies the change in the odds of the dependent event occurring for a one-unit increase in the predictor variable. Secondly, Chi-square tests were performed to analyse the relationship between the adoption of genetic tools and preferences regarding data recorded on farms; and reasons for using them within each group of farmers (commercial and seedstock). The survey data was managed in a Microsoft Excel spreadsheet. For logistic regression, Chi-square test and descriptive analysis, the Statistical Package for Social Sciences (SPSS version 28.0.0.0 (190)) was used.

RESULTS AND DISCUSSION

The survey was completed by 97 producers that met the conditions required for this study. A total of 116 postcodes were registered, distributed between QLD (86.2%), NT (12.9%) and north WA (0.9%). Of the participants, 55.67% answered that the main focus in their enterprise was commercial, followed by seedstock (44.33%). Therefore, the sample was biased towards seedstock, since in Northern Australia the proportion of seedstock producers is significantly lower than that of commercial farmers. The ratio between Adopters/Non-Adopters differed for each group of producers, being 69.77/30.23% and 48.15/51.85% for seedstock and commercial, respectively This study had a higher representation of commercial farmers who were classified as non-users of genetic tools and provides new information on potential users of genetic tools that could be used in the design of an extension activity.

Adoption of genetic tools. Results from logistic models considering demographic variables suggest that gender, level of education and the role on farm did not have any effect on the adoption of genetic tools. Age was the only demographic factor that had a significant effect on the odds ratio of adopting genetic tools. The fit quality of the models for age was similar across farmer groups, 14.9% and 14.1% of the variance in the dependent variables (Adopter/Non-Adopter) for seedstock and commercial producers respectively. The odds ratio *(Exp(B))* for age was 1.06 for seedstocks, indicating that for every one-year increase in age, the odds of adopting genetic tools increased by 6.4%. Conversely, the odds ratio *(Exp(B))* for the predictor was 0.95 for commercial producers, indicating that for every one-year increase in age, the odds of adopting genetic tools decreased by 5.2% (Table 1). Age is a factor that has shown inconsistencies in several studies of technology adoption processes (Montes de Oca Munguia and Llewellyn 2020). Furthermore, Martin-Collado *et al*. (2021) found that age did not generally affect farmers' attitudes towards breeding in all the breeds analysed, but it was significant for certain breeds and breeding systems. This is important to consider, as our analysis does not consider a breed effect.

The results of the logistic regression of productive variables suggest that on-farm data collection differs between producer groups. Data collection on commercial farms had a statistically significant effect (P=0.005) and explained 32% of the variation in genetic tool adoption levels in this group. The odds ratio (*Exp(B)*) indicates that the odds of the event occurring were 21.66 times higher when data was collected than when it was not. Sales weight, structure, conformation and appearance data are the phenotypic traits that commercial adopting breeders prefer to record. However, scrotal circumference, weaning weight and 200-400-600 daily weights are the traits that seedstock farmer adopters prefer to record. Data collection is crucial for identifying improvement opportunities and tracking progress (Bell and Sangster 2023). However, Northern Australia's challenging environment, including remote distances, large properties with low stocking rates, and poor connectivity, makes data collection difficult.

Table 1. Logistic regression parameters of age and data Collection on farm and their effect on the adoption of genetic tools (EBVs and genomic profile) by seedstock and commercial farmers

	n	Nagelkerke R^2	B	S.E.	Sig	Exp(B)	95% C.I. for EXP(B) Lower	Upper
Seedstock farmer								
Age	43	0.15	0.06	0.03	**0.05***	1.06	1.00	1.13
Data Collection	43	0.07	1.43	0.98	0.14	4.20	0.61	28.92
Commercial farmer								
Age	53	0.14	-0.05	0.02	**0.03***	0.95	0.90	0.99
Data Collection	54	0.32	3.08	1.09	**0.005***	21.67	2.57	182.74

*Level of significance: P<0.05

Table 2. Frequencies and Chi-square test results for the association between adopting genetic tools (Adopter; Non-Adopter) and reasons for using genetic tools (EBVs and genomic profile) by farmer's group

Variable, n (%) [a]	**A n(%)**	**NA n(%)**	**p-value**
Seedstock farmer	**(n =30)**	**(n = 12)**	
I understand the genetic tool	24 (85.7)	4 (14.3)	**0.009[2]***
It's easy to use	11 (73.3)	4 (26.7)	1.000[2]
I can see the impact of using it in my herd	25 (71.4)	10 (28.6)	1.000[2]
The cost/benefit ratio is appropriate	11 (78.6)	3 (21.4)	0.719[2]
Other farmers are using genetic tools	8 (88.9)	1 (11.1)	0.247[2]
It is important for the business	24 (88.9)	3 (11.1)	**0.001[2]***
I have received advice on how to use it	9 (100.0)	0 (0.0)	**0.041[2]***
Commercial farmer	**(n =25)**	**(n = 24)**	
I understand the genetic tool	20 (69.0)	9 (31.0)	**0.004[1]***
It's easy to use	12 (52.2)	11 (47.8)	1.000[1]
I can see the impact of using it in my herd	18 (58.1)	13 (41.9)	0.244[1]
The cost/benefit ratio is appropriate	11 (45.8)	13 (54.2)	0.572[1]
Other farmers are using genetic tools	2 (40.0)	3 (60.0)	0.667[2]
It is important for the business	22 (64.7)	12 (35.3)	**0.005[1]***
I have received advice on how to use it	8 (53.3)	7 (46.7)	1.000[1]

[a]Dummy variable. Only the values for which the option was selected are displayed. A: Adopter. NA: Non-Adopter. [1] Pearson's Chi-squared test, Level of significance: P<0.05; [2] Fisher's exact test, Level of significance: P<0.05.

Producer's preferences. For the seedstock farmer group, 88.37% of participants reported that they recorded data, however, only 42.1% selected recording birth weight. Associations were found between the adoption of genetic tools and on-farm recorded data such as scrotal circumference (P=0.031), birth weight (P=0.002), weaning weight (P=0.012) and 200/400/600-day weight (P<0.001). The reasons for using genetic tools that are related to the level of adoption (Adopters and Non-Adopters) are that they understand the genetic tools (P=0.009), that they are important for their business (P=0.001) and have received advice on how to use them (P=0.041). This could be an indication of the crucial and positive effect that extension activities and training systems have had in recent years in Northern Australia (MLA 2022).

For commercial famers, 74.1% indicated that they record traits on farm. However, only the structure/conformation and appearance record showed a relationship (P=0.01) with the adoption of genetic tools. For this group of farmers, the main reason for adopting genetic tools (Table 1) was that it was important for their business (69.4%, P=0.005), followed by they saw the impact on their herd (63.3%) and had an understanding of the genetics tool used (59.2%, P=0.004).

CONCLUSION

The results analysed showed that there are differences between groups of farmers in factors such as age and on-farm data collection in the probability of adopting genetic tools. Other variables analysed had no effect on the probability of adopting genetic tools. The results also show that the reasons for adopting genetic tools are associated with understanding economic motivations and the traits of the animals that farmers prefer. It is crucial to design extension strategies that consider the collection and interpretation of data on economically important traits, differentiating between commercial and stud farmers. Future research could be undertaken to determine the preferences regarding the training and extension systems preferred by farmers.

ACKNOWLEDGEMENTS

The authors express their acknowledgement to stakeholders and all the beef farmers of this study. This research received external funding from Meat & Livestock Australia (P.PSH.1186).

REFERENCES

Bell A. and Sangster N. (2023) *Anim. Prod. Sci.* **63**: 1.

Daniel W.W. and Cross C.L. (2018) Biostatistics: a foundation for analysis in the health sciences. Wiley.

Government of Western Australia. (2021). https://www.agric.wa.gov.au/export-services/western-australian-beef-industry?page=0%2C0#smartpaging_toc_p0_s0_h2

Greenwood P.L., Gardner G.E. and Ferguson D.M. (2018) *Asian-Aus. J. Anim. Sci.* **37**: 992.

Martin-Collado D., Díaz C., Benito-Ruiz G., Ondé D., Rubio A. and Byrne T.J. (2021) *Animal.* **15:** 100062.

Meat & Livestock Australia. (2022) Producer adoption outcomes report: 2021-2022. Meat & Livestock Austalia Limited, North Sydney.

Montes de Oca Munguia O. and Llewellyn R. (2020) *Appl. Econ. Perspect. P*42:, 80.

North Australia Beef Research Council (2022) Producer Identified RD&A priorities for 2023-24.

Nettle R., Major J., Turner L. and Harris J. (2022) *Anim. Prod. Sci.* **64**: AN22329.

Quddus M. A. (2022) *Proc. Natl. Acad. Sci., India, Sect. B Biol. Sci.* **92**: 461.

van der Werf J.H.J. (2023) *Anim. Prod. Sci.*, **63:** 925.

Proc. Assoc. Advmt. Anim. Breed. Genet: 26: 451-454

COMBINED LAMBPLAN: EXPANDING THE MULTI-BREED NATURE OF THE LAMBPLAN EVALUATION

S.F. Walkom[1], P. Alexandri[1], D.J. Brown[1], K.L. Bunter[1], N.K. Connors[1], S. de las Heras-Saldana[1], K.P. Gore[1], P.M. Gurman[1], L. Li[1], A.J. McMillan[1], S.P. Miller[1], P.E. Bradley[2] and A.A. Swan[1]

[1] Animal Genetics Breeding Unit*, University of New England, Armidale, NSW, 2351 Australia
[2] Meat and Livestock Australia, University of New England, Armidale, NSW, 2351 Australia

SUMMARY

The Australian sheep industry continues to mix breeds across all sectors, including ram breeding. Genetic evaluation needs to change with industry and provide relevant comparisons across breeds. This has prompted a development to combine the LAMBPLAN Terminal and Maternal evaluations. Combining analyses will provide a single breeding value 'language' across non-Merino breeds and continue to allow breeders to access selection tools to meet their breeding objectives irrespective of breed. This study summarises the key technical enhancements to deliver the Combined LAMBPLAN analysis, including: business decisions around data usage, formation of lifetime contemporary groups, genetic group structure and the use of a meta-founder model, fixed effects models, heterosis adjustments and parameter re-estimation. The combined evaluation will provide an improved utility of ASBVs for the Australian prime lamb industry.

INTRODUCTION

Sheep Genetics (SG) was established in 2005 to deliver a national across-flock genetic evaluation system to the Australian sheep industry for terminal, maternal and Merino sheep (Brown *et al.* 2007). Sheep Genetics provides a single national genetic evaluation "language" for Australian sheep breeders. The SG evaluations have focussed on industry-recorded data from three major breed types: 1) LAMBPLAN - terminal sire breeds, 2) LAMBPLAN - maternal/dual purpose breeds and 3) MERINOSELECT - Merino breeds. The two LAMBLAN evaluations represent across-flock multi-breed evaluations. Having a uniform language, Australian Sheep Breeding Values (ASBVs), to describe genetic evaluation for Australian sheep has proven extremely successful (Collison *et al.* 2018). This has, in part, enabled Australian sheep breeders to select for a diversity of breeding objectives irrespective of breed. Subsequently, a decline in the purity of the animals in the LAMBPLAN evaluations and the rise of composite animals has been observed (McMillan *et al.* 2023). This has also corresponded to an increase in common animals across LAMBPLAN analyses, with 35% of flocks represented in both, as breeders look to different genetic sources to achieve their desired genetic gains. An unintended consequence of sharing terminal and maternal genetics is that the common animals are provided with two ASBVs, which cannot be compared, exhibit a significant base difference and are not underpinned by the same information, ultimately creating confusion about how to interpret ASBVs across the analyses.

Some flocks, primarily the 'shedding' breeds (i.e. Dorpers, Australian White and Ultra Whites) who are breeding a self-replacing ewe wanted to shift to the Maternal analysis. This change would produce a substantial increase in the size of the LAMBPLAN – maternal analysis. Consequently, the decision was made to combine the LAMBPLAN evaluations and provide a single evaluation and common language from which all non-Merino-type sheep could be compared. This paper covers the historical technical advancements to the Sheep Genetic evaluation that enable the combined analysis

---

* A joint venture of NSW Department of Primary Industries and Regional Development and the University of New England

and how the evaluation is being changed to overcome the technical constraints associated with using genomics from diverse breed and composite populations.

TECHNICAL ADVANCEMENTS

Genomic reference flock. A significant feature of the Australian Sheep Cooperative Research Centre (CRC) program was the information nucleus flock (INF) (van der Werf *et al.* 2010). Via the INF and its later incarnations, the reference flock population has provided head-to-head comparisons of Merino, maternal and terminal sire types primarily for live animal and carcase traits. The reference flock was the initial industry investment into genomics and remains the cornerstone for the genomic reference flock for hard-to-measure traits such as intra-muscular fat, shear force, worm egg count and any future methane records. At the time of this paper approximately 1,700 sires of maternal and terminal origin have had progeny recorded.

Single-step genetic evaluation. The LAMBPLAN evaluations changed to a genomic single-step model in 2017 (Brown *et al.* 2018). The development of this model was associated with improvements in the accuracy and bias of the estimated breeding values in forward and cross-validation studies of the LAMBPLAN Terminal data, compared to the pedigree model (Gurman *et al.* 2018). The single-step GBLUP model further grew to a breed-adjusted GRM, which reduced the breed structure that was implicitly forming in the unadjusted GRM, resulting in an improvement in cross-validation accuracies and bias estimates (Gurman *et al.* 2019).

Sheep Genetics database re-development. Combining analyses was made possible due to the re-development and implementation of a single Sheep Genetics database, completed in 2022. This encapsulates the pedigree, phenotypic and genomic information for the MERINOSELECT and the two LAMBPLAN evaluations in one repository. The combined co-managed central database significantly reduces the associated risks for Sheep Genetics of maintaining consistency across multiple databases. This new platform provided the flexibility to allow the new analysis models and methods required for the development of the Combined LAMBPLAN analysis.

THE PROPOSED EVALUATION

A combined dataset. At the time of writing (January 2025) the LAMBPLAN terminal and maternal analysis were publishing ASBVs for 4.1m and 2.8m animals. Given the scale of these evaluations, the significant overlap of animals and to expedite timely delivery of analyses, several business decisions were made to develop the future Combined LAMBPLAN database. These decisions were focussed on data reduction including removal of fictional dams (remnant of old decommissioned software), New Zealand data (phenotypes and unlinked pedigree), multi-sire groups and historic Australian data from animals born before 2000 (phenotypes). Data reduction decisions were found to have negligible impact on the rank or spread of the breeding values of current selection candidates during initial testing. The Combined LAMBPLAN development discussed in this paper and the work presented by Alexandri *et al.* (2025), de las Heras-Saldana *et al.* (2025), Gurman *et al.* (2025) and McMillian *et al.* (2025) have been performed on a dataset of growth and development traits built by Sheep Genetics in November 2024 (Table 1). The pedigree included 4.8m animals, with 3.7m recorded for at least one trait presented in Table 1. The analysis also incorporates 246k genotypes imputed to 61.5k SNP density.

Contemporary group structure. The lifetime grouping of animals as part of contemporary group formation has been improved so that the contemporary group better reflects a group of animals that have been managed together. Contemporary group (CG) construction and age slicing of groups were moved out of the analysis software and applied within the SG database, providing greater flexibility and improved transparency for breeders. This reduced the number of small CGs in the later age stages and was made alongside the decision to remove data for single animal CGs.

Table 1: Summary of Combined LAMBPLAN research dataset, November 2024

Trait Description	Trait Code	Animals	Records	Mean (SD)	#cg	cg size
Birth Weight	BWT	2,262,969	2,262,969	4.82 (1.08)	55,629	40.7
Weaning Weight	WWT	2,932,332	3,327,446	36.53 (9.94)	105,564	31.5
Post-wean Weight	PWT	2,050,519	2,322,479	50.25 (12.24)	78,704	29.5
Yearling Weight	YWT	667,005	667,005	59.36 (15.42)	30,063	22.2
Hogget Weight	HWT	277,096	277,096	69.19 (17.40)	14,505	19.1
Adult Weight	AWT	120,529	185,917	70.88 (12.97)	16,052	11.6
Carcase weight	CWT	30,370	30,370	24.29 (4.15)	2,027	15.0
Weaning fat depth	WCF	268,023	268,023	3.27 (1.21)	12,324	21.7
Post-wean fat depth	PCF	1,586,894	1,586,894	3.52 (1.34)	54,394	29.2
Yearling fat depth	YCF	429,450	429,450	3.87 (1.63)	17,132	25.1
Carcase GR fat depth	CFAT	22,745	22,745	15.22 (6.09)	1,703	13.4
Carcase C-site fat depth	CCFAT	25,886	25,886	4.61 (2.47)	1,823	14.2
Intramuscular fat	IMF	27,297	27,297	4.25 (1.11)	1,947	14.0
Shear Force	SF5	24,910	24,910	32.46 (11.03)	1,757	14.2
Weaning eye muscle depth	WEMD	267,913	267,913	28.77 (4.72)	12,322	21.7
Post-wean eye muscle depth	PEMD	1,589,755	1,589,755	30.33 (5.10)	54,440	29.2
Yearling eye muscle depth	YEMD	429,341	429,341	31.57 (5.47)	17,123	25.1
Carcase eye muscle depth	CEMD	26,746	26,746	33.06 (4.92)	1,845	14.5
Dressing percentage	DRESS	18,570	18,570	46.23 (3.78)	1,388	13.4
Lean meat yield	LMY	8,707	8,707	55.84 (4.63)	567	15.4
Post-wean scrotal circumference	PSC	98,958	98,958	29.95 (3.98)	3,272	30.2
Yearling scrotal circumference	YSC	65,107	65,107	30.06 (3.39)	2,574	25.3

Multi-breed population. The Combined LAMBPLAN evaluation contains data from 76 breeds including but not limited to the major terminal (Poll Dorset, Texel, Suffolk, White Suffolk, Dorper and White Dorper), maternal (Border Leicester, Corriedale, Coopworth, Composite Maternal) and some small representation from Merino breeds. Leveraging the genomic data available, an unsupervised clustering method was utilised to improve the accuracy of breed composition assessment, which led to the development of a more inclusive multi-breed genetic evaluation (Alexandri *et al.* 2025), focussed on genomic population structure, not arbitrary breed allocation. The findings support a shift from relying solely on straight-bred reference populations, allowing for better representation of composite and smaller breeds in genetic evaluations.

Development of a new genetic grouping strategy was performed, taking the breeds and further dividing them into either 5 or 10-year time windows, producing a final count of 95 genetic groups. This has improved the accuracy of breeds across a range of traits (McMillian *et al.* 2025). Analysis of the genetic linkages between breeds confirmed that the Combined LAMBPLAN dataset enables effective comparisons across all the key breeds included in the evaluation.

OVIS solver modification. As a result of research on fixed modelling, an important change in the analysis methodology will be the transition from pre-adjusting the data for systematic fixed effects to directly fitting these effects in the solver (de las Heras-Saldana *et al.* 2025). This produces a better fit of the model to the data with little impact on computing performance. It also means that the effects are always current and relevant to the entire data set, in contrast to pre-adjustments which can remain unchanged over long periods. Data filtering restrictions around age have been adjusted to avoid extreme values and ensure more trustworthy estimates. Pre-adjustments are often based on subsets of data such as reference populations, where estimating effects can be problematic because of high levels of genetic variability included in the design. Not having to re-estimate adjustment factors represents an efficiency gain for research projects.

The decision to bring about significant change to the LAMBPLAN evaluation presents the opportunity to re-evaluate the most appropriate method for aligning the pedigree and genomic information in single-step genomic BLUP. Changing from a breed-adjusted GRM to a metafounder model produces better predictive ability (Gurman *et al.* 2025), indicating a further change to how breeds are modelled within the Combined LAMBPLAN analysis.

CONCLUSION

The Combined LAMBPLAN analysis represents a revolutionary step forward for the Australian sheep industry and is the result of several years of technical development. The new analysis will deliver more robust breeding values and allow breeders to further increase their rates of genetic progress. The Combined LAMBPLAN evaluation will be released in 2025 upon the completion of the research phase, the required implementation work for commercialisation and the relevant validation work to quantify the value of the new models. The evaluation will also be expanded to incorporate reproduction, health and wool traits throughout 2025. Combining LAMBPLAN provides a single breeding value 'language' across the non-Merino breeds and continues to allow breeders to access selection tools to meet their breeding objectives.

ACKNOWLEDGMENTS

This research was funded by Meat and Livestock Australia project L.GEN.2204. The authors gratefully acknowledge the teams behind the Sheep CRC for Sheep Industry Innovation Information Nucleus and MLA Resource flock as well as individual industry flock owners involved in flock management and data collection. The authors of the paper would personally like to acknowledge the legacy that Andrew "Swanny" Swan will leave behind within the national genetic evaluation and the broader Australian sheep industry.

REFERENCES

Alexandri P, Gurman P.M., Brown D.J., *et al.* (2025) *Proc. Assoc. Advmt. Anim. Breed. Genet.* **26**: *These proceedings.*

Brown D.J., Huisman A.E., Swan A.A., *et al.* (2007) *Proc. Assoc. Advmt. Anim. Breed. Genet.* **17**: 187.

Brown D.J., Swan A.A., Boerner V., *et al.* (2018) *Proc World Cong. Genet. Appl. Livest. Prod.* **11**: 460.

Collison C.E., Brown D.J., Gill J.S., *et al.* (2018) *Proc World Cong. Genet. Appl. to Livest. Prod.* **11**: 661.

de las Heras-Saldana, S., Brown, D.J., Gurman, P.M., *et al.* (2025) *Proc. Assoc. Advmt. Anim. Breed. Genet.* **26:** *These proceedings.*

Gurman P.M., Swan A.A., Boerner V. and Brown D.J. (2018) *Proc World Cong. Genet. Appl. Livest. Prod.* **11**: 362.

Gurman P.M., Bunter K.L., Boerner V., Swan A.A. and Brown D.J. (2019) *Proc. Assoc. Advmt. Anim. Breed. Genet.* **23**: 254.

Gurman P.M., Alexandri P, Brown D.J., *et al.* (2025) *Proc. Assoc. Advmt. Anim. Breed. Genet.* **26**: *These proceedings.*

McMillan A.J., Walkom S.F. and Brown D.J. (2023) *Proc. Assoc. Advmt. Anim. Breed. Genet.* **25**: 31.

McMillan A.J., Alexandri P, Gurman P.M., *et al.* (2025) *Proc. Assoc. Advmt. Anim. Breed. Genet.* **26**: *These proceedings.*

van der Werf J.H.J., Kinghorn B.P. and Banks R.G. (2010) *Anim. Prod. Sci.* **50**: 998.

EXPANDING AUSTRALIAN SHEEP GENOMIC REFERENCE POPULATION

P. Alexandri, P.M. Gurman, D.J. Brown, K.L. Bunter, K.P. Gore, S. de las Heras-Saldana, L. Li, A.J. McMillan, S.F. Walkom and A.A. Swan

Animal Genetics Breeding Unit*, University of New England, Armidale, NSW, 2351 Australia

SUMMARY

Allele frequencies play a crucial role in multi-breed genetic evaluations by enabling the construction of a breed-adjusted genomic relationship matrix (GRM), which can be derived from a genomic reference population. This study aimed to update the Australian sheep genomic reference population for a combined Terminal and Maternal LAMBPLAN genetic evaluation. The analysis incorporated genotypic data from sheep across maternal, terminal, shedding, and Merino breeds, as well as smaller breeds and composite populations. Population structure analysis identified 27 genetic clusters among terminal, maternal and shedding breeds and five clusters within Merino sheep. The findings demonstrate that genomic data and unsupervised clustering methods can improve breed composition accuracy and facilitate a comprehensive multi-breed genetic evaluation, reducing the need for strictly straight-bred reference populations.

INTRODUCTION

Genomic information has increased the accuracy of Australian Sheep Breeding Values (Brown *et al.* 2018). Genetic evaluation for the large and diverse Australian sheep population (Swan *et al.* 2015) requires individual sheep to be assigned breed proportions based on a genomic reference population used to identify breed content, for a breed adjusted GRM with a reduced breed structure to be constructed (Gurman *et al.* 2019). The Australian sheep genomic reference population included pedigree defined Border Leicester, Poll Dorset, Texel, Suffolk, White Suffolk and Merino animals (Gurman *et al.* 2017). Since then, the number of genotypes available for the LAMBPLAN genetic evaluation has vastly increased (Walkom *et al.* 2025). These genotypes come from both straight-bred individuals belonging to the above major sheep breeds, but also from smaller breeds and composite populations whose numbers in the genetic evaluations have been steadily increasing in the past few years (McMillan *et al.* 2023).

This work aimed to redefine the Australian sheep genomic reference population to be used for a breed adjusted GRM in preparation for a combined Terminal and Maternal LAMBPLAN genetic evaluation (Walkom *et al.* 2025) using genotype data from maternal, terminal, shedding and Merino sheep breeds, as well as from smaller breeds and composite populations. The latter were until recently, not included in the LAMPBPLAN single step genetic evaluation because reference populations for inclusion in breed adjusted relationship matrices had not been established. Two different datasets were used in a population genetic analysis to investigate population structure across different breeds and to calculate allele frequencies to define breed composition.

MATERIALS AND METHODS

Genotypes used in this analysis came from 260,200 sheep of different terminal and maternal breeds. A GRM was constructed within breed based on the method implemented by Yang *et al.* (2011). To avoid bias in the analysis that might be introduced by family structure, individual genotypes were filtered to represent different breeds but to be as distantly related as possible. Within breed, animals were retained using the following criteria: i) animals with no parents present in the

---

* A joint venture of NSW Department of Primary Industries and Regional Development and the University of New England

pedigree or no parents genotyped, ii) animals with an average pedigree Q matrix value > 0.9, and iii) average genomic relationship lower than 0.2. Because animals included in the data were genotyped with different single nucleotide polymorphism (SNP) panels, genotypes were filtered to include only the common SNPs (2,443) between these panels, resulting in a data set of 3,777 sheep from Terminal, Maternal and smaller breeds (Dataset 1, original genotypes). To account for Merino crosses in the LAMBPLAN analysis and to identify possible population structure within the Merino breed, a second data set of 20,501 unrelated Merino animals, genotyped in the same SNPs, was used (Dataset 2, original genotypes). To facilitate calculation of allele frequencies for all SNP data used in regular LAMBPLAN evaluations, all genotyped animals were imputed to 62,016 SNPs (imputed genotypes). The numbers of genotypes from each breed are presented in Table 1.

Table 1. Breeds, breed types and codes and number of unrelated animals (N) from each breed used in the analysis

Breed Code	Breed Name	Breed Type	N	Dataset
02	Border Leicester	Maternal	558	
03	Corriedale	Wool	26	
09	Wiltshire Horn	Shedding	32	
10	Polwarth	Wool	25	
11	Hampshire Down	Terminal	72	
12	Wiltipoll	Shedding	122	
14	Southdown	Terminal	77	
15	Coopworth	Maternal	72	
16	Poll Dorset	Terminal	896	
17	Texel	Terminal	59	
18	Romney	Wool	9	
19	Suffolk	Terminal	147	
22	Finnsheep	Maternal	20	
23	White Suffolk	Terminal	284	
24	Ultra White	Shedding	354	
25	Australian White	Shedding	181	1
26	Research	Composite	59	
31	English Leicester	Maternal	17	
38	East Friesian	Maternal	98	
40	Dorper	Shedding	132	
47	White Dorper	Shedding	147	
48	SAMM	Wool	65	
49	Damara	Shedding	29	
50	Merino	Wool	8,130	
51	Dohne Merino	Wool	19	
CM	Commercial Maternal	Maternal	25	
CS	Commercial Shedders	Shedding	43	
CT	Commercial Terminal	Terminal	145	
IF	Ile De France	Terminal	27	
VR	Van Rooy	Shedding	17	
60	Poll Merino	Wool	11,974	
74	NZ Merino	Wool	308	
76	NZ Lincoln	Terminal	20	2
85	NZ Poll Merino	Wool	89	
Total			24,278	

Population structure for Dataset 1 and 2 was inferred using the method implemented in STRUCTURE 2.3.4 software (Pritchard *et al.* 2000). STRUCTURE clusters individuals in groups minimising both linkage and Hardy-Weinberg disequilibrium (Falush *et al.* 2003) utilising a Bayesian approach. Individuals were assigned to a pre-determined number of clusters (K). For each cluster, allele frequencies were estimated, and individuals were re-assigned based on those frequency estimates until convergence was achieved. The existence of possible clusters within the Australian sheep population, was estimated using K values up to 40. Any previous breed information was discarded (unsupervised analysis), and the admixture model was used with correlated allele frequencies and eight independent runs of 10,000 iterations, after a burn-in of 4,000 iterations for each K. The optimal K value was informed based on the approach of Evanno *et al.* 2005. Final allele frequencies for the optimal K were estimated by aligning the replicates of each run and examining pairwise similarities between allele frequency matrices with the method described by Jakobsson and Rosenberg (2007). To avoid bias due to overrepresentation, for each of the clusters identified for Dataset 2, individuals were selected if their membership coefficient was greater than 90%. Since STRUCTURE analysis was performed for both the original and the imputed genotypes (results not shown), the consistency of estimated clusters was evaluated by calculating the correlation of the cluster membership Q matrices for each breed between the original and imputed data.

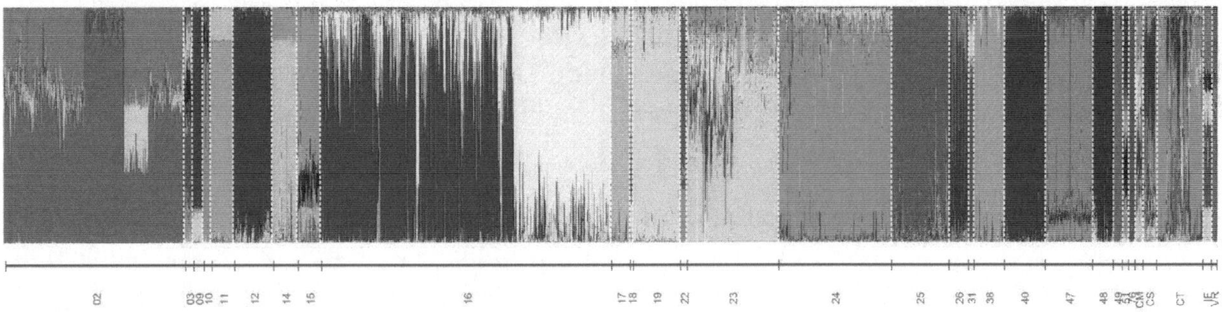

Figure 1. STRUCTURE analysis results for Dataset 1 with optimal number of clusters K=27. Each vertical line represents one individual and colours correspond to each of the identified clusters. Breed codes correspond to Table 1

RESULTS AND DISCUSSION

Analysis of the combined Terminal and Maternal breed genotypes (Dataset 1, original genotypes) defined 27 clusters (K) (Figure 1). The results showed several relatively homogeneous clusters, in some cases corresponding to different breeds like East Friesian (breed Code: 38), Hampshire Down (11), Southdown (14), Texel (17), Suffolk (19), Wiltipoll and Wiltshire Horn (12 and 09 respectively), Dorper (40) and Australian White (25). Several breeds were shown to include more than one cluster like Ultra White (24) which appear to share ancestry with White Dorper (47). This was also observed within the Commercial Maternal, Commercial Terminal and Commercial Shedder groups which contain different clusters corresponding to their composite origin. Analysis of Dataset 2 revealed the existence of five clusters within the Merino animals that seem to be more related to breeding objectives than geographical origin (Figure 2). This method identified 756 Merino animals from all five clusters to be included in the genomic reference population. Results were compared between original and imputed data for the same number of clusters, and mean cluster membership correlations across breeds was 0.93 (± 0.06) ranging from 0.77 to 1.00 for animals of different breeds. These high correlations demonstrate that similar results can be obtained when imputed data is used, and allele frequencies for each cluster calculated with this data correspond to those calculated with the original genotypes. Allele frequencies for the 62,016 SNPs in the imputed genotype data set were estimated for each of the clusters identified in the STRUCTURE analysis and used to calculate breed proportions for all genotyped animals used in the LAMBPLAN genetic evaluation.

The use of genomic data to identify genetic structure in livestock breeds had been previously used in beef and sheep (Sölkner *et al.* 2010) and have been used to assess breed diversity, determine breed composition and improve genetic evaluations (Gurman *et al.* 2019). A thoroughly selected set of genotyped animals combined with an unsupervised admixture approach can eliminate the need for straight-bred animals. In this light, smaller breeds and composite populations can be better accounted for and facilitate a combine Terminal and Maternal multibreed genetic evaluation.

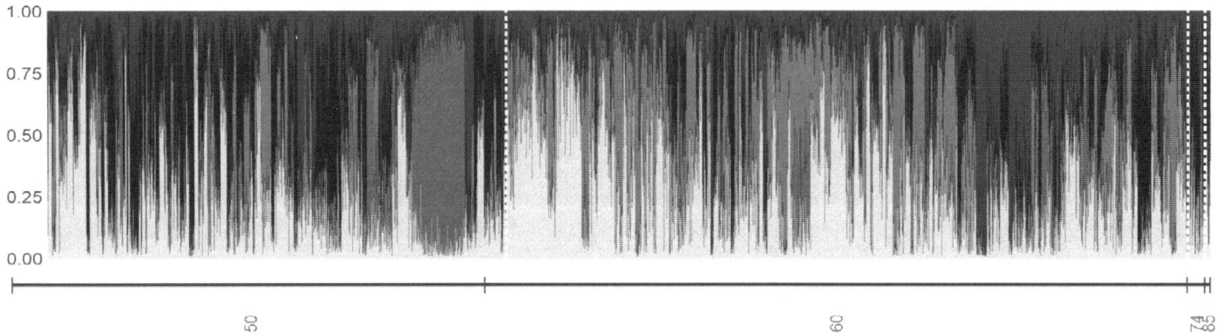

Figure 2. STRUCTURE analysis results for the Merino dataset with optimal number of clusters K=5. Each vertical line represents one individual and colours correspond to each of the identified clusters. Breed codes correspond to Table 1

CONCLUSION

The genomic reference population for Australian sheep has been successfully updated by incorporating genotypic data from maternal, terminal, shedding, and Merino breeds, and including smaller breeds and composite populations. By leveraging genomic data and unsupervised clustering methods, the accuracy of genomic breed composition assessment can be enhanced resulting in a more inclusive multi-breed genetic evaluation. The findings support a shift away from relying solely on straight-bred reference populations, allowing for better representation of composite and smaller breeds in genetic evaluations.

ACKNOWLEDGEMENTS

This research was funded by Meat and Livestock Australia project L.GEN.2204. The authors gratefully acknowledge the teams behind the Sheep CRC for Sheep Industry Innovation Information Nucleus and MLA Resource flock as well as individual industry flock owners involved in flock management and data collection.

REFERENCES

Brown D.J., Swan A.A., Boerner V., *et al.* (2018) *Proc. World Cong. Genet. Appl. Livest. Prod.* **11**: 16.
Evanno G., Regnaut S. and Goudet J. (2005) *Mol. Ecol* **14**: 2611.
Falush D., Stephens M. and Pritchard J.K. (2003) *Genetics* **164**: 1567.
Gurman P.M, Swan A.A. and Boerner V. (2017) *Proc. Assoc. Advmt. Anim. Breed. Genet.* **22**: 341.
Gurman P.M., Bunter K.L., Boerner V., *et al.* (2019) *Proc. Assoc. Advmt. Anim. Breed. Genet.* **23**: 254.
McMillan A.J., Walkom S.F. and Brown D.J. (2023) *Proc. Assoc. Advmt. Anim. Breed. Genet.* **25**: 31.
Jakobsson M. and Rosenberg N.A. (2007) *Bioinformatics* **23**: 1801.
Pritchard J.K., Stephens M. and Donnelly P. (2000) *Genetics*, **155**: 945.
Sölkner J, Frkonja A, Raadsma H.W, *et al.* (2010) Interbull Bulletin. **42**: 62.
Walkom S.F., Alexandri P., *et al.* (2025) *Proc. Assoc. Advmt. Anim. Breed. Genet.* **25**: *These proceedings.*
Yang J., Lee S. H., Goddard M. E. and Visscher P.M. (2011). *Am. J. Hum. Genet.* **88**: 76.

Proc. Assoc. Advmt. Anim. Breed. Genet: 26: 459-462

FIXED EFFECTS AND GENETIC PARAMETERS FOR THE COMBINED LAMPLAN EVALUATION OF AUSTRALIAN SHEEP

S. de las Heras-Saldana, D.J. Brown, P.M. Gurman, P. Alexandri, K.L. Bunter, K.P. Gore, L. Li, A.J. McMillan, S.F. Walkom and A.A. Swan

[1] Animal Genetics Breeding Unit[*], University of New England, Armidale, NSW, 2351 Australia

SUMMARY

The Australian sheep industry consistently improves performance through genetic evaluation provided by Sheep Genetics. Sheep Genetics is developing an enhanced evaluation system combining Terminal and Maternal breeds into a single LAMPLAN analysis. This improvement includes a review of fixed effects to ensure their suitability and an evaluation of alternative methods for their inclusion. This study outlines the steps taken to model fixed effects directly within the model, rather than through pre-adjustment, and to estimate genetic parameters. In the updated multi-trait analysis, two main changes were incorporated in the fixed effect model, 1) birth type and rearing type were combined into a single effect to account for their interaction, and 2) the age of the dam was fitted as class effect (rather than a quadratic polynomial). Including the effects directly in the model resulted in an improvement in the fit compared to using the current pre-adjustment methods. These enhancements should lead to more accurate breeding values for industry, and this will be the subject of further studies.

INTRODUCTION

In Australia, Sheep Genetics uses the OVIS (Brown *et al.* 2006) software to undertake the national genetic evaluation of sheep. Estimated breeding values (EBVs) are currently calculated by fitting a model with a single fixed effect of contemporary group using pre-adjusted phenotypes that account for systematic effects such as age (separated by sex), age of dam (linear and quadratic), birth and rearing type and liveweight (Brown *et al.* 2006). Although this approach is both accurate and computationally less demanding than fitting effects in the model, it requires estimating adjustment factors for each systematic effect on an ongoing basis (Brown and Reverter 2002). In most cases, adjustment factors remain unchanged over long periods of time and are often based on subsets of data such as well-recorded reference populations. However, this can be problematic because of high levels of genetic variability included in the design of reference flocks. Routinely revisiting fixed effects is necessary due to partial confounding (e.g. birth and rearing type) and the need to better account for any interactions. Importantly, given the decision to combine the terminal and maternal LAMBPLAN evaluations (Walkom *et al.* 2025), a meticulous assessment of the assumptions used for the national evaluation was required. This study describes the main steps followed to assess the modelling of fixed effects and estimation of genetic parameters for main traits in the Combined LAMBPLAN analysis.

FIXED EFFECTS MODELLING

All records from the Combined LAMBPLAN (Walkom *et al.* 2025) were used in the evaluation of fixed effects. To maximise the use of full dataset and to overcome the large number of records

[*] A joint venture of NSW Department of Primary Industries and Regional Development and the University of New England

available for weight traits such as birth weight (>2M), weaning weight (>4M) and post-weaning weight (>2M) required subsetting the data for analysis. A 'split-apply-combined' approach was applied in which the data for each trait was grouped by flock, year of birth, sex and breed type, with results combined for further analyses. This strategy maximised the use of all records to determine which effects were important across traits and identify where it would be necessary to fit interactions between the standard effects and factors such as breed and/or site.

The approach to evaluate the fixed effects of interest for each trait included fitting a linear model with fixed effects for the contemporary group (CG), birth (BT) and rear type (RT), or combined birth and rearing types (BRT), age of dam (AOD) and weight (WT; only for scan traits) [Model 1]. The adjustment factors were calculated as the ratio of the base level divided by the predicted mean. Further, multiple linear models were fitted to both the estimated adjustment factors (BT, RT, AOD) or the solutions (WT or age) to examine variation in effects of flock, sex, yob, breed [Model 2], site and their interactions. From each model, the variance explained (R^2) by the effect fitted in the model were compared. In general, for all the traits evaluated, the flock explained larger variance (R^2 from 0.06 to 0.71) compared to yob (R^2 from 0 to 0.05), sex (R^2 from 0 to 0.07) and breed (R^2 from 0.001 to 0.07). As expected more complex models that fitted all the effects or interactions resulted in larger variance explained (R^2 from 0.2 to 0.78).

Developing an approach that deals with the necessary interactions in a more parsimonious way while delivering accurate EBVs is important. Therefore, EBVs generated by the current model used in OVIS analysis were compared to those from models that fitted the fixed effects or their interactions. The correlations of EBVs generated from these models ranged from 0.9 to 0.999. These results suggest that the use of more complex models yield breeding values highly correlated with the current model used in OVIS. Therefore, the use of complex models would not be required.

A linear mixed animal model was fitted in ASReml v4.2 (Gilmour *et al.* 2015) to examine the importance of heterosis. In addition to the fixed effects mentioned above, heterozygosity coefficients were included in the models. Heterozygosity was calculated based on the pedigree and estimated both direct (DHB) and maternal (MHB) heterosis applying the classic "dominance" model described by Dickerson *et al.* (1973). A conditional Wald F statistic was applied to estimate the significance of fixed effects (p-value < 0.05).

In the final model, birth type (BT) was fitted only for birth weight (BWT) as class effect for single (1), twin (2), and triplet (3) levels. For traits where birth and rearing type were significant, a combined birth-rearing type (BRT) effect was fitted (six levels: 1-1, 2-2, 2-1, 3-3, 3-2, 3-1). The age at measurement was included as a linear covariate for females (ageF) and males (ageM) separately. Age of dams (AOD) was fitted as class effect with levels for two-years-old to 10 years-old-dams, dividing one-year-old dams into yearling and hogget at 420 days of age and an additional class of unknown age of dams. Heterozygosity, as DHB and MHB, were included only if significant in previous models. Weight (WT) was fitted as a covariate effect for the ultrasound scan, with both linear and quadratic terms. New contemporary groups were fitted, as defined by Walkom *et al.* (2025). The random effects consisted of genetic groups defined in McMillan *et al.* (2025), animal genetic, maternal genetic and permanent environmental.

A multi-trait analysis was performed using OVIS to compare results from the current model compared with fitting the effects directly in the model. A more complex model was tested in which the age solutions from Model 1 were split into categories based on six quantiles. Figure 1 shows the root mean square errors (RMSE) calculated on the solutions from three models: 1) using the preadjusted data in OVIS (preadjust model); 2) using age by sex interaction (age_MF model); and 3) fitting the age quantiles (age_Q6 model). In general, RMSE values were larger for solutions from the preadjusted model. Moreover, similar errors (RMSE) obtained in the other two models indicated that more complex models are not required to improve the efficiency of the analysis. The multi-trait analysis indicated that a change in approach from pre-adjusting data for systematic effects to directly fit the effects in the solver would be advantageous, given a better fit to the data with no impact on computing performance. Additionally, the effects are always current and relevant to the entire dataset. In contrast, pre-adjustments typically remain unchanged over long periods of time and are often based on subsets of data such as well-recorded reference populations where estimating effects can sometimes be problematic because of high levels of genetic variability included in the design. Moreover, avoiding the re-estimation of adjustment factors represents an efficiency gain for research projects in long term.

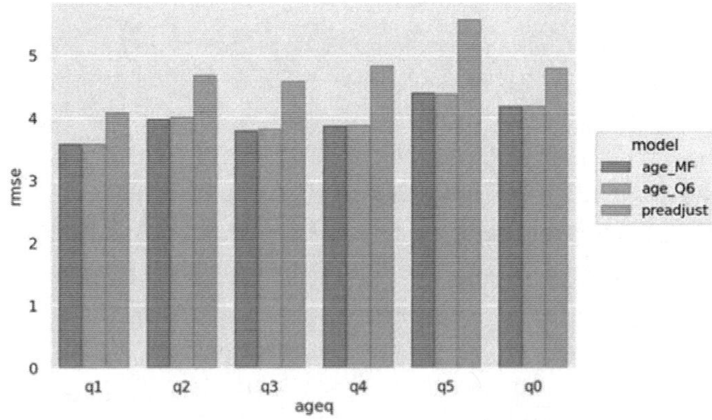

Figure 1. Root mean square error (rmse) for post-weaning weight fitting preadjusted data (green), age*sex interaction (blue) interaction or six age effects based on solution distributions (orange) in the models. The age solutions from the split-combine-apply were categorised into six quantiles (age_Q6)

GENETIC PARAMETERS

Initially, the genetic parameters for six terminal (Poll Dorset, Texel, Suffolk, White Suffolk, Dorper and White Dorper) and four maternal (Border Leicester, Corriedale, Coopworth, Composite Maternal) breeds were estimated and compared with the estimates obtained using all the breeds in the Combined LAMBPLAN. The records used in each dataset were from the most representative flocks selected, as indicated in de las Heras-Saldana *et al.* (2023). The final models were fitted in univariate model using the Combined LAMBPLAN dataset for birth weight (bwt), weaning weight (wwt), post-weaning weight (pwt), yearling weight (ywt), and adult weight (awt). In general, differences in genetic variances between the terminal and maternal breeds were not bigger than those already existing within breed type (Figure 2). Larger dispersion in ywt and awt reflects the lower number of records assigned to those traits for each breed, compared to a larger number of records in bwt, wwt and pwt (Walkom *et al.* 2025). Outliers corresponded to breeds with lower records, such as Texel, Suffolk, Dorper and White Dorper.

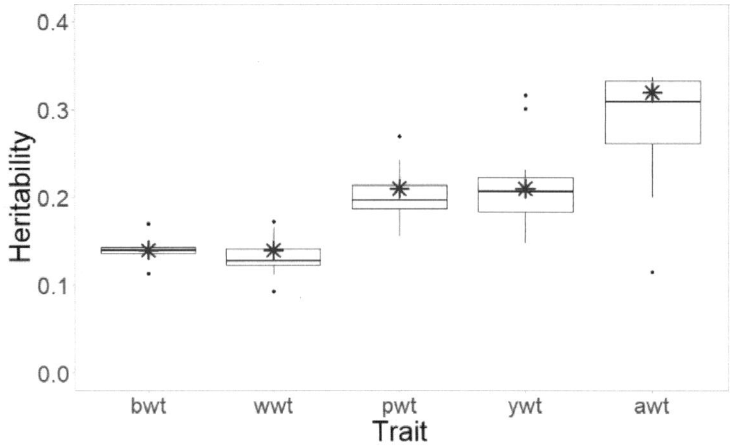

Figure 2. Distribution of heritabilities (h^2) estimates for the breeds in Terminals and Maternals (boxplots) and the current h^2 calculated in the combined LAMBPLAN dataset (indicated as '*')

CONCLUSIONS

Fitting fixed effects directly in the solver represents an efficiency gain for the national genetic evaluation of sheep since this approach is data-driven, always current and, avoids the maintenance of adjustment factors. It was also concluded that even with the combining of the two analyses, fitting global effects for BRT and age of dam remained the most worthwhile solution. These enhancements should lead to more accurate breeding values for the industry. In future analyses will consider a strategy to adjust weight for carcass traits and fitting a breed interaction.

ACKNOWLEDGMENTS

This research was funded by Meat and Livestock Australia project L.GEN.2204. The authors gratefully acknowledge the teams behind the Sheep CRC for Sheep Industry Innovation Information Nucleus and MLA Resource flock as well as individual industry flock owners involved in flock management and data collection.

REFERENCES

McMillan A.J., Brown D.J., Bunter K.L., de las Heras-Saldana S., Gore K.P., Gurman P.M., Li L, Alexandri P., Walkom S.F. and Swan A.A. (2025) *Proc. Assoc. Advmt. Anim. Breed. Genet.* **26***: These proceedings.*

Brown D.J. and Reverter A. (2002) *Livest. Prod. Sci.* **75**: 3.

Brown D.J., Ball A., Huisman A.E., Swan A.A., Atkins K.D., Graser H. U. and Woolaston R. (2006) *Proc.World Cong. Genet. Appl. Livest. Prod.* **9**: 5.

de las Heras-Saldana S., Gurman P., Swan A.A. and Brown D (2023) *Proc. Assoc. Advmt. Anim. Breed. Genet.* **25***:* 318.

Dickerson G.E. (1973) In *Proc. Anim. Breed. and Genet*. Symposium in Honor of Dr Jay L Lush.

Fogarty N.M. (1995) *Anim. Breed.* **63**: 3.

Gilmour A.R., Gogel B.J., Cullis B.R., Welham S.J. and Thompson R. (2015) ASReml User Guide Release 4.1, VSN International Ltd, Hemel Hempstead, UK.

Safari E., Fogarty N.M. and Gilmour A.R. (2005) *Livest. Prod. Sci.* **92**: 3.

Walkom S.F., Alexandri P., Brown D.J., Bunter K.L., de las Heras-Saldana S., Gore K.P., Gurman P.M., Li L., McMillan A.J. and Swan A.A. (2025) *Proc. Assoc. Advmt. Anim. Breed. Genet.* **26**: *These proceedings*.

GENOTYPING REDUCES PRESELECTION BIAS ON MERINOSELECT ASBVS IN MERINO FLOCKS

B.E. Clarke[1], D.J. Brown[2], P. Bradley[3] and A.A. Swan[2]

[1] Centre for Animal Production and Health, Murdoch University, Murdoch, WA, 6150 Australia
[2] Animal Genetics Breeding Unit*, University of New England, Armidale, NSW, 2351 Australia
[3] Meat and Livestock Australia, University of New England, Armidale, NSW, 2351 Australia

SUMMARY

Preselection of breeding animals to reduce the costs of phenotyping can lead to selection bias resulting in the loss of accuracy of Australian Sheep Breeding Values (ASBVs). This paper used a subset of data from the MERINOSELECT database to quantify the effects of selection bias in flocks with varying methods of preselection including random and selected phenotyping both with and without genotype information. Results show that genotyping all animals in combination with phenotyping the top 50% of animals based on genotype provided the most accurate individual and sire ASBVs. However, whilst this was the most accurate scenario, it still involved loss of ASBV accuracy and selection bias that varied according to trait when compared with the current model that includes genotyping and measuring all animals prior to selection.

INTRODUCTION

Accuracy of Australian Sheep Breeding Value (ASBV) prediction has increased with the implementation of single step genomic BLUP, with the largest increases in accuracies coming from genotyped animals (Brown *et al.* 2018). The MERINOSELECT analysis has seen an exponential increase in the number of genotypes since the introduction of single step genomic BLUP, with more than half of the 2023 born animals entering the analysis with a genotype. This increased uptake of genomics has led to higher costs for Merino ram breeders. To maintain a large selection pool, breeders produce and evaluate significantly more animals than are necessary for sale or breeding purposes. One approach to reducing both phenotyping costs and the costs associated with raising animals until phenotyping, is to preselect animals using either limited phenotype data or genomic information (Jibrali *et al.* 2020). While this preselection reduces the need to measure all animals from a cohort, it may also limit the information available for evaluating the preselected animals later, which can result in selection bias (Jibrila *et al.* 2021). The objective of this study was to establish the correlations between ASBVs calculated with different levels of phenotype and genotyping recording as well as to quantify the loss of selection advantage with differing levels of preselection within Merino flocks.

MATERIALS AND METHODS

Data. To quantify the effects of selection bias, a reduced dataset was extracted from the MERINOSELECT database. Extracted data came from 13 Merino flocks with a high proportion of genotyping. This resulted in a dataset of 273,896 animals of which 77,780 animals had genotypes.

Analyses. ASBVs were calculated for all animals in the extract using single-step genomic BLUP (ssGBLUP) methodology from the routine MERINOSELECT analysis (Brown *et al.* 2018). The dataset was then edited by taking out the 2023 drop of animals one flock at a time and removing

* A joint venture of NSW Department of Primary Industries and Regional Development and the University of New England

different combinations of phenotype and genotype data and rerunning the ssGBLUP. Analyses were carried out on a subset of 15 traits from which four key traits, with the most data and covering all trait categories, are reported in the results (yearling weight, yearling eye muscle depth, yearling greasy fleece weight and yearling fibre diameter).

The following scenarios examined included different levels of phenotypic recording both with (geno = 1) and without (geno=0) genotyping on the 2023 drop.

1. None – no phenotype recording of animals
2. Random – phenotype recording of a random 50% of animals
3. Selected – phenotype recording the heaviest 50% of animals at weaning
4. GenoSelect – phenotype recording the top 50% of animals based on their ssGBLUP midparent value for the Merino Production Plus index (MP+).

The full MERINOSELECT analysis of the 13 Merino flocks was assumed to provide the true breeding value (TBV) for these comparisons. Correlations were calculated between the TBV and estimated ASBVs for both individual animals born in 2023 and the sires of these animals. Comparisons were also made between the bias in mean ASBV compared to the TBV as well as the proportion of selection advantage (difference between average of flock and top 10% of flock) achieved relative to the full analysis across all flocks.

RESULTS AND DISCUSSION

Correlations. The correlations between ASBVs calculated with different levels of recording and an animals TBV varied from 0.70 to 0.98 depending on the level of recording and whether genotype information was included (Figure 1a) when averaged across all flocks and all traits. Correlations between individual ASBVs and TBVs varied by trait (Figure 2), but each scenario had a similar range. The impact of different levels of recording on the ASBVs of sires of the 2023 drop was less, with the range reducing to 0.93-0.99 (Figure 1b). This again varied by trait (Figure 2) with no phenotype recording having the largest impact. The impact of genomic information was also reduced on the sires ASBVs across all traits and within individual traits. There was an advantage by the addition of genotype data to all the phenotype only scenarios in correlations between individual TBVs and ASBVs for all 15 traits and for the four yearling traits (Figures 1a and 2). This advantage was not evident in the sire TBV and ASBV correlations (Figure 1b).

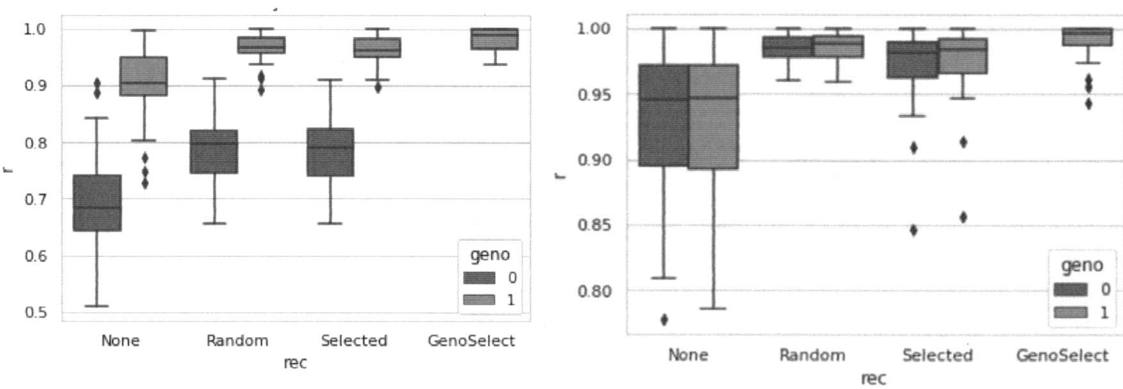

(a) **(b)**

Figure 1. Correlations (r), with standard errors, between ASBVs and True Breeding Values calculated with four different data recording scenarios (rec) that either include (geno = 1) or don't include (geno = 0) genotype information across 13 Merino flocks and 15 traits for both individuals (a) and sires (b)

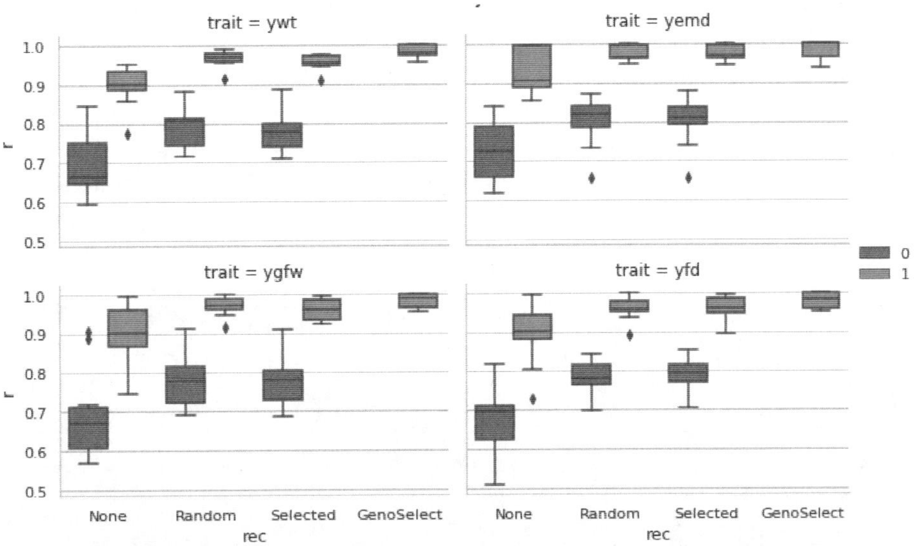

Figure 2. Correlations (r), with standard errors, between individual ASBVs and True Breeding Values calculated with four different data recording scenarios (rec) that either include (geno = 1) or don't include (geno = 0) genotype information across 13 Merino flocks for four yearling traits: weight (ywt), eye muscle depth (yemd), greasy fleece weight (ygfw) and fbre diameter (yfd)

Bias in mean ASBV. The bias in ASBVs estimated for individuals (Figure 3) varied by trait, with genotyping have a large impact on reducing both the range and average bias in individual ASBVs. This trend was the same for bias in mean sire ASBVs.

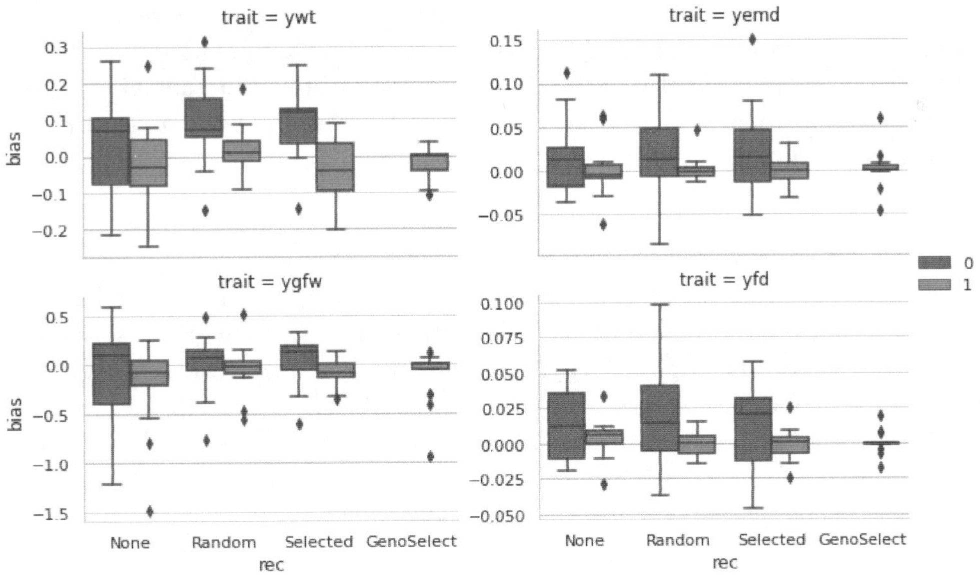

Figure 3. Bias in mean individual ASBVs from the full analysis across all flocks, with standard errors, between ASBVs and True Breeding Values calculated with four different data recording scenarios (rec) that either include (geno = 1) or don't include (geno = 0) genotype information across 13 Merino flocks for four yearling traits: weight (ywt), eye muscle depth (yemd), greasy fleece weight (ygfw) and fbre diameter (yfd)

Selection advantage. In all scenarios tested, genotyping allowed for above 90% of the selection advantage of the full analysis to be realised (Figure 4). The impact of genotyping was greater for the no phenotype scenario than for reduced phenotyping and no genotyping. Preselection based on genotype information (GenoSelect) achieved the highest proportion of selection advantage which has been shown to be due to the genotypes of preculled animals helping to more accurately reduce the bias caused by selected animals being a better-than-average subset (Jibrali *et al.* 2021).

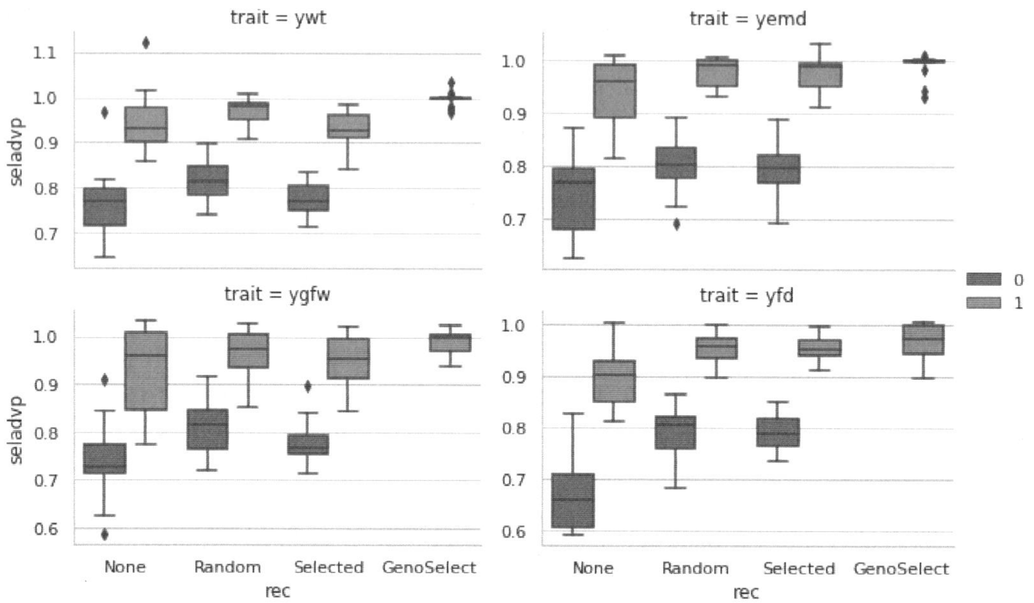

Figure 4. Proportion of selection advantage (seladvp) achieved relative to the full analysis across all flocks, with standard errors, for four different data recording scenarios (rec) that either include (geno = 1) or don't include (geno = 0) genotype information across 13 Merino flocks for four yearling traits: weight (ywt), eye muscle depth (yemd), greasy fleece weight (ygfw) and fibre diameter (yfd)

CONCLUSION

Genotyping and preselection based on genotype only ASBVs provides higher correlations with TBVs and less selection bias than preselection based on other variations of phenotyping examined in this paper. This study only looked at the effect of preselection on the ASBVs of the current generation. Further research is required to quantify the impact of the preselection on genetic gain in subsequent generations and the cost savings of this approach. This includes the long-term impact of selective recording on the reference population for individual flocks and the entire reference population. In addition, the impact of preselection on reproduction traits also needs to be examined.

REFERENCES

Brown D.J., Swan A.A., Boerner V., Li L., Gurman P.M., McMillan A.J., van der Werf J.H.J, Chandler H.R., Tier B. and Banks R.G. (2018) *Proc. World Cong. Genet. Appl. Livest. Prod.* **11**: 1.

Jibrali I.Ten Napel J., Vandenplas J., Veerkamp R.F. and Calus M.P.L. (2020) *Genet. Sel. Evol.* **52**: 42.

Jibrila I., Vandenplas J., Ten Napel J., Veerkamp R.F. and Calus M.P.L. (2021) *J. Anim. Breed. Genet.* **138**: 432.

IMPLEMENTATION OF GENOMIC MATE SELECTION AND OTHER NON-ADDITIVE ISSUES

B.P. Kinghorn[1,2] and A.J. Kinghorn[2]

[1] University of New England, Armidale, NSW, 2350 Australia
[2] BSK Software

SUMMARY

Increasing attention is being paid to our ability to influence mendelian segregation variance as a function of matched parental genetic markers. This results in 'Look Ahead Mate Selection' to target longer-term genetic gains. This paper addresses this and other aspects of genomic mate selection, in a new implementation method that can also accommodate other non-additive and compatibility issues, including animal behaviour compatibility, non-linear indices, transport costs and, in general, specification of must-have and must-not-have matings.

INTRODUCTION

The standard way of handling trait EBVs in a breeding program assumes that progeny genetic merit is predicted as the average of the parental EBVs. This is true for most traits. However, this is not appropriate for some traits and other factors – wherever we have knowledge that the progeny from a specific mating are predicted to deviate from their parental mean in some known way, eg. due to genetic dominance, or increased segregation variance.

This knowledge cannot be incorporated into any datafile that has one line of data per candidate. For this reason, we need to read these non-additive components from a separate file for each trait, which contains the predicted non-additive deviation for each possible pairing of a male candidate and a female candidate.

As some form of mate selection based on genomic information seems a most likely application, we refer to this approach as genomic mate selection (GMS), even though it can be applied without genomic information, and indeed without any genetic components. There is an increasing interest in GMS, with frequent new academic publications on how to use genomic information for increasing both short-term and longer-term genetic gains:

Short-term gains: Total genetic merit in progeny. Refined predictions of progeny merit can involve dominance components, and possibly some epistatic components, at nominated QTL. It could also involve genome-wide heterozygosity, although that may be less powerful. The Expected Cross Value of Ahadi *et al.* (2024) goes beyond this to consider the value of a mating pair as a function of allelic information and transmission probabilities for each partner, increasing the desirable allele frequency.

Long-term gains: Look Ahead Mate Selection (LAMS). "Genomic Mating uses information in a similar fashion to genomic selection but includes information on complementation of parents to be mated" (Akdemir and Sanchez 2016). This is in fact GMS, as there is selection of the parents as well as mate allocation among them. How these mate selections are chosen depends on several issues, including the number of generations ahead that we use to define as in our utility function. Niehoff *et al.* (2024a) illustrate scenarios in which different selection decisions should be made today according to the different objectives of aiming 1, 2 or 3 generations ahead. This is essentially driven by Mendelian Sampling Variance (MSV), which can be affected by:

- Selection decisions, where some individuals are expected to yield greater genetic variation among their progeny, even under random mating. An **Aa** heterozygote is expected to yield more genetic variation in progeny than offspring of **AA** or **aa** homozygotes.

- Mate selection decisions, where, in addition to selection effects, certain pairs synergise to give more variation among progeny. A cross **Aa** x **Aa** is expected to yield **AA**, **Aa** (or **aA**), and **aa** progeny in 1:2:1 ratio, which maximises genetic variation in the progeny.

Of course, mate selection has the most impact, and indeed when we use genomic information to increase genetic variation in the progeny, we are largely aiming at mating of heterozygotes genome-wide. This extra genetic variation leads to increased genetic gains in later generations, under Look Ahead Mate Selection.

Niehoff *et al.* (2024a) describe a number of objective functions that invoke LAMS, leading ultimately to increased genetic merit in great-grandprogeny. Tobias Niehoff has contributed notes in Kinghorn and Kinghorn (2025) that show how to invoke these objective functions in our software. Additionally, Niehoff *et al.* (2024b) generalise their approach to aim **n** generations ahead.

MATERIALS AND METHODS

As noted above, handling any type of non-additive effect usually requires more than the typical animal dataset that has one line of data per candidate. Where the dataset is to be used to make selection and mating decisions, these data lines typically include information on parent IDs, trait EBVs, genetic markers and usable information such as age and location. Non-additive effects that depend on mate allocation decisions require either storage of such information in files of dimension male candidates x female candidates, to cover all possible mate allocations, or calculation of non-additive effects on-the-fly as a function of information supplied in the main data file.

For the latter, calculation of predicted progeny merit is a simple function of the parents mated, either for each mating pair (eg. a non-linear selection index) or across all mating pairs (e.g. corrective mating about a target mean). These are simply covered by trait management tools (Kinghorn and Kinghorn 2025). In this paper we concentrate on non-additive effects that are not simply derived from additive effects, and require separate calculation and storage.

Classically, genetic merit in progeny is estimated as:

$$ProgenyMerit = \frac{A_{sire} + A_{dam}}{2} + D_{sire,dam}$$

where A and D are additive and dominance effects, respectively. Terms for epistatic effects can be added to this, but these are left out for simplicity. In general, we use this:

$$ProgenyMerit = \frac{EBV_{sire} + EBV_{dam}}{2} + Deviation_{sire,dam}$$

where the EBVs can be replaced by factors such as animal age, location etc, or more typically by 0's, where there is no additive effect involved. The *Deviation* can be any effect that can be calculated for each *sire* + *dam* combination. This could be a predicted progeny mean dominance deviation, a predicted MSV for the resulting full-sib family, or even the transport cost associated with implementing a mating between this sire and dam.

The three examples in the last sentence can all be included as non-additive 'traits' in a single analysis, with competition between them for emphasis in decision making, just as for normal additive traits.

Implementation in our software is simply achieved by making the name for such a trait start with "GMS" (Eg. GMS_BirthWt) and providing a file containing the non-additive effects for the male and female candidates for selection (in this case, named exactly "candGMS_BirthWt.txt").

For simple illustration, we here use progeny inbreeding from pedigree as the non-additive trait "GMS_PedInb". In the main datafile, the EBVs for this 'trait' are set to zero – as inbreeding is not heritable and is a function of parental pairs, not individual parents. The file candGMS_PedInb.txt looks like this, with males in rows and females in columns:

candGMS_PedInb	Lilianna	Guadalupe	Rayna	Chanel	Lyra	
Samir	0.155	0.050	0.063	0.059	0.054	...
Magnus	0.027	0.079	0.027	0.026	0.021	...
Jaziel	0.035	0.034	0.038	0.026	0.108	...
etc.						

RESULTS AND DISCUSSION

Figure 1 shows the results for the simple illustrative example. The 'trait' GMS_PedInb was chosen as the result can be compared to Progeny Inbreeding as reported separately in the application. GMS_PedInb has been managed to give a bimodal distribution in progeny, with 5 matings in a high mode around F=0.14 and 45 matings in a lower mode around F=0.06. As can been seen, the progeny distribution is the same as for progeny inbreeding, with a visual difference only because of the numbers of histogram bars differing between the graphs.

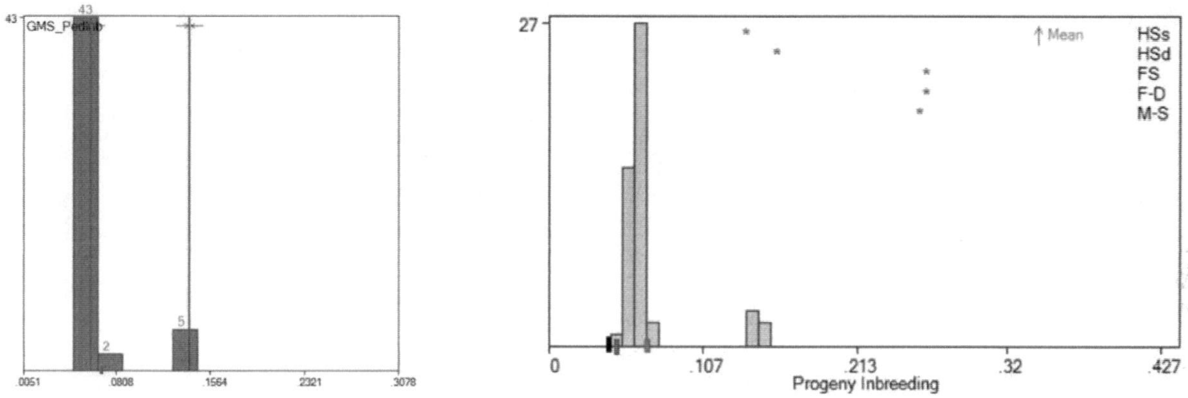

Figure 1. Results from targeting bimodality with 10% of families showing high values for the trait GMS_PedInb (left graph). As expected, the result is identical to Progeny Inbreeding (right graph), with difference in presentation only, due to the numbers of histogram bars used

The GMS approach can be used to accommodate many issues, some of which are demonstrated in this video https://youtu.be/XmFcKYBcJvc. Of course, the candGMS files of non-additive effects have to be generated in advance of the analysis, using the data resources available and probably appropriate scripting/coding to calculate each entry. Example applications include:

- LAMS, as described above.
- Phenotypic effects on mating/pregnancy/birth performance (eg. A small male might not reach a big female, or a big bull on a small cow could lead to dystocia).
- Animal behaviour compatibility scores for each possible male x female mating.
- The predicted transport cost associated with making each possible male x female mating.
- Imprinting effects, whereby some aspect of genetic merit depends on the gender of each parent involved – predicted merit is not simply an average of parental additive effects.
- Must-Have matings: In the file of effects for this 'trait' set all values to zero, except for the must-have matings, set these to 1. Then manage this trait to give the full number of must-have matings represented – or a lesser number as desired.
- Must-Not-Have matings: As above, except set value to 1 for the must-not-have matings, then manipulate to have that trait at value zero for all matings, or a greater number if desired.

Dynamic non-additive effects: Moeinizade *et al.* (2019) aimed a number of generations ahead using an approximate objective function of genetic merit in the last generation, involving simulation. However, for this method, "The contribution of a breeding parent is evaluated based on not only the favorable alleles that it carries but also the favorable alleles that it carries but are missing in other selected breeding parents." This means that the elements of the GMS matrix are dynamic, and cannot be read-in up-front. For such cases, further development would be required to give 'on-the-fly' calculation of the required elements, or indeed merit across all matings in light of the full mating set. Alternatively, direct calling of external routines that calculate such values can be developed.

CONCLUSION

The GMS approach described in this paper is simple to implement for most non-additive effects, and gives access to a wide range of technical, logistical and cost functions that involve a specific non-additive effect for each possible *sire x dam* pairing. Some of these functions extend the use of genomic information beyond genomic EBVs, giving some prospects to increase the returns from investment in genotyping animals.

REFERENCES

Ahadi P., Balasundaram B., Borrero J.S. and Chen C. (2024) *Heredity* **133**: 113.

Akdemir D. and Sánchez J.I. (2016) *Front. Genet.* **7**: 210.

Kinghorn B.P. and Kinghorn A.J. (2025) Instructions for Matesel. https://matesel.com/content/documentation/MateSelInstructions.pdf

Moeinizade S., Hu G., Wang L. and Schnable P.S. (2019) *G3-Genes Genom. Genet.* **9**: 2123.

Niehoff T.A.M., ten Napel J., Bijma P., Pook T., Wientjes Y., Hegedűs, B. and Calus M.P.L. (2024a) *Genet. Sel. Evol.* **56**: 41

Niehoff T.A.M., ten Napel J. and Calus M.P.L. (2024b) *G3-Genes Genom. Genet.* **14**: jkae205.

Proc. Assoc. Advmt. Anim. Breed. Genet. 26: 471-474

DATA QUALITY BY GEOGRAPHICAL LOCATION FOR AUSTRLIAN AND NEW ZEALAND MERINOSELECT FLOCKS

C.J. Bunter, P.E. Bradley, M.J. Hodge and K.L. Rummery

Meat and Livestock Australia, University of New England, Armidale, NSW, 2351 Australia

SUMMARY

The rate of genetic gain can be maximised by making selection decisions using accurate breeding values, underpinned by good quality data. This paper identifies and examines variation in data quality metrics in MERINOSELECT flocks across different geographic regions in Australia and New Zealand. Data quality metrics with significant variation across regions included pedigree completeness, variation in age and birth type of animals within a cohort, recording of fleece traits, and across flock linkage for reproduction traits. Improving such metrics via targeted adoption and extension messaging may lead to improved data quality, which may ultimately improve the rate of genetic gain resulting in more productive and profitable sheep.

INTRODUCTION

Maximising genetic gain in Merino flocks requires selection decisions to be informed by accurate breeding values, underpinned by high-quality data (Guy and Brown 2021). The Data Quality Score (DQS) report, was developed to help Sheep Genetics breeders identify key data quality metrics for improvement, and as a result, drive increased rates of genetic gain (Brown *et al.* 2022). Since the release of the DQS in 2022, anecdotal evidence suggests that there are regional differences in data quality (Sheep Genetics 2022).

The DQS presents as a score out of 100 for each flock and a star rating out of 5, considering the quantity and quality of data provided by the breeder over a 5-year rolling average. These metrics are weighted based on their relative importance (Guy and Brown 2021). This paper investigates the difference in individual DQS metrics for MERINOSELECT flocks between geographic regions, aiming to inform more targeted extensions and adoption campaigns.

MATERIALS AND METHODS

The 7th October 2024 MERINOSELECT run DQS metrics were used for all MERINOSELECT flocks. A flock's data was removed from the dataset if it did not have more than 3 years of current data, was non-active (such as research or legacy flocks), was from a non-Merino breed (such as Polwarth), or located outside of Australia and New Zealand. A total of 267 flocks were included in the final dataset.

DQS metrics were extracted and expressed as percentages including: i) proportion of full pedigree (FullPed); ii) average pedigree (AvgPed); iii) average recording of measurements of weight (RecWT), reproduction (RecRepro), wool (RecFW), carcase (RecCA) traits as percentage of the drop recorded; iv) proportion of syndicate pedigree (SynPed); v) proportion of age variation within a contemporary group (Ages); vi) proportion of birth type variation within a contemporary group (BT); vii) effective progeny as the percentage of progeny from a given sire relative to all progeny within a group (Brown *et al.* 2001) (Eff); and, viii) pedigree linkage to the analysis for weight (LinkWT), reproduction (LinkRepro), wool (LinkFW) and carcase traits (LinkCA) as a measure of the proportion of animals with a half sibling in another flock recorded for that trait group.

Flocks were assigned a region based on postcode as defined by the Australian and New Zealand governments (NSW Government 2025; NZ Immigration 2025; RDA 2025). Any regions with fewer than 3 flocks were combined into their relative states' 'other' category, allowing for better comparisons (Table 1). A series of one-way ANOVA tests were conducted to identify significant variation for data quantity and quality metrics between regions.

Table 1. Count of flocks per region, with region number as referenced in Figure 1

Region		Number of flocks	Region		Number of flocks
1	Barwon south west, VIC	22	13	Limestone coast, SA	13
2	Canterbury, NZ	15	14	Loddon Mallee, VIC	8
3	Capital Region, ACT	15	15	Marlborough, NZ	4
4	Central West, NSW	14	16	Murraylands and Riverland, SA	8
5	Queensland	3	17	New England and North West, NSW	18
6	Eyre Peninsula, SA	11	18	Otago, NZ	9
7	Far West and Orana, NSW	13	19	Riverina, NSW	10
8	Murray, NSW	10	20	Tasmania	9
9	Goldfields Esperance, WA	6	21	Wheatbelt, WA	27
10	Great Southern, WA	20	22	York and Mid north, WA	8
11	Hume, VIC	11	23	Western Australia other	4
12	Kangaroo Island, SA	5	24	Victoria other	4

RESULTS AND DISCUSSION

The DQS identified to have significant variation ($P<0.05$) between regions by one-way ANOVA are displayed in Figure 1. Greater levels of variation between regions were identified compared to variation within regions for metrics FullPed, AvgPed, SynPed, Ages, BT, RecFW and LinkRepro.

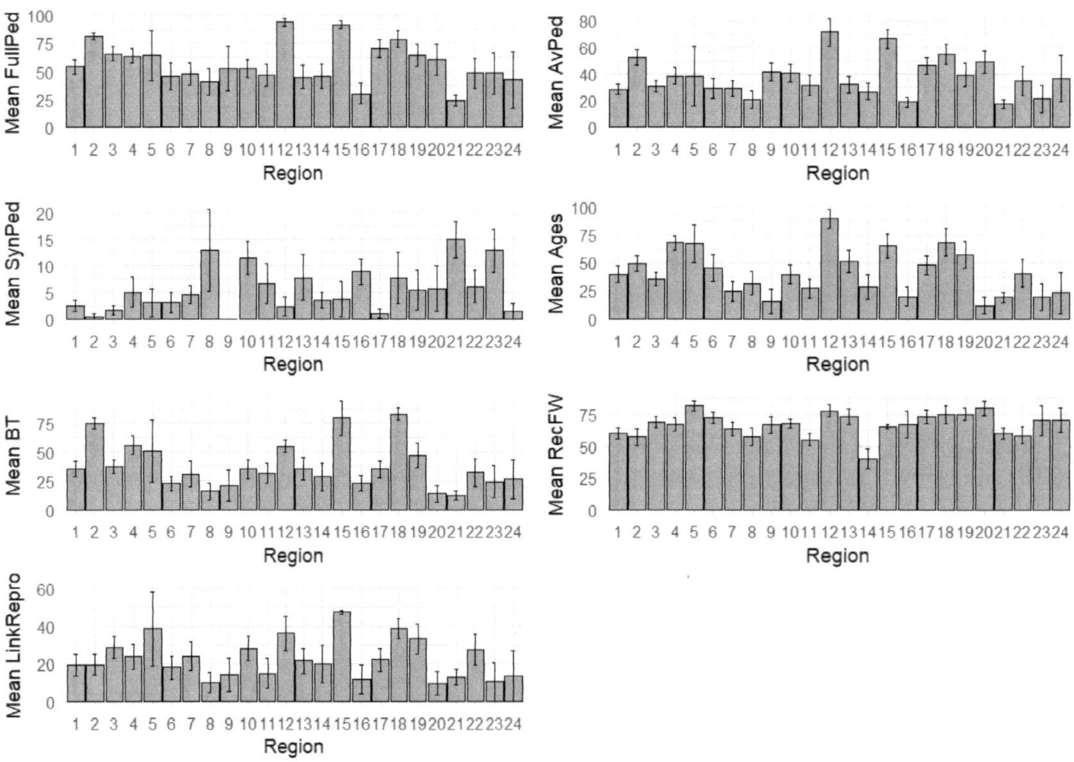

Figure 1. Average (box) and range (± standard error bar) for each data quality metric by region

Pedigree. Increased recording and depth of sire, dam and back pedigree provides more information for the calculation of breeding values, therefore increasing their accuracy. FullPed refers to the proportion of animals where both sire and dam are known. The average FullPed across regions was 56.6%, up on a previous study where Merino flocks averaged 43.6% in 2022 (Brown *et al.* 2022). Five regions averaged over 70% FullPed and two regions (Murraylands and Riverland, and Wheatbelt) averaged under 35% (Figure 1). AvgPed refers to the average depth of pedigree known. The average AvgPed was 37.4%, with the same 2 regions as FullPed averaging under 20% AvgPed (Figure 1). SynPed refers to the proportion of animals with syndicate sire pedigree, rather than an individual sire. The average SynPed was 5.5%, but 4 regions averaged greater than 10%: Murray, Great Southern, Wheatbelt and WA other (Figure 1).

Merino flocks historically have lower levels of pedigree recording compared to other breeds, but the increase in FullPed identified in this study could be attributed to the exponential uptake of genomic technology informing pedigree in recent years (Sheep Genetics 2024). While there has been an increase in pedigree recording, the previously mentioned regions still have low pedigree recording. Pedigree recording is integral to genetic evaluation and as a result, pedigree metrics are highly weighted in the DQS. This means that flocks that accurately provide full pedigree generally have a higher DQS. Pedigree is one of the most important components of genetic evaluations and as such, the Merino sheep industry could see some improvements by targeting resources towards improving pedigree recording and recording methods in Murraylands and Riverland, Wheatbelt, Murray, Great Southern, and WA other.

Fixed effects. Fixed effects are known non-genetic factors that have an impact on phenotypic expression and are therefore adjusted for or fitted in the model of the MERINOSELECT genetic evaluation. The accurate recording of fixed effects results in a more accurate breeding value calculation.

Ages refers to the proportion of age variation within a contemporary group, where a higher percentage indicates more variation in date of birth recorded. The average Ages was 39.4%, with 7 regions averaging under 25%, and 17 regions under 50% (Figure 1). This average is consistent with Brown *et al.* (2022), who found Merinos had significantly less age variation compared to LAMBPLAN flocks. Limited age information remains a difficulty, with little improvement from previous studies.

BT refers to the proportion of birth type variation within a contemporary group, where a higher percentage indicates more variation in birth type recorded. The average BT was 36.1% with 7 regions under 25% (Figure 1). BT may be supplied from pregnancy scan result, genomic testing identifying number of lambs per dam and count of lambs at birth. Evidence from the Sheep Genetics database suggests that in MERINOSELECT, BT is most commonly inferred from pregnancy scanning result. The differences in recording methods may explain some of the variation identified between regions. Targeting the nine regions that were below 25% for either Ages or BT with education specific to fixed effect recording would likely see an increase in the recording of Ages and BT, resulting in more correct breeding values and therefore a higher rate of genetic gain.

Phenotypic recording. RecFW refers to the proportion of animals recorded for fleece traits. The average RecFW was 66.8%, with only one region below 50%: Loddon Mallee (Figure 1). The recording of fleece traits is moderately weighted in the DQS, as phenotypic recording at a flock level provides information to the calculation of breeding values, increasing their accuracy, and it also helps inform the reference population for these traits. Phenotypic recording can vary across flocks based on the importance of that trait to each flock's breeding objective.

Linkage. LinkRepro refers to the proportion of animals that have a half sibling recorded for reproduction traits in another flock. The average LinkRepro was 22.6%, with Murray and Tasmania recording less than 10% LinkRepro, compared to Marlborough NZ averaging 47.5% (Figure 1). Linkage is integral to across flock genetic evaluation, allowing for the accurate comparison of animals in different environments. The variation in LinkRepro could be explained by differences in breeding objectives and therefore recording of reproduction traits. The current reproduction analysis (Weaning Rate analysis) used in MERINOSELECT was introduced in 2022 (Sheep Genetics 2022) and as the number of generations recorded for the Weaning Rate analysis increases (rather than the previously used NLB/NLW reproduction analysis), LinkRepro may increase as a result.

Targeting extension messaging. This analysis enables more targeted genetic extension activities, focusing on regions with lower performance in key data quality metrics, linked to genetic gain. By identifying which regions perform poorly in certain metrics, Sheep Genetics can prioritise targeting regions with content specific to their current weaknesses, increasing genetic gain. This allows for best use of time and resources, resulting in the largest impact. Additional work will need to be done to replicate this investigation with LAMBPLAN flocks, better identifying the variation between regions in data quality metrics that influence genetic gain across all Sheep Genetics evaluations.

CONCLUSION

Significant variation between regions was found for DQS metrics FullPed, AvgPed, SynPed, Ages, BT, RecFW and LinkRepro, suggesting there are region specific methods of data recording. The results from this paper may be used for region specific extension and education programs to improve data quality. This may produce a subsequent improvement in genetic gain for Australian and New Zealand Merino flocks. While genetic gain is not only impacted by data quality, this investigation provides Sheep Genetics with tangible outcomes to achieve better data quality in MERINOSELECT and therefore potentially increase future genetic gain.

REFERENCES

Brown D.J., Tier B. and Banks R. (2001) *Proc. Assoc. Advmt. Anim. Breed. Genet.* **14**: 441.

Brown D.J., McCrabb E.J., Bradley P.E., Rose I.J., Banks R.G. and Guy S.Z.Y. (2022). *Proc World Cong. Genet. Appl.Livest. Prod.* **12**: 743

Guy S.Z.Y and Brown D.J. (2021) *Proc. Assoc. Advmt. Anim. Breed. Genet.* **24**:143

New South Wales Government. (2025). *Smart and skilled regions.* https://www.nsw.gov.au/education-and-training/resources/smart-and-skilled-regions

New Zealand Immigration. (2025). *Regions and cities.* http://live-work.immigration.govt.nz/choose-new-zealand/regions-cities

Regional Development Australia. (2025). Regional Development Australia. https://www.rda.gov.au/

Sheep Genetics. (May, 2022). "Analysis Enhancements 2022". https://www.sheepgenetics.org.au/globalassets/sheep-genetics/updates/analysis-enhancements/analysis-enhancements-2022.pdf

Sheep Genetics. (2024). "Sheep Genetics Annual Outcomes Report". https://www.sheepgenetics.org.au/globalassets/sheep-genetics/about-us/annual-report-2023-24.pdf

Proc. Assoc. Advmt. Anim. Breed. Genet. 26: 475-478

SEVEN YEARS ON: WHAT HAS SINGLE-STEP BREEDPLAN MEANT FOR THE AUSTRALIAN BRAHMAN BREED?

C.A. Millen[1], B.J. Crook[1] and P.J. Williams[2]

[1] Agricultural Business Research Institute, University of New England, Armidale, NSW, 2351 Australia

[2] Australian Brahman Breeders' Association Ltd, Rockhampton, QLD, 4700 Australia

SUMMARY

A multi-trait single-step GBLUP (ssGBLUP) BREEDPLAN analysis that combined pedigree, performance and genomic data was first released for the Australian Brahman Breeders' Association (ABBA) in April 2017. In the seven years since, Australian Brahman breeders have embraced genotyping, with over 78,000 genotypes included in the analysis. The inclusion of this genomic information has improved the accuracy of BREEDPLAN Estimated Breeding Values (EBVs), although the improvement in EBV accuracy varies by trait and animal. Additionally, the ability to collect genomic information for use in the Brahman BREEDPLAN analysis has seen an increase in the number of young animals offered for sale with published EBVs and Selection Indexes, especially at large multi-vendor sales like the Rockhampton Brahman Week Sale.

INTRODUCTION

The BREEDPLAN analytical software (Graser *et al.* 2005) was developed by the Animal Genetics and Breeding Unit (AGBU) in the 1980s, with the first BREEDPLAN EBVs published by the Agricultural Business Research Institute (ABRI) in 1985. In the 40 years since, the BREEDPLAN model has been continually updated to incorporate new methodology and traits (Gudex *et al.* 2025). A recent development was the implementation of Single-Step BREEDPLAN evaluations in 2017 (Johnston *et al.* 2018), with over 20 Single-Step BREEDPLAN evaluations now routinely conducted by ABRI. This includes the Brahman Single-Step BREEDPLAN (Brahman ssBP) evaluation, which was the first multi-trait ssGBLUP BREEDPLAN evaluation to be released in 2017.

This paper reviews the size and composition of the Brahman Genomic Relationship Matrix (GRM) and how this has changed over the seven years since the release of Brahman ssBP in April 2017. The effect of including genomic information on BREEDPLAN EBV accuracy, and the practical implications for Australian Brahman breeders, is explored.

MATERIALS AND METHODS

The Brahman ssBP evaluation combines pedigree, performance and genomic information from ABBA in a complete multi-trait BREEDPLAN analysis of birth, growth, fertility, carcase and temperament traits. Monthly evaluations are run each year, with 18 trait EBVs reported. For the purposes of the current study, GRM data was extracted from the January 2018 and December 2024 Brahman ssBP evaluations.

The impact of genomics on EBV accuracy was examined by comparing results from the December 2024 Brahman ssBP evaluation with results from the same analysis run with genomic data excluded (BP). Data was analysed for all genotyped animals with an EBV accuracy of $\geq$5% in the Brahman ssBP evaluation.

To analyse the number of Brahman animals at the multi-vendor Rockhampton Brahman Week Sale (held annually in October) with BREEDPLAN EBVs available at the point of sale, the printed sale catalogues from the 2016 through to 2024 sales were examined. To avoid bias from the introduction of new EBVs during this time period (including shear force, flight time and percent normal sperm) or EBVs that were not reported in the hard copy catalogues across all years (e.g. mature cow weight), only a subset of EBVs were examined. These were birth weight, 200 day weight, 400 day weight, 600 day weight, scrotal size, days to calving, eye muscle area and rump fat.

RESULTS AND DISCUSSION

Brahman G-Matrix composition. The number of animals in the GRM has increased by 7.2 fold between January 2018 and December 2024, from 10867 animals in the January 2018 analysis to 78677 animals in the December 2024 analysis.

Interestingly, female animals are well represented in the GRM, representing 51% and 49% of animals in the GRM in January 2018 and December 2024, respectively. This is in contrast to what has been observed in other beef breeds such as Hereford, where only 16% of the animals in the GRM at the time of Single-Step BREEDPLAN implementation were female (Millen and Crook 2019). While strategies for building a reference population in other beef breeds have included a focus on strategic genotyping of influential sires (Millen and Crook 2019), the Australian Brahman breed has benefited from large-scale industry-funded projects, including the Brahman Beef Information Nucleus project and Repronomics™ project (Johnston *et al.* 2017), where large numbers of females were SNP genotyped and intensively recorded for a range of reproduction traits. The inclusion of data from these projects goes some way to explaining why female animals are well represented in the Brahman GRM across time.

One significant change in composition of the Brahman GRM over time is the percentage of genotyped animals without a phenotype. In January 2018, just 14% of genotyped animals had no phenotypes; by December 2024, this had risen to 49%. While the increase in the number of genotyped animals indicates that Brahman breeders have embraced genotyping since the release of Brahman ssBP, these results also indicate an increasing trend for Brahman breeders to genotype animals without collecting phenotypic information. Given the traditionally low uptake of BREEDPLAN performance recording amongst *Bos indicus* breeders in Northern Australia, the increase in genotyping has driven an increase in the numbers of young animals with BREEDPLAN EBVs available at the point of sale (as discussed further along in this paper). However, if the proportion of animals with genotypes only (and no phenotypes) continues, this could present problems for the longevity of Brahman ssBP. This highlights the importance of large-scale industry-funded projects where animals are intensively recorded and genotyped and/or the importance of encouraging phenotype collection through extension programs.

Effect of genomics on BREEDPLAN EBV accuracies. The inclusion of genomic data in Brahman ssBP has led to an increase in EBV accuracy. Across all traits, the greatest increase in EBV accuracy has been observed for animals that had a low EBV accuracy prior to the inclusion of genomic data, while those animals with a high EBV accuracy prior to the inclusion of genomic data have experienced little to no increase in EBV accuracy (Figure 1). However, the magnitude of this increase does vary by trait (Figure 1). This is further illustrated in Table 1, which shows that, for Brahman animals with an EBV accuracy of 30% prior to the inclusion of genomic data, the average increase in EBV accuracy following inclusion of genomic data ranged from 8% (Retail beef yield) to 37% (200 and 400 day weight).

Variation in the magnitudes of increase in EBV accuracy across different traits was also observed by Johnston *et al.* 2018 when the Brahman ssBP evaluation was first implemented. Similar results were also observed following the implementation of ssGBLUP for Angus (Johnston *et al.* 2018) and Hereford (Millen and Crook 2019) populations. The difference in magnitude of the accuracy increase across traits is likely explained by both heritability and the size of the reference population for each trait. Both are likely to be the case here; collection of additional phenotypes, particularly for lowly heritable and/or poorly recorded traits, are likely to drive larger increases in EBV accuracy than currently observed. Thus, the importance of performance recording remains a key extension message for both the Brahman breed in Australia and the wider beef industry.

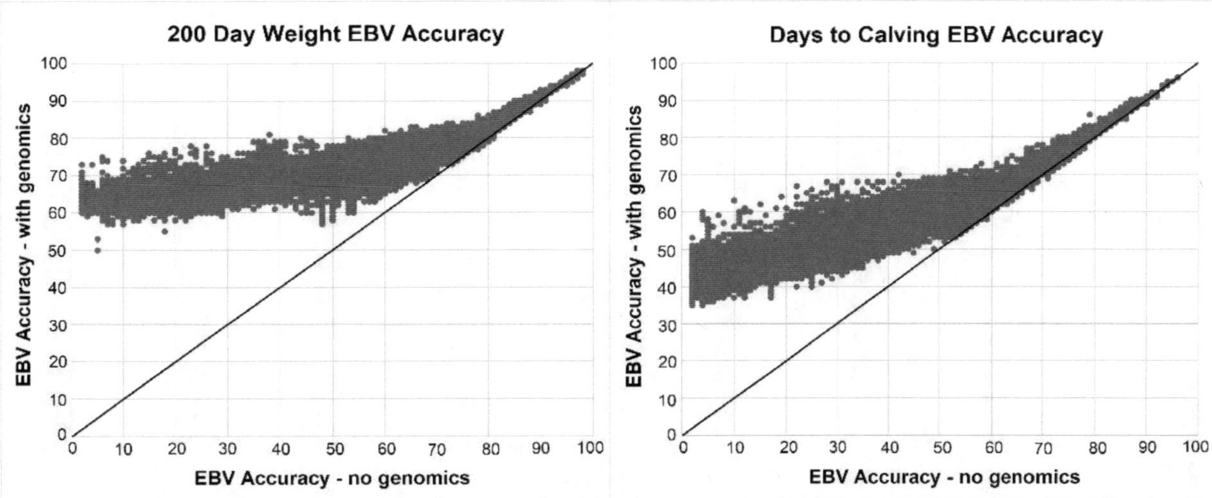

Figure 1. Changes in EBV accuracy for 200 Day Weight and Days to Calving when genomic information is included in the Brahman BREEDPLAN analysis

Table 1. Average increase in EBV accuracy when genomic information is included in the Brahman BREEDPLAN analysis, for animals that had a prior EBV accuracy of 30%

BREEDPLAN EBV	No. of animals with EBV accuracy of 30% (no genomics)	Avg. EBV accuracy (with genomics)	Avg. increase in EBV accuracy
Gestation length	2452	56	26
Birth weight	1056	63	33
200 day weight	755	67	37
400 day weight	982	67	37
600 day weight	793	66	36
Mature cow weight	1095	61	31
Milk	1726	50	20
Scrotal size	1031	62	32
Days to calving	1354	52	22
Carcase weight	1072	55	25
Eye muscle area	1156	50	20
Rib fat	1544	54	24
Rump fat	1759	54	24
Retail beef yield	2909	38	8
Intramuscular fat	2059	51	21
Percent normal sperm	1577	46	16
Flight time	783	54	24
Shear force	2301	44	14

Publication of EBVs for Rockhampton Brahman Beef Week sale animals. In 2016, the year prior to the release of Brahman ssBP, less than 20% of the sale animals (n=890) in the Rockhampton Brahman Beef Week sale catalogue had BREEDPLAN EBVs published (Figure 2). In 2024, this had increased to over 90% (n=838, Figure 2). This increase has largely been driven by the ability for Brahman breeders, following the implementation of Brahman ssBP, to conduct SNP genotyping for the purposes of obtaining Brahman ssBP EBVs. Indeed, of the 2024 sale lots, over 80% had genomic information recorded but no phenotypes. While the authors don't wish to downplay the importance of phenotypic recording for genetic evaluation, the availability of BREEDPLAN EBVs for an increased percentage of sale animals allows Brahman breeders opportunity to make more informed selection decisions and drive genetic progress in their operations.

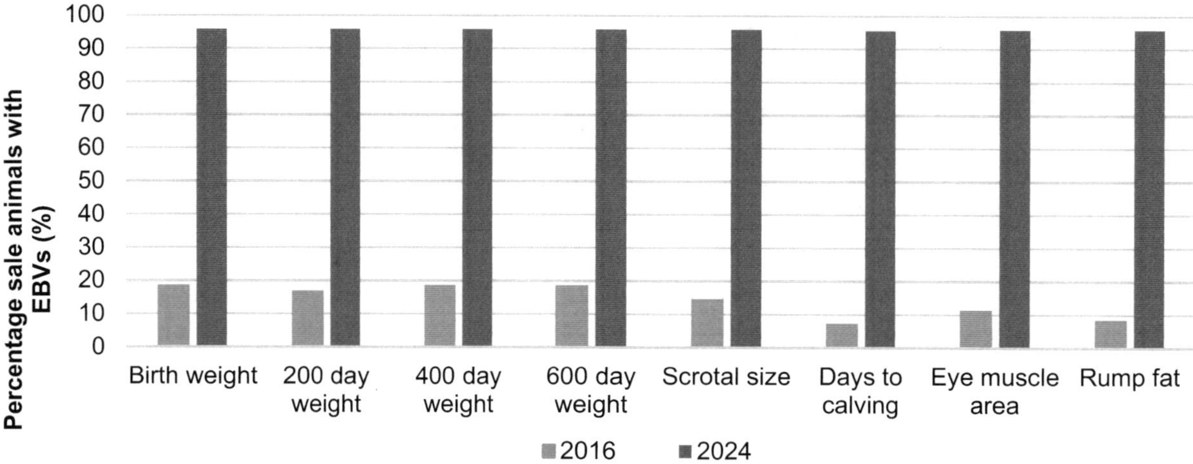

Figure 2. The percentage of animals in the Rockhampton Brahman Beef Week sale catalogues in 2016 and 2024 with birth weight, 200 day weight, 400 day weight, 600 day weight, scrotal size, days to calving, eye muscle area and rump fat EBVs available

CONCLUSIONS

Since Brahman ssBP was first released in April 2017, Brahman breeders have embraced genomic technology. There has been a rapid increase in the size of the genotyped population, which comprised 78677 animals in December 2024. The inclusion of genomic information has led to an increase in EBV accuracy, although the magnitude does vary by trait. Additionally, the ability for Brahman breeders, including those that have not traditionally collected performance information for BREEDPLAN, to conduct SNP genotyping, has led to an increase in the number of sale animals with BREEDPLAN EBVs available at the point of sale. The rapid rise in "genotyped only" animals highlights the on-going challenge of collecting phenotypic information and the importance of large-scale industry-funded research projects and extension programs for the future of the Brahman ssBP evaluation.

ACKNOWLEDGEMENTS

The authors wish to thank the Australian Brahman Breeders' Association Limited for supporting this work and providing access to their data.

REFERENCES

Graser H-U., Tier B., Johnston D.J. and Barwick S.A. (2005) *Aust. J. Exp. Agric.* **45**: 913.

Gudex B.W., Millen C.A., Johnston D.J. and Turner N. (2025) *Proc. Assoc. Advmt. Anim. Breed. Genet.* **26**: *These proceedings.*

Johnston D.J., Grant T.P., Schatz T.J., Burns B.M., Fordyce G. and Lyons, R.E. (2017) *Proc. Assoc. Advmt. Anim. Breed. Genet.* **22**: 385.

Johnston D., Ferdosi M., Connors N., Boerner V., Cook J., Girard C., Swan A. and Tier B. (2018) *Proc. World Cong. Genet. Appl. Livest. Prod.* **11**: 269.

Millen C.A. and Crook B.J. (2019) *Proc. Assoc. Advmt. Anim. Breed. Genet.* **23**: 536.

BREEDPLAN: 40 YEARS OF ACCELERATING GENETIC PROGRESS IN THE BEEF INDUSTRY

B.W. Gudex[1], C.A. Millen[1], D.J. Johnston[2] and N. Turner[1]

[1] Agricultural Business Research Institute, University of New England, Armidale, NSW, 2351 Australia
[2] Animal Genetics Breeding Unit*, University of New England, Armidale, NSW, 2351 Australia

SUMMARY

BREEDPLAN estimated breeding values (EBVs) were first published in 1985, the analysis having continuously evolved ever since. BREEDPLAN currently analyses over 15 million animals, including more than 1 million DNA profiles included in single-step BREEDPLAN genomic evaluations. This paper reviews some of the major enhancements to BREEDPLAN over the last 40 years, including the introduction of additional EBVs for a wide range of production traits, the introduction of BreedObject selection indexes, and the inclusion of genomic information culminating with the implementation of single-step BREEDPLAN.

INTRODUCTION

BREEDPLAN is a genetic evaluation scheme developed in Australia and used by beef cattle breeders worldwide. The BREEDPLAN BLUP within-herd multi-trait animal model was developed in 1984 and first implemented in 1985. This expanded upon the Australian National Beef Recording Scheme that had been in place since 1972. BREEDPLAN software development has been conducted by the Animal Genetics Breeding Unit (AGBU) with funding support from Meat & Livestock Australia (MLA). The commercial delivery of the technology has been the responsibility of the Agricultural Business Research Institute (ABRI) since 1985.

MATERIALS AND METHODS

Since the first BREEDPLAN run in March 1985, the number of records analysed has grown to over 15 million animals from approximately 6000 herds. These herds are from 80 breed organisations and non-breed affiliated groups (including large pastoral companies) located in 13 countries with some breed organisations representing multiple countries or having members outside of their 'home' country. Starting in 1994 with the first Trans-Tasman BREEDPLAN analysis, a feature has been the ability to combine data from more than one of these sources into a single analysis. Where possible, this feature enables comparable EBVs to be produced for each breed organisation, while preserving each organisations ownership and control of their data. In 1997 BREEDPLAN introduced genetic group methodology, and in 1998, the ability to import results from external genetic evaluations so that information from animals sourced from outside of the population (e.g. imported semen or embryos) could be used to inform their starting EBVs.

While the initial BREEDPLAN analyses in 1985 only produced EBVs for growth traits, subsequent research and data submission by users has allowed the development of new BREEDPLAN EBVs for male and female reproduction, calving ease, carcase, meat quality, temperament, net feed intake and structural soundness. Group BREEDPLAN was released in 1986 and enabled individual animals to have comparable information across herds. Crucial to the early Group BREEDPLAN development was the genetic linkage provided by breed organisation sire

* A joint venture of the NSW Department of Primary Industries and Regional Development and the University of New England.

referencing schemes and was supplemented by the linkage sires used in breeder herds. Other significant research that contributed to BREEDPLAN included, but is not limited to, the three Beef CRC projects, Repronomics, Southern Multi Breed project and various breed society trials. These projects also allowed the collection of genotypes and trait data, including 'hard to measure' traits (particularly fertility, carcase and meat quality). The combination of the data collected in these trials, and those collected by individual breeder herds, has enabled AGBU to publish an average of 12.8 scientific publications a year since 2006 that specifically reference BREEDPLAN or beef cattle breeding/genetics in the article title (http://agbu.une.edu.au/publications.html). Enhancements prior to 2006 were summarised in Graser *et al.* (2005) and these scientific publications are a key component of BREEDPLAN *modus operandi* as they allowed peer review/validation of BREEDPLAN methodologies by the wider scientific community to ensure that the methods were both valid and reflected the latest developments in genetic evaluation technology.

The evolution of the BREEDPLAN technology included the first use of BLUP methodology to calculate and publish EBVs and accuracies for beef cattle in 1981. BREEDPLAN also published the first multi-country multi-trait single-step genomic evaluation in 2017. One of the more advantageous enhancements was the inclusion of all quantitative traits into a single multi-trait BLUP analysis in 1989. The multi-trait BLUP analysis used the genetic relationships between all traits to both increase individual EBV accuracy by using correlated information and remove any selection bias. Research and development in variance component estimation has paralleled the major developments of BREEDPLAN. This included the implementation of new methods, models and software for the estimation of genetic parameters that underpin the BREEDPLAN BLUP analysis.

The development in 1995 of a threshold model allowed for the analysis of categorical data including calving ease, docility and structural soundness scores. EBVs were computed on the underlying scale and transformed to the observed scale for the published BREEDPLAN EBV. The ability to include crossbred animals in a single analysis, including accounting for heterosis, was first introduced into BREEDPLAN in 1997. Recent developments have included an increase in run frequency and developments in how DNA information is incorporated into the analyses. In 2010, DNA information was included using a 'blending' method, before upgrading to single-step methodology in 2017. The total number of DNA profiles included in the single-step BREEDPLAN genomic relationship matrices have rapidly grown since 2017 and exceeded 1 million for the first time in November 2024.

The addition of BreedObject selection indexes in 1996 enabled cattle producers to identify animals that were most profitable for particular production systems and markets. This was enhanced in 2001 with BreedObject on the web allowing breeders to customise selection indexes to suit their individual needs and production systems. The BreedObject software itself has been supported by over 40 scientific publications which can found at https://breedobject.com/page/About_BreedObject.html.

All of these developments in BREEDPLAN have come at the cost of more complex computing algorithms and strategies. These increases in complexity have also occurred concurrently with the ever increasing number of traits and records, and client desire for increased run frequency. Fortunately, advancements in computer software and hardware have kept pace and enabled these developments to be implemented.

RESULTS AND DISCUSSION

A survey of 639 Australian beef producers showed that the overall producer perception of BREEDPLAN is largely positive and is driven by sentiments of trust, scientific validity and a historical reputation for efficacy (Sloane and Walker 2023). These sentiments reflect the genetic progress that users of BREEDPLAN have been able to make. Figure 1 displays the average weighted selection index for all 41 breed organisations & groups around the world who currently have

BreedObject selection indexes supplied by BREEDPLAN. Overseas values were adjusted for their exchange rate to the Australian dollar in November 2024 and averaged across countries, breeds, all selection indexes within breed, and individual herds to give an indication of the average genetic progress that has been made using BREEDPLAN (equivalent to AU$2.82 per year). So, while there are individual herds and groups that perform better than the average (up to AU$8.90 average genetic progress per year for a breed society), there is little value of comparing breed groups that have different selection indexes. With one exception (Brahman in South Africa, Namibia and Zimbabwe), each country and breed combination in BREEDPLAN has unique selection indexes that are optimised for the specific production systems and markets that use their genetics. The effects of specific production systems, including the unique genetic parameters of each breed and other population specific parameters, local conditions and pricing, on selection indexes was demonstrated in Gudex *et al.* (2023).

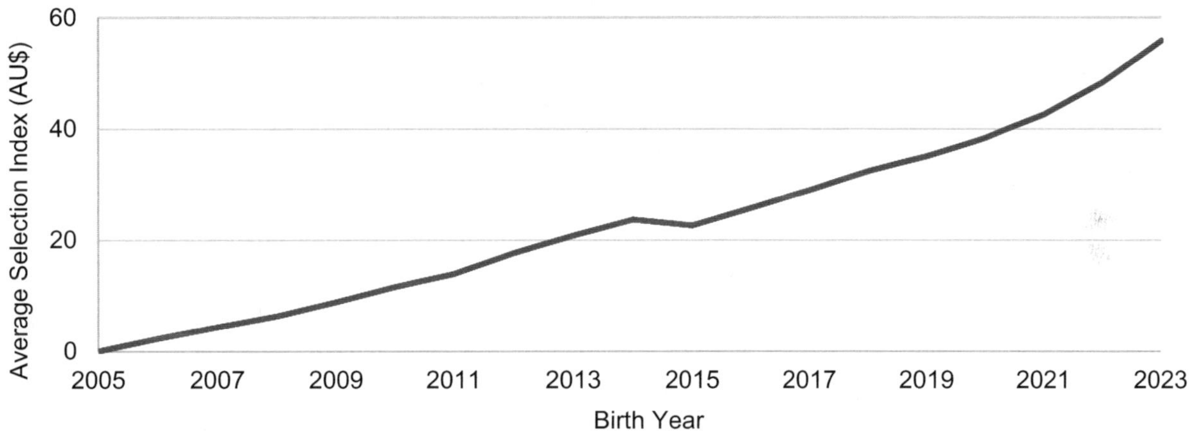

Figure 1. Average genetic progress since 2005 for breed organisations & groups with BREEDPLAN selection indexes

The staggered implementation of BREEDPLAN developments across different breeds has meant that any changes in the trend displayed in Figure 1 due to new developments is not obvious. However, such changes have been visible in studies conducted within individual breeds. Millen and Crook (2019) demonstrated an increase in genetic progress following the introduction of Hereford Single-Step BREEDPLAN for Australia, New Zealand and Namibia. The use of the single-step methodology in BREEDPLAN was validated by Johnston *et al.* (2023) using the forward cross-validation technique in their validation of the (separate) Angus, Brahman, Hereford, Santa Gertrudis and Wagyu single-step BREEDPLAN analyses. Millen *et al.* (2025, *these proceedings*) also reported that the use of single-step methodology in Australian Brahman has led to a large increase in the number of bulls being sold with BREEDPLAN EBVs.

Since 1985, there have also been numerous developments in how BREEDPLAN information is presented back to breeders. In 1989, ABRI developed and released its first iteration of herd recording software and an online sire selector in 1997. Today, websites are routinely used to view information (including pedigree and EBVs) on individual animals, sale, and semen catalogues, submit data, download reports and predict outcomes of matings, and 39% of Australian seedstock producers are using specialist herd management software (Sloane and Walker 2023). The websites also allow the results from companion products like GeneProb and Completeness of Performance to be viewed. GeneProb (Kinghorn 2000) was first provided by ABRI in 2009 and uses known gene test results and combines this with pedigree information to calculate the probability of alleles being present in untested animals. The Completeness of Performance product was first released in 2012 and provides a summary of the amount of data that each herd has submitted to BREEDPLAN. Concerns regarding the quality of data submitted and analysed was one of the biggest factors driving negative perceptions of BREEDPLAN in Australia (Sloane and Walker 2023). This perception remains despite the presence of several data checks that include identifying and omitting extreme records within a contemporary group until the records are verified by the breeder (outliers) and only accepting data for certain traits when it is recorded by an accredited professional. MateSel (Kinghorn 2011) is another companion product to BREEDPLAN that ABRI has offered since 2012. MateSel is a tool that enables breeders to optimise breeding outcomes by creating a suggested mating list based on a group of candidate sires and dams.

With genetic evaluations being an evolving and technical space, BREEDPLAN has always had a strong emphasis on extension services to support its clients. In Australia, this started with state department beef cattle officers, progressed through the industry funded Tropical Beef Technology Services (1995 to 2021) and Southern Beef Technology Services (2005 to 2021) projects and has continued with ABRI extension services today. One of the advantages of the current ABRI extension services model is its ability to work with and support BREEDPLAN's international clients. In 2024 alone, ABRI extension services delivered ~50 online workshops, webinars and podcasts, 40+ articles, 110+ social media posts, attendance at 11 training days, field days and youth events, and technical support to numerous individual herds around the world.

CONCLUSION

BREEDPLAN has a long and proud history of providing effective genetic evaluations to the beef industry around the world. This has been possible through the continuous incorporation of the latest developments in computing, methodology and scientific knowledge into the BREEDPLAN model. Additionally, the provision of companion products and extension services have supported BREEDPLAN clients to undertake a holistic approach to animal selection and breeding and drive genetic progress in their herds. As BREEDPLAN moves into the future, the implementation of new traits, products and services will continue to assist beef producers globally to improve productivity and profitability in their beef breeding businesses.

ACKNOWLEDGEMENTS

The authors would like to thank all the breeders, breed organisations and other breeding groups who have used BREEDPLAN over the last 40 years. We would also like to acknowledge all the staff at ABRI, AGBU, MLA and the breed associations who have contributed to the development and success of the BREEDPLAN system over this time.

REFERENCES

Graser H.U., Tier B., Johnston D.J. and Barwick S.A. (2005) *Aust. J. Exp. Agric.* **45**: 913.

Gudex B.W., Williams P.J. and Walmsley B.J. (2023) *Proc. Assoc. Advmt. Anim. Breed Genet.* **24**: 103.

Johnston D.J., Ferdosi M.H., Connors N.K., Cook J., Girard C.J. and Swan A.A. (2023) *Proc. Assoc. Advmt. Anim. Breed Genet.* **25**: 111.

Kinghorn B.P. (2000) In 'Animal Breeding – Use of New Technologies', pp. 112-117, editors B.P. Kinghorn, J.H.J. Van der Werf and M. Ryan, The Post Graduate Foundation in Veterinarian Science of the University of Sydney, Sydney.

Kinghorn B. P. (2011) *Gene. Sele. Evol.* **43**: 4.

Millen C.A. and Crook B.J. (2019) *Proc. Assoc. Advmt. Anim. Breed Genet.* **23**: 536.

Sloane B. and Walker (2023) L.GEN.2205 MLA Genetics Insights Report. https://www.mla.com.au/research-and-development/reports/2023/l.gen.2205---genetic-insights-report-2022/

GENETIC TRENDS IN AUSTRALIAN BEEF CATTLE AND SHEEP POPULATIONS, AND THEIR IMPACT

R.G. Banks, D.J. Brown, K.L. Moore, B.J Walmsley and A.A. Swan

Animal Genetics and Breeding Unit[1*], Armidale, NSW, 2351 Australia

SUMMARY

Since 2000, genetic improvement in the Australian beef and sheep industries has exhibited increasing rates of genetic progress expressed in economic index terms, increasing adoption of genetic technologies, and consequently, substantial and increasing economic impact. Economic benefits flow to both private and public beneficiaries, and imply substantial return on public and private investments. Monitoring and communication of ongoing progress will be important in maintaining investment, particularly as phenotyping increases in importance.

INTRODUCTION

Genetic evaluation at an industry level provides a basis for genetic improvement and for an informed market for genetic products (animals, semen, oocytes, embryos) and services (e.g. mating and breeding program design). These support public and private benefits, and attract both public (including both taxpayer and industry contributions) and private investment. Documenting and understanding the scale of these benefits is important for attracting and maintaining investment.

Genetic evaluation for the Australian beef and sheep industries is available through BREEDPLAN (Gudex *et al.* 2025) and Sheep Genetics (Brown *et al.* 2007), noting that some breeders access other systems. BREEDPLAN and Sheep Genetics have been available since the mid- and late 1980s respectively, and provide information on animals' genetic merit, which bull- and ram- breeders, and commercial cattle and sheep producers, can use to assist in selection.

These genetic evaluation systems routinely track information including numbers of animals entering the evaluation and average genetic merit for individual traits and economic indexes. This paper reports on genetic trends in BREEDPLAN and Sheep Genetics data, and using these derives estimates of economic impact.

MATERIALS AND METHODS

Data on genetic trends in BREEDPLAN and Sheep Genetics were provided by ABRI and AGBU, respectively. For beef cattle the indexes were developed using BreedObject v.6 (Barwick *et al.* 2018). Individual breed index trends were combined into aggregates for southern and northern breeds respectively by first averaging across the current indexes produced for each individual breed, and then combining into a weighted aggregate index using the numbers registered with each breed within birth year (Millen, *pers. comm.*; Gudex *et al.* 2025). The breed groups aggregated were:

- Southern: Angus, Charolais, Hereford, Limousin, Murray Grey, Red Angus, Shorthorn, Simmental, South Devon, Speckle Park and Wagyu.
- Northern: Belmont Red, Brahman, Brangus, Droughtmaster and Santa Gertrudis.

In sheep, trends were averaged across the main currently used indexes for each breed group; terminal sire breeds, maternal breeds and Merinos (https://www.sheepgenetics.org.au/getting-started/asbvs-and-indexes).

---

[*] A joint venture of NSW Department of Primary Industries and Regional Development and the University of New England

Estimates of annual economic impact were calculated as the product of:

- the difference between the index mean for each year and the mean for the 2000 drop)
- estimated number of sires entering industry annually as 0.55 of the number of animals enrolled in each year as male (Skinner and Brown *pers. comm.*), and 0.75 of those being used as sires
- estimated lifetime females joined as 100 times the estimated number of bulls entering industry and 125 times the number of rams
- number of animals enrolled in that year

Because the relevant indexes are all essentially a Net Present Value per female joined, the overall impact is therefore an estimate of Net Present Value added due to animals born in that year, compared to a 2000-drop base.

RESULTS AND DISCUSSION

Genetic trends for the breed groups, summarised by 5-year regressions for each period, are shown in Table 1. In all groups, the average rate of genetic progress in 5-year periods has increased through time. While unlikely to be the sole contributing factor, single step genomic evaluation was introduced to some extent in all 5 groups in 2016-2018, and may have contributed to the increases in genetic trend observed since 2015. Increase in genetic trend with the introduction of genomic prediction has been observed in other industries, most notably but not limited to dairy cattle (Guinan *et al.* 2022; Shaffer, *pers. comm.*, Knol *et al.* 2016).

Table 1. Breed group aggregated average genetic trends ($NPV per female joined per year)

Period	Southern Beef	Northern Beef	Merino	Terminal Sire	Maternal
2000-2005	$4.58	$0.75	$1.12	$1.22	$0.72
2005-2010	$4.86	$0.69	$1.13	$1.13	$2.56
2010-2015	$7.45	$0.67	$1.24	$1.00	$1.89
2015-2020	$8.73	$0.67	$1.53	$1.38	$2.14
2019-2023	$9.46	$0.83	$2.52	$1.99	$3.43

NB: the values here are the slope of the average genetic trend in that group for that period

Regardless of the extent to which these results reflect adoption of genomic prediction, they imply utilisation of genetic improvement technologies. Not shown here, 3 other trends contribute to this last observation:

- average index accuracy is rising steadily in all three sheep breed groups (data not available for beef at the time of writing), implying more and/or better recording practices. Sheep Genetics has provided metrics of data quality throughout the period of these results, and these coupled with general extension, would seem to be encouraging improved recording. The product of average index accuracy and number of animals evaluated, a measure of the informaton available to the market, has grown at about 7% per year in the 3 sheep groups (year data not shown).
- a steady genetic trend for improved fertility (expressed via the Days to Calving EBV) in Brahmans, with some suggestion of acceleration in the most recent years. Fertility is a major component of breeding objectives for northern breeds, so this trend is encouraging.
- and numbers of animals entering genetic evaluation are rising steadily through the period, most notably in sheep.

The estimates of economic impact across the whole industry are shown in Table 2.

Table 2. Breed group and industry total estimates of economic impact ($m)

Period	Southern Beef	Northern Beef	Merino	Terminal Sire	Maternal
2000-2005	$538.69	$18.06	$35.10	$42.10	$3.18
2005-2010	$1,220.91	$46.18	$113.89	$161.23	$55.70
2010-2015	$2,433.08	$73.56	$327.65	$313.45	$148.31
2015-2020	$3,679.22	$102.68	$721.87	$569.83	$322.59
2021-2023	$3,440.14	$86.81	$649.75	$517.35	$356.05
Total	**$11,312.95**	**$327.30**	**$1,848.27**	**$1,603.96**	**$885.83**

NB: economic impact for the last period is years 2021-2023. The last period for the genetic trend results is 2019-2023 (Table 1) – simply to have a 5-year trend rather than 3-year.

The estimated impacts in Table 2 are rising - in both species reflecting both increasing rates of genetic progress and increased numbers entering evaluation. Note that while the estimation method assumed constant proportion of rams entering the sheep industry to total animals evaluated, and constant lifetime joinings per ram, these parameters appear to be rising (Brown, *pers. comm.)*

The totals for the two species - $11.6bn and $4.3bn respectively – are approximately the current gross value of production of the beef and lamb and sheepmeats industries (Meat and Livestock Austalia 2025), meaning that gross margins have been growing at approximately 4% of gross value of production annually (which if gross margin averages 20% of gross value of production, equates to annual growth in real gross margin of 20%). These estimates suggest very substantial contribution of genetic improvement to industry economic outcomes, and steady and substantial increase in that contribution through the period 2000-2023.

Are there any caveats to these observations?

i) Average index change within a breed is not what every individual producer obtains by sourcing sires from a herd or flock making genetic progress, but the average trend describes progress in the population as a whole.

ii) Using the current indexes to define genetic trend does not mean that selection emphasis has been constant, as indexes have changed their trait coverage and weightings.

iii) The economic impact estimates depend on the scale of expression of genetic trend, which is here calculated using estimates of numbers of sires entering industry from herds or flocks per animal evaluated and their lifetime joinings. The values used here (41 and 52 for bulls and rams respectively) are similar to those derived by Fennessy *et al* (2014), but as noted above, appear to be rising, at least in sheep.

iv) The impact estimates are expressed in approximately current dollars – reflecting when the indexes were updated (through the period 2020-2024). Because nominal prices have risen during the period 2000-2025 in both species, the comparison with industry value should be done using current price and income levels (as was done here).

Other factors to consider include:

- counterfactual, including importation: importation has contributed to genetic improvement in the beef industry, but this fact does not lessen the value of that improvement, or the return on investment. Without local investment in genetic evaluation, importation would likely be less directed to local breeding objectives, and hence of lower value for the Australian industry, and also would be impossible to evaluate.
- the impact includes public, industry, and private benefits. The evaluation systems provide a platform that enhances each of them.
- the impact does not account for multiplication of these genetics such as sires bred from evaluated sires, which is a significant pathway especially in Northern Beef and Merino sheep.

- maintaining industry evaluation systems has to date depended on models of funding that were established in the 1980s, with understandably no appreciation of the ongoing challenge of attracting sufficient funds to collect phenotypes. Addressing this issue – funding phenotyping at sufficient scale and trait coverage – is urgent now.

CONCLUSIONS

The overall messages are clear:

- In both species, there have been significant increases in the rate and value of genetic change since 2000, with clear signs of acceleration in more recent years, which likely have made major contributions to maintain or improve profitability in value chains in both species.
- These trends reflect increases in adoption as well as improved recording resulting in increased accuracy, and implied increasing selection differential (as approximated by the ratio delta G/accuracy – not shown here).
- The economic impacts of the genetic change are substantial in both species.
- And while not estimated here, the return on investment ratios in both species, with investment including RDE and recording and selection costs, are approximately 33:1.

Genetic improvement in the Australian beef and sheep industries has generated susbtantial economic benefits, and there is good evidence that the scale of those benefits is increasing. Monitoring outcomes for these public-industry-private partnerships is essential, ideally tracking longer terms, and should include regular and open updating of economic indexes to ensure both their relevance to value chain businesses and to overal evaluation.

ACKNOWLEDGEMENTS

ABRI and Sheep Genetics are acknowledged for the provision of genetic trend data. Particular thanks to Catriona Millen (ABRI) and Peta Bradley (Sheep Genetics). This paper was initiated with inputs from Andrew Swan, and pays tribute to his contribution to beef and sheep genetic improvement in Australia.

REFERENCES

Barwick S.A., Henzell A.L., Walmsley B.J., Johnston D.J. and Banks, R.G. (2018) *J. Anim. Sci.* **96**: 1600.

Brown D.J., Huisman A.E., Swan A.A., Graser H.U., Woolaston R.R., Ball A.J., Atkins K.D. and Banks R.G. (2007) *Proc.Assoc.Advmt. Anim. Breed. Genet.* **17**: 187.

Fennessy P., Byrne T., Amer P. and Martin, G. (2014) Meat and Livestock Australia, Final Report B.EVA.001 and B. EVA.002 Meat & Livestock Australia, North Sydney.

Gudex B.W., Millen C.A., Johnston D.J. and Turner, N. (2025) *Proc. Assoc. Advmt. Anim. Breed. Genet.* **26**: *These proceedings.*

Guinan F.L., Wiggans G.R., Norman H.D., Dürr J.W., Cole J.B., Van Tassell C.P., Misztal I. and Lourenco D. (2023) *J. Dairy Sci.* **106**: 9911.

Knol E.F., Nielsen B. and Knap P.W. (2016) *Anim. Front.* **6**: 15.

Meat and Livestock Australia (2025) https://www.mla.com.au/about-mla/the-red-meat-industry/

BREEDING FOR FACIAL ECZEMA TOLERANCE – A SUCCESS STORY

P.L. Johnson[1] and R. Proffit[2]

[1] AgResearch, Invermay Agricultural Centre, Mosgiel, New Zealand
[2] Raupuha Stud, Pio Pio, New Zealand

SUMMARY

Facial Eczema (FE) or sporidesmin toxicosis is a significant animal health challenge that is set to worsen because of climate change. Tolerance to sporidesmin is measured as GGT21 (Gamma Glutamyl Transferase levels 21 days after exposure to a known dose of sporidesmin) is a highly heritable trait (~0.42). Genetic improvement in tolerance to the toxin has been achieved in many New Zealand maternal sheep flocks through the RamGuard programme in which animals are dosed with known amounts of the toxin. However, of interest is whether the observed RamGuard tolerance translates during a significant natural challenge of the toxin. 2016 was one of the worst years for FE in New Zealand's farming history. After exposure during the natural challenge, GGT data from 2014 born ewes from the Raupuha Stud was available. This stud consists of two maternal flocks enrolled in the RamGuard programme and two terminal flocks that are not. As anticipated the terminal flocks were highly susceptible to the natural challenge. The two maternal flocks had many more tolerant ewes. In support of the high heritability of the trait, there was strong evidence that how the Perendale sires responded themselves through RamGuard was reflected in their progeny. A total of 78% of the progeny from three sires that were very tolerant themselves (RamGuard Day 21 GGT less than 40 IU/L) had natural challenge GGT values less than 71 IU/L (the threshold used to indicate tolerance), compared to only 54% progeny from sires considered to be susceptible (RamGuard Day 21 GGT between 71 and 123 IU/L). However, noting that even these sires are on the tolerant end of the susceptible spectrum, as some of their contemporaries recorded RamGuard Day 21 GGT levels as high as 2000IU/L and as such the impact on progeny of highly intolerant sires was not observed. These results support that the RamGuard programme is identifying genetic variation in FE tolerance that results in tolerance when exposed to a natural challenge of sporidesmin and highlights the ongoing importance of breeding for genetic tolerance to FE as a tool to lessen its impact.

INTRODUCTION

Facial Eczema (FE) or sporidesmin toxicosis is an animal health challenge that results from grazing animals ingesting the toxin sporidesmin which is produced by spores of the fungus *Pseudopithomyces chartarum*. This fungus proliferates in late summer/autumn when conditions are warm and moist and pasture dead matter is present. The ingestion of the toxin causes liver damage initially producing sub-clinical effects but can result in clinical symptoms including photosensitivity and skin lesions. FE has long been a regular and significant animal health challenge in northern parts of New Zealand, but as a result of increased temperatures as a result of climate change, more areas of the North Island and areas of the South Island are more regularly observing elevated spore counts which means that animals are at risk of FE (Johnson *et al.* 2017a). In 2016 the weather conditions were such that it became one of the worst years reported for FE in New Zealand's history, both in terms of the geographical spread and also the severity of impact on animals (Johnson *et al.* 2017a).

Tolerance to FE is highly heritable (~0.42 – Morris *et al.* 1999; Morris *et al.* 2012; McRae *et al.* 2016) and maternal sheep studs who farm in or have clients in established FE areas have been breeding for increased tolerance to the toxin through the RamGuard programme (Amyes and Hawkes 2014). Raupuha Stud is one such stud and is home to four recorded sheep flocks, two maternal (Perendale and Romdale, a Perendale/Romney cross) and two terminal (Suffolk and SufTex, a Suffolk/Texel cross). Breeding for tolerance to FE has been taking place in the Perendale

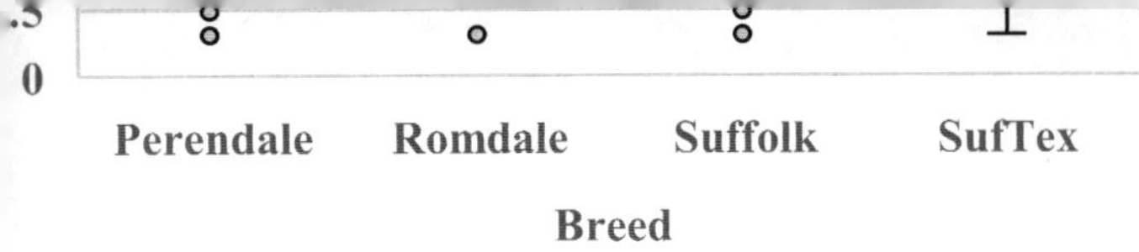

and Romdale flocks since 1999 and 2011 respectively. Because most terminal-sired lambs are expected to be processed by late summer, there has been limited genetic selection for tolerance to FE in terminal breeds, but it is generally thought that they are highly susceptible. The impact of the 2016 outbreak on liver damage (measured as blood gamma-glutamyl transferase (GGT) levels) and the productive outcomes of the Perendale and Romdale flocks were summarised by Johnson *et al.* (2017b) with significant impacts on all aspects of performance were observed. The purpose of this paper is to summarise the observed genetic differences in the four flocks but with emphasis on the Perendale flock for which more data was available.

MATERIALS AND METHODS

Blood GGT data was collected in 2016 during a significant FE outbreak on the Raupuha Stud in South Waikato of New Zealand (latitude –38, longitude 175). The farm recorded spore counts greater than 20,000 spores per gram, which is the level at which prevention is recommended, from early February through March, peaking as high as 145,000 in early March. Ewes born in from the Suffolk (n=72) and SufTex (n=41) flocks were blood sampled on the 11th of March and the Perendale (n=369) and Romdale (n=181) flocks on the 29th of March.

Given GGT values are non-normally distributed, the GGT values were log transformed for reporting. Within the RamGuard programme, GGT levels less than 71 IU/ml or logGGT 1.85 are used to indicate animals that are phenotypically tolerant to the toxin. The data are presented firstly considering the distribution in logGGT values at the breed level and then focusing on the Perendale flock results, presenting the logGGT distributions for progeny per sire and finally comparing the sire averages based on the 2016 data against their November 2015 breeding values.

RESULTS AND DISCUSSION

The distribution of logGGT values for the four breeds is in Figure 1. From this figure, despite the terminal breeds having been bled earlier, it can be observed that indeed the terminal breed animals were highly susceptible to the toxin with very few animals recorded as having low GGT levels. Conversely, a much larger proportion of the maternal breed animals exhibited tolerance to the natural challenge of sporidesmin.

Figure 1. Boxplot summary of logGGT values for born 2014 progeny of different breeds (maternal = Perendale and Romdale; Terminal = Suffolk and Suftex) exposed to a natural challenge of sporidesmin in 2016. Animals below the dotted line are considered to have exhibited tolerance to sporidesmin

Given tolerance to FE is a highly heritable trait, a single measurement on an individual animal has a moderate to high accuracy. Figure 2 shows the percentage of the born 2014 ewes within different GGT ranges, grouped according to how their sire had himself responded when he himself was assessed through RamGuard. The ewes were the progeny of fifteen sires selected for use in the breeding programme which came from a cohort of 118 rams that were measured through RamGuard over three years. Three of the sires represented were very tolerant to sporidesmin based on their RamGuard Day 21 GGT values (GGT < 40 IU/L), a further 6 sires were classed as tolerant (40 < GGT < 71 IU/L), with the remainder classed as susceptible (GGT> 70 IU/L). However, the most susceptible sire represented only had a RamGuard Day 21 value of 123 IU/L which is considerably lower than the maximum values for the contemporary groups from which they originated which included animals that had RamGuard Day 21 GGT values of nearly 2000 IU/L. As such the impact of using a highly susceptible sire was not observed in these ewes. That said, 78% of the progeny from the very tolerant sires were classed as tolerant, compared with only 54% of the progeny from the susceptible sires. Additionally, there were very few progeny from the very tolerant sires with extremely high GGT values.

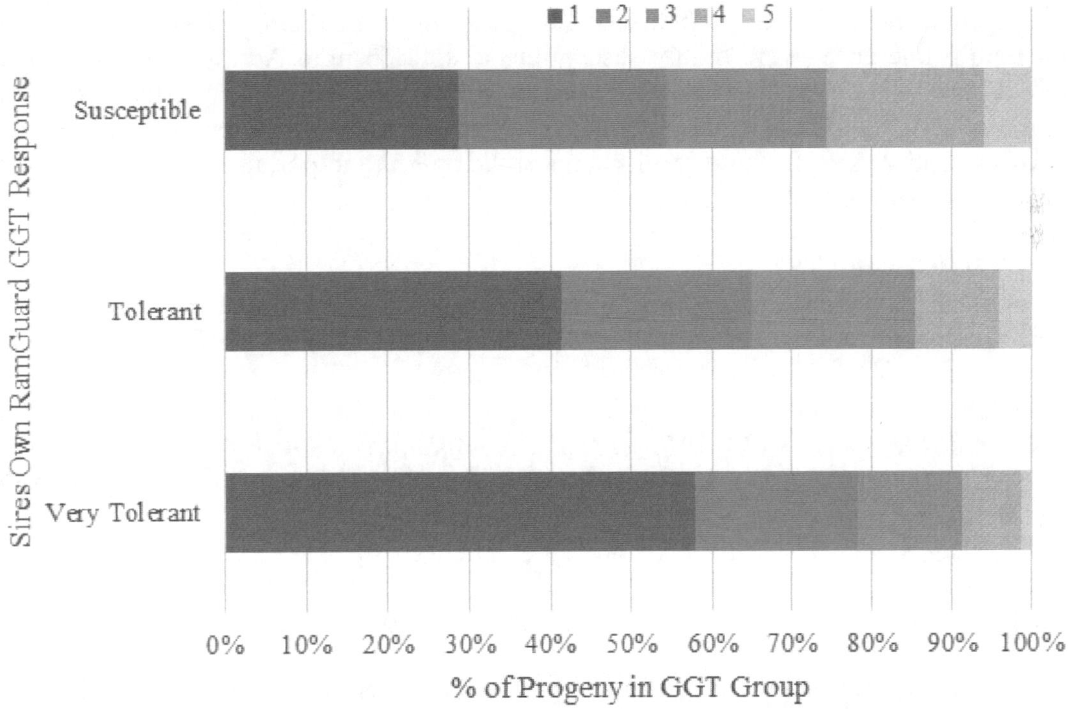

Figure 2. % Distribution GGT values for born 2014 progeny of Perendale sires exposed to a natural challenge of sporidesmin in 2016 according to their GGT level where 1= less than 40 IU/L; 2= 41 – 70 IU/L; 3=71-200 IU/L; 4=201-700 IU/L and 5 = greater than 700 IU/L. The progeny are grouped according to how their sires responded themselves to a targeted RamGuard Challenge:- Sire Very Tolerant (Sire RamGuard GGT <40 IU/L, n=69); Sire Tolerant (Sire RamGuard GGT between 41 and 70 IU/L, n=123); Sire Susceptible (Sire RamGuard GGT 71-123 IU/L, n=136).

One of the sires represented, Flock 1941 sire 120816/12, is of particular interest. This sire was one of the three that exhibited very high tolerance when he was assessed through RamGuard in 2013 at a dose rate of 0.43 mg/kg live weight (which was the highest dose rate for any of the sires represented), with the second lowest GGT Day 21 value (33 IU/L) for his contemporary group of 42 rams, with only one of his half-sibs having a lower Day 21 GGT value. Of his nineteen born 2014 ewe progeny, exposed to the 2016 natural challenge, the highest GGT value observed was only 156 IU/L, which was close to the average of all of the Perendale ewes (GGT average of 138 IU/L). The data reported on in this paper was entered into Sheep Improvement Limited (SIL), which when combined with this sires own previous data and that of his other progeny measured through RamGuard (n=2) led to him being the top-ranked sire in the SIL across flock FE evaluation for several years, and as of December 2024 still ranks in the top 2% of FE tolerant rams in SIL.

Based on the results of Johnson et al. (2017b) the tolerant progeny across all sires were significantly more productive throughout 2016 across aspects of reproduction, body condition and maternal ability, when compared to susceptible animals, and thus were more profitable and had improved welfare.

CONCLUSION

Facial Eczema is a significant animal health challenge that is only set to become even more significant as a result of climate change. It is known to be a highly heritable trait, and genetic improvement has been achieved by several breeders through the RamGuard programme, however, questions have always been raised about "how does this translate to a natural challenge scenario". This paper demonstrated that, as predicted, the progeny of terminal sires, who have not undergone selection for FE tolerance, were highly susceptible to sporidesmin. Additionally, that those sires who were highly tolerant themselves when measured through RamGuard produced progeny who when exposed to a significant natural challenge exhibited increased tolerance (compared to the progeny of susceptible sires). These results support the ongoing importance of breeding for genetic tolerance to FE as a tool to lessen its impact.

ACKNOWLEDGEMENTS

The authors acknowledge the support of Julia McCracken and the team at Pio Pio Vets. Financial support was obtained from the Farmer Initiated Technology Transfer Fund (FITT).

REFERENCES

Amyes N.C and Hawkes A.D. (2014) *Proc. N. Z. Soc. Anim. Prod.* **74**: 154.

Johnson P.L., Cameron C.A., Henderson H.V. and Cullen N.G. (2017a) *Proc. N. Z. Soc. Anim. Prod.* **77**: 107.

Johnson P.L., Henderson H.V. Proffit R. and Cullen N.G.(2017b) *Proc. N. Z. Soc. Anim. Prod.* **77**: 197.

McRae K.M., Cullen N.G., Amyes N.C. and Johnson P.L. (2016) *Proc. N. Z. Soc. Anim. Prod.* **76**: 43.

Morris C.A., Amyes N.C., Hickey S.M. and Binnie D.B. (1999) *Proc. N. Z. Soc. Anim. Prod.* **59**: 117.

Morris C.A., Phua S.H., Cullen N.G. and Towers N.R. (2012). *N. Z. J. Agric. Res.* **56**: 156.

AAABG – Association for the Advancement of Animal Breeding and Genetics
University of New England
Armidale NSW 2351
Australia

ISBN 979-8-3313-2668-5

9 798331 326685

CURRAN ASSOCIATES INC.
proceedings
.com